MW01518448

Understanding consumers of food products

Related titles:

Food product development, maximising success
(ISBN-13: 978-1-85573-468-5; ISBN-10: 1-85573-468-0)
Product development, from refining an established product range to developing
completely new products, is the lifeblood of the food industry. It is however a process
fraught with risk and often ends in failure. This book explains the means to making
product development a success. Filled with examples and practical suggestions, and
written by a distinguished team with unrivalled academic and industry expertise, *Food
product development* is an essential guide for R&D and product development staff, and
managers throughout the food industry concerned with this key issue.

Food process modelling
(ISBN-13: 978-1-85573-565-1; ISBN-10: 1-85573-565-2)
Food process modelling provides an authoritative review of one of the most exciting
and influential developments in the food industry. The modelling of food processes
allows analysts not only to understand such processes more clearly but also to control
them more closely and make predictions about them. Modelling thus aids the search
for improved and more consistent food quality. Written by a distinguished
international team of experts, *Food process modelling* covers both the range of
modelling techniques and their practical applications across the food chain.

Details of these books and a complete list of Woodhead titles can be obtained by:

- visiting our web site at www.woodheadpublishing.com
- contacting Customer Services (email: sales@woodhead-publishing.com; fax: +44
  (0) 1223 893694; tel.: +44 (0) 1223 891358 ext. 30; address: Woodhead Publishing
  Limited, Abington Hall, Abington, Cambridge CB21 6AH, England)

Understanding consumers of food products

Edited by
Lynn Frewer and Hans van Trijp

Sponsored by

CRC Press
Boca Raton Boston New York Washington, DC

WOODHEAD PUBLISHING LIMITED
Cambridge England

Published by Woodhead Publishing Limited, Abington Hall, Abington,
Cambridge CB21 6AH, England
www.woodheadpublishing.com

Published in North America by CRC Press LLC, 6000 Broken Sound Parkway, NW,
Suite 300, Boca Raton, FL 33487, USA

First published 2007, Woodhead Publishing Limited and CRC Press LLC
© 2007, Woodhead Publishing Limited
The authors have asserted their moral rights.

British Library Cataloguing in Publication Data
A catalogue record for this book is available from the British Library.

Library of Congress Cataloging-in-Publication Data
A catalog record for this book is available from the Library of Congress.

Woodhead Publishing Limited ISBN-13: 978-1-84569-009-0 (book)
Woodhead Publishing Limited ISBN-10: 1-84569-009-5 (book)
Woodhead Publishing Limited ISBN-13: 978-1-84569-250-6 (e-book)
Woodhead Publishing Limited ISBN-10: 1-84569-250-0 (e-book)
CRC Press ISBN-13: 978-0-8493-9144-6
CRC Press ISBN-10: 0-8493-9144-X
CRC Press order number: WP9144

The publishers' policy is to use permanent paper from mills that operate a sustainable
forestry policy, and which has been manufactured from pulp which is processed using
acid-free and elementary chlorine-free practices. Furthermore, the publishers ensure that
the text paper and cover board used have met acceptable environmental accreditation
standards.

Project managed by Macfarlane Production Services, Dunstable, Bedfordshire, England
(e-mail: macfarl@aol.com)
Typeset by Godiva Publishing Services Ltd, Coventry, West Midlands, England
Printed by TJ International Limited, Padstow, Cornwall, England

Contents

Contributor contact details

(* = main contact)

Editors and Chapter 30

Professor L. J. Frewer
Marketing and Consumer Behaviour
   Group
Wageningen University
Postbus 8130
6700 EW Wageningen
The Netherlands

E-mail: lynn.frewer@wur.nl

and

Professor Hans C. M. van Trijp
Marketing and Consumer Behaviour
   Group
Wageningen University
Postbus 8130
6700 EW Wageningen
The Netherlands

E-mail: hans.vanTrijp@wur.nl

Chapter 1

Professor Paul Rozin
Department of Psychology
University of Pennsylvania
3720 Walnut St
Philadelphia
PA 19104-6241
USA

E-mail: rozin@psych.upenn.edu

Chapter 2

Professor Cees de Graaf
Division of Human Nutrition
Wageningen University
P.O. Box 8129
6700 EV Wageningen
The Netherlands

E-mail: Kees.deGraaf@wur.nl

Chapter 3

Dr Herbert Meiselman
5 Harraden Avenue
Rockport
MA 01966
USA

E-mail:
   Herbert.Meiselman@natick.army.mil
   Herbert.L.Meiselman@us.army.mil

Chapter 4

Professor Egon Peter Köster*
Wildforsterweg 4A
3881 NJ Putten
The Netherlands

E-mail: ep.koster@wxs.nl

Dr Jozina Mojet
Wageningen UR-CICS
Building no. 118
Bornsesteeg 59
6708 PD Wageningen
The Netherlands

E-mail: Jos.Mojet@wur.nl

Chapter 5

Drs J. de Jonge*, E. van Kleef, L. J.
   Frewer and O. Renn
Wageningen University
Marketing and Consumer Behaviour
   Group
Hollandseweg 1
6706 KN Wageningen
The Netherlands

E-mail: Ellen.vanKleef@wur.nl
   lynn.frewer@wur.nl
   Janneke.deJonge@wur.nl
   ortwin.renn@soz.uni-stuttgart.de

Chapter 6

Ynte K. van Dam* and Hans C. M.
   van Trijp
Marketing and Consumer Behaviour
   Group
Wageningen University
PO Box 8130
6700 EW Wageningen
The Netherlands

E-mail: Ynte.vanDam@wur.nl
   hans.vanTrijp@wur.nl

Chapter 7

Professor Klaus G. Grunert
Department of Marketing and
   Statistics
MAPP
Aarhus School of Business
Haslegaardsvej 10
DK-8210 Aarhus V
Denmark

E-mail: klg@asb.dk

Chapter 8

Dr M. Buckley* and Cathal Cowan
Ashtown Food Research Centre,
   Teagasc
Ashtown
Dublin 15
Ireland

E-mail: ccowan@nfc.teagasc.ie
   Cathal.Cowan@teagasc.ie

Dr Mary McCarthy
Department of Food Business &
   Development,
University College Cork,
Ireland

E-mail: m.mccarthy@ucc.ie

Chapter 9

Gerrit Antonides*, Judith R.
    Cornelisse-Vermaat, Johan van
    Ophem*
Department of Social Sciences
Wageningen University
Hollandseweg 1
6706KN Wageningen
The Netherlands

E-mail: Gerrit.Antonides@wur.nl

Henriëtte Maassen van den Brink
'Scholar' Research Center for
    Education and Labor Market
Department of Economics and
    Econometrics
University of Amsterdam
Roetersstraat 11
1018WB Amsterdam
The Netherlands

Chapter 10

Michael Siegrist
University of Zurich
Department of Psychology
Binzmühlestrasse 14/15
CH-8050 Zürich
Switzerland

E-mail: siegrist@sozpsy.unizh.ch

Chapter 11

Professor Christopher Ritson* and
    Dr Elizabeth Oughton
School of Agriculture, Food and Rural
    Development
Faculty of Science, Agriculture and
    Engineering
Newcastle University
NE1 7RU
UK

E-mail: christopher.ritson@ncl.ac.uk
    e.a.oughton@ncl.ac.uk

Chapter 12

Rick Bell
Natick RD&E Center
Behavioral Sciences
Kansas Street
Natick 01760
Massachusetts
USA

E-mail: rickbell@post.harvard.edu

Chapter 13

Dr Einar Risvik*, Marit Rødbotten
    and Dr Nina Veflen Olsen
Matforsk AS
Osloveien 1
N-1430 Ås
Norway

E-mail: einar.risvik@matforsk.no
    marit.rodbotten@matforsk.no
    nina.veflen.olsen@matforsk.no

Chapter 14

Øydis Ueland
Matforsk AS
Osloveien 1
N-1430 Ås
Norway

E-mail: oydis.ueland@matforsk.no

Chapter 15

Sophie Nicklaus* and Sylvie
    Issanchou
Institut National de Recherche
    Agronomique (INRA)
Unité mixte de Recherche INRA/
    ENESAD FLAVIC
(Flaveur, Vision et Comportement du
    Consommateur)
17, rue Sully - BP 86510
21065 Dijon Cedex
France

E-mail:
    Sophie.Nicklaus@dijon.inra.fr
    Sylvie.Issanchou@dijon.inra.fr

Chapter 16

Dr David N Cox
CSIRO Human Nutrition
PO Box 10041
Adelaide BC
SA 5000
Australia

E-mail: David.Cox@csiro.au

Chapter 17

Dr David J Mela
Unilever Food and Health Research
    Institute
PO Box 114
3130 AC Vlaardingen
The Netherlands

E-mail: David.mela@unilever.com

Chapter 18

Dr Liisa Lähteenmäki*, Dr Marika
    Lyly and Dr Nina Urala
VTT Technological Research Centre
    of Finland
P.O. Box 1000
FIN- 02044 VTT
Finland

E-mail: Liisa.Lahteenmaki@vtt.fi
    Marika.Lyly@vtt/fi
    Nina.Urala@vtt.fi

Chapter 19

Dr Rohit Vaidya* and
    Dr Marcia Mogelonsky
HealthFocus International
449 Central Avenue, Suite 205
St. Petersburg
FL 33701
USA

E-mail: info@healthfocus.com (FAO
    Dr Rohit Vaidya)

Chapter 20

M. C. van Putten*, M. F. Schenck,
    Professor B. Gremmen and
    Professor L. J. Frewer
Wageningen University
Postbus 9101
6700 HB Wageningen
The Netherlands
E-mail: Martijn.Schenk@wur.nl
    bart.gremmen@wur.nl
    Margreet.vanPutten@wur.nl
    lynn.frewer@wur.nl

Chapter 21

Dr Elizabeth Redmond* and Professor
    Christopher Griffith
Food Research and Consultancy Unit
Cardiff School of Health Sciences
UWIC
Llandaff Campus
Western Avenue
Cardiff CF5 2YB
UK

E-mail: eredmond@uwic.ac.uk
    cgriffith@uwic.ac.uk

Chapter 22

Professor Annie S. Anderson
Centre for Public Health Nutrition
    Research
Department of Medicine
University of Dundee
Ninewells Hospital and Medical
    School
Dundee
UK

E-mail: a.s.anderson@dundee.ac.uk

Chapter 23

Dr Unni Kjaernes* and Lotte Holm
Postboks 4682
Nydalen
0405 Oslo
Norway

E-mail: unni.kjarnes@sifo.no
    loho@kvl.dk

Chapter 24

Marion Dreyer* and Ortwin Renn
Dr Marion Dreyer
DIALOGIK gGmbH
Seidenstr. 36
70174 Stuttgart
Germany

E-mail: dreyer@dialogik-expert.de

Chapter 25

Dr David Coles
Diedenweg 20b
6703GW Wageningen
The Netherlands
E-mail: david.coles@chello.nl
    david.coles@wur.nl

Chapter 26

Professor Johannes Brug
Erasmus MC
Department of Public Health
PO Box 1738
3000 DR Rotterdam
The Netherlands

E-mail: j.brug@erasmusmc.nl

Chapter 27

Dr Gene Rowe
Senior Research Scientist
Institute of Food Research
Norwich Research Park
Colney
Norwich NR4 7UA
UK

E-mail: gene.rowe@bbsrc.ac.uk

Chapter 28

Prof.dr. Frans W.A. Brom*, Tatjana
    Visak and. Franck Meijboom
Ethics of Life Sciences
Animal Sciences Group
Wageningen University
P.O. Box 387
NL-6700 AH Wageningen
The Netherlands

E-mail: Frans.Brom@wur.nl

Chapter 29

Professor Michiel Korthals
Professor Applied Philosophy
Social Sciences
Wageningen University
Hollandseweg 1
6706 KN Wageningen
The Netherlands

E-mail: Michiel.Korthals@wur.nl

Preface

In recent decades, research into, and knowledge about, understanding consumers of food products has grown and considerably expanded its scope. The field originated in the area of sensory science, where the aim was to identify factors which would inform food technologists and product development departments regarding quality control and (sensory) product optimisation. Today the field has expanded beyond a straightforward product focus to incorporate investigation of the key determinants of consumer decision-making and purchase behaviour in the market place. As a consequence, the field has incorporated other elements of relevance to consumer decision-making, including cognitive aspects of consumer behaviour, such as risk perceptions and consumer attitudes. Recently there has been increased emphasis on understanding the role of context, culture and environment, and how these interrelate to consumer behaviour. Research into the food consumer has developed from a focus on experience attributes (such as taste and flavour), to inclusion of usage aspects (such as convenience) and hygiene and credence attributes (such as safety and health). From this increased emphasis on external validity and market relevance, the study of understanding food consumer behaviour today has increased relevance to marketing and public policy. This is a response to recognition that their strategies are aligned to the requirements and preferences of the consumer and/ or citizen. At the heart of all these applications and sub-fields is fundamental understanding of the behaviour of food consumers. The field has also matured to become much more multi-disciplinary in nature where scientific disciplines such as psychology, economics, sensory science, sociology, marketing, public health, ethics and many others provide unique and complementary perspectives on food consumer behaviour. Increasingly, we also see collaboration between the natural and social sciences in understanding how food products, individual preferences, values and attitudes, and context and culture interact to produce specific food choices and behaviours.

The current volume reflects the state of the art in understanding consumers of food products. Part I of the book covers key influences of consumer food choices. Part II reflects the increasing importance of lifestyle choices of consumers which are grounded in cognitive beliefs and how lifestyles fit to particular food choices in addition to the intrinsic qualities of products *per se*. Part III builds on the increasing individualisation and globalisation of food choices, and the importance of understanding diversity in food choice patterns. Public health concerns, related to food and health, and how people respond to potential risk, are covered in Part IV, which focuses on the role of understanding the food consumer, and how choices relate to personal and public health. In today's society, it is important that public policy is grounded in good understanding of consumer behaviour, in order to safeguard societal priorities such as health problems associated with inappropriate health choices, or sustainable production. The need to understand societal issues is reflected in Part V of the book.

Part I is devoted to key influences on consumer food choice. After a general introduction by Paul Rozin, the role of the environment and eating context on food choices is discussed by Meiselman. De Graaf focuses on the important role of sensory qualities in determining food preferences and choice as well as how they may change according to context and external factors. Some of the less mainstream theories of food choice development are discussed by Köster and Mojet, who plead for more attention for non-cognitive and non-conscious effects in consumer choice behaviour. Ultimately, consumers in Western societies enjoy a great degree of freedom of choice and ultimately this choice will depend on how they trade off costs and benefits as part of their decision-making processes. However, many of the benefits remain largely hidden to the consumer at the moment of purchase and consumption, as in the case of food safety and health benefit associated with specific food choices. Perceptions of risk, benefit and trust are becoming increasingly important determinants of consumer choice, and the role of these factors is discussed by De Jonge, Van Kleef, Frewer and Renn in relation to food safety.

Part II of the book reflects the broader scope being developed within the field, and covers several of the fundamentals of consumer behaviour in relation to marketing activities within specific areas of application. Van Dam and Van Trijp discuss the fundamentals of branding and labelling, and how consumers understand the information these provide. Grunert discusses key aspects of food quality perception and their implications for product design. Other chapters deal with how consumers decide to outsource meal preparation, or not (Cornelisse-Vermaat, Antonides, Van Ophem and Maasen van den Brink) and consumer behaviour for specific subcategories of food products. These include consumer responses to new technologies such as those which are innovative and incorporate new technology (Siegrist), convenience foods (Buckley, Cowan and McCarthy) and organic foods (Ritson and Oughton). Many of these application areas have in common that their acceptance depends on consumer attitudes toward the issues in addition to the quality of the products being delivered.

Although it is easy to discuss understanding *the* food consumer, and many of the insights regarding key influences on consumer behaviour are generic in nature, it is important to realise that actual consumer behaviour is characterised by a high level of diversity. This is reflected in Part III of the book. Bell's chapter discusses the diversity of food choices as a result of individual life experiences and how these interact with socio-demographic variables, with a specific emphasis on the US market. Risvik, Rødbotten and Veflen Olsen discuss how and why food choices differ across Europe as a result of cross-cultural variability and Cox focuses specifically on understanding of Asian food consumers and food choices. Nicklaus and Issanchou consider issues associated with children and food choice, whereas Ueland discusses gender differences in relation to the same topic.

Part IV of the book reflects on the emerging importance of consumer understanding of food choices in the specific context of consumer health, and the relevance of research in this area to both public and commercial policies. The chapter by Vaidya and Mogelonsky presents the results of a commercial study on consumer attitudes towards health and nutrition, which aims to identify similarities and differences in attitudes worldwide. Mela addresses the importance of consumer beliefs, attitudes and understanding of the relevant health issues in the context of obesity, and emphasises the relative importance of consumer liking, wanting and eating different foods. Lähteenmäki, Lyly and Urala review the existing literature on consumer attitudes toward functional foods. Following on from the challenges posed in these chapters, Anderson focuses on mechanisms and approaches to change unhealthy food choices. Two other chapters focus on specific elements of consumer health and food choices. Van Putten, Schenk, Gremmen and Frewer review the state of the art in communication about food allergies, including problems associated with existing labelling practices, and issues associated with the adoption of novel low allergenic food products. Domestic food hygiene practices and food safety are discussed by Redmond and Griffith, where the focus is on improving domestic hygiene as potential determinants of food-related public health.

A chapter by Kjaerness and Holm on the social factors in food choice looking at the role of social practices introduces Part V of the book. Part V emphasises public policy and how societal objectives and preferences can be explicitly incorporated into the process of policy development. Dreyer and Renn discuss how this might be operationalised in the context of food safety policy development, while Brug and Wammes discuss the same issues in the context of public health policy. Rowe provides an overview of the role of public engagement in food policy, and how this might be effectively operationalised. The chapter by Coles discusses the interface between science, society and food policy in the context of EU regulatory practices. Brom, Visak and Meijboom describe how consumer values relate to consumer responsibility for ethical trading practices and ethical products. Korthals addresses the issue of ethics in food production and consumption and its implications for food product development and public policy. The final chapter looks to the future and outlines the challenges for food regulators and the new technologies to be taken account of.

Overall, by presenting a wide diversity of disciplinary approaches and applications to existing and emerging issues and problems related to food choices, this book provides a rich overview of the state of the art in understanding of food consumers. We sincerely hope that this effort may stimulate further scientific research activity into this domain both within and across the various disciplines which all have an important contribution to make to understanding the consumers of food products,

Lynn Frewer
Hans van Trijp

Part I

Key influences on consumer food choice

1

Food choice: an introduction

P. Rozin, University of Pennsylvania, USA

For the readers of this book, or even their acquaintances, human food choice probably brings to mind something like the following scenario: a person is faced with some foods, and has the option of trying them and choosing one or more to consume. In fact, that is a rare event in our species. A more generous set of images of food choice might be the following:

1. standing in front of an array of products in a supermarket, grocery store, or more traditional market
2. sitting at a restaurant and looking over the menu choices.

Note that in neither of these situations is a person actually *tasting* foods and deciding which one is to be selected. Tasting and choosing occurs in two other situations:

3. eating at home, if there is an array of food, one might taste a few of the options, and decide to consume more of some rather than others
4. in the laboratory or in communal food tastings, an individual may try two or more versions of a product, and report which one tastes best.

All four of these cases involve food choice. However, most food choice, and almost all food intake in the world occurs in situations other than these. Minimal amounts of food are eaten while one is shopping (1) or perusing a menu (2). At most home meals in the world (3), there is virtually no choice on the occasion about what is to be consumed. Laboratory food choices or communal tastings (4) are very rare events around the globe. In short, our vignettes of food choice are the tip of a food choice 'iceberg,' and an even smaller part of all of the occasions in which food is consumed in the world. We must remember that less than 20% of the people in the world live in what are often called 'developed' countries. Many people in the world, especially in less developed countries, grow or raise some of their own food. Eating at restaurants is a relatively rare event in the less developed world. And home meals in the less developed world do not usually include choices.

Even in the developed world, many situations in which food is consumed involve choice only in the sense that there is an option of whether to eat or not. One of the major aspects of food choice has to do with the number of alternatives. At the limits, there is one, with the choice of whether to eat/purchase or not, and at the other extreme are the 50+ flavors in some ice cream parlors, or the 50+ varieties of a particular type of product available in some supermarkets.

Consumption and purchase provide the two major frames that encompass most cases of explicit food choice. Purchase can be in a market or restaurant. In the purchase situation, choice is usually made without direct experience of the products. It may well be that true choice (more than one alternative) occurs more frequently in the purchase than the consumption situation.

Food choice presumes some sort of temporal and spatial unit (Rozin and Tuorila, 1993; Meiselman, 1996). The normal or basic unit is the dish or serving, as when we decide whether to have string beans or broccoli, a medium rare or well-done steak, tea or coffee. But in restaurants we often choose platters (a specified meat dish, with two designated vegetables), and we sometimes even choose a whole meal combination. The meal can be considered an alternative to the dish as the fundamental unit of consumption and food choice (Meiselman, 2000; Pliner and Rozin, 2000). Rarely do we actually choose which bite to take, as in a tasting or when offered a wide variety of foods. More critically, in the developed world, one might make 'indulgent' choices on weekends or special occasions, to be 'compensated' or 'neutralized' by more 'prudent' choices for the normal weekday meals (Sobal *et al.*, 2006). For the most part, food choices are made of what we will call dishes, and that is the framework we will presume for much of the treatment of choice in this chapter.

1.1   Intake versus preference

Our food choices, on the spot, as it were, are one of the determinants of what and how much food is eaten by our species. From the perspectives of economics, health, and commercial interests, the major question is, 'who eats what, and how much of it?' The relevant data come in a form such as 'the average Irish person consumes x kg of potatoes in one year.' A complete dietary survey, or a determination of all human food consumed by a group has almost all of the important economic information, and most of the health information. Worldwide, we find that by weight, rice and wheat are the two most widely consumed foods. Important as this information is, it raises few questions for the behavioral scientist; it says people eat what is available, what they have traditionally eaten, and what they can afford.

These are measures of intake. They are very relevant to health, and obviously of direct relevance to the study of obesity. Behavioral scientists are often more interested in preference, the choice of A rather than B when both are available. It is natural to presume, in the present context, that A and B are foods, like broccoli and asparagus, and that the choice has to do with which one to eat. But for some purposes, it is worth considering a broader frame for the preference for A over B. Suppose A is broccoli, and B is watching a favorite television program, or a two minute massage, or a bottle of body lotion. People often make this type of choice, but for convenience, we virtually always frame choices as between comparable entities, in this case, between things to eat.

1.2   Motivations, frames of reference, and the psychological categorization of potential foods

There is yet another limitation to what we might call the frame of reference of preference. Consider the choice between eating broccoli or paper. Now note that the answer here would be easy and virtually universal, but reversed if we were choosing what to write on, rather than what to eat. So, in framing preferences about food, we assume that the choices are among available and generally acceptable foods in the context in which eating is the issue at hand. But the paper example is instructive, because prior to making a judgment about what to eat, in a restaurant, or at home, we have already circumscribed the relevant domain of entities to what we might call 'edibles.' Now this *is* psychologically interesting, because it isn't obvious how any human being (or any animal) makes the fundamental distinction between the edible and the inedible. Human infants will put anything that fits into their mouth, and will swallow a wide range of things that adults would not consume (e.g., small coins, paper balls, feces) (Rozin *et al.*, 1986). So, in a sense, the first and most basic categorization that a human or other generalist animal makes, perhaps the most important categorization in early life, is what is potentially edible and what is not. We don't know how animals make this critical discovery. For humans it is easier, because the information is transmitted by parents and others, as a form of cultural wisdom. Paper, coins, wood and feces are just not food. The great majority of things we encounter in the modern world are inedible. We reject them as food not because we know they will taste bad, nor because we think they will harm us, but rather on ideational grounds; they are just not food. We can call this the fundamental distinction in food choice.

Our analysis of preferences led us to believe that there were three types of reasons for preferences and aversions (Rozin and Fallon, 1980). One was sensory/hedonic, most frequently, a reason based on flavor or texture, though sometimes on appearance. We classified foods that were accepted on these grounds as 'good tastes,' and those rejected on the same ground as 'distastes' (see Table 1.1). A clear example of a 'good taste' for Americans is chocolate. It is consumed almost entirely because of the sensory experience it provides. Common examples of 'distastes' for Americans include the frequent rejections by some of foods like Brussels sprouts, beer, anchovies or lima beans. A second

Table 1.1   Psychology taxonomy of food acceptance and rejection

	Sensory hedonic/ reasons	Anticipated consequences/ instrumental reasons	Ideational reasons	Examples
Rejection				
Distastes	+			Lima beans, beer
Dangers		+		Allergy foods, high fat foods
Inappropriates			+	Paper, sand
Disgust	(+)	(+)	+	Insects, snakes
Acceptance				
Good tastes	+			Chocolate
Beneficials		+		Health foods
Appropriates			+	?
Transvalued	(+)	(+)	+	Prasad

Source: modified and elaborated from Rozin and Fallon (1980).

reason for preference or aversion has to do with anticipated consequences. These often relate to health, but can also involve convenience or positive or negative relatively rapid postingestional consequences that may have minimal health influences. The category of food rejections based on anticipated consequences is particularly clear: these would be foods rejected because one might have an allergy to them, or based on beliefs about toxic effects of a particular food, or in a society in which obesity is frowned upon, foods that are high in fat and/or calories. We designated this category of foods as 'dangerous.' They are often appealing on sensory hedonic grounds, as for example, chocolate or ice cream for dieters. The opposite of 'dangerous' foods are 'beneficial foods,' consumed primarily because they are thought to have good consequences. Such foods are often appealing on sensory grounds, but for some, at least, things like whole wheat bread and many vegetables are accepted primarily because they are perceived to be healthy. Of course edible medicines, such as antacid pills, are the purest example of beneficials, items consumed only because of their anticipated consequences (Table 1.1).

The third grounds for accepting or rejecting a potential food is ideational. Something we know about the nature or origin of an entity determines whether it is edible or not. The great mass of things we deem inedible are just 'not food.' Things like paper, rocks, and metal. We do not reject them because they taste bad (we usually don't know) or because, in small amounts, they might be harmful. We call most things in the world 'inappropriate,' as food (Table 1.1). Some actual foods are also deemed 'inappropriate' for sociopolitical reasons, as for example, the meat of a species threatened with extinction, or an import from a disliked country.

There is a particularly interesting fourth category (after inappropriate, distasteful, and dangerous foods) of food rejection, which we have labeled 'disgust' (Rozin and Fallon, 1987). These involve an ideational rejection, like inappropriates. However, unlike inappropriates, disgusting foods are usually highly nutritive, and are responded to as if they are distasteful and dangerous. Thus, worms or caterpillars, for example, are potentially nutritious, and probably non-toxic, but people (in cultures that do not eat them) are inclined to think that they are both unhealthy and distasteful. Feces is the universal disgust (Rozin and Fallon, 1987). The category varies greatly across culture, but almost all of the entities that fall into it are of animal origin. Disgusting foods have the power of contamination; if they touch an otherwise edible food, they tend to render it disgusting and inedible.

Examining the taxonomy described above, and illustrated in Table 1.1, there are two 'open cells,' constituting the opposite, on the acceptance side, of 'inappropriate' and 'disgusting.' There are very few foods that fit uniquely into the 'appropriate' category. That is, all edibles are deemed 'appropriate' but are then either accepted or rejected principally on sensory or instrumental grounds into either of these categories within any culture/cuisine. The opposite of disgust, which we call 'transvalued' foods, are foods that are principally appealing, and elating, because of their nature. This ideational property imbues them with expected good taste and beneficial effects. An example might be prasad in Hindu India; food is given to priests in the temple, they remove some for the Gods, and return the rest. The returned food is blessed and thought to be superior on account of the sharing with the Gods. There are other occasional examples, such as perhaps the taking of the Host in the Catholic Mass.

Of course, many foods are consumed or rejected for multiple reasons; for many, milk is both beneficial and has a good taste. And some foods are conflicted, as for example, for a dieter, chocolate may be both a good taste and a danger.

Almost the entire psychology of food preference is based on foods accepted or rejected on some combination of sensory/hedonic and/or anticipated consequence reasons. The choice set rarely includes ideational options. This makes practical sense. However, there are exceptions, as when a person rejects sea bass because it is an endangered species, rejects beef on moral grounds, or prefers organic foods on the grounds that natural entities are inherently better.

1.3   Preference versus liking

Psychologically, there is a big difference between foods where acceptance/ rejection is based on anticipated consequences, and those for which the basis is primarily sensory/hedonic. In terms of psychological mechanism, it is much easier to explain a food avoidance based on fear of toxic effects ... of course it is avoided. And similarly, ingestion of a food or medicine because of anticipated positive consequences is also a straightforward matter of motivation or belief to action. To be sure, sometimes beliefs are incorrect, and sometimes there are impediments to action (like cost or bad taste), but the situation is still rather straightforward. But the situation is different for sensory/hedonic preferences, or what we could call 'likes.' It is again extremely clear why someone would accept something she likes, and reject something she dislikes. The puzzle for psychology is why some things are liked, and others disliked. Except for the case of rather few though important innate likes and dislikes, the problem is that we don't know much about how some things come to be liked and others disliked. Just as preference is a major but not total cause of intake, liking is a major but not exclusive cause of preference.

1.4   The opportunities for choice

As suggested in the introduction, most food 'choice' is actually made before the moment in which a person faces foods. The choices offered are invariably a small subset of all possible food choices, although the modern supermarket offers tens of thousands of options. What determines the 'choice set' is a set of prior contingencies and choices. First, it is generally the case that the choice is not usually between food and something else, such as a movie or making love. There are culturally designated occasions (such as meals, certain types of celebrations or other ritual events) at which one is offered food, possibly a choice of foods. If no choice is available, there is nominally the choice of whether to eat or not, but the not-eating choice is rarely opted for (except for some food phobic toddlers in the developed world, who may eat only peanut butter, milk and cookies). In the United States, what one is offered (or even what one considers eating) at 8am is quite a different set of foods, called breakfast foods, than what one is offered at 7pm, for dinner. The determinants of the choice set are principally economic factors (such as cost), geographical/ecological factors (what is available locally), and culturally-based practices, beliefs and attitudes, which determine the type of raw foods available, and the ways that they are combined into edible entities (that is, cuisine). Technology plays a large role, in producing palatable and inexpensive foods, such as frozen foods and canned goods, complex condiments, etc. Technology also allows for transporting foods from point of origin to consumer, often via intermediary food processors. At its simplest, this allows urban dwellers to consume fresh local farm products, and at its most complex, it allows these same urban dwellers to eat foods originating anywhere in the world. And, advances in agriculture and animal husbandry allow for a wide variety of inexpensive and often high quality foods to be made available.

We can refer to all of this critical background for food choice as the 'food system.' It is the hidden leviathan in food choice. It probably accounts for most of the 'variance' in what different people eat. Within culture variance in food choice is probably much higher in developed countries, where there are many options and where the cuisines of the world may mix, as opposed to a traditional culture situation, in which the set of food choices is narrower, composed of a set of dishes defined by the local cuisine.

1.5   Situating the person in the food choice situation: temporal perspectives

The context for food choice can conveniently be divided into simultaneous events (e.g., mood, physical and social surroundings) and events that either precede or are anticipated to follow a particular choice (Rozin and Tuorila, 1993). The temporal dimension, from past to present to future frames all preferences and choices (see Fig. 1.1 on page 18). This dimension is made explicit and important in the recent work of Daniel Kahneman (Kahneman *et al.*, 1997; Kahneman, 2000). There are three senses or dimensions of pleasure ('utility') described as experienced, remembered and anticipated. Experienced is pleasure as it is happening, remembered is the representation of that experience after the fact, and anticipated, refers to looking forward to some sort of food event. The important point made by Kahneman and his colleagues, coming largely from research on pain, is that the mapping of any of these on to any other is very complex. There are major differences between experienced and remembered pleasure, and major differences between anticipated and experienced pleasure, and these differences or if you will, distortions, have some lawful features. Experiences (at least of pain) are remembered principally in terms of their peaks and ends. As well, duration of episodes of pleasure, especially if these are rather uniform in their hedonic properties, are not represented in memory. A longer or shorter episode of rather uniform pleasure makes the same contribution to the total experience. This is called 'duration neglect.'

In the domain of anticipated pleasure, the basic finding is that people are very poor at predicting their own hedonic trajectories (Kahneman and Snell, 1992); that is, they cannot predict if and in what way their liking for an object will change on repeated experiences. In general, people seem to minimize the effect of adaptation or habituation, and presume that their current response to a new entity will accurately portray their future response. We now apply these important ideas to the domain of food and food choice.

When a person is facing a considered choice, either for purchase or consumption, there will almost always be some comparison of mental representations of the choices. In terms of the judgment about sensory pleasure, often the principal determinant of choice, it is rare that the chooser can actually sample the possibilities directly. Rather, the chooser has, principally, either verbal information (e.g., choice of coffee or tea), or visual information (the appearance of the food itself), or sometimes, olfactory or tactile information. If the foods in question are familiar, which is usually the case, the chooser has available his or her memories of past experiences (tastes) of the foods in question. In this case, it is largely reference to these memories that determines the choice. Insofar as the choice involves a sequence of experiences, such as a meal or a complex dish, the principles of peak, end and duration neglect might well hold in the food domain.

This has not been systematically investigated, but the only direct test of these ideas, in the context of a meal, suggests minimal evidence for peak or end effects, but strong evidence for duration neglect (Rode *et al.*, 2006). That is a subject's hedonic ratings from memory of two meals is identical, even though in one of the otherwise identical meals, the subject consumed twice as much of his or her favorite food. In the future, with further research on these processes, models of food choice at the moment will have to take into account the features of the memory representations of the foods that make up the choice set.

The issue of anticipated pleasure is extremely important in the domain of food for two reasons. First, people often make a committed food choice (for item and amount) before the act of consumption (e.g., by purchase or serving themselves a portion). Second, in the purchase situation, people often buy a large amount of a food, to be consumed over days or even months (as in buying a large box of breakfast cereal or a bottle of chocolate syrup or fruit preserves). Such an act presumes that the pleasure of consumption of a particular product will be maintained. If the product is familiar, and has been consumed hundreds or thousands of times, the person is probably at an hedonic asymptote. However, if it is new or relatively unfamiliar, errors in anticipatory judgments can be significant. This was shown initially by Kahneman and Snell (1992), who demonstrated that after sampling a new variety of yogurt or ice cream, people were unable to accurately predict how their liking would change after sampling it daily for a week. These findings were extended by Rozin et al. (2006b), to show the same inability for predicting the hedonic trajectory for two unfamiliar snack foods, and two unfamiliar body products, over the same period of sampling daily for one week. As well, these authors showed that older subjects (the sample consisted of college students and their parents) were no better at this than younger subjects, even though they had 20 or 30 more years to observe their own hedonic trajectories.

Memories of the recent past affect current choices in other ways. These memories may be explicit, that is conscious, or implicit. Barbara Rolls (2000) and her collaborators have documented sensory-specific satiety. The liking for a specific food declines within a session, after repeated sampling. This decline is usually greater for the specific food that has been repeatedly experienced, in contrast to other available foods. We don't know whether sensory specific satiety depends on explicit or implicit memory of the recent past, or some combination thereof.

It is certainly true that explicit recent memories may influence current intake, as if a person says he or she has had enough of X in a meal, and opts for Y. More dramatically, amnesics, who have no explicit memory for events that occurred more than a minute or so ago, will eat two or three consecutive meals (Rozin et al., 1998). When normals are offered a second meal, after the first has been cleared away, they typically say 'I just ate.' They are invoking a cultural norm which depends on explicit recent memory.

1.6   Biological, cultural, and psychological (individual) influences

It is fundamental to understand the origin (in evolutionary history or development) of the various forces that shape the person, the food, and the environment. These origins can be roughly categorized as biological (innate), cultural, or individual/psychological.

1.6.1 Optimal foraging

The environment in which humans accomplished almost all of their evolution is, of course, natural and 'biological,' in that it was minimally shaped by humans. Like all other animals, humans faced and still face the requirement of getting sufficient energy from food to maintain bodily functions, and to support survival activities. Most animals have some internal system that indicates a need for energy, expressed as what we call hunger in humans. Ironically, for many species, the most energy expending activity is the search for and capture of food. The more effort spent in food search and capture, the more food is required. In addition, food search and capture are activities which increase the risk of becoming prey to predators. Therefore, it is adaptive to spend as little energy as possible in the process of obtaining the food necessary for survival.

There is abundant evidence, from animals in many different phyla, that there is an exquisite system that discovers the way to get the maximum energy as food with minimal energy expenditure (Elner and Hughes, 1978; Krebs and Davies, 1984). This goes under the name of optimal foraging. In contemporary humans, this is expressed in terms of the importance of convenience in determining food choice. Many of the advances humans have made in cultural evolution (from tools, plant and animal domestication, to fast food and microwave ovens) were motivated by reducing the effort needed to obtain food. We are appropriately lazy and try to minimize the amount of energy we expend, in general, and in particular, the energy we spend in procuring, preparing, and consuming food.

1.6.2 Biological determinants of food choice in generalist animals

As generalist animals, humans face the problem already described, of distinguishing edible from inedible potential foods, avoiding toxins, and getting a balanced diet. On sensory grounds, it is very difficult, if not impossible, to detect all toxins and the presence of harmful microorganisms, as well as the presence or absence of a balanced set of nutrients. Unlike the case for specialist animals, it is not possible to evolve sense organs that can reliably detect the subset of entities that constitute food. As a result, for most generalist animals, and certainly for humans, food selection relies largely on the effects of experience (Mayr, 1974; Rozin and Schulkin, 1990). However, humans and many other generalists (including the common wild rat, *Rattus norvegicus*, the generalist most studied in the laboratory), do have certain biological predispositions that guide adaptive food selection.

- The need for food and water is signaled internally and innately by what we call hunger and thirst, and there exist systems for the regulation of food and water intake.

- There are a small number of innate taste biases that direct us towards entities that are likely to be foods, and energy sources in particular, and steer us away from entities that are likely to be toxic. There is good evidence for innate human preferences for sweet and avoidance of bitter (Steiner, 1979). It is probably also the case that there is an innate preference for fatty textures, associated in nature with the presence of a rich source of calories, and the existence of receptors that detect the presence of sodium ion is undoubtedly linked to the special importance of that mineral. Interestingly, in contrast to these biases in the taste system, it appears that there are minimal if any innate approach or avoidance biases in the olfactory system (Bartoshuk, 1990).
- There is a well studied and highly specialized system designed to evaluate potential foods in terms of their post-ingestive consequences. The system can learn the connection between foods and their consequences, even though these events are often separated by hours. The system is particularly effective in supporting the rejection of some categories of toxic foods (called conditioned taste aversions) but also effectively connects foods to their positive post-ingestive effects (Booth, 1982; Booth *et al.*, 1982; Garcia *et al.*, 1974; Rozin and Kalat, 1971; Sclafani, 1999).
- New foods offer both the possibility of new sources of nutrients and the possibility of toxins, harmful microorganisms, or nutritional imbalance. There is a delicate balance between interest in new foods and fear of them (Barnett, 1956), which I have described as the generalists' dilemma. The dimension of 'neophobia' is basic in both understanding humans as generalists, and as an individual difference variable across humans (Pliner and Hobden, 1992; van Trijp and Steenkamp, 1992).

1.6.3   Cultural determinants of food choice

There is a massive 'behind the scenes' effect of culture on food choice. All of the technologies that change the nature and availability of foods influence the opportunities for choice. All the accumulated wisdom and lore about what is edible and what is appropriate to eat are transmitted (as are the technologies) from one generation to the next. Culture-specific cuisines dictate what is to be eaten and in what forms and combinations. Elisabeth Rozin (1982) analyzes the food-specific part of cuisines into three basic components: the basic foods (e.g., rice for China), the characteristic flavors placed on the foods (which she calls flavor principles, e.g. soy sauce, rice wine and ginger for China), and typical modes of preparation of the foods (e.g., stir-frying for China). In a broader sense, we can include a variety of food-related practices as part of cuisine. These form the context of the meal. Schutz (1989) uses the appropriate term 'appropriatness' to encompass a whole set of culture-specific practices about what foods are served with or mixed with others, and about what foods are served when. For example, for Americans, ham is often served with something sweet, like pineapple, but beef is rarely served in such a manner. Frankfurters are often served with baked beans.

Cultures suggest or may require what is to be eaten when, and also how much. These can be called consumption norms. They are a major influence on portion size (Wansink, 2004). Consumption norms may vary with age, gender, and occasion, but there may also be some general norms; for example, at least in American culture, there is a norm that it is appropriate to eat no more nor less than *one* of many entities of moderate size; one bowl-full, one apple, one sandwich, one chicken breast. This norm may account for the fact that people eat less when given smaller portions, rather than correcting for this by eating extra portions (Geier *et al.*, 2006).

Cultures dictate food occasions. In many cultures, including many of the countries of Europe, the USA and Canada, breakfast is a distinct meal with specific foods (such as cereal). Special feast days involve more elaborate preparations, and sometimes specific dishes, such as the Thanksgiving turkey in the United States.

Cultures also dictate the importance and significance of food in life, more generally. The French are known to consider food particularly important. In India, food assumes a major social role, in that the meal enacts the social relations among family members, in terms of who eats with whom and in what order (Khare, 1976; Appadurai, 1981). This is greatly elaborated at events such as wedding celebrations. As well, in Hindu India, the caste structure is, in large part, maintained and enacted by food sharing rules which essentially prohibit consumption of food prepared by people of a lower caste than the person who will be the consumer (Marriott, 1968).

Finally, cultures dictate the manner of eating, what might be broadly called table manners. Appropriate behavior at the table, type of implements used (e.g., fingers, forks, or chopsticks). These practices have major social implications, but they also affect the experience of eating, itself.

1.6.4   Psychological (individual experience) determinants of food choice

Within the frame of the predispositions coming from genetically based universals, and culture-based rules, attitudes and practices, individuals develop particular sets of food-related beliefs, preferences, and practices. These have been aptly termed 'personal food systems' by Sobal and his colleagues (Sobal *et al.*, 2006; Furst *et al.*, 1996; Connors *et al.*, 2001). Individuals across cultures share the universal, genetically determined features of food choice, such as preferences for sweet and aversion to bitter. As the child grows up, its tendency to consume anything that goes in the mouth and does not have an innately aversive taste (Rozin *et al.*, 1986) is shaped and refined, by experience and instruction, so that the culture-wide practices are communicated and inter-nalized. But there are wide differences (at least in developed cultures, where they have been studied) in individual approaches to food and food preferences. Within-culture variation is due, to a small degree, to genetic differences in temperament and sensory function, but for the most part, it seems to be a result of the specific experience a child has with family, peers, and other cultural influences. Wide variation in aspects such as receptiveness to new foods (Pliner and Hobden, 1992) and preferences for particular foods (e.g., broccoli or lima beans for Americans) has been documented, as well as major attitudinal and behavioral differences expressed in terms of practices such as dieting or vegetarianism.

We know surprisingly little about how specific preferences or more general food attitudes are acquired. For the domain of liking for foods, there is evidence that exposure to a food is likely to produce increased liking (Zajonc, 1968; Pliner, 1982; Birch and Marlin, 1982). Through a variant of Pavlovian conditioning called evaluative conditioning, the consequences or contingent associates of consuming a food can influence liking for a food (deHouwer *et al.*, 2001; Rozin and Zellner, 1985). Taste aversion learning (conditioned taste aversions) are an example of this (Pelchat and Rozin, 1982). A third source of influence is social. Significant adults own likings as well as those of peers influence children (Birch *et al.*, 1980), and through evaluative conditioning, pairing of a food with a positive social signal (such as enjoyment or distress at eating a food) can transmit likings (Baeyens *et al.*, 1996). However, the powerful domain of social influence is, as yet, little understood. It is surely involved in the dramatic reversals of innate aversions that occur within specific cultures, as for example, the widespread liking for the irritant, chili pepper in many cultures (Rozin, 1990).

Any individual differences transmitted genetically, or by early experience with food, should reveal themselves in parent-child correlations in food preferences and attitudes. Although it is widely assumed that the early food environment has a powerful effect on later food preferences and attitudes, there is little evidence to support this view. Furthermore, given that milk, the first food of humans and other mammals, is unavailable as a food for adults in the precultural environment, it would be maladaptive for mammals to develop a strong preference for their first foods. What data is available suggest rather weak correlations, in the range of 0 to 0.30 between parent food preferences and those of their adult children (Rozin, 1991). This raises the 'family paradox'; if neither genetics nor parental influence are major influences on the development of preferences, what is the source for them? Candidates include peers as well as person-specific exposure to particular cultural influences.

Person, food, and environment
A number of authors, including Belk (1975), Bell and Meiselman (1995), Meiselman (1996), and Sobal *et al.* (2006) offer a framework for understanding food choice in terms of three components: the person, the product (food), and the environment (where the food products of choice are separated from the rest of the environment). This is a useful framework, so long as we understand that ultimately, the product and environment are filtered through the person: that is, it is the *perceived* product(s) and environment that influence choice. The following discussion uses this tripartite framework, with a focus on the interactions between them, and the fact that since the person ultimately makes the

food choice, the person is the final common path of action of all of the forces. A tentative model of food choice is presented in Figs 1.1(a) and 1.1(b) (see page 18). It is a composite of concepts and models developed by various students of food choice, but owes most to the work of Sobal and his collaborators (e.g., Furst *et al.*, 1996; Connors *et al.*, 2001; Sobal *et al.*, 2006), including their development of a concept of a 'personal food system.' It is important to note at the outset that virtually all models of food choice are about individuals in developed cultures, where tradition and necessity are less important forces, and number of alternatives and opportunities for choice are much higher.

Personal influences
We can divide the personal influences on food choices into those that are stable aspects of the individual, such as traits or demographic characteristics, and local, more immediate personal factors, that is, states as opposed to traits.

1.7   Stable features: demographic characteristics and traits

Although it has not been specifically demonstrated, it seems likely that culture is the most informative demographic feature about a person for prediction of food preferences or attitudes. There is probably no other predictor that compares in power. Age has predictive value if we include young children and the aged, but for the middle adulthood years (say 20–70), age has not proven to be a major predictor. Gender is of some importance, and is associated with modest differences in food preferences. A major exception is the large gender difference in concern about weight, which manifests in dieting and food choice, in the developed world (Rozin *et al.*, 2003a).

There are substantial differences across social class in both food preferences and attitudes. In the United States, it is primarily the upper middle and upper classes that have shifted to health and organic foods.

There are a few documentations of weak relations between general personality traits and food preferences and attitudes. Perhaps the most striking are links between sensation seeking and food neophobia. As well; there are links between general neophobia and food neophobia (Pliner and Hobden, 1992).

Beliefs about the natural world, the relation between diet and health, and trust in government and technology also influence food preferences (Frewer and Salter, 2003; Rozin *et al.*, 2003a; Siegrist, 1999). These beliefs tend to be stable across the lifetime. A striking example is vegetarianism, whether principally focused on its health features or the immorality of consuming animal foods (Amato and Partridge, 1989; Beardsworth and Keil, 1992). This can have a major effect on all aspects of eating, including the social context of eating. Vegetarianism is a moderately stable feature of an individual, but it does wax and wane in many individuals.

1.8   Momentary features: state variables

The obvious state variable that influences food choice is hunger. However, there are also demonstrable effects of mood on food choice (see Rogers, 1996), and in modern Western cultures, guilt about food indulgence, especially in females, can substantially affect choice and intake. A feeling that one has been indulging too much in recent meals, or even in the first part of a current meal, can lead to major changes in behavior, viewed as 'compensations,' in the present. Most critically, there are effects in both directions: food affects mood and mood affects food (Rogers, 1996).

1.8.1   Environmental/situational forces

Environmental or situational influences on food choice, other than properties of the food itself, are powerful determinants of choice. They have not been part of the major thrust of researchers in the field, although some, such as Belk (1975) and Meiselman (1996) have been calling attention to them for a long time. In recent years, the power of environmental influences on food intake and choice has become much more widely recognized (e.g., Bell and Meiselman, 1995; Hill and Peters, 1998; Rozin and Tuorila, 1993; Rozin et al., 2003b; Wansink, 2004). We do not have a good 'taxonomy' of environments, although different important aspects of environments have been discussed by a number of authors (e.g., Belk, 1975; Bell and Meiselman, 1995; Kahn and McAlister, 1997; Meiselman, 1996; Wansink, 2004). Important aspects of the environment that have been shown to influence intake by these authors include matters of location and accessibility (often, literally, the physical closeness to the food), effort to obtain the food, the number of distractions (alternate activities), the social context (both group size, e.g. de Castro, 1990, and the nature and relationships to eating partners e.g., Sobal et al., 2006), the culinary environment (e.g., home or restaurant, or type of restaurant), and the general atmosphere and mood at the site.

Belk (1975) notes that this type of analysis of the environment, into a number of distinct attributes, may not be the most appropriate way to parse it. He points out that a proper analysis of environments might be better accomplished by grouping together not environments that have many physical and social similarities, but rather environments or situations that tend to induce the same types of food choices. Thus, although a hurried breakfast and a hurried dinner at home may have little in common in terms of actual foods consumed, both bring forth eating rapidly, and choice of foods that do not take much preparation time. Belk examines a set of ten possible eating situations, and proposes a clustering of them based on the common behaviors they elicit. For example, the following situations both elicit similar behaviors as reported by his American respondents: 'You are too tired to cook dinner either because you have been cleaning the house all day or you had a very busy day at the office or you had been shopping all day or the children have given you a hectic day' and 'You are wondering what to serve yourself, and the children since your husband is not going to be home for dinner.' Belk factor analyzed the ten particular situations that he chose to explore into four types: variety seeking, entertaining guests, relaxing/informal and picnic. (Of course, any taxonomy of this sort is dependent on the exemplar situations, here only ten. But the point is the method or approach used here.)

1.9   Features of foods that influence food choice: the person–food interface

Although much research on food in relation to choice, often carried out by food producers, studies properties of foods (e.g., sugar content, saturated fat content, particular aspects of appearance and texture), the critical issue for acceptance is not these physical features, but the way they are perceived and interpreted by individuals. For this reason, this section combines the discussion of food properties with the processing of these properties by people. A wide range of research, much carried out and summarized by Bell and Meiselman (1995), Booth (1994), Cardello (1996), Kahn and McAlister, 1997; Pangborn, 1980; Rozin and Tuorila (1993), and Wansink (2004) suggest a number of features of the food–human interface as important influences. These include some features closely associated with food, but not technically properties of the food itself: information about the food that accompanies it (as on a label), salience of the food (including lighting), and packaging and portion size of the food (Wansink, 2004).

Generally, referring to the food itself, it seems that its sensory properties are the most powerful influence on choice, in most situations. These sensory properties include taste, smell, texture (sometimes combined into the term 'flavor') most prominently, but also visual and auditory qualities. These critical sensory features have been studied extensively by Booth (1994), Pangborn (1980) and others, and are well reviewed, in detail, by Cardello (1996). As well, variety of foods and their arrangement is a major influence. Kahn and Wansink (2004) have shown that a greater number of offered variants of a product (e.g., different colors/flavors of jelly beans) induces more intake, but only if the food is presented with each subtype segregated from the others, that is, in an organized presentation. Larger variety does not increase intake when the samples are presented in an unorganized mixture.

Other major influences on food choice include the perceived post-ingestional effects of the food (often summarized as 'health' effects). Sensory and 'health' effects are joined by a third major influence, convenience. People are much more likely to choose foods that require minimal effort investment between the point of selection and the point of consumption, in accord with the general biological bias to minimizing energy output. Convenience can manifest itself in varied ways, including avoiding elaborate preparation (even something such as detaching the sections of the grapefruit before eating it or shelling the peanut), favoring a food that is physically closer (even if the distances may be measured in inches!), or using an implement which maximizes energy efficiency.

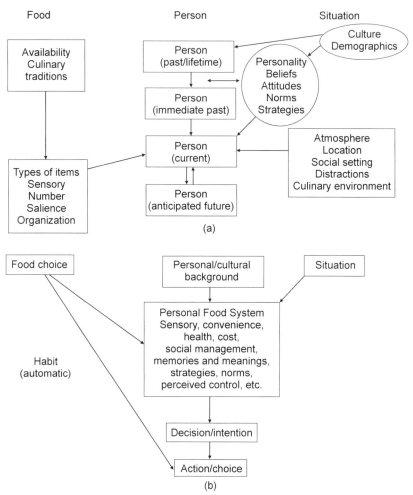

Fig. 1.1   (a) Schematic model of food choice, (b) enlargement of the area designated 'Person (current)' in Fig. 1.1(a), as well as the sequence from 'person' to 'action/choice' (based principally on a model developed by Sobal *et al.* (2006)).

Two other major influences are cost and managing social relations. Cost can be a major or minor influence, depending on both the situation (ordinary or celebration), the available choice set, and the income of the person involved. Managing social relations, brought to the fore by Sobal *et al.* (2006), based on interviews about food choice with a sample of adult Americans, can include many things. Relationships to co-eaters, degree of food sharing, expression of gratitude through food, and, critically, impression management. For example, some dieting American women feel uncomfortable being seen to eat abundant amount of foods or foods of high caloric content. As an example, in a survey of college undergraduates at six American universities, 13.5% of females reported being reluctant to *buy* a chocolate bar (Rozin *et al.*, 2003a).

These five forces (sensory, health, convenience, cost, and social management) are among the principal forces at work, but other influences, often related to beliefs, norms and values, also come to bear. These influences are represented in Figs 1.1(a) and (b), modified from a model presented by Sobal *et al.* (2006).

1.9.1   Food-situation-personal integrations: integrative models of food choice

A summary of integrative models, based primarily on the Personal Food System model (Sobal *et al.*, 2006), is presented in Fig. 1.1.

1.9.2   Attitudes, beliefs and norms to action

An important framing of food choice focuses on the link between attitudes and action. At the moment of choice, a confluence of mental influences, usually triggered by the food choice, arises, and resolves to an intention to make a particular choice, followed by the behavior that accomplishes this choice. The attitude to intention to action sequence, as well as the relations among influences on intentions has been part of the general approach to attitudes in social psychology. A major theoretical position on attitudes and action has been formulated by Ajzen and Fishbein, as the theory of reasoned action (Ajzen and Fishbein, 1980) or the later theory of planned behavior (Ajzen, 1988). This formulation has been systematically extended into the food domain, and elaborated appropriately by Richard Shepherd and his colleagues (Shepherd and Raats, 1996).

The theory of planned behavior indicates how beliefs and norms may interact to produce intentions and then actions. It considers intentions as a frequent mediator between thoughts or feelings and action, and seeks to predict intention and action from a measurement and appropriate weighting of three variables: attitude, subjective norm, and perceived control. Importantly, this approach makes a further distinction. Each of the three determinants are parsed into a belief component and a component that activates or magnifies that belief. Measurements of these two components are multiplied to result in the attitude, subjective norms, or perceived control. Thus (see Fig. 1.2), attitudes are composed of behavioral beliefs and outcome evaluations of the consequences of instantiation of the beliefs, subjective norms into normative beliefs and the motivation to comply with them, and perceived control into control beliefs and power components.

In later research applying the model to food choice, Shepherd and his colleagues (1996) have incorporated a fourth factor, personal identity, as an influence on intentions, using 'green identity' as a contributing component to intention to consume organic vegetables. These authors have also identified a fifth influence, which involves moral factors (Raats *et al.*, 1995). In the model of planned behavior, each of the relevant components (Fig. 1.2) is evaluated by questionnaire, and the linkages of the components are established with

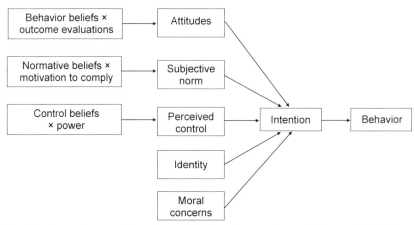

Fig. 1.2   Determinants of intentions and action in accordance with the modified Theory of Planned Action, as developed by Richard Shepherd and his colleagues (Shepherd and Raats, 1996).

correlations. Partial correlations or multiple regressions and path analysis allow for an evaluation of the importance and independence of the various components and pathways, and hence a quantitative model going from cognitions and feelings to action. Shepherd and his colleagues have used this approach successfully to explain a substantial amount of the variance in intention and action in food choice situations, involving such issues as reactions to genetically engineered foods (Sparks *et al.*, 1995). The work of Shepherd and his colleagues has magnified, quantified, and elaborated one segment of the complex total set of events involved in food choice.

1.10   An important and novel approach to the total food choice situation

Russell Belk (1975), a major student of consumer behavior in general, made a novel and insightful attempt to frame and explain food choice in 1975. He employed the person-food-environment triad, and attempted, with question-naires completed by American adults, to get information on the structure of each of these three domains, and the relations between these structures. The approach to discovering a behavioral taxonomy of environments is described above in the discussion of environments. Using a procedure called 'three-mode factor analysis,' factors are simultaneously extracted from data on persons, situations and food choices (responses). In this way, types of responses are related to types of situations, and both are related to types of persons, indicating what types of persons respond in what ways to each of a set of situations. This study is an illustration of a potentially very useful approach. Of course, this was a first study, and like any study, was constrained by the limited sets of responses, environments and persons sampled. But this is a procedure for generating three dimensional matrices of person/environment/food-response interactions, and exemplifies a disciplined approach to integrating influences on food choice.

1.10.1   The closest thing to a total model of food choice

The sociologist Jeffrey Sobal and his colleagues (e.g., Sobal *et al.*, 2006; Connors *et al.*, 2001; Furst *et al.*, 1996) have made the most complete attempt to capture the full context of the process of food choice, generating a model that is appropriately complex. The model is based principally on qualitative data, extended interviews with adult Americans. Figure 1.1(a) and (b), constructed by the present author, is heavily influenced by the Sobal model, and incorporates many of its features. It stands out among models for its richness, and for its emphasis on the role of social factors and social negotiations, and on the activation of strategies for food choice. The general view is that many categories of factors converge on choice, and that both the importance of the categories and the subfactors within them vary across individuals within culture (and of course, although this is not evaluated, they would differ substantially across cultures).

As Sobal *et al.* (2006) note: 'This model assumes that a key process in selecting foods is the construction of food choices based on cognitions and social negotiations. Overall, people are assumed to construct food choices in a variety of ways by actively selecting what, when, where, with whom, and how to eat.' Three very basic components of food choice are designated. The first, life course, essentially means relevant prior experiences. What emerges from this is the second major component, influences. The third component, the personal food system, which are 'are the mental processes whereby people translate influences upon their food choices into how and what they eat in particular situations. Personal food systems represent ways that options, tradeoffs, and boundaries are constructed in the process of making food choices.'

The major influences that contribute to the personal food system are sensory, convenience, instrumental consequences (usually health), cost, and social influences/management, with a significant 'other' category that includes moral issues and norms. This list of influences is similar to the lists of others, but significantly adds and emphasizes issues of social management, that are very important. This not only includes the general effects of the social influence of those with whom one is sharing a meal, but the social influences on shopping for food and preparing the food, customs about sharing (giving and receiving) food, and critically, consideration of the preferences, morals, sensitivities, and habits of one's eating partners.

The Sobal model incorporates factors already discussed, including the physiological state of a person, identities, financial resources, relevant skills, and influences of the eating environment.

What is particularly distinctive of the model is its description, based on interviews, of the way the many influences and factors are integrated into the 'personal food system.' These are described as value negotiations, balancing and integrating, and strategies. For simplicity, I think these important factors can be encompassed under the general term, 'strategies.' Sobal *et al.* (2006) define strategies as: 'the behavioral plans, routines, and rules that people develop for how and what they eat in recurring situations.' They include simplifying rules (heuristics), prioritization of values and influences, and contextual modifications of these prioritizations, such as limiting the food budget or the diet fat content on weekdays, but relaxing such limits on weekends. Notably, the time horizon for 'balancing' one's intake may often be one week. A set of specific strategies, designed to simplify food choices, are described, including: 1. focusing on only one value, 2. routinization, 3. elimination of a specific option, 4. restriction (as opposed to elimination) of a particular option. 5. substitutions, 6. additions, and 7. modifications (e.g., by changing foods). Although there are other strategies, this is the major attempt to describe them, for a Western adult consumer. The study of such strategies and their integration, and differences in them across cultures, is a major agenda for future research.

The inclusion of routinization as a strategy is critical. It no doubt accounts for behavior in many food choice situations, and more than many other strategies, may not even enter into conscious attention. The importance of this strategy is indicated in Fig. 1.1(b) by an arrow that by-passes the considered judgments included in the personal food system, and goes directly from the representation of the food choice to action (labeled as habit in Fig. 1.1(b)). One potentially important strategy, that lies 'between' habit and considered choices involves pre-commitment. That is, an individual makes a decision about how to resolve a choice situation (e.g., what brand of canned peaches to buy) and then follows that decision without further consideration in all future encounters with the relevant choice situation.

Two possible additional strategies that may be important are what I will call framing and coping with variety. Framing is a major concept in decision psychology, and refers to how one contextualizes a situation. For example, Debra Zellner and her colleagues (Zellner *et al.*, 2002, 2003) showed recently that how one categorizes a choice set may affect one's subsequent behavior. For example, standard coffee, beers, or fruit juices are rated as more desirable when they are separated categorically from high quality instances of the same products. When a choice set is categorized with the superordinate category (e.g., all beers), the less desirable category suffers by contrast; when separated, it is less susceptible. Framing is a strategy that is constantly, and often unconsciously applied. Without it, we would be comparing every eating episode to the best meal we ever had, with very negative consequences for day-to-day pleasure.

The explosion of microvarieties of foods (my local supermarket offers about 150 different kinds of yogurt, taking into account brand, flavor, container size, and degree of removal of sugar and fat) presents special problems, since a conscientious choice of each item would take an enormous amount of time, and make a supermarket visit a full-day activity. Barry Schwartz (2004) has addressed this problem as the 'Paradox of choice.' Too much choice can be aversive, and people develop strategies to handle such situations. One is to avoid situations in which there is too much choice, and another is a form of pre-commitment, in which one decides what particular product one wants, and always selects it from the array without considering others.

Of course, with enormous numbers of microvarieties, there is still the problem of *finding* the desired choice. Some individuals characteristically make narrow choices, of both product type and particular brands, but others seem to enjoy the variety, and so we have the important individual difference variables of variety seeking or neophobia (Pliner and Hobden, 1992; van Trijp and Steenkamp, 1992). This is as well a culture variable. Americans, in particular, seem to revel more than others with choices of 100 ice cream flavors, physically heavy restaurant menus that offer enormous numbers of choices, including almost limitless permutations of the side dishes to accompany the main (usually meat) item. And the American local restaurant table is typically decorated with salt, pepper, hot pepper, some type of oil, and various condiments, to allow further choices. Recent data suggests two things: first, even Americans are often overwhelmed and discomforted by choices (Iyengar and Lepper, 2000; Schwartz, 2004), and secondly, large numbers of options are less desirable in other cultures, particularly France and some other European countries (Rozin *et al.*, 2006a). American, perhaps Protestant, ideas of individually optimized foods, to meet the individual's unique preference functions, contrast with more collective food values in countries like France. After all, who should decide on what you eat, you or the chef?

The American food industry may be coming to realize that microvariety may have gone too far, especially since it increases product costs. Kahn and McAlister (1997) note that 'According to Durk Jager, president and COO of Proctor & Gamble: "[In 1996, the] average supermarket has about 31,000 SKUs [stock-keeping units, that is, specific products] and only about 500 – or less than 2% – move a case or more in a week (in an average store). The bottom 7,000 SKUs – almost 23% of the total – move less than one unit a week"' (p. 66).

1.11   Conclusion and future trends

The frightening complexity of food choice has motivated individual researchers to isolate and decontextualize particular aspects of food choice, and study these under controlled laboratory conditions. This is a viable strategy, so long as the decontextualization is done in a sensible way, and it is ultimately integrated back into the real world. Many investigators would now agree that decades of studying human responses to different sucrose concentrations in water was perhaps overdone, especially since preferred sugar levels vary markedly in terms of the context; we like more sugar in lemonade than in stringbeans. Recent research has generally become more context sensitive, more appreciative of the powerful role of the environment, more appreciative of the complexity of food choice, and more understanding that our biological regulation system does not play a dominant role in food choice and intake. Recently some more systematic attention has been paid to culture differences in the food world, with particular reference to the contrast between France and the United States (Stearns, 1997; Rozin *et al.*, 1999; Rozin, 2005).

We have yet to seriously take on the fact that the great majority of the world lives in less developed countries, with very strong cultural, environmental, and economic constraints on eating, and a lot less choice. Progress is being made. It is hard to give up the control of the laboratory to create a meaningful interface with the real world, and to give up the convenience of just working with the people next door, instead of the billions who live further away, but we are getting there.

1.12   Sources of further information

There are a few books that provide broad and integrative approaches to major issues in food choice and what we know about them. I single out three here: *The Psychobiology of Human Food Selection* (Barker, 1982), while old, includes a particularly broad array of edited papers including psychological, biological and cultural influences. Two more recent edited books that provide broad perspectives are *Food Choice, Acceptance and Consumption* (Meiselman and MacFie, 1996) and a book about to be published, *The Psychology of Food Choice* (Shepherd and Raats, 2006) (see also Maurer and Sobal, 1995; Meiselman, 2000; and Murcott, 1983). There are a few non-edited books that provide broad perspectives on the field, including *Psychology of Nutrition* (Booth, 1994), *The Psychology of Eating and Drinking* (Logue, 2004), *L'Homnivore* (Fischler, 1990), and *Sociology on the Menu* (Beardsworth and Keil, 1995). Two major, exhaustive, multiple volume reference books have appeared recently, the *Cambridge World History of Food* (Kiple and Ornelas, 2000) and *The Encyclopedia of Food* (Katz, 2004).

A number of journals publish many papers relevant to food choice, including *Appetite* and *Food Quality and Preference*. There are very useful articles that review important aspects of the field, including Birch *et al.* (1996), Booth (1982), Rozin and Tuorila (1993), Rozin (1976, 1982), Shepherd and Raats (1996), and Sobal *et al.* (2006). Finally, there are two books that constitute wonderful reading on the big culture-historical context of food in humans: *Guns, Germs and Steel* (Diamond, 1997), and *The Hungry Soul* (Kass, 1994) and four that focus on food, health, business and politics in the contemporary United States (Brownell, 2004; Nestle, 2002; Oliver, 2006; Schlosser, 2001).

1.13   References

AJZEN, I. (1988). *Attitudes, Personality and Behaviour*. Milton Keynes: Open University Press.
AJZEN, I. and FISHBEIN, M. (1980). *Understanding Attitudes and Predicting Social Behaviour*. Englewood Cliffs, NJ: Prentice Hall.

AMATO, P.R. and PARTRIDGE, S.A. (1989). *The New Vegetarians. Promoting Health and Protecting Life*. New York: Plenum Press.

APPADURAI, A. (1981). Gastro-politics in Hindu South Asia. *American Ethnologist*, **8**, 494–511.

BAEYENS, F., KAES, B., EELEN, P. and SILVERANS, P. (1996). Observational evaluative conditioning of an embedded stimulus element. *European Journal of Social Psychology*, **26**, 15–28.

BARKER, L.M. (ed.) (1982). *The Psychobiology of Human Food Selection*. Bridgeport, Conn.: AVI.

BARNETT, S.A. (1956). Behaviour components in the feeding of wild and laboratory rats. *Behaviour*, **9**, 24–43.

BARTOSHUK, L.M. (1990). Distinctions between taste and smell relevant to the role of experience. In E.D. Capaldi and T.L. Powley (eds), *Taste, Experience and Feeding* (pp. 62–72). Washington: American Psychological Association.

BEARDSWORTH, A. and KEIL, T. (1992). The vegetarian option: Varieties, conversions, motives and careers. *The Sociological Review*, **40**, 253–293.

BEARDSWORTH, A. and KEIL, T. (1995). *Sociology on the Menu*. London: Routledge.

BELK, R.W. (1975). The objective situation as a determinant of consumer behavior. *Advances in Consumer Research*, **2**, 427–437.

BELL, R. and MEISELMAN, H.L. (1995). The role of eating environments in determining food choice. In D. Marshall (ed.) *Food Choice and the Consumer* (pp. 292–310). London: Blackie.

BIRCH, L.L. and MARLIN, D.W. (1982). I don't like it; I never tried it: Effects of exposure on two-year-old children's food preferences. *Appetite*, **3**, 77–80.

BIRCH, L.L., FISHER, J.O. and GRIMM-THOMAS, K. (1996). The development of children's eating habits. In H.L. Meiselman and H.J.H. MacFie (eds) *Food Choice, Acceptance and Consumption* (pp. 161–206). London: Blackie Academic and Professional.

BIRCH, L.L., ZIMMERMAN, S.I. and HIND, H. (1980). The influence of social-affective context on the formation of children's food preferences. *Child Development*, **51**, 856–861.

BOOTH, D.A. (1982). Normal control of omnivore intake by taste and smell. In J. Steiner and J. Ganchrow (eds), *The Determination of Behavior by Chemical Stimuli. ECRO Symposium*. (pp. 233–243). London: Information Retrieval.

BOOTH, D.A. (1994). *Psychology of Nutrition*. London: Taylor and Francis.

BOOTH, D.A., MATHER, P. and FULLER, J. (1982). Starch content of ordinary foods associatively conditions human appetite and satiation, indexed by intake and eating pleasantness of starch-paired flavors. *Appetite*, **3**, 163–184.

BROWNELL, K.D. (2004). *Food Fight*. New York: McGraw-Hill.

CARDELLO, A. (1996). The role of the human senses in food acceptance. In: H.L. Meiselman and H.J.H. MacFie (eds) *Food Choice, Acceptance, and Consumption* (pp. 1–82). London: Blackie.

CONNORS, M., BISOGNI, C.A., SOBAL, J. and DEVNE, C.M. (2001). Managing values in personal food systems. *Appetite*, **36**, 189–200.

DE CASTRO, J.M. (1990). Social facilitation of duration and size but not rate of the spontaneous meal intake of humans. *Physiology & Behavior*, **47**, 1129–1135.

DEHOUWER, J., THOMAS, S. and BAEYENS, F. (2001). Associative learning of likes and dislikes. A review of 25 years of research on human evaluative conditioning. *Psychological Bulletin*, **127**, 853–869.

DIAMOND, J. (1997). *Guns, Germs, and Steel. The Fates of Human Societies*. New York: W.W. Norton.

ELNER, R.W. and HUGHES, R.N. (1978). Energy maximization in the diet of the shore crab, *Carcinus maenus*. *Journal of Animal Ecology*, **47**, 103–116.

FISCHLER, C. (1990). *L'Homnivore*. Paris: Odile-Jacob.

FREWER, L.J. and SALTER, B. (2003). The changing governance of biotechnology: The politics of public trust in the agri-food sector. *Applied Biotechnology, Food Science and Policy*, **1** (4) 199–211.

FURST, T., CONNORS, M., BISOGNI, C.A., SOBAL, J. and FALK, L.W. (1996). Food choice: A conceptual model of the process. *Appetite*, **26**, 247–266.

GARCIA, J., HANKINS, W.G. and RUSINIAK, K.W. (1974). Behavioral regulation of the milieu interne in man and rat. *Science*, **185**, 824–831.

GEIER, A.B., ROZIN, P. and DOROS, G. (2006). Unit bias: A new heuristic that helps explain the effect of portion size on food intake. *Psychological Science*, **17**, 521–525.

HILL, J.O. and PETERS, J.C. (1998). Environmental contributions to the obesity epidemic. *Science*, **280**, 1371–1374.

IYENGAR, S. and LEPPER, M. (2000). When choice is demotivating: Can one desire too much of a good thing. *Journal of Personality & Social Psychology*, **79**, 995–1006.

KAHN, B.E. and McALISTER, L. (1997). *Grocery Revolution. The new focus on the consumer*. Reading, MA: Addison-Wesley.

KAHN, B.E. and WANSINK, B. (2004). The influence of assortment structure on perceived variety and consumption quantities. *Journal of Consumer Research*, **30**, 519–533.

KAHNEMAN, D. (2000). Evaluation of moments: Past and future In: Kahneman, D. and Tversky, A. (eds) *Choices, Values, and Frames* (pp. 693–708). New York: Cambridge University Press.

KAHNEMAN, D. and SNELL, J. (1992). Predicting a change in taste: Do people know what they will like? *Journal of Behavioral Decision Making*, **5**, 187–200.

KAHNEMAN, D., WAKKER, P.P. and SARIN, R. (1997). Back to Bentham? Explorations of experienced utility. *The Quarterly Journal of Economics*, **12**, 375–405.

KASS, L. (1994). *The Hungry Soul*. New York: The Free Press.

KATZ, S.H. (ed.). (2004). *Encyclopedia of Food*. New York: Scribner.

KHARE, R.S. (1976). *The Hindu Hearth and Home*. Durham, N.C.: Carolina Academic Press.

KIPLE, K.F. and ORNELAS, K.C. (eds) (2000). *Cambridge World History of Food*. Cambridge: Cambridge University Press.

KREBS, J.R. and DAVIES, N.B. (eds) (1984). *Behavioral Ecology* (2nd edn). Oxford: Blackwell Scientific Publications.

LOGUE, A.W. (2004). *The Psychology of Eating and Drinking* (3rd edn). New York: Brunner-Routledge.

MARRIOTT, M. (1968). Caste ranking and food transactions: A matrix analysis. In: M. Singer and B.S. Cohn (eds), *Structure and Change in Indian Society* (pp. 133–171). Chicago: Aldine.

MAURER, D. and SOBAL, J. (eds) (1995). *Eating Agendas. Food and Nutrition as Social Problems*. Chicago: Aldine deGruyter.

MAYR, E. (1974). Behavior programs and evolutionary strategies. *American Scientist*, **62**, 650–659.

MEISELMAN, H.L. (1996). The contextual basis for food acceptance, food choice, and food intake: the food, the situation, and the individual. In: Meiselman, H.L. and MacFie, H.L.H. (eds) *Food Choice, Acceptance and Consumption* (pp. 239–263). London: Blackie.

MEISELMAN, H.L. (ed.) (2000). *Dimensions of the Meal. The Science, Culture, Business, and Art of Eating.* Gaithersburg, MD: Aspen.

MEISELMAN, H.L. and MacFIE, H.L.H. (eds) (1996). *Food Choice, Acceptance and Consumption.* London: Blackie.

MURCOTT, A. (ed.). (1983). *The Sociology of Food and Eating.* London: Gower.

NESTLE, M. (2002). *Food politics. How the Food Industry Influences Nutrition and Health.* Berkeley, CA: University of California Press.

OLIVER, J.E. (2006). *Fat Politics.* New York: Oxford University Press.

PANGBORN, R.M. (1980). A critical analysis of sensory responses to sweetness. In P. Koivistoinen and L. Hyvonen (eds), *Carbohydrate Sweeteners in Foods and Nutrition* (pp. 87–110). London: Academic Press.

PELCHAT, M. and ROZIN, P. (1982). The special role of nausea in the acquisition of food dislikes by humans. *Appetite,* **3,** 341–351.

PLINER, P. (1982). The effects of mere exposure on liking for edible substances. *Appetite,* **3,** 283–290.

PLINER, P. and HOBDEN, K. (1992). Development of a scale to measure the trait of food neophobia in humans. *Appetite,* **19,** 105–120.

PLINER, P. and ROZIN, P. (2000). The psychology of the meal. In H. Meiselman (ed.) *Dimensions of the Meal: the Science, Culture, Business, and Art of Eating* (pp. 19–46). Gaithersburg, MD: Aspen Publishers, Inc.

RAATS, M.M., SHEPHERD, R. and SPARKS, P. (1995) Including moral dimensions of choice within the structure of the Theory of Planned Behavior. *Journal of Applied Social Psychology,* **25,** 484–494.

RODE, E., ROZIN, P. and DURLACH, P. (2006). Experienced and remembered pleasure for meals: Duration neglect but minimal peak-end effects. Submitted manuscript.

ROGERS, P. J. (1996). Food choice, mood and mental performance: some examples and some mechanisms. In: Meiselman, H.L. and MacFie, H.L.H. (eds) *Food Choice, Acceptance and Consumption* (pp. 319–345). London: Blackie.

ROLLS, B. J. (2000). Sensory specific satiety and variety in the meal. In H. Meiselman (ed.) *Dimensions of the Meal: the Science, Culture, Business, and Art of Eating* (pp. 107–116). Gaithersburg, MD: Aspen Publishers, Inc.

ROZIN, E. (1982). The structure of cuisine. In L.M. Barker (ed.), *The Psychobiology of Human Food Selection* (pp. 189–203). Westport, CT: AVI.

ROZIN, P. (1976). The selection of foods by rats, humans, and other animals. In J. Rosenblatt, R.A. Hinde, C. Beer and E. Shaw (eds), *Advances in the Study of Behavior (Vol. 6)* (pp. 21–76). New York: Academic Press.

ROZIN, P. (1982). Human food selection: The interaction of biology, culture and individual experience. In L.M. Barker (ed.), *The Psychobiology of Human Food Selection* (pp. 225–254). Westport, CT: AVI.

ROZIN, P. (1990). Getting to like the burn of chili pepper: Biological, psychological and cultural perspectives. In B.G. Green, J.R. Mason and M.R. Kare (eds), *Chemical senses, Volume 2: Irritation* (pp. 231–269). New York: Marcel Dekker.

ROZIN, P. (1991). Family resemblance in food and other domains: The family paradox and the role of parental congruence. *Appetite,* **16,** 93–102.

ROZIN, P. (2005). The meaning of food in our lives: a cross-cultural perspective on eating and well-being. *Journal of Nutrition Education and Behavior,* **37,** S107–S112.

ROZIN, P. and FALLON, A.E. (1980). Psychological categorization of foods and non-foods: A preliminary taxonomy of food rejections. *Appetite,* **1,** 193–201.

ROZIN, P. and FALLON, A.E. (1987). A perspective on disgust. *Psychological Review,* **94,** 23–41.

ROZIN, P. and KALAT, J.W. (1971). Specific hungers and poison avoidance as adaptive specializations of learning. *Psychological Review*, **78**, 459–486.

ROZIN, P. and SCHULKIN, J. (1990). Food selection. In E.M. Stricker (ed.), *Handbook of Behavioral Neurobiology, Volume 10, Food and Water Intake* (pp. 297–328). New York: Plenum.

ROZIN, P. and TUORILA, H. (1993). Simultaneous and temporal contextual influences on food choice. *Food Quality and Preference*, **4**, 11–20.

ROZIN, P. and ZELLNER, D.A. (1985). The role of Pavlovian conditioning in the acquisition of food likes and dislikes. *Annals of the New York Academy of Sciences*, **443**, 189–202.

ROZIN, P., HAMMER, L., OSTER, H., HOROWITZ, T. and MARMARA, V. (1986). The child's conception of food: Differentiation of categories of rejected substances in the 1.4 to 5 year age range. *Appetite*, **7**, 141–151.

ROZIN, P., DOW, S., MOSCOVITCH, M. and RAJARAM, S. (1998). The role of memory for recent eating experiences in onset and cessation of meals. Evidence from the amnesic syndrome. *Psychological Science*, **9**, 392–396.

ROZIN, P., FISCHLER, C., IMADA, S., SARUBIN, A. and WRZESNIEWSKI, A. (1999). Attitudes to food and the role of food in life: Comparisons of Flemish Belgium, France, Japan and the United States. *Appetite*, **33**, 163–180.

ROZIN, P., BAUER, R. and CATANESE, D. (2003a). Attitudes to food and eating in American college students in six different regions of the United States. *Journal of Personality & Social Psychology*, **85**, 132–141.

ROZIN, P., KABNICK, K., PETE, E., FISCHLER, C. and SHIELDS, C. (2003b). The ecology of eating: Part of the French paradox results from lower food intake in French than Americans, because of smaller portion sizes. *Psychological Science*, **14**, 450–454.

ROZIN, P., FISCHLER, C., SHIELDS, C. and MASSON, E. (2006a). Attitudes towards large numbers of choices in the food domain: A cross-cultural study of five countries in Europe and the USA. *Appetite*, **46**, 304–308.

ROZIN, P. HANKO, K. and DURLACH, P. (2006b). Self-prediction of hedonic trajectories for repeated use of body products and foods: Poor performance, not improved by a full generation of experience. *Appetite*, **46**, 297–303.

SCHLOSSER, E. (2001). *Fast Food Nation*. New York: Houghton Mifflin.

SCHUTZ, H.G. (1989). Beyond preference: Appropriateness as a measure of contextual acceptance of food. In D.M.H. Thomson (ed.), *Food Acceptability* (pp. 115–134). Essex: Elsevier Applied Science Publishers.

SCHWARTZ, B. (2004). *The Paradox of Choice. Why More is Less*. New York: HarperCollins.

SCLAFANI, A. (1999). Macronutrient-conditioned flavor preferences. In: H.-R. Berthoud and R.J. Seeley (eds), *Neural Control of Macronutrient Selection* (pp. 93–106). Boca Raton, FL: CRC Press.

SHEPHERD, R. and RAATS, M.M. (1996). Attitudes and beliefs in food habits. In: Meiselman, H.L. and MacFie, H.L.H. (eds), *Food Choice, Acceptance and Consumption* (pp. 346–364). London: Blackie.

SHEPHERD, R. and RAATS, M. (eds). (2006). *The Psychology of Food Choice*. Oxfordshire, UK: CABI Press.

SIEGRIST, M. (1999). A causal model explaining the perception and acceptance of gene technology. *Journal of Applied Social Psychology*, **29** (10), 1093–2106.

SOBAL, J., BISOGNI, C.A., DEVINE, C.M. and JASTRAN, M. (2006). A conceptual model of the food choice process over the life course. In: R. Shepherd and M. Raats (eds), *The*

Psychology of Food Choice. Oxfordshire, UK: CABI Press.

SPARKS, P., SHEPHERD, R. and FREWER, L.J. (1995). Assessing and structuring attitudes toward the use of gene technology in food production: the role of perceived ethical obligation. *Basic and Applied Social Psychology*, **16**, 267–285.

STEARNS, P.N. (1997). *Fat History. Bodies and Beauty in the Modern West*. New York: New York University Press.

STEINER, J.E. (1979). Human facial expressions in response to taste and smell stimulation. In H.W. Reese and L.P. Lipsitt (eds), *Advances in Child Development and Behavior (Vol. 13)* (pp. 257–295). New York: Academic Press.

VAN TRIJP, H.C.M. and STEENKAMP, J-B., E.M. (1992). Consumers' variety seeking tendency with respect to foods: measurement and managerial implications. *European Review of Agricultural Economics*, **19**, 181–195.

WANSINK, B. (2004). Environmental factors that increase the food intake and consumption volume of unknowing consumers. *Annual Review of Nutrition*, **24**, 455–479.

ZAJONC, R.B. (1968). Attitudinal effects of mere exposure. *Journal of Personality and Social Psychology*, **9** (part 2), 1–27.

ZELLNER, D., KERN, B.B. and PARKER, S. (2002). Protection for the good: subcategorization reduces hedonic contrast. *Appetite*, *38*, 175–280.

ZELLNER, D., ROHM, E. A., BASSETTI, T. L. and PARKER, S. (2003). Compared to what? Effects of categorization on hedonic contrast. *Psychonomic Bulletin and Review*, **10** (2), 468–473.

2

Sensory influences on food choice and food intake

C. de Graaf, Wageningen University, The Netherlands

2.1 Introduction

Sensory preferences have a major impact on the food choice of the consumer. This is clear from different theories, models and a vast array of empirical data with respect to food choice and food intake. A good understanding of the food choice of the consumer requires a thorough understanding of the role that sensory factors play in food choice and food intake.

One of the earliest models about food choice was introduced by Pilgrim in 1957 from the US Army Quartermasters institute in Chicago (Pilgrim, 1957). This model distinguished between three major factors determining food choice, i.e. factors originating from internal physiology (hunger, satiety), factors originating from the interaction of the consumer with the environment (attitudes), and factors resulting from the interaction between the consumer and the product. The latter interaction determines the sensory perception of and preference for a particular food. Similar models focusing on the interaction between the consumer, the food and the environment were developed later on, and in all of these models, sensory preferences have a major role (e.g., Shepherd, 1988).

Food preferences/liking play an important role in Rozin's psychological taxonomy of food acceptance/rejection (e.g., Rozin and Vollmecke, 1986; Rozin, et al., 1986). The main dimension with respect to food choice in this taxonomy is like versus dislike. Liking results in the acceptance of a product and disliking leads to rejection. Other factors that are included in this model are anticipated consequences of consumption, ideas about appropriateness, and contamination issues. Positive anticipated consequences (feeling good after eating; e.g., medicines) increase acceptance, whereas negative anticipated consequences (feeling bad; e.g., allergic reactions) decrease acceptance. Ideas about appropriateness of food for certain occasions develop during childhood, and after some years we know that, for example, pizza and wine are not foods for early morning. Contamination issues refer the idea that foods can 'incorporate' properties of objects that have been in touch or have been associated with a certain food. For example, spitting saliva in another one's glass of orange juice will make the orange juice unacceptable for that person, although the taste won't be affected. For an extended discussion on this psychological taxonomy we refer to the original papers of Rozin and colleagues on this issue (Rozin and Vollmecke, 1986; Rozin et al., 1986).

Empirical data from surveys generally confirm the idea that liking is important in choice. Taste is one of the major factors in the food choice questionnaire of Steptoe et al. which was developed about 10 years ago (Steptoe et al., 1995). This questionnaire was meant to assess the most important motives in food choice. In a large EU survey ($n > 15\,000$) on influences on food choice, sensory factors appeared to be one of the mort important factors, along with the importance of health (Lennermas et al., 1997). The major importance of taste and health is also reflected in the development of the Taste and Health attitude questionnaire by Roininen et al. (1999). Roininen et al. (1999) showed that people differ in the extent to which they incorporate taste and health motives in their food choices. In another study, Stafleu et al. (2001) investigated the attitudes towards 20 fat-containing foods. It appeared that the attitudes and the intention to consume these 20 foods were mainly determined by beliefs about the sensory properties of foods, and much less by the health-related beliefs about these products. The affective dimension was more important than the cognitive component (Stafleu et al., 2001).

This chapter deals with sensory influences on food choice and food intake. The chapter starts with a general overview of empirical data on the relationship between rated liking and actual food intake and choice. In the next part of this chapter, the case is made that most food preferences are learned, this learning starts at a very early age (even before birth), and the early learned preferences may remain stable for years and years (until adulthood). Although preferences may be stable, they also depend on the nutritional status and the degree of exposure to sensory stimuli within a meal or across several days. For example, although one's favorite food may be beefsteak, having beefsteak every day may be too much of a good thing. This chapter ends with a discussion of empirical data on sensory preferences of children, the elderly, and the relationship between liking and intake in obese subjects.

Before we start with the discussion on the relationship between preferences and intake, it may be good to make a distinction between the various terms used throughout this chapter. In general liking and palatability refer to degree of pleasantness that subjects have when tasting a particular food. Liking, i.e. the pleasantness of the taste is different from 'wanting' which refers to the pleasantness to consume a food. Wanting may be measured by asking people for their desire to eat a particular food at a particular occasion. The term preference refers to the preference of one food over another.

2.2   Sensory perception, preference and food intake

2.2.1   Sensory perception

The sensory systems guide us through the outside world. In the framework of nutrition, they constitute the connecting principle between the food that we eat and the physiological processes of digestion and substrate utilization. The sensory perception of food involves vision, smell, taste, touch, audition, and the trigeminal system for sensing irritation (CO_2/bubbles in drinks, pepper) and temperature (Lawless and Heyman, 1998).

With our eyes we see the appearance, the color, size and shape of foods, and we recognize its identity. With our sense of smell we can 'smell' the volatile compounds of foods 'orthonasally' by sniffing with our nose just above the foods. When we move/break down the foods within our mouth we also perceive the volatile compounds within foods. These compounds are transported retronasally to the olfactory epithelium after swallowing parts of the food. The sense of taste mainly located at the tongue is responsible for the sensations sweet, salt, bitter, sour and umami (the taste of glutamate, Ve-Tsin). The texture of foods is perceived by a variety of senses, we perceive the hardness, the viscosity, the roughness and many other texture attributes. With some crispy foods, audition plays a role in the sensory perception. Another important attribute is the temperature.

2.2.2   Sensory preference, liking in relation to arousal

The sensory perception of foods is not the primary point of interest of most consumers. The most important characteristic of a food is whether we like it or not. Do we eat/drink more of it or not? The (dis)like for a food can be considered as the positive and/or negative evaluation of the sensory attributes of a particular food. With most sensory attributes, there is an optimal level of arousal (concentration, intensity), which is most liked. An arousal level high above the optimal levels may cause aversion. This optimal level of arousal is nicely reflected in the Wundt curve (Fig. 2.1), named after the nineteenth-century psychologist who discovered this principle (Wundt, 1907). Wundt curves are interesting, because they may differ from person to person, and, within subject, from moment to moment. Differences in sensory preferences between groups (e.g., between men and women, young and old, normal weight vs. obese) can be investigated through the comparison (concentration/arousal-liking) of Wundt curves for different groups (e.g., de Graaf *et al.*, 1996).

2.2.3   Sensory preference, liking and intake

Liking generally refers to the degree/amount of sensory pleasure that is derived from tasting or eating a particular product. Preference refers to a choice between two or more foods, where liking generally plays an important role. However, there are also many situations where people prefer a less liked food over a higher liked food because of various other motives that are involved, e.g. health, convenience, price, etc.

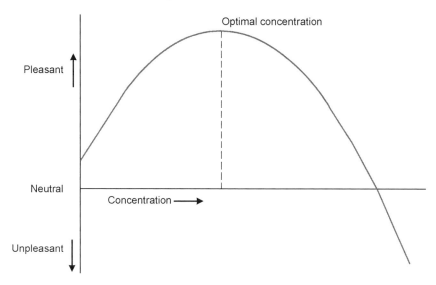

Fig. 2.1   Example of a Wundt curve, or a single peaked preference function, relating concentration to pleasantness. The concentration that concurs with the highest pleasantness is the optimal preferred concentration.

Liking has a positive effect on food intake; the more a food is liked, the more of it will be eaten. In many strictly controlled experimental studies it has been shown that ratings about the liking of a food are strongly correlated with the *ad libitum* intake of a food (e.g., Bellisle *et al.*, 1984; Bobroff and Kissileff, 1986; Guy-Grand *et al.*, 1994; Helleman and Tuorila, 1991; Zandstra *et al.*, 1999). For example, Zandstra *et al.* (1999) showed that the liking ratings during consumption of yogurt with various sugar concentrations had an average within subject correlation of 0.8 with actual *ad libitum* intake. A large number of other studies have found similar relationships between liking rating and intakes in laboratory conditions. According to Yeomans, liking affects food intake through the enhancement of the desire to eat a food, particularly at the start of the meal. He called this the appetizer effect (Yeomans, 1996; Yeomans *et al.*, 2004; see Fig. 2.2).

The relationship between liking and intake also holds outside the laboratory. In a recent paper on a series of four field studies with US Army men and women (de Graaf *et al.*, 2005), it was shown that the correlation between ratings on the 9-point hedonic scale and intake ranged between 0.22 and 0.62 for main dishes and between 0.13 and 0.56 for snacks. When the rating of a particular food was lower than 5 (is the neutral point; neither like, nor dislike) on the 9-point scale, subjects consumed on average less than 87% of the meal, and with ratings above 7 on this scale subjects consumed on average 100% of the provided main dishes (see Table 2.1).

Although liking has a strong positive effect on intake, the sensory hedonic dimension is not the only driver for intake. Mattes *et al.* (1990) showed that people with taste and smell disturbances still have a drive to eat, and do not

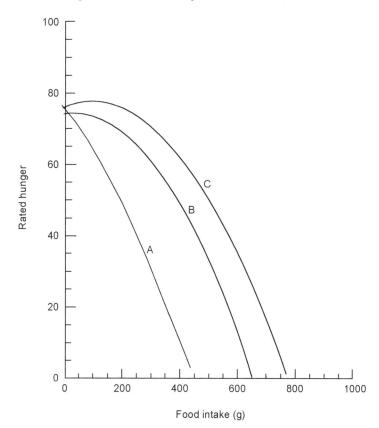

Fig. 2.2 Changes in rated hunger for normal weight men during eating a 'palatable' (A), bland (B), or overly strong flavored (C) test meal. The test meal was composed of pasta in sweetened tomato sauce. Palatability of the meal was manipulated by adding oregano. The bland condition did not contain added oregano, and had an average pleasantness rating of 46 on a 100 mm scale. The palatable meal contained 0.27% oregano, and rated 61 on pleasantness; the overly strong flavoured meal contained 0.54% and rated on average 40. Hunger was rated every 2 minutes, and the functions represent best fit quadratic functions (Source: Yeomans, 1996).

necessarily eat less than people with a normal sense of smell and taste. Without sensory stimulation, people still have a strong appetite, i.e. the internal drive to search, choose, and ingest foods.

2.2.4   Sensory preference and choice
If presented with two foods, people will generally choose the most liked food. This point also holds across foods and across people. In a canteen, the average market share of different products will tend to covary with the average liking scores of different products (Pilgrim and Kamen, 1963). In a recent paper with US Army men and women it was shown that the chance of selecting a meal a

Table 2.1 Relative amount eaten of main dishes in a series of field studies of the US Army, and the chance of selecting the meal at least a second time during the field test, as a function of acceptability ratings (1 = extremely unpleasant – 9 = extremely pleasant). (*n* refers to the number of observations/individual meals)

Acceptability rating	Amount eaten (1)	n (1)	Frequency rating First time (2)	Frequency of meals eaten > 1	Chance %
1	0.46	148	75	6	8
2	0.60	89	54	6	11
3	0.73	95	59	10	17
4	0.77	182	106	26	25
5	0.87	294	154	48	31
6	0.92	596	332	84	25
7	0.96	1168	665	213	32
8	1.00	1638	767	329	43
9	1.00	1497	601	315	52

Source: de Graaf *et al.* (2005)

second time was positively related to the hedonic rating on the first time that a product was tasted (de Graaf *et al.*, 2005) (Table 2.1).

2.2.5   The role of appropriateness (eating context)

Although it is clear that liking has a strong effect on intake and choice, we do not always eat the very best, the most liked food on each occasion. If someone's favourite food is pizza, he will not eat necessarily eat pizza every day. It is also improbable that he will eat pizza at breakfast or during the coffeebreak. This issue is related to the notion of 'appropriateness' (Schutz, 1988; Cardello and Schutz, 1996). We develop eating patterns, where many foods get a particular fitness for use for particular eating occasions. In the US many people eat turkey for Thanksgiving, and in The Netherlands, people consume 'oliebollen' at New Year's Eve. Appropriateness is linked to the various religious, symbolic, and emotional roles that foods can play. It is still an open question how foods acquire a fitness for use for particular eating occasions, and how food properties play a role in this.

Appropriateness interferes in the relationship between preference/liking and intake and choice. Some foods that are very special need special occasions on which they can be eaten; very common foods need common eating occasions.

2.2.6   The distinction between liking and wanting

Chocolate will generally get a higher rating on a sensory test than bread. Still most people will on average consume more bread than chocolate. So, the relationship between pleasantness and intake is not as straightforward as it seems to be at first sight. Part of this has to do with the appropriateness of certain foods for certain eating occasions. Another issue that plays a role in this respect is the distinction between 'liking' and 'wanting'. Liking refers to the pleasantness derived from sensory stimulation. Wanting refers to the motivation, or the desire to eat a particular food. With respect to actual food intake or choice, wanting seems to be more important than liking (Mela, 2006). In the last 10 years, Berridge has worked out the underlying neurophysiologic mechanisms. Liking and wanting have different neurochemical substrates in the brain, where liking seems to be more related to the opioid receptors, and wanting relates more to the dopaminergic receptors (Berridge, 1996, 2004).

The distinction between liking and wanting is important from a theoretical and a practical perspective. From a sensory testing point of view, liking is more stable than wanting, which makes liking much easier to measure. The wanting for a certain stimulus is much more dependent on external and internal factors than the liking, and therefore the question for wanting is much more difficult to measure in a practical way. However, as wanting is probably more relevant for choice and intake, it is necessary that we start working more on this aspect. Some investigators have started to work on this issue (e.g., Goldfield *et al.*, 2005; Yeomans *et al.*, 2004).

2.3   Development and stability of food preferences

Humans are born with a preference for sweet and a dislike for sour and bitter tastes. A preference for a salty taste develops within the first year of life. With respect to odors it is not clear whether or not there are inborn preferences, although it is clear that newborn infants detect and respond to certain odors. Recent work from Marlier and Schaal (2005) suggests that 3–4-day-old newborns prefer the odor of human milk over that of formula milk, irrespective of exposure to breast milk or formula milk. Apart from these few inborn preferences, the majority of the sensory preferences in humans are learned through repeated exposure to particular sensory events and their associated consequences. Basically, when an exposure to a certain stimulus is associated with positive consequences the preference goes up, and when the exposure with a certain stimulus is associated with negative consequences, the preference goes down.

There are three major operating mechanisms through which preferences are learned: exposure, post-ingestive consequences, and social interactions. The idea of exposure explains culturally mediated food preferences. It explains why many Dutch like cheese, and why Indian children like curry. Positive post-ingestive consequences explain why children very easily learn to like the taste of hamburgers but find it difficult to appreciate the taste of Brussels sprouts. Modeling and other social interaction help to explain the strong influences of family, friends, and commercials for specific food products.

2.3.1   The powerful role of exposure

One of the first experimental demonstrations of the effect of exposure on liking was a study by Pliner (1982), who exposed young adult subjects to different unfamiliar tropical fruit juices 0, 5, 10 or 20 times. 'The results showed a strong exposure effect such that the more frequently a juice had been tasted the better it was liked' (Pliner, 1982). In the same year, Birch and Marlin (1982) published a study with similar findings in 2-year-old children. The foods used in this study with two separate experiments were five unfamiliar fruits and five unfamiliar cheeses, to which the children were exposed 0–25 times. In a later study of Birch *et al.* (1987) it appeared that the exposure had to be in the form of really tasting the food in the mouth; just looking was not enough to produce a liking. In a more general analysis of food preferences in children, it was found that food preferences in children could be described by two main dimensions, the sweetness level and the familiarity to foods (Birch, 1979).

The role of exposure in the development of liking has later been confirmed in many other studies. The principle of exposure has also been applied to try to enhance the liking for fruits and vegetables in young primary schoolchildren. In two studies Wardle and colleagues (2003a,b) studied the effect of parent-led repeated exposure to unfamiliar vegetables in 2–8-year-old children. Children were exposed to vegetables 8–15 times, and they were asked to taste and ingest the vegetables. The results showed that the exposures led to higher rated liking or preference and higher *ad libitum* intakes of the vegetables. In a recently published study, Hendy *et al.* (2005) used a combination of exposure and social rewards to increase liking and intake of fruits and vegetables in 6–9-year-old children. This program was also successful in increasing both liking and intake in all three age groups for both fruit and vegetables. However, a six month follow up showed that the initial increments in preference were not maintained after this period of six months. In another recently published study on spinach it was found that repeated exposure to spinach resulted in slightly higher liking scores after repeated exposure, only in a group of initial spinach dislikers (Bingham *et al.*, 2005). These findings suggest that it is hard to produce a strong and long-lasting effect of exposure with respect to fruit and vegetable preference/consumption.

An interesting question with respect to the role of exposure in the formation in preferences is its timing within the life cycle. Pliner's study was done with adults and the studies of Birch *et al.* (1982) were done in 2-year-old children. In 2000, Schaal *et al.* (2000) published a study where they showed that newborns (3 hr after birth) of mothers who consumed anis-flavored food responded less negatively to anis odor than newborns of mothers who did not consume anis-flavored foods. Apparently, 'human fetuses learn odors from their pregnant mother's diet' (Schaal *et al.*, 2000). A study from Menella *et al.* (2001) showed that the mother's exposure to carrot juice in the third trimester of pregnancy and/or during lactation had a positive effect on the liking of carrot flavored cereals of weaned infants at the age of five months (Fig. 2.3). Another very early life study of Menella *et al.* (2004) showed that infants who were exposed to sour-bitter

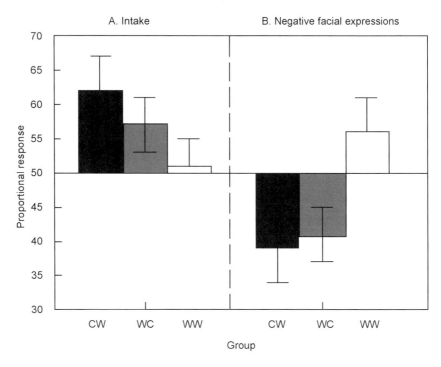

Fig. 2.3   The infants' relative acceptance of carrot-flavor cereal as indicated by display of negative facial expressions (right panel) and intake (left panel). There were three different experimental groups; the mothers in group CW drank carrot juice during the third trimester of pregnancy, and water during lactation; the mothers in group WC drank water during pregnancy and carrot juice during lactation. The control group WW drank water during both pregnancy and lactation (Source: Menella *et al.*, 2001). Reproduced with permission from *Pediatrics*, Vol. 107, page 88, Copyright 2001 by the AAP.

tasting formulas during the first 6 months of life were accepting this formula at 7 months of age, where children who were not exposed to this formula rejected this bad-tasting liquid. These studies show that the effects of exposure are already apparent very early in life. As is discussed later, these very early exposures may have long-lasting effects on later preferences.

One other interesting aspect in the relationship between exposure and preference is the effect of variety on the acceptance of new flavors. Breast-fed infants are more willing to accept a novel vegetable flavor than bottle-fed infants (Sullivan and Birch, 1994). In a later experimental study, newly weaned infants were exposed to either the same food (carrot) or three different foods (carrot, pea, squash) for 12 subsequent days (Gerrish and Mennella, 2001). In a test session with a novel food (chicken), the variety-exposed infants liked more of the chicken than the single-food-exposed infants. These effects of breast milk and variety exposure were very recently confirmed in a study of Maier *et al.* (2005).

2.3.2   Post-ingestive consequences

When a baby is hungry it starts crying. This is a powerful message to the mother, whose breasts (via the brain, of course) may respond by signaling that they want to get rid of the milk. After the consumption of milk by the baby, there is a sense of happiness. It is food that causes this dramatic change from the miserable state of hunger to the pleasant state of satiety. From this example it is clear that food has powerful reinforcing properties. Humans learn to associate the taste/flavor of a food with its metabolic consequences. Depending on the positive and/or negative consequences the preference goes up or down.

There is a difference between learned aversions and learned preferences. Most people have just a few learned taste aversions, which have been caused by the single coupling between the exposure to a particular food and subsequent nausea (Rozin and Vollmecke, 1986). This may occur after food poisoning, cancer radiation/chemotherapy treatment, and/or other types of sickness causing gastrointestinal discomfort (see, e.g., Bernstein, 1978). On the other hand, people have many food preferences, which are not formed after a single exposure but after the repeated consumption of a food.

It is clear that children learn to like those tastes/odors that are associated with carbohydrates and/or fats in food. In this way children learn to like foods that are high in energy density. The first demonstrations in this respect were three elegant, carefully designed studies by Birch and colleagues at the beginning of the 1990s (Birch et al., 1990; Johnson et al., 1991; Kern et al., 1993). Combining unfamiliar flavors with either carbohydrates or fats (ingestion of about 100–200 kcal) for 8–12 times caused an increase in liking for that particular flavor in 2–5-year-old children.

These energy-conditioned flavor preferences as found in children are difficult to replicate in adults. In a study of Stubenitsky et al. (see Zandstra, 2000) we tried to replicate the effect of the Birch studies in adults with novel tropical fruit flavors in 20 conditioning trials either with 200 ml, 67 kcal or 273 kcal yogurt drinks in 108 adults. Drinks were consumed mid-morning. There was also an exposure group ($n = 19$), who repeatedly consumed only small amounts (10 ml). The results showed a clear exposure effect on preference, but no energy-conditioning effect. This lack of effect may due to the age, or the more complex stimulation levels in adults compared to children, which may make it more difficult to 'learn/associate' the sensory signal with the energy signal.

The energy-conditioning effect has recently also been shown to operate in everyday life with adults in the study of Appleton et al. (2006). This study showed that energy conditioning of the liking of flavors was particularly produced when people consumed the energy-rich yogurt drink in a state of hunger. The finding is in line with the Darwinian point of view that it makes sense to learn to like those flavors/tastes that are associated with a high energy density. This idea explains why it is so easy to learn to like the taste of hamburgers or pizzas, but why it is so difficult to learn to like the taste of vegetables.

From the perspective of the energy-conditioning effect, one might expect that people will generally not like low energy versions of particular products, such as low fat, low sugar products. However, several studies have not shown any decrease in liking after repeated exposure to so called 'diet' products (e.g. Mela *et al.*, 1993). Low fat/low sugar versions of original products such as full-fat milk (→ semi-skimmed milk), full-fat margarine (→ semi-skimmed margarine), sugar-containing soft drinks (→ diet soft drinks) have now become widely available on the market place.

2.3.3   Social effects with respect to liking

Social mechanisms may produce strong effects on liking. These effects differ per age group. For example, infants are less affected by social pressures than adolescents. One of the major social effects on liking and intake is through modeling, imitation, and/or (un)conscious conformation to group pressure. Young children learn by observation what other respected people do, and what foods they prefer. Recent work of Hendy and Raudenbush (2000) showed that enthusiastic teacher and peer modeling may have a positive effect on the liking and intake for fruits and vegetables in children. An older study of Birch *et al.* (1980) showed that children adjust their preferences in a group to the preference of the majority in the group.

Apart from the effect of modeling, there are a number of rules on how social influences affect preferences. Giving food as a reward for good behavior increases the preference, e.g. giving a child a food as reward for playing in a cooperative way will enhance the preference for that food (Birch *et al.*, 1980). It must be noted, however, that not every food will be appreciated as a reward; a candy or a high energy dense snack may be viewed as a reward; this will not be true for a vegetable. Another rule is that presenting/giving food in combination with positive attention from respected others (e.g., adults) will lead to an increase in preference (Birch *et al.*, 1980). The consumption of a food in a pleasant social atmosphere may have a positive effect on its liking.

A third rule is more controversial; giving someone a reward for eating a particular food may lead to an increase in preference for that food, but may also be counterproductive, and lead to a decrease in liking. For example, the rule, 'first eat your vegetable, then you will get your dessert may lead to a decrease in preference for the vegetable and an increase for the preference of the dessert'. One early study of Birch *et al.* (1982) showed that using food as an instrument for getting a particular reward led to a decrease in preference for the particular food. However, in later studies, giving rewards led to positive effects on liking and intake (Horne *et al.*, 2004; Lowe *et al.*, 2004). One important aspect in this respect might be to what extent the rule is conceived as a kind of bribery (see also Hendy *et al.*, 2005 on overjustification).

2.3.4   Stability of food preferences

Preferences are remarkably stable, and especially early acquired preferences may have long-lasting influences. Recent work from Liem and Mennella indicates that children who were exposed to bitter-sour tasting protein hydrolysates during their first year of their life, had a preference for more sour tasting stimuli five or six years later (Liem and Mennella, 2002, 2003). A longitudinal study by Skinner *et al.* (2002) showed strong associations between the foods liked by a child at the age of 2–3 years, at the age of 4 years, and at the age of 8 years.

Results obtained by Nicklaus *et al.* (2004) suggest that preferences that are established by the age of 2–3 years are predictive for preferences in early adulthood. Nicklaus *et al.* (2004) measured food preferences (i.e. actual food choice) in 2–3-year-old children in a daycare center, and related these observed preferences to rated preferences later in life. Especially for vegetables, particular cheese varieties, and some types of meat, there were consistent positive associations. Results from a study of Haller *et al.* (1999) showed that neonatal experience with vanilla odor related to a preference for vanilla-flavored tomato ketchup during adulthood.

The results of these studies suggest that early established preferences are very important in predicting later preferences. Interventions to modify preferences may be most effective and long-lasting when carried out as early as possible.

2.4   Dynamics of liking: sensory specific satiety and boredom

2.4.1   Dynamics of liking

In the paragraph above, we have seen that preferences that are formed early in life may be stable until adulthood. However, this does not mean that the liking for a particular food is stable from day to day, and across the day or various situations. For example, the first cup of coffee in the morning tastes very pleasant, the second cup of coffee may still taste good, but after some more coffee, there will come a moment that the coffee becomes aversive. Preferences changes as a function of exposure, as a function of the time of the day, and as a function of nutritional status. When being full, we have different preferences to when we are hungry.

So, preferences vary. Within the scientific literature, the dynamics of liking has been studied from three perspectives:

- the change in liking for foods within an eating moment, i.e. sensory specific satiety
- the change in liking for foods across days, eating moments, i.e. boredom
- the change in liking as a function of nutritional status, i.e. alliesthesia (Cabanac, 1971; Appleton, 2005). This latter phenomenon has a physiological background, and is discussed in relation to food reward and weight status (Section 2.6.4).

2.4.2   Sensory specific satiety

Sensory specific satiety has been defined by Rolls *et al.* (1981a,b, 1982) as the decline in the reward value (liking, wanting) of an eaten food compared to the decline in reward value of uneaten foods. Originally it was discovered by Le Magnen in the 1950s in animal studies (Le Magnen, 1986). The experimental work with humans started in the early 1980s with a series of elegant studies by the Rolls family (Rolls, 1986). Sensory specific satiety can be considered as the basic driver for the search for variety in the diet, i.e. after eating a particular food with a particular taste, the reward value of that food will decline but the reward value of other foods will decline less. This process leads to the selection of a variety of foods with different sensory properties.

The early studies on sensory specific satiety showed that people can get satiated for cheese on a cracker and sausages, various flavors in yogurt, various types of sandwich fillings, or different pasta shapes. The sweet/savory dimension is important in sensory specific satiety, after eating a sweet food, appetite for something sweet will decline, and after eating a savory food, appetite for something savory will go down. Sensory specific satiety occurs within 2 minutes after eating a particular food and is at its peak 2–20 minutes after initial exposure. However, depending on the degree of exposure, it will last for at least 90 min or more (de Graaf *et al.*, 1993) (Fig. 2.4).

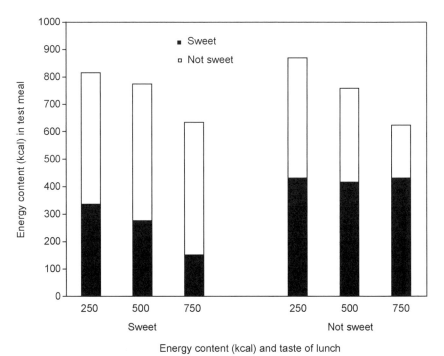

Fig. 2.4 Illustration of the effect of sensory specific satiety: energy intake at a test meal 2 h after consumption of a lunch (preload) that varied in energy content and taste (sweet and non-sweet). The preloads with equal energy content were matched with respect to weight, fat, protein, fiber and carbohydrate content. The energy intake in the test meal is divided into energy intake from sweet foods and the energy intake from non-sweet foods (Source: de Graaf *et al.*, 1993).

Texture is also important in sensory specific satiety. Guinard and Brun (1998) showed that after eating a hard food (e.g., an apple or a French 'baguette'), subjects expressed a lower liking for hard foods, but the liking for soft foods (e.g., apple sauce, soft bread) declined less. In that same study, the sweet/savory dimension seemed dominant over the hard/soft dimension (Guinard and Brun, 1998). Sensory specific satiety is dependent on the sensory properties of the food and not on the nutritional properties such as the energy density and/or the carbohydrate and fat levels (see, e.g., Rolls, 1986; Snoek et al., 2004).

Sensory specific satiety also affects food intake, and not only liking. When subjects are presented with a variety of foods, they will eat more than when they are presented with a single food. This has been shown repeatedly in the studies of Rolls et al. (1981a,b, 1982) with humans. This is also clear from a large number of animal studies (see Raynor and Epstein, 2001, for an overview).

2.4.3   Boredom

The definition of boredom, i.e. a decline in acceptance of a product after repeated exposure, is similar to the definition of sensory specific satiety. In the literature the phenomenon of boredom has been investigated across meals, and across days. The first studies on boredom were carried out in the 1950s by scientists from the US Army Quartermaster Institute. They showed that soldiers who were in training, or field test, did not meet their requirements for energy balance. When these soldiers were exposed to a limited diet with four different foods per day, they got bored of the foods, and their intake declined. In a later study (Schutz and Pilgrim, 1958; Siegel and Pilgrim, 1958), it was shown that a limited degree of variety may circumvent boredom or monotony effects.

Later studies confirmed monotony effects when people are exposed to a very limited number of foods. Cabanac and Rabe (1976) showed that subjects lose weight when they only consume vanilla flavored beverages, 3–4 times per day for a period of 3 weeks. Pelchat and Schaefer (2000) reported that young adults had cravings for savory, hard foods, when they were exposed to a nutritionally adequate diet composed of vanilla flavored beverages only for a period of 10 days. Rolls and de Waal (1985) showed within the context of an Ethiopian refugee camp, that subjects got bored of the few foods that were distributed over a six-month period. With study it should be kept in mind that the subjects were probably in a marginal nutritional status.

Meiselman et al. (2000) showed boredom to a lunch meal when subjects were exposed to the same lunch for five subsequent days. Boredom was more pronounced for the vegetable part (green beans) of this meal compared to the staple part (mashed potatoes). Hetherington et al. (2000, 2002) demonstrated that the daily consumption of chocolate led to a decline in the acceptance of chocolate, whereas daily consumption of French fries or bread with butter did not affect its acceptance. Zandstra et al. (2000) found that exposure to a meat sauce once a week was enough to produce a significant increase in boredom, and a significant decline in acceptance. Available variety in flavors of the meat sauces prevented this boredom.

2.4.4   Properties of food involved in sensory specific satiety and boredom

One of the most interesting questions within the field of sensory specific satiety and boredom is the nature of the food properties that are involved in these processes. In the early studies on boredom it was shown that subjects easily got bored of vegetables, fruits and canned meats, whereas staple foods were relatively resistant to boredom. These ideas were later confirmed by Meiselman *et al.* (2000) and Hetherington *et al.* (2000, 2002), who demonstrated that vegetables and chocolate produced strong boredom, whereas staple foods such as potatoes (French fries, mashed) and bread were resistant to boredom. In an interesting study on boredom responses to pasta sauces, Moskowitz (2000) found that the strongest boredom response was related to a sauce which had clearly identifiable visual characteristics (spikkles).

It has become clear that it is the sensory properties of the food and not the macronutrient composition that is responsible for sensory specific satiety. The energy density, fat content, and/or carbohydrate content have little to do with sensory specific satiety. This is also clear from the time frame of sensory specific satiety which is already apparent two minutes after sensory exposure. Within two minutes there is little nutrient adsorption in the gastrointestinal tract.

From the results of all the studies on boredom and sensory specific satiety, the picture emerges that these processes may be related to the sensory intensity/ impact of the taste. This idea concurs with the findings of Zandstra *et al.* (1999), who showed that appetite for something sweet was more suppressed after consumption of yoghurts with higher sweetness intensities. Vickers and Holton (1998) showed that stronger tasting tea produced more satiety/boredom than weaker tea. In a recent study Bell *et al.* (2003) showed that volume had a much stronger effect on sensory specific satiety than energy density. This finding is in line with the idea that the degree of sensory exposure is important in sensory satiety. Further confirmation of this hypothesis is required,

Another food dimension in this respect is the perceived complexity of a product. More complex products may give rise to lower sensory specific satiety and boredom. This idea originally comes from Berlyne (1970), who argued that simple stimuli will lead to earlier boredom than more complex stimuli. More complex foods (such as wine or cheese) may allow the consumer to focus on more different sensations during tasting. The effect of complexity on sensory specific satiety and boredom has been tested in various studies with mixed success. Some studies did not find any effect of stimulus complexity (Porcherot and Issanchou, 1998; Zandstra *et al.*, 2004). However a recent study by Russell *et al.* (2005) indicated that increasing the complexity of yogurt-like drinks prevented boredom.

2.5   Sensory preferences and food intake in children and the elderly

2.5.1   Sensory preferences and food intake in children

From the paragraphs on the development of food preferences it is already apparent, that children have an inborn preference for sweetness, and an aversion for sour and bitter tastes. Familiarity plays an important role in preferences of young children; 'I don't like it, I've never tried it' (Birch *et al.*, 1982). Children have higher optimal sucrose/sweetener concentrations in foods than young adults. They like very sweet tasting foods. This is clear from a number of cross-sectional studies (e.g., Desor *et al.*, 1975; de Graaf and Zandstra, 1999; Zandstra and de Graaf, 1998) and one longitudinal study (Desor and Beauchamp, 1987). The high optimal sucrose levels in foods for infants slowly goes down to adult levels (de Graaf and Zandstra, 1999). In the study of Desor *et al.* (1975) 9–15-year-old children also had higher optimal NaCl concentrations in water compared to adults. A substantial number (around 30%) of children like foods with a high sourness intensity (Liem and Mennella, 2003). Liem *et al.* (2004a) showed that there was a positive association between the liking for sour tastes and the degree of sensation seeking on other dimensions (bright colors; choosing a mystery candy over a familiar candy). There is very little information on odor preferences in children.

The higher liking for more intense stimuli might be related to a lower sensitivity in children compared to adults. Several studies have shown that children have a lower discriminatory ability for different sucrose concentrations compared to adults (Zandstra and de Graaf, 1998; de Graaf and Zandstra, 1999). It should be noted though, that it is difficult to disentangle sensory processes from cognitive processes in these age groups. For example, in a study of Liem *et al.* (2004a,b), it was clear that 4 and 5 year olds had consistent responses on preference tests, but only the 5 year olds were able to give consistent answers on questions related to sweetness intensity, (Which one of the two stimuli is sweeter?).

In their evaluation of products, younger children tend to focus more on one salient feature (e.g., the color) of a particular product than older children (e.g., Roedder-John, 1999). When children get older their evaluations take into account more dimensions, and also more abstract dimensions such as the health value, and/or appropriateness for certain use situations become more important. With respect to sensory modalities, it seems that younger children may focus more on visual and textural cues than older children (Rose *et al.*, 2004a). For example, in two studies on sensory preferences of meat, younger (6–7-year-old) children focused on the texture attributes, whereas older children focused more on taste and odor dimensions (Rose *et al.*, 2004a,b).

As noted in the paragraphs above, infants below the age of 7 months are very open to novel flavors. They very easily learn to like new tastes and flavors. However, older children from 2–5 years old are often neophobic and/or picky eaters, i.e. they do not readily accept novel foods. The willingness to try novel foods may be reduced by combining the novel food with familiar flavor combinations (Pliner and Stallberg-White, 2000) and/or acting as model to the child by tasting it (Hendy, 2002). A recent study with 2–5-year-old children, indicated that 'children are more likely to eat new food if others are eating the same type of food than when others are merely present or eating another kind of food' (Adessi *et al.*, 2005).

Regarding the effect of pleasantness on intake, it may be hypothesized that children are more responsive than adults. As we grow up, more issues such as appropriateness, health considerations, social pressures, social norms about portion size might interfere with the positive effect of liking on intake. However, there are little data on this issue, which may make it an attractive future research issue.

2.5.2   Sensory preferences and food intake in elderly

The elderly form a heterogeneous group with respect to preferences and nutritional status. As people get older, their average energy intake goes down, due to the so-called 'anorexia of ageing' (Morley, 2003). This low food intake results in high prevalences of malnutrition in elderly nursing homes in the industrialized world (e.g., Nijs *et al.*, 2006). One possible cause for this loss of appetite may be related to the loss of chemosensory sensitivity. From over 50 studies since the 1960s it is well known that there is a loss in sensitivity for taste and for smell (de Graaf *et al.*, 1994). The sense of smell is more affected than the sense of taste (de Graaf *et al.*, 1994, 1996). A nice example on the impairment in smell identification ability is given by results from a study by Thomas-Danguin *et al.* (2003) (see Fig. 2.5). This figure nicely shows that on average there is an impairment of the sense of smell, but there are still 90 year olds who perform better on the olfactory test than 40 year olds.

Several questions emerge from the observation that chemosensory sensitivity declines:

- Does a change in chemosensory sensitivity lead to changes in food/flavor preferences in the elderly?
- Is there a way through which we may compensate for the loss of sensitivity/ changes in preferences in order to stimulate food intake?
- Can the potential enhancement of intake be maintained for a longer term, so that it may have a positive effect on the nutritional status of elderly with a low food intake.

There have been about 20 studies on the question of whether or not there are changes in food preferences with age. Several of these studies showed that on average the elderly have higher optimal concentrations of tastants/odorants than young adults (de Graaf *et al.*, 1994, 1996; Griep *et al.*, 1997; Murphy and Withee, 1986, 1987; Kozlowska *et al.*, 2003). However, it must be acknowledged that not all studies show a similar effect, and also that the age changes in preferences may be different for different types of foods/flavors (e.g., Koskinen *et al.*, 2003; Mojet

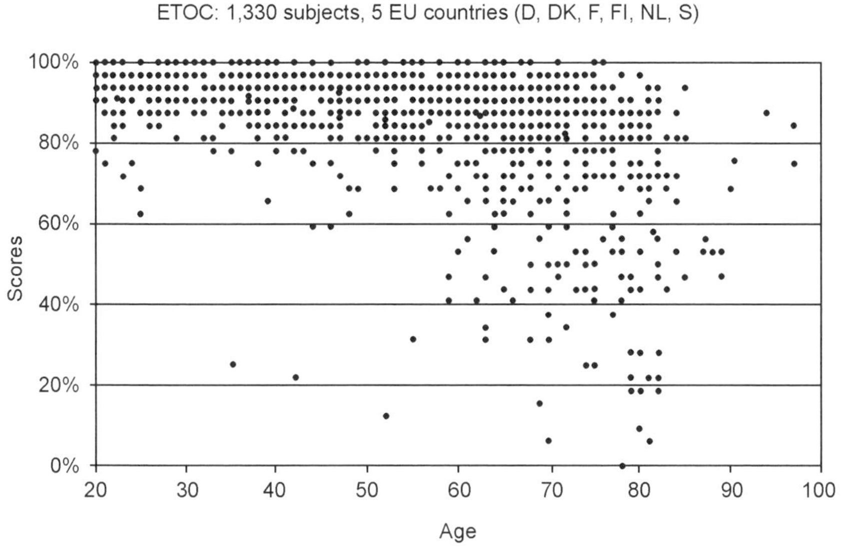

ETOC: 1,330 subjects, 5 EU countries (D, DK, F, FI, NL, S)

Fig. 2.5   Individual scores on the European Test on Olfactory Capabilities (ETOC) from 1330 people as a function of age (Source: Thomas-Danguin, 2003).

et al., 2005). The higher optimal concentration for the elderly compared to adults is more consistent across studies for sweeteners (e.g., de Graaf *et al.*, 1996; Mojet *et al.*, 2005) than for other tastes. The difference in findings between studies may also be attributed to the differences in population characteristics, where in older elderly populations (de Graaf *et al.*, 1994, 1996), there is more often a change in preference than in younger elderly (Mojet *et al.*, 2005).

In theory, the higher optimal flavor concentrations in the elderly relate to the lower chemosensory sensitivity. However, various studies failed to observe a relationship between a lower sensory performance and higher optimal flavor concentrations (e.g. Koskinen *et al.*, 2003). A study by Forde and Delahunty (2004) on preferences in orange juices distinguished between three groups of elderly, one group of elderly with similar preferences as young people, one group of elderly with a liking for more intense sensory stimuli, and a group of elderly which was indifferent to changes in sensory stimulation. The group of elderly which had a preference for higher sensory stimulation had an average lower sensory performance than the group of elderly who had similar prefer-ences as the young adults. This area still suffers from many methodological difficulties. For example, Koskinen and Tuorila (2005) showed that the scores on sensory performance tests had little predictive validity for the intensity ratings of flavors in real foods. Therefore, there is strong need for an improved methodology to assess/characterize chemosensory sensitivity and sensory preferences in elderly people.

The second question is whether or not flavor enhancement is a useful strategy to stimulate short-term food intake in the elderly. Results of studies on this question have been mixed. Studies by de Jong *et al.* (1996) and Koskinen *et al.* (2003) failed to find differences in intake in elderly with enhanced flavor/taste concentrations in foods, whereas in two other studies there was a higher intake in elderly at higher taste/flavor concentrations (Griep *et al.*, 2000; Kozlowska *et al.*, 2003). These studies require a good understanding and a precise estimate of optimal concentrations for different groups of elderly.

The third question deals with whether or not flavor enhancement can result in long term enhancement of food intake in elderly subjects. A first study on this issue was done by Schifmann and Warwick (1993) who showed that the elderly consumed more of flavor enhanced foods during a three-week intervention study in a nursing home. Total observed energy intake did not increase with flavor enhancement. In a second long-term flavor enhancement study of 16 weeks, Mathey *et al.* (2001) investigated the effect of the addition of MSG (monosodium glutamate = flavor enhancer) plus appropriate meat flavors (chicken, beef, fish, roast beef) to the hot meal on food intake and body weight. Mathey *et al.* (2001) found a higher food intake and body weight in the experimental (flavor enhancement) group of 36 elderly. In this study, changes in body weight were correlated with changes in food intake in the experimental group, which improved the credibility of the results of this study. After this study, Essed *et al.* (2006) tried to find out whether the observed effect in the Mathey study was due to the MSG or the flavors. This study was a parallel 16-week study, where there were four groups, placebo, flavor only, MSG only and flavor plus MSG. The results of this study were disappointing, and showed neither an effect of flavors nor of MSG on food intake and body weight (Essed *et al.*, 2006).

2.6   Sensory perception and preferences in relation to obesity

2.6.1   Sensory (intensity) perception and obesity

There are no indications that obese subjects have a different sensory perception of foods than normal weight subjects. Their threshold and supra-threshold sensitivity to various tastes (sweet, salt, bitter, sour) is not different from normal weight individuals. There have been some studies that have tried to link obesity with PROP sensitivity, but these attempts have led to mixed results, with the clear majority of the studies showing no relationship between PROP status and BMI (Mattes, 2004). There is little information available with respect to odor sensitivity and obesity, but a priori there does not seem a compelling reason why odor perception should be different in obese and normal weight individuals.

One area regarding sensory perception, which might be of interest to study in relation to obesity is fat perception. Fat is a texture attribute, and humans are not very well able to discriminate between small changes in fat concentrations in foods. People are much more sensitive to relative changes in sugar/salt concentrations than to fat. Mattes (2001, 2002b) showed that the oral exposure to free fatty acids in foods leads to elevated post-prandial triacylglycerol blood levels. These findings indicate that free fatty acids are 'sensed' on the tongue, and that they have metabolic consequences in terms of the postprandial lipid profile.

In relation to this observation, there are studies with rats by Gilbertson *et al.* (1998), who showed that specific strains of rats are more or less sensitive to the taste of linoleic acid. The insensitivity to taste of linoleic acid was inversely related to dietary preferences for fat. Kamphuis (2003) recently tried to distinguish between linoleic acid tasters and non-tasters in humans. They also tried to relate this sensitivity to food intake and fat preferences in humans. The relevance of this phenomenon for human taste perception remains to be elucidated (Kamphuis, 2003).

2.6.2   Sensory preferences and obesity

In the 1970s it was thought that obese individuals were characterized by a sweet tooth, i.e. a liking for high sweetness levels in foods. The rationale behind this idea is simple and attractive. Obese subjects ingest more energy than lean subjects; sweetness is the outspoken biological signal for the energy content of foods; a higher liking for higher sugar levels in obese individuals would then be a driver for a higher energy intake. Empirical studies on this hypothesis did not confirm this idea. This failure to find any effect of weight status on sweetness preferences is very consistent across a large number of studies Drewnowski (1987), Frijters (1984), Esses and Herman (1984), Thompson *et al.* (1977), Rodin *et al.* (1977), and Rissanen *et al.* (2002).

Although it is clear that on average obese and normal weight subjects do not differ in preferences for sugar, there are several studies that suggest that there is a difference in the preference for fat. Studies from Drewnowski (1987) on optimal sugar and fat levels in fat-sugar mixtures showed that obese subjects preferred higher fat levels than normal weight subjects. A study of Mela and Sacchetti (1991) showed a positive association between body fat percentage and the average optimal preferred fat levels in ten different foods across a group of 30 subjects. In a more recent study of Fisher and Birch (1995) with 18 3–5-year-old children it was shown that fat preferences as measured by a sensory test predicted fat intake from a standard menu. The fat preferences/intakes in the children were also positively associated with their parents BMI. In a large survey study with 428 4–5-year-old children, Wardle *et al.* (2001) observed that the children from the obese/overweight families had a higher preference for fatty foods in a taste test and a lower liking for vegetables.

Data from Rissanen *et al.* (2002) provide an interesting additional perspective. In a survey study with 23 pairs of monozygotic twins with discordant BMIs, the obese twins expressed a much higher preference for high fat foods than their lean counterparts. The obese twins also reported that they had a higher tendency to overeat from sandwiches, pastries and ice-cream, but not from sweets and soft drinks. The authors concluded that the acquired preference for fatty foods is associated with obesity.

In line with the data on a higher fat preference, there are results of studies that suggest that obese subjects have a relatively higher intake of foods with a higher energy density than normal weight subjects. Using data from a British national food survey MacDiarmid *et al.* (1996) showed that obese subjects had a higher consumption of high-fat/high-sugar food than subjects with a lower BMI. In a study with 41 lean subjects and 35 obese subjects, Cox *et al.* (1999) found that the obese subjects appeared to consume a diet higher in energy density, which was particularly associated with intakes of salty/savory food items. In a study with 34 obese and 34 normal weight subjects, Westerterp-Plantenga (2004) showed that the obese subjects had a relatively higher intake from energy dense foods (15–22.5 kJ/g), and a relatively lower intake from low energy dense foods (< 0–7.5 kJ/g). Le Noury *et al.* (2002) as cited by Yeomans *et al.* (2004) reported that obese subjects consumed greater amounts of high fat foods in a test-meal than their normal weight counterparts, and reported greater feelings of pleasantness and satisfaction with high-fat foods.

As far as the author is aware there are not many other data concerning differences in sensory preferences of obese and normal weight subjects. There seems to be no obvious reason why the sensory preferences of obese subjects are different from those of normal weight subjects. The issues around preferences for high-fat/high-energy dense foods in obese subjects warrant further study. One of the central questions in this respect is how these preferences develop, and how these preferences translate into actual eating behavior.

2.6.3   Responsiveness to palatability with respect to intake in obese consumers

In a fascinating paper in the journal *Science* in 1968, Schachter presented a number of new ideas about obesity and eating behavior (Schachter, 1968). He suggested that eating behavior of obese and normal weight subjects was differentially affected by internal and external cues. Schachter argued that obese individuals were more responsive to external cues not directly related to hunger, whereas normal weight subjects were more responsive to internal cues. External cues are signals related to emotional state (fear, stress, arousal), external environmental circumstances (e.g., easy-difficult access, time of day), and the palatability (unpleasant vs. pleasant) of food. In his paper, Schachter reported a series of experiments, the results of which supported the idea that obese subjects were more responsive to external signals whereas lean subjects 'listened' better to hunger signals. This theory was called the externality hypothesis.

After the presentation of the externality hypothesis a large number of studies were done to test this idea. Then, it appeared that there was not such a clear distinction between obese and normal weight subjects with respect to internal-external sensitivity. Obese subjects also responded to internal cues (such as hunger/deprivation time) and normal weight subjects also responded to external cues such as the palatability of foods. Also, the definition of an 'external signal' was not clear. Is arousal or anxiety an external signal? In addition, a large number of studies failed to replicate the internal-external difference between overweight and normal weight subjects. This was particularly the case for issues such as deprivation time, the effects of arousal, and cognitive and social cues (Spitzer and Rodin, 1981). These 'failures' to confirm the original ideas led in 1981 to a paper by Rodin (1981), with the title 'Current status of the internal-external hypothesis for obesity – What went wrong?'. After this paper, it became silent with respect to this idea for some time. However, the basic idea behind the externality hypothesis, i.e., that people are responsive to external cues, such as environmental circumstances/cues (social facilitation, portion sizes) and sensory stimulation is now at the center of many theories about overeating and the high prevalence of obesity (see also Herman *et al.*, 2005). Ideas with respect to the 'obesogenic environment' are a reflection of this concept (Finkelstein *et al.*, 2005).

In his paper Schachter also reported the results of two experiments (Hashim and van Itallie 1965; Nisbett, 1968) which were related to taste. In the study of Hashim and van Itallie (1965) it was shown that obese subjects showed a dramatic decrease in food intake when put on an unpalatable bland liquid diet regime, whereas normal weight subjects maintained their intakes at levels sufficient for energy balance. In the study of Nisbett (1968) two types of ice cream were used, one 'delicious' ice cream, and one ice cream with added quinine (called vanilla bitter). The results showed that the overweight subjects consumed more of the palatable ice cream, but not of the unpalatable ice cream. The results of these two experiments suggested that obese subjects eat less of unpalatable food than normal weight subjects, but eat more of palatable food, i.e. the hedonic value of a food had a stronger effect on food intake of obese individuals than of normal weight subjects.

Unlike with the other 'external' factors, this finding was later consistently replicated in a number of other studies (Hill, 1974; Hill and McCutcheon, 1975 (see Fig. 2.6); Rodin, 1975; Rodin *et al.* 1977; Spiegel and Steller, 1990). This finding was also replicated with normal weight and obese children (Ballard *et al.*, 1980).

In conclusion, the general hypothesis of a higher external responsiveness of obese subjects to external cues cannot be maintained in all its dimensions. However, there are quite a number of studies that confirmed the stronger effect of palatability on intake in obese subjects compared to normal weight subjects. This is one of the most consistently observed differences in eating behavior between normal weight and obese subjects. The higher sensitivity to palatability could be related to the idea that eating palatable foods is more rewarding (reinforcing) for obese subjects than for normal weight subjects.

2.6.4   Food reward and weight status

Ice-cream sales are higher in the summer than they are in the winter. In the ski-season in the Alps, many consumers show a high desire for a product called 'Glüh-wein', a product that makes you feel warm. The pleasantness of these sensations relate to the 'physiological usefulness' in relation to the setpoint for body temperature (Cabanac, 1975).

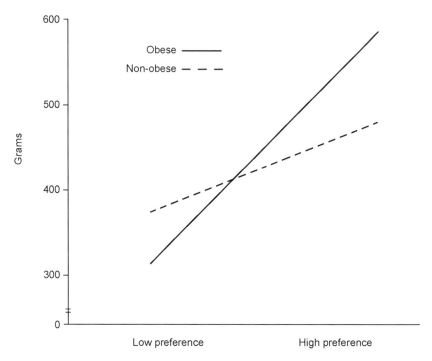

Fig. 2.6 Illustration of the externality effect. Mean grams of food eaten by seven obese and seven normal weight subjects during two low preference and two high preference dinner meals. Mean preference scores on an 11-point scale were 3.9 for the low preference meal and 8.6 for the high preference meals (Source: Hill and McCutcheon, 1975).

Cabanac (1971) argued that body weight also has its setpoint, and that the body defends this setpoint by changing the hedonics of foods, i.e. liking is modulated by the 'need-state' of the individual. The setpoint does not need to be on a body weight, which is optimal from an aesthetic or health point of view. The idea of a setpoint is widely supported by many animal studies (Woods and Seeley, 2002), showing that animals depending on genetic make-up and environmental circumstances defend a particular body weight. This idea also explains the observations in many human studies where after weight loss/starvation and/or short term food deprivation, most people return to their original body weight through increased energy consumption (Jeffery et al., 2004). The question within the framework of this chapter is whether or not the return to setpoint weight is guided by changes in the hedonic responses to food.

In one of Cabanac's original papers he showed that 'hungry' people before a preload of 50 g glucose showed an increased liking response to increasing sugar concentrations (Cabanac, et al., 1973; Duclaux et al., 1973). After the preload there was an optimal preferred sugar concentration, above which the pleasantness of the sugar concentration decreased. This latter observation is in line with the idea of the normal Wundt curve for sensory stimulation (see the section on liking in relation to arousal in the chapter). In the *Science* paper in 1971, Cabanac (1971) showed that people (three subjects) below their setpoint weight continued to like strong tasting/smelling solutions, even after a preload of 50 g glucose. The implication was that when people are below their setpoint weight, food intake during a meal will not satiate as quickly as when people are at or above their setpoint, i.e. food maintains its rewarding properties. This observation is in line with the idea that compensatory behavior after weight loss works through increases in meal size (Woods and Seeley, 2002).

Although Cabanac's idea is attractive and appealing, there have been only a very few studies that actually produced data from humans in line with this idea. Fantino *et al.* (1983) showed that overfed people (three North African women who were overfed before their marriage) expressed a lower liking for sweetness. A study from Raynor and Epstein (2003) found that the relative-reinforcing value of food increases after short-term (13 h) food deprivation. However, it is not clear whether or not the reinforcing value of food is directly/linearly related to the hedonic value of the food. Kleifeld and Lowe (1991) found no increase in sweetness liking after weight loss. Another study of Kaufman *et al.* (1995) showed that 20 h deprived subjects ate more of good tasting foods, but less of bad tasting foods, i.e. 'mild' food deprivation made subjects more finicky about the food they ate. This latter observation is not in line with the concept of liking as a straightforward function of physiological usefulness.

The reinforcing/rewarding value of foods has two components, which are sometimes difficult to distinguish in humans, i.e. 'liking', and 'wanting' (Berridge, 2004). Liking refers to the sensory-hedonic dimension, whereas wanting reflects the drive to ingest a particular food. A high wanting level does not necessarily coincide with a high liking level and vice versa. Pizza may be one's favourite food; most people do not want it for breakfast. Similarly, water is probably not the 'drink' with the highest hedonic value, but it is a drunk in large quantities. To summarize, it seems clear that liking in humans is not always dependent on need state (Yeomans *et al.*, 2004). It is also not clear that the hedonic value of a food increases when subjects are in a negative energy balance. More data and better theories are necessary to understand the effects of palatability on intake, and also to understand the effects of weight status on palatability and intake.

2.6.5   Sensory specific satiety, sensory cues and obesity

As discussed above, there are quite a number of studies that suggest that obese subjects are more responsive to the hedonic value of foods than normal weight subjects. One of the potential mechanisms that could explain this observation is the hypothesis that obese subjects are less sensitive to sensory specific satiety than normal weight subjects. This would manifest itself in a lower decline in the reward value of foods during consumption, i.e. obese subjects would continue to get reward from tasting a food compared to normal weight subjects. This leads to the postponement of meal termination, i.e. larger meals and higher energy intake. On the other hand, one could also make a case for the hypothesis that obese subjects are more sensitive to sensory specific satiety (Raynor and Epstein, 2001). A higher sensitivity may lead to earlier switching between foods/ flavors during and/or across meals. The resulting high variety of foods may lead to a higher energy intake.

There are some experimental data that support the hypothesis of a lower sensitivity to sensory specific satiety in obese subjects compared to normal weight subjects. In a study of Epstein *et al.* (1996) both 10 obese and 10 normal weight subjects were repeatedly (10 ×) stimulated to palatable food cues (lemon yogurt). The dependent variable in this study was the salivary response which can be considered as a measure for the desire to eat. The obese women showed a significantly slower decline in salivation than non-obese subjects. More recent data from Jansen *et al.* (2003), with obese and normal weight children, showed that after intense olfactory exposure, normal weight children decreased their intake during a test meal, whereas obese children overate. So, in both studies, the sensory exposure reduced the desire to eat in normal weight subjects but failed to do so in obese subjects.

The idea that obese subjects are less sensitive to sensory specific satiety was recently tested in a study with 21 obese and 23 normal weight women, matched for age, and restrained eating behavior (Snoek *et al.*, 2004). Food intake, appetite ratings and liking scores were measured before and after an *ad libitum* lunch. The experimental products differed in fat content (low, high) and taste (sweet, savory). The study comprised two experiments, one with sandwiches, and one with snacks. The results showed that the obese and non-obese subjects did not differ with respect to sensory specific satiety, i.e. the decline in liking ratings was about equal for obese and normal weight subjects. However, appetite ratings for something sweet and something savory after lunch were consistently higher for the obese than for the normal weight subjects. This finding shows that even after eating until satiation, obese subjects still expressed a higher wanting level to eat foods.

Apparently, obese subjects are not less sensitive to sensory specific satiety, but they do have a higher tendency to continue to eat. One hypothesis could be that obese subjects are more responsive to sensory cues, even after they are 'physiologically satiated'. This hypothesis is partly reflected in the theory of cue-reactivity of Nederkoorn and Jansen (2002) with respect to eating binges. 'This theory states that when a person regularly has eating binges, and these binges are reliably preceded by certain cues (e.g., the sight, smell and taste of the food, environment, cognitions, emotions), these cues becomes predictors of a binge' (Nederkoorn and Jansen, 2002).

The continued interest in sensory reward during a meal might be an attractive hypothesis why obese people have bigger meals than normal weight subjects. Much more work is needed in this area.

2.7   Discussion and conclusion

Sensory factors have a large effect on food intake and choice. Consumers choose foods that they like, and avoid foods that they do not like, and people eat more when the food is more palatable. When taking sensory factors into account with food choice and intake, it is helpful and important to distinguish between liking and wanting a particular food. Liking refers to pleasantness of the taste, whereas wanting refers to the desire to eat a particular food at a particular moment. It is important that the sensory field develops new methodologies that can distinguish between liking and wanting. These new methodologies refer to behavioral tests, but may also refer to methodologies that can measure motivational states on a more implicit level, such as physiological (e.g., skin conductance, brain imaging) signals related to the autonomic nervous system, or the central nervous system (see, e.g., Mela, 2006).

In relation to the point above, we can see a new research field emerging. It is becoming increasingly clear that sensing a food while we are eating has many more consequences than just the elicitation of taste and smell sensations. We know what to eat for breakfast in order to stay satiated until lunch. We know how filling chocolate is compared to chocolate milk. These observations imply that we have learned to associate the sensory signals of particular foods with the metabolic consequences. An excellent example of this type of work is from Mattes (2001, 2002b), who showed that oral exposure to free fatty acids in foods leads to elevated post-prandial triacylglycerol blood levels. In a similar way, it is clear that certain tastes/flavors are associated with certain emotional states, e.g. we know that sweetness has a calming, relaxing effect. These types of effect may also have a physiological background. These notions have far-reaching consequences for the field of sensory science. It means that sensory science will also get involved in the physiological consequences of sensory signals. This is an exciting development, which opens up a wealth of new research questions, which have great relevance for nutritionists, marketing researchers, public health institutions and food industries.

One example from the appetite area that relates to this new sensory field is the observation that solid foods have a higher satiating effect than liquid foods (Hulshof *et al.*, 1993; Mattes, 1996). People compensate much better after eating solid calories than after eating liquid calories (Mattes, 1996). One big distinction between solid and liquid foods is that they are processed differently in the mouth. It is probable that solid foods give rise to many more sensory signals than liquid foods. One exception to this rule is soup, which is a liquid which is eaten at a slow rate. Soups are also the only liquids that do have a higher satiating power (e.g., Mattes, 2005). It is hypothesized that solid foods are more satiating than liquids because solid foods produce sufficient sensory signals that are associated with the metabolic consequences. Therefore, we may learn from solid foods that they are satiating, while it seems more difficult to learn that from liquid foods. An interesting observation in line with this idea is that during evolution humans have never been exposed to liquid calories, except for the time just after birth, where babies get milk – this is a stage of rapid growth and a positive energy balance. This is a fascinating area for future research.

With respect to the development of taste preferences, it has become increasingly clear in recent years that many preferences are established at a very early age. Learning to like tastes is based on conditioning principles, and may start even before birth. This idea also applies to the concept of variety. Being exposed to a larger variety of tastes early in life may make children less neo-phobic and more ready to accept new flavors. Early established food preferences may have an impact that lasts for years. From this perspective it makes much sense to focus research on how early flavor preferences are established, and how long their impact lasts. It seems probable that this impact is different for different flavors. Which flavors from the mother's diet have an impact on the preferences of their children, how and which flavor enters into the mother's milk? If one gives a variety of vegetables to the infant after weaning, will this also help to form vegetable preferences later on?

During infancy, exposure and post-ingestive consequences form the basis of the establishment of stable food preferences. After infancy, preferences are also acquired through social processes like social rewards/punishments and modeling. One of the striking results in this area is that it is difficult to establish a longer lasting preference for vegetables (e.g. Hendy et al., 2005). This may have to do with the low energy content of vegetables, which results in little post-ingestive satiety. In this field it may help if we also take the cognitive development of children into account when designing social reinforcement strategies to increase vegetable preferences in liking. Eleven-year-old children need other reinforcement strategies than four year olds. Another strategy might be to investigate whether or not we can make vegetables more attractive by changing flavors and/or texture. We still know very little about which sensory properties of vegetables are disliked.

Early learned preferences have long term consequences. The liking for particular foods can be stable for years and years. However, the desire to eat (wanting) a particular product varies from moment to moment. We have seen that sensory specific satiety and/or boredom are very robust effects. People get satiated for the sensory properties of foods. The dimension sweet-savory is important in this respect, and it is also clear that nutritional properties are less relevant for sensory specific satiety or boredom. Further research is needed on the identification of the specific sensory properties that play a role in sensory specific satiety and boredom. It appears that the sensory intensity and the perceived complexity are relevant, but so far there have been few studies that have produced consistent effects.

Boredom/sensory specific satiety may actually play an important role in the working mechanisms of popular diets, like the Atkins diet (Atkins, 2002) or the Montignac diet (Montignac, 1997). In the Atkins diet, major sources of carbohydrates are removed from the diet. Major sources of carbohydrates are the staple foods such as bread, potatoes, pasta or rice, which, in general, have a neutral taste. In the Montignac diet, one is allowed to consume both fats and carbohydrates, but not combined in the same meal. In effect, this has the same consequences as the Atkins diet; it removes many neutral tasting foods from the meal. The result is that the meals only contain the more flavored items, such as the meat part and the vegetable part. These products cause earlier sensory specific satiety and/or monotony, probably because of their more intense taste. This early sensory satiety/boredom may result in a lower *ad libitum* food intake.

There is still much to learn in the field of sensory preferences of children. Most of the research on food preferences in children has been done with sweeteners. There is some work on salt, sour, bitter, and texture, and there little work on odors. The methodology of sensory research in children is also limited, with it strong reliance on verbal reports. We know that people do not always do what they say. Within this field it might be very helpful to make use of more advanced methods for the study of behavior, for example by making use of facial expression analysis, or making use of non-invasive measures that reflect autonomous nervous system activity (skin conductivity; heart rate variability, etc.).

This issue of changes in food preferences with age and their consequences for food intake and nutritional status remains a complex field. It is very clear that on average chemosensory sensitivity declines with age; however, it is also clear that the elderly is a heterogeneous group in this respect. The consequences of ageing for changes in food preferences are not well understood yet; it is clear that changes in preferences with age may be different for different flavors/tastants. The effects of changing preferences for food intake are also not well established. There seems a strong need to get a more thorough understanding of the underlying mechanisms that accompany chemosensory sensitivity. It may not be until we understand the underlying mechanisms, that we can say more about the consequences for preferences and intake (e.g., Mattes, 2002a).

Within in the field of sensory responses and obesity, it is clear that normal weight and obese consumers did not vary to a large extent in their perceptions or their sensory preferences. It seems that obese subjects have a larger responsiveness to palatability of foods than normal weight subjects. This relates more to 'wanting' of foods than to 'liking' of foods. This observation reinforces the urgency of finding good methods to measure 'wanting', and to find ways in which foods can effectively satisfy the wanting/desires without much energy.

In conclusion, the research area of relating sensory signals to food intake is a rapidly expanding area, with many exciting developments. The opening up of the sensory field to physiological responses is an important area that is highly relevant to the current major health problems in the industrialized world.

2.8   References and further reading

ADDESSI E, GALLOWAY AT, VISALBERGHI E, BIRCH LL (2005) Specific social influences on the acceptance of novel foods in 2–5–year–old children *Appetite*, 45, 264–271.
APPLETON KM (2005) Changes in the perceived pleasantness of fluids before and after fluid loss through exercise: a demonstration of the association between perceived

pleasantness and physiological usefulness in everyday life. *Physiol Behav*, 83, 813–819.

APPLETON KM, GENTRY RC, SHEPERD R (2006) Evidence of a role for conditioning in the development of liking for flavours in humans in everyday life. *Physiol Behav*, 87, 478–486.

ATKINS RC (2002) *Dr. Atkins new diet revolution.* Avon Books, New York.

BALLARD BD, GIPSON MT, GUTTENBERG W, RAMSEY K (1980) Palatability of food as a factor influencing obese and normal-weight children's eating habits. *Behav Res Therapy*, 18, 598–600.

BELL EA, ROE LS, ROLLS BJ (2003) Sensory-specific satiety is affected more by volume than by energy content of a liquid food. *Physiol Behav*, 78, 593–600.

BELLISLE F, LUCAS F, AMRANI R, LEMAGNEN J (1984) Deprivation, palatability and the microstructure of meals in human subjects. *Appetite*, 5, 85–94.

BERLYNE DE (1970) Novelty, complexity and hedonic value. *Percept Psychophys*, 8, 279–286.

BERNSTEIN IL (1978) Learned taste aversions in children receiving chemotherapy. *Science*, 200, 1302–1323.

BERRIDGE KC (1996) Food reward: brain substrates of wanting and liking. *Neurosci Biobehav Rev*, 28, 1–25.

BERRIDGE KC (2004) Motivation concepts in behavioural neuroscience. *Physiol Behav*, 81, 179–209.

BINGHAM A, HURLING R, STOCKS (2005) Acquisition of liking for spinach products. *Food Qual Prefer*, 16, 461–469.

BIRCH LL (1979) Dimensions of children's food preferences and consumption patterns. *J Nutr Educ*, 11, 189–192.

BIRCH LL (1980) Effect of peers models' food choices and eating behaviours on preschoolers food preferences. *Child Develop*, 51, 489–496.

BIRCH LL, MARLIN DW (1982) I don't like it; I never tried it: effects of exposure on two year old children's food preferences. *Appetite*, 3, 353–360.

BIRCH LL, ZIMMERMAN S, HIND H (1980) The influence of social-affective context on preschool childrens' food preferences. *Child Develop*, 51, 856–961.

BIRCH LL, BRICH D, MARLIN D, KRAMER L (1982) Effects of instrumental eating on children's food preferences. *Appetite*, 3, 125–134.

BIRCH LL, MCPHEE L, SHOBA BC, PIROK E, STEINBERG L (1987) What kind of exposure reduces children's food neophobia? Looking vs. tasting. *Appetite*, 9, 171–178.

BIRCH LL, MCPHEE L, STEINBERG, SULLIVAN S (1990) Conditioned flavor preferences in young children. *Physiol Behav* 47, 501–505.

BOBROFF EM, KISSILEFF HR (1986) Effects of changes in palatability on food intake and the cumulative food intake curve in man. *Appetite*, 7, 85–96.

CABANAC M (1971) Physiological role of pleasure. *Science*, 173, 1103–1107.

CABANAC M (1975) Temperature regulation. *Annu Rev Physiol*, 37, 415–439.

CABANAC M, RABE F (1976) Influence of a monotonous food on body weight regulation in humans. *Physiol Behav*, 17, 675–678.

CABANAC M, PRUVOST M, FANTINO M (1973) Negative alliesthesia for sweet stimuli after varying ingestions of glucose. *Physiol Behav*, 11, 345–348.

CARDELLO AV, SCHUTZ HG (1996) Food appropriateness measures as an adjunct to consumer preference/acceptability evaluation. *Food Qual Prefer*, 7, 239–249.

COX DN, PERRY L, MOORE PB, VALLUS L, MELA DJ (1999) Sensory and hedonics associations with macronutrient and energy intake of lean and obese consumers *Int J Obes*, 23, 403–410.

DE GRAAF C, ZANDSTRA EH (1999) Sweetness intensity and pleasantness in children, adolescents and adults. *Physiol Behav* 67, 513–520.

DE GRAAF C, SCHREURS A, BLAUW YH (1993) Short-term effects of different amounts of sweet and nonsweet carbohydrates on satiety and energy intake. *Physiol Behav*, 54, 833–843.

DE GRAAF C, POLET P, VAN STAVEREN WA (1994) Sensory perception and pleasantness of food flavours in elderly subjects. *J Gerontol: Psychol Sci*, 49, 93–99.

DE GRAAF C, VAN STAVEREN WA, BUREMA JA (1996) Psychophysical and psychohedonic functions of four common food flavours in elderly subjects. *Chem Senses*, 21, 293–302.

DE GRAAF C, BLOM WAM, SMEETS P, STAFLEU A, HENDRIKS HFJ (2004) Biomarkers of satiation and satiety. *Am J Clin Nutr*, 79, 946–961.

DE GRAAF C, KRAMER FM, MEISELMAN HL, LESHER LL, BAKER-FULCO C, HIRSCH ES, WARBER J (2005) Food acceptability in field studies with US Army men and women: relationship with food intake and food choice after repeated exposures. *Appetite*, 44, 23–31.

DE JONG N, DE GRAAF C, VAN STAVEREN WA (1996) Effect of sucrose in breakfast items on pleasantness and food intake in the elderly. *Physiol Behav*, 60, 1454–1462.

DESOR JA, BEAUCHAMP GK (1987) Longitudinal changes in sweet preferences in humans. *Physiol Behav*, 39, 639–641.

DESOR JA, GREENE LS, MALLER O (1975) Preferences for sweet and salty in 9–15-year-old and adults. *Science*, 190, 686–687.

DREWNOWKSI A (1987) Body weight and sensory preferences for sugar and fat. *Can Inst Food Sci Technol J*, 20, 327–330.

DREWNOWSKI A (1998) Energy density, palatability, and satiety: implications for weight control. *Nutr Rev*, 56, 347–353.

DUCLAUX R, FEISTHAUER J, CABANAC M (1973) Effects of a meal on the pleasantness of food and non-food odors in man. *Physiol Behav*, 10, 1029–1033.

EPSTEIN LH, PALUCH R, COLEMAN KJ (1996) Differences in salivation to repeated food cues in obese and nonobese women. *Psychosom Med*, 58, 160–164.

ESSED NH, VAN STAVEREN WA, KOK FJ, DE GRAAF C (2006) No effect of 16 weeks flavor enhancement on dietary intake and nutritional status of nursing home elderly. *Appetite*, in press.

ESSES VM, HERMAN CP (1984) Palatability of sucrose before and after glucose ingestion in dieters and nondieters *Physiol Behav*, 32, 711–715.

FANTINO M, BAIGTS F, CABANAC M, APFELBAUM M (1983) Effects of an overfeeding regimen – the affective component of a sweet sensation. *Appetite*, 4, 155–164.

FINKELSTEIN EA, RUHM CJ, KOSA KM (2005) Economic causes and consequences of obesity. *Ann Rev Public Health*, 26, 239–257.

FISHER JO, BIRCH LL (1995) Fat preferences and fat consumption of 3- to 5-year-old children are related to parental adiposity. *J Am Diet Assoc*, 95, 759–764.

FORDE CG, DELANHUNTY CM (2004) Understanding the role cross-modal interactions play in food acceptability in younger and older consumers. *Food Quality Prefer* 15, 715–727.

FRIJTERS JE (1984) Sweetness intensity perception and pleasantness in women varying in reported restraint of eating. *Appetite*, 5, 102–108.

GATENBY , TRUNCK F, AARON JI (1994) No effect of extended home use on liking for sensory characteristics of reduced-fat foods. *Appetite*, 21, 117–129.

GERRISH CJ, MENNELLA JA (2001) Flavour variety enhances food acceptance in formula-fed infants. *Am J Clin Nutr*, 73: 1080–1085.

GIBSON EL, WARDLE J (2003) Energy density predicts preferences for fruits and vegetables in 4-year-old children. *Appetite*, 41, 97–98.

GILBERTSON TA, LIU L, YORK DA, BRAY GA (1998) Dietary fat preferences are inversely correlated with peripheral fatty acid sensitivity. *Ann N Y Acad Sci*, 855, 165–168.

GOLDFIELD GS, EPSTEIN LH, DAVIDSON M, SAAD F (2005) Validation of a questionnaire measure of the relative reinforcing value of food. *Eat Behav*, 6, 283–292.

GRIEP MI, METS TF, MASSART DL (1997) Different effects of flavor amplification of nutrient dense foods on preference and consumption in young and elderly subjects. *Food Qual Prefer*, 8, 151–156.

GRIEP MI, METS TF, MASSART DL (2000) Effects of flavour amplification of Quorn and yoghurt on food preference and consumption in relation to age, BMI and odour perception. *Brit J Nutr*, 83, 105–113.

GRINKER J, HIRSCH J, SMITH DV (1972) Taste sensitivity and suspectibility to external influence in obese and normal weight subjects. *J Person Soc Psychol*, 22, 320–325.

GUINARD J-X, BRUN P (1998) Sensory-specific satiety: comparison of taste and texture effects. *Appetite*, 31, 141–157.

GUY-GRAND B, LEHNERT V, DOASSANS M, BELLISLE F (1994) Type of test-meal affects palatability and eating style in humans. *Appetite*, 22, 125–134.

HALL WL, MILLWARD DJ, LONG SJ, MORGAN LM (2003) Casein and whey exert different effects on plasma amino acid profiles, gastrointestinal hormone secretion and appetite. *Brit J Nutr*, 89, 239–248.

HALLER R, RUMMEL C, HENNEBERG S, POLLMER U, KOSTER EP (1999) The influence of early experience with vanillin on food preference later in life. *Chemical Senses*, 24, 465–467.

HASHIM SA, VAN ITALLIE B (1965) Studies in normal and obese subjects with a monitored food dispensing device. *Ann N Y Acad Sci*, 131, 654–691.

HELLEMAN U, TUORILA H (1991) Pleasantness ratings and consumption of open sandwiches with varying NaCl and acid contents. *Appetite*, 17, 229–238.

HENDY HM (2002) Effectiveness of trained peer models to encourage food acceptance in preschool children. *Appetite*, 39, 217–225.

HENDY HM, RAUDENBUSH B (2000) Effectiveness of teacher modelling to encourage food acceptance in preschool children *Appetite*, 34, 61–71.

HENDY HM, WILLIAMS KE, CAMISE TS (2005) 'Kids Choice' school lunch program increases children's fruit and vegetable acceptance. *Appetite* 45, 250–263.

HERMAN CP, POLIVY J, LEONE T (2005) The psychology of overeating. In Mela D (ed.), *Food, Diet and Obesity.* Cambridge, UK: Woodhead Publishing.

HETHERINGTON MM (1995) Sensory-specific satiety and its importance in meal termination. *Neurosci Biobehav Rev*, 20, 113–117.

HETHERINGTON MM, BELL A, ROLLS BJ (2000) Effects of repeat consumption on pleasantness, preference and intake. *Brit Food J* 102, 507–521.

HETHERINGTON MM, PIRIE LM, NABB S (2002) Stimulus satiation: effects of repeated exposure to foods on pleasantness and intake. *Appetite*, 38, 19–28.

HILL SW (1974) Eating responses of humans during dinner meals. *J Compar Physiol Psychol*, 86, 652–657.

HILL SW, McCUTCHEON NB (1975) Eating responses of obse and nonobese humans during dinner meals. *Psychosom Med*, 37, 395–401.

HORNE PJ, TAPPER K, LOWE CF, HARDMAN CA, JACKSON MC, WOLOLNER J (2004) Increasing children's fruit and vegetable consumption: a peer modelling and rewards-based intervention. *Eur J Clin Nutr*, 58, 1649–1660.

HULSHOF T, DE GRAAF C, WESTSTRATE JA (1993) The effects of preloads varying in physical state and fat content on satiety and energy intake. *Appetite*, 21, 273–286.

JANSEN A, THEUNISSEN N, SLECHTEN K, NEDERKOORN C, BOON B, MULKENS S, ROEFS A (2003) Overweight children after exposure to food cues. *Eating Behav*, 4, 197–209.

JEFFERY RW, KELLY KM, ROTHMAN AJ, SHERWOOD NE, BOUTELLE KN (2004) The weight loss experience: a descriptive analysis. *Ann Behav Med*, 27, 100–106.

JOHNSON SL, MCPHEE L, BIRCH LL (1991) Conditional preferences: young children prefer flavours associated with high dietary fat. *Physiol Behav*, 50, 1245–1251.

KAMPHUIS M (2003) The sense of dietary fat: food intake and body weight regulation. PhD thesis, University Maastricht, Datawyse, Universitaire Pers Maastricht, The Netherlands.

KAUFMAN NA, HERMAN CP, POLIVY J (1995) Hunger-induced finickiness in humans. *Appetite*, 24, 203–218.

KERN DL, MCPHEE L, FISHER J. JOHNSON S, BIRCH LL (1993) The postingestive consequences of fat condition preferences for flavours associated with high dietary fat. *Physiol Behav*, 54, 71–76.

KLEIFELD EI, LOWE MR (1991) Weight loss and sweetness preferences: the effects of recent vs. past weight loss. *Physiol Behav*, 49, 1037–1042.

KOSKINEN S, TUORILA H (2005) Performance on an odor detection and identification test as a predictor of ortho- and retronasal odor intensity ratings in the young and elderly. *Food Quality Prefer*, 16, 383–292.

KOSKINEN S, KALVIAINEN N, TUORILA H (2003) Flavor enhancement as a tool for increasing pleasantness and intake of a snack product among the elderly. *Appetite*, 41, 87–96.

KOZLOWSKA K, JERUSZKA M, MATUSZEWSKA I, ROSZKOWSKI W, BARYLKO-PIKIELNA N, BRZOZOWSKA A (2003) Hedonic tests in different locations as predictors of apple juice consumption at home in elderly and young subjects. *Food Quality Prefer*, 14, 653–661.

LAWLESS HT, HEYMAN H (1998) *Sensory evaluation of food: principles and practices.* Kluwer Academic Publishers, Dordrecht, the Netherlands.

LE MAGNEN J (1986) *Hunger.* Cambridge University Press, Cambridge.

LENNERNAS M, FJELLSTROM C, BECKER W, GIACHETTI I, SCHMITT A, REMAUT DE WINTER A, KEARNEY M (1997) Influences on food choice perceived to be important by nationally-representative samples of adults in the European Union. *Eur J Clin Nutr*, 51, Suppl 2, S8–S15.

LE NOURY JC, LAWTON C, BLUNDELL JE (2002) Food choice and hedonic responses: difference between overweight and lean high fat phenotypes. *Int J Obes*, 26, S125.

LIEM DG, MENNELLA JA (2002) Sweet and sour preferences during childhood; role of early experiences. *Develop Psychobiol*, 41, 388 395.

LIEM DG, MENNELLA JA (2003) Heightened sour preferences during childhood. *Chemical Senses*, 28, 173–180.

LIEM DG, WESTERBEEK A, WOLTERINK S, KOK FJ, DE GRAAF C (2004a) Sour taste preferences of children relate to preference for novel and intense stimuli. *Chemical Senses*, 29, 713–720.

LIEM DG, MARS M, DE GRAAF C (2004b) Consistency of sensory testing with 4- and 5-year-old children. *Food Quality Prefer*, 15, 541–548.

LOWE CE, DOWEY AJ, HORNE PJ (1998) Changing what children eat. In Murcott A (ed.), *The Nation's Diet: The Social Science of Food Choice*, London: Longman, pp 57–80.

LOWE CF, HORNE PJ, TAPPER K, BOWDERY M, EGERTON C (2004) effects of a peer modelling and rewards-based intervention to increase fruit and vegetable consumption in children. *Eur J Clin Nutr*, 58, 510–522.

MACDIARMID JI, CADE JE, BLUNDELL JE (1996) High and low fat consumers, their macronutrient intake and body mass index: further analysis of the National Diet and Nutrition Survey of British Adults. *Eur J Clin Nutr*, 50, 505–412.

MAIER AS, CHABANET C, ISSANCHOU S, SCHAAL B, LEATHWOOD P (2005) Exposure to different regimens of food variety influences the acceptance of new flavours by weanling infants. 6th Pangborn Sensory Science Symposium Oral Presentation no. 57.

MARLIER L, SCHAAL B (2005) Human newborns prefer human milk: conspecific milk odor is attractive without postnatal exposure. *Child Devel*, 76, 155–168.

MATHEY MF, SIEBELINK E, DE GRAAF C, VAN STAVEREN WA (2001) Flavor enhancement of food improves dietary intake and nutritional status of elderly nursing home residents. *J Geront; Biol Sci*, 56A, M200–M205.

MATTES RD (1996) Dietary compensation by humans for supplemental energy provided as ethanol or carbohydrate in fluids. *Physiol Behav*, 59, 179–187.

MATTES RD (2001) The taste of fat elevates postprandial triacylglycerol. *Physiol Behav*, 74, 343–348.

MATTES RD (2002a) The chemical senses and nutrition in aging: Challenging old assumptions. *J Am Dietet Ass*, 102, 192–196.

MATTES RD (2002b) Oral fat exposure increases the first phase triacylglycerol concentration due to release of stored lipid in humans. *J Nutr*, 132, 3656–3662.

MATTES RD (2004) 6-n-propylthiouacil taster status: Dietary modifier, marker or misleaders? In Prescott J, Bepper B, *Genetic Variation in Taste Sensitivity*. Marcel Dekker Inc, pp. 229–250.

MATTES RD (2005) Soup ans satiety. *Physiol Behav* 83, 739–747.

MATTES RD, COWART BJ, SCHIAVO MA, ARNOLD C, GARRISON B, KARE MR, LOWRY LD (1990) Dietary evaluation of patients of with smell and/or taste disorders. *Am J Clin Nutr*, 51, 233–240.

McCRORY MA, FUSS PJ, McCALLUM JE, YAO M, VINKEN AG, HAYS NP, ROBERTS SB (1999) Dietary variety within food groups: association with energy intake and body fatness in men and women. *Am J Clin Nutr*, 69, 440–447.

McKENNA RJ (1972) Some effects of anxiety level and food cues on the eating behaviour of obese and normal weight subjects: a comparison of the Schachterian and psychosomatic conceptions. *J Personal Soc Psychol*, 22, 311–319.

MEISELMAN HL, DE GRAAF C, LESHER LL (2000) The effects of variety and monotony on food acceptance and intake at a midday meal. *Physiol Behav*, 70, 119–125.

MELA DJ (2006) Eating for pleasure or just wanting to eat? Reconsidering sensory hedonic responses as a driver for obesity. *Appetite*, 47, 10–17.

MELA DJ, SACCHETTI DA (1991) Sensory preferences for fats: relationships with diet and body composition. *Am J Clin Nutr*, 53, 908–915.

MELA DJ, TRUNCK F, AARON JI (1993) No effect of extended home use on linking for sensory characteristic of reduced-fat foods. *Appetite*, 21, 117–129.

MENNELLA JA, JAGNOW CP, BEAUCHAMP GK (2001) Prenatal and postnatal flavor learning by human infants. *Pediatrics*, 107, 88–94.

MENNELLA JA, GRIFFIN CE, BEAUCHAMP GK (2004) Flavor programming during infancy. *Pediatrics*, 113, 840–845.

MOJET J, CHRIST-HAZELHOF E, HEIDEMA J (2005) Taste perception with age: pleasantness and its relationship with threshold sensitivity and supra-threshold intensity of five taste qualities. *Food Quality Prefer*, 16, 413–423.

MONTIGNAC M (1997) *Eat Yourself Thin*. Michel-Ange Publishing.

MORLEY JE (2003) Anorexia and weight loss in older persons. *J Geront A Biol Med Scie*, 58, 131–137.

MOSKOWITZ HR (2000) Engineering out food boredom: a product development approach that combines home use tests and time-preference analysis. *Food Quality Prefer*, 11, 445–456.

MURPHY C, WITHEE J (1986) Age-related differences in the pleasantness of chemosensory stimuli. *Psychol Aging*, 1, 312–318.

MURPHY C, WITHEE J (1987) Age and biochemical status predict preference for casein hydrolysate. *J Gerontol*, 42, 73–77.

NEDERKOORN C, JANSEN A (2002) Cue reactivity and regulation of food intake. *Eat Behav*, 3, 61–72.

NICKLAUS S, BOGGIO V, CHABANET C, ISSANCHOU S (2004) A prospective study of food preferences in childhood. *Food Quality Preference* 15, 805–818.

NIJS K, DE GRAAF C, VANNESTE V, KOK FJ, VAN STAVEREN WA (2006) Effect of family style meals on energy intake and risk of malnutrition in Dutch nursing home residents: a randomized controlled trial. *J Gerontol; Med Sci, BMJ*, 332, 1180–1184.

NISBETT RE (1968) Taste, deprivation, and weight determinants of eating behaviour. *J Personal Soc Psychol*, 10, 107–116.

PELCHAT ML, SCHAEFER S (2000) Dietary monotony and food cravings in young and elderly adults. *Physiol Behav*, 68, 353–359.

PILGRIM FJ (1957) The components of food acceptance and their measurement. *Am J Clin Nutr*, 5, 171–175.

PILGRIM FJ, KAMEN JM (1963) Predictors of human food consumption. *Science*, 355, 501–502.

PLINER P (1982) The effect of mere exposure on liking for edible substances. *Appetite*, 3, 283–290.

PLINER P AND STALLBERG-WHITE C (2000) 'Pass the ketchup please': familiar flavours increase children's willingness to taste novel foods. *Appetite*, 34, 95–103.

PORCHEROT C, ISSANCHOU S (1998) Dynamics of liking for flavoured crackers: test of predictive value of a boredom test. *Food Qual Pref*, 9, 21–29.

RAYNOR HA, EPSTEIN LH (2001) Dietary variety, energy regulation, and obesity. *Psychol Bull*, 127, 325–341.

RAYNOR HA, EPSTEIN LH (2003) The relative-reinforcing value of food under different levels of food deprivation and restriction. *Appetite*, 40, 15–24.

RAYNOR HA, JEFFERY RW, TATE DF, WING RR (2004) Relationship between changes in food group variety, dietary intake, and weight during obesity treatment. *Int J Obes Rel Met Disord*, 28, 813–820.

RISSANEN A, HAKALA P, LISSNER L, MATTLAR CE, KOSKENVUO M, RONNER T (2002) Acquired preference especially for dietary fat and obesity: a study of weight-disconcordant monozygotic twin pairs. *Int J Obes Relat Met Disord*, 26, 973–977.

RODIN J (1975) Effects of obesity and set point on taste responsiveness and ingestion in humans. *J Compar Physiol Psychol*, 89, 1003–1009.

RODIN J (1981) Current status of the internal-external hypothesis for obesity, what went wrong. *Am Psychol*, 36, 361–372.

RODIN J, SLOCHOWER J, FLEMING J (1977) Effects of degree of obesity, age of onset and weight loss on responsiveness to sensory and external stimuli. *J Compar Physiol Psychol*, 91, 586–597.

ROEDDER-JOHN D (1999) Consumer socialization of children: a retrospective look at twenty-five years of consumer research. *J Consum Res*, 26, 183–213.

ROININEN K, LATHEENMAKI L, TUORILA H (1999) Quantification of consumer attitudes to health and hedonic characteristics of foods. *Appetite*, 33, 71–88.

ROLLS BJ (1986) Sensory specific satiety. *Nutr Rev*, 44, 93–101.

ROLLS BJ, ROLLS ET, ROWE EA, SWEENEY K (1981a) Sensory specific satiety in man. *Physiol Behav*, 27, 137–142.

ROLLS BJ, ROWE EA, ROLLS ET, KINGSTON B, MEGSON A, GUNARY R (1981b) Variety in a meal enhances food intake in man. *Physiol Behav*, 26, 251–255.

ROLLS BJ, ROWE EA, ROLLS ET (1982) How sensory properties of food affect feeding behaviour. *Physiol Behav*, 29, 409–417.

ROLLS ET, DE WAAL WL (1985) Long term sensory-specific satiety: evidence from an Ethiopian refugee camp. *Physiol Behav*, 34, 1017–1024.

ROLLS ET, ROLLS JH (1997) Olfactory sensory-specific satiety in humans. *Physiol Behav*, 61, 461–473.

ROSE G, LAING DG, ORAM N, HUTCHINSON I (2004a) Sensory profiling by children aged 6–7 and 10–11 years, Part 1: a descriptor approach. *Food Quality Prefer*, 15, 585–596.

ROSE G, LAING DG, ORAM N, HUTCHINSON I (2004b) Sensory profiling by children aged 6–7 and 10–11 years. Part 2: a modality approach. *Food Quality Prefer*, 15, 597–606.

ROZIN P, VOLLMECKE TA (1986) Food likes and dislikes. *Ann Rev Nutr* 6, 433–456.

ROZIN P, HAMMER L, OSTER H, HOROWITZ T, MARMORA V (1986) The child's conception of foods: differentiation of categories of rejected substances in 16 months to 5 year range. *Appetite*, 7, 141–151.

RUSSELL K, DELAHUNTY C, JAEGER SR (2005) Monotony in food products: influence of food complexity on consumer liking and choice over time. 6th Panborn Sensory Science Symposium, Harrogate, UK, Oral Presentation no. 14.

SAELENS BE, EPSTEIN LH (1996) Reinforcing value of food in obese and non-obese women. *Appetite*, 27, 41–50.

SALBE AD, DELPARIGI A, PRATLEY RE, DREWNOWSKI A, TATARANNI PA (2004) Taste preferences and body weight changes in an obesity-prone population. *Am J Clin Nutr*, 79, 372–378.

SCHAAL B, MARLIER L, SOUSSIGNAN R (2000) Human foetuses learn odours from their pregnant mother's diet. *Chem Senses*, 25, 729–737.

SCHACHTER S (1968) Obesity and eating. *Science,* 161, 751–756.

SCHIFFMAN SS, WARWICK ZS (1993) Effect of flavour enhancement of foods for the elderly on nutritional status: food intake, biochemical indices, and anthropometric measures. *Physiol Behav*, 53, 395–402.

SCHUTZ HG (1988) Beyond preference: appropriateness as a measure of contextual acceptance of food. In Thomson DMH (ed.), *Food Acceptability*. Elsevier, New York, pp.115–134.

SCHUTZ HG, PILGRIM FJ (1958) A field study of monotony. *Psycholog Rep*, 4, 559–565.

SHEPHERD R (1988) Sensory influences on salt, sugar and fat intake. *Nutr Res Rev*, 1, 125–144.

SIEGEL PS, PILGRIM FJ (1958) The effect of monotony on the acceptance of food. *Am J Psychol*, 71, 756–759.

SKINNER JD, CARRUTH BR, BOUNDS W, ZIEGLER PJ (2002) Children's food preferences: a longitudinal analysis. *J Am Dietet Ass*, 102, 1638–1647.

SNOEK HM, HUNTJENS L, VAN GEMERT LJ, DE GRAAF, C, WEENEN H (2004) Sensory-specific satiety in obese and normal-weight women. *Am J Clin Nutr*, 80, 823–831

SORENSON LB, MOLLER P, FLINT A, MARTENS M, RABEN A (2003) Effect of sensory perception of foods on appetite and food intake: a review of studies on humans. *Int J Obes*, 27, 1152–1166.

SPIEGEL TA, STELLAR E (1990) Effects of variety on food intake of underweight, normal-weight and overweight women. *Appetite*, 15, 47–61.

SPITZER L, RODIN J (1981) Human eating behaviour: a critical review of studies in normal weight and overweight individuals. *Appetite*, 2, 293–329.

STAFLEU A, DE GRAAF C, VAN STAVEREN WA, BUREMA J (2001) Affective and cognitive determinants of intention to consume that contribute the fat intake. *Ecology of Food and Nutrition*, 43, 193–214.

STEPTOE A, POLLARD TM, WARDLE J (1995) Development of a measure of the motives underlying the selection of food: the food choice questionnaire. *Appetite*, 25, 267–284.

STUBBS RJ, JOHNSTONE AM, MAZLAN N, MBAIWA SE, FERRIS S (2001) Effect of altering the variety of sensorially distinct foods, of the same macronutrient content on food intake and body weight in men. *Eur J Clin Nutr*, 55, 19–28.

STUBBS RJ, WHYBROW S (2004) Energy density, diet composition and palatability: influences on overall food energy intake in humans. *Physiol Behav*, 81, 755–764.

STUBENITSKY K, AARON JI, CATT SL, MELA DJ (1999) Effect of information and extended use on the acceptance of reduced-fat product. *Food Quality Prefer*, 10, 367–376.

SULLIVAN SA, BIRCH LL (1994) Infant dietary experience and acceptance of solid foods. *Pediatrics*, 93, 271–277.

TEPPER BJ (1992) Dietary restraint and responsiveness to sensory-based food cues as measured by cephalic phase salivation and sensory specific satiety. *Physiol Behav*, 52, 305–311.

THOMAS-DANGUIN T, ROUBY C, SICARD G, VIGOUROUX M, FARGET V, JOHANSON A, BENGTZON A, HALL G, ORMEL W, DE GRAAF C, ROUSSEAU F, DUMONT JP (2003) Development of the ETOC: a Europoean test of olfactory capabilities. *Rhinology*, 41, 142–151.

THOMPSON DA, MOSKOWITZ HR, CAMPBELL RG (1977) Taste and olfaction in human obesity. *Physiol Behav*, 19, 335–337.

VICKERS Z, HOLTON E (1998) A comparison of taste test rating, repeated consumption, and postconsumption ratings of different strengths of iced tea. *J Sensory Stud*, 13, 199–212.

WARDLE J, GUTHRIE C, SANDERSON S, BIRCH LL, PLOMIN R (2001) Food and activity preferences in children of lean and obese parents. *Int J Obes*, 25, 971–977.

WARDLE J, HERRERA M-L, COOKE L, GINSON EL (2003a) Modifying children's food preferences: the effects of exposure and reward on acceptance of an unfamiliar vegetable. *Eur J Clin Nutr*, 57, 341–348.

WARDLE J, COOKE LJ, GIBSON EL, SAPOCHNIK M, SHEIHAM A, LAWSON M (2003b) Increasing children's acceptance of vegetables; a randomized trial of parent-led exposure. *Appetite*, 40, 155–162.

WESTERTERP-PLANTENGA MS (2004) Effects of energy density of daily food intake on long-term energy intake. *Physiol Behav*, 81, 765–771.

WOODS SC (1991) The eating paradox: how we tolerate food. *Psychol Rev*, 98, 488–505.

WOODS SC, SEELEY RJ (2002) Understanding the physiology of obesity: review of recent development in obesity research. *Int J Obes Rela Metab Disord*, 26 Suppl 4, S8–S10.

WUNDT W (1907) *Outline of Psychology.* Translated C.H. Judd. Leipzig: Engelsmann.

YEOMANS MR (1996) Palatability and the microstructure of feeding humans: the appetizer effect. *Appetite*, 27, 119–133.

YEOMANS MR, BLUNDELL JE, LESHEM M (2004) Palatability: response to nutritional need or need-free stimulation of appetite? *Brit J Nutr*, 92, Suppl. 1, S3–S14.

ZANDSTRA EH (2000) Preference and satiety; short- and long-term studies on food acceptance, appetite control and food intake. PhD thesis, Wageningen University: Grafisch Service Centrum, Wageningen, The Netherlands.

ZANDSTRA EH, DE GRAAF C (1998) Sensory perception and pleasantness of orange beverages from childhood to old age. *Food Qual Prefer*, 9, 5–12.

ZANDSTRA EH, DE GRAAF C, VAN TRIJP HCM, VAN STAVEREN WA (1999) Laboratory hedonic rating as predictors of consumption. *Food Qual Pref*, 10, 414–418.

ZANDSTRA EH, DE GRAAF C, VAN TRIJP HC (2000) Effects of variety and repeated in-home consumption on product acceptance. *Appetite*, 35, 113–119.

ZANDSTRA EH, STUBENITSKY K, DE GRAAF C, MELA DJ (2002) Effects of learned flavour cues on short-term regulation of food intake in realistic setting. *Physiol Behav*, 75, 83–90.

ZANDSTRA EH, WEEGELS MF, VAN SPRONSEN AA, KLERK M (2004) Scoring or boring? Predicting boredom thorugh repeated in-home consumption. *Food Quality Pref*, 15, 549–557.

3

The impact of context and environment on consumer food choice

H. Meiselman, US Army Natick Center, USA

3.1   Introduction: definition and conceptualization

This chapter will deal with how context and environment affect our choice of foods, our enjoyment of foods, and our intake of foods. It will deal with a number of eating events, especially meals, which have eluded careful definition and measurement. At the present time, there is no generally agreed upon definition of context, with some defining it narrowly and others defining it broadly. Context, in the narrow sense, could be the sensory stimuli which precede a test stimulus and thus provide a context for the test stimulus. At the other extreme, context could be all of those events and things which relate to a 'reference event but have some relationship to it' (Rozin and Tuorila, 1993). If the reference event were eating pizza at a particular time in a particular place, then hundreds if not thousands of variables might have some relationship to it.

The term environment is only slightly easier to define, but the definition becomes more complex as we add together the different parts of the environment, which include at least the following:

- food environment, referring to what other foods are present in a meal or which preceded the test food,
- physical environment, also referred to as location or setting, including all physical aspects within which the choice or consumption of food takes place,
- the social environment, involving the other people present or the people who influence the environment whether present or not,
- economic environment, involving both the cost and the perceived value of food in a situation,
- cultural environment, involving the traditions, patterns and beliefs handed down within a particular social group.

This chapter will deal mainly with the physical and social environments of meals, and to a lesser degree with the cultural, although it is not easy to separate different parts of the environment. The chapter will not deal with the strictly sensory environment, and the economic environment. The purpose of this chapter is to introduce the reader to eating environments; hopefully this will help place other chapters in this book into the context of actual eating. Our task is to move from the laboratory to natural eating environments, defined as places where people normally eat. Our task is also to move from isolated foods to natural meals. Natural eating places and meals bring with them a whole range of environmental variables, and as we shall see below, these environmental variables have powerful effects on food choice, enjoyment and intake. For readers who focus on food variables, our task will be to demonstrate that environmental variables often produce independent variables with larger effects on eating than food variables. Based on the powerful effects of environmental variables and the relatively large effects compared to the food itself, context must be considered when deciding what food people will choose, how much they will enjoy it, and how much they will eat.

There has been a gradually increasing interest in environment/context since the 1990s. Meiselman *et al.* (1988) discussed contextual variables at the Reading Conference on Food Acceptability (Thomson, 1988). At the same Reading meeting Schutz (1988) introduced the concept of appropriateness which tries to include situational concerns in food behavioral research. Meiselman (1992a,b) argued for greater use of natural contexts in research, suggesting that we, '... refocus human eating research towards greater use of real meals, served to real people (not subjects), in real eating situations' (p. 54) while acknowledging, 'This is not to suggest that all studies should be done in natural settings' (p. 54).

Rozin and Tuorila (1993) presented an organizational scheme for contextual variables which included a temporal dimension at the First Pangborn Conference. This temporal dimension was also used by Bell and Meiselman (1995) in their review of contextual variables. Meiselman (1996) organized context/environment into the food, the person and the environment. Most recently, Wansink (2004) and Stroebele and de Castro (2004) have reviewed work on the environment. Context and environment have been included in some models of food choice but context is not emphasized in current models.

Following this discussion of definitions and concepts of context and environment, the remainder of this chapter is devoted to describing research on the physical and social environments in which people eat. The chapter begins with a discussion of where research takes place, contrasting laboratory settings and natural settings, and the implications for both external and internal validity. The same theme of where research is conducted is picked up again in the discussion of meals at the end of the chapter. Next follows a survey of environmental variables, including powerful variables such as effort to obtain food, eating duration, and convenience. Following these more physical variables, the importance of the social environment of eating is presented. And finally, the important variable of choice is discussed, demonstrating the impact of providing diners a choice of foods when eating. All of the physical and social variables are combined in a discussion of meals, which is the natural context for most eating. The discussion of meals focuses on determinants of meal acceptability.

3.2   How context/environment is studied: laboratory vs natural studies

Most research on food choice can be conducted under laboratory control, in a naturalistic setting, or in a completely natural environment. While there seems to be an increase in the amount of food choice research on the effects of the environment, the vast majority of research continues to be conducted under controlled conditions such as exist in the laboratory, as was pointed out by Meiselman (1992b). The choice of environment in which to conduct context/ environment research is perhaps more critical when the subject of the research is the environment itself. Studies have begun to determine what might be the critical variables in natural environments. In other words, which variables differentiate laboratory and natural settings and which of those variables deter- mine any differences in outcomes between laboratory and natural or naturalistic studies. This is clearly a long-range issue to determine these critical variables, but early studies reported below suggest that meal context and having a choice about what to eat might be two of the critical variables (Hersleth *et al.*, 2003; King *et al.*, 2004, 2005). In the future, with increased knowledge of what are the critical variables, modifications of laboratory methods might be possible to improve prediction, and modifications of natural environments might be possible to increase the number of study designs.

There has been some confusion on what constitutes natural and naturalistic settings. Natural (research) settings are by definition places that exist in nature, in which people choose and purchase food and also places in which people eat. These include the supermarket, home, restaurant, cafeteria, and other food choice and eating locations. Naturalistic settings are settings designed to imitate or produce the effect or appearance of nature. When a laboratory is modified to make it 'like a restaurant' or 'like home', it is a naturalistic setting. When a natural setting is modified it may no longer be a natural setting.

Variables which contribute to environmental effects can be studied in both controlled settings such as the laboratory and in natural settings. The choice of setting is based on both practical and methodological considerations. Some have suggested that natural settings are more expensive and more difficult to manage. This is not always the case; in most natural studies the environment already exists and therefore nothing has to be designed and built. A test product (such as a new food) or a test procedure (such as price change) can simply be added into the existing environment and the effect noted. The dependent variable or measure might be the number of food portions purchased or consumed, or the waste of the test food left on the plates. As noted below, studies in natural settings have the advantage of higher external validity.

Such natural studies are often inexpensive to conduct, but they often lack the control which is present in the laboratory. The lack of trained research personnel in natural settings and the need to get a job done, such as feeding people in a restaurant, preclude complete control. This lack of control can be disturbing to laboratory-trained researchers, who might value control over realism. Both the advantage and the downside of natural settings is the lack of control. It is the lack of control in nature which we are seeking in natural settings, but the lack of control also means that things happen which we cannot control. A study might continue for one week, and during that week the weather might change, the air conditioning might break, and the serving personnel might change.

Do natural studies or laboratory studies produce higher external validity, the extent to which results relate to the real world? Schutz (1988) presented a model of external validity of eating research containing the variables of type of subjects, type of stimuli, and measurement procedure. More recently, van Trijp and Schifferstein (1995) also focus on type of respondents, type of stimuli, and scaling procedures as determinants of external validity. But they also consider test circumstances, including the context or location of research. Most current research on foods takes place at low levels of external validity, using expert or recruited laboratory subjects, simple substances and foods, hedonic/preference judgments, and laboratory settings. Van Trijp and Schifferstein point out that the traditional model of food sensory research emphasizes product attributes and sensory characteristics, producing higher internal validity but lower external validity. In this type of work, the focus is on the product. Natural subjects (consumers) and more natural stimuli both produce more externally valid results. In this type of work, the focus is on the consumer. Concerning the measurement procedure, Schutz suggests the measurement of use intentions, such as appropriateness, rather than affective/liking measures. One advantage of doing research in more natural locations is that the natural subjects and natural stimuli often come with the location – natural eating locations usually have real customers and real food, thus solving all three challenges.

Some people might argue that the lack of control in natural settings means that only observational studies are appropriate as opposed to manipulated studies. This is not the case. Both types of research can be conducted in natural settings, observation of an eating environment with no interference other than the presence of the observer (for example, Sommer and Steele, 1997; Bell and Pliner, 2003), or manipulated variables with clear independent and dependent variables and hypotheses of outcomes (for example, Meiselman *et al.*, 1994). Since manipulated natural studies might involve manipulation of the environment with resulting measurement of the effects of the manipulation, it can be debated whether these changes (independent variables) eliminate the claim of the studies as natural. But that claim stems from the fact that these are natural eating locations in which people eat regularly.

While natural settings are suitable for both observational and manipulated studies, some studies are better conducted in the laboratory. When the task is purely sensory, such as formulating a blend of ingredients, or selecting a substitute for an ingredient, or describing a product in agreed upon terms, then a laboratory with its appropriate controls is appropriate. Whenever one seeks to measure liking or actual consumption/intake, then one should at least consider a natural setting.

It is an empirical question whether controlled research settings or natural settings better predict the real world. Studies will be reported below which compare test results in laboratories, naturalistic settings, and natural settings, and later product use in other natural settings. The face validity of naturalistic studies is higher, and this is often important in commercial product development and in food service research.

3.3   Contextual variables

Until the 1990s there was very little research on contextual/environmental variables. An early publication from the US Army Natick Laboratories entitled 'Not Eating Enough' (Marriott, 1995) catalogued the findings to date on US Army studies on soldiers' eating in the field. These studies suggested some of the variables that might account for environmental effects. Other sources of information were the earlier studies, mainly from the Stunkard group at the University of Pennsylvania (Stunkard and Kaplan, 1977; Coll et al., 1979; Myers et al., 1980). These studies were conducted in the context of variables contributing to over- or under-eating as related to weight loss programs. This same orientation also produced many of the early food choice questionnaires, for example Stunkard and Messick (1985). What follows is a summary of some of the major environmental variables; the reader is also referred to earlier reviews (Meiselman, 1996; Bell and Meiselman, 1995), and more recent ones (Wansink, 2004; Stroebele and de Castro, 2004). First, four more or less physical environmental variables will be presented (effort, duration, convenience, and physical environment) followed by socialization and choice.

3.3.1   Effort to obtain food

Effort to obtain food is clearly one of the most important contextual/ environmental variables, because it is mentioned in all of the early approaches to environment. Among the early studies from the Stunkard group, Myers et al. (1980) examined the effect of product placement ('accessibility') in a cafeteria service line. Products placed with easier access, i.e. less effort, were selected more often. The US Army studies reported by Natick researchers in several publications (Marriott, 1995; Hirsch et al., 2005) mention the critical role of effort. The situations in which these studies were conducted, cafeterias and military field feeding, might exaggerate the role of effort in day-to-day household eating. But the importance of effort in human eating is consistent with its importance in animal eating, in which efficiency is a main driver.

Meiselman and colleagues (Meiselman *et al.*, 1994) conducted and reported the only study in which effort has been manipulated in a natural eating location, and food choice, acceptability and intake were measured and reported. The study took place in a student refectory or cafeteria where students ate daily, and paid for their food (as opposed to having meal cards which entitled them to food). To manipulate effort, one food item in each of two studies was moved from its usual location to a new location some distance away. In order to obtain this test food, the student had to obtain and pay for his meal in one meal line, and then go to the new line to obtain the test food. Both studies began with baseline periods in which regular eating was measured In the first study using chocolate the effort manipulation lasted one week, and in the second study using potato chips it lasted three weeks followed by a recovery phase in which the chips were returned to their former location.

While we expected the increased effort to reduce choice and intake of the test foods, we were unprepared for the dramatic effect, because increased effort reduced selection of the test foods to virtually zero (Tables 3.1 and 3.2). This was a strong and early indication that environmental variables can have very large effects. Research in product variables or psychological/physiological variables of the eater often have much smaller effects, yet they continue to receive more attention than environmental variables. This is likely because much product research is sponsored by the food industry which is interested in developing and improving its products. Another major source of attention to food choice and eating is the health care field which focuses on the psychological and physiological variables of the eater in addition to product variables. We might be able to produce greater effects related to healthy eating by manipulating the environment than by addressing the product or the eater.

Interestingly, the acceptability of the test foods did not vary with the effort manipulation, showing that choice, acceptance and intake are not always correlated. We have observed this same lack of correspondence between choice, acceptance and intake in many other studies, and these will be noted below.

Another interesting result from the two effort studies was the lack of full recovery of the earlier eating behavior when the chips were returned to their original location, even though this recovery period lasted three weeks (Table 3.2). We do not know whether the eating behavior would have returned to its beginning level, or whether we had introduced a very long-term change. These changes in eating pattern are of great concern to product manufacturers who do not want their regular customers to change their pattern of consuming their product, and thereby lose them.

3.3.2   Duration and temporal considerations

Eating duration has not been the subject of much eating research, but it might turn out to be a critical variable in eating. The Nordic study of 1200 consumers in each of four countries presented interesting data on meal duration based on people being asked how long they ate in 10 minute intervals of response (Holm,

Table 3.1 Selection rates for subjects who chose chocolate candy in the baseline period. Note that the candy selection rate drops from 0.39 in the baseline condition to 0.03 in the effort condition

Condition	Main dishes	Pizza	Alternatives	Salads	Sandwiches	Dessert	Fruit	Accessory foods	Candy	Total dessert fruit accessory foods candy	Total dessert fruit accessory foods
Baseline	0.471	0.157	0.020	0.059	0.294	0.078	0.098	0.177	0.392	0.608	0.353
Effort	0.380	0.085	0.051	0.098	0.366	0.192	0.216	0.255	0.031	0.580	0.549
	N.S.	N.S.	N.S.	N.S.	N.S.	N.S.	N.S.	N.S.	$p < 0.001$	N.S.	$p < 0.10$

Source: Meiselman *et al.* (1994).

Table 3.2 Selection rates for subjects who chose chips/crisps in the baseline period text. Chip/crisp selection dropped from 0.71 in baseline to 0.09 in effort and increased to 0.32 in recovery

Condition	Number of meals	Main meal	Pizzas	Starch items	Vegetables	Salads	Bread	Sandwiches	Desserts	Fruit	Crisps	Sweets/cakes	Sauces	Candy	Drinks
Condition	117	0.385	0.111	0.274	0.154	0.000	0.000	0.299	0.137	0.222	0.718	0.051	0.017	0.060	0.6
Effort	184	0.408	0.125	0.462	0.092	0.000	0.005	0.304	0.120	0.130	0.092	0.065	0.022	0.103	0.5
Recovery	171	0.398	0.088	0.398	0.123	0.003	0.006	0.333	0.152	0.135	0.322	0.070	0.029	0.140	0.5
Difference between the three periods		N.S.	N.S.	$p < 0.01$	N.S.	$p < 0.10$	N.S.	N.S.	N.S.	$p < 0.10$	$p < 0.001$	N.S.	N.S.	$p < 0.10$	N.S.
Contrast between baseline and manipulation		N.S.	N.S.	$p < 0.05$	N.S.	N.S.	N.S.	N.S.	N.S.	$p < 0.10$	$p < 0.001$	N.S.	N.S.	N.S.	N.S.

Source: Meiselman *et al.* (1994)

2002). The most frequent response for all meals in all countries was 10–20 minutes, with 21–30 minutes the second most frequent for three countries. The most infrequent response in all countries was 31–40 minutes. People probably think meals last longer than they actually do, although there is bound to be wide cultural variation.

Observational studies have documented eating durations in restaurants in the United States and correlated eating duration with the number of people present. Sommer and Steele (1997) observed eating in both American coffee shops and restaurants, and reported increased duration at the table for groups rather than individuals, and for those reading rather than non-reading. Being in a group added approximately 10 minutes to a meal, and reading added approximately another 10 minutes. Bell and Pliner (2003) observed eating duration and number of people at tables in thee types of eating establishments in the United States and found a good correlation between the two measures in all restaurant types. They also documented that people eat much longer in worksite cafeterias and moderately priced restaurants than in fast food restaurants (Table 3.3).

Waiting time in food service has also been studied. Edwards (1984) has distinguished waiting times before the meal (pre-process), during the meal (in-process) and after the meal (post-process). Some contributions to these three different waiting periods in food service are shown in Table 3.4. Waiting for food can affect food acceptability values as shown in Table 3.5. Waiting produced a monotonic decline in preference for almost all foods tested (except carrots).

The relationship between eating duration and intake might be critical based on recent data from Pliner *et al.* (2004). Their recent discovery that duration might underlie social facilitation of eating is presented below under socialization. Effort to obtain food and eating duration might be two of the most critical variables in controlling food intake, and could be used in an environmental program of weight control.

3.3.3 Convenience

Convenience is one of the major trends in eating in the past decades, with convenience food products and convenience (fast) food service. Steptoe *et al.*

Table 3.3 Commensality: People eat longer when eating with others; based on observational study in three settings. Results show increased meal times (in minutes) when eating with groups varying from 1 person (eating alone) to groups of 5; the study shows results for intermediate numbers of people from 2 to 4

• Fast food restaurant: groups from	1 (10.7 min) to 5+ (21.9 min)
• Worksite cafeteria: groups from	1 (12.6 min) to 5+ (44 min)
• Moderately priced restaurant: groups from	1 (27.6 min) to 5+ (58.5 min)

Source: Bell and Pliner (2003)

Table 3.4   Categories of waiting time

Pre-process: waiting to be seated; waiting to place an order:
- pre-schedule: waiting because the customer has arrived early
- delay: waiting because the 'table' is not yet ready
- queue: waiting to be attended to

In-process:
- waiting to be acknowledged
- waiting for the order to be taken
- waiting for the food to be served

and if in a hurry
- waiting for the second and subsequent course to be served

Post-process:
- waiting for the bill
- waiting for the waiter to return coats, etc.

Table 3.5   The effect of waiting in line on food preferences

Delay	Chicken	Roast potatoes	Boiled potatoes	Carrots
No delay	7.28	6.42	6.27	6.76
3 minutes	6.94	6.16	4.50	6.11
6 minutes	6.75	5.05	3.67	6.14
9 minutes	6.50	4.44	3.42	6.20
	$n = 62$	$n = 98$	$n = 32$	$n = 50$

Source: Edwards (1984)

(1995) developed a food choice questionnaire through factor analysis of questionnaires from a respondent group in the UK. The Food Choice Questionnaire contains 36 items on nine factors. Sensory appeal, health, convenience, and price were identified as the most important factors. The convenience factors dealt with the purchase and preparation of food, with the following items:

'It is important to me that the food I eat on a typical day':
Item 1. Is easy to prepare.
Item 15. Can be cooked very simply.
Item 28. Takes no time to prepare.
Item 35. Can be bought in shops close to where I live or work.
Item 11. Is easily available in shops and supermarkets.

Candel (2001) has suggested that meal preparation convenience has two key dimensions, time and effort, and he proposed a six-item rating scale to measure convenience orientation in food preparation. Jaeger and Meiselman (2004) argued that convenience needs to be considered throughout the entire food provisioning process, which includes acquisition, preparation, eating and cleaning up. They studied female US consumer perceptions of convenience, time and effort using a repertory grid analysis of responses to written scenarios. The scenarios included the elements of food acquisition/shopping, preparation, and cleaning up. They confirmed the importance of time and effort in the perception of convenience but noted that these two variables were highly interdependent, and not cleanly separated.

3.3.4   The physical environment

Where one eats is determined to some degree by the culture, but most people eat most of their food at home. Holm (2002) has pointed out that relatively few Nordic meals are consumed away from home. At the same time, almost 60% of Americans eat a meal away from home on any given day, and this figure is increasing. British and Swedish consumers eat out at about the same rate (about 22% weekly and about 38% monthly), lower than Americans and higher than other Nordic countries. The percentage of the American food dollar spent away from home is now near 50%. Eating away from home is an American pattern, and whether this pattern will be exported to other countries remains to be seen. It is also not clear what variables in the United States contribute to the pattern of eating meals away from home; greater spendable income, mobility, food available for long hours, the frequency of family oriented dining, and other variables are worth investigating.

There has been relatively little published research on variables of the physical setting, especially in the all-important home setting. This is in spite of the importance of this information for the food service industry and for the concern about weight control. There is a large business folklore on the proper environment for fast food restaurants, fine dining restaurants, and other eating settings. But very little of it appears to be based on sound data. For example, there is very little published information on the effects of lighting level; does less lighting promote longer meals with greater consumption? Does high lighting level promote faster eating? Bell and Meiselman (1995) have summarized some of the existing information on variables in the physical environment. Bell *et al.* (1994) studied restaurant decor on site with actual manipulations of a restaurant interior and demonstrated changes in consumer behavior.

Kimes and Robson (2004) note the lack of formal research on table configuration and table location in restaurants, and then present an analysis of such data from a single large chain restaurant in Phoenix USA. The restaurant seats 210 people at 65 tables seating 2, 4, and 5 people and is aimed at younger adults and families. Their dependent variables were duration and spending. Spending variables analyzed include spending per minute (SPM) and average check per person. Tables that offered more privacy regulation had both higher eating durations and higher checks. Customers in booth seating ate longer but spent more per check but not per minute (SPM). Customers in banquette seating stayed longest and had lower SPM. Interestingly, customers at seven less desirable tables (next to kitchen door, tables for two between larger tables with

Table 3.6   Effects of slow and fast music on foodservice variables

Variable	Slow music	Fast music	Significance
Service time	29 min	27 min	>0.05
Customer @ table	56 min	45 min	0.01
Food purchases	$55.81	£55.12	>0.05
Bar purchases	$30.47	£21.62	0.01
Estimated gross margin	$55.82	$48.82	0.05

Source: Milliman (1986).

Table 3.7   Effects of classical and pop music on meal expenditure (£)

Variable	Classical music	Pop music	No music
Total drink	£8.36	£7.55	£8.03
Total food	£24.13	£21.92	£21.70
Total spend	£32.52	£29.46	£29.73

Totals do not sum due to rounding.
Source: North *et al.* (2003).

Table 3.8   Effects of music

• Fast loud music increases the speed at which people eat:
4.40 bites per minute for fast loud music
3.83 bites per minute for slow soothing tunes

• Really loud 'techno' and 'disco' music:
deters the over 30s and drives people out

---

high traffic and noise) had shorter durations and therefore higher revenues (SPM). The authors do not discuss eating duration from the social facilitation viewpoint. Nor do they discuss whether higher income or higher spending customers might select or request certain (higher spending) table types. It would be interesting to combine such physical food service research with behavioral research on consumer attitudes and expectations.

There have been several published studies on sound level and the presence of music (Milliman, 1986, Table 3.6; North *et al.*, 2003, Table 3.7; Table 3.8). The results appear to support the hypothesis that fast music promotes faster eating, shorter table time and lower bar purchases, but not food purchases. Slower and classical music promote greater bar purchases. It would be interesting for health researchers to try to reverse these studies to determine if one can reduce food intake with music.

Several studies have sought to determine whether enhancing the food environment produces an enhancement in acceptance or intake. King *et al.* (2004) enhanced the laboratory setting in their study of contextual variables, and found that enhancing the physical environment had little reliable effect on product ratings (see Table 3.13, Test 4 on page 87). Hersleth *et al.* (2003) studied eight Chardonnay wines differing in three product characteristics served in four sessions (sensory laboratory or reception room, with or without food). The wines were rated on a 9-point scale by 55 consumers/wine users who completed all four sessions. The reception room had groups of eight in an enhanced social setting. Context effects were as large as product effects. The presence of food and the enhanced reception room raised hedonic scores 0.3–0.5 scale points (as did the product factors, one of which reduced liking). The presence of food was a more effective enhancer in the reception room than in the laboratory. Hoyer and de Graaf (2004) report an investigation with elderly diners in which enhancing the meal environment including flowers, tablecloths, acoustics, lighting increased food intake for males (4.9 vs. 4.4 MJ) and females (4.9 vs. 4.6 MJ) as compared to a more basic environment.

3.3.5   Socialization/commensality

Researchers who study eating in the controlled environment of the laboratory might succeed in dissociating eating from its social context. But most people eat food with other people, which is the definition of commensality, eating meals with others (Sobal, 2000). In fact, eating alone is devalued in many cultures, and it is not clear that eating alone in the laboratory is exempt from the lower status of eating alone. Do people eating alone tend to eat less, partially because the occasion seems less important?

Sobal and Nelson (2003) conducted a cross-sectional survey in one US county, yielding 663 usable questionnaires out of 1200 mailed. Meal partner data revealed that most respondents ate alone at breakfast, alone or with co-workers at lunch, and with family members at dinner (Table 3.9). Unmarried individuals more often ate breakfast and dinner alone and more often ate lunch

Table 3.9   Commensal patterns in one US community

- Over half (58%) ate breakfast alone, almost half (45%) lunch alone, but only 19% dinner alone (note: 25% live alone).
- 21% ate breakfast with partner/spouse, 13% lunch, and 37% dinner (note: 40% have a two-person household).
- 14% ate breakfast with family/children, 5% lunch, and 30% dinner.
- 24% ate lunch with co-workers and 13% with others.
- 2% skip breakfast, <1% lunch, and <1% dinner.
- 51% rarely or never eat at homes of family, 50% homes of friends, and 87% homes of neighbors.

Source: Sobal and Nelson (2003).

Table 3.10   Commensality: couples, families and adolescents sharing meals in selected studies

---

Couples reporting the number of shared meals per week:
US: 10 (Shattuck *et al.*, 1992)
Australia: 12 (Craig and Truswell, 1988)
Netherlands: 13 (Feunekes *et al.*, 1998)

Families sharing meals:
UK: 75% ate a meal together every day (Thomas, 1982; Warde and Martens, 2000)
US: 90% reported eating evening meal together

Adolescents reporting eating with parents:
US: 54% breakfast, 88% dinner (Hertzler *et al.*, 1976)
US: eating with family five or more times per week – 12% breakfast, 5% lunch, 69% dinner.
US: eating with all or most of family: 4.5 ± 3.3 meals per week

---

Source: Sobal and Nelson (2003).

with friends. Thus, work-oriented society leads people to eat alone during the day and with family in the evening. People maintain commensal relationships mainly with family. These data are consistent with other data on couples sharing meals (Table 3.10).

Demographic variables were not associated with commensal measures, and there were no gender differences. Further, living alone was not determinative of eating alone, so lone diners are not necessarily those living alone. It was expected that living and eating alone would characterize the elderly but this was not the case. Sobal and Nelson make the point that neighboring is important for the elderly but might not related to commensality. Neighboring and social interaction for the elderly need to be seen in a much broader context than just food intake.

The four-country Nordic study on eating patterns (Kjaernes, 2002) provides detailed data on commensal patterns (Holm, 2002) and confirms many of the observations of Sobal and Nelson. The study was based on 1200 surveys each in Norway, Sweden, Denmark, and Finland. People had eaten alone at least once the day before the survey for about 2/3 of people. The proportion of people eating alone and with family members was about the same, with the latter increasing in the evening. People living alone ate alone three times more often, and older people ate alone more than younger people. The chance of eating a full (proper) lunch or dinner did not vary whether eating alone or with others. Eating with colleagues peaked at midday, during the typical lunch time. Eating with friends and others was very infrequent and occurred on weekends. Family dining dominated weekends, weekdays and week evenings. Individual dining was more frequent on weekdays and weekend days.

The relationship between commensality and food intake was first raised by deCastro and colleagues in a series of papers based on the food diary method. DeCastro and deCastro (1987) trained people for one day on how to fill out a

Table 3.11   Social facilitation of eating: variables associated with higher food intakes are correlated with the number of people present

---

- Breakfast < Lunch < Dinner
  B < L < D #People B < #People L < #People D

- Restaurant meals > Home meals, other locations
  #People restaurant > #People home, (ns) other locations

- Meals > Snacks for all macronutrients
  #People meals = #People snacks

- Meals w/alcohol > Meals without alcohol
  #People w/alcohol > #People w/out alcohol

Source: DeCastro *et al.* (1990).

dietary diary and then had them take detailed records for the next week including what they ate and with whom. In many replications of the same pattern, deCastro and his associates found that people ate more meals socially than alone, and when eating socially, the amount consumed increased with the number of people present. The social facilitation of eating effects produced a large number of studies focusing on effects of other variables on intake. Variables which should increase eating also increased the number of people present (deCastro *et al.*, 1990), maintaining a strong correlation between the number of people present and how much is consumed (Table 3.11). Social facilitation of eating has recently been discussed in a lengthy review paper by Herman *et al.* (2003). Several authors have suggested that eating duration might be critical in social facilitation of eating effects. Pliner *et al.* (2004) have recently presented the first study which independently varied eating duration and group size. They found that the increased intake was related to eating duration and not to group size. This is an important effect which needs to be replicated in a variety of eating environments; it might present an important mechanism to increase or decrease eating for reasons of health. Social facilitation of eating remains an important phenomenon, but it might work through duration, and food intake might be especially sensitive to changes in eating duration.

Commensality involves other important aspects such as communication, both verbal and nonverbal. Edwards and Meiselman (2005) tested the impact of positive and negative verbal cues in a restaurant setting. Both male and female restaurant customers were offered a menu with a choice of five main courses. One of the main dishes was the target dish, and after customers were given menus, the waiter said either nothing or a positive or negative message. The positive message ('To assist in your selection, could I just say that the "target dish" (name) has been particularly popular this week/yesterday/last week') produced a 41% selection rate, while the negative comment ('To assist in your selection, could I just say that the "target dish" (name) has not been particularly popular this week/yesterday/last week') produced half the selection rate at 19%.

Interestingly, no comment yielded the same selection rate as the positive comment (40%). There were no significant differences among the acceptance ratings of the products in the different conditions; once the choice had been made the positive and negative comments had no effect on the perception of the product.

Whether social eating is healthier than solo eating is not yet clear. Sobal and Nelson suggest that commensal eating is healthier because of social facilitation (prevent under-consumption, risking over-consumption), social support (healthy food choices), and social control (healthy food choices). In many countries there is concern about the nutrition of the elderly and whether the higher percentage of elderly eating alone is a risk factor. But studies on the health of lone elderly diners give mixed results; male elderly might do worse. Perhaps the elderly are too heterogeneous for any generalization about eating alone and health.

3.3.6   Choice

Most dining environments present the diner with a number of choices. Should the diner eat at all, should they have a beverage (alcohol), should they have a full meal or a snack? Even at home, diners can politely refuse certain foods and select others. The variable of choice is one of the variables least studied and potentially most important, because choice varies tremendously across eating situations: the infant has very little choice, almost none, whereas the fine diner has a lot of choice. Unfortunately, the laboratory provides almost no choice, forcing the laboratory model of eating into an extreme situation. Most subjects in laboratory studies are expected to consume all of the samples offered to them, and very few if any research protocols ask the subject which sample they want or whether they want to skip samples. This is very unlike normal eating. While many human use regulations require that subjects can terminate a study, in fact most subjects feel pressured to conform to what is expected of them.

We are just beginning to demonstrate and measure the impact of choice. For example, one can easily demonstrate the impact of food monotony in a laboratory test. Providing laboratory subjects with the same daily lunch meal for a week produces reduced acceptance scores and reduced intake when the same meal is served every day, but not when the meal varies (Meiselman *et al.*, 2000a). The monotony effect disappears when the diners have choice (Kramer *et al.*, 2001). Soldiers who select the same meal every day, actually rate that product higher (Table 3.12). It makes sense, they like it more and they select it more frequently. The monotony phenomenon seen in the laboratory might not exist or rarely exist in the real world – why would people eat things they do not like? King *et al.* (2005) have demonstrated the criticality of choice in two successive tests. In both studies, providing choice enhanced acceptance scores. Choice had a bigger effect than the physical environment itself (see Tables 3.14 and 3.15 on pages 87 and 88).

Table 3.12 Kramer *et al.* data on the role of choice. Data are shown from two sites representing 2439 consumption incidents. The far left column shows the number of times any item was consumed: once (2159), twice (2159), twice (1037), etc. For example there were 35 times when an item was selected 10 times by the same person. The data from both sites show that there was an increase in acceptance ratings and percentage consumption for repeated consumption (a frequency of two or more times) as compared with single consumption

Number of times item was consumed	Site 1			Site 2		
	Frequency of consumption	Accept rating first/subsequent	Percentage consumed first/subsequent	Frequency of consumption	Accept rating first/subsequent	Amount consumed first/subsequent
1	2159	7.41/NA	95%/NA	1085	6.80/NA	84%/NA
2	1037			395		
3	538			162		
4	263			72		
5	140			46		
6	95	7.70/7.82	97%/98%	24	7.23/7.13	90%/88%
7	60			19		
8	45			14		
9	39			9		
10	35			0		
>10	187			0		
2 or more	2439			741		

** Significant at $\alpha = 0.01$
* Significant at $\alpha = 0.05$
N.S. Not significant

3.4   Meal context: putting the variables together

Those interested in environmental/contextual effects on food choice, acceptance and consumption inevitably find that they must address food as meals. The natural context for eating is the meal (however that might be defined) because most food is consumed as a part of a meal (for example, see Kjaernes, 2002). The decision to consider meals in research is a major one because meals are highly complex events comprising many dimensions including at least dietetic, culinary, sensory, social, nutritional, anthropological, cultural, health, temporal, economic and others. In fact the more one looks into what is involved in meals, the more complex this unit of eating is likely to appear.

3.4.1   Laboratory meals and natural meals

Unfortunately, too many researchers have focused on one or several dimensions of eating (for example, the sensory attributes of foods) and have ignored the meal context in which foods are consumed in combination with other foods and with other variables such as the social ones. Further, when researchers seek to serve a 'meal' to subjects in a laboratory setting, they have often served a single food or a combination of foods without appropriate consideration of what foods normally go together in that culture. This has resulted in unrealistic, and sometimes bizarre, meals served to subjects.

   This section of the chapter will deal with meal context, what we know about it, what we should be doing about it in research, and even how meals are evolving in the 21st century. The latter point is important because meals are constantly evolving. This is difficult for most students, laypeople, and researchers to grasp. What constitutes a meal changes at least once or twice a century, and probably more often, in our rapidly evolving cultures. Well into the 1900s most Western countries ate three hot meals per day (working people often went home to eat); we now eat one in many countries. In the 19th century, many working people ate up to five times per day or even more, a combination of meals and large snacks. This declined to three eating events, and then to between two and three. Not only the number of meals but what constitutes a meal has changed as will be noted below. These changing trends in eating, changing meals, are very important to understanding health issues in eating or product development needs of the future. The meal most changed from its traditional form is probably breakfast, and as Sobal and Nelson (2003) have pointed out: 'American breakfast is the most anomalous meal, more often small, short, skipped, and involving special foods.'

3.4.2   Past research on meals

Because of the many individual disciplines which undertake meal research, information on meals is scattered across a broad range of journals and books. One exception is the volume by Meiselman (2000) who attempted an inter-disciplinary approach to meals in *Dimensions of the Meal.* In addition to the problem of relevant meal literature residing in different sources, much of it is in the languages of individual countries, making it difficult to access without translation.

There are two main exceptions in English. There is a tradition in England of studying meals, beginning with Mary Douglas in the 1970s (Douglas, 1976; Douglas and Nicod, 1974), and carrying on to Anne Murcott in the 1980s. Other British investigators have built on these traditions, for example Marshall (1995, 2000). Douglas presented a framework for studying meals, within the social context, which remains a major influence to this day. Douglas emphasized that meals are highly structured events following a series of rules about where, when, and in what sequence foods could be served. Investigators who arbitrarily design laboratory meals often violate these rules.

The second is the recent study of eating and meals in the four Nordic countries of Denmark, Finland, Norway and Sweden. These countries participated in a survey of 1200 people in each of the four countries, and while many papers from this research were published in the local language of the countries involved, a book in English summarizing the work has been published (Kjaernes, 2002).

3.4.3   Daily meal patterns (past and present)

Rotenberg (1981) and others have pointed out that the three meal pattern which many people assume today was not present in the 19th century, when there were five or more daily eating events. At the beginning of the 20th century many (Western) cultures ate three hot meals per day. Some cultures have retained two hot daily meals, and others have moved to just one hot meal per day. Even within the Nordic countries the meal pattern of which meals are hot and which meal is the major meal varies from country to country.

The mean number of eating events in the Nordic survey is 3.9 eating events per day, with slightly less (3.7) in Denmark. Almost everyone (over 90%) eats at least three times per day. Denmark and Norway eat more cold meals, and Sweden and Finland eat more hot meals.

Is the meal pattern becoming more irregular? Are we becoming grazers? The Nordic study does not conclude that grazing typifies modern Nordic eating either based on meal pattern (Gronow and Jaaskelainen, 2001) or based on what is consumed at those meals (Makela, 2002).

3.4.4   Meal food combinations

Marshall and Bell (2003) asked students in both Scotland and Australia to provide hedonic ratings, frequency of use ratings, and appropriateness ratings for 51 common food names, and to then construct snack, lunch and dinner meals from the food names for 11 different physical locations. Through cluster analysis they identified six different meal types: main meal, light meal, fast food, snack, camping trip, and seafood snack. They emphasized that some foods are associated with specific meal types (hamburgers in fast food), and many food items belong in different meal types and in different locations (pasta). Fast food items fell into a clearly separate cluster. The main meals for both lunch and dinner were similar to the British 'proper meal' with a meat, vegetable and starch. Light meals contained the types of main dishes (pizza, pasta, sausage) not usually associated with main meals. Pizza fit into fast food, light meal, and snack depending on country and meal group. The effect of location on food choices was more important at lunch than dinner.

The Nordic study presents detailed data on what people eat at meal-times. Every meal pattern contains a main dish ('centre') by definition, and almost all (>95%) meals contain a beverage. Other meal components range from 10–65% present in meals. Hot meal patterns vary considerably with the pattern of main dish/starch/vegetable with or without sauces being the most common (Kjaernes, 2002, see CSV and CSVT in Table 4.2, p. 135). Considering all meals, most meals have 2 or 3 components, with 1 and 4 component meals about equal in frequency. This is similar to the British proper meal studied by Douglas and others.

3.4.5   Food acceptability within meals

The earlier work on food acceptability within meals produced a clear pattern of data, showing that the main dish within a meal contributes the largest portion to overall meal acceptability. This was assessed by asking respondents to rate individual food item acceptability and also overall meal acceptability for various combinations of foods, that is, meals. Respondents did not actually see or consume any food. Regression analysis on the individual foods and overall meals produced the following data:

Rogozenski and Moskowitz (1982)
Meal = 5.68 + 2.7 entrée + 0.53 starch + 0.42 veg + 0.25 salad + 0.57 dessert

Turner and Collison (1988)
Meal = 0.57 + 0.43 entrée + 0.21 sweet + 0.14 starter + 0.14 potato

Instead of modeling questionnaire meals, Hedderley and Meiselman (1995) modeled actual university cafeteria meals ($n = 309$) freely selected by university students in the UK. They found that the main dish accounted for varying amounts of overall meal acceptability depending on the type of meal and the type of main dish. Traditional meals with a main dish ($n = 175$) were the most frequent and the main dish accounted for 0.6 of overall meal acceptance. Sandwich meals ($n = 82$) and pizza meals ($n = 52$) were less frequent, and the main dish accounted for relatively more of the overall meal's acceptability, 0.7 and 0.9 respectively. For a pizza meal, the meal's acceptability is largely determined by the pizza itself.

King *et al.* (2004, 2005) have recently undertaken a series of studies on the effects of contextual factors on meal acceptability in both the laboratory and in restaurants. In the first series of six studies, they first studied meal acceptability in a standard central location/laboratory test. They then added in a series of

Table 3.13   Mean acceptance scores (scale 1–9) for overall meal and meal components in a sequence of tests with a different contextual variable added in each test

Meal component	Test 1 Traditional	Test 2 Meal	Test 3 Social	Test 4 Environment	Test 5 Choice	Test 6 Restaurant	P value
	($n = 104$)	($n = 93$)	($n = 106$)	($n = 106$)	($n = 101$)	($n = 35$)	
Overall	–	7.5a	7.3a	7.3a	7.3a	7.5a	
Salad	7.0c	7.5abc	7.6ab	7.1bc	7.7a	7.4abc	0.0021
Pizza	7.2ab	7.2ab	6.5c	6.9abc	6.7bc	7.4a	0.0032
Tea	5.9b	7.0a	6.8a	7.2a	7.1a	7.3a	<0.0001

Within row, means sharing letters are not significantly different.
Source: King *et al.* (2004).

contextual variables leading to an environment more and more similar to natural eating. The variables and the effects are shown in Table 3.13. Serving the foods as a meal and providing a choice of which foods to eat had the greatest effects on acceptance. When the results for Test 5, in which all four contextual variables are present, are compared with the laboratory test in Test 1, one can see that adding contextual variables increases acceptance scores for salad and tea. Testing the same foods in an actual restaurant also produces higher ratings than the laboratory. King *et al.* (2005) confirmed and extended these results in a second series of studies involving a national chain restaurant in the US. Once again laboratory ratings were lower that restaurant settings and mean context and food choice were again the critical variables (Tables 3.14 and 3.15). Meiselman *et al.* (2004) have concluded that laboratory testing underestimates true product acceptability and recommend adding 0.5 to 1.0 scale points to laboratory product ratings.

Meal acceptance ratings also vary depending on where the meal is served. Meiselman *et al.* (2000b) found lower ratings in institutional settings as com-

Table 3.14   Food acceptance scores in different test environments

Meal component	Test 1 Traditional CLT ($n = 74$)	Test 2 Restaurant CLT ($n = 83$)	Test 3 Restaurant survey ($n = 386$)	P value
Lasagna	4.9	4.7	5.0	0.2823
Cannelloni	3.7b	3.5b	4.7a	<0.0001
Iced tea	3.8b	3.7b	4.9a	<0.0001
Salad	4.1b	4.2b	5.2a	<0.0001
Breadsticks	3.8b	3.7b	5.0a	<0.0001

One-way analysis of variance was used to compare the effect of test for each meal component.
Within row, means sharing letters are not significantly different.
Overall rating score from 1 = poor, 2 = fair, ... 6 = excellent.
Source: King *et al.* (2005).

Table 3.15   Testing protocol. This protocol contains the context factor options included in each test

Context	Test 1 Traditional CLT	Test 2 Restaurant CLT	Test 3 Restaurant survey
Meal	Individual meal components	Individual meal components	Meal
Social	Self	Social	Social
Environment	Consumer testing facility	Restaurant	Restaurant
Choice	No choice	No choice	Choice

Source: King *et al.* (2005).

pared with restaurant settings, with laboratory ratings falling between. In a large follow-up study, Edwards *et al.* (2003) compared acceptability of the same prepared food in ten different locations, and observed the same distinction between institutional settings and restaurant settings. There was about a scale point difference between the highest and lowest ratings of the same product. They attributed these differences to differences in expectations which consumers have for products in these different locations (Cardello, 1994; Cardello *et al.*, 1996).

3.4.6   Meals of the future – are meals converging globally?

In my experience, no question excites a food lecture audience more than this question. People are passionate whether change lies ahead for foods and for eating, and that globalization will occur as it has in technology. Others are just as passionate on the need for traditionalism in foods and eating. The 2004 meeting of the European Sensory Network in Florence Italy focused on regional products ('typical products'), emphasizing both the cultural and economic aspects of the issue. The fact that some products have become global (for example, hamburger, cola, and pizza) does not mean that many others will follow. Perhaps a small number of foods will become global foods, while much larger numbers of foods will stay regional. This is an important issue for product developers, food service managers, and health officials. If what we eat is related to our health, then globalizing trends will affect all of these businesses. The food choice field has yet to undertake major studies of the trend toward globalization in food choices.

3.5   Future trends

The past twenty years have witnessed the introduction of environment as a major category of variables affecting food choice, and the identification of environmental variables. To a large degree studies have undertaken one or two variables at a time. However, any eating situation has many hundreds of variables if not more. The challenge is to study the interaction of environmental variables; do they add effects, multiply effects or cancel each others' effects? Another challenge is to study the interaction of environmental variables with product variables and consumer variables. The prediction of the interaction of product variables, consumer variables and environmental variables is really the goal of food choice research, and requires research of meals in natural eating locations.

3.6   Conclusions

This chapter has tried to introduce the reader to what is included under eating environment or eating context. In addition, it is hoped that the importance of the eating environment has become clearer. Research has identified a number of contextual/environmental variables; that list of variables can be expected to grow. Part of the task ahead is to identify the most important environmental variables. To date, meal context and food choice appear to be among those critical variables. Meal duration appears important by itself and as manifested through social facilitation. And effort to obtain food appears critical in increasing or decreasing intake. This suggests that laboratory research on eating should include meal considerations and choice considerations. Both are notably absent from much laboratory research on eating. As we learn more about environmental control of eating, it is hoped that this progress will produce better laboratory research and better natural research. In that way we have a better chance of producing acceptable food products and healthier eating situations.

3.7   References

BELL R and MEISELMAN H L (1995), 'The role of eating environments in determining food choice'. In D Marshall (ed.), *Food Choice and the Consumer*, Glasgow: Blackie Academic and Professional, pp. 292–310.

BELL R and PLINER P L (2003), 'Time to eat: the relationship between the number of people eating and meal duration in three lunch settings'. *Appetite*, 41, 215–218.

BELL R, MEISELMAN H L, PIERSON B and REEVE W (1994), 'The effects of adding an Italian theme to a restaurant on the perceived ethnicity, acceptability and selection of foods by British consumers'. *Appetite*, 22, 11–24.

CANDEL M J J M (2001), 'Consumers' convenience orientation towards meal preparation: conceptualization and measurement'. *Appetite*, 36, 15–28.

CARDELLO AV (1994), 'Consumer expectations and their role in food acceptance'. In MacFie H J and Thomson D M H (eds) *Measurement of Food Preferences*. London: Blackie Academic, pp. 253–297.

CARDELLO AV, BELL R and KRAMER F M (1996), 'Attitudes of consumers toward military and other institutional foods'. *Food Quality and Preference*, 7, 7–20.

COLL M, MYERS A and STUNKARD A J (1979), 'Obesity and food choices in public places'. *Archives of General Psychiatry*, 36, 795–797.

CRAIG P L and TRUSWELL A S (1988), 'Dynamics of food habits in newly married couples: food-related activities and attitudes towards food'. *Journal of Human Nutrition and Dietetics*, 1, 409–419.

DECASTRO J M and DECASTRO E S (1987), 'Spontaneous meal patterns of humans: influence of the presence of other people'. *American Journal of Clinical Nutrition*, 50, 237–247.

DECASTRO J M, BREWER E M, ELMORE D K and OROZOCO S (1990), 'Social facilitation of the spontaneous meal size of humans occurs regardless of time, place, alcohol and snacks'. *Appetite*, 15, 89–101.

DOUGLAS M (1976), 'Culture and food'. Russell Sage Foundation Annual Report 1976–77 (pp. 51–58). Reprinted in M Freilich (ed.) *The Pleasures of Anthropology*, New York: Mentor Books, 1983, pp. 74–101.

DOUGLAS M and NICOD M (1974), 'Taking the biscuit: The structure of British meals'. *New Society*, 19, 744–747.

EDWARDS J S A (1984), 'The effects of queuing on food preferences'. *International Journal of Hospitality Management*, 3(2), 83–85.

EDWARDS J S A and MEISELMAN H L (2005), 'The influence of positive and negative cues on restaurant choice and food acceptance'. *International Journal of Contemporary Hospitality Management*, 17, 332–344.

EDWARDS J S A, MEISELMAN H L, EDWARDS A and LESHER L (2003), 'The influence of eating location on the acceptability of identically prepared foods'. *Food Quality and Preference*, 14, 647–652.

FEUNEKES G I J, DE GRAAF C, MEYBOOM S and VAN STAVEREN W A (1998), 'Food choice and fat intake of adolescents and adults: association of intakes with social networks'. *Preventative Medicine*, 27, 645–656.

GRONOW J and JAASKELAINEN A (2001), The daily rhythm of eating. In Kjaernes U (ed.) *Eating Patterns: A Day in the Lives of Nordic Peoples*. Report No. 7-2001. Lysaker Norway: National Institute for Consumer Research, 91–124.

HEDDERLEY D and MEISELMAN H L (1995), 'Modeling meal acceptability in a free choice environment'. *Food Quality and Preference*, 6, 15–26.

HERMAN C P, ROTH D and POLIVY J (2003), 'Effects of the presence of others on food intake: a normative interpretation'. *Psychological Bulletin*, 129, 873–886.

HERSLETH M, MEVIK B-H, NAES T and GUINARD J-X (2003), 'Effect of contextual variables on liking for wine – use of robust design methodology'. *Food Quality and Preference*, 14, 615–622.

HERTZLER A A, YAMANAKA W, NENNINGER C and ABERNATHY A (1976), 'Iron status and family structure of teenage girls in a low-income area'. *Home Economics Research Journal*, 5, 92–99.

HIRSCH E, KRAMER F M and MEISELMAN H L (2005), 'Effects of food attributes and feeding environment on acceptance, consumption and body weight: lessons learned in a twenty year program of military ration research (Part 2)'. *Appetite*, 44, 33–45.

HOLM L (2002), The social context of eating. In Kjaernes U (ed.) *Eating Patterns: A Day in the Lives of Nordic Peoples*. Report No. 7-2001. Lysaker Norway: National Institute for Consumer Research, 159–198.

HOYER S and DE GRAAF C (2004), 'Workshop Summary: How do age-related changes in sensory physiology influence food liking and food intake. Influence of context and environment'. *Food Quality and Preference*, 15, 910–911.

JAEGER S R and MEISELMAN H L (2004), 'Perceptions of meal convenience: the case of at-home evening meals'. *Appetite*, 42(3), 317–325.

KIMES S E and ROBSON S K A (2004) 'The impact of restaurant table characteristics on meal duration and spending'. *Cornell Hotel and Restaurant Administration Quarterly*, Nov, 332–346.

KING S C, WEBER A J, MEISELMAN H L and LV N (2004), 'The effect of meal situation, social interaction, physical environment and choice on food acceptability'. *Food Quality and Preference*, 15, 645–654.

KING S C, MEISELMAN H L, HOTTENSTEIN A W, WORK T M and CRONK (2005), 'The effect of contextual variables on food acceptability: A confirmatory study'. Paper presented at *6th Pangborn Symposium, Harrogate UK.*

KJAERNES U (ed.) (2002), *Eating Patterns: A Day in the Lives of Nordic Peoples. Report No. 7-2001.* Lysaker Norway: National Institute.

KRAMER F M, LESHER L L and MEISELMAN H L (2001), 'Monotony and choice: repeated serving of the same item to soldiers under field conditions'. *Appetite*, 36, 239–240.

MAKELA J (2002), The Meal Format, In Kjaernes, U (ed.), *Eating Patterns: A Day in the Lives of Nordic Peoples.* Report No. 7-2001. Lysaker Norway: National Institute for Consumer Research, 125–158.

MARRIOTT B M (ed.) (1995), *Not Eating Enough.* Washington: National Academy Press.

MARSHALL D W (ed.) (1995) *Food Choice and the Consumer.* Glasgow: Blackie Academic and Professional.

MARSHALL D W (2000), British meals and food choice. In Meiselman H L (ed.), *Dimensions of the Meal.* Gaithersburg: Aspen, pp. 202–220.

MARSHALL D and BELL R (2003), 'Meal construction: exploring the relationship between eating occasion and location'. *Food Quality and Preference*, 14, 53–64.

MEISELMAN H L (1992a), 'Obstacles to studying real people eating real meals in real situations: Reply to Commentaries'. *Appetite*, 19, 84–86.

MEISELMAN H L (1992b), 'Methodology and theory in human eating research'. *Appetite*, 19, 49–55.

MEISELMAN H L (1996), The contextual basis for food acceptance, food choice, and food consumption: the food, the situation and the individual. In Meiselman H L and MacFie H J H (eds) *Food Choice Acceptance and Consumption.* Glasgow: Blackie Academic and Professional, pp. 239–263.

MEISELMAN H L (ed.) (2000), *Dimensions of the Meal.* Gaithersburg: Aspen.

MEISELMAN H L, HEDDERLEY D, STADDON S L, PIERSON B J and SYMONDS C R (1994), 'Effect of effort on meal selection and meal acceptability in a student cafeteria'. *Appetite*, 23, 43–55.

MEISELMAN H L, HIRSCH E S and POPPER R D (1988), Sensory hedonic and situational factors in food acceptance. In Thomson D M H (ed.) *Food Acceptability*, London: Elsevier Applied Science, 77–88.

MEISELMAN H L, DE GRAAF C and LESHER L (2000a), 'The effects of variety and monotony on food acceptance and intake at a mid-day meal'. *Physiology and Behavior*, 70, 119–125.

MEISELMAN H L, JOHNSON J L, REEVE W and CROUCH J E (2000b), 'Demonstrations of the influence of the eating environment on food acceptance'. *Appetite*, 35, 231–237.

MEISELMAN H L, KING S and HOTTENSTEIN A W (2004), 'Laboratory product testing produces an underestimation of true product acceptability'. Paper presented at *A Sense of Identity*, Florence, Italy.

MILLIMAN R E (1986), 'The influence of background music on the behavior of restaurant patrons'. *Journal of Consumer Research*, 13, 286–289.

MYERS A, STUNKARD A and COLL M (1980), 'Food accessibility and food choice'. *Archives of General Psychiatry*, 37, 1133–1135.

NORTH A C, SHILCOCK A and HARGREAVES D J (2003), 'The effect of musical style on restaurant customers' spending'. *Environment and Behavior*, 35(5), 712–718.

PLINER P, BELL R, KINCHLA M and HIRSCH E (2004) Time to eat? The impact of time facilitation and social facilitation on food intake. Presented at 5th Pangborn Sensory Science Symposium.

ROGOZENSKI J G JR and MOSKOWITZ H R (1982), 'A system for the preference evaluation of cyclic menus'. *Journal of Food Service Systems*, 2, 139–161.

ROTENBERG R (1981), 'The impact of industrialization on meal patterns in Vienna, Austria'. *Ecology of Food and Nutrition*, 11, 25–35.

ROZIN P and TUORILA H (1993), 'Simultaneous and temporal contextual influences on food acceptance'. *Food Quality and Preference*, 4, 11–20.

SCHUTZ H G (1988), Beyond preference: Appropriateness as a measure of contextual acceptance of foods. In D M H Thomson (ed.) *Food Acceptability*, London: Elsevier Applied Science, pp. 115–134.

SHATTUCK A L, WHITE E and KRISTAL A R (1992), 'How women's adopted low-fat diets affect their husband'. *American Journal of Public Health*, 82, 1244–1250.

SOBAL J (2000), Sociability and meals: Facilitation, commensality and interaction. In Meiselman H L (ed.) *Dimensions of the Meal*. Gaithersburg: Aspen, pp. 119–133.

SOBAL J and NELSON M K (2003), 'Commensal eating patterns: a community study'. *Appetite*, 41, 181–190.

SOMMER R and STEELE J (1997), 'Social effects on duration in restaurants'. *Appetite*, 29, 25–30.

STEPTOE A, POLLARD T M and WARDEL J (1995) 'Development of a measure of the motives underlying the selection of food: the Food Choice Questionnaire'. *Appetite*, 25, 267–284.

STROEBELE N and DE CASTRO J M (2004), 'Effect of ambience on food intake and food choice'. *Nutrition*, 20, 821–838.

STUNKARD A J and KAPLAN D (1977), 'Eating in public places: a review of reports of the direct observation of eating behavior'. *International Journal of Obesity*, 1, 89–101.

STUNKARD A J and MESSICK S (1985), 'The three factor eating questionnaire to measure dietary restraint, disinhibition and hunger. *Journal of Psychosomatic Research*, 29, 71–84.

THOMAS J E (1982), 'Food habits of the majority: evolution of the current UK pattern'. *Proceedings of the Nutrition Society*, 41, 211–217.

THOMSON D M H (ed.) (1988), *Food Acceptability*. London: Elsevier Applied Science.

TURNER M and COLLISON R (1988), 'Consumer acceptance of meals and meal components'. *Food Quality and Preference*, 1, 21–24.

VAN TRIJP H C M and SCHIFFERSTEIN H N J (1995), Sensory analysis in marketing practice: Comparison and integration. *Journal of Sensory Studies*, 10, 127–147.

WANSINK B (2004), 'Environmental factors that increase the food intake and consumption volume of unknowing consumers'. *Annual Reviews in Nutrition*, 24, 455–479.

WARDE A and MARTENS L (2000), *Eating Out: Social Differentiation, Consumption and Pleasure*. Cambridge: Cambridge University Press.

4

Theories of food choice development

E. P. Köster and J. Mojet, Wageningen University, The Netherlands

4.1 Introduction: the importance of models of food choice development

In humans, as in other omnivores, food choice is predominantly a learned behaviour. Very few aversions are inborn and even these few can be overcome by learning. This has given the human omnivore a large amount of freedom to live in very different circumstances. Eskimos and Papuans, Chinese, and South American Indians all have found places to live and find food.

Learning starts even before birth and carries on till the latest stages of life. It takes many forms, from completely unconscious conditioning and simple imitation to cognitive learning on the basis of reasoned argumentation. This means that food choice is a dynamic behaviour that is subject to almost continuous change and that can be influenced at very different levels. It varies not only from person to person and from situation to situation, but is also dependent on the type of food product. Preferences for staple foods, such as rice, cassava, bread or potatoes tend to remain quite stable in individuals belonging to a given culture, although even these preferences have become more variable due to the increase in travelling and the spreading of the major cuisines over the world. Liking for other food (vegetables, meats, bakery products, etc.) and condiments or spices not only varies much more between individuals, but also over the lifetime of each individual and varies depending on the eating situation and the frequency of the individual's previous exposure to the product. Phenomena such as 'learning to like through mere exposure' (Zajonc, 1968; Pliner, 1982) on the one hand and product boredom or slowly rising aversion on the other are well known. Underestimating the varied and dynamic nature of food preference and food choice behaviour is seen as one of the main causes of the many failures in introducing new food products in the market (Lévy and Köster, 1999; Köster *et al.*, 2003; Köster, 2003).

Most market and consumer research techniques are based on the implicit (and false) assumption that people do not change their preferences. Their results are like photographic stills that fixate the liking, preference or attitude of a subject for a food at a given moment in its development. The predictive value of such stills is very limited, because even if one collects them for many hundreds of people as is customary in most consumer research, one cannot construe the dynamic development of the preference in any of these individual subjects from them. Models of food choice must therefore incorporate insight in the role and relative importance of the different learning, motivational and situational factors that determine its development. Such insight is a necessary prerequisite for understanding food choice behaviour and finding effective ways of predicting and influencing it. It involves knowledge from a range of psychological theories that are involved in perception, learning and memory, motivation and emotion, decision making, cognition and social behaviour.

4.2   Learning theories

4.2.1   Learning and memory

Many different forms of learning and memory have been described. Traditionally, human learning and memory research was mainly concerned with explicit and active memorisation and explicit and conscious retrieval of the learned material (usually words or visual stimuli). However, over the last three decades special attention has been devoted to effortless and unconscious incidental learning and to implicit memory that shows itself in behaviour without the subjects' explicit awareness of its relation to the learning phase of the memory process. (Schacter, 1987, 1990, 1994; Schacter and Graf, 1986; Stadler and French, 1998).

In everyday life, these implicit learning and memory processes together with forms of simple associative or emotional conditioning are the predominant forms of learning where food-related sensations and expectations are concerned. Incidentally learned memory for taste, texture and flavour of food has been demonstrated in several studies (Mojet and Köster, 2002, 2005; Köster *et al.*, 2004). It has also been shown that, in contrast to intentionally learned memory, incidentally learned memory for these properties does not deteriorate with age (Møller *et al.*, 2004, 2006).

Other authors have shown that the perceptual impression and appreciation of attributes of food-related stimuli (odours and tastes) may be changed through simple conditioning (Baeyens *et al.*, 1990, 1995; Frank *et al.*, 1993; Stevenson *et al.*, 1995, 1998). Finally, food-related behaviour may also be influenced by learning at a cognitive and conscious level. Food labelling and promotion of healthy eating through publicity have had an effect on the consumption of products that are considered unhealthy, although they often do reach only part of the population. The extent to which each of these forms of learning determines food choice and eating and drinking behaviour and how they interact is still under discussion.

4.2.2   The role of perception and learning in early food choice development

Early perception and learning play an important role in establishing basic and probably very long-lasting preferences and in determining basic notions like the distinction between the edible and the non-edible. Non-cognitive learning mechanisms such as imprinting, affective and classical conditioning, and imitation are predominant in the formation of early food habits. There are recent indications that young children are particularly sensitive to certain odours and tastes and probably also have a very keen oral sense of touch related to sucking behaviour. This means that already in the transition from milk feeding to eating solid and varied food, many sensory experiences and preferences are incorporated in a fully non-cognitive way, which makes it difficult to change them later in life on rational grounds. This may also be the reason for the tenacity of culturally determined regional differences in staple food preferences. The infants' and young children's perceptual capabilities and the resulting perceptual experiences act as filters in food habit development. The five most important senses implicated in the acceptance of food in the oro-nasal area, olfaction, taste, touch, pain and kinaesthesia, seem to show only few signs of inborn preferences and aversions. There may exist an inborn aversion of odours of decay (rotten eggs, dead bodies), but an aversion like the one for faecal odours is certainly learned (Stein et al., 1958), as are probably most food odour preferences. In taste, there is an inborn aversion of bitter and perhaps sour and an inborn preference for sweet and probably umami (Steiner, 1977; Ganchrow et al., 1983), but the aversion for bitter can quite easily be overcome by learning (drinking beer, black coffee, etc.) and there are indications that some young children develop a strong liking for sour (Liem and De Graaf, 2004). With regard to tactile and kinaesthetic experiences, there may be an inborn mechanism that warns against the ingestion of particles in fluids and the swallowing of hard and sharp objects, although accidents with children show that it is not a very effective one. Pain provides a similar but also rather ineffective warning mechanism against the ingestion of very stinging and burning substances. Nevertheless, learning will turn even pain sensations into pleasurable experiences when people learn to appreciate carbonated drinks and to eat hot spices (Rozin and Schiller, 1980). As in olfaction, all preferences for taste, pain or texture, except the ones for sweet and perhaps for sour, seem to have been learned. Such food-related preference learning starts prenatally (Menella et al., 1995; Marlier et al., 1998; Ganchrow and Menella, 2003; reviews: Porter and Schaal, 2003), the amniotic fluid being a potential flavour carrier (Schaal and Orgeur, 1992) and continues in infancy via the mother's milk or formula feeding (Menella and Beauchamp, 1999; Menella et al., 2001). This type of learning has strong and long-lasting effects (Haller et al., 1999; Garcia et al., 2001; Nicklaus et al., 2004, 2005a,b,c) and resembles imprinting.

In a recent review on dietary learning in humans (Brunstrom, 2005), the possibility that early childhood might be one of the most important critical developmental periods in dietary learning is stressed. Thus, the kind of foods that an individual is exposed to in that period is a good predictor of habitual meal size (Brunstrom *et al.*, 2005) and of variety of food choice (Nicklaus *et al.*, 2005a) in adulthood. Furthermore, this critical period has also been mentioned in relation to the development of obesity in adulthood (Dietz, 1994, 1997).

Somewhat later in early childhood new forms of affective learning and conditioning, such as imitation and parental approval or punishment, become progressively more important. Birch and her co-workers documented many of the learning mechanisms that are functional in this period of life (Birch, 1980 and review Birch, 1999). According to them, setting an example by parents is much less effective – if not even counter-productive – than example setting by peers, and systems of punishment and reward by parents may work out in much more complicated ways than usually expected. Thus, promising rewards for finishing a certain dish will considerably reduce the liking for that dish, even if it was one that was initially well liked by the child. There are also indications that specific tastes (or brands) of products for which the consumption has been restricted by the parents during early childhood (peanuts, coca cola, ketchup in Europe) may become the object of lifelong craving and strong taste and brand fidelity, while for foods, the consumption of which has been encouraged by the parents (yoghurt, apple juice), the taste quality is later readily exchanged for another (Köster *et al.*, 2001). Furthermore, children learn quickly that healthy foods cannot be pleasant, when the parents insist that the child should eat a food it dislikes, because it is healthy. That this may have long-lasting effects was suggested by the results of a canteen study (Köster *et al.*, 1987). When presented during three introductory weeks as 'new', a healthy (low saturated fat and low sodium content) snack was chosen more than twice as often as when presented as 'healthy' by the large adult population of two sets of factory canteens ($N \sim 400$ each). Obviously, the word 'healthy' provoked negative feelings, but when in a later (unpublished) study three different versions of the healthy snack were presented, there was an almost equal acceptance of them under the 'healthy' as under the 'new' condition. Probably the explanation lies in the fact that as a child, people in those days never had a choice. If one disliked something, one had to eat it, because it was qualified as 'healthy'. Suddenly having a choice of healthy foods probably lifts these feelings of constraint and raises people's curiosity.

4.3   Motivation theories

4.3.1   Exploratory behaviour and need for activation
Curiosity, exploration and search for stimulation are at the basis of many learning processes and of modern motivation theory. Without them little development would take place. In contrast to the older motivation theories that

were based on notions of need- and drive-reduction (hunger leads to eating and eating leads to cessation of hunger), modern motivation theory is dominated by the idea that organisms actively look for stimulation and try to maintain an optimal level of activation (Hebb, 1946, 1949, 1966) or arousal (Berlyne, 1967). This means that the attractiveness of stimuli (e.g. music or food) depends on their arousing properties (e.g., intensity, novelty, complexity) in relation to the optimal level of arousal of the perceiving individual. Stimuli with a lower than optimal arousal potential will be considered boring and will be less liked, whereas stimuli with a much higher than optimal arousal potential (e.g., loud noises, very complex art forms or food tastes) will raise confusion, irritation and even fear (neophobia) or anger. Since people differ in the level of arousal that they prefer most, the same stimulus may be just right for some, while being either boring or too strong or complex for others. Furthermore, it will be clear that some arousing properties of the stimulus like, for instance, novelty and complexity will change with repeated exposure to the stimulus. Novelty will wear off over time, and music or foods that at first seemed very complex, will become simpler when we know them better. Also, our taste for aesthetic experiences (music, art, eating and drinking) seems to develop over our life time. Wines that we liked when we just started drinking wine, may later become too simple for our tastes. Theories about the dynamics of liking and preference have been developed by Berlyne (1960, 1970), Dember and Earl (1957), Zajonc (1968), and Walker (1980). The most important aspects of these theories are illustrated in Figs 4.1 and 4.2. The first figure shows the theory of Dember and Earl, which is based on the inverted U-shape postulated by Berlyne to describe the relationship between liking and arousal or perceived complexity of the stimulus (solid curve in Fig. 4.1), but adds a learning theory to it.

According to Dember and Earl, exposure to a stimulus with a somewhat higher complexity than the optimal one (B in Fig. 4.1) causes a shift of the optimum in the direction of this higher complexity, whereas exposure to stimuli of a lower than optimal complexity leaves the optimum unchanged. When such a shift takes place, it means that stimuli of lower complexity than the new optimum (including the previous optimum A) become less appreciated, whereas more complex stimuli gain in liking. This theory was verified both in animal and in human studies using visual stimuli (Dember et al., 1957; for review see Dember, 1970) and in human studies with foods and drinks (Léon et al., 1999; Köster et al., 2003; Lévy and Köster, 2001; Lévy et al., 2006).

Figure 4.2 shows the addition of Walker (1980) to the theory of Earl and Dember. Walker explained that experience with stimuli reduces their complexity and that this phenomenon leads to product boredom and reduced liking. This means that the general opinion in market research that one must choose the most liked product (A), leads automatically to product boredom, to reduced liking and to a short product life cycle. Choosing a less liked, but more complex, product (B) will lead to a gradual growth in liking with exposure and the product will stay in the market much longer even if the optimum of complexity shifts in the direction of more complexity as illustrated.

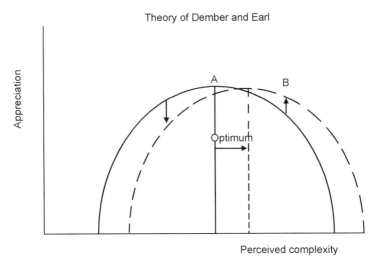

Fig. 4.1   The relationship between perceived complexity and appreciation according to Berlyne (solid line) and the shift of the curve (dotted line) and the optimum of perceived complexity (A) under the influence of the presentation of a 'pacer' (B) of a somewhat higher than optimal complexity according to Dember and Earl (1957).

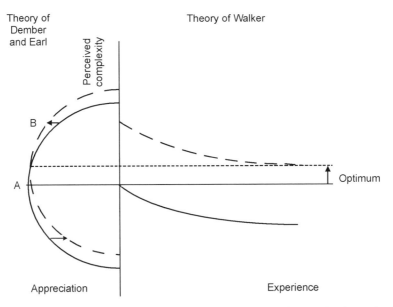

Fig. 4.2   The influence of repeated exposure to the stimulus on its perceived complexity and thus also on its appreciation according to Walker. Product boredom and mere exposure effects (Zajonc) can be explained as special cases of this theory. Their occurrence depends on the complexity of the presented stimulus relative to the optimal complexity of the subject.

Figure 4.2 also explains why the often cited 'mere exposure' theory of Zajonc (1968) that states that mere repeated exposure will lead to increased liking, has only limited value. In fact, this theory is a special case of the theory of Walker and will only hold for stimuli that are initially more complex than the optimally liked complexity (the dotted curve under Walker's theory), whereas for simpler than optimal stimuli Zajonc's theory will not hold and boredom prevails (solid line under Walker's theory). Experiments with repeated exposure to foods or drinks of different perceived complexities do indeed show both phenomena, the more complex ones being more and more liked and the simple ones being less and less liked (Léon et al., 1999; Köster et al., 2003).

Pliner (1982) exposed her subjects to novel drinks and ascribed the mere exposure effect that she found to the dissipation of food neophobia (fear for and dislike of new food), that may have been aroused by the novelty of her stimuli. It might be that people who are characterised as neophobic have a rather low optimal level of complexity and as a result find most novel stimuli too complex, but there is insufficient evidence to prove this.

4.3.2   Food neophobia and variety seeking

Food neophobia has often been described as a natural biological correlate of omnivorous exploratory behaviour. Omnivorous animals, that will try many food sources, will at the same time have to be cautious not to poison themselves (Rozin, 1976). There is good evidence, however, that neophobia is learned. Neophobia does not appear in children before the age of about three years (Cashdan, 1998; Pelchat and Pliner, 1995) and the extent to which it occurs seems to be strongly dependent on the reaction of the parents to the child's refusal to eat, a phenomenon that frequently occurs around this age. Anxious reactions by the mothers to this refusal seem quite strongly related to the development of severe neophobia (Hanse, 1994). Quite a bit of research has been devoted to the measurement of the stability of neophobic behaviour in children from 2–12 years. The results are rather variable and dependent on the measurement method employed. With questionnaire methods (Pliner and Hobden, 1992), both stability (Loewen and Pliner, 1999, 2000; Cooke et al., 2003) and reduction (Koivisto and Sjödén, 1996; Koivisto-Hursti and Sjödén, 1997) of neophobia have been found. With observational methods, only a decline of neophobic behaviour is found (Birch et al., 1987; Pelchat and Pliner, 1995; Pliner, 1994; Pliner and Loewen, 1997). On the basis of these results some authors speak of a passing phase of infantile food neophobia.

Of course, neophobia may have many other causes and is also often assumed to have been acquired later in life, as follows from the fact that elderly people sometimes score higher on a neophobia scale than do middle-aged adults (Tuorila et al., 2001). Such data have to be considered with much caution, however, since there is, as in the case of infantile neophobia, no conclusive evidence that such attitude measurements are related to actual behavioural differences in preference or in liking for novel products. In fact, the absence of conclusive behavioural results in studies with different types of products, may indicate that it is indeed the level of perceived complexity of the novel stimuli relative to the optimal complexity which determines whether the neophobic attitude will express itself in actual behaviour or not. Thus, Raudenbush and Frank (1999) found that responses of neophilics and neophobics were similar for familiar (less arousing and less complex) foods, but that they differed for unfamiliar (novel and complex) foods. Taking another approach, Pliner and co-workers showed that varying the arousal level in their adult subjects, by having them play different video games, influenced their willingness to accept novel foods. When manipulated arousal was low, subjects chose more novel foods than when it was high (Pliner and Melo, 1997). 'High sensation seekers' (Zuckerman, 1994) tried more novel foods than 'low sensation seekers' under conditions of low arousal. In a study with children, Pliner and Loewen (2002) showed that the willingness to eat moderately novel and very novel food was clearly diminished by creating a high level of arousal, while the willingness to eat familiar food remained unaffected by the arousal level. On the basis of these data, Pliner and Loewen suggest that the attractiveness of a food may also depend on the arousing properties of the situation in which it is consumed and thus that different types of arousal may complement each other in reaching the optimal level of arousal. Would this also mean that a decrease in psycho-physically or ecologically arousing properties (e.g., intensity of taste, quantity of food) could be compensated for by collative properties (e.g., perceived complexity)? There are indications that women, but not men, can use more complex food to compensate for a smaller quantity of food intake (Köster, 1988). This may be related to the fact that women do discriminate stimuli better in terms of complexity and do like more complex stimuli than men (Jellinek and Köster, 1983; Köster, 1985).

That the same food stimulus may be appreciated quite differently in different situations is of course well known and different forms of situational analysis are used in modern sensory consumer research (Köster 1996, 2003). However, the direct link proposed by Pliner and Loewen, between the arousing properties (relative to the optimal arousal level of the subject) of an eating situation and the arousal potential of the food, has not yet been investigated systematically. Even if, as suggested by Hanse (1994), neophobia is a learned personality trait that probably depends largely on the attitude of the parents during the young child's food refusal phase, it fits very well into the arousal theories discussed here. The initial complexity of very novel stimuli may be too perplexing for people that have a relatively low optimal complexity level.

Variety seeking, another personality trait that has received some attention in the last two decades (Van Trijp and Steenkamp, 1992) is also easily explained by these theories. People with a rather high level of optimal complexity may find many products boring and will look for temporal complexity to satisfy their needs, i.e. they will tend to be variety seekers, whereas people with a low level of optimal complexity will have a tendency to stick to the well known products they are familiar with. Like neophobia, variety seeking is usually measured by a questionnaire type scale. Measured in this way, neophobia seems to predict behaviour a bit better than variety seeking (Tuorila *et al.*, 2001; Van Trijp *et al.*, 1992). This might indicate that the negative sentiments expressed in neophobia have more predictive validity than the positive sentiments involved in variety seeking. Absence of variety seeking does not necessarily have to coincide with strong neophobia; it may simply reflect habit formation. Furthermore, variety seeking behaviour depends on the type of product involved. As one might expect, variety seeking is larger for products that are available in a greater variety and that evoke a rather high degree of involvement (Van Trijp, 1994).

Scales that explicitly measure the attitudes of people towards novelty and variety are in general rather poor predictors of behaviour. Nevertheless there are clear behavioural indications of differences among people in their reactions to new foods (Köster *et al.*, 2001) and in the stability of their likings and preferences (Köster *et al.*, 2003). Again, most of these differences can, perhaps, best be explained in terms of the interaction between individual differences in optimal arousal level of the subject and differences in the arousal potential that certain stimuli have acquired for different subjects. Thus, personal history and learning come into play and this may be the reason why 'personality traits' such as neophobia and variety seeking are only weakly expressed in behaviour.

All theories mentioned thus far share two characteristics. In the first place, they are 'descriptive' theories that explain the learning and motivation mechanisms functioning in the development of food choice. They are not especially devised for the development of strategies that lead to behavioural change, although effective ways of intervention to change behaviour can certainly be deduced from them. In their totality they also show that different approaches to influence choice, learning and habit formation may be needed at different stages in life. In the second place, the theories described above do rely on mechanisms that mostly function at a non-cognitive (and usually non-conscious) level.

Throughout life, arousal seeking and simple learning, based on sensory appreciation and on feelings of post-ingestive well-being or ill-feeling, remain probably the most important mechanisms in the formation and change of food preference and food choice. Nevertheless, the influence of other, more cognitive, social and cultural factors that may influence food choice decisions should also be discussed. These factors play a predominant role in a number of theories that are often invoked in the food domain to explain and predict food choice.

4.4 Cognitive theories

4.4.1 The theory of planned behaviour and related theories

The theory of reasoned action or later the theory of planned behaviour (Ajzen and Fishbein, 1970, 1980; Fishbein and Ajzen, 1975; Ajzen, 1985) is perhaps the most prominent of the many theories that are directly concerned with explicit cognitive factors and conscious strategies to obtain behavioural change. Here it will be discussed as an exemplar of these theories.

The theory is based on the idea that behaviour is directly provoked by intentions, which in turn are influenced by attitudes and beliefs about one's own values and control possibilities and about the judgements of others whose opinions seem important to the person involved. Responses to questionnaires, serial 'why-questions' as in the so-called laddering technique or self reports by the consumers are used to measure these attitudes, beliefs and intentions. Very often no observations of actual food choice behaviour are made to validate the results. This means that these methods tap only conscious processes and do take them at face value. The theory has been applied widely in the area of food choice, but in most cases only to show that certain beliefs and attitudes are highly correlated among each other and with measurements of intention. However, when the relation between intention and behaviour is investigated, the theory of planned behaviour and many similar theories based on conscious decision making have met with little success (for review see Baranowski *et al.*, 2003). The link between intention and actual behaviour is often weak. This has led some proponents of the theory to bypass the measurement of intentions and to try to link behaviour directly to attitudes (Shepherd and Stockley, 1985; Shepherd and Farleigh, 1986). Although in a number of instances improvement was obtained, the correlations between attitudes and behaviour still remain low. Moreover, they may actually be even weaker than they seem, because most authors do not verify whether the scores of both variables that they correlate are graded and normally distributed. If one correlates the attitudes toward eating meat with actual meat-eating behaviour and one does not exclude strict vegetarians, who are absolute in their refusal, one may get a spuriously high correlation, even if for all non-vegetarians the relation between attitude and behaviour is zero. It would at least be recommendable to show scatter-plots if one bases one's theory on correlation only.

In studies where other types of measurements than beliefs, attitudes and intentions are included, the factors food liking and past experience usually explain more of the variance in behavioural decisions than all hypothetical constructs (attitudes and beliefs) together (Koivisto and Sjödén, 1996; Rhodes and Courneya, 2003; Honkanen *et al.*, 2005; Verbeke and Vackier, 2005). This shows once more that memory and habit play a very important role in actual food choice.

4.4.2 Basic criticisms and the conscious-unconscious debate

Apart from its low predictive value, the theory of reasoned action and planned behaviour has been criticised on a more basic level. It assumes that people make rational decisions based on conscious deliberations. There is substantial evidence that casts doubt on both these pretensions. First of all, rationality in human decision making has come under attack since the publication of Kahneman and Tversky's famous papers on heuristics and biases in decision making under uncertainty (Kahneman and Tversky, 1972, 1973, 1979; Tversky and Kahneman, 1974, 1984). Their theory found application in many research areas (for reviews see Kahneman and Tversky, 2000; and Gilovich et al., 2002).

People are not only biased in a number of ways when they deal with probabilities, but their decisions are also strongly influenced by prior emotional experiences in comparable situations. Apart from gut feelings provoked by strong emotional signals that influence decision making consciously, Damasio (2003) describes two different ways in which emotions may bias the decision making process without any conscious awareness on the part of the subject. One of them implies the production of gut feelings by what he calls the 'as-if-body-loop', a mechanism, that also plays a role in empathy and in imitation behaviour, and in which the brain simulates certain emotional body states internally. Neural substrates involved in such an 'as-if-body-loop' have been described by a number of authors (Rizzolatti et al., 1996, 1999, 2001; Haari et al., 1998; Adolphs, 1999, 2002; Adolphs et al., 1999, 2000).

The second way in which the emotional signal may operate unconsciously on the decision-making process, is by 'producing alterations in working memory, attention and reasoning, so that the decision-making process is biased toward selecting the action most likely to lead to the best possible outcome, given prior experience' (Damasio, 2003). It has been shown in many instances (see, e.g., Bechara et al., 1997, 2000) that the individuals may probably remain completely unaware of the processes and strategies involved in their decision making, although recently doubt has arisen about the validity of this latter claim (Maia and McClelland, 2004). The evidence for the somatic marker hypothesis (Damasio, 1994), a model of how bodily states may provoke the selection of the optimal actions in decision processes has also been criticised in recent publications (Dunn et al., 2006; Rolls, 2005). These latter criticisms are directed at the physiological processes Damasio invokes as the determining factors of the decision making, rather than at the unconscious nature of it.

With regard to conscious deliberations – the second main pretension of the theory of planned behaviour – Wilson and his co-workers have shown that thinking about alternatives does not help people in making better decisions, but, on the contrary, may lead them astray (Wilson and Dunn, 1986; Wilson, 1990; Wilson and Schooler, 1991). Furthermore, in a large study on improving fruit and vegetable consumption, the so-called pre-contemplators, i.e. people who have never thought about changing their behaviour, were shown to change their behaviour at least as much as people in the so-called preparation stage, who are consciously considering change in the near future or have already begun changing (Resnicow et al., 2003). Apart from stressing once more the weakness of the link between intention and behaviour, this indicates that other than conscious deliberations may lead to the same changes and that the common belief that our actions are governed by our thinking may be false. In fact, according to many authors (Zajonc, 1984; Kihlstrom, 1987; Wilson, 2002; LeDoux, 1996; Kahneman, 2003) most of our fast decisions – and food choices fall in this category – are taken without any intervention at the conscious level and many of these decisions may even be contrary to what we think we ought to do. Wilson (2002) provides substantial evidence for the existence of an 'adaptive unconscious' that governs most of our daily doings and operates quite independently of our conscious convictions. This 'adaptive unconscious' is inaccessible to conscious inspection. It is an on-line pattern detector that functions automatically, fast, unintentionally and effortlessly, but with a certain rigidity that may lead to distortions based on bias or prejudice. It is concerned with the here-and-now and with immediate reactions.

According to other data (for review see Wegner, 2002), many of our everyday decisions to act have already been made before we come to realise our intention to carry them out. This may explain why the long-term goals that we have consciously set with regard to healthy eating or smoking behaviour are so easily overruled on the spur of the moment. The adaptive unconscious is also quite sensitive to negative information. Along similar lines, LeDoux (1996) suggests that humans and animals have a system of defensive behaviour that detects danger and produces responses that maximise the probability of surviving the dangerous situation. This system also operates fast and independently of consciousness and is part of the 'emotional unconscious', a concept mainly based on findings by Zajonc (1980, 1984) and others (Erdelyi, 1992; Bargh, 1990, 1992; Jacoby *et al.*, 1992) who showed that emotional processing can take place without being consciously experienced. And even if Zajonc's claim that emotion precedes cognition and is independent of it, is still debated after 20 years, there is no doubt that affective reactions can take place in the absence of conscious awareness (LeDoux, 1996). Is there no place left for conscious deliberation?

The adaptive unconscious has all the characteristics of the intuitive system (system 1) in the dual-process model of Kahneman (2003). Kahneman adds to the characteristics mentioned that the processes in system 1 function in parallel processing, are automatic, associative, implicit, emotionally charged and governed by habit, which makes them difficult to control or modify. In contrast, the operations of the reasoning system (system 2) are characterised by slow and effortful serial processing, that is more likely (not necessarily) to be consciously monitored and deliberately controlled. The system 2 processes are relatively flexible, potentially rule governed. They are also susceptible to interference by other effortful processes, whereas the effortless processes of system 1 neither cause nor suffer much interference when combined with other tasks. According to Kahneman the operating characteristics of system 1 are similar to the features of perceptual processes but are not restricted to the processing of current stimulation. Like the operations of system 2, the intuitive judgments of system 1 'deal with concepts as well as with percepts and can be evoked by language. In the model the perceptual system and the intuitive operations of system 1 generate *impressions* of the attributes of objects of perception and thought. These impressions are neither voluntary nor verbally explicit.' Since judgments are always intentional and explicit, system 2 is involved in all judgments, 'whether they originate in impressions or in deliberate reasoning'. Judgments that directly reflect impressions – i.e. are unaltered by system 2, that has as one of its tasks to monitor the activities of system 1 – are called *intuitive* judgments. Kahneman and Frederick (2002) suggest that the monitoring by system 1 is usually quite lax and that therefore many intuitive judgments are expressed. Some of these intuitive judgments may be erroneous, but intuitive thinking can also be powerful and accurate as has been demonstrated in many cases where high skill was acquired by prolonged practice, and performance is rapid and effortless (the performance of chess masters (Simon and Chase, 1973) and the detection of impending heart failure by experienced nurses (Gawande 2002; Klein 1998) are cited as examples).

Kahneman (2003) points out that since it is characteristic of many intuitive thoughts that they arise spontaneously and effortlessly under the proper circumstances, it is necessary to understand why some thoughts come to mind more easily than others and why some come spontaneously and others demand work. He uses the term accessibility as a common concept for the ease with which 'the different aspects and elements of a situation, the different objects in a scene and the different attributes of an object' come to mind and he points out that the determinants of accessibility subsume such notions as stimulus salience, selective attention, specific training, associative activation and priming. Motivationally relevant and emotionally arousing stimuli spontaneously attract attention, and accessibility reflects temporary states of priming and associative activation as well as enduring operating characteristics of the perceptual and cognitive systems. Uncertainty and doubt are phenomena of system 2. They seldom enter the scene in the minds of experienced decision makers that operate under pression when there is little time or in dealings with relatively unimportant decisions like in food choice. Priming and associative activation are also the main determinants of accessibility that together with the ideas of Bargh (Bargh, 1994; Bargh *et al.*, 2001) on the development of automatic behaviour and even automatic goal setting have entered the literature in the area of consumer psychology. Bargh and Chartrand (1999) define three forms of automatic self-regulation:

1. An automatic effect of perception on action, called the perception-behaviour link in which perceptions unconsciously give rise to our prime behaviours and ideas and stereotypes.
2. Automatic goal pursuit, which supposes that originally consciously set goals that are repeatedly invoked become 'mental representations' that can be automatically and unconsciously activated by situational features of the environment.
3. Continual automatic evaluation of one's experience, which states that emotional responses to events and moods are continuously and unconsciously evaluated, as was already postulated by Zajonc (1980) who claimed that 'preferences need no inferences'.

Like Kahneman (1973, 2003; Kahneman *et al.*, 1982; Kahneman and Tversky, 2000), from whom they are likely to have borrowed at least some of the ideas which they implemented in their interesting experiments, Bargh and Chartrand (1999), come to the conclusion that consciously and wilfully regulating one's behaviour requires considerable effort, is rather slow and 'requires a limited resource that is quickly used up'. This type of mental process can therefore only be used seldom and for a relatively short time. Non-conscious or automatic processes that are unintended, effortless and very fast can usually operate at any time and guide most of our behaviour. On one point they deviate from Kahneman and co-workers, however, when they introduce the notion of (re-)activation of 'mental representations' and knowledge, a notion which Kahneman and co-workers seem to have avoided, on the basis of their probably more dynamic view of perceptual processes. In current consumer psychology (Dijksterhuis *et al.*, 2005) that is strongly related to Bargh and Chartrands' ideas, this terminology leads to the danger that perception is seen as a passive receptive process that is only driven by priming from the environment (the perception-behaviour link), instead of as an active unconsciously driven searching process that conveys meaning to the environment on the organised basis of prior experience. A discussion organised around the paper by Dijksterhuis *et al.* (2005) on 'The unconscious consumer: Effects of environment on consumer behaviour', was thus unnecessarily limited to this passive view of the perception-behaviour link. In his most recent paper, Bargh (2006) seems to have become aware of the limitations of this approach and concludes in accordance with Cacioppio *et al.* (2000) that 'the mechanisms underlying mind and behaviour are not fully explicable by a biological or a social approach alone, but that rather a multilevel integrative analysis may be required'. Strangely enough, not a single mention of the more integrative theories of Damasio or LeDoux (see above) can be found in the article.

In all theories that deal with automatic behaviour and priming, learning and memory play a major role. Most of this learning is incidental, non-voluntary and non-conscious and the memories are implicit memories that probably take the form of unconscious situational expectations that direct attention rather than 'mental representations' of previous occurrences that are reactivated as such. In such a view, it is not that surprising that primes are multi-functional and have qualitatively different effects. Many of the questions that Bargh (2006) raises may perhaps be simplified if the rather rigid ideas about the nature of the primed 'content' are revised.

That habits and automaticity play a strong role in attitude-behaviour relations, has been stressed by many authors studying different behaviours (e.g., Aarts and Dijksterhuis, 2000; Ouellette and Wood, 1998; Verplanken and Aarts, 1999). The latter authors directly oppose habit to deliberate action as approached in the theory of planned behaviour and come to the conclusion that habits should be characterized as learned, goal-directed automatic responses and demonstrate how habits mark boundary conditions of planned behaviour. They furthermore show that there is something like 'a habitual mind-set which makes the individual less attentive to new information and courses of action and contributes to the maintenance of habitual behaviour'. This idea of a mind-set, which is here seen as a purely restrictive way in which learning through past experience influences attention, is clearly nearer to the concept of memory in the form of situational expectation, than the 'mental representations' of Dijksterhuis *et al.* (2005). That similar 'mind-sets' can also lead to positive and non-habitual results has been shown in the elegant research of Dijksterhuis and van Knippenberg (1998) in which they showed that overt complex behaviour such as answering diverse knowledge questions could be influenced through priming, by making the subject think about the stereotype of a professor. Subjects who had done so knew more answers to knowledge questions than subjects who had been in a control condition or had been thinking of the stereotype of a secretary.

Ouellette and Wood (1998) studied the influence of habits and past behaviour on future behaviour in a meta-analysis of 64 independent studies and came to the conclusion that past behaviour independently predicted future behaviour in studies that had conducted analyses controlling for the effects of intentions, attitudes toward the act, subjective norms and behavioural control, indicating that the effects of past behaviour cannot be attributed to these potential confounds. Although in unstable contexts (that were unfavourable to habit formation) the influence of past behaviour was mediated by intentions which then became the primary predictor of behaviour, past behaviour was a strong direct predictor in stable contexts (i.e. contexts that had favoured habit formation). These findings were once more confirmed in an independent experiment (Ouellette, 1996). They supported the idea of the authors that past behaviour guides future responses in two ways, either through the automaticity of the processing that initiates and controls their performance in constant contexts, or through the contribution of past behaviour to intentions in the conscious decision-making process necessitated by unstable contexts.

Ajzen (2002), in an effort to save the theory of planned behaviour, attacks the priority of past behaviour over reasoned action, first on the grounds that as long as the situation remains stable both theoretical perspectives would predict the same result. Although Ajzen admits that frequently performed behaviours can become habitual and can be enacted without much conscious attention, he questions whether the extra predictive value that is added to the predictive value of planned behaviour measures by including past behaviour as a factor (the 'residual effect') can be attributed to habit formation or 'habituation' as he calls it. In doing so he departs from the unproven premise that all behaviour is initially guided by explicit intentions. In view of what has been shown about unconscious motivation (Wilson, 2002; Damasio, 1994; LeDoux, 1996) and even goal setting (Aarts and Dijksterhuis, 2000; Bargh *et al.*, 2001) this seems an untenable position. Especially in a field like food choice where learning and preference formation start already before birth (see above) it becomes clear that much behaviour is not at all based on deliberation and reasoning. Based on his false premise, Ajzen then demands an explanation by the proponents of the habit formation approach for the fact that intentions lose their predictive validity when the behaviour becomes habitual. He argues that since intentions were there in the beginning and drove behaviour, there is no reason why they should stop doing so when the behaviour becomes habitual. In doing so he implies first of all that intentions always drive behaviour even when they become habitual, and he also tries to lay the burden of proof on his opponents. Ajzen continues to say that in order to have a residual effect on later behaviour, frequency of past behaviour *must (as if there were no other explanations!)* reflect the influence of factors not adequately captured in measures of the determinants of the behaviour. Obviously he overlooks that such supposed other determinants were also not foreseen in the theory of planned behaviour and that a much more parsimonious explanation of the fact that past behaviour is a better predictor was available. In fact, Ouellette and Wood, found that past behaviour is a better predictor of intention than intention is of past behaviour. They also indicate that this might mean that statements about intentions, and attitudes and beliefs are based on the notions that the subjects have about their past behaviour, as suggested by other authors (e.g., Bem, 1972; Cook *et al.*, 2005; see below). In fact, scrutinous contemplation shows that all but two of the examples cited by Ajzen (2002) in his strenuous efforts to reason away the priority of past behaviour over intention, could easily be explained in this way. The two exceptions are the post hoc analysis of two experiments by Ajzen and Madden (1986) on student course grade prediction based on past behaviour, or on intentions in groups of students that had respectively a low and a high accuracy in their grade expectations. Although past behaviour was clearly better than intention in predicting grade attainment in the low grade expectation accuracy groups, the reverse was true in the high accuracy groups where correlations between past performance and grade attainment were positive but too low to be significant, whereas the correlations between intention and grade attainment were rather high and significant.

Although this outcome seems to be a good counter example to the findings of Verplanken *et al.* (1998), who showed that intentions are only effective at low levels of habit, one must be very cautious in interpreting these results (of both groups of authors) because in these correlational data, no indication is given of the levels of performance in the groups and the dispersion in the results. In order to draw conclusions on such data, scatter plots should be given, especially when absence of correlation is used as an argument as in the case presented by Ajzen. The fact that Verplanken *et al.* (1998) also show the results for a middle group is more reassuring in this respect, but in the case of Ajzen and Madden (1986) it is possible that the high accuracy group had not enough variability in their past behaviour or in their final grades to produce a high correlation and the opposite may have been true of the intentions in the low opposite groups. All sorts of artefacts can lead to spurious correlations. This stresses once more that such analyses have no value as long as no complete insight in the data is given.

Since all other problems raised by Ajzen can indeed be solved by accepting Ouellette and Woods' suggestion and simply assuming that the subjects base their statements about attitudes beliefs and intentions on their past behaviour, it is proposed not to discuss here all of the rather tortuous arguments in defence of the planned behaviour approach in Ajzen's paper (Ajzen, 2002).

After all, according to the many authors mentioned earlier (Wilson; Damasio; LeDoux; Zajonc), people usually have no access to the motives in their adaptive unconscious and because questionnaires force them to come up with an answer, deducing an answer from their past behaviour may well be the only solution left to them. How trustworthy then is information based on questionnaires that explicitly ask for the reasons and motives of people's behaviour or for self-reports on attitudes, beliefs and intentions?

It was, of course, known for a long time that people say other things than they do, especially when an emotional and much debated topic like healthy eating is involved (Köster *et al.*, 1987; Köster, 1988, 2003), but the psychological evidence that explains this phenomenon has hardly ever entered the discussion on eating behaviour and food choice. Nevertheless, Nisbett and Wilson (1977) published their article entitled 'Telling more than we can know: Verbal reports on mental processes' a long time ago and since then there has been a continuous flow of papers that show that what people say about their beliefs, values, attitudes and intentions is often in discordance with their behaviour and their unconscious motives and prejudices (see Wilson, 2002).

Wilson *et al.* (1995) also explain the frequently made observation that thinking about reasons changes people's attitudes towards the object of their attitude by the fact that when thinking about reasons, people focus on attributes of the attitude object that are accessible in memory (1), that are plausible as causes for their feelings (2), and that are easy to verbalise (3). In other words, it is little wonder that past experience plays an important role in the expressed attitudes and intentions of people. All three elements mentioned rely heavily on it.

The reported reasons may not be the actual causes of their attitudes or intentions, however, because people do not have perfect access to the motives that are embedded in their adaptive unconscious and therefore have to invent plausible explanations for their behaviour. This is perhaps the most important and profound reason for the discrepancy between people's statements and their actual behaviour. It is a reason that can not easily be excluded or circumvented, precisely because access to the real reasons is not possible. Cook *et al.* (2005), while attacking the theory of planned behaviour on the basis of its misuse of causality notions, also criticise the use of questionnaires as the basis for measuring attitudes and beliefs. They cite Billig (1987) who explained that 'certain attitudes are put forward in particular contexts and are often accom-panied by particular argument and justification so that an attitude functions not as an "inner" state, but rather as a term used in relation to a particular conversation'. In other words: 'An attitude is an expression generated by a person for the occasion' (Cook *et al.*, 2005).

In the case of another common and often cited cause of the falsification of responses, social desirability, the subjects are at least usually aware of the fact that their statements do not correspond with their actual behaviour. This is not the case for 'why' questions about motives. Here, the responses are not white lies, but confabulations or narratives (Wilson, 2002) that, even to the person who utters them, seem the most correct answer. In many cases these responses do not transcend the level of 'common opinion' or hearsay.

4.5   Validity of measurement methods

4.5.1   Questionnaires and self reports

Does this mean that the responses to all questionnaires are at least doubtful and at most misleading? Much depends on the type of questions asked. As long as they are concerned with topics that are free from social desirability, responses to questions about the frequency of behaviour are usually more reliable and externally valid than questions about the reasons for the behaviour or about underlying hypothetical constructs such as values, beliefs, attitudes or intentions. The reason for this may be that frequency responses rely on past experience and past experience is the best predictor of actual behaviour. The same explanation might also be applied to the fact that, in almost all questionnaire research about beliefs, attitudes and intentions, some correlation with actual behaviour is found, notwithstanding the inaccessibility of the true motives for that behaviour. When indeed people deduce their statements about attitudes, beliefs and intentions from past experience and past experience is a good predictor of future behaviour, most of these statements will also show some relationship with actual behaviour. If this interpretation is correct, one might expect that the correlations found would grow with age – due to accumulation of past experience – and this is precisely what happens in the research of Olsen (2003). Of course there may be many other possible explanations of the difference in 'past experience-strength of attitude' relationship between different age cohorts. As Olsen himself admits, his findings are only based on correlational measures and thus have very limited value, especially because in no case do they explain more than 22% of the variance in self-reported consumption behaviour. Overlooking these data, an interpretation of the age dependency of the attitudes on past experience is also in accordance with the ideas of Daryl Bem, who proposed in his self-perception theory (Bem, 1972) that people use observation of their own behaviour in certain situations to infer conclusions about their feelings and to attribute their behaviour to internal causes.

Ninety years earlier, James (1884) had already proposed an interpretation of emotions – later known as the James-Lange emotion theory – along the same lines of reasoning. In contrast to the opinion of his contemporaries, he assumed that we feel sad because we are crying; we feel fear because we creep away. However, as Bem himself indicates, the applicability of his attribution theory is limited to situations in which emotions and motivational causes have to be named while it is unclear how we feel. Thus, it is not applicable to direct and clear feelings like hunger and thirst, but it might be perfectly relevant in the context of questionnaires about values, attitudes and intentions with regard to food choice, where the true reasons remain largely unconscious. For even our own basic values, attitudes and prejudices remain hidden to us in most cases and are often more easily visible to others than to ourselves (Wilson, 2002). And we even have to make up stories when someone asks us why we like mango or parmesan cheese, because we simply do not really know. In passing, Bem also introduces another important notion: the idea that our behaviour and our interpretation of it are dependent on the situation we are in.

4.5.2   Situational analysis

The present chapter is about the development of food choice and the dynamics of preferences. It has drawn attention to the importance of past experience in the prediction of present behaviour and to the role of arousal seeking in the evolution of food choice. Up till now, one thing has not been questioned: the integrity of the experiencing person over time. It may seem absurd, but in trying to understand our choices we should perhaps ask ourselves whether we are the same person in different situations.

Over the last two decades the influence of context on food appreciation has been studied extensively and it has been shown, for instance, that the same meal when presented in a student canteen-like surrounding is appreciated much less than when served in a luxury restaurant atmosphere (Meiselman, 1996; Edwards *et al.*, 2003). Interesting as they may be, these context effects are not exactly the same as the situational effects that are studied in situational analysis (Köster, 1996). Situations are not merely defined by the physical context, but by the meaning attributed to the surroundings by the 'intentionality' and resulting expectations of the individual subject. Why, once back at home, can we not touch the bottle of the ouzo we liked so much in Greece? The ouzo has not changed, but have we? The meanings of the things in our surroundings are as it were the crystallisations of our personal history. Our (mostly unconscious) intentions and expectations shape our surroundings. Hunger makes us see 'eatables', but hunger when sitting alone in front of the TV shows other 'eatables' than when at a dinner with friends at home or with a group of business relations at a restaurant.

The view that we attribute meaning to our surroundings fits well into a concept that considers perception to be an active instead of a passive and receptive process. According to this view we are never without (largely unconscious) expectations and hypotheses about what we will encounter and experience in our surroundings. These expectations are based on our 'intentions' and they are coloured by our personal history. Thus, the same physical context may have different meanings to people with different personal histories and to the same person under different 'intentions'.

Does this mean that all situations are highly individual and that no general conclusions can be drawn about their effects on people's choices? Not at all. As J.J. Gibson (1968, 1986) pointed out, our senses have been formed as an answer to the world we live in and this is the basis for a strong inter-subjectivity in our perception. Thus, gravity has made it necessary to develop muscles and kinaesthetic receptors to be able to stand and move. At the same time our senses make us see the meaning of 'under and above' and of horizontal versus vertical surfaces. We 'see' that we can not walk on the walls or on the ceiling.

Furthermore, members of the same culture and the same generation share an enormous commonality of meanings. They have learned to 'create' and understand situations in the same way. And such situational meanings may even transcend generations and cultures. If this was not the case there would be no film industry or literature. We immediately know what it is to be alone or in the company of family, friends or strangers. Many of the dishes we like most when in the family circle or when eating alone in front of the TV, we would not present to our guests when we organise an official dinner. We have summer and winter dishes, national and regional specialties and global products. We choose them according to the situation. This simply indicates that it is perhaps not wise to consider the consumer as the unity of research. Consumers behave quite differently in different situations in which they also have different wishes and expectations with regard to food. Funnily enough, most segmentations in consumer research are made on the basis of differences in age, gender, income, and in personality traits such as neophobia and variety seeking or in some of the more superficial 'life style' marketing typologies. All of these segmentations treat the consumer as if he/she were always the same person, whereas healthy people play many different roles in different situations and in doing so have different habits, wishes and expectations.

Thus, segmentation on the basis of eating and drinking situations seems to be a better concept for successful product development. This has been shown in a number of confidential experiments (summarised in Köster, 2003), and in a large scale study (Levy and Köster, 2001, unpublished report). Segmentation in age groups overlaps with situational segmentation to a certain extent – very few of the elderly go to discos – but it is much less specific. Further studies in situational analysis carried out with behavioural and observational methods would be very useful to improve consumer insight. For a correct understanding of the fact that situations are to such a large extent universal within a given culture, Bem's theory about the acquisition of beliefs and attitudes should be considered (Bem, 1970, Chapter 6). His self-perception theory is based on the idea that from early childhood on we learn from others (parents, peers) how to describe our internal states and perceptions on the basis of external cues. These others infer those internal states from our observable behaviour and then label the internal states that they assume to occur in us. Thus, they transmit a cultural pattern of interpretation of internal states very effectively from a very early age. In the beginning this learning takes place via conditioning. Bem furthermore proposes that 'in identifying his own internal states an individual partially relies on the same external cues that others use when they infer his internal states'. In other words, we read as it were our feelings and attitudes from our behaviour. This means that our conscious feelings, attitudes and intentions are the interpretation of behaviour rather than the cause of it. It also once more stresses the important influence of learning and past experience.

Situational analysis (Köster, 1996) is based on situational segmentation and on judgments of appropriateness, as in the work of Schutz (Schutz 1988; Cardello and Schutz, 1996). In such an analysis, people are first asked how often they are in certain situations (e.g., eating alone at home, in restaurants, in the car, or with family, friends, business partners, customers, etc.). They are also asked how well certain dishes or variants of these dishes would fit these situations. Subsequently, groups that are often experiencing certain situations are used in especially devised situational simulation experiments in which they evaluate products or in which their choice behaviour is observed (Köster, 1996, 2003; Henry and Köster, 1999, unpublished report), Levy and Köster (2001, unpublished report). Where tested, the method has led to a better prediction of actual choice behaviour in normal everyday life situations than hedonic tests, questionnaire methods, so called in-home-use tests or combinations of them. The reason is simple. People do not just eat products, they eat what they like in the meaningful situation that they forge themselves from the mixture of memory and intentionality.

4.6   Integration of the theoretical approaches

For a good understanding of the development of the determining factors of actual food choice behaviour it is necessary to accept the view that at different stages in life different combinations of the non-cognitive and cognitive mechanisms described in the theoretical approaches interact with each other in different ways. Nevertheless, as stated in the first sentence of this chapter, food choice is predominantly a learned behaviour and many different forms of learning do indeed play a significant role in all food choice behaviour. It also seems clear that throughout life there are sensitive periods for food choice habit formation and change. Different combinations of the types of learning described will probably be functional in these periods. Thus, in the perinatal period and very early childhood, only affective conditioning plays a role, but quite soon this is joined by learning through reward and punishment and basic emotional imitation learning. Since much of the habit formation based on this learning takes place at a pre-verbal and non-cognitive level, it is very difficult to change the habits formed in that period by cognitive influences later in life.

The imitation learning in this early phase is very different from the learning in late childhood and early puberty where imitation is based on social interactions, and social status plays the predominant role and where the intrinsic properties of the food itself are less important than their symbolic function. This may result in fads of a rather fleeting character. At the same time there is usually a period in late childhood, before the onset of puberty, that lends itself to cognitive learning of factual knowledge about food and eating and drinking behaviour and about the principles of good dietary behaviour. Whether such learning is integrated into the child's personal food habits is a different question. In most cases there is no immediate effect on the child's behaviour. Nevertheless, it may well be a good period to form conscious ideas about the relationship between health and eating and drinking behaviour, even if such ideas do not influence behaviour until much later in life.

At the end of puberty, when people pair off in couples, there is another important sensitive period for food habit formation and change. The different food habits of the pair members force them to change and new 'traditions' are formed that symbolise their unity. In this process a mixture of non-cognitive and cognitive arguments may be involved and it may lead to the exchange of past experiences and to a substantial amount of shared new ones. Most of these exchanges will be governed by the wish to find commonality on non-cognitive grounds. When couples stay together and children arrive there is a third major opportunity for food habit formation and change and it is one in which cognitive reasoning and planned behaviour play a role. The feelings of responsibility for the upbringing of the children come into play and may for a while dominate a part of the food choice behaviour. On the other hand, children's wishes and likes and dislikes often take command in many households. These wishes often form secret bonds with the old desires of one or both parents.

A fourth moment in life for the change of food habits is divorce. Although people often fall back on old habits in that situation, they may also make a radical change in their eating habits. Finally, retirement with its larger freedom is often another sensitive period for food habit change. Here health and convenience arguments are often in conflict with the desire to relive old pleasures.

4.7   Future trends

Research on the interaction of the different mechanisms at work in the development and change of food choice will be needed to fully understand the dynamics of preference and food choice. Such understanding will have to be tested by predicting food choice in real life situations. The growing insight into the unconscious origin of most of food choice behaviour will necessitate the development of new research methods that rely on observation of behaviour and simulation of situations rather than on questionnaires and self-reports. This research should go hand in hand with the development of insight into the role of the biological determinants of eating and drinking behaviour.

There also seems to be a need for the realisation that the most important food habits and food preferences are formed very early in life and that these are the ones least susceptible to being influenced by rational deliberations. This should lead to much more attention to the development of preventive educational programmes based on very early experiential learning with varied and healthy foods, instead of the current practice of convincing people at a stage where the basis for unhealthy eating habits has been laid and the chance of changing them is strongly reduced. At the same time, attention should be given to the adequate possibilities to counteract bad habits during the sensitive periods described above.

The construction of observational facilities to effectively study eating and drinking behaviour under normal everyday conditions has already started and methods for the study of simulated situational research have been developed. These new tools will help to understand the determinants of food choice and eating and drinking behaviour much better.

4.8   Sources of further information

Throughout the chapter a number of review articles and books that will provide insight have been mentioned already. Here we include them once more in an overview per subject.

4.8.1   Learning theories

Over the last three decades the views on the roles of learning and memory in everyday life have changed dramatically. Readers who are interested in these developments in general will find information in the *Oxford Handbook of Memory* edited by Tulving and Craik (2000). Accounts that deal more specifically with memory in the senses involved in food perception can be found in Herz and Engen (1996) and Köster (2004, 2005).

4.8.2   Motivational theories

The essential literature for the motivational theories described here is given in the text. Those interested, should perhaps devote special attention to the two books by Berlyne (1960, 1974) and to the chapters (9 and 10) on motivation in *Psychology of Perception* by Dember (1970). The relation between arousal and complexity has also received much attention in music psychology, but the extensive and interesting literature in this domain (that in many ways is comparable to food appreciation – difficult to verbalise, etc.) is rather dispersed.

4.8.3   Theories on decision and choice

The book edited by Shepherd (1989) provides an overview of the application of the theory of planned behaviour to food choice and eating and drinking behaviour. A critical evaluation of the many other conscious cognitive 'theories' (often not much more than overrated hypotheses) and their effectiveness in changing behaviour are discussed by Baranowski *et al.* (2003). Kahneman and Tversky's ideas on decision making and the biases that influence it can be found in Kahneman and Tversky (2000) and in Gilovich *et al.* (2002). Views on the unconscious and automatic nature of most decisions have been elegantly presented by Wilson (2002) and are supported by physiological underpinnings in the books of Damasio (1994, 1999, 2003) and LeDoux (1996, 2003), although these latter authors are criticised by Rolls (2005) for their views on the physiological mechanisms involved.

4.8.4   Active perception and expectations

Background literature on this topic is found in Gibson (1968) and more physiologically founded arguments for the viewpoint of active perception are discussed in Freeman (1999).

4.9   References and further reading

AARTS H and DIJKSTERHUIS A (2000), 'Habits as knowledge structures: Automaticity in goal-directed behavior', *J Personal Soc Psychol*, 78(1), 53–63.

ADOLPHS R (1999), 'Social cognition and the human brain', *Trends Cogn Sci*, 3, 469–479.

ADOLPHS R (2002), 'Neural mechanisms for recognizing emotion', *Current Opinion in Neurobiol*, 12, 169–178.

ADOLPHS R, TRANEL D and DAMASIO A R (1999), 'The human amygdala in social judgment', *Nature*, 393, 470–474.

ADOLPHS R, DAMASIO H, TRANEL D, COOPER G and DAMASIO A (2000), 'A role for somato-sensory cortices in the visual recognition of emotion as revealed by 3-D lesion mapping', *J Neurosci*, 20, 2683–2690.

AJZEN I (1985), 'From intentions to actions: A theory of planned behavior', in Kuhl J and Beckmann J, *Action Control; from Cognition to Behavior*, Heidelberg, Springer Verlag.

AJZEN I (2002), 'Residual effects of past on later behavior: Habiuation and reasoned action perspectives', *Personal Soc Psychol Rev*, 6(2), 107–122.

AJZEN I and FISHBEIN M (1970), 'The prediction of behavior from attitudinal and normative variables', *J Exp Soc Psychol*, 6, 466–487.

AJZEN I and FISHBEIN M (1980), *Understanding Attitudes and Predicting Behavior*, Englewood Cliffs NJ, Prentice Hall.

AJZEN I and MADDEN T J (1986), 'Prediction of goal-directed behavior: Attitudes, intentions, and perceived behavioral control', *J Exp Soc Psychol*, 22, 453–474.

BAEYENS F, EELEN P, VAN DEN BERGH O and CROMBEZ G (1990), 'Flavor-flavor and color-flavor conditioning in humans', *Learn Motiv*, 21, 434–455.

BAEYENS F, CROMBEZ G, HENDRICKX H and EELEN P (1995), 'Parameters of human evaluative flavor-flavor conditioning', *Learn Motiv*, 26, 141–160.

BARANOWSKI T, CULLEN K W, NICKLAS T. THOMPSON D and BARANOWSKI J (2003), 'Are current health behavioural change models helpful in guiding prevention of weight gain efforts?' *Obesity Res*, 11, 23S–43S.

BARGH J A (1990), 'Auto-motives: Preconscious determinants of social interaction', in Higgins T and Sorrento R M, *Handbook of Motivation and Cognition* (pp. 93–130), New York, Guilford.

BARGH J A (1992), 'Being unaware of the stimulus vs. unaware of its interpretation: Why subliminality per se does matter to social psychology', in Bornstein R and Pittman T, *Perception without Awareness*, New York, Guilford.

BARGH J A (1994), 'The four horsemen of automaticity: Awareness, intention, efficiency and control in social perception and cognition', in Uleman J S and Bargh J A eds, *Unintended thought*, New York, Guilford Press, pp. 3–51.

BARGH J A (2006), 'Agenda 2006: What have we been priming all these years? On the development, mechanisms, and ecology of nonconscious social behavior', *Eur J Soc Psychol*, 36, 147–168.

BARGH J A and CHARTRAND T L (1999), 'The unbearable automaticity of being', *Am Psychologist*, 54, 462–479.

BARGH J A, GOLLWITZER P M, LEE-CHAI A, BARNDOLLAR K and TRÖTSCHL R (2001), 'The automated will: Nonconscious activation and pursuit of behavioral goals', *J Personal Soc Psychol*, 81(6), 1014–1027.

BECHARA A, DAMASIO H, TRANEL D and DAMASIO A R (1997), 'Deciding advantageously before knowing the advantageous strategy', *Science*, 275, 1293–1295.

BECHARA A, DAMASIO H and DAMASIO A (2000), 'Emotion, decision making and the prefrontal cortex', *Cerebral Cortex*, 295–307.

BEM D J (1970), *Beliefs, Attitudes and Human Affairs*, Belmont CA, Brooks/Cole.

BEM D J (1972), 'Self-perception theory', in Berkowitz L (ed.), *Advances in Experimental Social Psychology Vol. 6*, New York, Academic Press, pp. 1–62.

BERLYNE D E (1960), *Conflict, Arousal, and Curiosity*. New York, McGraw-Hill.

BERLYNE D E (1967), 'Arousal and reinforcement', *Nebraska Symposium on Motivation*, University of Nebraska Press, pp. 1–110.

BERLYNE D E (1970), 'Novelty, complexity, and hedonic value', *Perc & Psychophys*, 8, 279–286.

BERLYNE D E (1974), 'The new experimental aesthetics', in Berlyne D E, *Studies in the New Experimental Aesthetics: Steps towards an Objective Psychology of Aesthetic Appreciation*. New York, John Wiley & Sons.

BILLIG M (1987), *Arguing and Thinking: a Rhetorical Approach to Social Psychology*, Cambridge, Cambridge University Press.

BIRCH L L (1980), 'The relationship between children's food preferences and those of their parents', *J Nutr Educ*, 12, 14–18.

BIRCH L L (1999), 'Development of food preferences', *Ann Rev Nutr*, 19, 41–62.

BIRCH L L, MCPHEE L, SHOBA B C, PIROK E and STEINBERG L (1987), 'What kind of exposure reduces children's neophobia?', *Appetite*, 9, 171–178.

BRUNSTROM J M (2005), 'Dietary learning in humans: Directions for future research', *Physiol Beh* 85, 57–65.

BRUNSTROM J M, MITCHELL G L and BAGULEY T S (2005), 'Potential early-life predictors of dietary behaviour in adulthood: a retrospective study, *Int J Obes*, 29(5), 463–474.

CACCIOPPO J T, BERNTSON G G, SHERIDAN J F and McCLINTOCK M K (2000), 'Multilevel integrative analyses of human behaviour: Social neuroscience and the complementing nature of social and biological approaches', *Psychol. Bull*, 126, 829–843.

CARDELLO A V and SCHUTZ H G (1996), Food appropriateness measures as an adjunct to consumer preference/acceptability evaluation, *Food Qual Pref*, 7, 239–249.

CASHDAN E (1998), 'Adaptiveness of food learning and food aversions in children', *Information pour les sciences sociales*, 37, 613–632.

COOK A J, MOORE K and STEEL G D (2005), 'Taking a position: a reinterpretation of the theory of planned behaviour', *J Theory Soc Beh*, 35(2), 143–154.

COOKE L, WARDLE J and GIBSON E L (2003), 'Relationship between parental report of food neophobia and everyday food consumption in 2–6-year-old children', *Appetite*, 41, 205–206.

DAMASIO A R (1994), *Descartes Error: Emotion, Reason and the Human Brain*, New York, Avon.

DAMASIO A R (1999), *The Feeling of What Happens*, San Diego, Harcourt, Inc.

DAMASIO A (2003), *Looking for Spinoza*. London, Harcourt Inc.

DEMBER W N (1970), *The Psychology of Perception* (2nd edn), London, Holt, Rinehart and Winston, pp. 341–375.

DEMBER W N and EARL R W (1957), 'Analysis of exploratory, manipulatory and curiosity behaviors', *Psychol Rev*, 64, 91–96.

DEMBER W N, EARL R W and PARADISE N (1957), 'Response by rats to differential stimulus complexity', *J Comp Psychol*, 49, 93–95.

DIETZ W H (1994), 'Critical periods in childhood for the development of obesity', *Am J Clin Nutr*, 95, 955–959.

DIETZ W H (1997), 'Periods of risk in childhood for the development of adult obesity – what do we need to learn?', *J Nutr*, 127, S1884–1886.

DIJKTERHUIS A and VAN KNIPPENBERG A (1998), 'The relation between perception and behavior, or how to win a game of trivial pursuit', *J Person Soc Psychol*, 74(4), 865–877.

DIJKSTERHUIS A, SMITH P K, VAN BAAREN R B and WIGBOLDUS H J (2005), 'The unconscious consumer: Effects of environment on consumer behavior', *J Consum Psychol*, 15, 193–202.

DUNN B D, DALGLEISH T and LAWRENCE A D (2006), The somatic marker hypothesis: a critical evaluation, *Neurosi Biobehav Rev*, 30, 239–271.

EDWARDS J S A, MEISELMAN H A, EDWARDS A and LESHER L (2003), 'The influence of eating location on the acceptability of identically prepared foods', *Food Qual Pref*, 14, 647–652.

ERDELYI M H (1992), 'Psychodynamics and the unconscious', *Am Psychol*, 47, 784–787.

FISHBEIN M and AJZEN I (1975), *Belief, Attitude, Intention and Behavior: An Introduction to Theory and Research*, Reading MA, Addison-Wesley.

FRANK R A, VAN DER KLAAUW N J and SCHIFFERSTEIN H N J (1993), 'Both perceptual and conceptual factors influence taste-odor and taste-taste interactions', *Perc Psychophys*, 54, 343–354.

FREEMAN W J (1999), *How Brains Make up their Minds*. London, Orion Books Ltd.

GANCHROW J R, STEINER J E and DAHER M (1983), 'Neonatal facial expressions in response to different qualities and intensities of gustatory stimuli', *Infant Behav Dev*, 6, 189–200.

GANCHROW J R and MENELLA J A (2003), 'The ontogeny of human flavour perception', in Doty R L *Handbook of olfaction and Gustation* (2nd edn), New York, Marcel Dekker, pp. 823–846.

GARCIA P, SIMON C, BEAUCHAMP G K and MENELLA J (2001), 'Flavor experiences during formula feeding are related to childhood preferences', *Chem Senses*, 26, 1039.

GAWANDE A (2002), *Complications: A Surgeon's Notes on an Imperfect Science*, New York, Metropolitan Books.

GIBSON J J (1968), *The Senses Considered as Perceptual Systems*, London: George Allen & Unwin, pp. 116–129.

GIBSON J J (1986), *The Ecological Approach to Visual Perception*, Hillsdale NJ, Lawrence Erlbaum.

GILOVICH T, GRIFFIN D and KAHNEMAN D (eds) (2002), *Heuristics and Biases: The Psychology of Intuitive Judgement*, Cambridge, Cambridge University Press.

HAARI R, FORSS N, ARVIKAINEN S, KIRVESKARI E, SALENIUS S and RIZZOLATTI G (1998), 'Activation of the primary motor cortex during action observation: A neuromagnetic study', *Proc Nat Acad Sci*, 95, 15061–15065.

HALLER R, RUMMEL C, HENNEBERG S, POLLMER U and KÖSTER E P (1999), 'The influence of early experience with vanillin on food preference later in life', *Chem Senses*, 24, 465–467.

HANSE L (1994), *La néophobie alimentaire chez l'enfant* (Food neophobia in children) Doctoral thesis, Université de Paris 10, Nanterre, France.

HEBB D O (1946), 'Emotion in man and animal: an analysis of the intuitive processes of recognition', *Psychol. Rev.*, 53, 88–106.

HEBB D O (1949), *The Organization of Behavior*, New York, Wiley.

HEBB D O (1966), *A Textbook of Psychology*, 2nd edn, Philadelphia, Baunders,

HENRY S and KÖSTER E P (1999) A comparison of three ways to simulate eating situations, Rapport Confidentiel, CESG, Dijon.

HERZ R S and ENGEN T (1996), 'Odor memory: Review and analysis', *Psychonomic Bull Rev*, 3, 300–313.

HONKANEN P, OLSEN S O and VERPLANKEN B (2005), 'Intention to consume seafood – the importance of habit', *Appetite*, 45, 161–168.

JACOBY L L, TOTH J P, LINDSAY D S and DEBNER J A (1992), 'Lectures for a layperson: Methods for revealing unconscious processes', in Bornstein R and Pittman T, *Perception without Awareness*, New York, Guilford, pp. 81–120.

JAMES W (1884), 'What is an emotion', *Mind*, 9, 188–205.

JELLINEK J S and KÖSTER E P (1983), 'Perceived fragrance complexity and its relation to familiarity and pleasantness II', *J Soc Cosmet Chem*, 34, 83–97.

KAHNEMAN D (1973), *Attentionand Effort*, Englewood Cliffs, NJ, Prentice-Hall.

KAHNEMAN D (2003), 'A perspective on judgement and choice: Mapping bounded rationality', *Am Psychologist*, 58(9), 697–720.

KAHNEMAN D and FREDERICK S (2002), 'Representativeness revisited: Attribute substitution in intuitive judgment', in Gilovitch T, Griffin, D and Kahneman D (eds), *Heuristics and Biases*, New York, Cambridge University Press.

KAHNEMAN D and TVERSKY A (1972), 'Subjective probability: A judgment of representativeness', *Cogn Psychol*, 3, 430–454.

KAHNEMAN D and TVERSKY A (1973), 'On the psychology of prediction', *Psychol Rev*, 80, 237–251.

KAHNEMAN D and TVERSKY A (1979), 'Prospect theory: An analysis of decision under risk', *Econometrica*, 47(2), 263–291.

KAHNEMAN D and TVERSKY A (eds) (2000), *Choices, Values and Frames*, Cambridge, Cambridge University Press.

KAHNEMAN D, SLOVIC P and TVERSKY A (eds) (1982), *Heuristics and Biases*, NewYork, Cambridge University Press.

KIHLSTROM J F (1987), 'The cognitive unconscious', *Science*, 237, 1445–1452.

KLEIN G (1998), *Sources of Power: How People make Decisions*, Cambridge, MA, MIT Press.

KOIVISTO U and SJÖDÉN P (1996), 'Reasons for rejection of food items in Swedish families with children aged 2–17', *Appetite*, 26, 89–103.

KOIVISTO-HURSTI U K AND SJÖDÉN P O (1997), 'Food and general neophobia and their relationship with self-reported food choice: familial resemblance in Swedish families with children of ages 7–17 years', *Appetite*, 29, 89–103.

KÖSTER E P (1985), 'The importance of stimulus complexity in the analysis of sensory preferences', in Adda J, *Progress in Flavour Research*, Amsterdam, Elsevier Science Publishers, pp. 15–27.

KÖSTER E P (1988), 'Problems in consumer research with health products', in Manley C H and Morse R E, *Healthy Eating. A Scientific Perspective*, Chicago, Allured Publ Corp, pp. 191–225.

KÖSTER E P (1996), 'The consumer? The quality?', in *Production industrielle & qualité sensorielle Agoral 96. Huitièmes rencontres scientifiques et technologiques des industries alimentaires 2-3 Avril 1996, Dijon*, Paris, LavoisierTec & Doc, pp. 11–19.

KÖSTER E P (2003), 'The psychology of food choice: some often encountered fallacies', *Food Qual Pref*, 14, 359–373.

KÖSTER E P (2004), 'Perception and incidental memory in three "lower" senses', in Oliveira A M, Teixeira M, Borges G F and Ferro M J (eds), *Fechner Day 2004*, Coimbra, pp. 102–107.

KÖSTER E P (2005), 'Does odour memory depend on remembering odours?' *Chem Senses*, 30, (suppl.1), i236–i237.

KÖSTER E P, BECKERS A W and HOUBEN J H (1987), 'The influence of health information on the acceptance of a snack in a canteen test', in Martens M, Dalen G A and Russwurm H, *Flavour Science and Technology*, New York, John Wiley & Sons, pp. 391–398.

KÖSTER E P, RUMMEL C, KORNELSON C and BENZ K H (2001), 'Stability and change in food liking: food preferences in the two Germanys after the reunification', in Rothe M, *Flavour 2000: Perception, Release, Evaluation, Formation, Acceptance, Nutrition and Health*. Bergholz-Rehbrücke, Rothe, 237–253.

KÖSTER E P, COURONNE T, LÉON F, LÉVY C and MARCELINO, A S (2003), 'Repeatability in hedonic sensory measurement: a conceptual exploration', *Food Qual Pref*, 14, 165–176.

KÖSTER M A, PRESCOTT J and KÖSTER E P (2004), 'Incidental learning and memory for three basic tastes in food', *Chem Senses*, 29, 441–453.

LÄHTEENMÄKI L and VAN TRIJP H C M (1995), 'Hedonic responses, variety-seeking tendency and expressed variety in sandwich choices', *Appetite*, 24, 139–152.

LEDOUX J (1996), *The Emotional Brain*, 3rd edn, London, Phoenix Orion Books Ltd.

LEDOUX J (2003), *Synaptic Self*, London, Penguin Books.

LÉON F, COURONNE T, MARCUZ M C and KÖSTER E P (1999), 'Measuring food liking in children: a comparison of non verbal methods', *Food Qual Pref*, 10, 93–100.

LÉVY C M and KÖSTER E P (1999), 'The relevance of initial hedonic judgements in the prediction of subtle food choices', *Food Qual Pref*, 10, 185–200.

LÉVY C M and KÖSTER E P (2001), 'Validation d'une méthode de simulation situationelle du comportement consommatoire dans des discotèques', *Rapport confidentiel, Pernod Ricard*.

LÉVY C M, MacRAE A W and KÖSTER E P (2006), 'Perceived stimulus complexity and food preference', *Acta Psychologica*, 123, 394–413.

LIEM D G and DE GRAAF C (2004), 'Sweet and sour preferences in young children and adults: role of repeated exposure', *Physiol Behav*, 83, 421–429.

LOEWEN R and PLINER P (1999), 'Effects of prior exposure to palatable and unpalatable foods on children's willingness-to-taste other novel foods', *Appetite*, 32, 351–366.

LOEWEN R and PLINER P (2000), 'Development of the Food Situations Questionnaire; a self-report measure of food neophobia in children', *Appetite*, 35, 239–250.

MAIA T V and McCLELLAND J L (2004), 'A re-examination of the evidence for the somatic marker hypothesis: what participants really know in the Iowa gambling task, *Proc Nat Acad Sci USA*, 101(45), 16075–16080.

MARLIER L, SCHAAL B and SOUSSIGNAN R (1998), 'Neonatal responsiveness to the odor of amniotic and lacteal fluids: A test of perinatal chemosensory continuity', *Child Dev*, 69, 611–623.

MEISELMAN H L (1996), 'The contextual basis for food acceptance, food choice and food intake: the food, the situation and the individual', in Meiselman H L and Macfie H J H, *Food Choice, Acceptance and Consumption*, London, Blackie Academic & Professional, pp. 239–263.

MENELLA J A and BEAUCHAMP G K (1999), 'Experience with a flavor in mother's milk modifies the infant's acceptance of flavored cereal', *Dev Psychobiol*, 35, 197–209.

MENELLA J A, JOHNSON A and BEAUCHAMP G K (1995), 'Garlic ingestion by pregnant women alters the odor of amniotic fluid', *Chem. Senses*, 20, 207–209.

MENELLA J A, JAGNOW J A and BEAUCHAMP G K (2001), 'Prenatal and postnatal flavor learning by human infants', *Pediatrics*, 107, E88.

MOJET J and KÖSTER E P (2002), 'Texture and flavour memory in foods: An incidental learning experiment', *Appetite*, 38, 110–117.

MOJET J and KÖSTER E P (2005), 'Sensory memory and food texture', *Food Qual Pref.*, 16, 251–266.

MØLLER P, WULFF C and KÖSTER E P (2004), 'Do age differences in odour memory depend on differences in verbal memory?', *Neuroreport*, 15(5), 915–917.

MØLLER P, MOJET J and KÖSTER E P (2006), 'Incidental and intentional flavour memory in young and older subjects', *Chem. Senses*, in press.

NICKLAUS S, BOGGIO V, CHABANET C and ISSANCHOU S (2004), 'A prospective study of food preferences in childhood', *Food Qual Pref*, 15, 805–818.

NICKLAUS S, BOGGIO V, CHABANET C and ISSANCHOU S (2005a), 'A prospective study of food variety seeking in childhood, adolescence and early adult life', *Appetite*, 44, 289–297.

NICKLAUS S, BOGGIO V and ISSANCHOU S (2005b), 'Food choices at lunch during the third year of life: high selection of animal and starchy foods but avoidance of vegetables', *Acta Paediatr.*, 94, 1–11.

NICKLAUS S, CHABANET C, BOGGIO V and ISSANCHOU S (2005c), 'Food choice at lunch during the third year of life: increase in energy intake but decrease in variety', *Acta Paediatr.*, 94, 12–18.

NISBETT R E and WILSON T D (1977), 'Telling more than we can know: Verbal reports on mental processes', *Psychol Rev*, 84, 231–259.

OLSEN S O (2003), 'Understanding the relationship between age and seafood consumption: The mediating role of attitude, health involvement and convenience', *Food Qual Pref*, 14, 199–209.

OUELLETTE J A (1996), 'How to measure habits? Subjective experience and past behaviour'. Unpublished Doctoral Thesis, Texas A & M University, College Station.

OUELLETTE J A and WOOD W (1998), 'Habit and intention in everyday life: The multiple processes by which past behavior predicts future behavior', *Psychol Bull*, 124, 54–74.

PELCHAT M and PLINER P (1995), '"Try it, you'll like it". Effects of information on willingness to try novel foods'. *Appetite*, 24, 153–166.

PLINER P (1982) 'The effects of mere exposure on liking for edible substance', *Appetite*, 3, 283–290.

PLINER P (1994) 'Development of measures of food neophobia in children', *Appetite*, 23, 147–163.

PLINER P and HOBDEN K (1992), 'Development of a scale to measure the trait of food neophobia in humans', *Appetite*, 19, 105–120.

PLINER P and LOEWEN R (1997), 'Temperament and food neophobia in children and their mothers', *Appetite*, 28, 239–254.

PLINER P and LOEWEN R (2002), 'The effects of manipulated arousal on children's willingness to taste novel foods', *Physiol Behav*, 76, 551–558.

PLINER P and MELO N (1997), 'Food neophobia in humans: effects of manipulated arousal and individual differences in sensation seeking', *Physiol Behav*, 61, 331–335.

PLINER P, PELCHAT M and GRABSKI M (1993), 'Reduction of neophobia in humans by exposure to novel foods', *Appetite*, 20, 111–123.

PORTER R H and SCHAAL B (2003), 'Olfaction and the development of social behaviour in neonatal mammals', in Doty R L, *Handbook of Olfaction and Gustation* (2nd edn), New York, Marcel Dekker, 309–327.

RAUDENBUSH B and FRANK R A (1999), 'Assessing food neophobia: The role of stimulus familiarity', *Appetite*, 32, 261–271.

RESNICOW K, MCCARTHY F and BARANOWSKI T (2003), 'Are precontemplators less likely to change their dietary behavior? A prospective analysis', *Health Educ Res*, 18, 693–705.

RHODES R E and COURNEYA K S (2003), 'Modelling the theory of planned behaviour and past behaviour', *Psychol Heath Med*, 8, 57–69.

RIZZOLATTI G, FADIGA L, GALLESE V and FOGASSI L (1996), 'Premotor cortex and the recognition of motor actions', *Cogn Brain Res*, 3, 131–141.

RIZZOLATTI G, FADIGA L, FOGASSI L and GALLESE V (1999), 'Resonance behaviors and mirror neurons', *Arch Italiennes de Biologie*, 137, 85–100.

RIZZOLATTI G, FOGASSI L and GALLESE V (2001), 'Neurophysiological mechanisms underlying the understanding and imitation of action', *Nature Rev Neursi*, 2, 661–670.

ROLLS E T (2005), *Emotion Explained*, Oxford, Oxford University Press.

ROZIN P (1976), 'The selection of food by rats, humans and other animals', in Rosenblatt J, Hinde R A, Beer C and Shaw E, *Advances in the study of behaviour, Vol. 6*, New York, Academic Press, 21–76.

ROZIN P and SCHILLER D (1980), 'The nature and acquisition of a preference for chilli pepper by humans', *Motiv Emot*, 4, 77–101.

SCHAAL B and ORGEUR P (1992), 'Olfaction in utero: can the rodent model be generalized?', *Quart J Exp Psychol*, 44, 254–278.

SCHACTER D L (1987), 'Implicit memory: History and current status', *J Exp Psychol: Learn, Mem Cogn*, 13, 501–518.

SCHACTER D L (1990), 'Perceptual representation systems and implicit memory: Toward a resolution of the multiple memory system debate', *Ann NY Acad Sci*, 608, 543–571.

SCHACTER D L (1994), 'Priming and multiple memory systems: Perceptual mechanisms of implicit memory', In D. L. Schacter and E. Tulving (eds), *Memory Systems*, Cambridge, MA, Bradford Book, pp. 233–268.

SCHACTER D L and GRAF P (1986), 'Effects of elaborative processing on implicit and explicit memory for new associations', *J Exp Psychol:Learn, Mem Cogn*, 12, 432–444.

SCHUTZ H G (1988), 'Beyond preference: Appropriateness as a measure of contextual acceptance of food', in Thomson D M H, *Food Acceptability*, New York, Elsevier, pp. 115–134.

SHEPHERD, R (1989), 'Factors affecting food preference and choice', in Shepherd R, *Handbook of the Psychophysiology of Human Eating*, London, Wiley, pp. 3–24.

SHEPHERD R and FARLEIGH C A (1986), 'Attitudes and personality related to salt intake', *Appetite*, 7, 343–353.

SHEPHERD R and STOCKLEY L (1985), 'Fat consumption and attitudes towards food with a high fat content', *Human Nutr: Appl Nutr*, 39A, 431–442.

SIMON H A and CHASE W G (1973), 'Skill in chess', *Am Scientist*, 61, 394–403.

STADLER M A and FRENSCH P A (eds) (1998), *Handbook of Implicit Learning*, Thousand Oaks, Sage Publications.

STEIN M, OTTENBERG P and ROULET H (1958), 'A study of the development of olfactory preferences', *Arch Neurol Psychiatr*, 15, 201–213.

STEINER J E (1977), 'Facial expressions of the neonate infant indicating the hedonics of food-related chemical stimuli', in Weiffenbach J M *Taste and Development: The Genesis of Sweet Preference*, Bethesda, MD, US Dept H E W Publications, pp. 173–188.

STEVENSON R J, PRESCOTT J and BOAKES R A (1995), 'The acquisition of taste properties by odors', *Learn Motiv*, 26, 433–455.

STEVENSON R J, BOAKES R A and PRESCOTT J (1998), 'Changes in odor sweetness resulting from implicit learning of a simultaneous odor-sweetness association: an example of learned synesthesia', *Learn Motiv*, 29, 113–132.

TULVING E and CRAIK I M (2000), *The Oxford Handbook of Memory*, Oxford, Oxford University Press.

TUORILA H, LÄHTEENMÄKI L, POHJALAINEN L and LOTTI L (2001), 'Food neophobia among the Finns and related responses to familiar and unfamiliar foods', *Food Qual Pref*, 12, 29–37.

TVERSKY A and KAHNEMAN D (1974), 'Judgment under uncertainty', *Science*, 185, 1124–1131.

TVERSKY A and KAHNEMAN D (1984), 'Extensional versus intuitive reasoning: The conjunction fallacy in probability judgment', *Psychological Review*, 91, 293–315.

VAN TRIJP H C M (1994), 'Product-related determinants of variety seeking behavour for foods', *Appetite*, 22, 1–10.

VAN TRIJP H C M and STEENKAMP J E B M (1992), 'Consumers's variety seeking tendency with respect to foods: measurement and managerial implications', *Europ Rev Agricult Econ*, 19, 181–195.

VAN TRIJP H C M, LÄHTEENMÄKI L and TUORILA H (1992), 'Variety seeking in the consumption of spread and cheese', *Appetite*, 18, 155–164.

VERBEKE W and VACKIER I (2005), 'Individual determinants of fish consumption: application of the theory of planned behaviour', *Appetite*, 44, 67–82.

VERPLANKEN B and AARTS H (1999), 'Habit, attitude, and planned behaviour: Is habit an empty construct or an interesting case of goal-directed automaticity?', in Stroebe W and Hewstone M (eds), *European Review of Social Psychology Vol. 10*, New York, John Wiley & Sons, pp. 101–134.

VERPLANKEN B, AARTS H, VAN KNIPPENBERG A and MOONEN A (1998), 'Habit versus planned behaviour: A field experiment', *Br J Soc Psychol*, 37, 111–128.

WALKER E L (1980), *Psychological Complexity and Preference: a Hedgehog Theory of Behavior*, Belmont, Wadsworth.

WEGNER D M (2002), *The Illusion of Conscious Will*, Cambridge, MA, MIT Press.

WILSON T D (1990), 'Self-persuasion via self-reflection', in Olson J M and Zanna M P (eds), *Self-inference Processes: the Ontario Symposium*, Vol. 6, Hillsdale NJ, Erlbaum, pp. 43–67.

WILSON T D (2002), *Strangers to Ourselves: Discovering the Adaptive Unconscious*, Cambridge, The Belknap Press of Harvard University Press.

WILSON T D and DUNN D S (1986), 'Effects of introspection on attitude-behavior consistency: Analyzing reasons versus focussing on feelings', *J Exp Soc Psychol*, 22, 249–263.

WILSON T D and SCHOOLER J W (1991), 'Thinking too much: Introspection can reduce the quality of preferences and decisions', *J Pers and Soc Psychol*, 60, 181–192.

WILSON T D, HODGES S D and LAFLEUR (1995), 'Effects of introspection about reasons: inferring attitudes from accessible thoughts', *J Pers Soc Psychol*, 69, 16–28.

ZAJONC R B (1968), 'Attitudinal effects of mere exposure', *J Pers Soc Psychol*, Monograph suppl, 9 (part 2), 65–74.

ZAJONC R B (1980), 'Feeling and thinking: Preferences need no inferences', *Am Psychol*, 35, 151–175.

ZAJONC R B (1984), 'On the primacy of affect', *Am Psychol*, 39, 117–123.

ZUCKERMAN M (1994), *Behavioral Expressions and Biological Bases for Sensation Seeking*, Cambridge, Cambridge University Press.

5

Perceptions of risk, benefit and trust associated with consumer food choice

J. de Jonge, E. van Kleef, L. Frewer and O. Renn, Wageningen University, The Netherlands

5.1 Introduction

The concept of consumer risk perception is important for understanding and explaining consumers' food choices (Conchar *et al.*, 2004). Risk is an inherent part of food choice as consumers continuously make trade-offs between exploring new foods and avoiding unsafe foods, a phenomenon referred to as the omnivore paradox (Fischler, 1990; Rozin, 1976). Consumer perceptions of risks associated with specific foods or food-related hazards have been extensively investigated (see, among others, Dosman *et al.*, 2001; Frewer *et al.*, 1994; Rosati and Saba, 2004), as well as the consequences of perceived risk for consumption intentions (see, among others, McCarthy and Henson, 2005; Pennings *et al.*, 2002; Verbeke and Viaene, 1999; Verbeke, 2001). In addition, it has been recognised that consumers' ethical concerns or other core values such as preferences for specific production methods may interact with risk perceptions in terms of influencing consumer food choices (Dreezens *et al.*, 2005). Besides consumer perceptions of risk, recent research activities have focused on the potential impact of consumer trust (in both food safety information sources, and institutions with responsibility for consumer protection) on consumer acceptance of food products and public responses to regulatory activities in the food safety area (Berg *et al.*, 2005; Siegrist, 2000).

Consumers are frequently confronted with food risk information via a range of different sources. Commonly cited examples include news media reports on food safety, for example from newspapers, television and radio, or from the internet. Prominent recent examples include that of Bovine Spongiform Encelophathy (BSE) (Frewer and Salter, 2002), and dioxin contamination of

European food chains (Verbeke, 2001). In addition, extensive negative media attention has been paid to the introduction of emerging food technologies (for example, genetic modification applied to the agri-food sector), which may result in increased risk perceptions and public concerns (Frewer *et al.*, 2002b; Verbeke, 2001; Hampel *et al.*, 2000). Disagreement regarding the extent and consequences of particular food hazards between different actors and stake-holders, should the associated risks occur, has likewise focused public attention on issues of food safety.

In designing and implementing appropriate risk management strategies, it is valuable to understand how consumers and experts differ in their risk perceptions (Frewer, 2001; Hansen *et al.*, 2003; Renn, 2004a; Slovic, 1987). The traditional approach to managing food risks has focused on translating the outcomes of technical risk assessments into both risk management activities and communication to consumers. Good risk assessment, risk management, and risk communication practices are needed in order to maintain and increase consumer confidence in the safety of food. As part of the process of risk analysis, it is important to take the actual concerns of consumers into account when developing both mitigation strategies and risk governance structures. Effective assessment and handling of risks may prevent the occurrence of substantial economic costs associated with a food safety crisis, such as the destruction of animals used for food production, product recalls, and managerial costs associated with tracking the origin of the product and identifying all potential contamination. In addition, good risk assessment, risk management, and risk communication practices may limit the frequently observed temporary or permanent reductions in consumer consumption levels of foods of which consumption is considered to be risky (Pennings *et al.*, 2002; Verbeke, 2001; Verbeke and Van Kenhove, 2002; Verbeke and Viaene, 1999).

Slovic and colleagues (see, for example Fischhoff *et al.*, 1978; Slovic, 1987, 1992; Slovic *et al.*, 1982) conducted the seminal research which demonstrated that consumer risk psychology is of direct relevance to understanding best practice in risk management and risk communication. Application of psycho-metric techniques has demonstrated that consumers are not only influenced in their assessment of the impact of potential hazards by technical assessments alone. They are also influenced by psychological factors such as whether they perceive that they are involuntarily exposed to a hazard, and the extent to which they believe a particular hazard is potentially catastrophic and uncontrollable. These perception patterns describe properties of risks or risky situations based on which people judge risks, namely beyond the two classical factors of risk analysis, i.e. level of probability and degree of possible harm. Here psychologists differentiate two classes of qualitative perception patterns: on the one hand risk-related patterns, which are based on the properties of the source of risk; on the other hand situation-related patterns, based on the idiosyncrasies of the risky situation (Renn, 2004a). They all contribute to the cognitive belief system which people include in their judgments of the seriousness of the risk as well as the acceptability of such a risk to themselves and society (Fischhoff, 1985).

Building on these early studies that focused on public perceptions of risk, a body of research has examined the role of consumer perceptions of both risk and benefit associated with different food hazards (see, for example, Saba and Messina, 2003; Siegrist and Cvetkovich, 2000). In addition, many studies have focused on consumer attitudes to emerging food technologies such as genetically modified foods (see, for example, Bredahl, 2001; Frewer et al., 2003b; Siegrist, 2000). Other research has focused on lifestyle hazards such as inappropriate dietary choices (see, for example, Anderson et al., 2000; Brug et al., 1995; Dibsdall et al., 2002; Steptoe et al., 2004) or microbiological risk (see, for example, Fischer et al., 2005; Gordon, 2003; Griffith and Worsfold, 1994; Griffith et al., 1998). The focus of research has tended towards an examination of the interrelationships between perceptions of risk, benefit, and trust, and how these relate to consumer attitudes towards specific food hazards, as well as food risk management practices more generally.

Another stream of research that evolved from the psychometric studies, explicitly focused on the role of affect (or human emotion) in guiding judgments of risk and benefit (Finucane et al., 2000b; Loewenstein et al., 2001; Slovic et al., 2002, 2004). The basic premise of this research is that affect, i.e. a readily available impression that something is good or bad, may serve as an important cue for making judgments of and responding to risk. Using a mental shortcut can be more efficient and easier than analytic reasoning about advantages and disadvantages of engaging in a particular behaviour, and can be helpful when mental resources are limited (Slovic et al., 2004).

Consumer risk perceptions have also been extensively examined in a marketing context (for a review of this literature see Conchar et al., 2004; Mitchell, 1998, 1999; Stone and Grønhaug, 1993). In this area, not only food risks with (potential) negative health consequences are studied, but other negative consequences relating to the consumer purchase process as well, such as the risk of losing valuable time in shopping or the potential harm to one's social standing that may arise from buying a potentially risky product.

A final stream of research has arisen which relates to the issue of increasing societal trust in food risk governance through involving stakeholders in the process of food safety risk analysis in a meaningful way (for example, see, Renn et al., 1995; Renn, 1999; Renn, 2004b for discussion of participatory processes in general; Renn, 2003; Rowe et al., 2004; Rowe and Frewer, 2005 for discussion of such processes focusing on food risks).

Several decades of research in different disciplines, such as economics, psychology, sociology and marketing has shown that each discipline uses a different approach and focuses on different aspects of risks. We believe, however, that an interdisciplinary approach is the most useful when trying to understand consumer behaviour with respect to food safety. The aim of this chapter is to provide an overview of consumer research into risk perception in relation to food choice behaviour. First, the range of methodological approaches that have been applied to study consumer risk perceptions will be reviewed. Next, the focus will be on what knowledge about consumer risk perceptions means for effective risk communication strategies and ways to involve the public in policy development.

5.2   Research into consumer food risk perceptions

In this section, four different approaches to investigate consumer risk perceptions related to food and food production will be reviewed. The methodological approaches adopted by the various studies within each research stream will be considered, and the key findings will be described.

5.2.1   Psychometric hazard classification studies

Research in the psychometric tradition focused on systematic assessment of how the risks associated with a range of potentially hazardous activities (for example, implementation of technologies such as nuclear power, human activities such as skiing, and methods of transportation such as railroads) were perceived by lay people, as opposed to experts (see, among others, Fischhoff et al., 1978). Examining public perceptions of risk and benefit was new in the sense that decision-making regarding the acceptance of technological developments in society used to be based on an economic cost analysis, where a trade-off was made between economic risks and benefits associated with a certain technology or activity (Starr, 1969). In the economic cost approach of Starr (1969), risk was measured as the probability of fatalities per hour of exposure of the individual to the activity considered, and benefits were represented by a monetary value, that is the average annual benefit per person involved. Risk and benefit were assumed to be positively correlated, as society would not accept high risk technologies with few societal benefits. Major drawbacks of this approach were that the results, to a large extent, depended upon the particular measures of risk and benefit used, and that decisions were limited only to economic aspects. However, the research highlighted the importance of making decisions based on risk-benefit trade-offs.

Inspired by and extending on the study by Starr (1969), Slovic and colleagues developed a psychometric approach which aimed at understanding the different perceptions that people associate with different types of hazards. They showed that public perceptions of risk are multidimensional, which indicates that risk perceptions may be more complex than a trade-off between perceptions of risk and benefit per se (Fife-Schaw and Rowe, 1996, 2000; Fischhoff et al., 1978; Slovic, 1987; Sparks and Shepherd, 1994). In research applying the psychometric approach, lay people rated a range of different hazards on different risk characteristics (e.g., the catastrophic potential and controllability of the hazard), which facilitated characterisation of public perceptions (Slovic, 1987). One of the most important findings of these studies was that the public perceived a specific hazard to vary along various cognitive and affective psychological dimensions, such as perceived knowledge and perceived dread. Although not all studies have used the same set of risk characteristics, frequently used aspects of risk were the voluntariness of exposure to the hazard, knowledge of the hazard by science and by those who are exposed, control, catastrophic potential, the degree of dread, the number of people exposed, the newness of the risk, and the seriousness of the risk for future generations. It was found that these risk dimensions were correlated and could be represented by two main principal components, namely the extent to which the hazards were dreaded and known by those exposed to them. By plotting the hazards in a two-dimensional space, defined by the two components, hazards could be described and classified according to similarities and differences in risk perceptions. For example, the public considered risks associated with genetic modification of food products as unknown and dreaded, whereas high fat diets were perceived to be known and not much dreaded (Kirk *et al.*, 2002; Sparks and Shepherd, 1994). A drawback of the psychometric studies was that most of the inferences drawn from the data were based on mean values while the variance in individual differences in risk perceptions was not taken into account. That is, respondent ratings of hazards on a number of risk characteristics were aggregated in order to perform factor analysis. However, many studies have shown that risk perceptions differ across individuals (Gould *et al.*, 1988; Barnett and Breakwell, 2001; Baron *et al.*, 2000; Burger *et al.*, 2001; Parry *et al.*, 2004). For example, increased perceptions of risk have been found in relation to personal prior experience with food poisoning (Parry *et al.*, 2004), whereas lower risk perceptions were observed for people that indicated high perceived personal control over potential hazards (Baron *et al.*, 2000). Interestingly, an analysis of individual perceptions of hazards on the risk dimensions distinguished by Fischhoff *et al.* (1978), where ratings were not aggregated across individuals, resulted in the same well-known two dimensions underlying public risk perceptions (i.e., unknown risk and dread risk) as were obtained with the psychometric studies that used aggregated data (Siegrist *et al.*, 2005). However, it was found that the extent to which hazards were perceived to be unknown and dreaded was dependent upon individual perceptions, indicating that individual differences do exist in the ratings of different hazards on several risk characteristics.

5.2.2   Research into risk and benefit perception in relation to the acceptance of food technologies and attitudes toward food-related hazards

Developing results from the psychometric studies in which food-related hazards were characterised in terms of a range of risk dimensions, research evolved to an examination of the influence of people's perceptions of risk and benefit on their acceptance of food technologies as well as their attitudes toward food-related hazards (Bredahl, 2001; Eiser *et al.*, 2002; Frewer *et al.*, 2003b; Saba and Messina, 2003; Siegrist, 2000; Siegrist and Cvetkovich, 2000; Siegrist *et al.*, 2000; Williams and Hammitt, 2001). In general, technology-related hazards include food processing and production-related activities, such as genetic modification, the use of pesticides, or food irradiation. An important finding was that consumers' response to genetic modification of food products was affected by perceptions of both risks and benefits associated with the technology, as well as concerns about their impact on the integrity of nature, indicating that values represent an important component of consumer decision making (Bredahl, 2001; Siegrist et al., 2000).

In empirical studies, perceived risk and perceived benefit were consistently found to be inversely related (Alhakami and Slovic, 1994). This is in contrast to what would be expected in the real world, where high risks are only acceptable when they are offset by high benefits (see Slovic et al., 2004). Different explanations have been proposed for this phenomenon. One is that people approach the judgment task analytically and produce a net riskiness and a net benefit judgment (Alhakami and Slovic, 1994). This suggests that people do not evaluate risks and benefits independently from each other. In particular, when net risk is high, net benefit is low, and vice versa. Further, it has been suggested that people perceive hazards in terms of general attitudes; when general attitudes are positive people tend to give low risk and high benefit judgments, and vice versa (Alhakami and Slovic, 1994). This idea has been confirmed in further empirical research, where general evaluative judgments with regard to particular food technologies had a high impact on individual perceptions of risk (Eiser et al., 2002; Poortinga and Pidgeon, 2005; Frewer et al., 2003b). A third explanation is that people have a tendency to avoid cognitive dissonance and a need for consistency in beliefs (Alhakami and Slovic, 1994), which makes it difficult for people to perceive high risks and high benefits associated with the same hazard simultaneously. In addition, it has been proposed that the inverse relationship between risk and benefit perception results from not taking into account the role of trust, which is proposed to influence risk and benefit perceptions (Siegrist, 2000; Siegrist and Cvetkovich, 2000; Siegrist et al., 2000, but see also Eiser et al., 2002). The influence of consumer trust in scientists, authorities, and the industry on consumer perceptions of risk and benefit was found, for example, with respect to the use of pesticides in agriculture (Saba and Messina, 2003). The importance of trust has also been stressed by Slovic (1993, 1999), who contends that risks are accepted when the public trust expert knowledge, regulators and risk managers in being able to control risks. However, the converse has also been observed. If people have a strongly held attitude about a potentially hazardous activity, they will confer trust upon a source which provides a risk message congruent to their attitude, but distrust a source which provided a dissonant message (Frewer et al., 2003b).

Finally, it has been suggested that the negative correlation between perceived risk and perceived benefit results, because people use emotional cues or affect as a heuristic when judging risks and benefits (Finucane et al., 2000b; Slovic et al., 2002). The role of affect in relation to individual risk perceptions will be discussed in the next section.

5.2.3   Research into risk, benefit and affect

Alkahami and Slovic (1994) found that the inverse relationship between perceptions of risk and perceptions of benefit was related to the individual's feelings about the technology. Favourable feelings about the technology were associated with low ratings of risk and high ratings of benefit, whereas unfavourable feelings resulted in high perceptions of risk and low perceptions of benefit. Finucane *et al.* (2000b) found empirical support for the contention that affect comes prior to, and influences, judgments of risk and benefit. The results of the study indicated that under time pressure, that is, when opportunity for analytic deliberation was limited and people had to rely on affective judgments, the strength of the inverse relationship between perceptions of risk and benefit increased. In addition, providing people with information about either the degree of benefit or risk influences their subsequent perceptions of both risk *and* benefit. For example, information indicating that benefit was high led to a higher judgment of perceived benefit as well as reduced perceptions of risk (Finucane *et al.*, 2000b). So, manipulation of the degree of benefit influences perceptions of risk, which shows that benefits and risks are not judged independently from one another. The authors argue that this effect is due to the fact that people use the affect heuristic for making judgments about risk and benefit, that is, manipulating one attribute (for example, benefit) causes an affectively congruent but inverse effect on the non-manipulated attribute (for example, risk). Johnson and Tversky (1983) found similar results in experimental studies where respondents had to estimate the frequency of fatalities of a range of risks and undesirable events. When negative affect was increased by letting respondents read a newspaper report of a tragic event, frequency ratings of various undesirable events were systematically higher compared to the ratings made by the control group, even when the news report and the estimated risk were unrelated to each other (Johnson and Tversky, 1983). This tendency for overall affect to serve as a cue for making judgments is termed the *affect heuristic* (Finucane *et al.*, 2000b; Slovic *et al.*, 2002, 2004). Loewenstein *et al.* (2001) proposed the *risk-as-feelings* hypothesis, which is very similar to the affect heuristic as it 'postulates that responses to risky situations (including decision making) result in part from direct (i.e., not cortically mediated) emotional influences, including feelings such as worry, fear, dread, or anxiety'.

5.2.4   Studies focusing on consumer risk perceptions in relation to purchase and consumption behaviour

Research reported in the economic and marketing literature has focused on the perceived risks in relation to purchase and consumption of (food) products. The level of consumer confidence in the safety of food and the extent to which risks are perceived in relation to the consumption of food products, may be related to product choice (i.e., does the consumer avoid certain products), brand choice, retail choice, and preferences for distinct product types (e.g., organic products) (see, among others, Mitchell, 1998; Pennings *et al.*, 2002; Saba and Messina,

2003; Verbeke, 2001; Verbeke and Van Kenhove, 2002; Verbeke and Viaene, 1999). For example, two consecutive studies on consumer perceptions of meat indicated that consumers' claimed past and future meat consumption was related to, among others, consumer perceptions regarding the presence of hormones or harmful substances in meat, the safety of meat, the trustworthiness of meat, and perceptions of whether meat was produced in an environmentally friendly way (Verbeke, 2001; Verbeke and Viaene, 1999). The more consumers believed that meat containing hormones or harmful substances, was unsafe, not trustworthy, and not produced in an environmentally friendly way, the more they indicated having decreased consumption as compared to the previous year, and the more they intended to decrease meat consumption in the future. In addition, McCarthy and Henson (2005) found that in the context of purchasing beef, purchase location, colour of the meat, country of origin, and quality marks were the most important aspects that were considered by consumers in order to reduce perceived risks associated with beef products.

Mitchell (1998) argues that consumers are often more motivated to avoid mistakes than to maximise utility in purchase behaviour, a tendency which has been referred to as loss aversion (Rabin, 1998). In this context, risk is equated with perceived uncertainty attributable to possible negative outcomes associated with purchasing the product. Consumers may be imperfectly informed and thus uncertain about product attribute levels, such as the level of product quality (Erdem *et al.*, 2004). In particular, two aspects of risk that are distinguished in this literature are *uncertainty*, i.e. the chance that a negative outcome will result, and *consequences*, i.e. the expected magnitude or seriousness of the loss in the case of a negative outcome. The extent to which consumers realise that they may not attain all of their goals associated with buying the product will result in perceived risk (Mitchell, 1998).

Various types of perceived risk have been distinguished in the literature: performance risk (e.g., will the product deliver what the advertisement promises), physical risk (e.g., to what extent is a product safe to eat), financial risk (e.g., is the product value for money), psychological risk (e.g., does the product fit with a person's self-image), social risk (e.g., the extent to which the product influences the way others think of a person) and time risk (e.g., how long does it take to learn how to use the product) (see, for example, Conchar *et al.*, 2004). Every product may have several of these perceived risks associated with its purchase and each consumer has an individual risk tolerance for them (Mitchell, 1998). In the context of food safety, performance risk and physical risk are most important to consumers, as 'regarding food safety, the goal is to acquire food products which have the desired consumption attributes, are safe to eat, and are free of contamination and therefore free of worry to the consumer' (Yeung and Morris, 2001).

Another classification that has been made regarding the concept of perceived risk is the distinction between inherent risk and handled risk (Chaudhuri, 1998). The former refers to the risk inherent in a product class or product choice risk (e.g., should I eat beef, should I buy a new computer?), while the latter refers to the risk in the buying situation or brand choice risk within a product class (should I choose organically produced beef, do I buy a desktop or a laptop computer?).

Several authors (e.g., Cooper *et al.*, 1988; Pennings *et al.*, 2002) argue that it is necessary to distinguish risk perception from risk attitude. In contrast to more general risk perception research, these authors argue that *risk perceptions* are consumers' interpretation of the chance to be exposed to the content of the risk. Consequently, risk perceptions may be defined as a consumer's assessment of the uncertainty of the risk content inherent in a particular situation. In addition, *risk attitude* is considered to reflect a consumer's general predisposition to risk in a consistent way, that is, the willingness to take a particular risk (Pennings *et al.*, 2002). In their study, Pennings *et al.* (2002) modelled consumers' reactions to the BSE crisis as a combination of risk perceptions, risk attitudes and the interaction between them. They contend that the best way of dealing with a risk-related crisis depends on whether public responses to a crisis are driven by risk perceptions or risk attitudes. When consumer behaviour is driven by risk *perceptions*, effective risk communication should be combined with efforts to reduce the risk. The assumption here is that providing the true probabilities of being exposed to the risk will be a useful way to respond to consumer concerns, and that consumers believe the messages communicated to them, as well as the provider of the information. When risk *attitudes* mainly influence consumer behaviour, the best solution is to eliminate the risk. In this case it is important for risk managers to focus on consumers' emotional responses to the potential risk (Mitchell, 1999).

Risk perceptions and risk attitudes differ across individuals (Bouyer *et al.*, 2001; Verbeke and Van Kenhove, 2002). In the context of a potential hazard, some people might actively try to reduce the risk or mitigate the exposure to the risk, whereas others may try to avoid exposure altogether. Another way to deal with potential risks is to ban any information about the risk and to deny its existence (Renn, 2005). Zwick and Renn (2002) observed individual differences in consumers' responses to the BSE crisis in Germany. Around 20% of the respondents changed their diets, asked for local meat, and became more suspicious of food products where the country of origin was not known to them. They kept basic patterns of their behaviour even after the crisis was over. Around 15% of the respondents avoided beef for several months and resumed their dietary habits after they felt the crisis was over. Their consumption patterns then remained identical to the pre-BSE crisis. Almost 25% of the respondents hardly changed their diet at all but expressed a strong belief that the media was amplifying the risk and that they tried to avoid reading more about it. It should be noted, however, that 40% of the respondents could not be easily grouped into one of the three categories (Zwick and Renn, 2002). Individual differences in consumer concerns and behaviour in response to the BSE crisis were also observed in other European countries (Berg, 2004; Verbeke and Viaene, 1999). Besides different responses to hazards on the individual level, several studies have indicated that risk perceptions also differ across cultures and are related to predominant worldviews and socio-political factors (Finucane *et al.*, 2000a; Flynn *et al.*, 1994; Palmer, 2003).

Research on public perceptions of risk over recent decades has provided valuable insights. It has resulted in increased knowledge about how the public conceptualise risk, how the concept of risk is related to other concepts, such as trust, benefit, and acceptance of technologies in food production, and how risk perceptions influence food choice behaviour. The next section deals with the issue of risk communication. Potential underlying causes for difficulties with respect to developing effective risk communication and risk management strategies will be discussed, in particular the role of public trust in information sources and regulatory institutions will be considered.

5.3   Implications for food risk communication and public involvement in policy development

A substantive body of literature has focused on experimental approaches to developing effective risk communication. The goal of risk communication is to assist stakeholders and consumers in understanding the rationale of risk assessment results and risk management decisions, and to help them arrive at a balanced judgment that reflects the factual evidence about the matter at hand in relation to their own interests and values (OECD, 2002). Good practices in risk communication help stakeholders and consumers to make informed decisions about self-protective behaviour (for example, healthy eating, or purchasing food products produced by application of emerging or controversial technologies).

5.3.1   Public trust in information sources and regulatory institutions

Increasing public distrust in science, regulation, and associated institutions, and the information provided by them, has, not unsurprisingly, been of concern to those responsible for managing food risks. This increased distrust has been linked to a public 'decline in deference' or a 'crisis in confidence' associated with scientific and regulatory institutions (Laird, 1989). Uncertainty and public dispute in many areas of science have rendered the automatic belief in the accuracy of scientific conclusions less tenable. Indeed, this longstanding decline in the public's trust in science has resulted in the legitimacy of scientific judgment being regularly questioned by the public (House of Lords Select Committee on Science and Technology, 2000).

Frewer and Salter (2002) have observed that various factors have contributed to this process. These include the rise of the 'consumer citizen' and informed choice making at the level of consumer choice, the diminished role of the 'expert' as a result of the wide availability of specialist information, and broad shifts in the national (and in some cases international) political culture towards more open forms of government. Increased transparency in regulation is often presented as a 'trust increasing' factor (Lang, 1998; HM Government, 2001).

However, public distrust in risk assessment is likely to arise under circumstances where uncertainties in risk assessment become open to public scrutiny through increased transparency, but are not explained explicitly as part of the risk communication process (Frewer *et al.*, 2002a).

Broadly speaking, research into public trust and distrust in risk management has focused on trust in information sources (which has theoretical origins in social psychology and communication studies) and trust in regulatory institutions (which has origins in socio-political theoretical approaches). The former primarily has implications for the practice of risk communication, the latter primarily for the structure of institutions and the development of risk management policy, although the two are not independent of each other. Each will now briefly be considered.

Trust in information sources
McComas and Trumbo (2001) provide an overview of research that has attempted to develop methods to assess the drivers of trust and distrust in risk information sources. Typically, research in risk communication has theorised that trust and credibility were multi-dimensional (Renn and Levine, 1991). In general terms these dimensions comprised items assessing different information source characteristics. That is, the extent to which the source is perceived to possess knowledge and expertise, to be open and honest with the information it provides to the public, and primarily concerned about public welfare (see also Covello, 1992; Kasperson *et al.*, 1992; Peters *et al.*, 1997). Alternative approaches to understanding the drivers of trust and distrust have employed combined qualitative and quantitative methodologies to generate respondent (rather than experimenter) generated credibility constructs (Frewer *et al.*, 1996). In this research, truthfulness, honesty and concern for public welfare were associated with trust, but distrust was associated with concern about the motives of the information sources in providing the information to the public.

A key question pertinent to the development of effective risk communication is whether trust in information actually influences risk perceptions and associated behaviour. The elaboration likelihood model, or ELM, has been proposed as a model that can predict whether persuasion or attitude change will occur following a communication about a particular issue (Petty and Cacioppo, 1986). The impact of source characteristics on responses to risk messages provides a framework in which such effects might be systematically investigated. The ELM proposes that individuals may adopt attitudes for reasons other than their understanding and evaluation of persuasive arguments contained in a particular message. In other words, attitude change may occur in the absence of argument scrutiny (for example, because of an individual's beliefs about information source characteristics), particularly under circumstances where personal involvement or interest in the subject of the message is also low (Priester and Petty, 1995).

Research conducted within the framework of the ELM has been inconsistent. People are likely to utilise trust as an important cue to evaluate risk communications about lifestyle hazards (such as microbiological contamination of food). This is because they personally believe that they are at low personal risk from the hazard, a phenomenon known as 'optimistic bias' or unreal optimism (Miles and Scaife, 2003). People receiving the risk communication are more likely to change their risk attitudes if the information source providing the message is trusted (Frewer et al., 1997). For technological hazards, the results are equivocal. For example, in the case of genetic engineering applied to food and agriculture, trust appears to co-vary with attitude rather than to predict it (Eiser et al., 2002). Furthermore, there is evidence to suggest that, if the message contained in the information does not align with already well-established attitudes held by the message recipient, the result will have negative impacts on both the perceived honesty and expertise of the information source (Frewer et al., 2003b). Of course, stigmatisation effects may also result from increased distrust on the part of the recipient – in other words, once the source is distrusted, this may have subsequent impacts on all of the messages provided by the same source.

Risk communication messages vary in the way they are communicated to consumers. The *framing* of a risk communication message influences its evaluation, even when two message frames describe objectively equivalent situations (Tversky and Kahneman, 1981; Levin et al., 1998). For example, the most common finding in the literature is that, in the context of attribute framing (e.g., 75% lean versus 25% fat), people respond more favourably to positive than negative framing (Krishnamurthy et al., 2001). In addition, persuasive messages can have different appeal depending on whether they stress positive consequences of performing the act to achieve a particular goal (e.g., taking folic acid supplements will lead to a healthier heart function) or negative consequences of not performing the act (e.g., not taking folic acid supplements will increase the risk of cardiovascular diseases) (see, among others, Levin et al., 1998). With respect to risk communication, consumer concerns and risk perceptions tend to be biased towards negativity: information sources providing bad news (e.g., indicating potential health risks for consumers) are perceived to be more credible than information sources providing good news (e.g., indicating no health risks for consumers) (Siegrist and Cvetkovich, 2001). Together with the tendency that negative events also receive more attention than positive events, it is often assumed that trust is easier to destroy than to create (Siegrist and Cvetkovich, 2001; Slovic, 1993, 1999).

Although there is some evidence that public risk perceptions exhibit a basic universality across cultures and with the passage of time, which may facilitate a collective focus on risk and provides a basis for communication (Rohrmann, 1995; Renn and Rohrmann, 2000), it has also been found that individual differences exist in consumer preferences for risk communication messages and responses to information. For example, Kornelis et al. (submitted) surveyed Dutch consumers regarding their intended use of a range of food safety information sources, their interest in food safety information, the extent to which

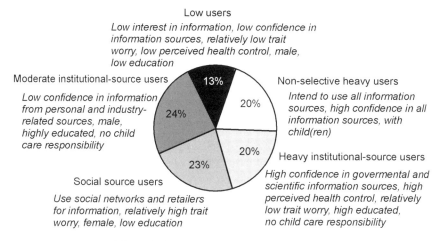

Low users

Low interest in information, low confidence in information sources, relatively low trait worry, low perceived health control, male, low education

Moderate institutional-source users

Low confidence in information from personal and industry-related sources, male, highly educated, no child care responsibility

Non-selective heavy users

Intend to use all information sources, high confidence in all information sources, with child(ren)

Heavy institutional-source users

High confidence in govermental and scientific information sources, high perceived health control, relatively low trait worry, high educated, no child care responsibility

Social source users

Use social networks and retailers for information, relatively high trait worry, female, low education

Fig. 5.1   Consumer segment characteristics.

they perceived information sources to be of high quality, and the extent to which they perceived that they were personally knowledgeable about food safety issues. Data were also collected on several personality characteristics, as well as socio-demographic variables. The application of a cluster analysis indicated that five distinct consumer segments could be identified on the basis of individuals intended use of food safety information sources (Fig. 5.1).

Some individuals, the 'non-selective heavy users', actively seek food safety information from a variety of sources. 'Heavy' and 'moderate' institutional-source users, and 'social source' users also actively seek food safety information, but the former two tend to use institutional sources, the latter their own social networks and retailers. 'Low users' are the most problematic in terms of food risk communication, as they tend to possess a relatively high external and low internal locus of control, reinforcing their perception that they have little personal control over their own health. Results of this study indicate that individual differences exist in terms of searching for and processing of food safety information. These differences should be taken into account in developing risk communication strategies targeted at distinct consumer groups.

Different issues arise when communicating about the adoption and application of emerging food technologies, where the debate focuses on developing adequate information to enable consumers to make their own decisions about the acceptability of the products resulting from the application of new technologies. Even as late as 2001, researchers were advocating a comparative risk approach as a method to 'correct' the 'erroneous' perceptions of the public. In this approach, the level of risk associated with a particular hazard was expressed in relation to another hazard or person in order to create more objective perceptions of risk, for example negative health consequences of unhealthy dietary behaviour are 40 to 100 times bigger than negative health consequences of unsafe food (Rijksinstituut voor Volksgezondheid en Milieu, 2004, pp. 33). Williams and Hammitt (2001), for example, report that North American consumers perceive high levels of risk to be associated with conventionally grown products compared to those produced using organic production methods. The researchers recommended that relative risk estimates of the impact of consumption on human health be used in order to 'correct' these perceptions. No assessment was made of perceptions of the risks of unintended effects to either health or to the environment, which may have contributed to consumer fears, nor of other factors (for example, green values or self identity, Von Alvensleben, 2001) which may be influencing consumer risk estimates.

Garvin (2001) has argued that there are important issues of epistemology inherent in the process and practice of risk communication. She specifies that there are three groups of key players in understanding risk – scientists, policy makers and the public. Arguably 'each group employs different, although equally legitimate, forms of rationality'. For scientists, rationality is constructed in *scientific terms*. For members of the policy community, the basis of rationality is *political* and *expedient*. For the third group, the public, rationality is essentially social, taking account of risk context, cultural factors and local conditions. Garvin notes that each form of rationality has a legitimate input into how risk management is operationalised, and increased institutional recognition of this may reduce the polarity of view observed between the different sectors. Indeed, most risk communication research has tended to assume that people are passive risk perceivers, who need to be informed about the implications of a particular hazardous event. However, Lion *et al.* (2002) have found that people prefer information which they can use to determine the personal relevance of the risk confronting them.

It is, of course, important to take into account the fact that, in terms of developing effective risk communication, the safety of food cannot always be judged directly by consumers (e.g., Nelson, 1970; Renn, 1997, 2005). For example, it is difficult to infer from the physical product whether it is con-taminated with pesticide residues, dioxins, or BSE. Evaluating the safety of food often has a strong *credence* component. That is, consumers make inferences about the quality of the product based on external attributes or beliefs. However, these cannot be verified by direct experience either prior to, or after, purchase and consumption, because it is often difficult to establish with certainty that adverse health effects are a direct consequence of consuming a particular food item. Since many of the food risks that people are concerned about relate to chemical or other artificial ingredients in food, neither exposure nor dose-effect are detectable by human senses. Moreover, these risks are highly complex, i.e. there are usually many years of latency between consumption and effect. When food is consumed with a high concentration of pesticide residues health symptoms may be visible only many years later, if at all (Böcker and Hanf, 2000). Therefore consumers depend on information provided by third parties, because they cannot judge the seriousness of the risks to which they are exposed

(Green *et al.*, 2003). Green *et al.* (2003) found that consumers use their trust in salespersons, especially the ones they personally know, and regulatory institutions as a heuristic to assess perceptions of safety, which shows that consumers compensate for the lack of knowledge about the food they eat by conferring trust in actors in the food chain and regulating authorities (Berg, 2004; Siegrist and Cvetkovich, 2000). If people trust the responsible institutions to provide the necessary information, they are willing to use a balancing approach between risks and benefits and to assign trade-offs between the two. If they do not trust the institutions, they are more likely to demand 'zero risk'. For if one is dependent on information provided by third parties for the assessment of such risks and these third parties are not considered trustworthy, then one does not accept any cost-benefit calculation. If the answer is 'maybe', external cues as postulated by the ELM described above become major indicators for assigning credibility. The scope of external cues ranges from personal appeal to assumed hidden interests (Renn, 1997, 2005).

Communicating scientific uncertainties
Risk management decisions may be particularly difficult under conditions of ambiguity, uncertainty about the occurrence of risks and potential consequences of risks should they occur (Kunreuther, 2002). Perhaps it is the difficulty associated with risk management under conditions of uncertainty that has led expert groups to assume that lay people cannot conceptualise uncertainty in risk assessment or risk management (Frewer *et al.*, 2003a). However, there is evidence that elite groups in the scientific and policy community have underestimated the ability of non-experts to understand uncertainty (Gigerenzer and Selten, 2001).

   Groenewegen (2002), in his analysis of Dutch toxicology, has observed that scientists who are also advisors to the policy community are frequently involved in the definition of the social problem as well as the research agenda put into place to counter it. He concludes that, in early toxicology activity (for example, that arising in the early 19th century as a response to the public health problems associated with industrialisation) research, advice and policy developments were closely linked. More latterly, advice, policy and research activities were much more clearly demarcated, which has resulted in differentiation of agendas within the risk management process. Historically, many scientists thought that providing the public with information about uncertainty would increase distrust in science and scientific institutions, as well as cause panic and confusion regarding the extent and impact of a particular hazard on human health, the economy and the environment. However, there is evidence that members of the public (drawn from different social backgrounds) are very familiar with the concept of uncertainty (Frewer *et al.*, 2002a). The failure of institutional actors to communicate about uncertainty was actually increasing public distrust in institutional activities designed to manage risk. Furthermore, the conclusion that participants in the research indicated that they would also like to be provided with information about risk uncertainty is likely to become particularly important as increased transparency in risk management processes means that scientific uncertainties associated with risk analysis become more likely to be the subject of public scrutiny and debate.

One conclusion that can be drawn at this point is that the most important factor in developing effective risk communication appears to be developing messages that are personally relevant to the recipient (e.g. Lion *et al.*, 2002). If, for example, research into risk perceptions associated with a particular pesticide indicates that people are concerned about unintended environmental effects, and the risk message is actually about effects on human health, the receiver of the message may dismiss the information as irrelevant to their concerns. When communicating about health-related behaviour, it is important to target information to groups of individuals affected (using multiple communication strategies to target different groups if necessary). This may be particularly relevant in the context of health risks associated with particular food choices, where it is important to target those individuals who actively avoid health-related information. It is important to target vulnerable groups by selecting their preferred information channels. Scientific uncertainty, what is being done by the scientific community to reduce this uncertainty, and the development of effective targeted information directed towards vulnerable groups are becoming increasingly relevant.

Trust in institutions and consulting stakeholders and the public
The issue of trust in institutions represents a complex issue. Trust is likely to be particularly important under circumstances where people feel that they have very little personal control over potential hazards. For example, Siegrist (2000) reports that the extent to which an individual distrusts an institution with specific responsibility for regulating gene technology and its products has a direct relationship with the extent to which perceived risk is increased and perceived benefit decreased. Societal responses to emerging and/or potentially transformative technologies (nanotechnology, for example) may reflect increased public distrust unless institutions and organisations act to develop and maintain public confidence in their risk management practices.

Zwick and Renn (2002) have demonstrated that public perceptions of the performance of risk management institutions were related to perceptions of risk. This effect was particularly important when the hazard associated with the potential risk was seen as controversial and the evidence about potential harm was perceived to be ambiguous (Zwick and Renn, 2002). In a multiple regression, perceived institutional trust was highly influential for evaluating the acceptability of electromagnetic field (EMF) base stations (13% of the explained variance) and genetically modified food (29% of the explained variance), but had no effect on the acceptability of smoking and climate change, and only limited effect on acceptance of nuclear energy (9%) and BSE (8%). It may be surprising that trust was less important for nuclear energy in spite of its controversial nature and a visible dissent among experts about its acceptability.

The main reason here was that most people were sceptical about nuclear power and found the risks very serious independent of whether they trusted authorities or not. It interesting to note, however, that those who had more trust in regulatory authorities were more willing to accept the statement that nuclear power has distinct benefits to society (Zwick and Renn, 2002, p. 24).

While public trust in institutions is, to some extent, contingent on making institutional decisions transparent and open to public scrutiny, other factors, such as institutional reactivity to public concerns, and involvement of the public in the risk management decision-making process itself, are also likely to increase public trust in the extent to which lay-people perceive institutions take public concerns and values into account. If people perceive that they are empowered in the decision-making process through consultation or actual direct involvement in final decisions, this may result in greater acceptance of both process and outcome. There is an extensive literature on how best to involve the public in the decision-making process – public trust in public participation may actually decrease if the activity is not independently evaluated as to its effectiveness in terms of how the exercise is conducted and its subsequent impact on policy development (Rowe and Frewer, 2000). Furthermore, the way in which the results of the exercise are incorporated into the policy process needs to be communicated in an effective way to both the participants in the public participation exercise and the general public. Again, the process assumes that societal concerns be included explicitly in the whole risk management process. If the results are not used to develop policy, the reasons must be made clear to participants and the general public (Renn, 2004b).

5.4   Implications for risk management

The issue of reduced public confidence in risk management practices, in part, originates in the exclusion of societal concerns and values from the risk management framework. These concerns might usefully be included in risk assessment ('what is assessed?', 'how is uncertainty dealt with?', 'how safe is safe enough?') and risk management ('how can socially inclusive risk management practices be developed?'), as well as risk communication ('what information is needed and by whom?'). There is a need to more efficiently integrate societal concerns and values into risk assessment and risk management procedures, as well as incorporate risk perceptions into risk communication. At present, failure to do so is one of the causative factors associated with the decline in public confidence in risk assessment and risk management.

5.5   Conclusions

Risk assessment as developed within the natural sciences is a beneficial and necessary element regarding the development of an effective risk policy.

Arguably it is the only means by which relative risks can be compared, and risk mitigation options with the lowest statistical expectations identified. However, it cannot and should not be used as the only relevant guide for policy development. That is, it is essential to also take into account context and situation-specific circumstances. Food is associated with more than just safety aspects: it relates to concerns about nutrition, about well-being, naturalness, and eating culture. Consumers expect from public authorities that they consider these concerns when regulating food safety. In addition, it is important to take account of public priorities and values in the handling of different risks, as differences between the technical and consumer communities have resulted in differences in perceptions of how the policy community should handle different existing and emerging food risks. Among others many consumers are concerned about fairness with respect to those who impose risks on others. If people feel that they are used as guinea pigs for serving the interests of very few they will reject the implicit risks of such food even if these risks are regarded as minute. When these aspects are not part of the decision-making process, decisions will not meet the requirement of achieving collective objectives in a rational, purposeful and value-optimising manner.

It is useful to systematically identify the various dimensions of consumer risk perceptions and to measure those dimensions against characteristics derived from technical risk assessments. In principle, the extent to which different technical options distribute risk across the various groups of society, the extent to which institutional control options exist and to what extent risk can be accepted by way of voluntary agreement can all be measured using appropriate research tools.

In order to develop effective risk communication, it is important to investigate dynamic changes in both the extent and nature of public perceptions associated with specific hazards. Greater understanding of individual differences in perceptions will facilitate information delivery. However, communication is a two-way process, and institutions need to learn how to internalise public views and societal values into risk analysis. Both the food industry and the policy community need to understand how to communicate about credence characteristics associated with particular products or production processes (which are likely to be highly influenced by consumer trust in both information sources, institutions with responsibility for consumer protection, and all of the different actors in the food chain). Understanding consumer responses to a food scare, and maintaining consumer confidence in food safety during a crisis, is all contingent on understanding consumer perceptions of risk associated with different food hazards.

Of course, increasingly transparent risk assessment and management activities means that the effective communication of uncertainty and variability associated with risk assessment is increasingly important, as information about uncertainty and variability is available in the public domain but hitherto may not have been explicitly communicated as such. In itself, such communication may not increase trust. However, it behoves risk communicators to develop effective ways to communicate about these issues, as they are in the public domain as a result of increased communication, as failure to so do may be trust destroying.

Simply understanding consumer perceptions of risk in isolation of other factors in itself is unlikely to be the only influential factor relating to developing effective communication about food risks, and healthy food choices. We have already noted that one institutional response to increased consumer distrust in the process of risk analysis in the food area has been to increase transparency in the risk analysis process. As a consequence, risk uncertainty and variability about the potential impact or extent of a particular hazard are placed in the public domain and become open to public scrutiny. Lay people also recognise that further research may be needed in order to reduce the uncertainty, and acknowledgment of this need may, in turn, be trust inducing. Indeed, the public appears to be more accepting of uncertainty resulting from shortfalls in scientific processes than to uncertainty associated with the failure of institutions to reduce scientific uncertainty through conducting appropriate empirical investigation. This all serves to confirm the recommendation of *The National Research Council* (1994) that risk communication should focus on the sources of uncertainty as well as the magnitude of uncertainty associated with a particular hazard.

Other factors associated with risk assessments also influence risk management decisions (for example, the severity and immediacy of the potential risk, the cost and side effects of mitigation options, and the cost and time required for research). Uncertainty associated with risk assessments, risk management, and the link between risk assessment and risk management should be communicated to the public and other key stakeholders as well as to decision-makers if there is to be an informed public debate about how risks should be handled.

Another risk assessment issue that must be disseminated to all interested parties, including the public, is that of risk variability, when the risk varies across a population but the distribution is well known. Increasingly sophisticated understanding of whom is at risk (for example, through knowledge about human genomics) coupled with improved risk assessment methodologies (for example, probabilistic approaches which can take account of risk uncertainties and population level variability) means that population level communication strategies are becoming increasingly irrelevant. Vulnerable groups may be identified, who merit targeted communication about the personally specific effects of making different food choices. Understanding variability about both the risks and benefits of particular food choices may also have implications for the allocation of resources to risk mitigation activities, another potential focus of public debate. Discussion of how such resources are allocated is important in development of public confidence in risk management and, ultimately, *risk–benefit* assessment.

At present, however, there is insufficient knowledge about how to develop best practice in risk communication about uncertainty and variability. The former is contingent on developing ways to discuss different kinds of uncertainty; the latter may entail methodological development in targeting information to 'at risk' populations. Both uncertainties associated with the risks of consuming a particular food, as well as the potential benefits, have profound implications for consumer decisions associated with different food choices. Risk management decisions associated with resource allocation (for example, how research funds are distributed across hazards in order to reduce uncertainties, or how risk mitigation activities are prioritised for risks which differentially affect different sub-populations) are also affected. The selection of communication messages about food consumption under circumstances where there is both risk and benefit associated with consuming a particular food is also increasingly a subject of research. For example, consumers in Northern European countries are being encouraged by experts in nutrition to consume more fish to prevent the occurrence of cardio-vascular diseases. However, scientists with expertise in toxicology recommend limitations in the amount of fish consumed by an individual, owing to the accumulation of potentially toxicological substances in the product. Public trust in these processes is likely to be low unless there is informed public debate regarding both risk management and risk assessment procedures, which permits the inclusion of wider societal values and priorities into decision-making processes.

One question which frequently arises relates to who has the authority to make decisions, and how the decision-making process can be justified in terms of prevalent societal values? Greater involvement of those affected by a particular risk, greater transparency in decision-making, a non-hierarchical discourse between different stakeholders and end-users, and the risk analysis community all represent potential solutions that have the potential to bridge the gap between assessment and perception and assist risk managers to design management options that meet the protective goals and the perceptions of the affected consumers, in the food area and beyond.

5.6   References

ALHAKAMI, A.S. and SLOVIC, P. (1994). A psychological study of the inverse relationship between perceived risk and perceived benefit. *Risk Analysis*, **14**(6), 1085–1096.

ANDERSON, E.S., WINETT, R.A. and WOJCIK, J.R. (2000). Social-cognitive determinants of nutrition behavior among supermarket food shoppers: A structural equation analysis. *Health Psychology*, **19**(5), 479–486.

BARNETT J. and BREAKWELL, G. (2001). Risk perception and experience: Hazard personality profiles and individual differences. *Risk Analysis*, **21**(1), 171–177.

BARON, J., HERSHEY, J.C. and KUNREUTHER, H. (2000). Determinants of priority for risk reduction: The role of worry. *Risk Analysis*, **20**(4), 413–427.

BERG, L. (2004). Trust in food in the age of mad cow disease: a comparative study of consumers' evaluation of food safety in Belgium, Britain and Norway. *Appetite*, **42**, 21–32.

BERG, L., KJAERNES, U., GANSKAU, E., MININA, V., VOLTCHKOVA, L., HALKIER, B. and HOLM, L. (2005). Trust in food safety in Russia, Denmark and Norway. *European Societies*, 7(1), 103–129.

BÖCKER, A. and HANF, C.-H. (2000). Confidence lost and – partially – regained: Consumer response to food scares. *Journal of Economic Behavior & Organization*, **43**, 471–485.

BOUYER, M., BAGDASSARIAN, S., CHAABANNE, S. and MULLET, E. (2001). Personality correlates of risk perception. *Risk Analysis*, **21**, 457–465.

BREDAHL, L. (2001). Determinants of consumer attitudes and purchase intentions with regard to genetically modified foods – results of a cross-national survey. *Journal of Consumer Policy*, **24**, 23–61.

BRUG, J., LECHNER, L. and DE VRIES, H. (1995). Psychosocial determinants of fruit and vegetable consumption. *Appetite*, **25**(3), 285–296.

BURGER, J., GAINES, K.E. and GOCHFIELD, M. (2001). Ethnic differences in risk from mercury among Savannah river fishermen. *Risk Analysis*, **21**(3), 533–544.

CHAUDHURI, A. (1998). Product class effects on perceived risk: The role of emotion. International *Journal of Research in Marketing*, **15**, 157–168.

CONCHAR, M.P., ZINKHAN, G.M., PETERS, C. and OLAVARIETTA, S. (2004). An integrated framework for the conceptualization of consumers' perceived-risk processing. *Journal of the Academy of Marketing Science*, **32**(4), 418–436.

COOPER, A.C., WOO, C.Y. and DUNKELBERG, W.C. (1988). Entrepreneurs perceived chances for success. *Journal of Business Venturing*, **3**(2), 97–108.

COVELLO, V.T. (1992). Trust and credibility in risk communication. *Health and Environment Digest*, **6**(1), 1–3.

DIBSDALL, L., LAMBERT, N. and FREWER, L.J. (2002). Using interpretative phenomenology to understand the food related experiences and beliefs of a select group of low-income UK women. *Journal of Nutrition Education and Behavior*, **34**(6), 298–209.

DOSMAN, D.M., ADAMOWICZ, W.L. and HRUDEY, S.E. (2001). Socioeconomic determinants of health- and food safety-related risk perceptions. *Risk Analysis*, **21**(2), 307–317.

DREEZENS, E., MARTIJN, C., TENBÜLT, P., KOK, G. and DE VRIES, N.K. (2005). Food and values: an examination of values underlying attitudes toward genetically modified- and organically grown food products. *Appetite*, **44**, 115–122.

EISER, J.R., MILES, S. and FREWER, L.J. (2002). Trust, perceived risk, and attitudes toward food technologies. *Journal of Applied Social Psychology*, **32**(11), 2423–2433.

ERDEM, T., ZHAO, Y. and VALENZUELA, A. (2004). Performance of store brands: A cross-country analysis of consumer store-brand preferences, perceptions, and risk. *Journal of Marketing Research*, **41**(1), 86–100.

FIFE-SCHAW, C. and ROWE, G. (1996). Public perceptions of everyday food hazard: A psychometric study. *Risk Analysis*, **16**(4), 487–500.

FIFE-SCHAW, C. and ROWE, G. (2000). Extending the application of the psychometric approach for assessing public perceptions of food risk: Some methodological considerations. *Journal of Risk Research*, **3**(2), 167–179.

FINUCANE, M., SLOVIC, P., MERTZ, C.K., FLYNN, J. and SATTERFIELD, T.A. (2000a). Gender, race, and perceived risk: the 'white male' effect. *Health, Risk and Society*, **2**, 159–172.

FINUCANE, M.L., ALHAKAMI, A., SLOVIC, P. and JOHNSON, S.M. (2000b). The affect heuristic in judgments of risks and benefits. *Journal of Behavioral Decision Making*, **13**(1), 1–17.

FISCHER, A.R.H., DE JONG, A.E.I., DE JONGE, R., FREWER, L.J. and NAUTA, M.J. (2005). Improving Food Safety in the Domestic Environment: The Need for a Transdisciplinary Approach. *Risk Analysis*, **25**(3), 503–517.

FISCHHOFF, B. (1985). Managing risk perceptions. *Issues in Science and Technology*, **2**(1), 83–96.

FISCHHOFF, B., SLOVIC, P., LICHTENSTEIN, S., READ, S. and COMBS, B. (1978). How safe is safe enough? A psychometric study of attitudes towards technological risks and benefits. *Policy Sciences*, **9**, 127–152.

FISCHLER, C. (1990). *L'homnivore. Le gout, la cuisine et le corps.* Odile Jacob: Paris.

FLYNN, J., SLOVIC, P. and MERTZ, C.K. (1994). Gender, race, and perception of environmental health risks. *Risk Analysis*, **14**, 1101–1108.

FREWER, L.J. (2001). Environmental risk, public trust and perceived exclusion from risk management. *Environmental risks: Perception, Evaluation and Management*, **9**, 221–248.

FREWER, L.J. and SALTER, B. (2002). Public attitudes, scientific advice and the politics of regulatory policy: the case of BSE. *Science and Public Policy*, **29**, 137–145.

FREWER, L.J., SHEPHERD, R. and SPARKS, P. (1994). The interrelationship between perceived knowledge, control and risk associated with a range of food-related hazards targeted at the individual, other people and society. *Journal of Food Safety*, **14**, 19–39.

FREWER, L.J., HOWARD, C., HEDDERLEY, D. and SHEPHERD, R. (1996). What determines trust in information about food-related risks? Underlying psychological constructs. *Risk Analysis*, **16**(4), 473–486.

FREWER, L.J., HOWARD, C., HEDDERLEY, D. and SHEPHERD, R. (1997). The Elaboration Likelihood Model and communication about food risks. *Risk Analysis*, **17**(6), 759–770.

FREWER, L.J., MILES, S., BRENNAN, M., KUSENOF, S., NESS, M. and RITSON, C. (2002a). Public preferences for informed choice under conditions of risk uncertainty. *Public Understanding of Science*, **11**(4), 1–10.

FREWER, L.J., MILES, S. and MARSH, R. (2002b). The media and genetically modified foods: Evidence in support of social amplification of risk. *Risk Analysis*, **22**(4), 701–711.

FREWER, L.J., HUNT, S., KUZNESOF, S., BRENNAN, M., NESS, M. and RITSON, C. (2003a). The views of scientific experts on how the public conceptualize uncertainty. *Journal of Risk Research*, **6**(1), 75–85.

FREWER, L., SCHOLDERER, J. and BREDAHL, L. (2003b). Communicating about the risks and benefits of genetically modified foods: The mediating role of trust. *Risk Analysis*, **23**(6), 1117–1133.

GARVIN, T. (2001). Analytical paradigms: The epistemological distances between scientists, policy makers and the public. *Risk Analysis*, **21**(3), 443–455.

GIGERENZER, G. and SELTEN, R. (2001). Rethinking rationality. In G. Gigerenzer and R. Selten (eds), *Bounded rationality: The adaptive toolbox. Dahlem Workshop Report* (pp. 1–12). MIT Press: Cambridge, Mass.

GORDON, J. (2003). Risk communication and foodborne illness: Message sponsorship and attempts to stimulate perceptions of risk. *Risk Analysis*, **23**(6), 1287–1296.

GOULD, L.C., GARDNER, G.T., DELUCA, D.R., TIEMANN, A., DOOB, L.W. and STOLWIJK, J.A.J. (1988). *Perceptions of technological risks and benefits.* Russell Sage Foundation: New York.

GREEN, J.M., DRAPER, A.K. and DOWLER, E.A. (2003). Short cuts to safety: risk and 'rules of thumb' in accounts of food choice. *Health, Risk and Society*, **5**, 33–52.

GRIFFITH, C.J. and WORSFOLD, D. (1994). Application of HACCP to food preparation practices in domestic kitchens. *Food Control*, **5**, 200–204.

GRIFFITH, C.J., WORSFOLD, D. and MITCHELL, R. (1998). Food preparation, risk communication and the consumer. *Food Control*, **9**(4), 225–232.

GROENEWEGEN, P. (2002). Accommodating science to external demands. The emergence of Dutch toxicology. *Science, Technology & Human Values*, **27**, 479–498.

HAMPEL, J., KLINKE, A. and RENN, O. (2000). Beyond 'red' hope and 'green' distrust. Public perception of genetic engineering in Germany. *Politeia*, **16**(60), 68–82.

HANSEN, J., HOLM, L., FREWER, L., ROBINSON, P. and SANDØE, P. (2003). Beyond the knowledge deficit: Recent research into lay and expert attitudes to food risks. *Appetite*, **41**, 111–121.

HM GOVERNMENT (2001). The Phillips Report. The BSE Inquiry: The inquiry into BSE and CJD in the United Kingdom. London: The Stationery Office: London.

HOUSE OF LORDS, SELECT COMMITTEE ON SCIENCE AND TECHNOLOGY (2000). Science and Society, HL38, Session 1999–2000. Stationery Office: London.

JOHNSON, E.J. and TVERSKY, A. (1983). Affect, generalization, and the perception of risk. *Journal of Personality and Social Psychology*, **45**(1), 20–31.

KASPERSON, R.E., GOLDING, D. and TULER, S. (1992). Social distrust as a factor in siting hazardous facilities and communicating risks. *Journal of Social Issues*, **48**, 161–187.

KIRK, S.F.L., GREENWOOD, D., CADE, J.E. and PEARMAN, A.D. (2002). Public perception of a range of potential food risks in the United Kingdom. *Appetite*, **38**(3), 189–197.

KORNELIS, M., DE JONGE, J., FREWER, L. and DAGEVOS, H. (submitted). Consumer selection of food safety information sources. *Risk Analysis*.

KRISHNAMURTHY, P., CARTER, P. and BLAIR, E. (2001). Attribute framing and goal framing effects in health decisions. *Organizational Behavior and Human Decision Processes*, **85**(2), 382–399.

KUNREUTHER, H. (2002). Risk Analysis and Risk Management in an uncertain world. *Risk Analysis*, **22**(4), 655–664.

LAIRD, F.N. (1989). The decline of deference. The political context of risk communication. *Risk Analysis*, **9**(4), 543–550.

LANG, T. (1998). BSE and CJD: recent developments. In S. Ratzan (ed.), *The Mad Cow Crisis: Health and the Public Good*. UCL Press: London.

LEVIN, I.P., SCHNEIDER, S.L. and GAETH, G.J. (1998). All frames are not created equal: A typology and critical analysis of framing effects. *Organizational Behavior and Human Decision Processes*, **76**(2), 149–188.

LION R., MEERTENS, R.M. and BOT, I. (2002). Priorities in information desire about unknown risks. *Risk Analysis*, **22**(4), 765–776.

LOEWENSTEIN, G.F., WEBER, E.U., HSEE, C.K. and WELCH, N. (2001). Risk as feelings. *Psychological Bulletin*, **127**(2), 267–286.

McCARTHY, M. and HENSON, S. (2005). Perceived risk and risk reduction strategies in the choice of beef by Irish consumers. *Food Quality and Preference*, **16**, 435–445.

McCOMAS, K.A. and TRUMBO, C.W. (2001). Source credibility in environmental health risk controversies: Application of Meyer's credibility index. *Risk Analysis*, **21**(3), 467–480.

MILES, S. and SCAIFE, V. (2003). Optimistic bias and food. *Nutrition Research Reviews*, **16**, 3–19.

MITCHELL, V.-W. (1998). A role for consumer risk perceptions in grocery retailing. *British Food Journal*, **100**(4), 171–183.

MITCHELL, V.-W. (1999). Consumer perceived risk: Conceptualisations and models. *European Journal of Marketing*, **33**(1/2), 163–195.

NATIONAL RESEARCH COUNCIL (1994). *Science and Judgement in Risk Analysis.* National Academy Press: Washington DC.

NELSON, P. (1970). Information and consumer behaviour. *Journal of Political Economy*, **78**, 51–57.

OECD (2002). *Guidance document on risk communication for chemical risk management.* OECD Publication: Paris.

PALMER, C.G.S. (2003). Risk perception: another look at the 'white male' effect. *Health, Risk and Society*, **5**, 71–83.

PARRY, S.M., MILES, S., TRIDENTE, A., PALMER, S.R. and SOUTH AND EAST WALES INFECTIOUS DISEASE GROUP (2004). Differences in perception of risk between people who have and have not experienced salmonella food poisoning. *Risk Analysis*, **24**(1), 289–299.

PENNINGS, J.M.E., WANSINK, B. and MEULENBERG, M.T.G. (2002). A note on modeling consumer reactions to a crisis: The case of the mad cow disease. *International Journal of Research in Marketing*, **19**, 91–100.

PETERS, R.G., COVELLO, V.T. and McCALLUM, D.B. (1997). The determinants of trust and credibility in environmental risk communication: An empirical study. *Risk Analysis*, **17**(1), 43–54.

PETTY, R.E. and CACIOPPO, J.T. (1986). *Communication and Persuasion: Central and Peripheral Routes to Attitude Change.* Springer-Verlag: New York.

POORTINGA, W. and PIDGEON, N.F. (2005). Trust in risk regulation: Cause or consequence of the acceptability of GM food? *Risk Analysis*, **25**(1), 199–209.

PRIESTER, J.R. and PETTY, R.E. (1995). Source attributions and persuasion: Perceived honesty as a determinant of message scrutiny. *Personality and Social Psychology Bulletin*, **21**, 637–654.

RABIN, M. (1998). Psychology and economics. *Journal of Economic Literature*, **36**(1), 11–46.

RENN, O. (1997). Mental health, stress and risk perception: Insights from psychological research. In J.V. Lake, G.R. Bock and G. Cardew (eds), *Health impacts of large releases of radionuclides.* pp. 205–231. Ciba Foundation Symposium 203. Wiley: London.

RENN, O. (1999). A model for an analytic deliberative process in risk management. *Environmental Science and Technology*, **33**(18), 3049–3055.

RENN, O. (2003). Acrylamide. Lessons for risk management and communication. *Health Communication*, **8**(5), 435–441.

RENN, O. (2004a). Perception of risks. *The Geneva Papers on Risk and Insurance*, **29**(1), 102–114.

RENN, O. (2004b). The challenge of integrating deliberation and expertise: Participation and discourse in risk management. In T. McDaniels and M.J. Small (eds), *Risk Analysis and Society. An Interdisciplinary Characterization of the Field*, pp. 289–366. Cambridge University Press: Cambridge, Mass.

RENN, O. (2005). Risikokommunikation – der Verbraucher zwischen Information und Irritation. In Office of Technology Assessment at the German Bundestag (ed.), *Risikoregulierung bei unsicherem Wissen: Diskurse und Lösungsansätze*, pp. 51–72. Discussion Paper No. 11 (March 2005). TAB: Berlin.

RENN, O. and LEVINE, D. (1991). Trust and credibility in risk communication. In R.E. Kasperson and P.J. Stallen (eds), *Communicating Risks to the Public: International Perspectives*, pp. 175–218. Kluwer: Dordrecht, Boston, London.

RENN, O. and ROHRMANN, B. (2000). Cross-cultural risk perception research: State and challenges. In O. Renn and B. Rohrmann (eds), *Cross-cultural Risk Perception. A Survey of Empirical Studies*, pp. 211–233. Kluwer: Dordrecht, Boston, London.

RENN, O., WEBLER TH. and WIEDEMANN, P. (1995). The Pursuit of Fair and Competent Citizen Participation. In O. Renn, Th. Webler and P. Wiedemann (eds), *Fairness and Competence in Citizen Participation. Evaluating New Models for Environmental Discourse*, pp. 339–368. Kluwer: Dordrecht, Boston, London.

RIJKSINSTITUUT VOOR VOLKSGEZONDHEID EN MILIEU (2004). *Ons eten gemeten. Gezonde voeding en veilig voedsel in Nederland [Our food measured. Healthy nutrition and safe food in the Netherlands]*. Rijksinstituut voor Volksgezondheid en Milieu: Bilthoven.

ROHRMANN, B. (1995). Technological risks: Perception, evaluation, communication. In R.E. Mechlers and M.G. Stewart (eds), *Integrated Risk Assessment. Current Practice and New Directions*, pp. 7–12. Balkema: Rotterdam.

ROSATI, S. and SABA, A. (2004). The perception of risks associated with food-related hazards and the perceived reliability of sources of information. *International Journal of Food Science and Technology*, **39**, 491–500.

ROWE, G. and FREWER, L.J. (2000). Public participation methods: An evaluative review of the literature. *Science, Technology & Human Values*, **25**, 3–29.

ROWE, G. and FREWER, L.J. (2005). A typology of public engagement mechanisms. *Science, Technology, & Human Values*, **30**(2), 251–290.

ROWE, G., MARSH, R. and FREWER, L.J. (2004). Evaluation of a deliberative conference using validated criteria. *Science, Technology & Human Values*, **29**(1), 88–121.

ROZIN, P. (1976). The selection of foods by rats, humans, and other animals. In J.S. Rosenblatt, R.A. Hinde, E. Shaw and C. Beer (eds), *Advances in the study of behavior, Vol 6*, pp. 21–76. Academic Press: New York.

SABA, A. and MESSINA, F. (2003). Attitudes towards organic foods and risk/benefit perception associated with pesticides. *Food Quality and Preference*, **14**, 637–645.

SIEGRIST, M. (2000). The influence of trust and perceptions of risks and benefits on the acceptance of gene technology. *Risk Analysis*, **20**(2), 195–203.

SIEGRIST, M. and CVETKOVICH, G. (2000). Perception of hazards: The role of social trust and knowledge. *Risk Analysis*, **20**(5), 713–719.

SIEGRIST, M. and CVETKOVICH, G. (2001). Better negative than positive? Evidence of a bias for negative information about possible health dangers. *Risk Analysis*, **21**(1), 199–206.

SIEGRIST, M., CVETKOVICH, G. and ROTH, C. (2000). Salient value similarity, social trust, and risk/benefit perception. *Risk Analysis*, **20**(3), 353–361.

SIEGRIST, M., KELLER, C. and KIERS, H.A.L. (2005). A new look at the psychometric paradigm of perception of hazards. *Risk Analysis*, **25**(1), 211–222.

SLOVIC, P. (1987). Perception of risk. *Science*, **236**, 280–285.

SLOVIC, P. (1992). Perception of risk: Reflections on the psychometric paradigm. In S. Krimsky and D. Golding (eds), *Social Theories of Risk*, pp. 117–152. Westport: Praeger.

SLOVIC, P. (1993). Perceived risk, trust, and democracy. *Risk Analysis*, **13**, 675–682.

SLOVIC, P. (1999). Trust, emotion, sex, politics, and science: Surveying the risk-assessment battlefield. *Risk Analysis*, **19**, 689–701.

SLOVIC, P., FISCHOFF, B. and LICHTENSTEIN, S. (1982). Facts versus fears: Understanding perceived risk. In D. Kahneman, P. Slovic and A. Tversky (eds), *Judgment under Uncertainty: Heuristics and Biases*, pp. 463–489. New York: Cambridge University Press.

SLOVIC, P., FINUCANE, M., PETERS, E. and MacGREGOR, D.G. (2002). The affect heuristic. In T. Gilovich, D. Griffin and D. Kahneman (eds), *Heuristics and Biases: The Psychology of Intuitive Judgment*, pp. 397–420. New York: Cambridge University Press.

SLOVIC, P., FINUCANE, M., PETERS, E. and MacGREGOR, D.G. (2004). Risk as analysis and risk as feelings: Some thoughts about affect, reason, risk, and rationality. *Risk Analysis*, **24**(2), 311–322.

SPARKS, P. and SHEPHERD, R. (1994). Public perceptions of the potential hazards associated with food production and food consumption: An experimental study. *Risk Analysis*, **14**(5), 799–806.

STARR, C. (1969). Social benefit versus technological risk. *Science*, **165**, 1232–1238.

STEPTOE, A., PERKINS-PORRAS, L., RINK, E., HILTON, S. and CAPPUCCIO, F.P. (2004). Psychological and social predictors of changes in fruit and vegetable consumption over 12 months following behavioral and nutrition education counseling. *Health Psychology*, **23**(6), 574–581.

STONE, R.N. and GRØNHAUG, K. (1993). Perceived risk: Further considerations for the marketing discipline. *European Journal of Marketing*, **27**(3), 39–50.

TVERSKY, A. and KAHNEMAN, D. (1981). The framing of decisions and the psychology of choice. *Science*, **211**(30), 453–458.

VERBEKE, W. (2001). Beliefs, attitude and behaviour towards fresh meat revisited after the Belgian dioxin crisis. *Food Quality and Preference*, **12**, 489–498.

VERBEKE, W. and VAN KENHOVE, P. (2002). Impact of emotional stability and attitude on consumption decisions under risk: The coca-cola crisis in Belgium. *Journal of Health Communication*, **7**(5), 455–472.

VERBEKE, W. and VIAENE, J. (1999). Beliefs, attitude and behaviour towards fresh meat consumption in Belgium: Empirical evidence from a consumer survey. *Food Quality and Preference*, **10**, 437–445.

VON ALVENSLEBEN, R. (2001). Beliefs associated with food production methods. In L.J. Frewer, E. Risvik and R. Schifferstein (eds). *Food, People and Society. A European Perspective of Consumers' Food Choices*, pp. 381–399. Springer-Verlag: Berlin.

WILLIAMS, P.R.D. and HAMMITT, J.K. (2001). Perceived risks of conventional and organic produce: Pesticides, pathogens, and natural toxins. *Risk Analysis*, **21**(2), 319–330.

YEUNG, R.M. and MORRIS, J. (2001). Food safety risk: Consumer perception and purchase behaviour. *British Food Journal*, **103**(3), 170–187.

ZWICK, M. and RENN, O. (2002). *Perception and evaluation of risk. Findings of the Baden-Württemberg Risk Survey 2001*. Working Report No. 203. Centre of Technology Assessment: Stuttgart.

Part II

Product attributes and consumer food choice

6

Branding and labelling of food products

Y. K. van Dam and H. C. M. van Trijp, Wageningen University, The Netherlands

6.1 Introduction

Brands and quality labels are important communication vehicles to consumers. They are often seen as the most valuable asset of companies. This also holds for brands in the food category. However, brands and branding have not received much specific research attention within the food consumer science literature (Jaeger, 2006) although there have been notable exceptions (e.g., Bredahl, 2003; Akbay and Jones, 2005). In the scientific marketing and consumer behaviour literature brands and branding strategies have been heavily researched in terms of their value both to consumers and to (food) companies. To stimulate this cross-fertilisation, the aim of this chapter is to familiarise food researchers with some of the marketing thinking on branding, brand equity, and brand management from different viewpoints, in order to derive implications for branding and labelling of food products. In addition to branding we will also discuss aspects of food labelling more generally with special emphasis on nutrition and health labelling.

6.1.1 What is a brand?

A brand is defined by the official American Marketing Association (AMA) as 'a name, term, sign, symbol, design or a combination of them intended to identify the goods or services of one seller, or a group of sellers and to differentiate them from those of competitors' (AMA, 2006; Aaker, 1991; Blois, 2000). The legal term for brand is trademark. Brands as a reference to the maker have a long history. For example, the medieval guilds required that craftspeople put trademarks on their products to protect themselves and consumers against inferior quality (Kotler and Keller, 2006), or at least against unlicensed competitors.

Brands may be owned by the manufacturer (known as A-brands or manufacturer brands) but also by a trading house or the reseller of the product (private labels). In many cases these private labels are owned by retailers in which case they are known as store brands (e.g., Tesco, Carrefour). With many manufacturers outsourcing all their production and with trading houses extending the range of products carrying their private label, the distinction between brands and private labels is rapidly fading. What remains important these days is a distinction between store brands and others. The distribution of store brands is restricted to the outlets of that store, whereas the other brands do not have this restriction. Increasingly retailers are using their store brands in their marketing strategies as a means to differentiate themselves from other retail chains (e.g., Ailawadi and Keller, 2004).

A brand may identify one item, a family of items, or many or even all items of that seller. As a result, brands come in different forms and formats (e.g., Keller, 2003). Some companies carry one brand name and one visual style in different product groups or product classes. Examples include Philips (light bulbs and CD-players), Yamaha (motorcycles and pianos), and Peugeot (cars, bicycles, pepper grinders) in which case the brand is referred to as an umbrella brand or family brand. Sometimes two brand names are used on the same article. For example, companies like Nestlé and Unilever carry their corporate brand name on the pack together with the individual brand name with the aim to benefit both from the corporate image and the individual brand name image. Sometimes this combination of corporate and individual brand names is even more subtle such as in the Nestlé brands Nestea, Nesquick and Nescafe or the McDonald's brand names McDrive and McChicken. New brands may also be combined with existing individual brand names (as in Becel/Flora ProActive) in which case the brand value of the parent brand (Becel/Flora) extends into the sub-brand (Pro-Active). Brand combinations may also occur in the case of ingredient branding and co-branding.

From a consumer perspective, brands extend well beyond their pure descriptive information content on what is the source or the maker of the brand. Companies produce and name products, but brands are made in the minds of the consumer. An example of this consumer basis of a brand is the following: Blaupunkt was founded in 1923 under the name *Ideal*. The core business was the manufacturing of headphones. If the headphones came through quality tests, the company would give the headphones a blue dot. The headphones quickly became known as the *blue dots* or [in German] *blaue Punkte*. The quality symbol would become a trademark, and the trademark would become the company name in 1938. A brand is 'an identifiable product or service augmented in such way that the buyer or user perceives relevant unique added values that match their needs most closely' (De Chernatony, 1992). Fundamental to this definition is the implicit assumption that the branded product or service delivers the functional and psychological benefits that the customer has paid for and has a right to expect (Hankinson, 2000).

6.1.2 Branding in the food domain

Within foods, it is important to recognise the wide diversity in branding practice. On the one hand some of the world's most valuable brands are from food companies as are some of the most trusted brands worldwide. For example, brands like Coca Cola and McDonald's feature in the Top 10 of the world's most valuable brands (Berner and Kiley, 2005) and brands like Coca Cola and Spa are identified as the most trusted brands in several countries (Reader's Digest, 2005). On the other hand, within the food category a substantial part of the fresh produce is still sold as generic products without any branding or at best under the implicit 'brand' of the retailer that sells it. Increasingly also, within foods and other fast moving consumer goods there has been a rise of store brands in recent years, and national manufacturer brands have lost market share to retailer brands (AC Nielsen, 2005). More often than not both branded, private labels and generic food products are sold through the same retail outlet. The resulting competition between brands, private labels and generic products within one and the same supermarket has not yet been studied in the literature.

There seems to be a persistent belief that branding is not feasible for fresh produce. For example, Riezebos (1994) claims that to some product categories, like fresh vegetables and potatoes, branding may be less applicable. Opposed to this, the Chiquita brand shows that fresh fruits are also differentiable.

6.2   The role of brands in the consumer decision process

Historically brands can be seen as carriers of information that were born out of the necessity of a time in which markets grew faster than communication lines (Mitchell, 2001). Once the personal feedback between producer and customer becomes impossible, the producer needs a marker for the quality that prospective buyers could expect of their products. Likewise the consumer needs a marker to identify products and producers that reliably match their expectations (*cf.* Domizlaff, 1939). In terms of learning theory (Van Osselaer and Alba, 2003), brands are markers of intrinsic product attributes. This value of the brand is something that consumers must learn from their interactions with the brand. Some of this learning can be based on advertising, some can be based on word-of-mouth, and some of this will be based on personal experiences with branded products.

For the consumer the brand name and its appearance is an information stimulus. The brand is a 'bundle of information' (Riezebos, 1994) representing a cluster of knowledge, experiences, and emotions that is stored in memory and that can be triggered or accessed through the association with the brand name. The Human Associative Memory (HAM) theory (Anderson, 1983) and schemata theory (Neisser, 1976) assume that human declarative knowledge is stored in memory as a network of interlinked concept nodes. The link between two concepts is strengthened every time the concepts (e.g., brand and product attribute) are co-occurring (but see Janiszewski and Van Osselaer, 2000 and Van Osselaer and Janiszewski, 2001 for complementary insights). Brand names are conceived of as a node in the network interconnected with a variety of associations including facts about the brand, but also thoughts, feelings, perceptions, images, experiences, usage situations, etc. (Keller, 2006). A strong brand is one which has strong, unique and positive associations attached to it (Krishnan, 1996).

In the following sections we will use the combined approaches of information cue and associative processing to discuss the role of brands in the different stages of consumer decision making, in particular arising from brand awareness and brand associations (e.g., Aaker, 1991).

6.2.1   Brand awareness

Consumers use brands as heuristics in their search for products that provide optimal catering to their needs. Consumers associate brand names to facts, thoughts, feelings, perceptions, images, and experiences. Associations also combine the brand to various usage situations. Brands may also signal social dimensions of the product, as brand awareness implies that the brand has a reputation within the consumer's social network or within society as a whole. Through its various associations, brand name awareness may render the particular brand salient in the face of a consumer decision problem. This means that brands come to mind at the very moment of product choice and that they are more likely to feature in the consumer's consideration set, out of which a final product choice is being made (Keller, 2006).

Brand awareness is likely to guide decision making in the first trial of a new product. Consumers may reduce their uncertainty by relying on brand awareness. For habitualised or routinised purchases the associations between usage context and brand name may be so strong, that need recognition is immediate 'brand need recognition'. If the brand Nescafe generates high awareness when the consumer runs out of coffee, it would put the Nescafe brand in a favourable position to enter (consideration set) or determine (routinised choice) the decision process.

Brand awareness is characterised in terms of depth and breadth. Depth relates to the likelihood that the brand is recalled or recognised. Breadth is the variety of purchase and consumption situations in which the brand is recalled or recognised (Keller, 2006). The brand awareness of some brands is so high that they become synonymous to category choice as in the case where a kid wants to go to McDonald's, rather than 'a fast-food restaurant', or in the way that every pain-killer is referred to as Aspirin. Brand awareness also strongly biases product choice towards the known brand in repeat purchases, even when the quality of the well-known-branded product is significantly lower than other brands that have been sampled (Hoyer and Brown, 1990).

6.2.2   Associations in information processing and evaluation

In addition to serving as memory tags in themselves, brand names also serve an important role as an information cue that retrieves or signals product attributes, benefits, affect or overall quality (Warlop *et al.*, 2005). The brand name is a marker for the product quality that the consumer has experienced or that the consumer expects to experience. The strength of a brand name is related directly to the information that is contained in the brand name. In this light we can look at the study of Bello and Holbrook (1995), who found no brand equity over and above the quality rating of products. For those products it would appear that the brand name is a nearly perfect indicator of product quality, and therefore carries a lot of information that is reinforced by product experiences.

In terms of information processing models (e.g., Steenkamp, 1990), the brand name serves as an extrinsic attribute of the product offering which may play an important role in the consumer's quality perception (e.g., Bredahl, 2003) and decision-making process. Brand name cues trigger associations related to the performance and psychosocial meaning of the product offering. Studies in this field aim at understanding the associations that consumers have with information stimuli such as brand cues, and they illustrate that information from the specific cue may spread easily within the consumer memory to expectations on specific product attributes, product benefits, and affective associations up to the level of consumer values (as often measured through means-end-chain analysis). Because the primary associations with the brand name are part of the associative network as well, information may spread beyond the primary brand association (e.g., Pepsi and Michael Jackson) into the associations that consumers hold with that primary association. The latter are called secondary associations and over time they may transfer to the brand name (Keller, 1993). This pattern of primary and secondary associations that arises from the brand cue is known as brand image. If the expectations generated by the brand cue are reinforced at the level of purchase and consumption, the value of the brand cue as a source of information will be reinforced. This simplifies the consumer's next purchase and consumption decision. Strong brands are those that help simplify the consumer decision process by triggering (and delivering) truly relevant benefits in a consistent and distinctive manner (Keller, 2000). By delivering up to their promise, they increase the level of consumer satisfaction as well as confidence in choice.

For many products, and especially for food products, brand awareness may also guide information processing in a way that can be explained by vicarious learning (*cf.* Bandura, 1977) or even conformity (Asch, 1956). This is particularly true when product quality is difficult to judge by the consumer, as in the case of experience or credence goods such as foods with characteristics that can be partly evaluated in use and partly cannot be evaluated at all (*cf.* Darby and Karni, 1973). In the absence of intrinsic search attributes, brand awareness can be used as a proxy for quality, as the best known brand (or the brand with the largest market share) apparently is the socially accepted ideal point of quality. Also, in complex choice situations where the consumer is confronted with a wide assortment – as in front of the supermarket shelves – brand recognition can be an important factor in simplifying the choice task, and choice can be biased towards the brand with the higher awareness. Even when consumers are experienced with a product, having sampled various different brands and claiming to base their judgment on taste and quality, this perceived quality is likely to be mainly dependent on brand awareness (Hoyer and Brown, 1990). Over the past 40 years research has consistently shown that in a blind taste test consumers are unable to detect their preferred brand, that in a blind taste test they rate their preferred brand lower than in an identified taste test, and that in a blind taste test they may rate other brands higher than their preferred brand. This implies that strong brands actually shape and change the consumer's perception of the product, improving taste and quality in a way that no product innovation could achieve. In the next subsections we will go deeper into these processes.

6.3   The brand and its sources of value

Brands perform important functions within the firm in terms of providing a corporate culture, countervailing power to the retailers and other customers, internal administrative processes as well as legal protection (e.g., Kotler and Keller, 2006). Brand value to companies arises to some extent from these factors internal to the company, but primarily from the consumer franchise or loyalty that they help to generate. It is important to recognise that much – if not all – of the value of brands originates from the consumer trust and the confidence that consumers experience from these brands. This is the element of brands and branding that we will focus on in the remainder of this chapter. The concept of 'the value of a brand' or brand equity is central to much of the branding literature. Ailawadi *et al.* (2003: 1) among others define brand equity more precisely as 'the marketing effects or outcomes that accrue to a product with its brand name compared with those that would accrue if the same product did not have the brand name'. Much effort has been devoted to understanding and measuring both the sources and the consequences of brand equity (e.g., Keller, 2003).

One of the challenges in brand equity research is to identify and quantify the associations in brand knowledge as a source of brand equity. From a consumer perspective, brand equity is based on consumer attitudes about positive brand attributes and favourable consequences of brand use (AMA, 2006). In other words, for consumers brands have value which extends beyond the purely objective value of what the brand delivers.

Throughout the marketing literature several models have been proposed for the added value of brands. One of the early models on brand equity was put forward by Aaker (1991) who distinguishes between sources and consequences of brand equity at the company level and at the consumer level. In the marketing and consumer behaviour literature these two levels of brand value have largely developed separately into what might be called the financial-managerial approach and the behavioural approach (Keller, 2001). Riezebos (2003) extended Aaker's model to further detail out the components of brand equity (see Fig. 6.1).

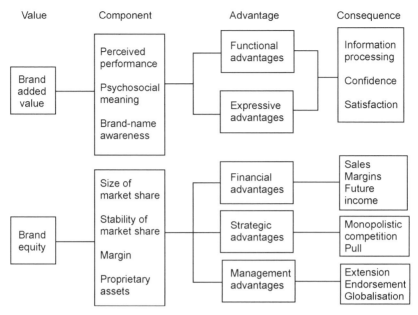

| Value | Component | Advantage | Consequence |

Fig. 6.1 Components of brand equity (Riezebos, 2003).

6.3.1   Consumer-based brand equity

Ultimately most of the company value arises from consumer behaviour with respect to its brands. As summarised by Keller (2001) consumer perceived brand equity has been studied from four different streams of academic research. The consumer psychology approach, which we emphasise here, builds largely on associations with the brand as represented in consumer memory. Prominent models within this approach are developed by Aaker (1991) and Keller (1993, 1998). The information-economics perspective builds on market imperfection and information asymmetry, arguing that brand equity largely arises from brand credibility which reduces consumer uncertainty and lowers the information costs and perceived risk (e.g., Erdem and Swait, 1998). The sociology-based approach emphasises the cultural meaning of brands and products (e.g., McCracken, 1986) and the biology-based approach has emphasised the way brand information becomes integrated into memory processes even generating subconscious effects on consumer behaviour (e.g., Zaltman and Coulter, 1995).

From a consumer orientation, brands are the platform for managing consumer relations and building customer equity (Rust et al., 2004). The customer equity of the brand is defined as the lifetime value of the firm's customers. The word *lifetime* deserves special attention, as customer equity is based on the cumulative value of a lasting relationship between the company and its brand loyal customers, rather than being a snapshot in time.

Others have defined specific elements of brand equity at the consumer level which make the concept measurable. For example, Aaker (1996) defines brand awareness, perceived quality, specific brand associations, brand loyalty and other proprietary assets (such as patents, trademarks and channel relationships) as key assets of the brand. He also identifies the unique set of brand associations that represents what the brand stands for and promises to customers (perceived brand identity) as the cornerstone. The 12 dimensions of brand identity are organised around four perspectives: brand-as-product (product scope, product attributes, quality/value, uses, users, country of origin), brand-as-organisation (organisational attributes, local vs global), brand-as-person (brand personality, brand-customer relationships) and brand-as-symbol (visual imagery/metaphors and brand heritages). Keller's (2000; 2001) brand resonance model identifies six components organised in four levels of increasing consumer-intimacy: (1) brand identification – operationalised in brand salience, (2) brand meaning – operationalised in brand associations at the functional (performance) and psychological and social (imagery) level, (3) response – operationalised as positive functional evaluations (judgments) and emotional reactions (feelings) toward the brand, and (4) the relationships that consumers have with the brand (resonance). Srinivasan et al. (2005) review previous measures for brand equity and develop a modelling framework for measuring brand equity incorporating the components of brand awareness, performance associations and imagery associations with the brand. Some authors (e.g., Ailawadi et al., 2003) have questioned the use of survey-based 'consumer mindset' measures to quantify brand equity and favour outcome-based so-called product-market measures. Ailawadi et al. (2003) argue that these product-market measures allow more direct interpretation in terms of real life earnings in the market place due to brand equity (see also Srivastava and Reibstein, 2005).

6.3.2   Measurement of consumer-based brand equity

One of the key challenges for research on brands and branding is to identify and quantify the sources of brand equity, for which a wide variety of methodologies is available (see Agarwal and Rao, 1996; Chandon, 2003 and Keller, 2006 for overviews). We will discuss the measures of brand equity under three headings here, being measures of brand awareness, measures of brand image, and outcome-based measures.

Brand awareness can be identified through a variety of aided and unaided memory measures to test brand recall and recognition. Qualitative techniques (see also Supphellen, 2000) involve free association in which the consumer is asked 'what comes to mind when you think of [this product category or purchase situation]?'. Respondents may then be further probed to express thoughts about the positivity, uniqueness and strengths of these associations. More quantitative approaches to brand recognition can be based on the simple question whether consumers have seen, heard of and used the brand before. Usually, fake items are included to correct the data for yes-saying tendencies. A more advanced method is where the brand is masked or distorted to assess whether consumers would still correctly identify the brand. Recall is measured through more abstract questions such as which brands come to mind when you think of a particular product category (e.g., drinks), a particular usage situation (e.g., drinks when you do sports or eat breakfast) or even more specifically (e.g., soft drinks). The appropriateness (e.g., Cardello and Schutz, 1996) measure can meaningfully be applied here when recall is measured from a usage situation perspective.

Brand image measures aim at identifying and quantifying the network of primary and secondary brand associations that underlie brand equity. In addition to free elicitation technique, projective techniques can be applied as an indirect way to identify brand associations. Again a variety of methods is available such as completion and interpretation tasks (e.g., with empty bubbles in brand related cartoons asking the consumer to fill in the text) and comparison tasks (e.g., 'if the brand were a person or an animal what would it be and why?'). Means-end chain analysis can be applied to the elicited associations in order to identify the higher order (more abstract) meanings of the brand associations in terms of consumer benefits and value delivery. In brand personality methods, the respondent is asked to express the human characteristics that can be attributed to the brand. Often these associations are explored in qualitative tests as in the case where we would ask the respondent: 'if the brand would come alive as a person, what would it be like?'. A very comprehensive approach to brand equity measurement is Zaltmann's Metaphor Elicitation Technique (ZMET) in which consumers collect images that they consider representative for the brand. From these images the deeper meaning of the brand is further explored. Based on qualitative research, Fournier (1998) identified six dimensions of consumer-brand relationships: (1) self-concept connection, (2) commitment or nostalgic attachment, (3) behavioural interdependence, (4) love/passion, (5) intimacy and (6) brand partner quality. Aaker (1997) developed a brand personality question-naire to quantify five dimensions of brand personality: (1) sincerity, (2) excitement, (3) competence, (4) sophistication, and (5) ruggedness.

Quantitative measurement of brand image (see also Agarwal and Rao, 1996), is usually conducted in rating tasks in which consumers rate one or more brands on a number of items that represent their perceptions. In terms of brand performance, these items may be very specific such as the perceived sensory features/expectations but also more abstract in terms of brand reliability, durability and service. Brand imagery extends beyond pure brand performance to include aspects of how the brand is being used. Categories of items on which brand image is assessed typically include user profiles, usage situations, personality and values of the brand as well as history, heritage and experiences. Brand judgment is sometimes defined as the more abstract, integrative personal opinions and evaluations of the brand expressed in terms of brand quality, brand credibility, brand consideration and brand superiority. Brand feelings represent the more emotional responses and reactions to the brand and are sometimes expressed as (Keller, 2003) warmth, fun, excitement, security, social approval and self-respect. Elements of brand image can be assessed through the conven-tional perceptual mapping techniques (e.g., Steenkamp *et al.*, 1994) which may be compositional (as in Factor Analysis), decompositional (as in Multi-dimensional Scaling) or a combination of both (as in Free Choice Profiling with Generalised Procrustes Analysis). Several authors have proposed specific brand equity measurement tools. Keller (2001) and Rust *et al.* (2004) both suggest a set of candidate measures for the constructs in their brand equity models. Netemeyer *et al.* (2004) validated a set of survey measures to quantify the dimensions of perceived quality, perceived value for cost, uniqueness and willingness to pay a price premium. Yoo and Donthu (2001) developed survey measures for overall brand equity as well as its components: brand loyalty, perceived quality, brand awareness and brand associations. Whereas these measures focus on the more functional or tangible aspects of brand equity, Aaker (1997) developed a specific scale for measuring an important brand intangible: that of brand personality, the human characteristics or traits that can be attributed to a brand. Her measure is composed of five brand personality dimensions: sincerity, excitement, competence, sophistication and ruggedness.

Outcome measures of brand equity quantify the extent to which brand names deliver the customer-related benefits to the company. One prominent method is conjoint analysis (see Carneiro *et al.*, 2005; Enneking *et al.*, 2005 for recent food-related applications), in which the brand name features as a design factor. By comparing the part worth values for the different brand name level, the relative importance of branding in overall evaluation can also be quantified in relation to other marketing factors such as price, information and design features. Arguing that brand equity is what remains after the consumer ratings have been corrected for the differences in preference due to physical product features, the residuals approach (e.g., Park and Srinivasan, 1994; Dillon et al., 2001) attempts to separate consumer attribute ratings for a brand into two components: (1) brand-specific associations and (2) general brand impressions. Valuation approaches attempt to put a financial figure on brand equity. However, as argued by Keller (2006), there is no conventional accounting approach available for doing so. See Keller (2006) for further description of these financial measures.

6.3.3   Finance-based brand equity

For many firms brands are the most valuable asset, being valued much higher than physical assets. Recognising that brands hold value to the company, it has become increasingly popular during the 1980s to express the value of brands in financial terms (e.g., Barwise, 1993; Simon and Sullivan, 1993; Ailawadi *et al.*, 2003; Chandon, 2003) either through a cost approach (amount of money that would be required to reproduce or replace the brand) or a market approach (the present cash flow derived from the brand's future earnings). Whereas some 50 years ago, 80% of a typical firm value was made up of tangible assets such as its plant, equipment, inventory and land, today on average nearly 50% of a firm's value (and even 70% of Fortune 500 companies (Keller, 2003)) is determined by intangible assets such as intellectual property, brands and the firm's customers. Brands are often the largest of these assets (Srivastava and Reibstein, 2005). For example the Coca Cola company recently valued its physical assets at around $6 billion, whereas the value of the Coca Cola brand was valued at around $67 billion (Berner and Kiley, 2005) hence making up over 90% of its total assets.

Brands provide advantages to the firm at the financial, strategic and management level (Riezebos, 2003). Strong brands that communicate and reinforce consumer value are likely to generate higher consumer clientele and therefore put the company in a competitive advantage compared to competition. Also, because brands reinforce consumer decisions, they are likely to generate repeat purchases among consumers and thereby brand loyalty over time. This brand loyalty is an important consequence of brand equity (Aaker, 1991) and may result in more stable market shares over time. Strong brands with high consumer loyalty will also generate consumer pull at the retail level. Consumers expect the brand to be present on the shelves at their supermarket. As a result, retailers may be willing to accept a lower trade margin for strong brands, which in turn puts the branding company in an advantageous position in negotiations with retailers. Also strong brands may reduce the company costs in terms of economy of scale and because brands help to streamline the internal organisation processes. All of these benefits help to increase the profit margin that the company realises on the branded product. Finally brands can also add value through proprietary brand assets related to patents, and legal protection of the brand. These company-level components of brand equity translate into a number of specific advantages at the financial, strategic and management level. Financial advantages accrue from the fact that a brand strategy provides higher rewards in the long run than a generic product strategy. In the short run costs are higher for brands, due to packaging and advertising, but in the long run brands are believed to generate higher return on investment than unbranded products. Eventually, strong brands generate higher sales and higher profits thus adding to the value of the company. Also, brands are important to the continuity of firms as they may generate a certain guarantee for future income.

At the strategic level brands strengthen the position of the company in relation to potential competition. Strong brands generate consumer loyalty, which gives the company some influence over the market. Brand loyalty raises barriers to other brands attempting to capture a share of the market. A market with branded products is in a state of monopolistic competition, as each brand monopolises a differentiated proposition, allowing firms to maximise economic profit. Brands are also important in relationships with the retailer, as mentioned before. Strong brands facilitate acceptance of the brand by retailers, implying wider distribution and better trade margins. In addition, brands may also have strategic value internally to the company to provide a sense of direction and pride to its employees as well as in attracting new and best employees in the labour market.

At the management level brands are very important because strong brands provide a platform from which the company can reach larger markets through extension and endorsement as well as geographic reach in the form of global branding. The concept of brand extension will be discussed in more detail later in this chapter.

6.4   Brand management

As argued above strong brands hold a number of advantages for the company. In marketing terms (Keller, 2001) strong brands deliver positive effects on all marketing mix elements. Product-related effects are due to strong brands being linked to more positive evaluations, quality perceptions and purchase rates. Also, the sheer familiarity that strong brands have increases consumer confidence, brand attitudes and purchase intention and also mitigates the potential impact of negative product experience. At the price level, strong brands are able to command higher prices, their sales are more immune to price increases as well as to price competition from small share brands. Also, strong brands tend to have a loyal clientele which is less price-sensitive. At the channel level, strong brands have a higher chance of channel acceptance and shelf space and are more likely to feature in higher quality image stores. In terms of communication-related effects it has been found that positive feelings for the brand can positively bias the evaluation of brand advertising, lower the negative reaction to advertising repetition, better withstand competitive ad interference and make it better able to withstand a product-harm crisis. To many companies it is worthwhile to invest in the building and maintenance of strong brands, which is the field of brand management that we will discuss next.

6.4.1   Managing the brand

The brand name is a separate asset, and to the firm it has become an independent entity that has an economic value of its own. Brands are bought and sold or licensed. This economic value is the yardstick for contemporary brand management: the evaluation of the effectiveness of marketing decisions and brand extensions, the competitive strength of the brand, is judged from the contribution to this financial value of the brand. Brand value and brand equity are a result of marketing mix decisions, of the corporate organisation, and external factors (cf. Porter, 1985). The 'house of quality' approach (Hauser and Clausing, 1988) acknowledges that the brand may further benefit from the synergy among tactical marketing decisions. The objective of brand management is investing in brand equity by optimising these corporate and external elements through marketing mix decisions.

It was stated in the previous paragraph that brand equity is based on consumer responses to the brand, like brand loyalty, brand awareness, perceived quality and positive brand associations. Furthermore brand equity can be enhanced by proprietary assets, like patents, trademarks and channel relationships. Brand management therefore is managing consumer responses and proprietary assets in order to increase the value of the brand equity.

6.4.2   Managing consumer responses

Consumer-based components of brand equity are brand loyalty, brand awareness, perceived quality and positive brand associations (Aaker, 1991). It should be noted that these are components of the management view of the brand image. To the consumer brand equity denotes the superior performance of the brand, and the consumer's 'total understanding of the brand', which may cover elements like product hierarchy, performance, value for money, attitude, recognition, trustworthiness, confidence, satisfaction, social image, values and identification (Howard, 1994; Lassar *et al.*, 1995). Chaudhuri and Holbrook (2001) note that consumer-based brand equity hinges on the psychological associations with the brand. Brand associations can be combined into a 'brand personality'. Packaging and other physical cues, advertising, and even other users of the brand create an understanding of the brand that can be described by personality dimensions, such as, for example, 'old-fashioned', 'intelligent', 'sexy', 'athletic', 'glamorous' or 'rugged'. Likewise marketing actions may lead to personality inferences like 'schizophrenic' for repeated repositioning, 'comfortable' for a continuing character in advertising, or 'sophisticated' for exclusive distribution (Solomon, 2002).

Brand associations are not merely verbal associations. A majority of the brand associations are likely to be based on visual or other sensory impressions, without a corresponding verbal description (Supphellen, 2000). Brand associations may also be emotional impressions, affective responses to the brand that are also coded non-verbally. Consumers can have strong emotional attachments to brands (Thomson *et al.*, 2005), and therefore Tsai (2005) urges brand managers to enhance the full spectrum of consumer experiences and to create a holistic brand value structure, which can unite the consumer's sensory, emotional, social, and intellectual experiences in a new and positive way. Lindstrom (2005) notes that brand managers focus on only sight and sound, but that 'emotional connections are effectively made with a synergy of all five senses, and as such those brands that are communicating from a multi-sensory brand platform have the greatest likelihood of forming emotional connections between consumers and their product'. This is especially important in the branding of food products, as food products by their very nature are experienced and evaluated in use by a synergy of all the human senses.

In their study among numerous food and non-food products Chaudhuri and Holbrook (2001) find that brand loyalty is among others based on trust in the brand and affect to the brand. By separating brand loyalty into purchase loyalty and attitudinal loyalty they show differential effects on markets share and relative price of the brand. Purchase loyalty, which is mostly dependent on brand trust, is predictive of market share. Attitudinal loyalty, which is equally dependent on brand trust and brand affect, is predictive of relative price. Managing consumer responses then boils down to building consumer confidence and positive affect towards the brand. Strong brands are brands that are trusted by the consumers and that generate positive affect in consumers. It should be noted that being a strong brand does not imply being the market leader, as Howard (1994) reports that the [attitudinal] brand loyalty to the share leader in a product category is consistently lower than the loyalty to the most loyal brand. Conversely this implies that having the most loyal consumers does not mean having the most consumers, but it does mean having customers who consistently are willing to pay the full price for their brand.

Managing a brand name is managing a symbol and a promise of quality and consistency (Kardes, 2001). The perceived quality of the brand is affected by many elements, the most important of which are objective quality and consistency. Consistency refers to both consistent quality, and image consistency. Kardes (2001) especially warns against inconsistency due to complex and multidimensional brand images as a result of uncontrolled brand extensions.

6.4.3   Brand extensions

Recognising that brands are valuable assets, many companies are leveraging those assets by introducing new products or product ranges under their strongest brand names. When an existing brand name is used for a new product this is called a brand extension, and the existing brand is called the parent brand. Applying an existing brand name to new products in the same product category as the parent brand is called a line extension. Line extensions are usually introduced to target a new market segment, or to increase the variety offered to the market segment that is already served. The brand extends into a new market segment within the product category that is currently catered for by the existing brand (e.g., Diet Coke). Using an existing brand name for new products in a different product category is called a category extension. In category extension the existing brand is used to enter a different category not currently served by the parent brand as in the case of Virgin entering the cola market. And sometimes it is a matter of interpretation whether an extension is to be considered a line extension or a category extension, as in the case of Mars extending from candy bars into ice-cream and drinks. This is a category extension from chocolate bars into ice-cream and drinks, but a line extension from chocolate snacks to frozen snacks and liquid snacks.

Other types of brand extensions are ingredient branding, and companion brands. In co-branding (also known as dual branding or brand bundling) two brands are combined (e.g., Lays crisps with Heinz ketchup) in order to benefit from the virtues of both brands. Ingredient branding is a special case of co-branding where the purpose is to create brand equity for materials or components that are necessarily contained within other branded products. The famous example is, of course, 'Intel-inside' but in food we also see similar strategies for artificial sweeteners (NutraSweet) and other ingredients (Toblerone in Hertog Ice cream).

Advantages of brand extensions over the introduction of a new brand are numerous. The brand extension can benefit from the existing brand name awareness and the brand image of the parent brand. The reputation of the parent brand may reduce the perceived risk that is experienced with the new product. These factors facilitate acceptance of the product both by retailers and by consumers. Furthermore the costs of developing, introducing, advertising, and supporting a new brand with its packaging and labelling, which may easily run into tens of millions, are dramatically reduced. Advertising and promotion for a brand extension may benefit the parent brand as well, thereby increasing efficiency even further. The parent brand can benefit from the extension as well, as the extension can support the image of the parent brand, but a brand image can be seriously damaged by inconsistent brand extensions. Klein (2000) goes as far as stating that by now the majority of brand extensions are developed mainly to support the image of the brand and to increase the brand equity. In the past decades she notices a reversal process: where brand names used to support products in the market, now products are supporting brand names in the market.

Successful brand-extensions rely on similarity between the parent brand and the extension for their success (Aaker and Keller, 1990; Boush and Loken, 1991; Keller and Aaker, 1992). It is, however, not always obvious what causes this similarity. Especially because extensions that seem consistent and coherent to the brand manager may seem inconsistent and incoherent to consumers (Kardes, 2002). Similarity may be based on common attributes among parent brand and extension, or on both products having the same image (Bhat and Reddy, 2001). Similarity may be based on the parent brand being associated with benefits that are also valued in the extension category (Meyvis and Janiszewski, 2004). Similarity may also be based on two products being part of the same or similar product categories (Boush and Loken, 1991), but even so it makes a difference whether categories are defined taxonomically (e.g., fruit juices) or functionally (e.g., breakfast drinks). Depending on which kind of category is considered, orange juice could be similar either to tomato-juice or to tea.

Perceived similarity organised around shared goals facilitates the transfer of knowledge and affect from a parent brand to an extension of that brand (Martin and Stewart, 2001; Martin et al., 2005). The availability of well-formed, goal-derived categories associated with a parent brand establishes an organising framework for consumers' assessments of similarity that facilitates the transfer of consumer knowledge and attitude from the parent brand to a brand extension in another product category. This facilitating effect of similarity does not occur in the absence of goal-derived categories (Martin et al., 2005).

Wrapping up the brand
In summary, company-derived value from brand equity is largely based on the fact that brands add value for consumers in terms of ease and promotion of information processing, increased confidence in choice and increased level of satisfaction. By providing consumers with information on the product's origin, brand names can reduce the imperfection and asymmetric information structure between consumers and the supply chain (e.g., Erdem and Swait, 1998), provided that the information contained or implied by the brand is correct and manageable for the consumer (see also Verbeke, 2005).

Brands provide an important tool to differentiate products from the generic category on the basis of its seller or source, which 'loads' the brand with information. And conversely the products that belong to an extended brand family help to differentiate the brand from other brands and contribute to the brand equity. Strong brands have strong, unique and positive associations attached to them (Krishnan, 1996). Strong brands help simplify consumer decision making by consistently and distinctively triggering and delivering truly relevant benefits (Keller, 2000). Strong brands also provide a platform for brand extension. But only brand extensions that consistently add to the positive associations and distinctive relevant benefits of the brand, contribute to the customer equity of the brand name.

6.5   Labelling

Brands in a descriptive sense as a reference to the product source or maker are a label, defined by AMA as 'the information attached to or on a product for the purpose of naming it and describing its use, its dangers, its ingredients, its manufacturer, and the like. A label is usually thought of as printed material, but labelling in the broader sense has been ruled to include spoken information and separate promotional pieces, if they serve the information purpose and are closely allied to the product' (AMA, 2006). To some extent these labels follow the same logic as branding. For example, Van Trijp and Steenkamp (1997) used Aaker's brand equity scheme to analyse the consumer value of Integrated Quality Control labels. However, whereas labels are often purely informative to describe certain objective qualities of the product (such as country/region of origin or nutritional quality) and apply to a range of different products that conform to a certain criterion or certification scheme, brands are often designed to communicate more specific, competitive and more implicit or less tangible information to the consumer.

In many instances, and particularly when observable product differentiation is low and mainly based on so-called credence attributes (those that cannot be verified by the consumer even after normal consumption) consumers may have a particular need for information to reduce their uncertainty and to allow informed choice. This is especially applicable to the food market as 'most foods products can be classified as credence goods' (Anania and Nistico, 2004). In such instances, consumers have to trust the information that is provided and credible information is required. Quality labels can be an effective way to provide transparency, reduce information asymmetry between consumers and supply chains and enhance informed choice (Caswell, 1998). Similar to brands, quality labels are also being used as an information cue by the consumer and can be meaningfully analysed with brand equity models (Van Trijp and Steenkamp, 1997). However, quality labels differ from brands in two important ways. They do not signal the source, maker or seller but rather that the product conforms to specific criteria often related to the way the product is produced (e.g., EKO, ISO, Marine Stewardship Council), the country or region of origin (e.g., Made in ...) or the specific content of the product (as in health labelling). Secondly, food labels are often regulated and controlled by (independent) certification organisations, with induced costs to the producer (see, e.g., Cheftel, 2005).

But like brands, quality labels can help consumers as an information cue, as a means to simplify the choice process and as a means to re-identify a product for re-purchase (Grunert, 2005). Labels can be awarded by manufacturers, groups of manufacturers, government bodies and independent organisations. The criteria for the labels can range from very strict to virtually non-existent and can be quite specific (such as organic means of production) to very general (as in country-of-origin) (Grunert, 2005).

Labels only reduce market imperfection from information asymmetry if they provide honest information, are properly understood/interpreted on the part of the consumer and are used by consumers (Verbeke, 2005). As summarised by Grunert (2005) and Verbeke (2005) many of these assumptions are questionable in the case of food labelling as the empirical evidence suggests that labels are frequently not understood, misinterpreted and over-generalised and not used intensively by the consumer. This explains why labelling debates are being dominated by the (perceived) information content and the processing and use of these labels by consumers (Teisl and Roe, 1998).

6.5.1 'Brand equity' of labels

Research confirms a willingness to pay premium prices for eco-labelled and Fair Trade products (e.g., Bjørner et al., 2004; Jaffry et al., 2004; Loureiro and Lotade, 2005; Veisten and Solberg, 2004), as well as a shift in consumer preference and choice. Apparently eco-labels and Fair Trade labels carry 'brand equity' as well. For consumers that value the ethical label (Fair Trade, eco) in their preference formation and choice, the brand name is of minor importance (de Pelsmacker et al., 2005). From the 'Fair Trade' label we can also learn that ethical labels allow for 'brand extensions'. The 'Max Havelaar' label was originally developed for Fair Trade coffee, but has been extended into new Fair Trade product categories like bananas and cocoa.

Ethical labels can support the products of new entrants and weak brands in a market, as the label may attract a consumer segment that the brand could not reach. For the same reasons ethical labels may weaken brand equity. There is a growing segment of consumers that value the ethical label, and to them the label carries more weight than a brand name. Therefore a weak brand or a generic product can boost its sales by adopting an ethical label, and it may actually gain market share against a strong competitor that does not carry the label. Consumer trust and confidence, however, is gained by the label more than by the brand, and therefore does not necessarily add to the brand equity. According to the information processing approach of Van Osselaer (2004) the attribution of the perceived product value is divided between the brand and the label, implying less information (and therefore less value) being uniquely awarded to the brand. However, Van Osselaer (2004) also shows that the equity of existing strong brands need not be harmed by the adding of additional attribute information, as when an ethical label is applied to a strong brand. Applying an ethical label to a strong brand could be comparable to co-branding, like the earlier example of

Lays crisps with Heinz ketchup, benefiting from the virtues of both brand and label.

Even if several brands carry the label, for the consumer the label reduces the complexity in the market, because the label effectively splits the market supply in two subsets, one certified and one not. Preference formation and choice can be limited to the products within either subset. In a generic market, like fresh produce, introduction of an ethical label creates differentiation and enables a choice where none existed.

6.5.2   Nutrition and health labels

A particular hot issue in food is related to the use of nutrition and health labels on food products. Nutritional information, including health claims, is regulated on a country-by-country basis (see Hawkes, 2004 for an overview). For nutrition labelling, the US has adopted mandatory labelling on almost all pre-packaged foods since the introduction of the 1990 Nutrition and Labelling and Education Act in 1990 (revised in 1994), and the same holds for Australia and New Zealand, although the format and content varies between countries. In the EU, nutrition labelling is not compulsory, and required only if a nutrition claim is being made, but also across Europe large differences exist in the regulation of health claims (see, e.g., Hill and Knowlton, 2000). Also for health claims, regulatory schemes vary considerably across the globe. Out of the 74 countries reviewed by Hawkes (2004), about half had no regulation of health claims, 30 do not allow any reference to disease in claims, 23 allowed nutrient function and other claims and only 7 would allow disease risk reduction claims (Williams, 2005). Furthermore three countries permit product-specific health claims within a specified framework, as, for example, some form of self-regulatory system.

Currently, the EU is in the process of developing pan-European legislation in which nutrient content claims and health claims based on existing science will be allowed provided that they are in the pre-defined list of allowed nutrition and health claims. For health claims based on 'new science' a more central role of EFSA will be introduced. EFSA will conduct a pre-market authorisation in which they assess the scientific substantiation as well as the degree to which the proposed health claim can be understood by and is meaningful to the consumer (EU, 2003). In Europe, the PASSCLAIM project has specifically focussed on the level of scientific substantiation required for different levels of health claims (e.g., Richardson et al., 2003) and consensus seems to be emerging in many different health benefit areas (see EJN, 2003). Some of the other restrictions in the new regulation, such as the application of positive nutrient profiles as a prerequisite for allowing health claims in the first place, are still under discussion

Generally speaking, food and nutrition labels serve at least three purposes (e.g., Przyrembel, 2004) relating to providing information to the consumer, protection of the consumer from potentially misleading information and stimulating competitiveness through fairness in trade. These are not necessarily fully compatible goals and as Hill and Knowlton (2000: 443) conclude: 'The US legislation on nutrition and health claims is targeted toward striking the balance between the need to ensure a high level of consumer information and protection on the one hand and competitiveness of the food and dietary supplements industry on the other'. Przyrembel (2004) concludes that over time attention has tended to shift from pure consumer protection against fraud more in the direction of consumer information and education through food labels. The latter goal can only be achieved if consumers understand and use nutritional information. Keller *et al.* (1997) apply an information processing perspective to consumer perceptions of health claims, arguing that health claims can only exert a positive effect on consumer behaviour to the extent that: (1) consumers are aware of the claim; (2) consumers understand the claim; (3) consumers make appropriate inferences from the claim; (4) consumers consider the claim credible; (5) consumers attach attitudinal relevance to the claim; and (6) consumers translate the claim into action tendency (purchases).

Several studies on consumer understanding and use of nutritional labelling (see European Heart Network, 2003 and Cowburn and Stockley, 2005 for a recent review) and health claims (see Williams, 2005 for a recent review) have challenged several of these assumptions. Cowburn and Stockley (2005) summarise the evidence that in the context of nutrition labelling those who look at labels understand some of the information and are confused by other information. Use of nutrition labelling seems to be limited. Despite the fact that in surveys and focus groups consumers claim to read labels, studies that use verbal protocol analysis seems to suggest that this information is not being processed in great depth. Reasons for not using labels include lack of time, lack of understanding and concerns about the accuracy of the information. Jacoby *et al.* (1977) already established that the vast majority of consumers neither understands nor uses nutrition information as presented in 'back-of-pack' labels, though negative information (Russo *et al.*, 1986) and information that contains arousing and specific consequence information (Moorman, 1990) does influence consumers. Consumer groups that are most likely to use the nutrition labelling information are women and consumers with higher education and income and those who already have a special interest or positive attitude to diet and health. Consumers seem able to execute simple calculations and comparisons for nutrition information but many have difficulty in translating that information to the total diet context. Consumers claim to look at nutritional information to support their purchase decision, but observational studies indicate that consumers make very fast choices at the shelf due to time pressure and lack of involvement with the food products, suggesting that they use the information selectively at best (Williams, 2005). Similarly, a recent review on consumer understanding and use of health claims for food (Williams, 2005) concludes that there is still a low level of consensus. Comparing evidence from survey and focus groups with that obtained from stronger research design focussing on experimental work and outcome studies, Williams (2005) concludes that there are some common findings, though. Health claims are generally seen as useful and consumers will perceive products with health claims as healthier and they express higher purchase intent for them. But at the same time, consumers are sceptical about commercial health claims and they want them approved by the government.

Many countries allow both nutrient-content claims (e.g., 'low sodium') and structure-function claims (e.g., 'contains folic acid: folic acid contributes to the normal growth of the foetus'), even if they prohibit health claims. However, consumers do not make clear distinctions between different claim levels such as nutrient-content claims, structure-function claims and health claims (Van Trijp and Van der Lans, 2006a). Also for nutrition and health claims there are substantial differences across countries and socio-demographic groups in how consumers evaluate and appreciate these claims (Williams, 2005; Van Trijp and Van der Lans, 2006b). In terms of format, consumers prefer short and simple claims over long, complex and scientifically worded claims, and express a preference for split claims consisting of front-of-pack succinct claims with supporting evidence on the back of the pack.

This raises a number of issues as the more succinct information might allow for more interpretational freedom on the part of the consumer. Results on the use of front label health claims are far from consistent, but an early study by Roe *et al.* (1999) identified four potential sources of bias in consumer inference making from health claims. A *positivity bias* may occur when consumers evaluate the product more positively due to the mere presence of a health claim. A *halo* effect is said to occur when consumers generalise positive attribute perceptions (e.g., low cholesterol) to imply other positive attributes (e.g., low fat) even though that is not justified by the claim. A *magic bullet* effect occurs if a consumer attached inappropriate health benefits from the attribute perception of the product (e.g., low fat to imply the product to help against cancer). Finally, an *interactive* effect may occur when the presence of a health claim affects consumer processing or use of nutritional facts information as available on the back of the pack. Mazis and Raymond (1997) find that especially informationally disadvantaged consumers, who do not have access to government information on health and diet, respond more to (front-of-pack) health claims. At the same time they refer to Brucks *et al.* (1984) who concluded that only consumers with a high level of nutritional knowledge were able to interpret and use the (back-of-pack) information.

An important discussion in current and future nutrition and health labelling will be how to balance the objectives: consumer protection, optimal consumer information and competitiveness and fairness in trade. Consumer protection is best served by availability of detailed nutritional information as is already available on the back of many food products. However, from a consumer information point of view, more information does not necessarily mean accessibility of information, as it would implicitly assume that consumers pay attention to this information, understand the nutritional information and can process the information for the benefit of their decision processes.

A consumer protection perspective should focus on the credibility of claims ensuring that the information is correct and not misleading. This objective extends into understanding and inference making from the claim to ensure that the consumer does not over-generalise what is stated, implied or possibly suggested by the claim. From a consumer information perspective the focus should be on information transparency such that the consumer is aware of this source of information, is armed to understand the information and able to interpret the claim correctly (inference making).

Lawrence and Rayner (1998) go as far as stating that from a public health point of view a general prohibition on health claims should be maintained, if only to adhere to the principle that health is related to dietary patterns rather than to specific food products. They also notice that scientific substantiation of health claims would require longitudinal studies of the effect of food or food-components as part of a normal diet by the target population that is specified in the health claim.

From a competitiveness of the food industry perspective the focus is on the attitudinal relevance of the claim and the use of this information for product choice. In this way, the claim becomes a competitive advantage vis-à-vis products not carrying that claim. From a branding and labelling perspective the key question is whether those competitively advantageous claims have an informational effect as suggested by the labelling literature or an emotional effect as suggested by the branding literature, and if the effect is mainly emotional, whether they still serve consumer protection.

6.6   Conclusion

This chapter aimed at introducing the food research community to some of the state of the art thinking on branding and brand management as it is available in the marketing literature. Also it drew a parallel between branding and labelling as processes that follow a similar logic in the consumer information processing area. Branding aims at building strong equity for the brand as evidenced by higher awareness, positive perception of overall quality and more specific brand associations in the mind of the consumer. As such, strong brands provide a competitive advantage in terms of coming to mind more easily when the consumer is faced with a purchase or consumption decision and by providing a cue from which the consumer infers specific product qualities. Brands elicit a consumer expectation which can bias actual consumer perception of the product even when it is actually consumed, as is evident from studies in the expectation (dis-)confirmation literature. However, brand images need to be nurtured and reinforced as brand associations need to be confirmed, and because lack of reinforcement of the brand promises may lead to disconfirmation and potentially may backfire on the brand image. But as long as they are built on distinctive quality and managed consistently strong brands will return their investments both in processed and fresh food products.

Whereas brands are designed to communicate and convince about one particular source, maker or company, food labels are designed to apply to a broader set of competitive products. Food labels may take many different forms, informing on the region or country of origin, production process, nutritional value and the like. Food labelling presupposes a set of criteria against which the product performs. Labels are not designed to favour one particular product but to favour a class of products, often from a social responsibility or public health perspective.

This makes the situation more complex as a balance needs to be found between public interest, consumer interest and commercial interest. First, consumers have the right to be informed in order to choose in full knowledge of the facts. Also, from a consumer protection point of view consumers should be protected against incorrect and misleading information. For this reason, many of the quality labels are regulated and controlled by independent regulatory bodies.

Of particular concern, currently, are the food and nutritional labelling and the new EU legislation that is being prepared. We have reviewed some of the rationale and consumer research behind nutrition and health labelling. This research is still in rudimentary stage and several authors have pointed out shortcomings and biases in existing research on consumer evaluation of nutrition information. For example, the vast majority of this research has been conducted in the US, the UK and some other north western European countries. There is a strong need for more research in a wider variety of countries and also of a comparative nature as consumer understanding, perception and preference for nutrition information is like to vary with culture (Williams, 2005; Cowburn and Stockley, 2005). Also, much of the research does not use very strong research designs. For example, of the 103 papers on consumer understanding of nutrition labelling reviewed by Cowburn and Stockley (2005) only 9% were judged to be of high or medium-high quality. Many findings are based on surveys and focus groups. Although these may provide important insights, it is important that such findings are complemented with experimental studies and outcome studies. Cowburn and Stockley (2005) further indicate a research need for better methodology that would allow us to study consumer understanding and use of nutrition information in real-life situations and also to quantify with the use of more objective methods. Individual and cross-cultural differences should be an important focus when the purpose is to let consumers benefit from nutrition information which would typically require the use of larger and more representative samples. As a final research gap, Cowburn and Stockley (2005) argue that more research is needed on consumer motivation to use and understand food labels and on interventions that would enhance understanding and use also in relation to diet quality. Research on framing of nutrition information would add to an understanding of the consumer responsiveness to nutrition and health claims and nutrition information more generally (Van Trijp and Van der Lans, 2006a; Van Kleef et al., 2005). Finally the literature suggests that the effect of claims and labels are dependent on level of education, with higher educated consumers benefiting from the informational label and lower educated consumers responding to the claim.

6.7   References

AAKER, D.A. 1991, *Managing Brand Equity.* New York, Free Press.

AAKER, D.A. 1996, *Building Strong Brands*. New York, Free Press.

AAKER, J. 1997, Dimensions of brand personality. *Journal of Marketing Research*, XXXIV (August): 342–352.

AAKER, D.A. and K.L. KELLER, 1990, Consumer evaluations of brand extensions. *Journal of Marketing* 54: 27–40.

AC NIELSEN 2005, *The Power of Private Labels in Europe: an insight into consumer attitudes*: www.acnielsen.com.

AGARWAL, M.K. and V. RAO, 1996, An Empirical Comparison of Consumer-Based Measures of Brand Equity, *Marketing Letters*, 7(3): 237–247.

AILAWADI, K.L. and K.L. KELLER, 2004, Understanding retail branding: conceptual insights and research priorities. *Journal of Retailing* 80 (4): 331–342.

AILAWADI, K.L., D.R. LEHMANN and S.A. NESLIN, 2003, Revenue Premium as an outcome measure of brand equity. *Journal of Marketing* 67 (October), 1–17.

AKBAY, C. and E. JONES, 2005, Food consumption behavior of socio-economic groups for private labels and national brands. *Food Quality and Preference* 26, 621–631.

AMA 2006. The Marketing Dictionary on-line [http://www.marketingpower.com/mg-dictionary.php].

ANANIA, G. and R. NISTICÒ, 2004, Public regulation as a substitute for trust in quality food markets: what if the trust substitute cannot be fully trusted?   *Journal of Institutional and Theoretical Economics* 160 (4): 681–701.

ANDERSON J.R. 1983, A spreading activation theory of memory. *Journal of Verbal Learning and Verbal Behavior*, 22 (June): 261–295.

ASCH, S.E. 1956. Studies of independence and conformity. A minority of one against a unanimous majority. *Psychological Monographs*, 70(9), Whole no. 416.

BANDURA, A. 1977. *Social Learning Theory*. Prentice Hall. Englewood Cliffs, NJ.

BARWISE, P. 1993, Brand equity: Snark or Boojum? *International Journal of Research in Marketing*, 10(1), 93–104.

BELLO, D.C. and M.B. HOLBROOK, 1995, Does an absence of brand equity generalize across product classes? *Journal of Business Research* 34: 125–131.

BERNER, R. and D. KILEY, 2005, Global brands: Interbrand annual report. *Business Week* August 1, 2005, 86–94. See also interbrand.com.

BHAT, S. and S.K. REDDY, 2001, The impact of parent brand attribute associations and affect on brand extension evaluation. *Journal of Business Research* 53 (3): 111–122.

BJØRNER, T.B., L.G. HANSEN *et al.*, 2004, Environmental labelling and consumers' choice – an empirical analysis of the effect of the Nordic Swan. *Journal of Environmental Economics and Management* 47 (3): 411–434.

BLOIS, K. 2000, *The Oxford Textbook of Marketing*. Oxford, Oxford University Press.

BREDAHL. L. 2003, Cue utilisation and quality perception with regard to branded beef. *Food Quality and Preference* 15: 65–75.

BOUSH D.M. and B. LOKEN, 1991, A process-tracing study of brand extension evaluation. *Journal of Marketing Research* 28(1): 16–28.

BRUCKS, M., A.A. MITCHELL and R. STAELIN, 1984, The effect of nutritional information disclosure in advertising: an information processing approach. *Journal of Public Policy and Marketing* 3: 1–27.

CARDELLO A.V. and H.G. SCHUTZ, 1996, Food appropriateness measures as an adjunct to consumer preference/acceptability evaluation. *Food Quality and Preference* 7(3–4):

239–249.

CARNEIRO, J.D.S., V.P.R. MINIM, R. DELIZA, C.H.O. SILVA, J.C.S. CARNEIRO, O. and F.P. LEAO, 2005, Labeling effects on consumer intention to purchase for soybean oil. *Food Quality and Preference.*, 16(3) 275–282.

CASWELL, J.A. 1998, How labelling of safety and process attributes affects markets for food. *Agricultural and Resource Economics Review* 27: 151–158.

CHANDON, P. 2003, *Note on measuring brand awareness, brand image, brand equity and brand value.* INSEAD Working paper, Fontainebleau.

CHAUDHURI A. and M.B. HOLBROOK, 2001. The chain of effects from brand trust and brand affect to brand performance. The role of brand loyalty. *Journal of Marketing* 65(2): 81–93.

CHEFTEL, J.C. 2005, Food nutrition labelling in the European Union. *Food Chemistry*, 93: 531–550.

COWBURN, G. and L. STOCKLEY, 2005, Consumer understanding and use of nutrition labelling: a systematic review. *Public Health Nutrition.* 8(1): 21–28.

DARBY, M.R. and E. KARNI, 1973, Free competition and the optimal amount of fraud. *Journal of Law and Economics*, XVI(1): 67–88.

DE CHERNATONY, L. 1992, *Creating powerful brands.* Oxford: Butterworth/Heinemann.

DE PELSMACKER, P., W. JANSSENS *et al.*, 2005, Consumer preferences for the marketing of ethically labelled coffee. *International Marketing Review* 22(5): 512–530.

DILLON, W.R., T.J. MADDEN, A. KIRMANI and S. MUKHERJEE, 2001, Understanding what's in a brand rating: A model for assessing brand and attribute effects and their relationship to brand equity. *Journal of Marketing Research* 38, 415–429.

DOMIZLAFF, H. 1939, *Die Gewinnung des offentlichen Vertrauens.* Hamburg: HVA .

EJN 2003, PASSCLAIM Process for the assessment of scientific support for claims on food. *European Journal of Nutrition* 42, Supplement 1, 1/1–1/119.

ENNEKING, U., C. NEUMANN and S. HENNEBERG, 2005, How important intrinsic and extrinsic product attributes affect purchase decisions. *Food Quality and Preference.* doi: 10.1016/j.foodqual.2005.09.008.

ERDEM, T and J. SWAIT, 1998, Brand equity as a signaling phenomenon. *Journal of Consumer Psychology.* 7 (April): 131–157.

EU 2003, Draft proposal for a regulation of the European Parliament and the council on nutritional and health claims made on foods. Commission of the European Communities. Brussels: June 2003: http://europe.eu.int/eur-lex/en/com/pdf/2003/com2003/0424en01.pdf. [2005 update not yet public].

EUROPEAN HEART NETWORK 2003, Systematic Review on the research of consumer understanding of nutrition labelling. see: *http://www.ehnheart.org/files/consumer%20nutrition-143058A.pdf.*

FOURNIER S. 1998, Consumers and their brands: developing relationship theory in consumer research. *Journal of Consumer Research*, 24(4): 343–373.

GRUNERT, K.G. 2005, Food quality and safety: consumer perception and demand. *European Review of Agricultural Economics* 32(3) 369–391.

HANKINSON, G. 2000, Brand management. In: K. Blois, *The Oxford Textbook of Marketing.* Oxford, Oxford University Press, pp. 481–499.

HAUSER J.R. and D. CLAUSING, 1988, The House Of Quality. *Harvard Business Review* 66(3): 63–73.

HAWKES, C. 2004, Nutrition Labelling and Health Claims: the global regulatory environment. Geneva: World Health Organization. http://whqlibdoc.who.int/publications/2004/9241591714.pdf.

HILL and KNOWLTON, 2000, Study on nutritional, health and ethical claims in the European Union. Brussels: Hill and Knowlton. Prepared for the European Commission Directorate General for Health and Consumer Protection. April 2000. http://europa.eu.int/comm/consumers/cons_int/safe_shop/fair_bus_pract/ green_pap_comm/studies/nutri_claim_en.pdf

HOWARD, J.A. 1994, *Buyer Behaviour in Marketing Strategy* (2nd edn). Englewood Cliffs, NJ: Prentice Hall.

HOYER, W.D. and S.P. BROWN, 1990, Effects of brand awareness on choice for a common, repeat purchase product. *Journal of Consumer Research* 17(2): 141–148.

JACOBY, J., R.W. CHESTNUT and W. SILBERMAN, 1977, Consumer use and comprehension of nutrition information. *Journal of Consumer Research* 4 (2): 119–128.

JAEGER, S.R. 2006, Non sensory factors in sensory science research. *Food Quality and Preference* 17, 132–144.

JAFFRY, S., H. PICKERING *et al.*, 2004, Consumer choices for quality and sustainability labelled seafoods in the UK. *Food Policy* 29(3): 215–228.

JANISZEWSKI CHR. and S.M.J. VAN OSSELAER, 2000, A Connectionist Model of Brand-Quality Associations. *Journal of Marketing Research* 37 (August): 331–350.

KARDES, F.R. 2001, *Consumer Behavior and Managerial Decision Making*. Upper Saddle River, NJ: Prentice Hall.

KARDES, F.R. 2002, *Consumer Behaviour and Managerial Decision Making* (2nd edn). Upper Saddle River, NJ: Prentice Hall.

KELLER, K.L. 1993, 'Conceptualizing, measuring, and managing customer-based brand equity', *Journal of Marketing*, 57 (January), 1–22.

KELLER, K.L. 1998, *Strategic Brand Management: Building, Measuring and Managing Brand Equity*. Upper Saddle River, NJ: Prentice Hall.

KELLER, K.L. 2000, The brand report card. *Harvard Business Review* (Jan–Feb): 147–157.

KELLER, K.L. 2001, *Building customer-based brand equity: a blueprint for creating strong brands.* MSI working paper 01-107.

KELLER, K.L. 2003, *Strategic Brand Management*. Upper Saddle River, NJ: Prentice Hall.

KELLER, K.L. 2006, Measuring brand equity. In: Grover, and M. Vriens (eds), *Handbook of Marketing research: uses, misuses and future advances*. Thousand Oaks, CA: Sage Publications (forthcoming).

KELLER, K.L. and D.A. AAKER, 1992, The effects of sequential introduction of brand extensions. *Journal of Marketing Research* 29 (Feb): 35–50.

KELLER, S.B., M. LANDRY, J. OLSON, A.M. VELLIQUETTE, S. BURTON and J.C. ANDREWS, 1997, The effects of nutrition package claims, nutrition facts panels, and motivation to process nutrition information on consumer product evaluations. *Journal of Public Policy and Marketing* 16(2): 256–269.

KLEIN, N. 2000, *No Logo, No Space, No Choice, No Jobs*. London: Flamingo.

KOTLER, P. and K.L. KELLER, 2006, *Marketing Management*, 12th edn. Upper Saddle River, NJ: Pearson Education Inc.

KRISHNAN, H.S. 1996, Characteristics of memory associations: a consumer-based brand equity perspective. *International Journal of Research in Marketing* 13: 389–405.

LASSAR, W., B. MITTAL and A. SHARMA, 1995, Measuring customer-based brand equity. *Journal of Consumer Marketing* 12(4): 11–19.

LAWRENCE, M. and M. RAYNER, 1998, Functional foods and health claims: a public health policy perspective. *Public Health Nutrition* 1(2): 75–82.

LINDSTROM, M. 2005, Broad sensory branding. *Journal of Product and Brand Management* 14(2): 84–87.

LOUREIRO, M.L. and J. LOTATE, 2005, Do fair trade and eco-labels in coffee wake up the consumer conscience? *Ecological Economics* 53(1): 129–138.

MARTIN, I.M. and D.W. STEWART, 2001, The differential impact of goal congruency on attitudes, intentions, and the transfer of brand equity. *Journal of Marketing Research* 38(4): 471.

MARTIN, I.M., D.W. STEWART and S. MATTA, 2005, Branding strategies, marketing communication and perceived brand meaning: The transfer of purposive, goal-oriented brand meaning to brand extensions. *Journal of the Academy of Marketing Science* 33(3): 275–294.

MAZIS, M.B. and M.A. RAYMOND, 1997, Consumer perceptions of health claims in advertisements and on food labels. *The Journal of Consumer Affairs* 31(1): 10–26.

McCRACKEN, G. 1986, Culture and consumption: a theoretical account of the structure and movement of cultural meaning of consumer goods. *Journal of Consumer Research* 13 (1: June), 71–84.

MEYVIS, T. and C. JANISZEWSKI, 2004, When are broader brands stronger brands? An accessibility perspective on the success of brand extensions. *Journal of Consumer Research* 31(2): 346–357.

MITCHELL, A. 2001, The camel, the cuckoo and the reinvention of win-win marketing. *The Journal of Brand Management* 8(4/5): 255–269.

MOORMAN, CH. 1990, The effects of stimulus and consumer characteristics on the utilization of nutrition information. *Journal of Consumer Research* 17 (3): 362–374.

NEISSER, U. 1976, *Cognition and Reality: Principles and implications of cognitive psychology.* San Francisco: Freeman.

NETEMEYER, R.G., B. KRISHNAN, C. PULLIG, G. WANG, M. YAGCI, D. DEAN, J. RICKS and F. WIRTH, 2004, Developing and validating measures of facets of customer-based brand equity. *Journal of Business Research.* 57: 209–224.

PARK, C.S. and V. SRINIVASAN, 1994, A survey-based method for measuring and understanding brand equity and its extendibility. *Journal of Marketing Research* 31, 271–288.

PORTER, M.E. 1985, *Competitive Advantage: Creating and Sustaining Superior Performance.* New York: The Free Press.

PRZYREMBEL, H. 2004, Food labelling legislation in the EU and consumers information. *Trends in Food Science and Technology* 15: 360–365.

READER'S DIGEST 2005, *Reader's Digest European trusted brands 2005.* [http://www.rdtrustedbrands.com/index.shtml].

RICHARDSON, D.P., T. AFFERTSHOLT, N-G. ASP *et al.,* 2003, PASSCLAIM – synthesis and review of existing processes. *European Journal of Nutrition* 42 [Suppl 1], 1/96–1/111.

RIEZEBOS, H.J. 1994, *Brand Added Value: Theory and Empirical Research about the Value of Brands to Consumers.* Delft: Eburon.

RIEZEBOS, R. 2003, *Brand Management: a theoretical and practical approach.* Harlow, Essex: Pearson Education Limited.

ROE, B.E., A.S. LEVY and B.M. DERBY, 1999, The impact of health claims on consumer search and product evaluation outcomes: results from FDA experimental data. *Journal of Public Policy and Marketing* 18(1), 89–115.

RUSSO, J.E., R. STAELIN, C.A. NOLAN, G.J. RUSSELL and B.L. METCALF, 1986, Nutrition information in the supermarket. *Journal of Consumer Research* 13(1): 48–70.

RUST R.T., V.A. ZEITHAML and K.N. LEMON, 2004, Customer-centered brand management. *Harvard Business Review* 82(9): 110.

SIMON C.J. and M.W. SULLIVAN, 1993, The measurement and determinants of brand equity – a financial approach. *Marketing Science* 12(1): 28–52.

SOLOMON, M.R. 2002, *Consumer Behavior*. Upper Saddle River, NJ: Prentice Hall.

SRINIVASAN, V., CH.S. PARK and D.R. CHANG, 2005, An approach to the measurement, analysis, and prediction of brand equity and its sources. *Management Science* 51(9): 1433–1448.

SRIVASTAVA R. and D.J. REIBSTEIN, 2005, *Metrics for linking marketing to financial performance*. MSI special report 05-200.

STEENKAMP J.B.E.M. 1990, Conceptual model of the quality perception process. *Journal of Business Research* 21(4): 309–333.

STEENKAMP J.B.E.M., H.C.M. VAN TRIJP and J.M.F. TEN BERGE, 1994, Perceptual mapping based on idiosyncratic sets of attributes. *Journal of Marketing Research* 31(1): 15–27.

SUPPHELLEN, M. 2000, Understanding core brand equity: guidelines for in-depth elicitation of brand associations. *International Journal of Market Research* 42(3): 319–338.

TEISL, M.F. and B. ROE, 1998, The economics of labelling: an overview of issues for health and environmental disclosure. *Agricultural and Resource Economics Review* 27: 140–149.

THOMSON, M.D., D.J. MACINNIS, *et al.*, 2005, The ties that bind: measuring the strength of consumers' emotional attachments to brands. *Journal of Consumer Psychology* 15(1): 77–91.

TSAI, S-P. 2005, Utility, cultural symbolism and emotion: A comprehensive model of brand purchase value. *International Journal of Research in Marketing* 22(3): 277–291.

VAN KLEEF E., H.C.M. VAN TRIJP and P. LUNING, 2005, Consumer research in the early stages of new product development: a critical review of methods and techniques. *Food Quality and Preference* 16(3): 181–201.

VAN OSSELAER, S.J.M. 2004, *Of rats and brands: a learning-and-memory perspective on consumer decisions*. Inaugural address Erasmus University Rotterdam.

VAN OSSELAER, S.J.M. and J.W. ALBA, 2003, Locus of equity and brand extensions. *Journal of Consumer Research* 29: 539–550.

VAN OSSELAER, S.J.M. and CHR. JANISZEWSKI, 2001, Two ways of learning brand associations. *Journal of Consumer Research* 28 (September): 202–223.

VAN TRIJP, H.C.M. and J.B.E.M. STEENKAMP, 1997, Quality labelling as instrument to create product equity: the case of IKB in the Netherlands. In: Wierenga, B., A. van Tilburg, K.G. Grunert, JB-EM Steenkamp and M. Wedel (eds), *Agricultural Marketing and Consumer Behaviour in a Changing World*. Boston: Kluwer Academic Publishers.

VAN TRIJP, H.C.M. and I. VAN DER LANS, 2006a, Consumer perceptions of nutrition and health claims. *Appetite* (in press).

VAN TRIJP, H.C.M. and I. VAN DER LANS, 2006b, *Individual differences in consumer perceptions of nutrition and health claims*. Wageningen University, Marketing and Consumer Behaviour Group working paper.

VEISTEN, K. and B. SOLBERG, 2004, Willingness to pay for certified wooden furniture: a market segment analysis. *Wood and Fiber Science*. 36(1): 40–55.

VERBEKE, W. 2005, Agriculture and the food industry in the information age. *European Journal of Agricultural Economics* 33(3): 347–368.

WARLOP, L., S. RATNESWAR and S.M.J. VAN OSSELAER, 2005, Distinctive brand cues and memory for product consumption experiences. *International Journal of Research in Marketing* 22: 27–44.

WILLIAMS, P. 2005, Consumer understanding and use of health claims for foods. *Nutrition Reviews* 63(7): 256–264.

YOO, B. and N. DONTHU, 2001, Developing and validating a multidimensional consumer-based brand equity scale. *Journal of Business Research* 52: 1–14.

ZALTMAN, G. and R.H. COULTER, 1995, Seeing the voice of the customer: Metaphor-based advertising research. *Journal of Advertising Research* 35(4): 35–51.

7

How consumers perceive food quality

K. G. Grunert, Aarhus School of Business, Denmark

7.1 Introduction

Quality is a ubiquitous term when talking about food and consumers. Most food manufacturers usually maintain that they produce food of good quality. At the same time, manufacturers sometimes complain that consumers are not willing to pay for good quality. Consumers sometimes complain that the quality of food is not what it used to be, that the quality of products is not good enough in relation to the price, that really good quality is hard to obtain.

Everybody agrees that quality is a good thing. We want quality. If we hold price constant, more quality is always better. Quality is also a matter of degree. We talk about more or less quality, about good quality and bad quality. In other words, quality is a central term in the *evaluation* of food products.

But quality is also an evasive concept. If manufacturers complain that consumers don't appreciate the quality of their products, and consumers at the same time say that good quality is hard to obtain, this could indicate that they are not talking about the same thing. Quality is not self-evident. We cannot look at a food product and immediately conclude whether it is high or low quality. Just by looking at the product, we will often be uncertain about the quality. And different consumers may have different opinions about the quality of the same product.

When quality is a central term in the evaluation of food products, and at the same time is an evasive concept that may be viewed differently by different people in different situations, understanding the way in which consumers perceive quality becomes an important topic. Only when we understand how consumers perceive quality in food products can we direct food production into directions consumers will appreciate. The high failure rate of new food products introduced in the market shows that many manufacturers find it difficult to understand consumer food quality perception. A good understanding of consumer quality perception is therefore a prerequisite for successful product development in the food sector. But understanding consumer food quality perception is also of relevance from other perspectives. From a health policy perspective, where we are interested in getting consumers to make healthier choices, it is important to understand how the health component enters consumer quality perception. From an environmental perspective, it is interesting to understand why some consumers see an additional quality in that some food is organically produced, while others don't. For regulators working with questions of food labelling, it is interesting to understand how food labels affect consumer quality perception.

This chapter deals with consumer quality perception. It is based on research that has been done trying to understand the role of quality in consumer choices. This research is spread across different areas, noticeably marketing, psychology, and agricultural economics. We start with a brief introduction to the term 'quality', settling for a comprehensive approach. We then go through five basic propositions on the way in which consumers perceive food quality. We summarize them in a model, the Total Food Quality Model, that we regard as a useful framework for analysing consumer food quality perception. We close with some speculation about future trends.

7.2   Defining food quality

There is an abundance of ways in which the term *quality*, both in food and otherwise, has been defined (see the special issue of *Food Quality and Preference* in 1995 for a broad range of proposals). There is general agreement that quality has an *objective* and a *subjective* dimension. Objective quality refers to the physical characteristics built into the product and is typically dealt with by engineers and food technologists. Subjective quality is the quality as perceived by consumers. The relationship between the two is at the core of the economic importance of quality: only when producers can translate consumer wishes into physical product characteristics, and only when consumers can then infer desired qualities from the way the product has been built, will quality be a competitive parameter for food producers.

In the subjective realm we can, as a gross simplification, distinguish between two schools of thought about quality. The first one, which we can term the *holistic* approach, equates quality with all the desirable properties a product is perceived to have. The second, which we can term the *excellence* approach, suggests that products can have desirable properties which consumers, in their own language, may not view as part of quality. In food, convenience is sometimes named as an example: consumers may say that 'convenience goods are generally of low quality', even though they regard convenience as a desirable property of food products (see, e.g., Olsen, 2002; Zeithaml, 1998). In the following, we will use the holistic approach, i.e. we regard quality as encompassing all the desirable properties a consumer expects from a food product. Framed in another way, quality is what the consumer wants to *get* out of buying a food product – as opposed to what the consumer has to *give* in order to get, like paying a price.

7.3   Five propositions on how consumers perceived food quality

In the following, we present five propositions that we believe characterize the way in which consumers perceived food quality. These are synthesized from a broad range of research on quality perception, both in food and in other areas.

7.3.1   Quality perception is based on inferences

When asking consumers in an open-ended interview what they regard as food products of good quality, the answers always radiate around four central concepts: taste (and other sensory characteristics), health, convenience, and – for some consumers – process characteristics like organic production, natural production, animal welfare, GMO-free, etc. Let us assume that a consumer wants to evaluate a main meal ingredient on these major quality dimensions while shopping. How can the consumer do this? If the product is a new one, the consumer cannot draw on previous experience, and hence all aspects of quality are *a priori* unknown.

Quality dimensions are thus characterized by *uncertainty*. This is the rule more than the exception. But since a purchase decision has to be made nevertheless, the consumer forms expectations about quality based on the information available at the time of purchase. The quality itself is not known, but the consumer *infers* it from the information available. In the literature on quality perception, information used to infer quality is usually called *cues*.

Inference making in the quality perception process is one of the more mysterious areas of consumer behaviour, and the literature abounds with more or less well-documented cases of bizarre inferences: consumers seem to use the smell of stockings as a cue for inferring longevity, the viscosity of a cleaning product to infer cleaning power, and the strength of the speeder spring of a car to infer engine power (see Peter *et al.*, 1999). In the food area, consumers are known to use colour and fat content of meat as an indicator of taste and tenderness, organic production as an indicator of superior taste of vegetables, and animal welfare as an indicator of more healthy products – all inferences which are, from an objective point of view, at least questionable.

Two basic principles are useful in trying to understand the way in which consumers make inferences. They go back to the Sorting Rule Model which Cox published back in 1967 (Cox, 1967), and they say that consumer prefer cues (a) which they believe to be *predictive* of the quality they want to evaluate, and (b) which they feel *confident* in using.

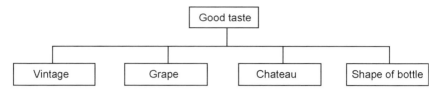

Fig. 7.1   Inferences when buying wine – example.

Figure 7.1 shows a constructed example which can illustrate these principles. We are looking at a consumer who is buying a bottle of wine; the quality he is interested in is a good taste, based on enjoyment as the major purchase motive. The taste of the wine is unknown at the time of purchase (unless one has bought the same wine before), but lots of cues are available on the bottle, and in a wine shop most salespeople will be happy to provide additional ones. Figure 7.1 shows only a small selection of these cues: vintage, grape, chateau and shape of bottle. Which of them will the consumer use in forming an expectation about taste?

Most wine drinkers will have an idea about the relative predictive ability of these four cues. They will probably believe that of the four, the grape is most predictive of taste, the shape of the bottle is least predictive, and the other are in between. So based on predictive value, we would expect the consumer to make an inference mainly based on the cue *grape*.

However, the consumer may not feel confident in making an inference based on this cue. There is a Merlot and a Shiraz, both of which the consumer has heard about before, and the consumer may remember that one is round and mellow and the other strong and fruity, but which was which? There is also a Pinotage, a Zinfandel and a Lemberger, grapes which the consumer has never heard about. Thus, the consumer is not *confident* in his ability to make an inference based on this cue.

The concept of confidence is thus strongly linked to knowledge and expertise of the consumer (Selnes and Troye, 1989). A real wine expert will have no problems in making inferences based on grape, vintage, and chateau. But many other wine drinkers will, and they may end up making an inference based on the shape of bottle, because they feel confident in making an evaluation of the shape – knowing that the predictive ability of shape with regard to taste is quite limited. The basic principle is that consumers prefer cues which they believe have high predictive power, but that lack of confidence in using the cue can veto its use – and the consumer scans cues of subsequently less and less predictive power until one is found that he is confident in using. This principle can explain, if not all, then at least a good deal of the strange cases we have heard about cue inference making.

One particular cue that warrants special mention is the brand. Not all food products are branded, but even fresh produce like meat and vegetables are branded to an increasing extent, and in addition to manufacturers' brands there are other brand-like symbols which are issued by groups of manufacturers, regions, countries, independent bodies, etc. These are sometimes called quality labels or generic trade marks. Brands and related symbols can be powerful cues in making quality inferences. They are powerful cues for consumers to the extent that consumers actually find them predictive of the quality of the product, and once a brand has been well-established, consumers usually also feel confident in using it. To the extent that a brand is widely used for quality inference, it accumulates *brand equity*, i.e. becomes a valuable asset for the manufacturer owning it. There is a sprawling mass of literature on branding and brand equity (two good introductory sources are Erdem and Swait, 1998; Keller, 1993), but the main conclusions there are in line with what we stated above: that brands are valuable to the extent that consumers confidently use them to make predictions about quality of the product.

Inference making characterizes quality perception not only at the purchase stage. Once the product has been bought and taken home, more inferences about quality can take place. When the product is being prepared, more cues may become available – like the touch of a piece of meat, the smell of a cheese, the pouring characteristics of sea salt, all of which may lead to new inferences about quality. And while some quality aspects, noticeably the taste of the food, eventually become amenable to direct experience, many others do not: whether the product is healthy or not will even after consumption be a matter of inferences, and perception of healthiness may, for example, have been affected by the experience that the taste was good, because healthy products are not expected to have a good taste.

7.3.2   Quality perception is related to underlying values and attitudes

Quality is something that is desirable, but why? With some aspects of quality, the answer seems so self-evident that the question is rarely asked. Of course everybody prefers a good taste to a bad taste, and of course a healthy product is better than a non-healthy one – or is it? Here we already run into exceptions, because there are clearly cases where consumers *prefer* the unhealthy alternative, like in cases where one wants to *indulge* in a cream cake, and the fact that it is unhealthy and expensive adds to the attraction. When we get to the quality dimensions, convenience and process characteristics, the answers are not self-evident at all, since convenience – saving time and effort in preparing meals – is attractive for some and not for others, and may even for the same person be attractive in some situations but not in others. Can we explain why and when certain dimensions of quality become attractive to consumers?

There are streams of research that have attempted to come up with answers to this question. One is the *means-end theory* of consumer behaviour (Gutman, 1982; Mulvey *et al.*, 1994; Peter *et al.*, 1999; Pieters *et al.*, 1995; Walker and Olson, 1991; Zeithaml, 1998). The basic assumption of means-end theory is that consumers are not interested in products per se, but in what the product is doing for them – in the self-relevant consequences of the product, in the way the product helps them attain their life values. Whether a consumer finds a product attractive will therefore depend not only on whether the consumer infers that the product has desirable qualities, but also on the extent to which these qualities are perceived to contribute to the attainment of that consumer's life values. In a means-end chain context, food quality, as perceived by consumers, is a *bridging concept* – by forming impressions of the quality of a product, consumers form a judgement on whether the characteristics of the product, as they have been perceived, will help in attaining that consumer's life values.

The means-end approach has been widely used in studying consumer food choice (e.g., Bech-Larsen, 2001; Bredahl, 1999; Fotopoulus *et al.*, 2003; Grunert and Grunert, 1995; Grunert et al., 2001b; Jaeger and MacFie, 2000; Miles and Frewer, 2001; Nielsen et al., 1998; Russell et al., 2004; Valette-Florence *et al.*, 2000). Results from studies employing this approach are usually presented in so-called *hierarchical value maps*, and Fig. 7.2 shows an example from a study on consumers' choice or non-choice of fresh fish as the mainstay of an evening meal (from Nielsen *et al.*, 1997). If we look at the middle part of the diagram, we can see that quality perception of fresh fish centres around three major dimensions: enjoyment of eating (mainly related to taste), health aspects (mainly related to the content of vitamins and minerals), and a perceived lack of convenience (because the fish is difficult to prepare and has to be bought at a fishmonger). At the bottom of the diagram we see the concrete product characteristics from which these quality dimensions are inferred. At the top we see the basic life values which motivate consumers' choice or non-choice of fresh fish.

There is a good deal of general research on life values that can be brought to the area of quality perception. Life values have been defined, by two of the most prominent researchers in the field, as: (1) concepts or beliefs (2) about desirable end states or behaviours (3) that transcend specific situations, (4) guide the selection or evaluation of behaviour and events, and (5) are ordered by relative importance (Schwartz and Bilsky, 1987). Life values thus provide motivation to select and choose between different options across a range of different life situations. Various researchers have tried to come up with universally valid catalogues of people's life values (the most prominent examples being Kahle, Rokeach, and Schwartz; see Kahle *et al.*, 1986; Rokeach, 1973; Schwartz, 1992). Figure 7.3 shows the system of life values as proposed by Schwartz: based on comprehensive research in many countries of the world, he concludes that people's life values can be structured into ten domains, and that these domains are organized as in Fig. 7.3, namely along two dimensions: tradition vs. innovation and individual vs. collective. While the ten domains claim to be universal, their relative importance will differ among individuals.

Life values can be viewed as the most abstract concepts in a system of attitudes explaining what a consumer regards as desirable qualities of a food product and what not. Such systems of attitudes have been investigated especially in the context of consumers' interest in process characteristics – characteristics of the way in which a food product has been produced, such as organic production, production including or excluding genetically modified organisms, fair trade products, production with regard to animal welfare. For

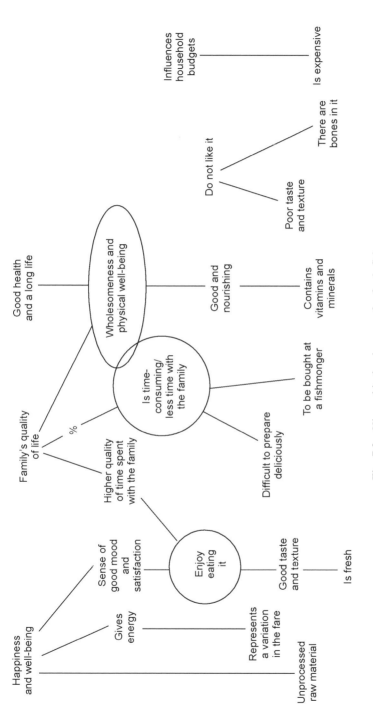

Fig. 7.2 Hierarchical value map for fresh fish.

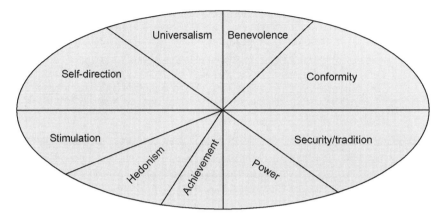

Fig. 7.3   Life value domains according to Schwartz (1992).

example, it has been shown that consumers critical with regard to the use of genetic modification have positive attitudes to environment and nature and negative attitudes to technological progress; such attitudes are linked both to general life values like universalism and benevolence and to the perception of quality of products where genetic modification has been involved (Bredahl, 2001; Grunert *et al.*, 2003; Søndergaard *et al.*, 2005).

7.3.3   Quality perception is an expression of lifestyle

The four basic dimensions of quality mentioned earlier – sensory characteristics, health, convenience and process characteristics – are probably quite universal, but otherwise the process of quality perception is characterized by individual differences: not only will there be differences in the relative importance of the quality dimensions, but also in the way they are inferred from available cues, in the way consumers shop and are thus exposed to various kinds of quality cues, and in the way they prepare and eat their meals, with resulting differences in the quality experienced during consumption. Furthermore, the systems of life value and attitudes driving the food choice and quality perception process will differ between consumers.

These differences can be viewed as part of different *lifestyles*. Lifestyle has been a popular concept in segmenting consumers, and with increasing fragmentation of consumer lifestyles, one has increasingly adopted domain-specific approaches to lifestyle, where lifestyle is analyzed only with regard to a certain life domain, like food (van Raaij and Verhallen, 1994). There has been a good deal of work trying to categorize consumers according to their *food-related lifestyle* (Brunsø and Grunert, 1998; Grunert *et al.*, 1997, 2001a), which is defined as the general pattern of how consumers use food to fulfil basic motives or attain life values, and of which quality perception is an important component. Food-related lifestyle can be measured by means of a questionnaire that has been extensively tested for cross-cultural validity, i.e. for its ability to obtain results which can be compared even though respondents come from different countries, cultures, and language areas (Scholderer *et al.*, 2004). Extensive research on consumers' food-related lifestyle in a number of European countries (Grunert *et al.*, 2001a), and also some countries outside Europe (Askegaard and Brunsø, 1999; Reid *et al.*, 2001), has established a number of basic food consumer segments:

- *The uninvolved food consumer.* For these consumers, food is not a central element in their lives. Consequently, their purchase motives for food are weak, and their interest in food quality is limited mostly to the convenience aspect. They are also uninterested in most aspects of shopping, don't use specialty shops, and don't read product information, limiting their exposure to and processing of food quality cues. Even their interest in price is limited. They have little interest in cooking, tend not to plan their meals, and snack a great deal.
- *The careless food consumer.* In many ways, these consumers resemble the uninvolved food consumer, in the sense that food is not very important to them, and, with the exception of convenience, their interest in food quality is correspondingly low. The main difference is that these consumers are interested in novelty: they like new products and tend to buy them spontaneously, at least as long as they don't require a great effort in the kitchen or new cooking skills.
- *The conservative food consumer.* For these consumers, the security and stability achieved by following traditional meal patterns is a major purchase motive. They have a major interest in the taste and health of food products, but are not particularly interested in convenience, since meals are prepared in the traditional way and regarded as part of the woman's tasks.
- *The rational food consumer.* These consumers process a lot of information when shopping; they look at product information and prices, and they use shopping lists to plan their purchases. They are interested in all aspects of food quality. Self-fulfilment, recognition and security are major purchase motives for these consumers, and their meals tend to be planned.
- *The adventurous food consumer.* While these consumers have a somewhat above-average interest in most quality aspects, this segment is mainly characterized by the effort they put into the preparation of meals. They are very interested in cooking, look for new recipes and new ways of cooking, involve the whole family in the cooking process, are not interested in convenience and reject the notion that cooking is the woman's task. They want quality, and demand good taste in food products. Self-fulfilment in food is an important purchase motive. Food and food products are an important element in these consumers' lives. Cooking is a creative and social process for the whole family.

7.3.4   Quality perception changes over time

It was noted above that the evaluation of food quality at the point of purchase is characterized by uncertainty, and that quality is therefore inferred by cues. We also noted that these inference processes do not stop once the product has been bought – during preparation new cues may become accessible, and even once the product has been consumed new information may affect the perception of, for example, the product's healthiness.

Some quality dimensions, however, are amenable to experience once the product has been bought and prepared. This goes for the sensory characteristics, and for the convenience aspects. This means that the quality expectations that have been formed based on inferences from quality cues can be confirmed or refuted. Confirmation or non-confirmation of expectations is a major determinant of consumer satisfaction and of consumer intent to repurchase the product or not. Especially for new products, which have been bought for the first time and where the formation of expectations at the point of purchase cannot be based on own previous experience, the confirmation or non-confirmation of expectations is a crucial factor in the success or otherwise of the product.

Whether quality expectations will be confirmed or not is, of course, a question of how good the consumer was in predicting the quality based on the cues at hand. Given the discussion about inference making above, it may not come as a surprise that very often consumers are not especially good at predicting the quality, with non-confirmation of expectations and dissatisfaction as a consequence. Figure 7.4 shows an example of this. Consumers evaluated three types of steak based on their visual appearance, as they would when the meat is in a cooling disk. They then received samples of all three types of steak to take home, prepare, and eat, on consecutive days. The steaks differed in the degree of fattening up of the animals before slaughtering (details are in Brunsø et al., 2005). As the figure shows, the quality experience points in the opposite direction to the expectations: the steak with the highest expectations was actually the least liked after consumption. The discrepancy can be traced back to the way cues were used for inference making: consumers took visible fat as the main cue for evaluating quality, inferring that more visible fat means lower quality. Actually, higher degrees of intramuscular fat lead to more taste and tenderness, which explains the opposite results.

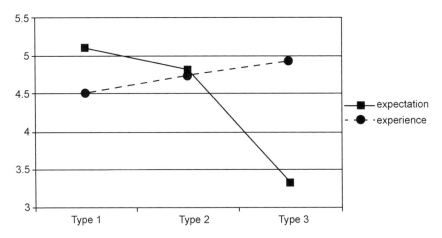

Fig. 7.4   Quality expectations and experience for three meat types.

We have thus seen that quality perception changes over time. It changes most notably from the pre-purchase phase to the post-consumption phase. But the distinction of only two phases – before and after purchase – is a simplification. Quality perception can change continuously over time, as new cues become available and as new experiences are being made. We may smell or feel a cheese on the way home, which we did not dare to do in the shop. We may experience how the appearance of the product deteriorates during storage at home. During preparation we will experience the degree of convenience in assembling the product into a meal, and there may be new sensory cues in terms of smell and texture. Even after eating, the quality perception may change, like when we get sick and attribute it to the product, or when we are confronted with new information about the healthiness of the product.

In addition, a single purchase does not stand alone. Most food products are bought several times, and many are bought continuously over extended periods. Quality perception may change across the whole range of purchases. We usually assume, though, that the biggest change occurs in connection with the first purchase, because in the first purchase – the *trial purchase* – quality expectations are necessarily based on informational cues only, not on own experience, and the first purchase will then lead to the first actual experience with the product, which may lead to fundamental changes in the perception of quality. When purchasing the product for the second and subsequent times – the *repeat purchases* – own previous experiences will play a role in forming the quality expectations, which will therefore be more accurate. But changes may still occur, for a variety of reasons. There may occur *learning* with regard to how to handle the product, resulting in better quality experiences. The *situation* in which the product is consumed may change, which may have an impact on the experienced quality. When consumers are variety-seeking and like stimulation and change, a positive quality perception may *wear off* over time. For credence qualities, the quality perception can always change when new information about the quality becomes available.

7.3.5   Perceived quality is actionable

The more we understand how consumers perceive quality, the more we can use that knowledge in developing and marketing new food products. By the way we design a new product, and by the way we communicate about the product, we can *influence* how consumers will perceive the quality of this new product. Therefore, we propose that perceived quality is *actionable*. More specifically, we propose that a good understanding of consumer quality perception will allow a food manufacturer to achieve *positioning for perceived quality, confirmation of expectations*, and *competitive differential* (see also Grunert, 2005).

Positioning for perceived quality
A new food product is only successful to the extent that consumers see some desired qualities in it. More importantly, unless the new product is just a low

cost copy of an existing product, it has to be a quality that goes beyond what is currently on the market. We have identified four major dimensions of quality in food products: sensory quality (mostly taste, but also appearance, smell, texture), healthiness, convenience, and process characteristics, where the latter covers various aspects of the production process (e.g., organic production) that may be of interest to certain consumer segments.

These four dimensions, especially when seen in conjunction, open endless possibilities for product innovation. With regard to product innovation related to taste and other sensory characteristics, consumers' increasing interest in variety and new stimulation creates possibilities for product differentiation. Healthiness covers both food safety and nutrition, but with the rise of the functional food category is also increasingly covering selected positive health benefits, like prevention of cardiovascular diseases. Convenience relates not only to convenience in preparation, but also in shopping, storage, eating, and disposal. As far as consumer interest in process characteristics is concerned, we believe there will be a rising interest of consumers in these matters which will go far beyond narrow concepts like organic production and animal welfare (see also Section 7.5).

It is not enough that a product has certain qualities; consumers must also be able to perceive them. Successful product development means, therefore, also developing the right set of *cues* that consumers can use to infer the presence of the quality. Cues, as we have seen, can be intrinsic (part of the physical product) and extrinsic. The more abstract and intangible the qualities become, as in health- and process-related benefits, the more the consumer is forced to rely on extrinsic, i.e. information cues, and the more the *credibility* of information provided will play a role.

Confirmation of expectations

Consumers may be successfully prompted to try a new product once, but whether they will buy it again, or even become loyal customers, will depend heavily on the post-purchase experience. When expectations which were formed before and during the purchase are not confirmed, consumers will be disappointed and not buy again. Designing new products therefore entails designing them both for the *creation* and for the *confirmation* of experience. This goes both for the physical product and for the communication about it. Special care has to be taken when aspects of the physical product and/or the communication may have different, perhaps even opposing, effects on expectations and actual quality experience.

Competitive differential

Perception of quality and of cost will in most cases be relative concepts anchored by the perception of other products currently on the market. A new product will only be successful when it is perceived as having an advantage compared to known alternatives, in terms of better quality or new types of quality, and only when the trade-off between these qualities and the perceived price is regarded as attractive compared to existing products. Successful new product development is, therefore, also to a large extent a question of identifying the right competitors which will function as benchmarks in the mind of consumers, in order to find out whether the relationship between perceived quality and perceived price is likely to result in a purchase.

7.4   The Total Food Quality Model

The Total Food Quality Model (TFQM), originally proposed by Grunert *et al.* (1996), integrates the five propositions discussed in the previous section into an overall framework for analysing consumer food quality perception. The model is shown in Fig. 7.5.

First of all, the TFQM distinguishes between 'before' and 'after' purchase evaluations. As already mentioned, the quality perception processes before and after the purchase are fundamentally different – before the purchase basically all quality dimensions are uncertain, and consumers infer quality from the cues at hand. After the purchase, some aspects of quality become amenable to experience, so that expectations formed in the prepurchase phase can be confirmed or refuted. The distinction between before and after purchase thus forms the basis of the TFQM.

In the before purchase part, the model shows how quality expectations are formed based on the quality cues available. Cues are defined as pieces of information used to form quality expectations (Steenkamp, 1990). The intrinsic quality cues cover the physical characteristics of the product, and are related to the product's technical specifications, which also include its physiological characteristics, i.e. characteristics which can be measured objectively. The extrinsic quality cues represent all other characteristics of the product, such as brand name, price, distribution, outlet, packaging, etc. Of all the cues consumers are exposed to, only those which are perceived will have an influence on expected quality. The cues consumers are exposed to and those they perceive are affected by the shopping situation: the amount of information in the shop, whether purchases are planned or spontaneous, the pressure of time while shopping, etc.

According to the TFQM, and in line with our discussion about the role of values and attitude systems, quality is not an aim in itself, but is desired because it helps to satisfy purchase motives or values. The model therefore includes motive or value fulfilment, i.e. how food products contribute to the achievement of desired consequences and values. Extrinsic cues such as a label and its content may, for example, generate expectations about exceptionally high eating quality – giving the consumer a feeling of luxury and of pleasure in life. The values sought by consumers will, in turn, have an impact on which quality dimensions are sought and how different cues are perceived and evaluated. The sequence from cues, through quality, to purchase motives forms a hierarchy of increasingly abstract cognitive categories. In this way, the TFQM integrates the means-end model of consumer behaviour.

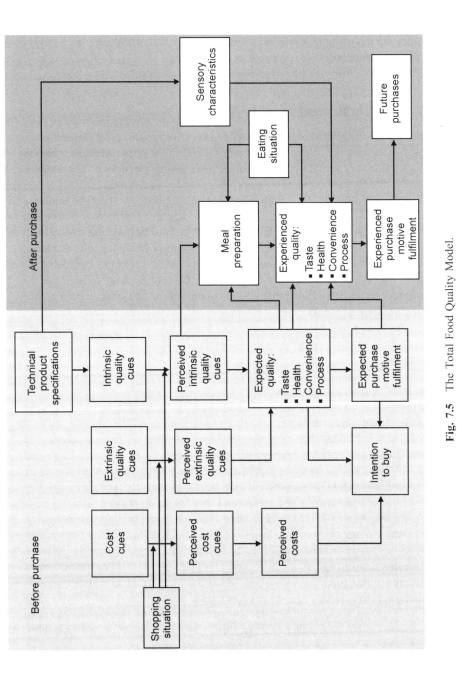

Fig. 7.5 The Total Food Quality Model.

Expected quality and expected fulfilment of purchase motives constitute the positive consequences consumers expect from buying a food product, and are offset against the negative consequences in the form of various (mostly monetary) costs. The trade-off determines the intention to buy.

After the purchase, the consumer will be able to experience some aspects of quality, notably the sensory and convenience aspects, and these experiences will often deviate from expected quality, especially when it is based on quality cues with a low degree of predictive power. The experienced quality is influenced by many factors. The product itself, especially its sensory characteristics (in an objective sense, as measured by a sensory panel), is obviously one determinant, but there are many others: the way the product has been prepared, situational factors such as time of day and type of meal, the consumer's mood, previous experience, etc. And the expectation itself may also be an important variable in determining the experienced quality of the product (Deliza and MacFie, 1996; Oliver, 1993; Schifferstein, 2001). The relationship between quality expectation and quality experience (e.g., before and after purchase) is commonly believed to determine product satisfaction, and consequently the probability of purchasing the product again.

7.5   Future trends

It is a widely held belief in the food sector that the way in which consumers perceive quality in food has become more complicated over the past decades. New quality dimensions have been added or have become more important – health is many times as important as good taste now, the convenience dimension has risen dramatically in importance, and process characteristics like organic production have been added. In addition, the way these general quality dimensions are perceived and inferred has changed as well – for example, consumers' subjective health theories have changed, and while some years ago most consumers simply believed that fat is bad for you, many have now started to understand the intricate distinctions between saturated and unsaturated fatty acids.

We would like to close, therefore, with some speculation about trends in the way consumer food quality perception will develop in the coming years. We mention four such trends: *healthy living, variety seeking, convenience as way of life*, and *sympathetic food production.*

Healthy living
Health is already an important component of quality perception, but the way in which it enters quality perception may change considerably. This will be driven both by increased consumer learning about healthy eating, and by new products coming on the market. Regarding the latter, it is reasonable to expect that there will be an increasing stream of so-called functional foods – products with built-in specific health properties – and research on nutrigenomics carries the promise that these products will be increasingly tailored to specific consumer segments.

In this situation, consumers will have to develop new views with regard to what constitutes healthy eating, and the extent to which they will come to attach credibility to the promises of functional foods and be able to perceive the new health-related qualities will have a major impact on the success (or lack of it) of this new stream of products.

Variety seeking
Seeking stimulation and more variety has been a trend affecting many areas of life, including food, and the ability of food products to live up to the demand for variety may come to play a bigger part in quality perception. Variety can be related to the sensory quality of the food, but may also be related to meal composition, eating occasions, ways of cooking, and ways of eating out.

Convenience as way of life
Convenience is one of the strongest ongoing trends in the food area. Whereas convenience food earlier was mainly something for people not very interested in food, and consequently was usually at the lower end of the quality spectrum, all types of consumers, including those interested in gourmet food, now demand convenience. Convenience thereby gets a much broader meaning, where it includes parameters like processing shortcuts, meal components, intelligent storing devices, and new ways of shopping.

Sympathetic food production
The promotion of specific ways of food production, like organic production, together with a stream of food scandals has resulted in a generally increased interest in the way food is being produced. The 'process characteristics' dimension of quality perception may thus develop from consisting of a few distinct aspects – organic production, GMOs, animal welfare, fair trade – into a more holistic evaluation of food production, where consumers form opinions about whether they like what they have learned about the production process or not. We presently know very little about which types of production processes consumers will find more sympathetic than others, beyond the specific examples mentioned, but one likely development is that consumers will find production processes that can be framed more sympathetically as industrializations of traditional craftsmanship-type processes.

7.6   Sources of further information

A more in-depth treatment of many of the issues dealt with here can be found in:

GRUNERT, K. G. (2005). Food quality and safety: Consumer perception and demand. *European Review of Agricultural Economics*, **32**, 369–391.

A thorough treatment of inference processes in consumer behaviour can be found in:

KARDES, F. R., POSAVAC, S. S. and CRONLEY, M. L. (2004). Consumer inference: A review of processes, bases, and judgment contexts. *Journal of Consumer Psychology*, **14**, 230–256.

More information on food-related lifestyle is available in:

GRUNERT, K. G., BRUNSØ, K., BREDAHL, L. and BECH, A. C. (2001). Food-related lifestyle: A segmentation approach to European food consumers. In L. J. Frewer, E. Risvik, H. N. J. Schifferstein and R. von Alvensleben (eds), *Food, People and Society: A European perspective of consumers' food choices* (pp. 211–230). London: Springer.

7.7   References

ASKEGAARD, S. and BRUNSØ, K. (1999). Food-related life styles in Singapore: Preliminary testing of a Western European research instrument in Southeast Asia. *Journal of Euromarketing*, **7**(4), 65–86.

BECH-LARSEN, T. (2001). Model-based development and testing of advertising messages: A comparative study of two campaign proposals based on the MECCAS model and a conventional approach. *International Journal of Advertising: The Quarterly Review of Marketing Communications*, **20**, 499–519.

BREDAHL, L. (1999). Consumers' cognitions with regard to genetically modified food. Results of a qualitative study in four countries. *Appetite*, **33**, 343–360.

BREDAHL, L. (2001). Determinants of consumer attitudes and purchase intentions with regard to genetically modified foods: Results of a cross-national survey. *Journal of Consumer Policy*, **24**, 23–61.

BRUNSØ, K. and GRUNERT, K. G. (1998). Cross-cultural similarities and differences in shopping for food. *Journal of Business Research*, **42**, 145–150.

BRUNSØ, K., BREDAHL, L., GRUNERT, K. G. and SCHOLDERER, J. (2005). Consumer perception of the quality of beef resulting from various fattening regimes. *Livestock Production Science*, **94**, 83–93.

COX, D. F. (1967). The sorting rule model of the consumer product evaluation process. In D. F. Cox (ed.), *Risk Taking and Information Handling in Consumer Behaviour* (pp. 324–369). Boston: Graduate School of Business Administration, Harvard University.

DELIZA, R. and MacFIE, H. J. H. (1996). The generation of sensory expectation by external cues and its effect on sensory perception and hedonic ratings: A review. *Journal of Sensory Studies*, **11**, 103–128.

ERDEM, T. and SWAIT, J. (1998). Brand equity as a signaling phenomenon. *Journal of Consumer Psychology*, **7**, 131–158.

FOTOPOULUS, C., KRYSTALLIS, A. and NESS, M. (2003). Wine produced by organic grapes in Greece: using means-end chains analysis to reveal organic buyers' purchasing motives in comparison to the non-buyers. *Food Quality & Preference*, **14**, 549–566.

GRUNERT, K. G. (2005). Consumer behaviour with regard to food innovations: Quality perception and decision-making. In W. M. F. Jongen and M. T. G. Meulenberg (eds), *Innovation in Agri-food Systems* (1st edn, pp. 57–85). Wageningen: Wageningen Academic Publishers.

GRUNERT, K. G. and GRUNERT, S. C. (1995). Measuring subjective meaning structures by the laddering method: Theoretical considerations and methodological problems. *International Journal of Research in Marketing*, **12**, 209–225.

GRUNERT, K. G., LARSEN, H. H., MADSEN, T. K. and BAADSGAARD, A. (1996). *Market Orientation in Food and Agriculture*. Boston: Kluwer.

GRUNERT, K. G., BRUNSØ, K. and BISP, S. (1997). Food-related lifestyle: Development of a cross-culturally valid instrument for market surveillance. In L. R. Kahle and L. Chiagouris (eds), *Values, Lifestyles, and Psychographics* (pp. 337–354). Mahwah, NJ: Lawrence Erlbaum.

GRUNERT, K. G., BRUNSØ, K., BREDAHL, L. and BECH, A. C. (2001a). Food-related lifestyle: A segmentation approach to European food consumers. In L. J. Frewer, E. Risvik, H. N. J. Schifferstein and R. von Alvensleben (eds), *Food, People and Society: A European Perspective of Consumers' Food Choices* (pp. 211–230). London: Springer.

GRUNERT, K. G., LÄHTEENMÄKI, L., NIELSEN, N. A., POULSEN, J. B., UELAND, O. and ÅSTRÖM, A. (2001b). Consumer perceptions of food products involving genetic modification: Results from a qualitative study in four Nordic countries. *Food Quality and Preference*, **12**, 527–542.

GRUNERT, K. G., BREDAHL, L. and SCHOLDERER, J. (2003). Four questions on European consumers' attitudes to the use of generic modification in food production. *Innovative Food Science and Emerging Technologies*, **4**, 435–445.

GUTMAN, J. (1982). A means-end chain model based on consumer categorization processes. *Journal of Marketing*, **46**(2), 60–72.

JAEGER, S. and MacFIE, H. J. H. (2000). Incorporating 'health' into promotional messages for apples: A means-end theory approach. *Journal of Food Products Marketing*, **6**(2), 57–78.

KAHLE, L. R., BEATTY, S. E. and HOMER, P. (1986). Alternative measurement approaches to consumer values: The List of Values (LOV) and Values and Life Style (VALS). *Journal of Consumer Research*, **13**, 405–409.

KELLER, K. L. (1993). Conceptualizing, measuring, and managing customer-based brand equity. *Journal of Marketing*, **57**(1), 1–22.

MILES, S. and FREWER, L. J. (2001). Investigating specific concerns about different food hazards. *Food Quality and Preferences*, **12**, 47–61.

MULVEY, M. S., OLSON, J. C., CELSI, R. L. and WALKER, B. A. (1994). Exploring the relationships between means-end knowledge and involvement. *Advances in Consumer Research*, **21**, 51–57.

NIELSEN, N. A., SØRENSEN, E. and GRUNERT, K. G. (1997). Consumer motives for buying fresh or frozen plaice: A means end chain approach. In J. B. Luten, T. Børresen and J. Oehlenschläger (eds), *Seafood from Producer to Consumer: Integrated Approach to Quality* (pp. 31–43). Amsterdam: Elsevier.

NIELSEN, N. A., BECH-LARSEN, T. and GRUNERT, K. G. (1998). Consumer purchase motives and product perceptions: A laddering study on vegetable oil in three countries. *Food Quality & Preference*, **9**, 455–466.

OLIVER, R. L. (1993). Cognitive, affective, and attribute bases of the satisfaction response. *Journal of Consumer Research*, **20**, 418–431.

OLSEN, S. O. (2002). Comparative evaluation ad the relationship between quality, satisfaction, and repurchase loyalty. *Journal of the Academy of Marketing Science*, **30**, 240–249.

PETER, J. P., OLSON, J. C. and GRUNERT, K. G. (1999). *Consumer Behaviour and Marketing*

Strategy – European edition. Maidenhead: McGraw-Hill.

PIETERS, R., BAUMGARTNER, H. and ALLEN, D. (1995). A means-end chain approach to consumer goal structures. *International Journal of Research in Marketing*, **12**, 227–244.

REID, M., LI, E., BRUWER, J. and GRUNERT, K. G. (2001). Food-related lifestyles in a cross-cultural context: Comparing Australia with Singapore, Britain, France and Denmark. *Journal of Food Products Marketing*, **7**(4), 57–75.

ROKEACH, M. (1973). *The Nature of Human Values*. New York: Free Press.

RUSSELL, G. C., FLIGHT, I., LEPPARD, P., VAN LAWICK VAN PABST, J. A., SYRETTE, J. A. and COX, D. N. (2004). A comparison of paper-and pencil and computerised methods of hard laddering. *Food Quality & Preference*, **15**, 279–291.

SCHIFFERSTEIN, H. (2001). Effects of product beliefs on product perception and linking. In L. Frewer, E. Risvik and H. Schifferstein (eds), *Food, People and Society* (pp. 73–97). Heidelberg: Springer Verlag.

SCHOLDERER, J., BRUNSØ, K., BREDAHL, L. and GRUNERT, K. G. (2004). Cross-cultural validity of the food-related lifestyles instrument (FRL) within Western Europe. *Appetite*, **42**, 197–211.

SCHWARTZ, S. H. (1992). Universals in the content and structure of values: Theoretical advances and empirical tests in 20 countries. In M. P. Zanna (ed.), *Advances in Experimental Social Psychology* (vol. 25, pp. 1–65). San Diego, CA: Academic Press.

SCHWARTZ, S. H. and BILSKY, W. (1987). Toward a universal psychological structure of human values. *Journal of Personality and Social Psychology*, **53**, 550–562.

SELNES, F. and TROYE, S. V. (1989). Buying expertise, information search, and problem solving. *Journal of Economic Psychology*, **10**, 411–428.

SØNDERGAARD, H. A., GRUNERT, K. G. and SCHOLDERER, J. (2005). Consumer attitudes to enzymes in food production. *Trends in Food Science &Technology*, **16**, 466–474.

STEENKAMP, J.-B. E. M. (1990). Conceptual model of the quality perception process. Journal of Business Research, **21**, 309–333.

VALETTE-FLORENCE, P., SIRIEIX, L., GRUNERT, K. G. and NIELSEN, N. A. (2000). Means-end chain analyses of fish consumption in Denmark and France: A multidimensional perspective. *Journal of Euromarketing*, **8**, 15–27.

VAN RAAIJ, W. F. and VERHALLEN, T. M. M. (1994). Domain-specific market segmentation. *European Journal of Marketing*, **28**(10), 49–66.

WALKER, B. A. and OLSON, J. C. (1991). Means-end chains: Connecting products with self. *Journal of Business Research*, **22**, 111–119.

ZEITHAML, V. A. (1998). Consumer perceptions of price, quality, and value: A means-end model and synthesis of evidence. *Journal of Marketing*, **52**(3), 2–22.

8

Consumer attitudes towards convenience foods

M. Buckley and C. Cowan, Ashtown Food Research Centre, Ireland and M. McCarthy, University College Cork, Ireland

8.1 Introduction

New product development (NPD) is crucial to the long term survival and profitability of operators within the food-manufacturing sector. Consumers change, markets change and companies need to proactively develop new products to satisfy the needs of their consumers. In a food and drink industry characterised by low overall volume growth, increasing consolidation and competition, efficient NPD is essential to gaining competitive advantage. For NPD to be successful, consumer research is essential and should be carried out from the earliest stages of the NPD process, to ensure that products are developed with the consumer foremost in mind.

Clearly market segmentation is a fundamental step in the NPD process as it provides a better understanding of the market by providing information about the motives and needs of different consumers. Segmentation also allows for behaviour to be predicted with greater accuracy and aids in the identification and exploitation of potential market opportunities (Kotler, 1991). Furthermore, it provides the necessary information on which to arrange all other marketing strategies, including product development, pricing decisions and communication (Elmore-Yalch, 1998).

However, the process of identifying meaningful segments has become more complex and researchers such as Boedekker and Marjanen (1993) note that traditional methods of segmentation, such as those based on demographics, are becoming less practical in the analysis and prediction of consumer behaviour. Thus it is not surprising that marketers are increasingly using consumer life-styles, which are seen as a more multi-dimensional basis, for explaining behaviour. Indeed the impact of changing consumer lifestyles is most evident in food markets. In high income countries with significant female participation in the workforce, solutions to time and energy deficits are sought and new food lifestyles emerge. When considering this, Grunert *et al.* (1993) developed an instrument specifically tailored to segment food markets, the Food-Related Lifestyle (FRL) instrument. The FRL instrument, which has been cross-culturally validated, measures consumers' attitudes towards the purchase, preparation, and consumption of food products. Different segments are found in different European countries but the instrument is cross-culturally valid in terms of factor loadings, factor covariances and factor variances, although item specific means and item reliabilities are biased across cultures (Scholderer *et al.*, 2004). Interestingly, for Ireland and Great Britain the measurement charac-teristics are completely invariant when applied to consumer populations (O'Sullivan *et al.*, 2005). In fact the instrument has identical measurement characteristics in both populations.

The theme of this chapter is the investigation of consumer attitudes towards convenience foods and the importance thereof. Following detailed consideration about the definition of convenience, the market trends influencing demand for convenience are reviewed. As the demand for convenience is a lifestyle issue, some background information on the FRL instrument is provided. Following this, the views and attitudes of identified food-related lifestyle segments in Ireland and Great Britain with regard to convenience and convenience food are presented. Convenience food lifestyle (CFL) segments in the British population are then examined. Selected segments are discussed with a view to contrasting the varying requirements of different more convenience-oriented segments as well as those of the less convenience-oriented segments. Some remarks on the value of both the FRL and CFL approaches, in the context of convenience in food choice, conclude the chapter.

8.2   Definitions of convenience and convenience foods

Convenience can be of substantial importance in deciding consumer behaviour towards food products (Candel, 2001). Convenience is a term that is frequently used, but is also a word that is not fully understood or operationalised by marketers (Gofton, 1995). According to Yale and Venkatesh (1985), conveni-ence is an important concept on both a utilisation behaviour level and a product characteristic level. It is necessary for marketers to be aware of the complex nature of convenience in order to determine convenience seeking consumer segments and to formulate products and marketing designs for these segments, which emphasise consumer perceived convenience attributes (Yale and Venkatesh, 1985).

Initially when examining the concept of convenience, one must consider the orientation of the individual toward demanding such an attribute. In 1972, Anderson suggested that convenience orientation is driven by the motives of fulfilling some instantaneous want or need and freeing time and/or energy for other purposes. This is not dissimilar from Candel's (2001, p. 17) view that convenience orientation is 'the degree to which a consumer is inclined to save time and energy as regards meal preparation'. More recently, Scholderer and Grunert (2005) defined convenience orientation as a positive outlook towards time and energy saving aspects of home meal production. Convenience orientation is considered by many as a critical factor determining the inclinations of consumers in their food-related behaviours. Anderson (1972, p. 50) argues that convenience orientation is of interest to marketers as it serves as a basis 'for market segmentation and more effective allocation of marketing effort'. This view is reinforced by Candel (2001), who suggests that by identifying the extent to which consumers are convenience-orientated, the behaviour of food consumers in relation to their food preferences may be better understood. In fact Madill-Marshall *et al.* (1995) provide direction on how to predict the level of convenience food usage when they suggest examining convenience orientation towards food-related activities (e.g. food shopping and cooking). These views clearly suggest that convenient food lifestyle segments are prevalent in the food market.

Much research on the concept of convenience tended to focus on the notion of time. Candel (2001), for example, spoke about the use of time-buying and time-saving strategies being used to resolve the time constraints resulting from the increased participation by women in the labour force.

However, it was recognised by authors such as Gofton and Marshall (1988) that convenience involved more than the quality of 'time-saving' in food activities. Convenience is a concept to which there are multiple dimensions (Brown, 1989; Gofton and Marshall, 1992). Gofton (1995, p. 178) points out 'convenience eating is a complex and contentious object of analysis. To assume it can be dismissed as time saving is simply misleading and truly a waste of time'.

Brown (1989) proposed that convenience consisted of five dimensions – time, place, acquisition, use and execution. According to Brown (1989) the ultimate convenience product would be available to the individual on a continuous basis, ubiquitously, and would require virtually no effort to acquire or use. Furthermore, the consumer may choose as much physical or mental effort as he or she wishes to expend in acquiring the product. Man and Fullerton (1990, p. 75) defined convenience foods as products 'in which all or at least one significant part of the preparation process has been transferred from the kitchen to the factory. In general, convenience foods offer the consumer easier preparation, faster cooking, portion control, a variety of choices, less cleaning, and a minimum of waste'.

Brown and McEnally (1993) argued that convenience should be considered at all stages in the process of food consumption and to determine the proportionate importance that consumers attach to time and energy use in acquisition, consumption and disposal. Thus it is not surprising that convenience definitions such as IGD (2002, p. 1) embraced these concepts '[convenience is] associated with reducing the input required from consumers in either food shopping, preparation, cooking or cleaning after the meal'. Similarly, Darian and Cohen (1995) referred to the idea of convenience being important at one or more of a number of stages including when deciding what to eat, purchasing, preparation, consumption and when clearing up.

8.2.1   The food provisioning process

As suggested earlier, convenience is associated with reducing effort, time, and/ or skill during the food provisioning process, a process that follows each of the stages from shelf to stomach (Gofton 1995; Marshall, 1995). Given the import-ance of the food provision process in the definition of convenience we will briefly consider the meaning of convenience at the acquisition, meal preparation/cooking, consumption and post consumption stages.

With regard to convenience in acquisition, the discussion tends to focus on the time and effort expended in the purchase outlet, storage characteristics of the product; location of the purchase outlet (time and effort used getting to the outlet), and time and effort spent planning purchases. Berry (1979) referred to a study carried out by the Food Marketing Institute that identified a number of time-saving strategies employed by consumers while shopping for food. Such tactics included: not demanding special cuts of meat, bulk buying, shopping in less-congested stores, visiting convenience stores, and purchasing pre-packaged foods. Gofton and Ness (1991) referred to convenience in acquisition in terms of products being widely available and ease of storage and thus availability for use at any time. For Gofton and Marshall (1988), convenience was related to food shopping and acquisition and how the food is to be stored. McMillan (1994) suggested that many individuals seek to minimise the time spent on food shopping, while Darian and Cohen (1995) add that the time spent on planning is an important component of the shopping process.

Umesh et al. (1988) referred to the concept of convenience in relation to shopping activities, proposing that convenience shopping offered products to the consumer in a location that required very little time to make purchases, as well as offering products and services that were designed to save time. Swoboda and Morschett (2001) considered it useful to examine convenience in terms of the actual purchasing behaviour of consumers with regard to convenience-orientated shopping.

Meal preparation is a time-consuming activity and accounts for much of the time spent in household production (Davies and Madran, 1997; Sloan, 1997; Capps et al., 1985). Thus at the most basic level the importance of convenience reflects the propensity of consumers to try and minimise the time that is spent on meal preparation (Verlegh and Candel, 1999). In fact, the purchase of con-venience foods removes some of the need for a certain amount of the household manager's time, effort and cooking expertise (Capps et al., 1985). Similarly, Costa et al. (2001) related convenience attributes to effective resource employment in the process of food preparation and argued that the time strains experienced by consumers resulted in less time for cooking. Gofton and Ness (1991) argued that convenience involved simplicity in the cooking process, speedy cooking, being able to cook without special utensils, and the ability to serve without cooking or without special tableware.

Darian and Cohen (1995) referred to the mental effort involved in planning ahead and argued that the importance of this aspect has increased due to the individual food preferences of family members, requiring that a variety of foods be prepared for dinner. The attempts to satisfy individual preference within a family unit highlights the importance of culinary skills (or rather lack thereof) in the demand for convenience foods. Interestingly, Candel (2001) suggested that the transference of culinary skills can be seen as ultimately leading to time and energy saving in meal preparation. He noted that expertise/skills result from time and energy spent on learning a particular job. When food products offer convenience in the sense that particular skills are not required, this implies that no time/energy is needed to learn to prepare the product successfully.

Gofton and Ness (1991) referred to convenience in the process of consumption as suitability of a product for use as a lap or TV meal. Costa *et al.* (2001) stated that the time pressures being experienced by consumers resulted in individuals not having enough time to eat. Consumption of convenience foods has increased significantly, as has the propensity of individuals to eat out (Davies and Madran, 1997). Darian and Cohen (1995) referred to the mental effort involved in the planning and preparation of meals and argued that the different schedules of family members has made it difficult to eat dinner at the same time each day or even to eat dinner together as a family.

With regard to convenience and post-meal activities, according to Darian and Cohen (1995), having very little or no clearing up after eating is very important to time-pressured consumers. Research carried out by IGD in 2002 found that 22% of the respondents would choose a meal that did not require much washing up as a convenient meal solution (IGD, 2002). When consumers prepare a meal from scratch, they are often faced with the task of disposing of food ingredients that were not used in the food preparation process. This is particularly relevant in the context of UK, where there is an increasing number of one- and two-person households. Convenience foods are often seen as offering 'convenience' as they remove the need to dispose of ingredients, since exact quantities of ingredients are provided (IGD, 2002).

8.2.2   Categorisation of convenience foods

Based on the review of the literature, it is feasible to conclude that convenience can be defined in terms of the time and effort savings that it offers to the consumer in food-related activities. Various authors have proposed many different convenience food groupings. From the definitions and categorisations, it is clear that consumer understanding of convenience has altered, and that it is important to have an adequate understanding of convenience in order to develop

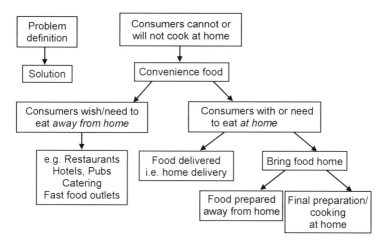

Fig. 8.1 Categorisation of convenience foods.

marketing strategies that meet consumer demands. Convenience orientation is a measure of the extent to which the consumer is likely to seek time and energy savings in relation to his/her food-related activities.

Based on the premise that consumers are incapable of or reluctant to prepare and cook a full meal from scratch at home, a categorisation of convenience foods was modelled (Fig. 8.1), similar to the categorisation developed by Costa *et al.* (2001), who developed a meal solutions categorisation. Convenience foods may be subdivided into two categories based on where ultimate consumption takes place, i.e. at home or away from home. The 'at-home' category may be further segregated into the 'food delivered' and 'bring food home' classes. The latter category may be further divided on the basis of whether the food is prepared in the home or away from home.

8.3   Consumer forces driving the convenience food market

In order to understand the behaviour of consumers with regard to convenience foods, it is necessary to understand what is motivating people's behaviour and what needs are being fulfilled by this behaviour. Several factors, particularly changing consumer trends and lifestyles, have contributed to the growth of the convenience food market. These are discussed with a particular focus on Great Britain, but similar trends are apparent in many European markets.

8.3.1   Demographic changes

Populations right across Europe are ageing, thus one can anticipate a larger percentage of the EU population falling into the 65 and over category in the coming decades (Mintel, 2000; Bass *et al.*, 1999). It is anticipated that there will be changes in food demand from this growing cohort of the population as the life experiences of the 'new old' will be very different from their predecessors. These future retirees will demand convenience as they are current consumers of partly prepared meals, takeaways or meals eaten away from home (IGD, 2001). In fact future retirees gave a 10% higher value rating for convenience foods than current retirees (IGD, 2001). According to IGD (1998), the 60 year old in 2021 will be 'more accustomed to grabbing a quick meal and eating out than cooking at home' (p. 111). Future retirees will also demand single or smaller serving packages because of the convenience offered and the minimisation of food waste (Senauer et al., 1991).

The average size of the household is decreasing across Europe with more households made up of one or two people (IGD, 2001; Mintel, 2000). The implications of declining household size include a greater demand for smaller pack sizes and single servings, for ready prepared food and for eating out (IGD, 1998). As early as 1976 Kahn (1976) stated that single person households might use more convenience foods since the former desire foods that are simple to prepare and require little preparation and cleanup time. Thus, it is not surprising that the increasing number of smaller households is increasing the demand for smaller pack sizes and individual portioning of products, in addition to ease and speed of preparation (Traill, 1997; Kahn, 2000).

8.3.2   Work and income

A number of changes pertinent to the demand for convenience food are linked to the labour force activity and household incomes. These influences include female participation in the labour force, more use of household technologies, longer working hours and increasing consumer incomes.

The increase in the number of women in the workforce has been identified by various authors as one factor driving the demand for convenience foods (e.g., Stafford and Wills, 1979; Somogyi, 1990; Senauer et al., 1991; Traill, 1997). The increase in the number of families with both partners participating in paid employment has resulted in a situation whereby they have less leisure time and are therefore more time sensitive. These have become important factors influencing shopping behaviour and demand for convenience foods (Traill and Harmsen, 1997; Umesh et al., 1988). McKenzie (1986) cited a number of reactions to the increase in the number of women in the workplace including more consumption away from home and greater use of convenience foods. Traill (1997) asserted that the increased participation of women in the labour force has contributed to the end of the family meal in favour of snacking.

McKenzie (1986) and Traill (1997) have argued that the increased participation by women in the workplace has increased the use of such appliances as microwave ovens, freezers and food processors, dishwashers and deep freezers. According to IGD (1998, p. 171), 'new technologies have had considerable impact on the type of food consumed, and how and where it is being purchased and eaten'. When one looks at the adoption of various technologies, it becomes clear that there has been an uptake of time and effort saving technologies. For example, the microwave saves time in preparation and cooking, while the dishwasher saves time and effort in the clearing up process. The freezer stores time by freezing food for later consumption (Shove and Southerton, 2000).

British people work longer hours than their European counterparts (Geest, 2001); the average working week for the British worker was 44 hours in 1998 (IGD, 1998). The result of a longer working week has been that workers experience greater time pressures. Household duties such as shopping, cooking and cleaning are still required to be performed and thus extra pressures are placed on leisure time. Therefore individuals seek out means of easing such pressures, for example by seeking out convenient methods of meal preparation and cooking. Furthermore, growth in personal incomes due to such factors as longer working hours has resulted in increased demand for convenience foods (Traill, 1997).

According to Bonke (1992), income encourages the consumption of convenience foods, independently of the time commitment required. Incomes in Great Britain have been increasing due to the increased participation of females in the labour force and the practice of having children later in life (Mintel, 2000). Consequently people are eating out more (IGD, 2001) and 'prepared foods command premium prices over just buying the ingredients because the consumer pays the manufacturer for the time taken to prepare the product' (Geest, 2001).

8.3.3   Breakdown of traditional mealtimes

The structure of meals is becoming more fragmented (IGD, 2000) and increasing in informality (Mintel, 2000) and the traditional family meal is disappearing. In Britain eating in front of the television is now the norm for 43% of adults and 37% snack between meals (Mintel, 2000). The breakdown of mealtimes leads to an increase in the demand for convenience foods to meet the individual demands resulting from families not eating together (IGD, 2000). Related to this there has been a declining number of people who have and use cooking skills (Geest, 2001; IGD, 1998).

According to Senauer (2001, p.1) there has been a breakdown in 'traditional, culturally determined food preferences', and individuals exhibit very dissimilar food consumption patterns. This is apparent in Great Britain, where there has been a shift towards a more individualistic eating behaviour (Gofton, 1995). The individual, rather than the family, has become the decision-maker in terms of food choice. In addition, children are given more responsibility at an earlier age, which leads to a situation within a household whereby different family members eat different things on the same eating occasion (IGD, 1998).

8.3.4   More than convenience required

Many consumers interested in convenience foods also have a desire for new food experiences. Consumers are increasingly looking for novelty and excitement in the foods that they eat (Leatherhead Food RA, 2001). A number of influences may be responsible for this including cheaper air travel, which means that consumers can easily travel to various destinations around the world (IGD, 2001). The ethnic composition of the population also influences food demand (IGD, 2001). In fact, many foreign-cuisine restaurants have opened throughout Great Britain. The experience of unusual foods while eating away from home and on holiday has inspired individuals to imitate these types of meals in the home (Mintel, 2000).

Consumers believe that they 'are what they eat' (Khan, 2000, p. 16) and there is demand amongst consumers for meal solutions that address many health concerns (Reuters, 2002). The association between food and health has been accepted by consumers (Geest, 2001; IGD, 2000; Mintel, 2000). There has been a shift in food consumption towards lower fat products such as pasta, rice, and some fruit and vegetables. A number of barriers to healthy eating have been identified. These include lack of time, increase in snacking and stressful lifestyles (Promar, 1997). Concerns about healthy eating play an important role in the selection of convenience foods: 'Convenience in the mind of the consumer is no longer just about food that is simple and easy to prepare, but also about food which is healthy. The rise in information available about what we eat is clearly having a growing impact on our food buying habits as a nation' (i.e. of British consumers) (TNS, 2000).

8.4   Usefulness of lifestyles in understanding demand for convenience: food-related lifestyle

While the trends highlighted above indicate a general move towards increased demand for convenience, this may not apply to the population as a whole. It has long been accepted that the public are not homogenous with regard to their demands and in fact there are many heterogeneous groups within any population. Thus, an examination of consumer segments with regard to demand for convenience foods is useful. One characteristic that may be used to segment a population, which is very relevant to demand for convenience, is lifestyles. In fact, the evolving and dynamic lifestyle of today's consumer makes it essential to use lifestyle variables in the segmentation process. Of particular relevance to this discussion is food lifestyle. In an attempt to understand these lifestyles Grunert *et al.* (1993) developed the Food-Related Lifestyle (FRL) instrument. The FRL instrument is a measurement instrument that collects consumer information about attitudes and behaviour relating to the purchase, preparation and consumption of food products (Grunert *et al.*, 1993). The concept of food-related lifestyle endeavours to explain the manner in which people use food to achieve basic life values.

In the food-related lifestyle approach, lifestyle is regarded as a mental construct that explains, but is not identical to behaviour. The concept takes its roots in a hierarchy of cognitive categories and based on cognitive psychology, the instrument aims to relate lifestyle to other cognitive categories, and also shows how they are related to behaviour (Brunsø *et al.*, 1996). The FRL

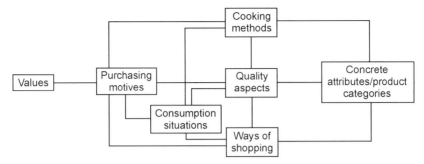

Fig. 8.2 A model of food-related lifestyle.
Source: Kluwer Academic Publishers Massachusetts, *Market Orientation in Food and Agriculture* (1996), page 48, Chapter 3, Analysing consumers at the aggregate level, Grunert K G, Baadsgaard A, Larsen H H and Madsen T K, figure 3.6; Copyright © 1996 by Kluwer Academic Publishers, with kind permission of Springer Science and Business Media.

instrument is based on means end-chain theory and was developed from the idea that consumers perceive a product to hold value to the extent that its use will convey self-relevant consequences (Grunert *et al.*, 1996). The means-end chain model for the FRL instrument is shown in Fig. 8.2. It illustrates the system of cognitive categories, scripts, and their associations, which relate food products to values. The FRL instrument has also been applied widely and has been shown to be cross-culturally valid (Brunsø *et al.*, 1996; O'Sullivan *et al.*, 2005).

8.4.1 Components of food-related lifestyle

The FRL instrument is a combination of five inter-related elements that cover shopping, meal preparation and consumption of meals (Grunert *et al.*, 1993). These comprise 69 items that measures 23 lifestyle dimensions in the five major life elements. The FRL embraces all of the food processing stages that are referred to as important when assessing demand for convenience.

- The first element is '*ways of shopping*'. This reflects consumers' shopping behaviour for food, with regards to whether they read the labels, if they are reliant on the advice of others, their attitude towards advertising and whether they shop for themselves or for others.
- Consumers '*cooking methods*' examines such aspects as the manner in which the products are transformed into meals, the length of time taken to prepare them, if any time at all, and whether the meals are planned or spontaneous.
- '*Quality aspects*' refers to attitudes to health, nutrition, freshness and the luxury attributes of a product.
- '*Purchasing motives*' explores what consumers expect from a meal and the importance of these expectations. For example, the tradition of a meal can mean more to one person than to another.
- '*Consumption situations*' refer to where the meal takes place, and whether the meal is thought of differently when eaten alone, with family or with friends.

Table 8.1   FRL segments identified in five European countries

	Germany	France	Great Britain	Denmark	Ireland	Great Britain
	1993	1994	1994	1995	2001	2002
Conservative	18	13	19	11	21	9
Rational	26	35	33	11	–	26
Uninvolved	21	18	9	11	16	14
Adventurous	24	–	12	25	8	17
Careless	11	–	27	23	–	14
Eco-moderate	–	–	–	20	–	–
Moderate	–	16	–	–	13	–
Hedonistic	–	18	–	–	28	–
Enthusiastic	–	–	–	–	14	–
Snackers	–	–	–	–	–	20

The FRL instrument has been applied in nationally representative surveys in a number of countries. FRL segments were identified in Denmark, France, Germany, Great Britain (Brunsø et al., 1996; Buckley et al., 2005), Spain (MAPP, 1996) and Ireland (Ryan et al., 2004). The segments derived in each country are shown in Table 8.1, including the segments identified for Ireland and Great Britain in 2001 and 2002 respectively (Ryan et al., 2004; Buckley et al., 2005).

Food-related lifestyle segments (in the case of both Ireland and Great Britain) were identified using hierarchical cluster analysis and employing Ward's method. The identified segments were profiled using the 23 dimensions of the food-related lifestyle instrument using a procedure that had been followed in previous FRL studies: by '*comparing the mean scores on the dimensions with the population mean scores on the same dimension*' (Bredahl and Grunert, 1997, p. 12).

8.5   Food-related lifestyle research in Ireland and Great Britain

As the focus of this chapter is convenience we will consider two studies, one in Ireland and one in Great Britain, that set out to look at food-related lifestyles and convenience orientation. In these two studies of consumers a number of dimensions that explicitly looked at convenience orientation were developed. Such dimensions resulted in a better profiling of the FRL segments in the context of convenience orientation than the FRL dimensions alone. The approach differed for Ireland and Britain.

For Ireland, two convenience elements were developed following a review of the Irish convenience food market, which highlighted a number of convenience-related food issues driving this market. The drivers of convenience were categorised under 'lifestyle issues that drive the convenience food market' and 'attitude towards convenience food'. The former included five dimensions: time pressures, stress levels, breakdown of mealtime, eating alone and individuality/

family. 'Attitude towards convenience food' included three dimensions: convenience food value for money, health value of convenience food and time and convenience food. In the Irish study, the convenience dimensions were used solely as descriptive variables to explain the FRL segments.

In the case of both of these studies the FRL and convenience food attitudinal components of the questionnaire are explicated by reference to the dimensions whose mean scores are at least $+1$ greater than or -1 lower than the sample mean score.

8.5.1   Irish segments

Of the six FRL segments identified in Ireland three had a convenience orientation namely: the hedonistic (28%), the extremely uninvolved (16%) and the adventurous segments (8%) (Ryan *et al.*, 2002). The other three segments, the conservative (21%) the enthusiastic (14%) and the moderate consumers (13%) are less interested in convenience foods and are among the lowest purchasers of these foods. This convenience orientation was defined based on the respondent's lifestyle, attitudes toward convenience, and their reported purchase behaviour for a range of convenience foods. It should be noted that the convenience orientation measures developed during the research were not used in the process of identifying the FRL segments. The measures merely served as descriptive variables to further describe each of the segments. Table 8.2(a) highlights the mean scores of segments on the lifestyle issues and attitudes toward convenience.

Hedonistic consumers are particularly interested in the more pleasurable aspects of food and they enjoy seeking out new ways of cooking foods. The conservative consumers are quite traditional in their shopping and cooking behaviour, they do not like change, and security is a strong purchasing motive. The view of the conservative consumers is that cooking and shopping is very much the women's responsibility. The extremely uninvolved segment is the least interested segment of all in every aspect of food. Food does not have a role other than a basic functional one. They do not enjoy shopping or cooking and show a lack of interest in the quality aspects of food elements. The enthusiastic food consumers are very interested in every aspect of food shopping, preparation and cooking. In addition, more than any other group they are interested in health, value for money, freshness, taste and organic food. They are more likely to plan their meals and preparation involves the entire family. The moderate segment do not have very strong characteristics, and are neither particularly interested nor uninterested in all aspects of food. The adventurous consumers are very involved with food. For them food personifies novelty, and a way to socialise and develop relationships (Ryan *et al.*, 2002).

The demand for convenience differed between the segments. To illustrate, hedonistic food consumers perceived greater time pressures and were also more likely than consumers in other segments to believe that convenience foods were good value for money. They were the most frequent purchasers for several

Table 8.2(a) Comparison of the Irish FRL segments' mean scores with the sample mean scores for the eight convenience-related issues

	Mean score of sample	Hedonistic consumer	Conservative consumer	Extremely uninvolved consumer	Enthusiastic consumer	Moderate consumer	Adventurous consumer
Lifestyle issues which drive the convenience food market							
Time pressures	12.85	+1.07	−1.50	+0.48	+0.98	−1.37	–
Stress levels	8.68	+0.92	−0.50	+0.61	+0.39	−1.37	−1.42
Breakdown of mealtimes	11.67	+0.55	–	+0.99	+0.98	−2.28	−1.46
Eating alone	12.97	+0.79	−0.95	+0.70	–	−0.79	−0.49
Individuality	12.77	–[1]	–	+0.56	+0.70	−0.94	−0.55
Attitude towards convenience food							
Convenience food value for money	9.51	+1.20	−0.68	–	–	−0.94	−0.43
Health value of convenience food	11.43	+0.80	−0.57	–	+1.12	−0.78	−2.19
Time and convenience food	14.90	+0.29	−0.52	+0.51	+1.10	−1.11	−0.79

Note: [1] There is no difference between the segment mean score and the population mean score.

Table 8.2(b) Comparison of the British FRL segments' mean scores with the sample mean scores for the 16 convenience-related issues

	Mean score of sample	Snacking	Careless	Uninvolved	Rational	Adventurous	Conservative
Forces driving the convenience food market							
Time pressures	12.43	−0.20	+1.33	+0.77	−0.19	−1.04	−0.30
Stress levels	9.41	+1.90	−0.12	+0.84	−0.38	−1.78	−0.71
Breakdown of mealtimes	11.20	+0.74	+0.15	+0.96	−0.32	−1.36	+0.26
Eating alone	12.56	−0.24	+0.92	+1.34	+0.13	−1.80	+0.23
Individuality	11.75	+0.08	+0.13	+0.48	+0.10	−0.78	+0.16
Time and ways of shopping	12.24	+0.06	+2.00	+1.40	−0.76	−1.44	−0.31
Enjoyment of meal preparation	10.18	+1.47	+0.97	+4.42	−1.94	−3.39	+0.78
Skills requirement	9.48	+1.89	−0.39	+2.49	−0.61	−2.78	−0.09
Beliefs about convenience food							
Convenience food value for money	10.53	+0.94	+0.12	+1.05	−0.30	−1.31	−0.49
Health value of convenience food	10.95	+1.14	+0.16	+0.93	−0.12	−1.68	−0.65
Time and convenience food	13.30	−0.44	+0.58	+1.47	+0.28	−1.25	−0.73
Involvement with convenience food	9.62	+2.21	+0.23	+1.36	+0.03	−2.99	−1.78
Post-meal convenience							
Clearing up	8.57	+2.26	+0.25	+0.95	−0.10	−2.65	−1.32
Disposal of waste ingredients	9.77	+1.64	+0.48	+1.30	−0.30	−2.43	−0.78
Convenience consumption							
Propensity towards convenience products	9.64	+1.75	+0.65	+1.86	−0.76	−2.68	−0.25
Propensity towards convenience processes	9.44	+1.78	−0.38	+1.18	−0.40	−2.27	−0.72

convenience food products and services, including frozen pizzas, prepared dinners, prepared sauces and prepared vegetables. They were the second most frequent purchasers of frozen chips, prepared meal centres and for eating in public houses. Additionally, these consumers snacked more than the other segments. In terms of product development as these consumers bought more convenience foods than any other group, they are an excellent segment for companies to target. Factors like avoidance of waste, reduction in washing up and the ability to microwave foods could be important buying motivations for this segment, who have little time to eat and probably even less time for shopping and cooking. Companies could also consider the possibility of developing snacks for these consumers.

The adventurous food consumers also used a number of convenience food products, but their motivations, needs and demands differed from the hedonistic food consumers. Tasting and preparing new foods was extremely important to them. They also did not believe that convenience foods are healthy. Price was relatively less important to these consumers and they are among the highest income earners. However, product information was important to this group. Therefore, companies could develop convenient higher priced food products which are considered healthy, new and/or ethnic meals that are restaurant quality ready meals. Information and promotion campaigns should emphasise health, quality, and the variety of new and ethnic convenience food products on offer.

8.5.2   British FRL segments and their convenience orientation

In 2002, six FRL segments were identified in Britain: the snacking food consumer (20% of consumers), the careless food consumer (14%), the uninvolved food consumer (14%), the rational food consumer (26%), the adventurous food consumer (17%) and the conservative food consumer (9%). Table 8.2(b) shows the mean scores of the segments on the convenience dimensions. The former three, the snacking, careless and uninvolved were identified as convenience-seeking segments. They represented 52% of the sample. The rational food consumers, adventurous food consumers and conservative food consumers were more concerned with the quality attributes of foods, including health, taste and freshness. Inspection of segment sizes shows that British consumers have become more convenience-oriented in their attitudes to food. In an earlier application of the FRL, one segment, the snackers, did not exist. This illustrates the importance and relevance of using segmentation to understand changing consumer requirements. The proportion of convenience-oriented consumers rose from 36% of consumers in 1994 to 48% in 2002. The size of the adventurous segment has also increased. In parallel, the proportions of rational and conservative consumers decreased over the eight-year period, indicating that British consumers have become less traditional in their food habits. The segments are described by Buckley *et al.* (2005) and as the approach was broadly the same as for the Irish study they are not further discussed. Instead convenience-specific lifestyle segmentation is discussed in the next section.

8.6   Convenience food lifestyle segmentation in Great Britain

For Great Britain, in addition to the FRL segmentation and convenience profiling on the FRL segments, a further segmentation was completed. (The profiling was similar but not identical with that carried out in Ireland in 2001.) In the latter case, only the convenient dimensions developed, based on an analysis of trends driving the market, were used.

In order to further investigate consumer attitudes to convenience, Buckley (2003) segmented the British market based on convenience-specific food life-styles (CFL). Four CFL segments were identified namely, the food connoisseurs (26% of consumers), the home meal preparers (25%), the kitchen evaders (16%) and the convenience-seeking grazers (33%). Twenty convenience-specific dimensions were identified, some of which are also part of the FRL tool. Thus where appropriate, relevant FRL variables were used. Table 8.3 compares the mean score of the sample with the mean scores of the CFL segments identified for each of the convenience food lifestyle factors. Statistical tests between the mean score of the particular segment and the mean score of the remaining segments combined ascertained where significant differences existed for each of the CFL dimensions.

Table 8.3   Mean scores of the convenience food lifestyle (CFL) segments in Great Britain compared with the population mean score across the CFL factors

	CFL sample	Food connoisseurs	Home cookers	Kitchen evaders	Convenience-seeking grazers
Convenience food choice	3.26	−0.50	−0.77	+0.74	+0.65
Convenience in meal preparation and cooking	3.47	−0.29	−0.72	+0.97	+0.33
Neophilia	4.48	+0.30	+0.10	−0.57	−0.05
Freshness versus convenience	5.39	+0.22	+0.53	−0.12	−0.54
Convenience in shopping	3.92	−0.20	−0.62	+0.85	+0.23
Time pressures	4.13	−0.46	−0.84	+1.13	+0.48
Individualism	3.93	−0.18	−0.62	+0.69	+0.31
Price check	4.47	−0.61	+0.35	−0.27	+0.37
Shopping list	4.26	−1.33	+1.49	−1.88	+0.84
Disposal of waste ingredients	3.25	−0.42	−0.83	+1.04	+0.49
Information check	3.45	+0.04	+0.34	−0.81	+0.10
Eating out	4.26	+0.18	–	+0.13	−0.20
Whole family	4.06	+0.14	+0.24	−0.43	−0.08
Woman's task	3.24	−0.33	+0.03	−0.02	+0.25
Stress levels	3.12	−0.21	−0.70	+0.47	+0.47
Propensity towards convenience processes	3.12	−0.60	−0.77	+0.33	+0.92
Planning	3.72	−0.18	+0.53	−0.80	+0.12
Breakdown of mealtimes	3.74	−0.35	−0.66	+0.75	+0.42
Snacking	3.23	−0.43	−0.38	+0.18	+0.55
Eating alone	4.51	−0.46	−1.06	+1.46	+0.48

Two of the four CFL segments were identified as being particularly convenience oriented, the kitchen evaders and the convenience-seeking grazers. Of all the segments, the kitchen evaders are the most likely to select convenience foods to make their lives easier in terms of washing up and preparing something they wouldn't otherwise know how to cook from raw ingredients. They place the highest emphasis on convenience in meal preparation and shopping. Of all segments, the kitchen evaders experience the highest level of perceived time pressures. In line with this, and along with the convenience-seeking grazers, the kitchen evaders encounter the highest perceived stress levels compared with the other segments. Thus for a manufacturer of ready meals, the major benefit of these products should be emphasised as the speed and ease with which they can be cooked.

The convenience-seeking grazers use the microwave a lot and they have the highest tendency to snack between meals. They are more frequent purchasers of takeaway meals to eat both at home and away from home. Hand held snacks, hot or cold, for immediate consumption and suitable for eating on the move might provide a lucrative opportunity for manufacturers targeting the segment of convenience-seeking grazers.

Less convenience-oriented segments may also offer opportunities. The food connoisseurs may be characterised as a group that takes the most pleasure from new foods and experimenting in the kitchen. They feel less time pressured and stressed than most other groups and saving time and energy in meal preparation is less important to these consumers than to some of the other groups. Thus, for the food connoisseurs, communication strategies could focus particularly on fun and enjoyment elements in the context of meal preparation. These consumers enjoy cooking, and food has an important role to play from an entertainment perspective. An opportunity for food manufacturers might present itself in the merchandising of primary meal ingredients with recipe cards.

8.7   Segmentation and product development for convenience foods

The value of the FRL lies in it providing a complete approach to understanding consumers in the context of looking at all the important factors influencing food choice. It gives a general understanding of consumer motivations and cooking and eating perceptions. As a measurement instrument it has a sound theoretical base in the form of the means-end chain theory (Grunert *et al.*, 1996). The linking of the consumers' value perception from the FRL with demographic data provides further valuable information on particular groups to target with existing and new food products. In terms of providing understanding on a specific market such as convenience foods discussed here, this was only possible because additional information on the convenience-related attitudes and reported purchase behaviour of many convenience foods was collected. This enabled detail profiling of the segments from a convenience viewpoint. Such an approach, while very useful, is expensive as additional data had to be collected. Thus the idea of a specific segmentation instrument for convenience was conceived. The instrument developed provided segmentation based on convenience perceptions and this is more directly useful for convenience food companies. However, while the CFL is more specifically suited to the understanding consumers of convenience foods it does not deal in overall terms with general factors influencing the consumer and has not been tested for cross-cultural validity.

Although FRL results can form an important part of the analysis for any food company in a market, it is not sufficient in itself to be the only basis of decisions on product development and communication. Specific research relevant to the product area is also essential in deciding on new product development or modifications of the promotional programme for existing products. In Denmark a commercial company (Jysk Analyseintsitut, Denmark) undertakes regular FRL surveys for a consortium of food companies and it is suggested that this approach has considerable merit as it allows access to very useful consumer information inexpensively on a confidential basis. As these surveys are repeated at regular intervals, long-term trends can be identified and appropriate marketing strategies put in place.

The complex nature of convenience needs to be understood in order to determine convenience-seeking consumer segments. Convenience in food-related activities deals with all stages of the food provisioning process from shopping and meal preparation to consumption and post meal activities. Considering convenience in the context of market segments, FRL and CFL segmentation provides a useful means of discussing homogenous food groups. The segmentation studies presented illustrated there are opportunities for food manufacturers to develop specific product offerings and develop communication strategies for each of the identified segments.

8.8   Sources of further information and advice

Readers interested in the FRL instrument are referred to the MAPP website http://www.asb.dk/centres/mapp.aspx. For more information on applications in Ireland and Great Britain by the Irish team from Ashtown Food Research Centre (formerly The National Food Centre) Teagasc, UCC and Bord Bia, readers are referred to the publications cited by Ryan *et al.* (2002, 2004) and Buckley *et al.* (2005) and also to the Teagasc website http://www.teagasc.ie where working papers presented to industry are available.

8.9   Acknowledgements

The Irish and British FRL studies were carried out with the financial support of the Irish Government under the National Development Plan, 2000–2006, Food Institutional Research Measure. The authors would like to thank Dr Karen

Brunsø and Dr Klaus Grunert, Centre for research on customer relations in the food sector (MAPP Institute), The Aarhus School of Business, Denmark for their advice and support and Mr Julian Smith, Bord Bia for his contribution to the project.

8.10   References and further reading

ANDERSON B B (1972), 'Working women versus nonworking women: a comparison of shopping behaviour', in Boris W Becker and Helmut Becker (eds), *Combined Proceedings, Chicago*, American Marketing Association, 335–359.

BASS E, VAN BATTUM S, VOORBERGEN M and ZWANENBERG A (1999), 'The fight for stomach share', Special Report, Food and Agribusiness Research, Rabobank International.

BERRY L L (1979), 'The time-buying consumer', *Journal of Retailing*, 55 (4), 58–69.

BOEDEKER M and MARJANEN H (1993), 'Choice orientation types and their shopping trips, 7th International Conference on Research in the Distributive Trades', *Conference Proceedings, Stirling*, Institute for Retail Studies, University of Stirling, 59–67.

BONKE J (1992), 'Choice of Foods – allocation of time and money, household production and market services', MAPP, Working Paper no. 3.

BREDAHL L and GRUNERT K G (1997), 'Food-related lifestyle trends in Germany 1993–1996', MAPP Working Paper no. 50.

BROWN L G (1989), 'The strategic and tactical implications of convenience in consumer product marketing', *The Journal of Consumer Marketing*, 6 (3), 13–19.

BROWN L G and McENALLY M R (1993), 'Convenience: definition, structure, and application', *Journal of Marketing Management*, 2 (2), 47–56.

BRUNSØ K, GRUNERT K G and BREDAHL L (1996), 'An analysis of national and cross-national consumer segments using the food-related lifestyle instrument in Denmark, France, Germany and Great Britain', MAPP, Working Paper no. 35.

BUCKLEY M (2003), 'Lifestyle segmentation of food consumers in Great Britain, with specific reference to their convenience orientation', MSc thesis, Cork, UCC.

BUCKLEY M, COWAN C, McCARTHY M and O'SULLIVAN C (2005), 'The convenience consumer and food-related lifestyles in Great Britain', *Journal of Food Products Marketing*, 11 (3).

CANDEL M J J M (2001), 'Consumers' convenience orientation towards meal preparation: conceptualisation and measurement', *Appetite*, 36, 15–28.

CAPPS JR O, TEDFORD J R and HAVLICEK JR J (1985), 'Household demand for convenience and nonconvenience foods', *American Journal of Agricultural Economics*, November, 861–869.

COSTA A I A, DEKKER M, BEUMER R R, ROMBOUTS F M and JONGEN W M F (2001), 'A consumer-oriented classification system for home meal replacements', *Food Quality and Preference*, 12, 229–242.

DARIAN J C and COHEN J (1995), 'Segmenting by consumer time shortage', *Journal of Consumer Marketing*, 12 (1), 32–44.

DAVIES G and MADRAN C (1997), 'Time, food shopping and food preparation: some attitudinal linkages', *British Food Journal*, 99 (3), 80–88.

ELMORE-YALCH R (1998), 'A Handbook: Using Market Segmentation to Increase Transit Ridership', Report 36, Transportation Research Board NRC, National Academy Press, Washington, DC.

GEEST PLC (2001), 'Market Trends', www.geest.co.uk/html/profile/markets/market_trends.htm, 18th September, 2001.

GOFTON L (1995), 'Convenience and the moral status of consumer practices', in Marshall DW (ed.), *Food Choice and the Consumer*, UK, Chapman and Hall, 152–181.

GOFTON L and MARSHALL D W (1988), *A Comprehensive Scientific Study of the Behavioural Variables Affecting the Acceptability of Fish Products as a Basis for Determining Options in Fish Utilisation Research and Development at Torry Research Station*, University of Newcastle Upon Tyne.

GOFTON L and MARSHALL D W (1992), *Fish: A Marketing Problem*, Horton Publishing, and Horton.

GOFTON L and NESS M (1991), 'Twin trends: health and convenience in food change or who killed the lazy housewife', *British Food Journal*, 93 (7), 17–23.

GRUNERT K G, BRUNSØ K and BISP S (1993), 'Food-related lifestyle: Development of a cross-culturally valid instrument for market surveillance', MAPP, Working Paper no. 12.

GRUNERT K G, BAADSGAARD A, LARSEN H H and MADSEN T K (1996), *Market Orientation in Food and Agriculture*, Massachusetts, Kluwer Academic Publishers.

IGD (1998), *Food Consumption '98: the one stop guide to the food consumer*, IGD Business Publications.

IGD (2000), *Catering for the Consumer*, IGD Business Publications.

IGD (2001), *Meal Solutions: Simplifying food choice and provision*, IGD Business Publications.

IGD (2002), Fact Sheet, UK Foodservice Market Overview, 21 August 2002, www.igd.com.

KAHN P (1976), 'One- and two-member household feeding patterns', *Food Product Development*, 10 (8), 22–23, 30.

KHAN Y (2000), 'Meal solution trends', *The World of Food Ingredients*, April/May, 16–23.

KOTLER P (1991), *Marketing Management: Analysis, Planning, Implementation and Control*, New York, Prentice-Hall.

LEATHERHEAD FOOD RA (2001), *The UK Food and Drinks Report, April 2001*, 16, Leatherhead Publishing.

MADILL-MARSHALL J J, HESLOP L and DUXBURY L (1995), 'Coping with household stress in the 1990s: who uses "convenience foods" and do they help?', *Advances in Consumer Research*, 22, 729–734.

MAN D and FULLERTON E (1990), 'Single Drop Depositors B An Aid to Production of Chilled Ready Meals'. In: *Process Engineering in the Food Industry, Convenience Foods Quality Insurance*, Field R W and Howell J A (eds), UK, Elsevier Science Publishers Ltd.

MAPP (1996), *Newsletter No. 4, October 1996*, MAPP.

MARSHALL DW (1995), *Food Choice and the Consumer*, Chapman and Hall.

McKENZIE J C (1986), 'An integrated approach – with special reference to changing food habits in the UK', in Ritson C, Gofton L and McKenzie J (eds), *The Food Consumer*, London, John Wiley and Sons.

McMILLAN N H (1994), 'Mall of America: can a big time entertainment venue win by fighting customers?', *International Trends in Retailing*, 11 (2), 87–91.

MINTEL INTERNATIONAL GROUP LIMITED (2000), *The Evening Meal*.

OFFICE FOR NATIONAL STATISTICS (2002), *Social Trends, No. 32*, Jill Matheson and Penny Babb (eds), London, The Stationery Office.

O'SULLIVAN C, SCHOLDERER J and COWAN C (2005), 'Measurement Equivalence of the Food-Related Lifestyle (FRL) Instrument in Ireland and Great Britain', *Food Quality and Preference*, 16, 1–12.

PROMAR INTERNATIONAL (1997), *Convenience Plus: New Approaches and Opportunities for Food Companies in Europe: Strategy to 2000 and Beyond*, UK.

REUTERS BUSINESS INSIGHT (2002), *The Meals Solutions Outlook to 2007: Increasing Market Share in Frozen, Chilled and Ambient Foods*.

RYAN I, COWAN C, McCARTHY M and O'SULLIVAN C (2002), 'Food-related lifestyle segments in Ireland with a convenience orientation', *Journal of International Food and Agribusiness Marketing*, 14 (4): 29–48.

RYAN I, COWAN C, McCARTHY M and O'SULLIVAN C (2004), 'Segmenting Irish food consumers using the food-related lifestyle instrument', *Journal of International Food and Agribusiness* 16 (1): 89–114.

SCHOLDERER J and GRUNERT K G (2005), Consumers, food and convenience: The long way from resource constraints to actual consumption patterns, *Journal of Economic Psychology*, 26, 105–128.

SCHOLDERER J, GRUNERT K G, BREDAHL L and BRUNSØ K (2004), Cross-culturally validity of the food-related lifestyles instrument (FRL) within Western Europe, *Appetite*, 42 (2): 197–211.

SENAUER B (2001), 'The food consumer in the 21st century: New research perspectives', Working Paper 01-03, The Food Industry Center, University of Minnesota.

SENAUER B, ASP E and KINSEY J (1991), *Food trends and the changing consumer*, St. Paul, Egan Press.

SHOVE E and SOUTHERTON D (2000), 'Defrosting the freezer: from novelty to convenience. A story of normalization', *Journal of Material Culture*, 5 (3), 301–320.

SLOAN A E (1997), 'What's cooking?', *Food Technology*, 51 (9), 32.

SOMOGYI J C (1990), 'Convenience foods and the consumer – current questions and controversies', *Nutritional Adaptation to New Lifestyles*, (45), 104–107.

STAFFORD T H and WILLS J W (1979), 'Consumer demand increasing for convenience in food products', *National Food Review*, (13), 15–17.

SWOBODA B and MORSCHETT D (2001), 'Convenience-oriented shopping: a model from the perspective of consumer research', in Frewer L, Risvik E and Schifferstein H (eds), *Food, People and Society: A European Perspective of Consumers' Food Choices*, Berlin, Heidelberg, Springer-Verlag, 177–196.

TNS (2000) Press Release, *'Healthy Eating' is Key in Britons' Choice of Convenience Foods – Survey Findings*, 10 November 2000.

TRAILL B (1997), 'Structural changes in the European food industry: consequences for innovation', in Traill B and Grunert K G (eds), *Product and Process Innovation in the Food Industry*, UK, Chapman & Hall, 38–60.

TRAILL B and HARMSEN H (1997), 'Pennine Foods: always prepared for a new ready meal', in Traill B and Grunert KG (eds), *Product and Process Innovation in the Food Industry*, UK, Chapman & Hall, 187–199.

UMESH U N, PETTIT K L and BOZMAN C S (1988), 'Shopping Model of the Time-Sensitive Consumer', *Decision Sciences*, 20, 715–729.

VERLEGH P J W and CANDEL M J J M (1999), 'The consumption of convenience foods: reference groups and eating situations', *Food Quality and Preference*, 10, 457–464.

YALE L and VENKATESH A (1985), 'Toward the construct of convenience in consumer research', *Proceedings*, Association for Consumer Research, 402–408.

9

Outsourcing meal preparation

**J. R. Cornelisse-Vermaat, G. Antonides and J. van Ophem,
Wageningen University, The Netherlands and H. Maassen van den
Brink, University of Amsterdam, The Netherlands**

9.1   Introduction

Due to the increased female labour participation in the past decades, households are short of time to perform all household and care activities. There are three strategies to solve this problem: (1) outsourcing household and care activities; (2) substitution of household and care tasks by domestic appliances; and (3) time arrangement (adjusting working hours or shop opening hours) (SCP, 2000; Van Ophem and De Hoog, 1995; and Van Dam *et al.*, 1994). In addition, households may simply reduce their level of household and care activities. In this chapter, we will focus on the outsourcing of household and care activities and its determinants, in particular with respect to meal preparation.

Outsourcing is defined as an arrangement for a particular service outside the household (either privately or subsidised) to take care of household activities. Since the end of the 1990s in western European countries, US, Canada, and Australia, outsourcing household and care activities has become more and more common (see, for example, Kim, 1989; Bittman *et al.*, 1998, Cohen, 1998; Spitze, 1999; Mihalopoulos and Demoussis, 2001; RIVM, 2004; SCP, 2000; and Tijdens *et al.*, 2000). This trend was enabled by increased female market labour supply which in many households has increased the financial budget. Over the years, the number of outsourcing possibilities has grown. Still, until now, little research has been performed on the determinants of outsourcing.

Becker (1965) gives an important theoretical argument for outsourcing. Households can be considered as small production companies that try to maximise their output, restricted by their time and budget. If time spent on the labour market is more valuable than time spent on home production, it could be profitable to outsource home production (of course, this also depends on the price of outsourcing). It is therefore more likely for working wives to outsource household activities, since they will place a higher marginal value on their household production time than non-working wives (Kim, 1989).

Like in other Western cultures, at present Dutch households can outsource home cleaning to a cleaning lady/man, cooking to restaurants (or people can eat ready-to-eat-meals, takeaway food or delivery food), and childcare to day-care centres. The increased household income, due to higher female labour participation, gives more possibilities to outsource domestic work to others. A total of 82 percent of the Dutch dual earner households eat takeaway food more than once per month, as compared with 62 percent of the single-income households.

Until now, not much has been known about the outsourcing behaviour of immigrants in the Netherlands and about outsourcing differences between immigrants and the native Dutch. Outsourcing behaviour is not only determined by socio-economic and demographic variables, but culture (or ethnicity) is also important in explaining outsourcing within households. For example, ethnic groups may choose different (cheaper) outsourcing possibilities than the native Dutch. Earlier research shows that immigrants in the Netherlands use formal childcare less frequently than the native Dutch (NIBUD, 2004).

This chapter aims at: (1) studying the determinants of the demand for outsourcing meal preparation in the Netherlands within the conceptual context of household production theory; and (2) investigating differences in outsourcing meal preparation between non-Western immigrants and the native Dutch.

This chapter is structured as follows. Section 9.2 provides an overview of the theory and earlier empirical results on outsourcing meal preparation and con-structs a conceptual framework. Section 9.3 describes our data and the methods used. In Section 9.4, the estimation results on outsourcing of meal preparation in different household types are given. Section 9.5 concludes and discusses some issues of this chapter.

9.2   Literature and conceptual model

Earlier research shows that households with higher-educated wives are more likely to use outsourcing opportunities than households with lower-educated wives (Bellante and Foster, 1984; Soberon-Ferrer and Dardis, 1991). With respect to outsourcing household activities, both the household's income and the wife's income are important. Soberon-Ferrer and Dardis (1991) found that unearned income (non-labour income), wives' wages, wives' education, and being white are significant factors in outsourcing home cleaning. Spitze (1999) and Oropesa (1993) found that in the United States, higher-income households receive more paid household help than lower-income households. This is also true in Dutch households (Van der Lippe et al., 2004; Lambriex and Siegers, 1993). In the United States, the top-income households spend more than twelve times as much on housekeeping services and four times as much on food away from home as the lowest income group (Cohen, 1998).

With respect to outsourcing meal preparation in Western countries, the following can be found in the literature. Outsourcing meal preparation is positively related with income, employment status, urban location, and the number of people in the household aged over 14 (Heiman *et al.*, 2001; Mihalopoulos and Demoussis, 2001; and Manrique and Jensen, 1998). For example, US-data from Soberon-Ferrer and Dardis (1991) show that the wage rate of both male and female in the household are significantly positively related to the expenditure on food-away-from-home. The same is true when the wife is higher educated and when the family owns the home.

Greek data shows that the expenditure on food-away-from-home has increased from 18 percent of the household budget in 1982 to 25 percent of the food budget in 1994. (Mihalopoulos and Demoussis, 2001). Mihalopoulos and Demoussis found that singles and fully employed people spend more on food-away-from-home, whereas people aged above 55, married couples, and households with young children spent less on food-away-from-home. In addition, a Spanish study concluded that households with children aged above 14 and households with fewer children were more likely to outsource meal preparation (Manrique and Jensen, 1998). The increase in expenditure on outsourcing meal preparation in households with children aged above 14 was also found in an Australian study (Bittman *et al.*, 1998).

Several studies mention that outsourcing meal preparation is more common in urban areas (Mihalopoulos and Demoussis, 2001; Manrique and Jensen, 1998; Cohen, 1998; and Soberon-Ferrer and Dardis, 1991). An explanation could be that the possibilities to outsource meal preparation (especially fast food) are more established in the urban areas compared to rural areas.

Research on outsourcing behaviour of immigrants is scarce. A few US studies show that blacks, Hispanics, and Asians (although the question arises whether these people should be considered as immigrants or not) spend less on food away from home and domestic services, but more on clothing care than whites (Cohen, 1998; Soberon-Ferrer and Dardis, 1991; Bellante and Foster, 1984). In Switzerland, after correcting for wage rate, immigrants spend more time on household activities, which could imply that they outsource less (Sousa-Poza *et al.*, 2001).

Data from the Dutch Social and Cultural Planning Bureau (SCP) show that meal preparation is mainly outsourced by two-earner households: 60 percent eat out more than once per month, and 82 percent eat takeaway food more than once per month (35 percent even more than once a week). For one-earner households these figures are 26 percent for restaurant visits more than once per month, respectively 62 percent for takeaway food more than once per month (SCP, 2000). Visiting restaurants is highly positively correlated with income and negatively correlated with having children.

Takeaway food is cheaper and less time-consuming, which explains why, in particular, middle-income households with children eat takeaway food (Van der

Lippe *et al.*, 2004; SCP, 2000). Between 1980 and 1999, the expenditure on meal-preparation outsourcing in the Netherlands increased from 3.1 to 4.2 percent of the total household budget (CBS, 2001). Time spent on cooking and dishwashing significantly decreased with the consumption of takeaway food and increased with the number of children living at home, cohabiting without children, and having children aged 15 years and older (Tijdens *et al.*, 2000). Time spent on cooking also decreased with the working hours of both males and females (Labriex and Siegers, 1993).

Using household production theory of Becker (1965) and Gronau (1986, 1977) as a linear specification of the demand function for outsourcing meal preparation (O) can be constructed. Let Y represent the household income per month and N the weekly working hours, the demand function for outsourcing meal preparation can be represented as follows for a one-person household:

$$O = \beta_0 + \beta_1 Y + \beta_2 \bar{N} + \beta_3 D + \epsilon \tag{9.1}$$

with β_0 the constant term, and β_1–β_3 coefficients to be estimated. D is a vector of socio-economic and demographic variables like age, level of education, children at home, ethnicity, etc. ϵ is a stochastic disturbance term with normal distribution and zero mean.

We are aware of the fact that the number of working hours (N) is endogenous. However, in real-life, it is difficult to adjust the time spent on market labour (at least in the short run) when compared with household and care time or leisure. Therefore, we consider N as given (see also Van Ophem and De Hoog, 1995; Lambriex and Siegers, 1993; and Homan, 1988), and equal to $\bar{N}$. This has the additional advantage of avoiding the problem of over-determination.

With model (9.1) we are able to estimate the demand for takeaway food, delivery food, and eating out in restaurants and determine which factors are of importance with the respect to outsourcing meal preparation.

In this chapter, we apply the conceptual framework of the demand for outsourcing meal preparation as described to the Dutch situation. We have adapted the model into a model for two-person households. The following demand function was constructed for takeaway food, delivery food, and eating out (O, measured in expenditures per month[1]) for two-person households (couples):

$$O = \beta_0 + \beta_1 Y + \beta_2 \bar{N}_f + \beta_3 \bar{N}_m + \beta_4 D + \epsilon \tag{9.2}$$

where β_0 represents the constant term, and β_1–β_4 are the coefficients to be estimated. Y equals the household income per month (net income of each partner, and/or other income like social security or children's allowance). $\bar{N}_f$ and $\bar{N}_m$ equal the weekly market working hours for females and males, respectively.

---

1. These were measured as expenditures directly, rather than as P_oO, because our cross-section data did not give enough information about the prices. For simplicity reasons we write O in equation (9.1).

D is a vector of socio-economic and demographic variables and includes dummies for ethnicity (with the native Dutch as reference for Surinamese/ Antilleans, Moroccans, and Turks[2]), age of each partner, level of education of each partner, children living at home, living in an urban area, being a home-owner, religious affiliation, and health. ϵ is a stochastic disturbance term with normal distribution and zero mean. The appendix comprises a list of the variables used in the estimations and shows the sample distributions of the variables used.

A high number of working hours is normally correlated with a high household income. In households with a high number of working hours, household and care activities may be outsourced to 'buy' time for activities with their children (Hallberg and Klevmarken, 2003). Therefore, working hours are expected to have a positive relationship with outsourcing. Education is also an indication of high working hours and a high income and therefore higher expenditure on outsourcing.

Immigrants are expected to spend less on outsourcing meal preparation than the native Dutch, partly because of their lower income levels, but also because of cultural differences (Cohen, 1998; Soberon-Ferrer and Dardis, 1991; Bellante and Foster, 1984). Immigrants may run their households in a relatively traditional way and, as in traditional households, they prefer to do more of the cooking themselves instead of outsourcing it (Oropesa, 1993).

Younger people may outsource more, because they work more hours on the labour market and have less time for household and care tasks (Mihalopoulos and Demoussis, 2001; Manrique and Jensen, 1998; Cohen, 1998). On the other hand, middle-aged people usually have more money available for outsourcing household and care tasks and may have a higher need for it (Bittman et al., 1998).

We assume that people owning their home have a higher income in general; therefore, a positive effect of home ownership on outsourcing meal preparation is hypothesised as is also found in the literature (Cohen, 1998; Bellante and Foster, 1984; Soberon-Ferrer and Dardis, 1991). A positive relation between living in an urban area and outsourcing is expected, since living in an urban area will give more opportunities for outsourcing meal preparation (Manrique and Jensen, 1998; Cohen, 1998; and Soberon-Ferrer and Dardis, 1991).

People with a religious background may have a more traditional time allocation within the household and make less use of outsourcing facilities because, in general, they favour traditional values. Literature also suggests that traditional women are less likely to have meals delivered, because they like to cook more, or feel guilty when they serve convenience foods (Soberon-Ferrer and Dardis, 1991). The effect of health on outsourcing meal preparation may be negative, since people with poor health may be less able to prepare meals themselves, whereas people with better health will be more able to cook their own meals.

---

2. We assumed that both partners in each household had the same ethnicity, which was true for about 80 percent of the households in our sample.

9.3   Data and methods

In cross-section data from the Dutch population, the number of immigrants is usually low. Therefore, a stratified sample was needed in order to compare the outsourcing behaviour of immigrants with that of the native Dutch in an efficient way. In order to obtain enough information to compare both sexes and different household types, we aimed at obtaining 700 respondents per group.[3]

The data was collected in the Netherlands between September and November 2001 by a Dutch organization for market research. Interviewers who could speak both Dutch and Moroccan, or both Dutch and Turkish were hired to conduct bilingual interviews.[4]

The Dutch sub-sample was drawn randomly from the total pool of phone numbers (about 6.8 million) administered by the Dutch Telephone Company in 2001. The immigrant sub-samples were drawn from a sample of about 80,000 names owned by the market research company. The immigrants were selected on the basis of their names (indicating their ethnicity).[5] The Surinamese and Antilleans were considered as one group, since they are from comparable origin. Each group consisted of about 700 respondents, except for the Moroccan group in which only 449 respondents participated within the data collection time period.

The respondents were asked about their behaviour concerning outsourcing meal preparation. They were asked which types of outsourcing meal preparation they used, how much money they spent on it, and how frequently they used these outsourcing opportunities. Three types of outsourcing meal preparation were questioned: takeaway food (like Chinese food, food from snack bars, or kebab), delivery food (like pizza-delivery at home), and eating in (fast food) restaurants. Also, socio-economic and demographic questions concerning income, level of education, age, and children living at home were asked.

Both bivariate and multivariate analyses have been performed to examine determinants of the demand for outsourcing meal preparation in the Netherlands and to investigate differences in outsourcing meal preparation between non-Western immigrants and the native Dutch. Analysis of variance was performed to study outsourcing differences between immigrants and native Dutch. To estimate the demand for outsourcing meal preparation, equation (9.2) was regressed using the Ordinary Least Squares method. Elasticities have been calculated to find out whether outsourcing meal preparation is considered as a normal good or a luxury good.

---

3. In 2003, the percentage of non-Western immigrants (including Moroccans, Turks, Surinamese and Antilleans) within the Dutch population was 10 percent, amounting to about 1,483,000 immigrants within the total population of about 16 million. The share of Moroccans, Turks, Surinamese, and Antilleans within the non-Western immigrants was about 70 percent (CBS, 2003).
4. This was not needed for the Surinamese/Antillean group, which speak Dutch as their mother language.
5. An immigrant is defined as a person having at least one parent who was born abroad (CBS, 2000).

9.4   Results

Table 9.1 gives an overview of the monthly frequencies of using outsourcing methods by the native Dutch, Surinamese/Antillean, Moroccan, and Turkish households. The percentages using these categories are stated between parentheses. Analysis of variance was accomplished to investigate the differences in outsourcing frequencies across the ethnic groups.

The analyses of variance show that the outsourcing use frequencies differed significantly across the groups in all three meal preparation outsourcing categories. The Turks outsourced meal preparation the most frequently in all three outsourcing categories. Compared to the other groups, Moroccans seemed to outsource meal preparation the least frequently. All three immigrant groups used delivery food more as compared with the native Dutch, who in turn preferred eating out in restaurants. Takeaway food was most often used for outsourcing meal preparation in all groups.

Table 9.2 shows the average expenditure on takeaway food, delivery food, and eating out among the native Dutch and immigrants in the Netherlands. Moroccans spent the most on takeaway food and delivery food. Although the Turks ate takeaway food the most frequently, they chose cheaper takeaway food than the Moroccans. The relatively large standard errors for almost all categories indicate that expenses on outsourcing facilities differed widely among the households.

The native Dutch and Surinamese/Antilleans spent the most on eating out. Although the groups differed in their frequency of outsourcing meal preparation,

Table 9.1   Average outsourcing frequencies within ethnic groups (percentage of use between parentheses)

	Native Dutch ($N = 701$)	Surinamese/ Antill. ($N = 701$)	Moroccans ($N = 449$)	Turks ($N = 700$)
Takeaway food***	2.13 (61.5%)	2.73 (64.1%)	2.65 (50.6%)	3.12 (61.3%)
Delivery food***	0.62 (14.8%)	0.62 (22.3%)	0.61 (19.4%)	0.80 (20.6%)
Eating out***	1.41 (54.6%)	1.22 (45.1%)	1.14 (37.3%)	1.68 (38.1%)

*** $p < 0.01$

Table 9.2   Average monthly expenditure* on meal preparation outsourcing per ethnic group (standard errors in parentheses)

	Dutch	Surinamese/ Antilleans	Moroccans	Turks
Takeaway food	39 (1.8)	48 (2.3)	52 (4.7)	49 (3.9)
Delivery food	34 (2.8)	39 (5.5)	55 (11.0)	49 (6.4)
Eating out	88 (6.5)	88 (10.2)	69 (4.4)	72 (6.6)

* rounded off to the nearest Euro

their total monthly expenditure was not significantly different across the groups.

In order to enable estimation of the demand for outsourcing meal preparations, we only included the respondents in the sample who reported use of the outsourcing options and who made expenditure on outsourcing (this implies that we left out 381 people who made no expenditure on outsourcing). We estimated equation (9.1) by Ordinary Least Squares regression. Table 9.3 shows the results for outsourcing meal preparation in two-person households (couples), either with or without children.

Table 9.3 Demand for outsourcing meal preparation for couples (t-values in parentheses)

Constant	4.022	(10.958)***
Native Dutch	ref. group	
Surinamese/Antilleans	−0.570	(−3.818)***
Moroccans	−0.749	(−4.078)***
Turks	−0.569	(−3.558)***
Working hours female	−0.004	(0.576)
Wrk hrs fem. not observed	−0.140	(0.693)
Working hours male	0.002	(0.475)
Wrk hrs male not observed	−0.385	(−1.519)
Net monthly household income	0.000	(4.478)***
Children at home 0–3	−0.320	(−2.467)**
Children at home 4–11	−0.070	(−0.587)
Children at home 12–15	0.319	(2.229)**
Children at home 16–25	0.261	(1.754)*
Children at home > 25	reference group	
Low level education female	reference group	
Medium level education female	0.316	(2.444)**
High level education female	0.421	(2.412)**
Low level education male	reference group	
Medium level education male	0.323	(2.557)**
High level education male	0.295	(1.881)*
Living in an urban area	0.125	(1.065)
Religious affiliation	−0.138	(−1.247)
Health	0.064	(1.201)
Homeowner	−0.013	(−0.115)
Age female 18–34	reference group	
Age female 35–44	0.011	(0.066)
Age female 45–64	−0.164	(−0.649)
Age female ≥65	−0.674	(−1.503)
Age male 18–34	reference group	
Age male 35–44	−0.568	(−3.678)***
Age male 45–64	−0.873	(−3.784)***
Age male ≥65	−0.750	(−1.973)**
# Observations	1452	
Adjusted R^2	0.136	
F statistic	9.822	

* $p < 0.10$
** $p < 0.05$
*** $p < 0.01$

For one-person households (including single parents with children), these results may be different, because in such households there is only one person taking care of all household and care activities and of generating income. Therefore, we have repeated the estimation for singles (including both one-person households and one-parent families). In this case only the working hours, level of education, and age of the respondent were included in the demand function (9.1). The results are shown in Table 9.4.

The results in Table 9.3 show that all immigrant groups spent less on outsourcing meal preparation than the native Dutch. This could be explained by a lower socio-economic status and different culture of the ethnic groups as compared with the native Dutch. For the Turks and Moroccans this could be explained by their lower income, too. However, the analyses were controlled for socio-economic status by income, level of education, and working hours. This indicates that the culture effect might have been stronger than the effect of socio-

Table 9.4   Demand for outsourcing meal preparation for singles and single parents (t-values in parentheses)

Constant	4.576	(12.489)***
Female	−0.092	(0.633)
Native Dutch	reference group	
Surinamese/Antilleans	−0.022	(−0.121)
Moroccans	−0.441	(−1.889)*
Turks	0.210	(0.877)
Working hours	−0.009	(−1.263)
Wrk hrs not observed	−0.521	(−2.001)**
Household income	0.000	(3.040)***
Children at home 0–3	−0.839	(−2.394)**
Children at home 4–11	−0.809	(−2.901)***
Children at home 12–15	0.651	(1.972)**
Children at home 16–25	−0.091	(−0.304)
Children at home >25	reference group	
Low level of education	reference group	
Medium level of education	0.150	(0.812)
High level of education	0.243	(1.111)
Living in an urban area	0.156	(1.094)
Religious affiliation	−0.290	(−1.842)*
Health	−0.042	(−0.590)
Homeowner	0.365	(1.982)**
Age 18–34	reference group	
Age 35–44	−0.678	(−3.157)***
Age 45–64	−1.157	(−5.222)***
Age ≥65	−2.341	(−7.792)***
# Observations	715	
Adjusted R^2	0.202	
F statistic	10.031	

* $p < 0.10$
** $p < 0.05$
*** $p < 0.01$

economic status. This finding will be discussed in more detail in Section 9.5. The analysis for singles shows that only Moroccans spent less than the other groups on outsourcing meal preparation. This result shows that outsourcing behaviour between couples and singles (with or without children) was very different.

The results show that single men spent more on outsourcing meal preparation than single women. This may imply that single men either bought more expensive meals, or outsourced meal preparation more frequently than single females. The expenditure on outsourcing varied significantly with the working hours of each partner for both singles and couples. Singles who own their home spent more money on outsourcing meal preparations compared to singles who rented a home. Of course, owning a home could be considered as an indication of wealth.

Household income was significantly positively related to outsourcing expenditure for both couples and singles, which was also found in earlier research (Lambriex and Siegers, 1993). The level of education of both females and males positively affected the total amount spent on outsourcing meal preparation. A social-class effect could explain this result, since in higher social classes it is more common (and more accepted) for females to work and more household and care activities need to be outsourced.

No significant effect of health on outsourcing expenditure for meal preparation was found. This indicates that people with poorer health did not spend more money on outsourcing than healthy people. Although expected, people living in urban areas did not spend more on outsourcing than city dwellers.

Expenditure on outsourcing meal preparation was significantly higher for households with older children than for households with young children. Obviously, having children younger than 12 years implied that more food was prepared at home. When children grow older, meal preparation was outsourced more. This could be due to a generation-effect since children aged between 12 and 15, generally like takeaway food (like McDonald's) a lot.

In all households the expenditure on outsourcing meal preparation decreased with age. The expenditure on outsourcing decreased significantly with the male's age (in many cases the main provider of household income). Of course, this could be explained partly by the fact that older people have fewer working hours (or do not work at all), which decreases the need for outsourcing in general. Also, older men may generally have a relatively traditional household where the wife does not do paid labour and takes care of (almost) all household and care activities. The results indicate that outsourcing meal preparation (probably, mostly in the case of delivery food and takeaway food) is mainly accomplished by younger people.

In order to calculate the elasticity of outsourcing meal preparation, the regression on outsourcing expenditure was repeated for the four groups taking the logarithm of both outsourcing expenditure and household income in Table 9.5. The elasticity shows the percentage change in outsourcing expenditure which is due to a percentage increase in income. The analyses exclude working hours, level of education, and own home (which are all related to household income).

Table 9.5   Elasticity of expenditure on outsourcing for the native Dutch and immigrant couples (standard errors in parentheses)

Native Dutch	Surinamese/ Antilleans	Moroccans	Turks
0.441 (0.144)	0.887 (0.217)	0.738 (0.314)	0.871 (0.194)

Although outsourcing meal preparation may be considered a luxury good (with elasticities greater than 1), the results of our data show that it was more like a normal good, since the elasticities were between 0 and 1. The elasticities indicate that for both the native Dutch and the non-Western immigrant groups, the expenditure on outsourcing meal preparation changed less than proportionally with the income.

9.5   Conclusions

In this chapter, we have studied: 1) the determinants of the demand for outsourcing meal preparation; and 2) differences between non-Western immigrants and the native Dutch regarding their outsourcing behaviour. Generally, the results of our study in the Netherlands correspond to the results of earlier research on outsourcing meal preparation in Western countries, although our study reveals additional information on the specific situation of immigrants in the Netherlands. Table 9.6 shows the expected and confirmed effects of the estimations.

As was found in other western European countries, the US, and Australia, in the Netherlands we found that having children, level of education, and household income are important determinants of the demand for outsourcing meal preparation (Mihalopoulos and Demoussis, 2001; Spitze, 1999; Manrique and Jensen, 1998; Bittman *et al.*, 1998; Cohen, 1998; Soberon-Ferrer and Dardis, 1991; and Bellante and Foster, 1984).

As expected and in confirmation with the literature, all immigrant groups spent less on outsourcing than the native Dutch (Cohen, 1998; Soberon-Ferrer and Dardis, 1991; and Bellante and Foster, 1984). Since the analyses were corrected for socio-economic status, there should be another factor explaining these differences. The analyses showed that the outsourcing behaviour differed significantly among the ethnic groups. The native Dutch and Surinamese/ Antilleans spent the most on eating out in restaurants, whereas the Moroccans and Turks spent more on delivery food and takeaway food. The results suggested that Turks and Moroccans preferred to eat at home, even if they do not cook by themselves. The native Dutch, on the other hand, were more likely to visit restaurants than the immigrants.

When people grow older, generally their expenditure on outsourcing meal preparation decreases. In particular, people younger than 35 years outsource

Table 9.6 Expected and confirmed (significant) effects of the estimations

Variable	Expected	Confirmed
Surinamese/Antilleans	−	Yes
Moroccans	−	Yes
Turks	−	Yes
Gender (female)	+/−	No
Working hours female	+	No
Working hours male	+	No
Household income	+	Yes
Children living at home	+	Yes
Level of education female	+	Yes
Level of education male	+	Yes
Living in an urban area	+	No
Religious affiliation	−	No
Health	−	No
Homeowner	+	No

+ positive effect, − negative effect, +/− either positive or negative effect expected

meal preparation, which confirms the earlier literature (Manrique and Jensen, 1998 and Cohen, 1998). Females do not spend less on outsourcing meal preparation than males as we had hypothesised. Moreover, working hours of both partners do not affect the outsourcing expenditure. Apparently, long working days are less important than income when it comes to the decision to outsource meal preparation.

People with a religious background (possibly being somewhat more traditional) do not spend less on outsourcing meal preparation compared to people who are not religiously affiliated. Obviously, also in these families outsourcing the preparation of meals has become quite common although one might expect that in traditional households, women prefer to do the cooking by themselves (Oropesa, 1993).

The calculated elasticities show that for immigrants as well as for the native Dutch outsourcing meal preparation is a normal good, indicating that outsourcing meal preparation is not considered as a luxury.

In this study the frequency and the expenditure on outsourcing meal preparation have been studied in relation to socio-economic and demographic variables. In order to have a more in-depth insight in the cultural differences in food habits, it may be of interest to include the types of food people eat when they outsource their meal preparation.

9.6 References

BECKER, G. S. (1965). A theory of the allocation of time. *Economic Journal* 75, 493–517.
BELLANTE, D. and FOSTER, A.C. (1984). Working wives and expenditure on services. *Journal of Consumer Research* 11, 700–707.

BITTMAN, M., MEAGHER, G. and MATHESON, G. (1998). The changing boundary between home and market. Australian trends in outsourcing domestic labour. *Social Policy Research Centre Discussion Paper* 86, 1–29.

CENTRAAL BUREAU VOOR DE STATISTIEK (CBS) (2000). *Standaarddefinite allochtonen* [Standard definition of immigrants]. Webmagazine: www.cbs.nl.

CENTRAAL BUREAU VOOR DE STATISTIEK (CBS) (2001). *Allochtonen in Nederland 2001* [Immigrants in the Netherlands 2001]. Centraal Bureau voor de Statistiek. Voorburg/Heerlen.

CENTRAAL BUREAU VOOR DE STATISTIEK (CBS) (2003). *Statistich jaarboek* [Statistics annual]. Centraal Bureau voor de Statistiek. Voorburg/Heerlen.

COHEN, P.N. (1998). Replacing housework in the service economy: gender, class, and race-ethnicity in service spending. *Gender & Society* 12, 219–231.

GRONAU, R. (1977). Leisure, home production, and work – theory of the allocation of time revisited. *Journal of Political Economy* 85, 1099–1123.

GRONAU, R. (1986). Home production – a survey. In: Ashenfelter, O. and Layard, R. (eds). *Handbook of Labour Economics* Volume I. Elsevier Science Publishers BV.

HALLBERG, D. and KLERMARKEN, A. (2003). Time for children: a study of parent's time allocation. *Journal of Population Economics*, 16, 205–226.

HEIMAN, A., JUST, D.R., McWILLIAMS, B. and ZILBERMAN, D. (2001). Incorporating family interactions and socio-economic variables into family production functions: the case of demand for meats. *Agribusiness* 17, 4, 455–468.

HOMAN, M.E. (1988). *The allocation of time and money in one-earner and two-earner families: an economic analysis*. PhD dissertation. Erasmus Universiteit. Rotterdam

KIM, C. (1989). Working wives' time-saving tendencies: durable ownership, convenience food consumption, and meal purchases. *Journal of Economic Psychology* 10, 391–409.

LAMBRIEX, G.E.E.M. and SIEGERS, J.J. (1993). *Een geintegreerde analyse van tijds- en inkomenbesteding* [Integrated analysis of time allocation and income allocation]. SWOKA. Instituut voor consumentenonderzoek. Onderzoeksrapport 136. Den Haag.

MANRIQUE, J. and JENSEN, H.H. (1998) Working women and expenditures on food-away-from-home and at-home in Spain. *Journal of Agricultural Economics* 49, 3, 321–333.

MIHALOPOULOS, V.G. and DEMOUSSIS, M.P. (2001). Greek household consumption of food away from home: a microeconometric approach. *European Review of Agricultural Economics* 28, 4, 421–432.

NATIONAAL INSTITUUT VOOR BUDGETVOORLICHTING (NIBUD) (2004). *De inkomsten, uitgaven en het financieel beheer van alochtone huishoudens* [Income, expenditures, and financial management in immigrant households]. NIBUD. Utrecht.

OROPESA, R.S. (1993). Using the service economy to relieve the double burden. Female labor force participation and service purchases. *Journal of Family Issues* 14, 3, 438–473.

RIJKSINSTITUUT VOOR VOLKSGEZONDHEID EN MILIEU (RIVM) (2004). *Ons eten gemeten. Gezonde voeding en veilig voedsel in Nederland* [Measuring our food. Healthy and safe food in the Netherlands]. Van Kreijl, C.F. and Knaap, A.G.A.C. (eds). Bohn Stafleu Van Loghum. Houten.

SOBERON-FERRER, H. and DARDIS, R. (1991). Determinants of household expenditures for services. *Journal of Consumer Research* 17, 385–397.

SOCIAAL EN CULTUREEL PLANBUREAU (2000). *De kunst van het combineren. Taakverdeling onder partners* [The skill to combine. Division of tasks among partners]. Keuzekamp, S. and Hooghiemstra, E. (eds). SCP. Den Haag.

SOUSA-POZA, A., SCHMID, H. and WIDMER, R. (2001). The allocation and value of time assigned to housework and childcare: analysis for Switzerland. *Journal of Population Economics* 14, 599–618.

SPITZE, G. (1999). Getting help with housework. *Journal of Family Issues* 20, 6, 724–745.

TIJDENS, K., LIPPE, T. VAN DER and RUIJTER, E. (2000). *Huishoudelijke arbeid en de zorg voor kinderen: herverdelen of uitbesteden?* [Household labour and childcare: redistribution or outsourcing?]. Serie wetenschappelijke Publicaties. Elsevier Bedrijfsinformatie b.v. 's Gravenhage.

VAN DAM, Y.K., HOOG, C. DE and OPHEM, J.A.C. VAN (1994). Ten geleide: reflecties op gemak bij voeding [Reflections on convenience and food]. In: Van Dam, Y.K., De Hoog, C., and Van Ophem J.A.C. (eds). *Eten in de jaren negentig. Reflecties op gemaksvoeding* [Eating in the 1990s. Reflections on convenience food]. Landbouw Universiteit Wageningen

VAN DER LIPPE, T., TIJDENS, K. and DE RUIJTER, E. (2004). Outsourcing of domestic tasks and time-saving effects. *Journal of Family Issues* 25, 2, 216–240.

VAN OPHEM, J.A.C. and DE HOOG, K. (1995). Monetarisering in de huishouding: een historische beschouwing [Monetarisation in households: a historic consideration]. *Huishoudstudies* 5, 2, 1–8.

9.7   Appendix: distribution of the variables *N* = 2170

(ln) Expenditure per month on outsourcing meal preparation	4.10 (2.00)
Working hours female	12.92 (16.11)
Working hours male	24.75 (20.89)
Children at home 0–3 y/n	21.2%
Children at home 4–11 y/n	31.3%
Children at home 12–15 y/n	15.5%
Children at home 16–25 y/n	14.0%
Children at home ≥25 y/n	3.3% (reference group)
Level of education female*	
Low	24.5% (reference group)
Medium	39.6%
High	17.5%
Level of education male	
Low	21.6% (reference group)
Medium	33.9%
High	20.3%
Age female	
18–34	47.2% (reference group)
35–44	26.9%
45–64	19.0%
≥65	5.4%
Age male	
18–34	39.3% (reference group)
35–44	30.7%
45–64	22.5%
≥65	5.8%

Surinamese/Antilleans	28.0%
Moroccans	15.9%
Turks	27.6%
Living in an urban area	30.3%
Religious affiliation**	30.4%
Health***	2.69 (0.98)
Homeowner	40.1%
Net monthly household income (in Euro)	1613.50 (930.25)

* Low level of education: primary school and vocational education. Medium level of education: lower and higher secondary education, pre-university education, and intermediate vocational education. High level of education: people holding a bachelor's or a master's degree.
** Going to a church, temple, mosque, or synagogue more than once per month.
*** Scale: 4 = excellent–0 = poor.
Standard deviations in parentheses.

10

Consumer attitudes to food innovation and technology

M. Siegrist, University of Zürich, Switzerland

10.1   Introduction

Development of new products is expensive and risky for the food industry. Most new food products are failures in terms of consumer acceptance and disappear from the market shortly after their introduction. The failure rate for new food products is between 60 and 80% (Grunert and Valli, 2001). It is important, therefore, to know what factors facilitate and what factors hinder consumer acceptance of new food products.

The sensory qualities of foods are important to their success in the market place, and the price must also be right, of course. One must not ignore the possibility, however, that other factors, such as the technologies used in processing novel foods, may affect their acceptance. The introduction of GM (genetically modified) food, for example, has not been successful in Europe. In most European countries, consumers are opposed to GM foods (Gaskell *et al.*, 2000). Nanotechnology will be the next innovation that will be important in the food sector. This technology can be used to alter foods or to create innovative packaging materials. It seems possible, for example, to change the texture of certain foods utilizing nanotechnology. Or the technology can be used to produce new packaging materials that have, for example, anti-bacterial coatings (Kaiser and Tang, 2004). Survey results suggest that Europeans are more sceptical regarding this new technology than people in the US (Gaskell *et al.*, 2004). How the public will react when nano-food is introduced in the market place remains to be seen.

Attitudes toward food technology and food innovations may play an important role in the acceptance of novel foods. The present chapter reviews the research that has examined the influence of attitudes on the acceptance of food innovations and technology. Consideration of this knowledge at an early stage of product development may help reduce the failure rate of new food products.

10.2   Methods and models for analysing consumer attitudes to food innovation and technology

Psychological research on influence has employed the concept of attitudes to explain public reactions toward new technologies (Frewer *et al.*, 2004). Attitudes are evaluations of objects in our environment. Attitudes present a summary evaluation of an object (Ajzen, 2001). These evaluations can vary from positive to negative, and they are experienced as affect. Typical evaluative dimensions are good-bad, pleasant-unpleasant, or likable-dislikable (Ajzen, 2001). Positive attitudes are associated with approaching behaviour and negative attitudes are associated with avoidance behaviour.

The expectancy-values model is the most popular conceptualization of attitude (Fishbein and Ajzen, 1975). According to this model, readily accessible beliefs or attributes associated with an object determine the attitude toward the object. The subjective value of the attribute is multiplied by the strength of the association between the object and the attribute. The products for all accessible attributes are summed. This summative index is directly proportional to a person's attitude (Ajzen, 1991). The theory of planned behaviour, or variations of it, have been widely used to explain people's intentions to buy new foods (Cook *et al.*, 2002; Saba and Vassallo, 2002).

In recent social psychological and cognitive models, two distinct processing modes have been identified (Smith and DeCoster, 2000). Based on these dual-mode models, Slovic and colleagues (Finucane *et al.*, 2000; Slovic *et al.*, 2002, 2004) draw a distinction between the experiential system and the analytic system. The analytic system uses probabilities or formal logic in making decisions. The experiential system, on the other hand, has a strong affective basis. It is an intuitive, fast, mostly automatic system. These intuitive feelings are our primary means of evaluating risks (Slovic *et al.*, 2004). The experiential system helps us to quickly decide whether something is good or bad. Slovic and colleagues assume that the affective reactions evoked by stimuli serve as cues for judgments. According to this view, perceived benefits and perceived risks are shaped by the affect associated with a technology. This phenomenon is known as the 'affect heuristic.' Slovic and colleagues use affect as it is employed in the concept of attitude (e.g., Ajzen, 2001), to mean overall degree of positivity or negativity toward the attitude object.

Slovic's (1987) psychometric paradigm has been widely used to study why people perceive various hazards differently (Slovic, 1987; see also the chapters by de Jonge *et al.*, this volume; Dreyer and Renn, this volume). Results of this research suggest that feelings of dread are the major factor affecting public perception and acceptance of risk for a broad range of hazards (Slovic *et al.*, 2004). Food-related hazards, like BSE or pesticide residuals, are perceived as dreaded risks, while food colouring or saturated fats are perceived as non-dreaded risks (Fife-Schaw and Rowe, 2000; Kirk *et al.*, 2002). Recently, Slovic *et al.* (2004) suggested that the importance of the dreadfulness of a hazard for perceived risks can be viewed as evidence of 'risk as feelings'. Affect or attitudes seem to determine risk perception.

Attitudes help us to make sense of and give meaning to our experiences. It has been shown that existing attitudes can affect the evaluation of new information (Prislin *et al.*, 1998). The influence of existing beliefs on the meaning of new information was demonstrated in a study by Eiser *et al.* (1995). Participants were asked about global warming and about the cause of an oil tanker collision in the English Channel. Although these two topics were unconnected, answers to questions about them were closely linked. Thinking about one of the issues primed people to think about the second issue in ways that differed from non-primed conditions.

Information conveyed by risk communication is, therefore, mediated by the attitudes people hold. Scholderer and Frewer (2003) examined the effects of various information strategies on consumer attitude change. Results indicated that the information strategies used by the researchers decreased consumer's acceptance of GM foods compared with the control group. The authors concluded that the information material was more likely to activate pre-existing attitudes than the no-additional-information condition in the control group. The activation of the pre-existing attitudes resulted in an increased consistency of the beliefs and choices expressed by the participants. People's attitudes toward GM foods seem to be so strong that new information is overridden. Informing the public about new technologies may often fail to increase acceptance unless other factors (such as personal or societal benefits, and the values placed on these) are also addressed.

Methods for measuring implicit attitudes have recently been developed (Greenwald *et al.*, 1998). In almost all studies examining attitudes toward foods, however, explicit measures have been used. Various instruments to measure attitudes toward foods have been proposed. Roininen *et al.* (1999) describe a scale that measures the importance of health and taste characteristics. Other scales measure attitudes toward new foods and food technology (Huotilainen and Tuorila, 2005). Most of these scales are not pure attitudinal measurements. They include mixtures of attitudinal items, behavioural intentions, and beliefs.

People's attitudes toward food are related to their other attitudes and beliefs. The dichotomy between nature and technology, for example, is important for a better understanding of the acceptance of food innovations. People tend to have confidence in natural food and the way it is produced, but they are suspicious toward new foods and new food technologies (Huotilainen and Tuorila, 2005). Assessments of the naturalness of foods seem to be correlated with sensory appeal (Steptoe *et al.*, 1995). Natural food is associated with better looks and better taste compared with foods containing additives or artificial ingredients.

Attitudes toward GM technology are influenced by more general environmental attitudes (Siegrist, 1998; Sparks *et al.*, 1995). The attitude of favouring the protection of nature because of its intrinsic value had a negative impact on acceptance of GM technology. Valuing nature because of its usefulness and benefits to humans, however, had a positive influence on acceptance of GM technology. In a similar study, general attitudes or world views had an important influence on the perception of GM technology (Siegrist, 1999).

The concept of attitudes is a psychological approach toward a better understanding of the acceptance or non-acceptance of novel food. However, a psychological view may be too narrow. Attitudes toward a new food technology will not only be influenced by the innovation itself but also by the surrounding social, economic, and political environments (Henson, 1995). Various dynamic social processes may generate public concern about hazards that are judged as low risks by experts, to the neglect of hazards that they judge as high risks (Kasperson *et al.*, 2003). Such a process of the social amplification of risk perceptions can be observed in the domain of GM foods in Europe.

10.3   Outline of consumer attitudes to food innovation and technology

Novel foods and new food technologies may be more acceptable to the public if there are tangible benefits to the consumer (Frewer *et al.*, 2003). A Swiss study (Siegrist, 2000) examined lay people's perceptions of GM applications in the domains of food and medicine. Results suggested that acceptance of GM products was largely determined by perceived risk and perceived benefit. Standardized path weights show that perceived benefits are much more important for the acceptance of GM products than perceived risks. A Swedish study reported similar findings (Magnusson and Hursti, 2002). Tangible benefits – products that are better for the environment, for example, or products that are healthier – increased peoples stated willingness to purchase GM products. The importance of perceived benefits for the acceptance of GM food was also demonstrated in experimental studies. In one study, participants received information about genetically engineered soybeans (Brown and Ping, 2003). Two groups received information that differed in the presence or absence of a consumer benefit. Results showed that participants who were informed about a GM application with a consumer benefit perceived lower personal risks compared with participants who were informed about an application without a consumer benefit.

Recent studies suggest, however, that benefit alone does not guarantee acceptance. Cox *et al.* (2004) observed a low intention to consume GM food, even though they communicated clear benefits to the consumer. It should also be emphasized that consumers are not a homogenous group. In other words, consumers differ in what they perceive as benefits. Organic food, for example, may constitute a benefit for one segment of consumers but not for others. In sum, results of these studies suggest that perceived benefits may have an impact on how GM food is assessed. However, the acceptance of novel food cannot be reduced to perceived risks and perceived benefits.

10.3.1   Attitudes toward specific ingredients

Consumers may hesitate to purchase food products because they contain certain food additives or food colourings. Little research has examined attitudes toward food additives and food colourings, however (for an exception see Kajanne and Pirttilä-Backman (1996)). There are a few studies in which the psychometric paradigm was utilized to examine how people perceive various food hazards. In these studies, participants used a variety of rating scales to evaluate a set of hazards (Fischhoff et al., 1978). They assessed, for example, how well the hazard is known to science and the degree of dread associated with it. In the studies focusing on food hazards, a very heterogeneous set of hazards was presented, ranging from GM food to salmonella.

Based on the studies utilizing the psychometric paradigm, it can be concluded that food colourings and food additives are perceived as unknown risks and as hazards with low dreadfulness (Fife-Schaw and Rowe, 2000; Sparks and Shepherd, 1994). Results of another study suggest that growth hormones are perceived in a similar way (Kirk et al., 2002). Low severity suggests that food additives and food colourings are not perceived as a source of concern or a problem for future generations. These ingredients, therefore, may not pose a serious problem for the acceptance of new foods.

10.3.2   Attitudes toward new processes

New processes enable innovations in the food sector. Processes like food irradiation or high-pressure processing are methods for food preservation. Recombinant DNA technology is used to create new varieties – of plants, for example – such as golden rice, a variety with improved nutritional value (Ye et al., 2000). Some of these new processes are not well accepted by consumers. Attitudes toward these new processes may help to explain why this is the case.

Food irradiation offers a number of benefits for consumers (Henson, 1995). This method kills micro-organisms in food, and it is a method of food preservation. A number of countries have approved the use of irradiation of specific doses on certain foods (Henson, 1995). However, there are still countries in which food irradiation is not approved.

The benefits associated with food irradiation are not tangible to the consumer; they must be explicitly communicated. This may not be an easy task since radiation is strongly associated with nuclear power, a technology that tends to evoke negative associations and images (Slovic et al., 1991). These negative attitudes may shape attitudes toward food radiation, helping to explain why a number of consumers perceive food irradiation as a risky technology (Bord and

O'Connor, 1990). Furthermore, consumers must trust that the food industry uses food irradiation properly. The public may fear that this technology lowers the quality of foods (e.g., contaminated foods are irradiated and resold).

In an experimental study, utilizing a student sample from Brazil, the effects of a video about food irradiation on attitudes toward this technology were examined (Oliveira and Sabato, 2004). Results suggest that people hold more positive attitudes toward food irradiation after receiving information about it. Utilizing correlational data, Bord and O'Connor (1990) found that knowledge is positively correlated with acceptance of irradiated food. However, greater fear of radiation resulted in less acceptance. In a similar vein, Bruhn (1998) concluded that, when provided with science-based information, a high percentage of consumers favour irradiated food.

High-pressure processing is a new method for increasing food safety with minimum quality loss (Ozen and Floros, 2001). This processing technology was developed to meet consumer demands for fresh products with reduced microbiological contamination. It is a non-conventional and new technology since high-pressure processing does not use heat to preserve food (Deliza et al., 2005). Results of focus group interviews showed that use of this new technology had a positive impact on the perception of the product (Deliza et al., 2003). Information about this new processing technology, emphasizing its benefits, had a positive influence on purchase intention (Deliza et al., 2005). Future studies must show whether these results from Brazil can be generalized to developed countries.

In sum, knowledge seems to have an impact on the acceptance of food irradiation and high pressure processing. However, these technologies may not evoke strong feelings. Risk communication may have an effect when people do not hold strong convictions related to the technology (Earle and Siegrist, 2006). As a consequence of GM technology being likely to evoke more affective responses by consumers, risk communication and knowledge may not positively affect its acceptance. Results of surveys examining public perception of biotechnology suggest that more knowledgeable persons tend to have more extreme attitudes than less knowledgeable persons (Durant et al., 1998). Those attitudes, however, may be positive or negative.

A good deal of research regarding new processes related to food has focused on GM foods. Frewer et al. (1997a), for example, used conjoint analysis to examine attitudes toward various processing technologies. Conjoint analysis is a statistical method that is based on multi-attribute decision theory. The results showed that genetic modification was the least acceptable production method; the traditional method was most acceptable. Additional results showed that the benefits of the product could compensate for the fact that it was produced by a less preferable method. Consumers may accept a food-processing technology, even though they have negative attitudes toward it, when the product is associated with tangible benefits. The study focused on genetic manipulation of micro-organisms used for the production of cheese. Manipulation of micro-organisms is perceived as less problematic than other applications, such as GM animals (Siegrist and Bühlmann, 1999). Therefore, tangible benefits may not result in higher acceptance for all food products; instead, it is contingent on consumer acceptability of specific applications.

Attitudes toward new technologies are shaped by the perceived benefits associated with them. In addition, consumers are susceptible to framing effects. The use of new technologies is accepted for some products, but it is not accepted for other products. Such effects have been demonstrated in various studies in the domain of GM foods. Perception and acceptance varies according to the type of application (Frewer *et al.*, 1997b; Gaskell *et al.*, 1999). Results have clearly indicated that people in Europe and in the US have more positive attitudes toward medical applications than toward agricultural applications (Gaskell *et al.*, 1999). Furthermore, differing applications in the food domain are perceived completely differently (Siegrist and Bühlmann, 1999). In this study, several scenarios described various applications of gene technology drawn from the domains of agriculture, food, drugs, and medicine. Participants rated the similarity of the different applications. Results of multidimensional scaling showed that two dimensions were relevant for the perception of gene technology. The first was related to the nature of the application (food related/ medical application). Medical applications were viewed more positively than food applications. The second dimension was related to the organisms involved (animals, plants/micro-organisms). The golden rice application (a rice variety with an enhanced level of vitamin A) was located between the medical applications and the agricultural applications. These results suggest that framing applications in a certain way may alter attitudes in a more positive or a more negative direction.

Various factors seem to affect attitudes toward gene technology. People who trusted institutions involved in using or regulating gene technology judged the benefits to be greater and the risks lower for this technology (Siegrist, 2000). Since most people possess only limited knowledge of gene technology (Durant *et al.*, 1998), the importance of trust should be of no surprise. One way people cope with a lack of knowledge is to rely on trust to reduce the complexity of risk management decisions (Earle and Cvetkovich, 1995). A causal model that has been proposed to explain acceptance of gene technology and other technologies is shown in Fig. 10.1. This model has been successfully tested (Siegrist, 1999, 2000; Siegrist *et al.*, 2000). Based on the results, one can conclude that trust in institutions, or in persons doing genetic modification research or using modified products, is an important factor influencing the perception of gene technology. Trust has an impact on perceived risk as well as on perceived benefit. Acceptance of, or willingness to buy, GM foods is directly determined by the perceived risk and the perceived benefit. In other words, trust has an indirect impact on the acceptance of GM foods. Perceived value similarity seems to be an important antecedent of trust (Siegrist *et al.*, 2000). People tend to trust persons who share their values, and people tend to distrust people who do not share their values. If the value similarity approach of trust is correct, trust in gene technology can be increased if a technology is framed to

Fig. 10.1   Model explaining acceptance of GM foods.

reflect the public's salient values (e.g., medical application and not food application).

An important segment of consumers are those who are willing to buy more expensive organic foods (see also Chapter 11 by Ritson). There are at least two motives that can be identified to explain why some people show a preference for organic foods. Self-reported purchase of organic foods was related to perceived benefit for human health and to environmental concern (Magnusson *et al.*, 2003). The results of a recent study suggest, however, that even when the healthfulness of natural and artificial foods is specified to be equivalent, most of the people with a preference for natural food continue to prefer it (Rozin *et al.*, 2004). Perceived naturalness or lack of naturalness seems to be a factor that influences attitudes toward genetic engineered foods. Results of a study by Tenbült and colleagues suggest that the more a product is seen as natural, the less acceptable will be a genetic engineered version of that product (Tenbült *et al.*, 2005). Similarly, a study by Siegrist (2003) showed that consumers considered it more important to have baby food and unprocessed food free of gene technology than to have processed foods, such as chocolate, frozen foods or convenience food, free of gene technology.

10.3.3   Attitudes toward new foods

New varieties of foods have been introduced in the market recently. The most important new categories are probably functional food and convenience food. So-called functional foods are products that promise consumers improvements in targeted physiological functions (Diplock *et al.*, 1999). So-called convenience food saves or reduces some kind of effort (Scholderer and Grunert, 2005). Time, physical energy or mental energy can be saved, and the saving can occur at different stages of home food preparation.

Consumers cannot directly experience the benefits of functional food. Producers must explicitly communicate the benefits. This makes trust crucial for the acceptance of functional food (Siegrist and Cvetkovich, 2000) because consumers must trust the health claims provided by the producers (Verbeke, 2006). Participants with stronger health benefit beliefs in functional food products showed more acceptance for functional food than participants with weaker health benefit beliefs (Verbeke, 2005).

Urala and Lähteenmäki (2004) measured attitudes toward functional food. The authors analyzed responses to 42 functional-food-related statements. Results of a factor analysis showed that the following seven factors account for consumers' attitudes toward functional foods:

- reward from using functional food
- confidence in functional food
- necessity for functional food
- functional food as medicine
- functional food as part of a healthy diet
- absence of nutritional risk in functional food
- taste of functional food.

Perceived reward from using functional food was the best predictor of consumers' stated willingness to use functional food. Perceived risks had no effect, indicating that perceived benefits are more important than perceived risks. However, different results were observed for various types of functional food (e.g., probiotic juice, energy drinks). Consumers do not perceive functional foods as a homogenous food category (Urala and Lähteenmäki, 2003, 2004). Further, results from Urala and Lähteenmäki (2004) suggest that attitudes toward functional food are different from general health interests.

Attitudes influence acceptance of functional food. However, other factors such as price or taste are important as well. Several studies suggest that consumers are not willing to compromise taste for possible health benefits (Tuorila and Cardello, 2002; Verbeke, 2006). Consumers no longer believe that good taste and healthiness are mutually exclusive.

Natural foods are valued more than functional foods. A majority of US consumers prefers to eat more fruits and vegetables, as opposed to functional foods, in order to obtain phytochemical health benefits (Childs and Poryzees, 1997). Not all food carriers are comparably compelling to consumers. Carriers with a good health image (e.g., yoghurt) are more attractive than carriers lacking such an image (e.g., chewing gum) (van Kleef *et al.*, 2005). This study further suggests that consumers have more positive attitudes toward physiology-related health benefits (e.g., osteoporosis) than toward benefits that are psychology-related (e.g., stress).

The markets for functional foods and convenience foods do not overlap (Shiu *et al.*, 2004). An individual is not likely to be a heavy user of both convenience and functional foods. Further, the consumption of convenience food is influenced by situational factors (Verlegh and Candel, 1999) as well as by socio-demographic factors (Shiu *et al.*, 2004). Little is known, however, about the influence of attitudes on the consumption of convenience food.

10.4   Understanding consumer choice

Most studies measured the intention or the willingness to buy food products. It is not the intention to buy a certain product that is of interest, however; it is the actual purchasing behaviour. Unfortunately, purchasing behaviour cannot be investigated before a product is introduced in the marketplace. In the developmental phase of food innovations one therefore has to rely on data about willingness to purchase such new products. There are other reasons why, in the domain of GM foods, for example, researchers are forced to measure intentions to buy instead of actual purchasing behaviour. In many countries, GM products are not labelled. As a consequence, even consumers with no intention to buy GM foods may end up purchasing them. In other countries, GM foods are not available to consumers because grocery stores do not stock such items. Stated intentions, unfortunately, are not a very good proxy for actual behaviour. It can be expected that attitudes toward food innovations are much better predictors for reported 'willingness to buy' new products compared to actual purchase behaviour. The reason for this is that not only consumer attitudes, but also situational factors, determine consumer purchasing behaviour.

10.4.1   Impact of attitudes to food innovation and technology on purchasing behaviour

Frewer *et al.* (1996) measured the likelihood of purchase for different product categories. A container of yoghurt, a tomato and a chicken drumstick were presented as real products. Products were shown in pairs of photographs, one was labelled as a GM product; the other was labelled as conventional product. Results showed that participants were significantly less likely to purchase GM products compared to conventional equivalents.

Social and situational factors must not be neglected in studying food choice, preferences and purchasing behaviour. It has been shown, for example, that social situations are a major influence on the intent to consume TV dinners (Verlegh and Candel, 1999). In some situations (e.g., dinner alone), consuming convenience food is perceived as appropriate; few people, however, serve TV dinners when entertaining friends. Attitudes toward new foods may be moderated by social or situational factors.

10.4.2   Impact of attitudes to food innovation and technology on willingness to try new foods

Much research on acceptance of new food products has focused on abstract situations (e.g., description of new products). Only few studies have examined people's reactions toward realistic products and their willingness to taste such products. Townsend and Campbell (2004) conducted an experiment in which participants were asked to taste three apples and to decide which of them was traditionally grown, organically grown or GM. In fact, all apples had been traditionally grown. About half of the participants indicated that they would buy

GM food, the other half of the respondents stated that they would not buy GM food. Purchasers and non-purchasers differed significantly in their attitudes toward GM food. However, 86% of the non-purchasers were happy to taste an apple that was labelled as a GM product. Due to this ceiling effect, attitudes were of little value in determining the willingness to taste new foods.

In a study with participants from several Scandinavian countries, participants were offered five pieces of cheese as a reward for taking part in the study (Lähteenmäki et al., 2002). They could choose among cheese labelled as a GM product and cheese labelled as a traditional product. Two thirds of the respondents took at least one piece of GM cheese. Attitudes toward the use of gene technology were the best predictor of participants' choice behaviour.

The rated willingness to try different new foods does not seem to be a one-dimensional construct (Backstrom et al., 2004). The willingness to try modified dairy products (e.g., functional yogurts, fat-free yogurts) loaded on a different factor than the stated willingness to try GM products. Adherence to technology was a strong predictor of willingness to try GM foods. It was a weak predictor, however, for willingness to try functional dairy products.

Attitudes toward new foods may not only have an impact on willingness to try new foods, but they may also influence the actual liking of a product. Caporale and Monteleone (2004) examined the effect of specific product information on the perceived quality of the product. In an experiment, participants received information about selected aspects of the manufacturing process (i.e., use of GM yeast, organic methods, traditional technology). Results suggested that information about the manufacturing process can have an impact on the perception of a product. GMOs are seen as unnecessary in food manufacturing, and these negative attitudes toward GMOs decreased the actual liking of the beers compared with identical beers that were labelled as being produced by traditional technology.

Consumers, in general, are more positive about familiar foods as compared to unfamiliar foods (Raudenbush and Frank, 1999). Personality variables may also shape consumers' attitudes towards new foods. It has been shown that some people have a stronger tendency to avoid new foods than other people, a phenomenon that has been labelled food neophobia. Results of several studies suggest that food neophobia negatively influences consumers' willingness to try new foods (Raudenbush and Frank, 1999; Tuorila et al., 1998, 2001).

In sum, a number of studies examined the influence of attitudes toward new food technologies on the willingness to try new foods. Attitudes toward food innovations not only influenced the likelihood of trying new foods, but they also influenced how much consumers liked the taste of products.

10.5   Understanding consumer attitudes to innovation and technology for food product development

A number of factors must be taken into account for successful product development. Sensory quality and price are important factors, of course. However, socio-demographic variables like income, household size, education, age, gender, and cultural-background influence food-related behaviour (Axelson, 1986), and may, therefore, affect the success of new food products. In other words, attitudes toward food technology are one factor among many that may have an impact on the success of novel products.

Results of the research examining consumer attitudes toward new food technologies may contribute to a better understanding of which factors negatively influence acceptance of new food products. New processes, such as food irradiation or genetic modification, seem to be the least acceptable technologies. In Europe, especially, it seems very difficult to market such food products. Specific ingredients and new foods like functional food, however, are easier to market. Nevertheless, attitudes may help to explain why some consumers buy new food products, and why, at the same time, other consumers are hesitant to buy the very same food products.

Results of surveys about GM foods conducted in European countries and in the US suggest that there are considerable differences in public opinion on this subject (Gaskell et al., 1999). Cultural differences must, therefore, be taken into account when consumer attitudes towards new food technologies are examined. According to Gaskell et al. (1999) two explanations may account for the lower consumer acceptance of GM foods in Europe compared with the US. First, people in the US have more trust in regulatory authorities than in Europe. As a consequence, US consumers are less concerned about new and unfamiliar technologies than European consumers. Second, there seems to be a greater prevalence of menacing food images (e.g., eating GM foods may result in genetic infection) in Europe than in the US. This may be the result of recent food safety scares in Europe (e.g., BSE or 'mad cow disease').

It is common sense that GM foods with tangible benefits for the consumer will be easier to market than GM foods without obvious consumer benefits. For the industry it might be tempting to assume that attitudes toward GM foods will be more positive if a GM product with a desirable benefit is on the market. However, novel foods that have clear health benefits may not be appealing to all consumers. Thus, introducing such novel foods is unlikely to result, generally, in more positive attitudes toward GM food (Frewer et al., 2004). It is more likely that GM food is accepted for some products, but not for other products.

10.6 Future trends

In most of the reviewed studies, beliefs about, and not attitudes toward, different foods were measured. Few researchers clearly distinguished between attitudes and beliefs. In one such study, beliefs about genetic engineered food influenced attitudes toward genetic engineered food (Dreezens et al., 2005). However, based on this study, not much can be concluded about the causal relationship between attitudes and beliefs. Beliefs could be post-hoc rationalizations of decisions and attitudes. The affect dimension, good-bad, is probably the primary factor that influences purchasing behaviour in the food domain. In many cases, food purchasing can be viewed as an automatic process. Once people have established an attitude toward certain foods, it will be difficult to change these attitudes. Childhood experiences may be important, because it is very likely that in this phase of life attitudes toward foods are formed that are difficult to change later on. Future studies should focus more on the emotional aspects of attitudes and less on the beliefs. These studies should also examine factors that influence attitudes toward food innovation and technology.

In most studies, explicit measures have been used. However, explicit attitudes toward foods may be distorted by social desirability or self-presentation biases. In recent years, indirect measures have been used as an alternative to direct measures. A very popular indirect measure is the implicit association test (IAT), which has been used successfully in numerous studies (Greenwald *et al.*, 1998). The IAT was used to predict brand preferences, product usage, and brand recognition in a blind taste test (Maison *et al.*, 2004). Other methods, such as affective priming, were used to indirectly measure food attitudes (Lamote *et al.*, 2004). It is premature to conclude whether these indirect measures will be valuable for a better understanding of attitudes toward food innovation and technology. The results so far are promising, however, and future research should address the question of whether indirect measures can better predict food choice behaviour than direct measures of food attitudes.

10.7   Sources of further information and advice

The Journals *Appetite* and *Food Quality and Preference* regularly publish articles related to attitudes to food innovation and technology (http://www.sciencedirect.com/).

The Institute of Food Technologists provides information about new developments and acceptance of new technologies. The institute publishes the journal *Food Technology* (http://www.ift.org/cms/).

Articles about consumer choice, preferences, concerns and related topics can be found in the *British Food Journal* (http://www.emeraldinsight.com).

Articles related to consumer attitudes to food innovation and technology are occasionally published in journals focusing on risk perception and risk assessments (e.g., *Risk Analysis*, *Journal of Risk Research*).

The European Commission is monitoring public opinion in the Member States. Some of the surveys are related to food issues (e.g., attitudes towards gene technology) (http://europa.eu.int/comm/public_opinion/index_en.htm).

10.8   References

AJZEN, I. (1991). The theory of planned behavior. *Organizational Behavior and Human Decision Processes*, **50**, 179–211.

AJZEN, I. (2001). Nature and operation of attitudes. *Annual Review of Psychology*, **52**, 27–58.

AXELSON, M. L. (1986). The impact of culture on food-related behavior. *Annual Review of Nutrition*, **6**, 345–363.

BACKSTROM, A., PIRTTILA-BACKMAN, A. M. and TUORILA, H. (2004). Willingness to try new foods as predicted by social representations and attitude and trait scales. *Appetite*, **43**(1), 75–83.

BORD, R. J. and O'CONNOR, R. E. (1990). Risk communication, knowledge, and attitudes: Explaining reactions to a technology perceived as risky. *Risk Analysis*, **10**, 499–506.

BROWN, J. L. and PING, Y. C. (2003). Consumer perception of risk associated with eating genetically engineered soybeans is less in the presence of a perceived consumer benefit. *Journal of the American Dietetic Association*, **103**(2), 208–214.

BRUHN, C. M. (1998). Consumer acceptance of irradiated food: Theory and reality. *Radiation Physics and Chemistry*, **52**(1–6), 129–133.

CAPORALE, G. and MONTELEONE, E. (2004). Influence of information about manufacturing process on beer acceptability. *Food Quality and Preference*, **15**(3), 271–278.

CHILDS, N. M. and PORYZEES, G. H. (1997). Foods that help prevent disease: Consumer attitudes and public policy implications. *Journal of Consumer Marketing*, **14**, 433–447.

COOK, A. J., KERR, G. N. and MOORE, K. (2002). Attitudes and intentions towards purchasing GM food. *Journal of Economic Psychology*, **23**(5), 557–572.

COX, D. N., KOSTER, A. and RUSSELL, C. G. (2004). Predicting intentions to consume functional foods and supplements to offset memory loss using an adaptation of protection motivation theory. *Appetite*, **43**, 55–64.

DELIZA, R., ROSENTHAL, A. and SILVA, A. L. S. (2003). Consumer attitude towards information on non conventional technology. *Trends in Food Science & Technology*, **14**(1), 43–49.

DELIZA, R., ROSENTHAL, A., ABADIO, F. B. D., SILVA, C. H. O. and CASTILLO, C. (2005). Application of high pressure technology in the fruit juice processing: benefits perceived by consumers. *Journal of Food Engineering*, **67**(1–2), 241–246.

DIPLOCK, A. T., AGGETT, P. J., ASHWELL, M., BORNET, F., FERN, E. B. and ROBERFROID, M. B. (1999). Scientific concepts of functional foods in Europe: Consensus document. *British Journal of Nutrition*, **81**(4), S1–S27.

DREEZENS, E., MARTIJN, C., TENBULT, P., KOK, G. and DE VRIES, N. K. (2005). Food and values: an examination of values underlying attitudes toward genetically modified- and organically grown food products. *Appetite*, **44**(1), 115–122.

DURANT, J., BAUER, M. W. and GASKELL, G. (eds). (1998). *Biotechnology in the public sphere*. London: Science Museum.

EARLE, T. C. and CVETKOVICH, G. T. (1995). *Social trust: Toward a cosmopolitan society*. Westport, CT: Praeger.

EARLE, T. C. and SIEGRIST, M. (2006). Morality information, performance information, and the distinction between trust and confidence. *Journal of Applied Social Psychology*, **36**, 383–416.

EISER, J. R., REICHER, S. D. and PODPADEC, T. J. (1995). Global changes and local accidents: Consistency in attributions for environmental effects. *Journal of Applied Social Psychology*, **25**, 1518–1529.

FIFE-SCHAW, C. and ROWE, G. (2000). Extending the application of the psychometric approach for assessing public perceptions of food risks: Some methodological considerations. *Journal of Risk Research*, **3**, 167–179.

FINUCANE, M. L., ALHAKAMI, A., SLOVIC, P. and JOHNSON, S. M. (2000). The affect heuristic in judgments of risks and benefits. *Journal of Behavioral Decision Making*, **13**, 1–17.

FISCHHOFF, B., SLOVIC, P., LICHTENSTEIN, S., READ, S. and COMBS, B. (1978). How safe is safe enough? A psychometric study of attitudes towards technological risks and benefits. *Policy Sciences*, **9**, 127–152.

FISHBEIN, M. and AJZEN, I. (1975). *Belief, attitude, intention, and behavior*. Reading, MA: Addison-Wesley.

FREWER, L. J., HOWARD, C. and SHEPHERD, R. (1996). The influence of realistic product exposure on attitudes towards genetic engineering of food. *Food Quality and Preference*, **7**(1), 61–67.

FREWER, L. J., HOWARD, C., HEDDERLEY, D. and SHEPHERD, R. (1997a). Consumer attitudes towards different food-processing technologies used in cheese production – The influence of consumer benefit. *Food Quality and Preference*, **8**(4), 271–280.

FREWER, L. J., HOWARD, C. and SHEPHERD, R. (1997b). Public concerns in the United Kingdom about general and specific applications of genetic engineering: Risk, benefit, and ethics. *Science, Technology, and Human Values*, **22**, 98–124.

FREWER, L., SCHOLDERER, J. and LAMBERT, N. (2003). Consumer acceptance of functional foods: Issues for the future. *British Food Journal*, **105**, 714–731.

FREWER, L., LASSEN, J., KETTLITZ, B., SCHOLDERER, J., BEEKMAN, V. and BERDAL, K. G. (2004). Societal aspects of genetically modified foods. *Food and Chemical Toxicology*, **42**(7), 1181–1193.

GASKELL, G., BAUER, M. W., DURANT, J. and ALLUM, N. C. (1999). Worlds apart? The reception of genetically modified foods in Europe and the US. *Science*, **285**, 384–387.

GASKELL, G., ALLUM, N., BAUER, M., DURANT, J., ALLANSDOTTIR, A., BONFADELLI, H., *et al.* (2000). Biotechnology and the European public. *Nature Biotechnology*, **18**, 935–938.

GASKELL, G., TEN EYCK, T., JACKSON, J. and VELTRI, G. (2004). Public attitudes to nanotechnology in Europe and the United States. *Nature Materials*, **3**, 496.

GREENWALD, A. G., McGHEE, D. E. and SCHWARTZ, J. L. K. (1998). Measuring individual differences in implicit cognition: The implicit association test. *Journal of Personality and Social Psychology*, **74**, 1464–1480.

GRUNERT, K. G. and VALLI, C. (2001). Designer-made meat and dairy products: Consumer-led product development. *Livestock Production Science*, **72**, 83–98.

HENSON, S. (1995). Demand-Side Constraints on the Introduction of New Food Technologies – the Case of Food Irradiation. *Food Policy*, **20**(2), 111–127.

HUOTILAINEN, A. and TUORILA, H. (2005). Social representation of new foods has a stable structure based on suspicion and trust. *Food Quality and Preference*, **16**, 565–572.

KAISER, H. and TANG, X. (2004). *Nanotechnology 2015 and converging markets*. Tübingen: HKC22.com.

KAJANNE, A. and PIRTTILÄ-BACKMAN, A. M. (1996). Toward an understanding of laypeople's notions about additives in food: Clear-cut viewpoints about additives decrease with education. *Appetite*, **27**(3), 207–222.

KASPERSON, J. X., KASPERSON, R. E., PIDGEON, N. and SLOVIC, P. (2003). The social amplification of risk: Assessing fifteen years of research and theory. In N. Pidgeon, R. E. Kasperson and P. Slovic (eds), *The Social Amplification of Risk* (pp. 13–46). Cambridge: Cambridge University Press.

KIRK, S. F. L., GREENWOOD, D., CADE, J. E. and PEARMAN, A. D. (2002). Public perception of a range of potential food risks in the United Kingdom. *Appetite*, **38**, 189–197.

LÄHTEENMÄKI, L., GRUNERT, K., UELAND, O., ASTRÖM, A., ARVOLA, A. and BECH-LARSEN, T. (2002). Acceptability of genetically modified cheese presented as real product alternative. *Food Quality and Preference*, **13**, 523–533.

LAMOTE, S., HERMANS, D., BAEYENS, F. and EELEN, P. (2004). An exploration of affective priming as an indirect measure of food attitudes. *Appetite*, **42**, 279–286.

MAGNUSSON, M. K. and HURSTI, U. K. K. (2002). Consumer attitudes towards genetically modified foods. *Appetite*, **39**(1), 9–24.

MAGNUSSON, M. K., ARVOLA, A., HURSTI, U. K. K., ABERG, L. and SJODEN, P. O. (2003). Choice of organic foods is related to perceived consequences for human health and to environmentally friendly behaviour. *Appetite*, **40**(2), 109–117.

MAISON, D., GREENWALD, A. G. and BRUIN, R. H. (2004). Predictive validity of the implicit association test in studies of brands, consumer attitudes, and behavior. *Journal of Consumer Psychology*, **14**, 405–415.

OLIVEIRA, I. B. and SABATO, S. F. (2004). Dissemination of the food irradiation process on different opportunities in Brazil. *Radiation Physics and Chemistry*, **71**(1–2), 495–499.

OZEN, B. F. and FLOROS, J. D. (2001). Effects of emerging food processing techniques on the packaging materials. *Trends in Food Science & Technology*, **12**, 60–67.

PRISLIN, R., WOOD, W. and POOL, G. J. (1998). Structural consistency and the deduction of novel from existing attitudes. *Journal of Experimental Social Psychology*, **34**, 66–89.

RAUDENBUSH, B. and FRANK, R. A. (1999). Assessing food neophobia: The role of stimulus familiarity. *Appetite*, **32**, 261–271.

ROININEN, K., LÄHTEENMÄKI, L. and TUORILA, H. (1999). Quantification of consumer attitudes to health and hedonic characteristics of foods. *Appetite*, **33**(1), 71–88.

ROZIN, P., SPRANCA, M., KRIEGER, Z., NEUHAUS, R., SURILLO, D., SWERDLIN, A., *et al.* (2004). Preference for natural: Instrumental and ideational/moral motivations, and the contrast between foods and medicines. *Appetite*, **43**, 147–154.

SABA, A. and VASSALLO, M. (2002). Consumer attitudes toward the use of gene technology in tomato production. *Food Quality and Preference*, **13**(1), 13–21.

SCHOLDERER, J. and FREWER, L. J. (2003). The biotechnology communication paradox: Experimental evidence and the need for a new strategy. *Journal of Consumer Policy*, **26**, 125–157.

SCHOLDERER, J. and GRUNERT, K. G. (2005). Consumers, food and convenience: The long way from resource constraints to actual consumption patterns. *Journal of Economic Psychology*, **26**(1), 105–128.

SHIU, E. C. C., DAWSON, J. A. and MARSHALL, D. W. (2004). Segmenting the convenience and health trends in the British food market. *British Food Journal*, **106**, 106–127.

SIEGRIST, M. (1998). Belief in gene technology: The influence of environmental attitudes and gender. *Personality and Individual Differences*, **24**, 861–866.

SIEGRIST, M. (1999). A causal model explaining the perception and acceptance of gene technology. *Journal of Applied Social Psychology*, **29**, 2093–2106.

SIEGRIST, M. (2000). The influence of trust and perceptions of risks and benefits on the acceptance of gene technology. *Risk Analysis*, **20**, 195–203.

SIEGRIST, M. (2003). Perception of gene technology, and food risks: Result of a survey in Switzerland. *Journal of Risk Research*, **6**, 45–60.

SIEGRIST, M. and BÜHLMANN, R. (1999). Die Wahrnehmung verschiedener gentechnischer Anwendungen: Ergebnisse einer MDS-Analyse. *Zeitschrift für Sozialpsychologie*, **30**, 32–39.

SIEGRIST, M. and CVETKOVICH, G. (2000). Perception of hazards: The role of social trust and knowledge. *Risk Analysis*, 20, 713–719.

SIEGRIST, M., CVETKOVICH, G. and ROTH, C. (2000). Salient value similarity, social trust, and risk/benefit perception. *Risk Analysis*, **20**, 353–362.

SLOVIC, P. (1987). Perception of risk. *Science*, **236**, 280–285.

SLOVIC, P., FLYNN, J. H. and LAYMAN, M. (1991). Perceived risk, trust, and the politics of nuclear waste. *Science*, **254**, 1603–1607.

SLOVIC, P., FINUCANE, M., PETERS, E. and MacGREGOR, D. G. (2002). The affect heuristic. In T. Gilovich, D. Griffin and D. Kahneman (eds), *Heuristics and biases: The psychology of intuitive judgment* (pp. 397–420). Cambridge: Cambridge University Press.

SLOVIC, P., FINUCANE, M. L., PETERS, E. and MacGREGOR, D. G. (2004). Risk as analysis and risk as feelings: Some thoughts about affect, reason, risk, and rationality. *Risk Analysis*, **24**, 311–322.

SMITH, E. R. and DeCOSTER, J. (2000). Dual-process models in social and cognitive psychology: Conceptual integration and links to underlying memory systems. *Personaliy and Social Psychology Review*, **4**, 108–131.

SPARKS, P. and SHEPHERD, R. (1994). Public perceptions of the potential hazards associated with food production and food consumption: An empirical study. *Risk Analysis*, **14**, 799–806.

SPARKS, P., SHEPHERD, R. and FREWER, L. J. (1995). Assessing and structuring attitudes toward the use of gene technology in food production: The role of perceived ethical obligation. *Basic and Applied Social Psychology*, **16**, 267–285.

STEPTOE, A., POLLARD, T. M. and WARDLE, J. (1995). Development of a measure of the motives underlying the selection of food: The food choice questionnaire. *Appetite*, **25**, 267–284.

TENBÜLT, P., DE VRIES, N. K., DREEZENS, E. and MARTIJN, C. (2005). Perceived naturalness and acceptance of genetically modified food. *Appetite*, 45, 47–50.

TOWNSEND, E. and CAMPBELL, S. (2004). Psychological determinants of willingness to taste and purchase genetically modified food. *Risk Analysis*, **24**, 1385–1393.

TUORILA, H. and CARDELLO, A. V. (2002). Consumer responses to an off-flavor in juice in the presence of specific health claims. *Food Quality and Preference*, **13**, 561–569.

TUORILA, H., ANDERSSON, A., MARTIKAINEN, A. and SALOVAARA, H. (1998). Effect of product formula, information and consumer characteristics on the acceptance of a new snack food. *Food Quality and Preference*, **9**, 313–320.

TUORILA, H., LÄHTEENMÄKI, L., POHJALAINEN, L. and LOTTI, L. (2001). Food neophobia among the Finns and related responses to familiar and unfamiliar foods. *Food Quality and Preference*, **12**, 29–37.

URALA, N. and LÄHTEENMÄKI, L. (2003). Reasons behind consumers' functional food choices. *Nutrition & Food Science*, **33**, 148–158.

URALA, N. and LÄHTEENMÄKI, L. (2004). Attitudes behind consumers' willingness to use functional foods. *Food Quality and Preference*, **15**(7–8), 793–803.

VAN KLEEF, E., VAN TRIJP, H. C. M. and LUNING, P. (2005). Functional foods: Health claim-food product compatibility and the impact of health claim framing on consumer evaluation. *Appetite*, **44**, 299–308.

VERBEKE, W. (2005). Consumer acceptance of functional foods: socio-demographic, cognitive and attitudinal determinants. *Food Quality and Preference*, **16**(1), 45–57.

VERBEKE, W. (2006). Functional foods: Consumer willingness to compromise on taste for health. *Food Quality and Preference*, **17**, 126–131.

VERLEGH, P. W. J. and CANDEL, M. J. J. M. (1999). The consumption of convenience foods: reference groups and eating situations. *Food Quality and Preference*, **10**(6), 457–464.

YE, X., AL-BABILI, S., KLÖTI, A., ZHANG, J., LUCCA, P., BEYER, P., *et al.* (2000). Engineering the provitamin A (β-carotene) biosynthetic pathway into (carotenoid-free) rice endosperm. *Science*, **287**, 303–305.

11

Food consumers and organic agriculture

C. Ritson and E. Oughton, Newcastle University, UK

11.1   Introduction

This chapter differs from most of the rest of this book in that it is defined by a
food product category, rather than the personal or social characteristics of
consumers, or the food choice environment. The reason for its inclusion must
therefore be that understanding why some people choose to consume organic,
and others do not, can provide additional insight into food consumer behaviour.
Thus the purpose of this chapter is not to describe 'the organic consumer' per se,
but to explore which, and to what extent, factors underlying food choice
influence consumption of organic products.

A number of features of organic products (known in some European countries
as ecological or bio products) suggest that they have the potential to provide a
valuable case study for food choice.

First, although historically, organic production has been associated with fruit
and vegetables, today it is possible to buy an organic version of virtually any
food product, from milk to wine, eggs to bread, bottled baby foods to chocolate.
(In the EU of 15 countries, organic beef, milk and sheep have as high a share of
their markets as do organic fruit and vegetables of theirs (Hamm and Gronefeld,
2004)). Equally, there is always an alternative – known in the organic literature
as 'conventional' – product available. So although it is quite possible that an
organic consumer may, say, be more likely to be a vegetarian, or perhaps less
likely to purchase processed products, neither 'conventional' nor 'organic'
constrains the product choice range available.

Organic consumption therefore resembles a real world 'food choice
laboratory' in which in almost all cases the choice to purchase the organic

product will be a consequence of the perceived attributes of organic versus the conventional alternative (and nothing else).[1]

The second interesting feature of organic from the perspective of consumer food choice is, paradoxically, the lack of choice. Lampkin and Measures (2001) describe organic farming as:

> . . . an approach to agriculture where the aim is to create integrated, humane, environmentally and economically sustainable agricultural production systems. Maximum reliance is placed on locally or farm-derived, renewable resources and the management of self-regulating ecological and biological processes and interactions in order to provide acceptable levels of crop, livestock and human nutrition, protection from pests and diseases, and an appropriate return to the human and other resources employed. Reliance on external inputs, whether chemical or organic, is reduced as far as possible (2001, p. 2).

All this is backed up by a complex set of rules relating to farm production, and to some extent food processing, under the umbrella of two EU regulations, 2092/91 for plants, and 1804/99 for animals. National certification bodies implement and monitor the regulations, sometimes imposing 'stronger' rules. In some countries there is a single body/organic label; in others competing bodies/labels. Thus an organic product comprises a set of attributes, and the consumer buys a prescribed package and is not in a position to choose a variety of different 'quantities' of organic product attributes.

The third feature of the organic product category of interest from the perspective of food choice is that most of the product attributes which distinguish the organic product from the conventional alternative can only be imparted by the primary producer, rather than the food manufacturer, distributor, or retailer, although of course there must be conformity along the food chain with organic regulations. This provides a link between consumer food choice and farmer decision-making lacking in much of the modern food system.

Moreover, most of these product characteristics which distinguish organic from conventional are 'credence attributes' (Ritson and Mai, 1998); that is, not attributes that can be identified before purchase ('search') or ascertained after consumption ('experience'), but ones that require 'belief'. This therefore involves trust on the part of the consumer in the behaviour of the primary producer and other actors in the organic food chain, and the mechanism for achieving that trust will be an important factor in food choice.

Finally, organic consumption represents an excellent real world empirical base for understanding aspects of food choice because it is a dynamic market. Real world purchase data can help to explain the factors underlying food choice in one of two ways: cross-sectionally, that is, what differences between consumers lead one to choose to purchase a product and the other not; or time-

---

1. Of course, occasionally the organic purchase may be an 'accident' or a consequence of temporary availability.

related, what is it that has changed about an individual to lead them to purchase a product this year, but not last year. To provide useful data, the latter requires *change*. The limitation to this type of analysis is that as yet there is no systematic collection of organic market data, which makes it difficult to generalise across countries.[2] The current chapter therefore draws on a number of different studies and data collections in order to illustrate the arguments put forward.

A notable feature of the market for organic products over the past 10 to 15 years has been rapid growth world wide with global sales reaching US$25 billion in 2003. Of this total Western Europe accounts for 51% and North America 45%. Sales in other regions similarly continue to grow, although from a very small base (Willer and Yussefi, 2005). Market growth in Europe has slowed from 8% in 2002 to approximately 5% in 2003, with a total value estimated at 10.5–11 billion Euro. (Richter and Padel, 2005). Rapid growth has continued in US markets. The slower growth in Europe in general masks considerable national differences: Spain, the UK and many of the new EU accession countries have shown annual increases of more than 10% per annum whereas there has been slower growth in Denmark, Germany, Austria and Switzerland. In terms of the absolute market size Germany is the largest market in Europe with sales of organic food over €3 billion in 2003, followed by Great Britain €1.6 billion, France €1.5 billion, Italy €1.4 billion and Switzerland €0.74 billion. However, average consumer expenditure in Switzerland is highest in Europe at €103 per consumer per year, double the second highest Denmark at €51 per year. Per capita consumption on organic food in the USA is closer to the UK, France or Netherlands at €30 per head per year (Richter and Padel, 2005).

Across Europe the relative significance of organic foods within different food groups varies. The most recent comprehensive analysis of these patterns is provided by Hamm and Gronefeld (2004). In the EU as a whole the most significant share of organics (by volume) in total consumption is for cereals (1.8%), beef (1.6%), eggs (1.3%), vegetables (1.3%), fruit and nuts (1.3%) and milk and milk products (1.2%). These relatively low European averages mask some very high shares within countries: for example, in Denmark 8.4% of cereals, 8.8% of potatoes, 8.8% of eggs and 10% of milk and milk products consumed are organic. Similarly, in Switzerland, organic sales by volume account for 8.9% of cereals, 7.3% of oilseeds, 3.7% of milk and milk products and 3.5% of vegetables. There is also evidence for the increasingly rapid growth of prepared organic foods. For example, data collected by Mintel in Britain show very large increases in prepared organic foods and baby and infant foods between 1998 and 2003 (Mintel, 2003).

Against this brief background description of organic markets we now look more carefully at the factors accounting for these changes.

---

2. The situation in Europe is being addressed by the activities of EISfOM. See, for example, Recke *et al.* (2004).

11.2 The expanding organic market: consumer led or producer driven?

It is tempting to assume, particularly in a book concerned with consumer food choice, that the rapid growth in consumption of a particular kind of food must represent incontrovertible evidence of a fundamental change in consumer attitudes, perceptions, and beliefs about that kind of food; and that an end to that growth in turn indicates a stabilising of those changing attitudes. But the organic market is more complicated than that.

In simple market economics, a distinction can be made between a demand led, and supply driven, growth in market size. Changing consumer attitudes and perceptions towards the food product could indeed lead to market growth. Changing tastes, lifestyles, meal patterns, environmental or health concerns, but also income growth, can all lead consumers to wish to purchase more of the product, causing a shortage at existing price levels. Production responds to market signals (higher prices – more profitable) and more is produced and consumed.

But an increase in consumption can also be supply driven. The usual reason for this is that a new, cost-reducing, technology is adopted which makes production more profitable at existing prices. Supply increases, prices fall and consumption increases. In the case of organic agriculture, there are two further, and rather peculiar, reasons why the increase in consumption could be supply driven. The first is what one might term 'producer values'. The early producers of organic products did so almost exclusively because 'they believed in it'. They were not particularly market orientated or market aware and their produce found consumers in a rather haphazard way. Second, conversion of land to organic production has been subject to substantial government incentives. When the government subsidises the production of anything – more is produced, and more consumed.

It is possible in principle for a market to grow due to the interaction of independent demand-led and supply-driven factors which, if balanced, allow consumption to increase at constant prices (or in the case of the organic market, at what are usually referred to as price premiums over 'conventional' produce – see below).

In addition, there is a particularly interesting way in which a supply driven market growth can lead to an increase in consumption. If a particular product simply becomes more available, consumers become more aware of its existence, qualities and attributes, and more is purchased without the incentive of cheaper prices, or any fundamental change in consumer attitudes, tastes and preferences for the product.

Ritson (1993) argues that there was an element of this in the rapid growth in consumption of farmed salmon during the late 1980s and early 1990s. The new, sophisticated, technology of salmon farming reduced costs. Salmon farming was profitable at prevailing prices, attracted substantial investment and supply increased rapidly. But this coincided with a period in which the product itself was an ideal match for a demand trend towards more 'healthy' and convenient products. In addition, though, a product that had previously been mostly associated with restaurant meals became increasingly available in supermarkets, product awareness increased, and more was purchased. Marketing specialists at the time indicated that the most important reason for increased consumption was 'availability'. The conditions were therefore in place for a rapid expansion of the market at constant prices. Quite suddenly, though, the 'supply drive' overtook the demand pull, and prices collapsed: that is, the continual increase in supply could only be converted into an equivalent growth in consumption by falling prices.

Many of the same forces appear to have been present in the case of the recent rapid rise in the consumption of organic food. To oversimplify a little, it is probably the case that the growth in the organic market in most European countries will have displayed an early, supply-driven period, followed by a demand-led phase, with supply growth again taking over more recently.

The substitution of a supply-driven growth by a demand-led one is illustrated by the Danish experience. In the early 1990s in Denmark, most farmers who converted to organic did so because of a belief in the environmental superiority of organic production. In response to survey questions nearly 90% said that they converted because they 'disagreed with conventional agriculture' or 'had environmental concerns'. By the end of the decade this had been replaced by the market-orientated lure of the organic price premiums prevailing, with more than half quoting 'higher incomes' as their reason for conversion.

Similarly, the rapid growth in the UK at the end of the 20th century appears to have been demand led. The best evidence of this is the degree to which the UK was reliant on imports for its organic supplies (see Table 11.1). Domestic production was failing to meet the growth in demand and high price premiums were pulling in imports.

The UK House of Commons Agriculture Committee (2001) observed that the expansion in organic production was having to race to keep up with the growth in customer demand. At the same time, A German academic commented: '... the market growth during the last years has not primarily been driven by the demand side, but was mainly caused by activities on the supply side' (Alvenslaben, 2001, p. 388).

Table 11.1   UK self sufficiency (%)

Product	2000	2001	Product	2000	2001
Beef	77	60	Cereals	19	28
Sheep	97	94	Potatoes	63	66
Pork	66	34	Vegetables	40	55
Poultry	46	67	Fruit	16	4
Eggs	93	90	Milk	80	97

Source: based on data in Hamm and Gronefeld (2004)

Table 11.2   Share of organic production sold as organic in the EU: 2001

Animal products		Vegetable products	
Milk	68	Cereals	93
Beef	69	Oilseeds	91
Sheep	54	Olive oil	73
Pork	94	Potatoes	96
Poultry	99	Vegetables	95
Eggs	97	Fruit	84
		Wine	61

Source: based on data in Hamm and Gronefeld (2004)

The main explanation for this was that generous government incentives for conversion to organic were stimulating the increase in production in some countries. This was evidenced by the proportion of organic production which was sold into conventional marketing channels (that is, not sold to consumers as being organically produced). This is illustrated in Table 11.2. It should be noted though that the gap between the production and consumption of organic food may reflect a failure in the institutions governing the processing, transport and retailing of organics, i.e. structural failures in the link between producers and consumers.

It is important to point out here the very important characteristic of organically produced food that distinguishes it from conventionally produced and which raises complications for market analysis. The European Action Plan for Organic Food and Farming (SEC2004 739) argues that the environmental benefits of organic farming are important public goods and should therefore be financed by public means:

> Both roles of organic farming contribute to the income for farmers ... In order to achieve the objectives of consumers, producers and the general public, organic farming should develop a balanced approach to these societal roles. It should offer a fair and long-term support for public goods, and at the same time foster the development of a stable market (2004, p. 6).

They note, for example, that in Sweden farmers are encouraged to produce organic for its public good attributes even though they sell into a conventional food chain. The private benefits are reaped by consumers who have organic foods available to them but this aspect of production should be subject to market rules. Given that any organic product embodies both those benefits, analysis of the market becomes very complex indeed.

In summary, it is clear that a substantial part of the increase in consumption of organic products has been demand led, the consequence of a positive shift in consumer attitudes to organically produced food.

But part has also been supply driven, with consumer reaction to more competitive prices and increasing availability the main vehicle for increasing consumption. It is to these two features of organic consumer behaviour that we now turn.

11.3   Factors influencing organic purchase[3]

In the previous section we examined the development of the European market for organic agricultural products and the inter-relationship between supply and demand factors within the market. In this section we examine more closely the consumer of organic foods and ask: what factors influence the consumers to choose organic rather than conventional foods? This is:

> ... a potentially complex task in which many different aspects might need to be considered. Health, environmental concern, ethics, authenticity and taste, and concerns about the relations between people and nature are examples of broad themes that recur in the literature (Torjusen *et al.*, 2004, 39).

There are a large number of studies of European organic consumers but it is difficult to generalise the findings across countries or to untangle the complex inter-connections because of the different methodologies and conceptual models that have been used. As Ritson and Kuznesof (2006) note in their study of consumption and alternative production technologies, a number of models of food choice have emerged drawing on contributions from different academic disciplines and including: economic factors, sensory aspects of eating, perceptions relating to health, nutrition and well-being, lifestyle factors and beliefs about production technologies. Approaches to the study of food consumers can be split broadly into those taking a cognitive or behavioural approach and those with a more socially or culturally determined view of behaviour.[4] Cognitive approaches depend on psychological models that explore the consumers' knowledge and perceptions of the characteristics of the food in relation to the needs that they are trying to satisfy through their purchase. Within this approach differing emphases are place upon the consumers' values, beliefs and attitudes, their intentions to act and their actions. Social and cultural studies of organic food, on the other hand, emphasise the many symbolic meanings of food and the activities surrounding its purchase, preparation and consumption. Both approaches show consumers concerned with quality and safety aspects of organic food but these concerns are constructed in different ways.

---

3. The following section draws on current work being carried out on EU integrated project No. 506358 *Improving quality and safety and reduction of cost in the European organic and 'low input' food supply chains.* In particular deliverable 1.2 Subproject 1: Determining consumer expectations and attitudes towards organic/low input food quality and safety, Midmore *et al.* (2005).
4. See chapters in the current volume and for particular reference to organic foods Torjusen *et al.* (2004).

As Midmore *et al.* (2005) note:

From the point of view of the organic consumer, 'organic' implies 'quality' in itself, and support for organic agriculture and 'safe' food-processing techniques. The use of wholesome, unadulterated ingredients contributes not only to the individual good, in terms of healthy eating, but also to broader environmental and social goals, which benefit the community as a whole through fundamentally sustainable and 'caring' production methods.

Just as in the case of conventional foods, differences in organic food choices are found according to socio-economic and demographic factors. Other significant factors affecting organic food choice include whether consumers are traditional and heavy, medium or light consumers of organic foods. There is also an interesting pattern emerging that shows consumers changing across the life cycle; for example families with children in the 15–20 age group living at home having lower consumption than those with younger children. The type of distribution channel that the consumer chooses to use is similarly significant; 'heavy' users of organic food frequently buy though alternative, small scale channels. However, over 80% of organic food purchased in Britain, Denmark, Finland and Sweden is from supermarket and conventional channels, whereas in Belgium, Germany, Spain, Greece, Portugal and Norway the majority of organic food sales is through organic or wholefood shops or through direct sales from farmers (Hamm and Gronefeld, 2004).

Table 11.3 provides an overview the reasons that consumers have given for purchasing organic foods across a number of European countries. Clearly, the table does show some general patterns emerging across the eight countries.[5] In all of the countries studied, health, either for self and or family, appears as an important factor for consumers. Health benefits may be associated with the idea: (a) of fewer additives and chemicals in the food – that it is produced 'naturally' and (b) of healthy eating, which in turn helps in avoiding health problems. Furthermore, there is a strong association between health, well being and quality of life in general. Health associations derive not only from what is absent from the food but also the belief that it contains higher nutrient values. This view is particularly true for fruit, vegetable and cereal products, which it is believed contain more vitamins and minerals, and contribute to a more wholesome meal.

The 'health' attribute of organic food is not just a reflection of a positive 'pull' feature of organic products, but reflects a 'push' from the conventional food market. Consumers have become anxious following food scares, such as those associated with BSE and Salmonella, and uncertainty about the effects of novel technologies such as genetic engineering of food and increasingly seek 'safe' food. Organic food is perceived as one of the ways of dealing with the

---

5. For details of consumer organic preferences based on laddering interviews and focus groups across a range of product categories, see Zanoli (2004) and for an additional review of Denmark, Italy, UK and Hungary see Torjusen *et al.* (2004).

Table 11.3   Reasons that consumers purchase organic foods

Country	Reasons for buying organic food	Reasons for not buying organic food
Austria	• Own health (improvement, avoidance of risk) • Responsibility for children • Contribution to regional development	• Price, habit, mistrust and lack of motivation, poor availability and product range
Switzerland	• Better taste • Health, especially for mothers and people with illness • Altruistic motives; environment, animal husbandry, remuneration of farmers	• Price, low perception of difference between organic and conventional production, mistrust in organic standards
Germany	• Own or children's health (avoidance of harmful ingredients) • Support of organic shops and farmers in their aspiration • Occasional consumers mention taste as a buying motive more often than regular consumers	• Price, poor availability, shopping habits, doubts about quality, lack of interest, taste
Denmark	• Motives reflect a lifestyle choice: environmental protection • Own health • Support of, and contribution to, a better world	• Poor quality, no perceptible difference between organic and conventional food
Finland	• Motives reflect a lifestyle choice: environmental protection • Health (products are pure and contain no residues) • Conscience	• Price conscious consumer affected by unreliable quality
France	• Healthy nutrition (healthy, nutritious, unaltered) • Taste • Respect for the living world	• Lack of information, large number of different labels
Italy	• Health (safety) • Taste	• Availability, lack of trust in standards, product quality, price for regular users
UK	• Own health (no chemicals, purity) • Local farming and fair trade • Environmental protection	• Product related (price, appearance, availability, quality, variety, taste) information about the product (confusion, habit, trust, information)

Adapted from Zanoli (2004), various sources.

Table 11.4 Association with the stimulus 'bio-products'

Association	
1. Without chemicals	10. Expensive
2. Natural products	11. No pesticides
3. Without artificial fertiliser	12. Controlled farming
4. 'Biological' farming	13. Not containing noxious agents
5. Healthy	14. Not genetically modified
6. 'Ecological' farming	15. Natural manure
7. Caring animal husbandry	16. Free range animals
8. Not sprayed	17. Negative associations
9. Environmentally friendly	

Source: based on data in Alvensleben, 2001

anxieties associated with conventional food production and processing systems (Alvensleben, 2001).

This feature of the organic food consumer – the belief that organic products lack the negative credence attributes associated with conventional agriculture and food production, is illustrated by Tables 11.4 and 11.5. Two thousand consumers in Germany were asked what they most associated with Bio (organic) products. The various responses are shown in Table 11.4, the responses ranked from most frequently mentioned association.

Second, in a survey of 1000 British consumers, respondents were asked 'how worried' they were about a series of potential food safety issues previously

Table 11.5 UK public concerns about food

Concern
1. The use of hormones in animal production
2. The use of antibiotics in food production
3. The use of pesticides in food production
4. Animal welfare standards in food production
5. Eating genetically modified food
6. Safety of meat products produced by intensive farming methods
7. The use of additives in food
8. Quality of food using intensive farming methods
9. Conflicting information on food safety
10. Lack of information about food from government
11. Hygiene standards in the food industry
12. Hygiene standards in restaurants and take-aways
13. Being able to afford good quality food
14. Amount of fat in your diet
15. Information about what foods are good for you keeps changing
16. Knowing what to do when there is a food scare
17. Getting food poisoning
18. Hygiene standards in your home

Source: Miles *et al.* (2004)

identified from focus groups as things which concerned consumers about food consumption. In Table 11.5 the 'worries' are now ranked from most to least worried (percentage of the sample which said they were either highly or extremely worried). The striking observation is that many of features of food consumption which seem to cause most concern to consumers – pesticides, hormones, antibiotics, additives, intensive farming and poor animal welfare – represent negative characteristics thought to be absent from organic products (without chemicals, without artificial fertilisers, no pesticides, not sprayed, caring animal husbandry).

A second frequently mentioned characteristic of organic food is the taste. Taste is a sensory or organoleptic attribute of food that may be experienced directly by the consumer and may be compared directly to the physical aspects of other foods. Sensory analysis of organic foods has shown that there is no consistent difference across different product categories of the taste of organic and conventional foods (Fillion and Arazi, 2002) Particular foods may have a measurably better taste when organic but this will depend on a range of other production conditions apart from whether the food is just organic or not. As Midmore et al. (2005) point out, these physical factors can be viewed as an effect of the organic production process as well as hedonistic characteristics. Taste is very subjective, and positive feelings about taste tend to be linked to the authenticity of the organic product. Consumers describe the taste as being 'real' or 'genuine' and as in the case of health there is an association with 'naturalness'. However, taste is also given as a reason for non-purchase of organic foods where there may be no discernable difference in taste between conventional and organic foods or where the freshness or 'look' of the food suggests that it will not taste good.

A range of ethical issues are given for the consumption of organic foods including animal welfare – natural rearing, humane slaughter techniques, protection of the environment, fair trade, local production and the reduction of food miles as well as broader economic and social impacts. An interesting aspect of this 'ethical' group of issues is the number of characteristics that are not 'organic' (according to the EU regulations defining organic production and processing) but are associated with organic in the consumer's mind: fair trade, food miles, small scale production, origin labels, regional images, etc. These issues are complicated by the number of different labels under which organic foods are marketed – some being more inclusive of broader social values than others. Further systematic work is required on the ways in which the differing cultural values across Europe may relate to different ethical concerns.

Tables 11.3 and 11.4 illustrate clearly that many of the reasons that con-sumers offer for purchasing organic foods are unseen positive credence charac-teristics. Even after eating the product the consumer may not be certain that it was organic. Thus the degree of knowledge on the part of the consumer and the amount of information on the production, and processing techniques that the food has undergone, play an important role in the decision to purchase. It is for this reason that the certification and labelling of food plays such an important part in organic food choice. It is not surprising then that 'trust' or 'lack of trust' is mentioned frequently as a feature of the purchasing decision.[6] Many of the credence characteristics associated with the positive decision to purchase organic food are not required by the formal EU regulations governing its production and processing. For example, 'small scale' and 'local' are not organic attributes, but valued by organic consumers and associated by them with organic agriculture.

Whereas many of the positive characteristics associated with the decision to buy organic foods are credence characteristics, many of those given for not purchasing are more directly 'experience' characteristics.[7] For example: price, poor availability, limited product range, too many different labels, poor taste. The implications of this are discussed in more detail below.

The attributes that consumers attach to organic food can be split into those with private use values – such as health, taste and freshness; and public use values – for example, animal welfare, environmental conservation. Private use values are consumed by the individual, whereas public good values are shared and consumption by any one person does not exclude consumption by another. Historically, the decision to buy organic produce has been associated more strongly with public use values, with a concern for the environment. However, more recently, as consumers become more concerned with food safety and health, the significance of private values has increased. It is interesting to note, therefore, that all the reasons – experience characteristics – given for not buying organic described above are private use values. The public good attributes of organic foods are not mentioned. Is it then the private good aspects of organic foods that fail to satisfy consumer needs?

Work being carried out in Denmark is revealing an interesting and dynamic relationship between the values that consumers hold with respect to organic foods and their purchasing patterns (Wier et al., 2005). Denmark has the highest consumption per capita of organic food in Europe and government support has emphasised the public good aspects of organic farming. Wier et al. show that public good attributes are widely acknowledged by the Danish respondents in the study, and that over one quarter would be willing to pay extra taxes to support the future of organic farming. However, the study of household purchasing behaviour shows that although public-good values appear to be a prerequisite for purchasing organic foods, it is private good attributes that determine the actual degree to which purchases take place.

The paradox generated by these findings is further illustrated by a comparative study of Denmark and Britain, a country with the fastest growth of organic food consumption in Europe (Wier et al., 2005). Demand in both markets is shown to be sustained primarily by the private good attributes, health

---

6. Trust illustrates well the differences that cognitive/behavioural and socio/cultural theories offer in understanding and analysing organic food choices. Whereas in the former lack of trust is seen as being something that may be remedied by education and the provision of information; in the latter in may be regarded as a positive communication within the development of the food system. (See Kjærns, 2003, quoted in Torjusen et al., 2004.)
7. The authors wish to thank Hans van Trijp for bringing this to their attention.

and safety, of organic foods. However, in both countries much of organic food is produced and handled in concentrated and industrialised sectors characteristic of the conventional food systems that consumers are trying to avoid. These results indicate that there is still much to learn about the development of the organic food market and what it is about it that affects consumer choice.

11.4   The price premium

One of the most fundamental and durable things that we know about food consumers (Ritson and Petrovici, 2001) is that, in almost all cases, when price falls (and nothing else changes) more is purchased. The increase in consumption will be a combination of existing consumers increasing their frequency/amount of consumption; and lower prices inducing new consumers of the product. Almost all of the increase in consumption will involve substitution. In the case of organic produce, most of the substitution is likely to be between the organic food product and a similar 'conventionally produced' one. This in turn has led much of the debate over organic prices, and such data as is available, to be characterised in the form of the 'organic price premium' – the percentage excess of the organic product price over the conventionally produced alternative.

Table 11.6 provides estimates from different sources for selected products in selected EU Member States. It is tempting to conclude that these, often substantial, organic price premiums provide us with monetary estimates of the value consumers attach to the attributes of organic products relative to conventionally produced ones. Up to a point, this is correct, but subject to a number of important qualifications.

First, the price premium may reflect attribute differences other than these specifically associated with organic production. It is clearly difficult in many cases to establish a conventional benchmark when a range of quality exists for

Table 11.6   Consumer price premiums (%) for organic products in selected EU Member States (2002)

Product	Denmark	UK	France	Germany	Italy	EU (15) Average
Bread	13	34	106	25	38	41
Potatoes	56	128	140	83	101	94
Tomatoes	85	118	126	110	68	102
Apples	36	53	41	98	53	75
Milk	19	38	64	42	117	50
Yoghurt	8	8	85	25	63	37
Eggs	17	56	51	53	121	54
Chicken	91	138	64	181	107	129
Steak	46	70	56	74	70	59

Source: based on data in Hamm and Gronefeld (2004)

the conventional product, for example, with chicken, eggs and tomatoes. Hill and Lynchehaum (2002, 5), claim that, 'The primary reason that organic is more expensive is simply because it is good quality food. Organic is not expensive when compared to other quality foods'.

Second, the pricing policy adopted by retail stores may incorporate elements of consumer value additional to those tied to organic production. In a study of retail store pricing policies for organic fruit and vegetables in France, Germany, Spain and the UK, La Via and Nucifora (2002) found that the organic price premium had a basic component of about 40%, which appeared to reflect the higher production and marketing costs of organic products. Beyond that, there could be an additional mark-up of up to another 40%, reflecting store location and additional services and display information associated with the organic sections of the store.

Third, the price premiums are unstable over time and very sensitive to changes in the balance of supply and demand in a dynamic market. As indicated earlier, the time taken for farms to convert to organic can lead the growth in demand to outstrip supply, and price premiums rise. Equally, when produce from the converted land comes onto the market, supply can overshoot demand and price premiums are squeezed (and some organic produce goes into conventional marketing channels). In a more stable market one would expect the price premium to reflect the extra organic production and marketing costs.

Clearly these fluctuations in price premiums do not of themselves indicate major shifts in the value attached by consumers to organic attributes. Rather consumers with lower valuations are varying their levels of consumption, or coming in and out of the market, in response to varying price premiums.[8] Thus, what the price premium indicates is the lower boundary of consumer valuation, *at the prevailing levels of consumption*. It does not tell us the extent to which 'committed consumers' value organic attributes – what they would be 'willing to pay'.

There are three ways in which to explore this issue: analysis of the impact of changing prices on purchases; 'stated preference' interviews; and 'choice experiments'. In the case of the first of these – price analysis – broadly speaking, the less sensitive is the level of organic consumption in response to a price increase, then the more this indicates the presence of 'committed consumers', valuing organic attributes at more then the prevailing price premium.

Wier and Smed (2000) used data for 2000 Danish households in 1997–98 to estimate the response of organic consumption to price changes. Their results suggest that the demand for organic products is much more sensitive to price than the demand for conventional food. More formally, the price elasticity of demand – the percentage change in consumption for a given percentage change

8. The growth in organic consumption has typically been associated with the view that it indicated 'extra' consumers joining the existing market. However, analysis of panel data in Denmark (Wier *et al.*, 2005) clearly shows existing consumers 'dropping out' of organic consumption, as well as new consumers joining.

in price – is greater for organic products. This is to be expected, if variation in price premiums cause consumers to switch between organic and conventional consumption. For example, the price premium could increase either because of a rise in organic prices or a fall in conventional prices. If this increase in premiums leads to, say, 10% less organic consumption, then the corresponding increase in conventional consumption will be much lower when expressed as a percentage of total non-organic consumption. Wier and Smed estimated that a decrease of 20% in the organic price premium would increase the market share of organic consumption for dairy products from 10 to 15%, for bread and cereals from 5 to 7%, and for meat from 1 to 2%.

Another way of exploring the sensitivity of organic consumption to organic prices, *relative to* benchmark conventionally produced product prices, is by the concept of cross-price elasticity of demand – the percentage change in purchases of one product relative to a given percentage change in the price of another product. If food products are substitutes for each other – that is consumers are willing to switch consumption in response to changes in relative prices – then one would expect a rise in the price of one product to lead to an increase in the consumption of the substitute product. This is shown in Table 11.7 for organic and conventionally produced dairy products in Denmark.

The estimates are intuitively very credible. They suggest, for example, that a 10% increase in the price of organic dairy products (with no change in conventional prices) would decrease organic consumption by 22.7%, and *increase* consumption of conventional products by 1.3%. However, an increase in the price of conventional products by 10% (with no change in organic prices, and thus narrowing the price premium) would decrease the consumption of conventional products by 11.3% and *increase* organic consumption by 12.7%. Again the fact that organic consumption seems to be much more sensitive to changes in conventional prices, than conventional consumption is to organic prices, reflects the much lower market share (about 10% in this case) for organic dairy products.

Table 11.8 provides a more comprehensive set of estimates of own-price elasticities of demand for organic products, compared to the conventional equivalent, this time for the UK. In all cases the response of consumption of organic products to change in price is about double that for conventional produce.

Table 11.7   Own and cross price elasticities of demand for organic and conventionally produced dairy products in Denmark

	Elasticity with respect to price	
	Organic	Conventional
Organic	−2.27	1.27
Conventional	0.13	−1.13

Source: Wier *et al.* (2001)

Table 11.8 Own price elasticities of demand for organic and conventional products in the UK

	Conventional	Organic
Dairy	−0.57	−1.14
Milk	−0.76	−1.54
Eggs	−0.26	−0.52
Cheese	−0.34	−0.67
Meat	−0.95	−1.91
Beef	−1.64	−3.28
Lamb	−0.52	−1.05
Pork	−1.87	−3.74
Chicken	−1.37	−2.75
Vegetables	−0.31	−0.62
Processed vegetables	−0.67	−1.34
Potatoes	−0.21	−0.43
Green vegetables	−0.47	−0.94
Fruit	−0.21	−0.43
Bananas	−1.31	−2.62
Apples	−0.49	−0.97
Citrus fruits	−0.46	−0.92

Source: ADAS (2004)

The second way of attempting to estimate the extent to which consumers value organic attributes, in the sense of being willing to pay a price premium, is simply to ask them. Wier and Calverley (2002) report on a range of such studies and a synthesis of these results, for the studies carried out in Sweden, Norway, Denmark and the UK, is presented in Table 11.9.

The two notable features of Table 11.9 are, first, the substantial proportion of consumers willing to purchase organic at low price premiums – in general much lower premiums than those shown as prevailing in Table 11.6. Second is that the proportion indicating a willingness to pay seems to level off, once premiums exceed 30%, suggesting a core of 'committed' organic consumers willing to pay the higher price premiums that tend to apply.

Wier and Calverley also note that studies in the Netherlands and Germany indicated higher proportions of consumers willing to pay high price premiums.

The strength of the 'willingness to pay' approach is that it allows insight into

Table 11.9 European consumers' willingness to pay for organic food

Price premium (%)	5–10	10–20	20–30	30–40	40–50	50–60
Proportion of consumers willing to buy (%)	45–80	20–50	10–25	5–20	3–18	3–15

Source: based on data collected by Wier and Calverley (2002)

potential consumer behaviour lying outside the range of price premiums provided by market data. There are, however, serious doubts over the reliability of individual consumers' own estimates of how they would behave in different market circumstances, and the willingness to pay technique (sometimes described as 'contingent valuation' or 'stated preference') is more commonly used to attempt to value goods for which markets do not exist, in particular environmental goods.

Some experts argue that, for products which do possess markets, a more reliable method for predicting consumer behaviour outside the range of observed market prices is by 'choice experiments', in which consumers use real money and real products under laboratory conditions. Soler and Gil (2002) used an experimental auction market to attempt to elicit consumer willingness to buy organic olive oil. They did this first on its own, and then, as a reference point, provided information relating to prevailing prices of conventional olive oil. They found that only 5% of consumers were willing to pay the prevailing organic price premium in Spain, but that up to 70% were willing to pay some premium.

11.5   Conclusions

The past 10–15 years have seen rapid growth in the consumption of organic products in Europe and North America. Not all of this consumption growth can, however, be attributed to a fundamental shift in consumer attitudes towards organic products. Part of the growth in consumption has been supply driven by government support and because of the environmental goals leading some producers to convert to organic production.

Many studies of organic consumption indicate that health reasons underpin much of the consumer motivation to purchase organic. The 'health' attribute of organic food is not just a reflection of positive 'pull' factors such as perceived higher nutrient value, but also 'push' factors associated with the absence of negative associations of conventionally produced food.

For markets to function efficiently it is assumed that buyers and sellers have complete information. However, the organic food market shows not just that consumers lack information but that in some cases they are misinformed about organic foods. This may be particularly true for 'occasional' consumers who are significant for the further development of the organic market. Many of the positive attributes associated with organic foods are strongly linked to attributes associated with other 'alternative' production and food handling systems, e.g. fair trade, small scale, local production, low travel miles, etc. The reasons that consumers give for purchasing organic foods reflect their beliefs about 'organic' and what it means, not the formal, regulated, definition of organic. Until consumers are much better informed about the meaning of organic it is difficult to be sure about their attitudes; currently we are really looking at attitudes towards a bundle of non-conventional food production characteristics.

Similarly, price analysis suggests that organic consumption appears to be very sensitive to changes in the price premium over conventional produce with, however, a core of 'committed consumers' willing to pay the substantial premiums which usually prevail; but, there is a much larger pool of potential consumers at more modest price premiums. The analysis of the organic food market is complicated, however, by the fact that organic foods embody both public and private goods and these are intimately bound together. Whereas it may be argued that private goods should be subject to market behaviour, the situation is more complex with public goods. Subsidy of the public good – the environmental benefits of organic farming – is thus also a subsidy of the private good. Moreover, emerging evidence seems to suggest that although consumers list the environment as a significant and positive attribute of organic food, their market behaviour seems to indicate that it is the private good attributes – health and taste – that determine purchase. Public good characteristics may form a necessary but not sufficient condition for private action.

11.6   References

ADAS (2004) *Evidence Assessment to Inform the Review of the Organic Farming Scheme – November 2003, http://statistics.defra.gov.uk/esg/evaluation/ofs/*.

ALVENSLEBEN, R. VON (2001) 'Beliefs Associated with Agricultural Production Methods', in Frewer, L., E. Risuik and H. Schiferstein (eds) *Food People and Society*, Springer Verlag, Berlin Heidelberg.

EUROPEAN COMMISSION (2004) *European Action Plan for Organic Food and Farming* Commission Staff Working Document SEC (2004) 739.

FILLION, L. and S. ARAZI (2002) Does organic food taste better? A claim substantiation approach, *Nutrition and Food Science*, Vol 32, No 4: 153–157.

HAMM, U. and F. GRONEFELD (2004) *The European Market for Organic Food: Revised and Updated Analysis*, Organic Marketing Initiatives and Rural Development Vol. 5, The University of Wales, Aberystwyth.

HILL, H. and F. LYNCHEHAUM (2002) 'Organic Milk: Attitudes and Consumption Patterns', *British Food Journal*, Vol 104, No 7: 526–542.

HOUSE OF COMMONS (2001) *Select Committee Report on Organic Farming*, HMSO. London

LAMPKIN, W. and M. MEASURES (2001) *Organic Farm Management Handbook,* University of Wales, Aberystwyth.

LA VIA, G. and A.M.D. NUCIFORA (2002) 'The Determinants of the Price Mark-Up for Organic Fruit and Vegetables in the European Union', *British Food Journal*, Vol 104, No 3/4/5: 319–336.

MIDMORE, P., S. NASPETTI, A-M. SHERWOOD, D. VAIRO, M. WEIR and P. ZANOLI (2005) *Consumer attitudes to quality and safety of organic and input foods: a review*. September. EU Intrgrated Project 506358. Quality Low Input Food, pp67.

MILES, S., M. BRENNAN, S. KUZNESOF, M. NESS, C. RITSON and L.J. FREWER (2004), 'Public Worry about Specific Food Safety Issues', *British Food Journal*, Vol 106, No 1: 9–22.

MINTEL (2003) *Organic Foods – UK – November 2003*, Mintel International Group Ltd.

RECKE, G., H. WILLER, N. LAMPKIN and A. VAUGHAN (2004) *Development of a European Information System for Organic Markets – Improving the Scope and Quality of Statistical Data.* Proceedings of the 1st EISfOM Seminar, Berlin Germany 26–27 April 2004.

RICHTER, T. and R. PADEL (2005) 'The European market for organic foods'. In Willer and Yussefi *The World of Organic Agriculture Statistics and Emerging Trends 2005* IFOAM Berlin.

RITSON, C. (1993) *The Behaviour of the Farmed Salmon Market in Europe: A Review*, Centre for Rural Economy, University of Newcastle-upon-Tyne.

RITSON, C. and S. KUZNESOV (2006) 'Food consumption, risk perception and alternative production technologies', Chapter 3 in Eilenbery, J. and H. M. T. Hokkanen (eds) *An Ecological and Societal Approach to Biological Control*, Springer, Dordrecht.

RITSON, C. and L.W. MAI (1998) 'The economics of food safety', *Nutrition and Food Science*, No 5: 253–259.

RITSON, C. and D. PETROVICI (2001) 'The economics of food choice: is price important?' in Frewer, L., E. Risuik and H. Schifferstein (eds) *Food People and Society*, Springer.

SOLER, F. and J.M. GIL (2002) 'Consumers' acceptability of organic food in Spain: Results from an experimental auction market', *British Food Journal*, Vol 104, No 8: 670–687.

TORJUSEN H., L. SANGSTAD, K. O'DOHERTY JENSEN and U. KJÆRNES (2004) 'European Consumers' Conceptions of Organic Food: A Review of Available Research', Professional Report no. 4-2004, *National Institute for Consumer Research*, Oslo Norway.

WIER, M. and C. CALVERLEY (2002) 'Market potential for organic foods in Europe', *British Food Journal*, Vol 104, No 1: 45–62.

WIER M. and S. SMED (2000), reported in Wier and Calverley, *Modeling demand for organic foods.* Paper presented at The 13th International Scientific IFOAM Conference, Basel, Switzerland.

WIER, M., L.G. HANSEN and S. SMED (2001) Explaining Demand for Organic Foods. Paper presented at 11th annual EAERE Conference, Southampton, UK, June 2001.

WIER, M., L.M. ANDERSON and K. MILLOCK (2005) 'Information Provision, Consumer Perceptions and Values – The Case of Organic Foods', Forthcoming in: Russell, C. and S. Karup: *Environmental Information and Consumer Behaviour.* New Horizons in Environmental Economics Series. Edward Elgar.

WILLER H. and M. YUSSEFI (eds) (2005) *The World of Organic Agriculture Statistics and Emerging Trends 2005*, IFOAM, Bonn.

ZANOLI, R. (ed.) (2004) 'The European Consumer and Organic Food', *Organic Marketing Initiatives and Rural Development, Vol 4,* University of Wales, Aberystwyth.

Part III

Diversity in consumer food choice: cultural and individual difference

12

Life experience and demographic variables influencing food preferences: the case of the US

R. Bell, Natick RD&E Center, USA

12.1 Introduction

This chapter was originally conceived as a basic discussion of demographic variables affecting food preferences in the US, and this is a noble cause for a book chapter. However, as I looked through the literature I began to question whether demographic data are the most valid and useful variable for understanding food preferences and consumer behavior regarding food and for guiding product developers, and I further questioned the myriad of definitions we seem to have for the concept of 'food preferences.' These questions motivated me to revise the content of this chapter as one that does explore the relationship between demographics and preferences, but spends significant time discussing the appropriateness of these variables for truly understanding consumers and their eating behavior.

In the fields of demography and consumer research, scientists have accepted the definitions of the constructs of demographic variables and food preferences for the past fifty years. It is true that in regard to food preferences we are continually developing new methods and underlying frameworks for analyzing these data (see Gains, 1994; Greenhoff and MacFie, 1994); and new psychophysical approaches to asking and answering the same basic questions (see Conner, 1994; Cardello, 1994), and these methods and measures have been validated and proven reliable during this time, and our understanding of how these variables influence people's attitudes, knowledge and capabilities has progressed. But what we are measuring and its predictability for behavior have not progressed as far as the methods and measures, and given the quickly changing nature of society and civilization over the past fifty years, one could argue that if what we are measuring is not predictive of behavior, then the measures and definitions, as currently utilized, are limited in terms of helping us to truly understand consumers in regard to eating behavior. If food preferences are to be a part of our understanding consumer behavior, and if they are to suggest future trends in food product development, then we need to update our definition of both demographics and food preferences, and how they relate to each other.

This chapter will present information on how demographics influence food preferences, but it will go beyond that by questioning the use of demographics as a valid variable for understanding consumers and by suggesting the use of life experience variables as a research approach that could better elucidate an understanding of consumer behavior in regard to food preferences. It will also look at the often-unclear definition of the term 'food preferences,' and will make arguments for taking a more 'whole-diet' view of preference, rather than viewing preference in the context of single food ratings or as the preferred option among two or more items at a single food choice event. In addition to these definitions, the methods for collecting data on each will be briefly described, as will the importance of collecting these data. Following this, US demographic data will be presented, as will implications for how these data relate to food preferences across the diet and how they suggest directions for future trends in product development. Among these implications will be a re-examination of food preferences in the more salient context of life experiences, rather than as numerical or categorical demographic data.

12.2   Measuring and defining demographics and life experiences

The demographics and eating patterns of the United States are best examined when looking at the US Census, which is conducted once every ten years, and includes a complete count of the persons and households in the US. This is an incredibly expensive way to do research, primarily creates cross-sectional data, does not take into account interactions, and it cannot examine all possible relationships. In addition, the data tend to group individuals and lead to an assumption that all who belong to a demographic group possess similar beliefs, attitudes and knowledge. This would have made more sense in the 1950s, when homogenous neighborhoods were the norm and cultural diversity was hard to find in the US. The growth of individualism, the peace and freedom movements of the 1960s, the economic opportunities since the 1980s in the US that has brought millions of new immigrants from countries in Central America, South America, Asia and Europe, the explosion of communication and advertising and communication networks such as the Internet, the increased prosperity and income inequality in developed countries, have all contributed to more diverse beliefs, attitudes, norms and behavior in the US. No longer can 13-year-old

Caucasian males from an urban community such as New York City be considered to behave in similar ways; too many subcultures have been created, leading to a wide range of diversity, even within regional socio-demographic groups, within age groups and youth subcultures, and within affluent and less affluent working subcultures. This is a critical factor that should make us question the adequacy of demographics for providing predictive validity when these simple demographic variables are used as independent or dependent variables in food preference studies.

Researchers who collect data from convenient samples of only a few hundred will tend to collect demographic information from individuals or groups in a background questionnaire and then use these variables to stratify outcomes, thereby being able to explain a finding based on demographics; at other times, the demographics are the subject of the study, providing the independent variable predicting the dependent variable. However, it can be questioned whether these individual or group studies, where the total sample number is in the few hundreds, are representative samples for those in these demographic categories. And if they are not, then trying to explain the effects of age or gender or race on food preferences, based on a convenient sample of several hundred, could provide invalid and misleading results. Although there might be some merit in showing a statistical effect, the issue is whether or not to make changes in product development based on this small sample, as opposed to looking at much larger numbers.

As an example, Logue and Smith (1986) examined food preferences in over 300 subjects stratified by gender and age, body mass index, sensation-seeking propensity, and ethnicity. Their findings were consistent with other smaller studies: females reported higher preferences for low-calorie foods, candy and wine, and lower preferences for milk, meat, beer, and spicy foods. Those subjects who were younger reported higher preferences for sweet foods and lower preferences for chili pepper. Lower BMI subjects have lower preferences for sweet foods and meat. Sensation seekers have higher preferences for spicy foods and lower preferences for sweet or bland foods. Those who grew up on Asian cuisine had higher preferences for alcoholic beverages and lower preferences for Asian food compared with subjects who did not grow up on Asian cuisine. Logue and Smith suggested that predictors of food preferences could improve research on the determinants of food preferences, and it can be implied, could guide product developers in creating new products that tap into those food preferences. However, Logue and others argue that much of the variance in food preferences remains to be explained.

Large epidemiological studies using national data sets, such as NHANES (e.g., Block *et al.*, 2004), and the Nurses Health Study (e.g., Liu *et al.*, 1999), use sampling procedures that give a valid representation of the population and the findings can be generalized to the entire nation. However, owing to the costs of studies of this nature, the smaller studies will always continue to be present and necessary. And we find that even in these larger epidemiological studies the data do not often explain more of the variability in the outcome of interest than do the smaller studies.

When demographics are considered in large-scale sociological, nutritional, epidemiological or public health studies, demographics frequently are focused on the nature of the entire household, rather than on specific individuals. The data can be either cross-sectional (wherein the demographic and the preference data are population-based, and the two variables are not necessarily retrieved from the same individual) or within-subject in nature, whereas most of the smaller studies are within-subject designs. This latter design-type is one of the strengths of the smaller studies, although the inability to generalize to the greater population is a weakness. The component characteristics of the household that are generally collected when looking at large-scale demographics are household size, age distribution, and marital status. But also included in many demographic data collections, in both large and small studies, are categorizations of socio-economic status, including education, race and occupation, and an individual's living location (urban, rural, suburban, and exurban classifications). The debate as to whether smaller data sets provide accurate assessments of demographic variables will continue; but the debate could include whether or not the demographic data collected in the large-scale data sets are valid and useful at all for making comparisons on outcomes such as food preferences.

12.2.1   Segmentation strategies

If demographics are not the most valid or useful variable for segmenting the population, there is still the need for segmentation. In the place of demographics, we would need other actionable variables that would allow marketing managers to target specific groups with advertising and communications, and from a research perspective there is still a need to segment consumers in order to provide variables by which we can understand different aspects of life that might influence the food consumer. At present, socio-demographics are really the principal, and for some, the only set of variables we have to judge the representative nature of a sample for making an inference back to the entire population, and therefore, they still play an important role in positioning and targeting of food strategies.

In the marketing literature, segmentation strategies that go beyond demographics have received a good deal of attention. Marketers have argued that it is difficult to validly profile market segments using finite statistic models that mix traditional descriptor variables based on demography (Wedel and Desarbo, 2002). These researchers, and others, have proposed nested types of statistical models that provide more predictability of demographic data for product usage outcomes, but regardless of the modeling approach, demographic data are still the variables used as predictors. Other marketing researchers have argued for usage-based segmentation, stating that goods that are purchased based primarily on hedonic value are different from other types of consumable goods and that segmentation should be based on categories of purchase incidence (Boter and Wedel, 1999).

There is agreement among marketers that preference for a product is an important variable to measure, but it is segmentation that permits a more focused understanding of the consumer and an ability to execute the marketing mix. The most common approach to segmentation since the 1950s has been dividing consumers based on descriptive information about benefits sought, attitudes and beliefs about a product category, purchase history and styles, purchase channels used by self and family. This segmentation basis is then either defined by group membership – like heavy, medium, and light buyers or older versus younger consumers – or defined by groups that are embedded in the data and uncovered by statistical analysis – benefits segments, attitude segments, or psychographic segments. These segments would then be cross-tabulated against the remaining questions in a particular research project to profile each group and hypothesize the characteristics besides the segmentation base that might distinguish one group from another.

Data from these analyses frequently distinguish obvious differences, for example that higher income consumers buy more goods and services than those in lower income groups, and men buy and use certain products more than do women. However, when buying motivations, benefits derived from products, and their sensitivity to such marketing constructs as price, promotions, and communication channel strategies are examined, members of groups segmented based on buying behavior are often found to be indistinguishable from one another. Because of this, marketing segmentation has focused more recently on product benefits, consumer psychographics (lifestyle factors), needs and wants, and marketing elasticities (Haley, 1985) and this approach has become the mainstay of many market segmentation studies. In short, product benefits are measured and then people with similar sets of benefits are termed 'benefit segments.' This post hoc approach has gained favor with most marketing strategists.

A tandem segmentation method is carried out to derive segments (Haley, 1985; Myers, 1996): rating-scales are administered to consumers eliciting perceptions of benefits and expected or experienced deliverables of a particular product, and factor analysis is used to reduce the data to a smaller number of underlying dimensions. Cluster analysis is then applied and profiled to describe the various types of consumers the data suggest are underlying the population data (Punj and Stewart, 1983).

However, this approach is not without its criticisms, and because of the items and the wording in surveys, results often show a small range of mean item scores that hover towards the top-end of the scale. And despite attempts to counteract these limitations, these rating methods still lead to scale use bias and have limitations that question the validity of these approaches for providing actionable data for executing marketing strategies (Baumgartner and Steenkamp, 2001).

Successful segmentation is usually considered to have been achieved when six criteria are satisfied:

1. identifiability – the extent to which we can truly identify segments;
2. substantiality – are identified segments large enough to warrant separate marketing targeting?;

3. accessibility – the extent to which we are able to reach the customers in our segments via communication channels;
4. stability –do these segments we have identified remain stable over a certain period of time?;
5. responsiveness – the extent to which different market segments respond uniquely to the marketing efforts directed at them; and
6. actionability – the extent to which the identified market segmentation provides direction for guiding marketing efforts.

A specific market segmentation strategy will be a function of the variables used to segment the market and the methods or procedures used to arrive at a certain classification. Given the number of failed products compared with successful products, it is obvious that these criteria, although derived from the best thinking of its time, have a long way to go until we truly understand our food consumer and can identify how to target appropriate segments within the population.

Moskowitz (1985) proposed sensory segmentation for understanding the food consumer and for directing marketing strategies. He argued that by using a large number of questions based on critical variables about a range of products that are produced within a product category (a technique called 'category appraisal'), the data could be analyzed in a way that elucidated the 'sensory' consumer segments for particular products. For example, there might be a segment of consumers who prefer sweet, crunchy pickles and another segment that prefers sour, less crunchy ones. By the use of Conjoint analytic procedures, these segments could be uncovered from the data. Moskowitz and colleagues (Moskowitz and Rabino, 1994) have also proposed sensory and consumer segmentation based on other product-centered consumer variables.

These approaches to segmentation, both marketing and sensory, and choice of segmentation variables can produce categorizations of large or small consumer segments. Some of these segments might be easy to target; others might be difficult. They might even indicate that consumers respond in a homogenous fashion within segments; and they might even be more predictive of food preferences than would be demographics. However, most are based around product characteristics, usage or beliefs, or buying behaviors in general, and not necessarily around the variables that influence an individual's life. Demographics are numbers that attempt to categorize individuals based on what is going on in their lives; however, they do not describe the experience of the individual's life within that demographic segment or how individuals within a segment might differ and why. This is an area that neither marketing researchers nor psychology researchers have spent enough time exploring. Specifically, we have not spent enough effort examining the influence of life experiences on food preferences, and these variables might prove to be the most useful for categorizing individuals and for understanding the food consumer and explaining his or her behavior.

12.2.2   Life experiences

I would argue that demographic variables are not appropriate variables for predicting food preferences. It is true that from a product development perspective and for understanding the food consumer, we need some variable by which we can segment consumers and by which we can reach those segments with communications and marketing concerning new and existing products, but just because we need one does not mean we should use the one that is most available, if it is not providing the most valid segmentation by which to understand the consumer. Socio-demographics are a mechanistic variable comprising numbers reflecting group membership, but those across groups have different types of life experiences, and even those within each group have different life experiences. The mechanistic aspect of demographics does not adequately reflect an individual's life experiences, which are variables that are more proximal to the behavior of interest – eating patterns – than are demographic variables, which are more distal to that behavior. It seems more appropriate to try to understand food preferences based on a variable that is temporally and experientially closer in relationship, and reflects a wider range of emotional and cultural contexts, than does the classification in a demographic category. All of those people living in New York City might not share the same experiences, and it is these experiential variables that are more proximal to eating behavior.

This suggests that with more salient variables to examine, demographics could be relegated to static numerical or categorical entities that do not, by themselves, reflect what is going on in the mind and heart of a consumer; they are merely numbers. What is more important than membership in these categories is the effect these demographic variables have on the lives of the individuals in those categories, how the individuals experience the demographic attributes they possess, and then how those experiences might relate to food preferences. Sociologist Ned Block (1994) describes a concept known as 'qualia.' Qualia include the unintentional perceptions of things around you, as well as the properties of sensations, feelings, thoughts, desires, and pain. There are debates as to the existence of qualia, because most perception is assumed to be intentional, and it is unproven whether the content of an experience needs to be intentional (like the content of cognitive thought), or unintentional, in order to have an effect on the individual. But whether the perceptions are intentional or unintentional, it is likely that perceptions, being real to the perceiver, can produce changes in physiological states, and that their scientific nature is in large part related to our life experiences.

In a similar vein, public health research in the US has recently focused beyond the numerical differences between people from different demographic categorizations (e.g., socio-economic condition, gender, age, race, etc.), and instead has moved toward the assessment of physiological and psychological differences between people who have had different life experiences. For example, compared with White Americans, Black Americans have higher blood pressure levels that are associated with a higher incidence of cardiovascular ischemic events. There are those who would argue that this effect is a biological one, but in the past decade, data have clearly indicated that the effect is, to a large extent, a social one. In fact, Krieger (2000) and colleagues (e.g., Krieger *et al.*, 1993; Williams, 1997, 1999; Williams and Collins, 1995) argue that it is the *experience* of Black Americans living their lives in certain demographic conditions that has contributed greatly to this epidemiological difference, not the mere categorization of their being Black and with membership in certain demographic categories. The reason this is an important distinction is that there are those within these demographics who are resilient, a concept known as positive deviance (see Palmer and Humphrey, 1990; Spreitzer and Sonenshein, 2004), who do not suffer the poorer health outcomes despite possessing the demographics of those who are more inclined to them. Conversely, there are those not within these demographics who do suffer the poorer health outcomes. This suggests that there is a sense of susceptibility, independent of demographic categorization, that leads certain individuals to react to the conditions in which they live – the sensations, the feelings, the pain … the qualia – that have contributed to these poorer health outcomes. So here, it is not enough to look at the numbers of those in demographic conditions, but rather how the individuals within the demographic experience being in their demographic condition. This implies that looking only at demographic numbers can result in a poor estimation of the true effects of these demographics, since we would be bypassing the perceived experience and the unintentional perceptions – the qualia – resulting from being a part of these demographics. Therefore, perceptions of life experiences seem to be a more specific, proximal, and appropriate variable to investigate.

I am not positing that the quantitative measure of a variable is a faulty measurement, but rather, that we tend to assume the number derived or the categorization for an individual represents the same set of life experiences for each individual who possesses that particular number or categorization. When a study finds an effect of a demographic variable, what are we to infer from the finding? How should it be explained? Is it fair to assume that the majority of individuals within a demographic categorization would react similarly to the sample tested?

As an example, age is a variable that, when considered in conjunction with gender, has been shown to discriminate food preferences. Logue and colleagues (1988) studied 241 subjects, including 77 students, their siblings and parents, and found that females had higher preferences for low-calorie foods when they were older; males had higher preferences for alcohol when they were older; regardless of gender, individuals had a higher preference for coffee when they were older. But in this example, is it their ages that we are truly concerned with, or are we more concerned with what is in the mind and experiences of a 19 year old? Wouldn't the low-calorie food preference be predicated on some notion of restricted eating in the family or among the students' friends? And have all 19 year olds experienced their age in the same way? Should we not attempt to focus more on the experience of being 19, and let that drive our understanding of how these consumers behave, what they want and what types of foods to then provide for them?

In another example of using demographic data and having little ability to interpret the data in a way that would add to our understanding of consumers, Turrell (1998) showed that food preferences could be influenced by socio-economic status (income, education, occupation). In a sample of 403 individuals in Australia, food choices of respondents in lower socio-economic groups were the least consistent with dietary guideline recommendations. In addition, individuals in this group reported liking fewer healthy foods; and overall, socio-economic condition explained more than 10% of the variability in healthy food purchasing behavior. The author concluded: 'Whilst it is not clear why socio-economic groups differ in their food preferences, possible reasons include: differential exposure to healthy food as a consequence of the variable impact of health promotion campaigns, structural and economic barriers to the procurement of these foods, and subculturally specific beliefs, values, meanings, etc.' So within the conclusion, Turrell is suggesting that it is the differences in experiences, and the subsequent beliefs, values and meanings, within and across socio-economic groups that could explain the data. I agree; and I am suggesting we focus our data collection on these experiential measures and not merely on demographics.

Therefore, rather than focusing on demographic numbers and categories, I will provide interpretations of demographic data, where appropriate, in the context of the experiential properties of these demographics – the qualia resulting from belonging to or possessing these demographics, and in doing so, will attempt to broaden the definition of demographics and to urge a re-examination of the type of data we should be collecting in the field of consumer food behavior. And as part of these life experiences, I am choosing to include other critical variables that individuals have experienced that might be a part of the unintentional perceptions associated with the way they have lived and experienced their demographic conditions. Fortunately, over the past twenty years, we have begun to create measures to examine many of these other life experience variables and have assessed their relationship to food preferences, but we have not placed them into the context of a demographic/experiential duality. Constructs such as perceived meal and dietary variety (Bell and Meiselman, 1995), perceptions of food combinations as meals or not (Pliner *et al.*, 2004), perceived complexity of foods and meals (Bell and Ueland, 2005), prior eating behaviors (e.g., Rolls *et al.*, 1981), contextual aspects of eating situations (Edwards *et al.*, 2003; King *et al.*, 2006), food involvement (Bell and Marshall, 2003), food neophobia (Pliner and Hobden, 1992), and expectations (Cardello, 1994) are just a few of the variables that should be considered as part of life experience mix, as it is these experiences that drive our choices. And each new choice and each new life experience can further shape our perceptions and our evaluations of these experience constructs. And if perceptions are to be included, so must the coping strategies for living in these conditions, such as overeating, restrictive eating (Stunkard and Messick, 1985), binge eating, and cravings. And if these are to be included, then the results of eating in these manners must also be included; therefore, body mass index is another important variable that should be collected. It is all of these life experiences, and our qualia related to the demographic categories of which we have been and are members, not merely our categorizations, that we bring with us when we approach a new eating or food choice event.

So when in this chapter you are presented data related to trends in demographic characteristics and their effects on food preferences, I ask that you take these terms and their associated data cautiously and skeptically; and to consider how these demographics might affect the life experience of individuals – their thinking, their attitude, their unintentional and intentional perceptions – and ultimately, how these life experiences might translate to effects on consumer behavior, with food preferences as a part of that behavior.

12.3   Measuring and defining food preference

In the field of consumer research in relation to food product development, we have been using the phrase 'food preferences' for more than fifty years. But the definition of food preference has not been clearly differentiated from acceptability or choice. Some use the terms interchangeably, while others have argued for a difference. For example, Rozin (1979, 1990) has argued that food preferences, which are determined by a desire to maximize pleasure, imply choice and that by preferring one food means that you are choosing it over another food or some other activity. He described it as a descriptor of behavior, not a mechanistic variable. He went on to suggest that 'liking' or acceptability is a mental descriptor, and is probably the major, though not the only, cause of preference when other variables such as health and social factors and satiety and prior eating are taken into consideration. The two terms, liking and preference, would be equivalent only when these other variables were controlled.

Cardello (1996) described preference as an attitudinal response to the *names* of foods, or in essence, to the thought of the food rather than the eating of the food, while acceptance or acceptability is a response to a direct and immediate eating of or sensory experience with a product. Drewnowski (1997) used preference in regards to tastes, rather than foods, and linked these taste preferences, along with taste perceptions, food preferences to food choices and volume of food consumed. Drewnowski went on to argue that food preferences and food choices of populations are dependent upon attitudinal, social, and economic variables, including income. Other researchers have suggested that a stated preference indicates a choice of one item over another or over an array of others (e.g., Shepherd and Sparks, 1994). These concepts, however interchangeable in how we have used them in the field, are assumed to reflect how much an individual is said to like a product, whether it being an immediate experience or the thought of having an experience with that food product. Even the first psychophysical measure to examine the liking of a product, the 9-pt hedonic scale (Peryam and Pilgrim, 1957), referred to this as a measurement of food preferences. Psychologists have often stated that the phenomenological issues of food ratings are among the biggest challenges in understanding human behavior regarding food, but what seems to occur in the field is that we add a new elemental variable that then relates to acceptance, e.g. expectations (Cardello, 1994), or a new method for examining comparative preferences, e.g. preference mapping (Greenoff and MacFie, 1994), and these have been excellent advances in our understanding of the phenomenological and mathematical aspects of food ratings. But although these elements and techniques have improved our ability to optimize acceptance, they do not suggest product development directions that will match consumer choices over the course of an individual's diet. In essence, they remain psychophysical exercises that, although they do explicate human perception issues, they do not always easily translate to actionable product development and marketing directions or to an understanding of long-term eating behavior.

12.3.1   Preference and choice

Based on the reasoning just presented, preference and acceptability might or might not be assumed to relate to food choices across an individual's diet, because at present, there is no conclusive evidence that choice is associated with stated preferences or acceptability beyond single choice events. Some studies have shown isolated choice events where these variables are related (e.g., Hirsch and Kramer, 2002), while others have shown that their relationship is dependent on certain meal occasions and eating locations (Marshall and Bell, 1996; 2003) or on repetitive patterns of eating specific foods and associated cravings (Bell, 2006). Although it is intuitively pleasing to think that preference and choice across the diet are highly correlated, preferences for food options will likely change from choice event to choice event as a result of many variables, including prior choices, as has been described in the concept of sensory specific satiety (Rolls *et al.*, 1981). It is more likely that intervening experiences and our perceptions of those experiences will shape our desire for any food product at any given time more than would general preferences for a food.

Therefore, since I do not feel that single event preferences are an important consideration for guiding product development or marketing strategies, or for truly understanding the long eating patterns of the food consumer, I will not focus on preference as a rating of names or of products or as a preferred choice between two or more items. Instead, I will interpret preference in this paper as being the choices made over time in the diet in general, or, eating patterns. And it is these eating patterns, culled from large datasets from agricultural and government sources that will be examined in the context of demographic and life experience differences in the US population. The importance of examining this relationship is self-evident: without understanding it, rather than targeting specific segments of the population we would have to try to guide the development of a product that tries to be all things to all people – something that is not likely to produce success in the market place, as many segments of the market want different things.

12.4   US demographics and food preferences: past, current and projected trends

Up until now I have been critical of the use of demographic data as a variable that differentiates food preferences and that provides direction for the foods to produce. But because for decades these data have been the gold standard for understanding categories of consumers, there are large volumes of demographic data available; and as they are the 'best approach we have' at this point in time, these will provide a parsimonious starting point for our current understanding of the characteristic composition of those individuals residing in a particular region of a country or in the entire country.

12.4.1   Demographic trends

In the United States, the Department of Commerce, Bureaus of Census has as its primary purpose to count, and then to project future trends for, the population of the US by age, race, and sex. A full census is conducted every ten years. This set of data provides information, as close as possible, to actual numbers of individuals in the US, and they serve as the best representation of demographics for the population. There are limitations and assumptions that are made when considering the nation's demographics as well as demographic projections for the future, both of which will influence food preference trends over time. If we combine the US Census Bureau's projections of demographic change over the next twenty years with the variations found in food expenditures by household income, age composition, places of residence, race, and diet-health knowledge, the results will show that household food expenditures are also likely to change accordingly. But this relationship assumes that the relationships of income and demographics to food expenditure will be maintained, implying that relative prices and alternative opportunities for food choices, as well as tastes and preferences, remain unchanged. Additionally, as their economic and demographic circumstances change, consumers are assumed to acquire the expenditure patterns of individuals already observed in those circumstances.

For example, a family that moves from the West to the Southeast of the US acquires the expenditure characteristics of households in the Southeast. And from an age perspective, a 17 year old in 2020 is assumed to have the same food expenditure pattern as a 17 year old from the most recent Census. And finally, projection models are driven by projected changes in demographics and projected income growth, even though it is known by demographers that these are subject to differential changes that could alter actual outcomes.

The projections from these data sets and final population counts from the Census suggest some rather large differences in the US population in the year 2000, in comparison to various times in the 1900s. Even though the total number of people – 295,734,134 – is at the highest level ever in the US, the rate of population growth in the US has been steadily declining over the past 20 years. During the past 15 years, within-country migration has resulted in only three states having the majority of the country's population growth during that time:

California, Texas and Florida. In addition to moving to the South and the West, the existing population is growing older, is living longer, is residing in smaller households, is becoming more ethnically diverse, is increasingly exposed to a world-wide cuisine, and is becoming more health-conscious.

In particular, the median age in the US in 2005 is around 37 years, up from 32 years in 1990. The population of individuals who are between the ages of 30 and 50 has increased by more than 40 million over the past 25 years. The Department of Commerce projects that the number of individuals over the age of 65 will more than double by the year 2040.

The average household size is around 2.5 individuals, compared with nearly 3.5 in the early 1990s. The number of single individuals living alone, and this includes both young and old, comprises more than 25% of all US households, and more than 50% of households have only one or two people living in them.

The Hispanic and Asian populations are growing rapidly in the US. It is projected that by around 2030, at least half of the population of the US will be from one of these ethnic categorizations. There are more families with two wage earners today compared with 1950, and these families also have two or fewer children. In the 1950s, the typical household was comprised of one wage earner, a stay-at-home-mom, and more than two children. Today, this particular typical household demographic comprises less than 10% of US households. The difference is due to the changing desire, and for some families the economic need, for having two wage earners. More than 80% of women aged 35–44 are working, and even more than 65% of women with children are working. Family units are less stable than they once were, with nearly 43% of first marriages ending in separation or divorce within 15 years (US Census Bureau, 1998). This is in comparison with 1950, when the divorce rate was 2.8% (US Center for Disease Control, 2001).

12.4.2   Where Americans spend their food money

There are food trends that are also tracked by the Census and other researchers. A trend prevalent in the US, as well as in many developed countries, is the increasing propensity to consume food away from home. In 1970, 10.2% of disposable income was spent on food in the home, while 3.6% was spent on food eaten away from home. This can be compared with 1995, when only 6.7% of disposable income was spent on food eaten in the home, while 4.3% was spent on food eaten away from home (Putnam and Allshouse, 1997). The proportions of food taken in and out of the home are getting closer to each other. As a percentage of the consumer's total food budget, the share spent for food away from home has grown from roughly 26 to 39% between 1960 and 1995, and it is estimated that more than one-third of this expenditure is devoted to fast-food consumption (Manchester and Clauson, 1996). This latter trend is likely due to higher disposable incomes and a growing demand for convenience and value-added food products.

On average, Americans spend about 11% of their disposable income on snacks, meals and beverages, compared with 24% in the middle of the last century. Since less disposable income today is spent on food, consumers can be more selective in what they spend it on. In order to compete, each of the three major categories in the food industry – packaged food manufacturers, restaurants and retail distributors – have had little choice but to attempt to lure business away from the other two. This goes some of the way toward explaining changes in delivery of food items, packaging of food items, and combinations of food items, all of which have had to address the varied tastes, changing needs, and mobile lifestyles of US consumers.

In 2002, analysts, consultants, researchers, economists, marketers and other industry experts were asked to identify which societal shifts would have the most salient effects on the food industry (Gardyn, 2002). Each of the three major categories in the food industry faced different obstacles: Packaged foods companies have experienced flat sales growth for more than a decade because, as more women work full-time, they cook fewer meals at home. So these demographic implications for product developers in the packaged foods area are to provide more food products that offer the ease, taste and convenience that is found in restaurant and fast food establishments. Restaurants, which are projected in the US to grow to 1 million by 2010, up from 858,000 today, according to the National Restaurant Association (2003), base their future success on how well they can distinguish themselves through new and innovative dining concepts, especially those that attempt to reach specific targeted consumer segments, such as the Baby Boomers. The retail food markets have to compete with warehouse clubs and other retailers that offer more food in their aisles and convenient, lower-cost, one-stop shopping options. And a repetitive circle is created: demographic and life experience trends set the agenda for product developers who provide products to support the trend, thereby encouraging the trend to continue. These potential changes in foods that are available could have substantial influences on the proximal behaviors and experiences of the food consumer, changing our eating patterns and preferences for the future, including the types of ingredients available, and the amount of time we might spend in the kitchen, which could further influence the amount of time we spend with our families, leading to an effect on life experiences and, distally, on demographics. There is a circular dynamic relationship between life experiences, demographics and food preferences, whereby a change in one can have repercussions for another. And once a change is set in motion, there is inertia generated by the relationship that takes extraordinary changes in demographics and life experiences to alter it.

12.5   Implications for food preferences and product development

The increase in single living has led to a higher demand for smaller packaging of products, and single meals marketed to these individuals. In addition, data suggest that individuals who live alone or in smaller families tend to eat more foods away from home, leading to changes in preferences for fast food and convenience food establishments.

One of the biggest changes in life experiences resulting from two-income families is the difficulty in managing time. In addition, the increase in cultural pressures to achieve, and to buy the latest popular consumer goods, other than food, also has increased the need to earn more money and thereby has enhanced the burden on people's time. The resultant change in food patterns is that breakfasts are eaten more quickly, less socially, and often on the move; lunches are taken more quickly, and fast food and convenient food outlets are filling the need. Less time is available, or made, for cooking. The idea of three square meals a day seems reserved for the traditional-type eaters and the Army (though data suggest that time constraints have even shattered the 'three squares a day' motto of the US military). Grazing has become the norm, and the explosive growth of coffee shops and high-end, high-priced coffees are the result. The family meal is a rare occasion, and single serving packaged products are everywhere in stores.

12.5.1   Ethnic and cultural diversity in population, food and eating behavior

Migration and immigration patterns within the US have implications for food preferences and product development. As people move, they tend to abandon their older ways of eating and adopt the local cuisine and habits. This phenomenon is referred to as 'situational ethnicity' (Stayman and Deshpande, 1989), and it suggests, 'When in Rome, eat as the Romans do.' This change in eating patterns shows that food preferences are not necessarily stable. The migration of people to the South and Southwestern US has produced a boom in the Tex-Mex, Mexican and Southern cuisines in the US. Food companies and restaurants have sped to market with these items in response to the population's demographic changes.

The combination of the increase in Asian and Hispanic populations in the US, coupled with the increase in two-income families, has created a completely different set of consumer challenges for companies to consider when marketing. This diversity has coincided with increases in target marketing, and simultaneously, with the rapid diffusion of use of the Internet as well as mass communication strategies. It is unclear if the diversity of the population led to the need to market in a different way, which led to the development of the Internet, or vice-versa, but it is clear that the relationship has allowed for the exploitation of the Internet and electronic media for marketing purposes. It is possible that demographic data has forced companies not only to change products, but also to create new ways to market these products; and from this need, mass communication mechanisms changed accordingly. But the causal nature of this relationship is questionable.

12.5.2   Future segmentation strategies

With the need to target market and to get specific food products to those who have a preference for these products, food companies had to take a different tact than merely demographics. In the 1970s, they began looking less at demographics, and more at lifestyle variables. These variables have been used to classify people into such categories as inner-directed vs. outer-directed, status-oriented versus action-oriented versus strugglers. These types of descriptors have been examined in the market research, consumer behavior and psychology literature (e.g., Rokeach, 1973; Mitchell, 1983), and they were also potential predecessors for some of the constructs that were later explored in relation to experiences with food, such as food neophobia, food involvement, variety-seeking, dietary restraint, expectations, and associated behaviors.

In 1982, Leonard sought to categorize individuals regarding their eating behavior. Rather than doing so by demographics, or even lifestyle, he hypothesized that a better way of guiding product development and marketing was to look at how consumers might approach food choices and use food, over time. Perhaps in response to this, companies began to alter the naming and conceptualizing of consumer categories to suit their own needs, and the needs of their product developers. Although much of this corporate literature is proprietary and unpublished, Senauer *et al.* (1991) did publish a study wherein they revealed the names of the target groups to which they had marketed their various products over the prior several years. One group, though called many different names, was basically what has been referred to as the Yuppies, or what Pillsbury called, the 'Chase and Grabbits.' This group was described as being from households with higher incomes, highly mobile and active, generally two income earners, who were neophilic (willing to try, and often seeking out, new things). Products and marketing efforts were aimed directly at the lifestyle and life experiences of individuals within this group.

A second group, a more traditional category, was called the 'Down Home Stokers.' This group, usually households with only one wage earner and with lower incomes, comprised primarily immigrated ethnic groups. They tended to maintain their traditional eating patterns of the culture from where they or their older generation community members emigrated, rather than adopt a situational ethnicity pattern of eating behavior. These individuals had completely different preferences than other groups and food companies needed to provide what they desired and to market directly to them, without necessarily marketing these items to other groups.

A third group, which they called the 'Careful Cooks,' would today likely be referred to as the 'Health and Food Conscious.' These individuals tend to have more years of education, to be older, to be retired, have higher income, and to be healthy; but they also are appreciative of food, and experience joy in eating – a group I like to call 'Foodies,' whose members place a high importance on the role of food in their life. This construct has also been researched recently and defined as food involvement (see Bell and Marshall, 2003; Marshall and Bell, 2004). This group lives to eat.

A fourth categorization defined by Pillsbury was the 'Functional Feeders' – those who tended to be older, male, often conservative in their thinking, and who do not place a high value on the joy of eating. This group tends to eat to live. They called the final category, 'Happy Cookers.' These individuals love to cook, and think of cooking meals first rather than going out to eat. They have a joy of cooking, not just eating. Regardless of which names are given to these groups by corporations or by researchers, or the number of groups derived, the categories tend to fall more in line with a combination of demographics, life experiences and with qualia than solely with demographics. With the psychological and sociological examination of lifestyle variables in the 1970s, one would have assumed these variables would have usurped demographic data in regard to food preferences; but this has not been the case. We have not truly translated this framework to our consumer research efforts over the past thirty years in an effort to better understand consumer behavior regarding food preferences.

12.5.3 Future food trends: based on past trends or projected demographics?

The following are food preference trends in the US over the past 50 years. Most of the data are from the following sources (Day, 1996; Putnam and Allshouse, 1997; Hollman *et al.*, 2000). The consumption of chicken and fish is increasing, while the consumption of beef is decreasing. Consumption of fresh and frozen foods is increasing, while cured foods are decreasing and canned foods remain stable. Frozen dairy consumption (ice cream and frozen yogurt) and cheese consumption are increasing, while animal fats and butter consumption is decreasing. But in general, all fat and oil consumption in increasing. Consumption of legumes is increasing, while flour and cereal products consumption is decreasing, perhaps owing to the recent wave of low-carbohydrate diets. Frozen juice consumption is increasing, while canned and chilled juices remain stable. Fresh fruits are slightly increasing, while processed fruits and total fruit consumption has remained stable. Vegetable intake has seen large increases recently, but consumption of potatoes is decreasing. And non-calorie sweets are increasing, while sugared sweets have been decreasing.

Total food consumption is decreasing, probably owing to the aging population, who require less total caloric intake in their diets. Eating out is increasing, while eating in is decreasing. Consumer spending at full-service and fast food restaurants will continue to grow over the remainder of this decade and the next. However, the larger increase is predicted to occur at full-service restaurants. Simulations assuming modest growth in household income plus expected demographic developments show that per capita spending could rise by 18% at full-service restaurants and by 6% for fast food between now and 2020. The assumed increase in income alone could cause such spending to rise by almost 15% and 7% at full-service and fast food restaurants, respectively. The increasing proportion of households containing a single person or multiple adults without live-at-home children will cause per person spending to rise by another

1 to 2% in each of these segments. However, the aging of the population will decrease spending on fast food by about 2% per capita (Stewart *et al.*, 2004).

Rapid changes in nutrition recommendations by the USDA and other public health and medical sources have been conflicting, and messages have confused the population who want to eat for health. But it is likely that the concern for health has aided the organic food market to grow, and it will likely to continue to do so, even though the primary improvement provided by organic food is for the soil and adjacent crops, not for individual biochemistry, though organic eating is more advantageous for humans to eat, due to decreased frequency of exposure to pesticides.

So for health-conscious reasons, we tend to monitor our calories and fat intake during the main meal, but we save room for the high-fat, high-sugar (or alternate sweetener) dessert. This is reflected by the success of both types of foods in the marketplace. These are not always meant to satisfy one target market or another; but rather, the 'forbidden' foods are the reward for the calorie-restricting main course. Whether these preferences are a function of physiology, advertising, marketing, the explosion of brands, or psychology is unclear; but their presence is overwhelming, and could help explain how preferences and eating patterns change in the short term in response.

In contrast to the consumer buying behavior, attitudes, and demographic trends that lead to increases in food consumed away from home, the use of processed foods, and the movement toward larger supermarkets, there are some changes in local and national agriculture that could provide enough force to alter the inertia of existing larger forces. For example, the number of local farmers' markets has increased nationally (Johnson *et al.*, 1996). This has been accompanied by a recent increased demand for fresh fruits and vegetables in the US, primarily due to the health conscious consumer and the growing obesity problem and associated diseases. The increase in 'green consumerism' – those who are concerned about the sustainability of the environment – has influenced the number of environmentally friendly products and locally grown products (Hartman, 1996). This has changed the nature of retail stores, many of whom now provide natural food sections and organic foods in their markets. Hartman and others indicate that these agricultural and green trends will likely continue, leading to a further 'fragmentation' of the food market into diverse segments. Due to their own economic demands, farmers are selling off their farmland, either getting out of the farming industry or dividing up land and selling it for housing developments. This has also been induced by larger market demands, in which commodities are bringing lower prices than they cost to produce.

Another example of the 'changing forces of change' is that in the last twenty years, Americans have purchased more salsa each year than catsup, reflecting a growing trend toward the widespread acceptance of international foods in the US diet. And one of the most interesting changes brought on by the faster pace of life and the need for convenience in food items is the increase in liquid food items, including caloric and non-caloric liquids, and most surprisingly, meal replacement drinks. Americans spend some $821 billion on food today, from supermarket produce to restaurant meals to snack foods at vending machines. The US Department of Agriculture reports this figure will grow to $1.2 trillion over the next decade. Much of that growth will be hard earned by the food industry.

12.6   Future trends

For the future, there are increases expected in the elderly population over the next decade that will likely lead to more home delivery services and direct or farmers' markets located within close proximity of elderly residential areas. In addition, the number of available health foods designed and positioned for the growing elderly market segment will likely increase.

The greater proportion of one- and two-person households is expected to continue, increasing the demand for smaller serving sizes, perhaps in greater varieties. The rural aging and single-parent households with children comprise the limited-resource consumer segment of the population: Welfare and Assistance programs, such Women, Infants, and Children (WIC) and Food Stamp Voucher programs, will increasingly be utilized by each state's direct market operators. Already in several states in the US, many farmers' markets use this payment system. At the same time, higher-income, well-educated individuals will become a growing market segment for specialty produce and niche food producers.

One critical, unexpected issue facing food developers in the future is how demographics affect people's feelings of security. The life experience associated with a demographic (e.g., living in New York City) in the present is associated with the fear of another terrorist attack. This can lead to a change in how individuals feel about eating out or eating with family. This could bring about more traditional eating, as people choose to spend more time 'nesting' with their families. This could, in turn, lead to a return to the nostalgic appeal of very specific direct markets, such as local farmers' markets, which will likely be in greater need in growing urban areas, small towns, as well as tourist centers.

It is likely that women will still be the primary food purchasers and preparation decision-makers in most households. But with increasing numbers of women employed outside of the home, marketing strategies that address time constraints and convenience will continue to be an important market niche. In addition, the African-American, Asian, and Hispanic ethnic groups will also continue to grow in the US, providing further niche food market opportunities.

12.7   Summary

In this chapter, I have offered a challenge to our current use of preference measures as the key measure for trying to predict future trends in consumer behavior and to guide food product development. Yes, it is a simple and easy measure to collect, but that does not make it predictive in the context of the world of life experiences and the world of food in which we live today – a world that forces us to cull through the effects of our living conditions, sensory-overloading advertising, psychology, economics, physiology, biology and chemistry – when a judgment about a food is made. I am suggesting we go beyond 'preference' as a single measure of understanding how consumers relate to food products, because the future of understanding consumers and their relationship with new product development should be based on far more complex issues, including meals, diets, location of choice and consumption, prior experiences, social context – the qualia of how we have lived and are living at present.

Additionally, I have argued that a food choice event should be considered in the context of all choices that have come before. And this suggests that to evaluate preference or acceptability on its own and for only one single eating event might be demonstrating a phenomenon, but does not translate well into durable food choice predictions and therefore, provides less understanding of consumer eating behavior.

I am not suggesting that we throw out the response of food preference; but the notion of what comprises 'preference' must advance, and it must include other variables that relate to consumer behavior in regard to food; and our notion of categorical demographic variables must advance. Otherwise, the relationship between these variables will remain a scientific exercise and not progress our understanding of consumer behavior regarding food.

Perhaps by replacing 'demographics' with the broader term 'life experiences' and the myriad of variables that might comprise this term, making it more comprehensive, more specific to how consumers experience their living situations, we could, in multivariate models, make preference a more predictable variable. And if we broaden the predictors for understanding consumer behavior, we can explain more of the variability in consumer behavior regarding food, and give much greater direction on how to target market food items and suggest future trends for food product development.

12.8   References and further reading

BAUMGARTNER, H. and STEENKAMP, J. B. E. M. (2001). Response styles in marketing research. *Journal of Marketing Research*, 38, 143–156.

BELL, R. (2006). Brief review: the impact of repeatedly eating the same foods on food cravings and weight status. *Appetite*, In Press.

BELL, R. and MARSHALL, D. W. (2003). The construct of food involvement in behavioral research: scale development and validation. *Appetite*, 40(3), 235–244.

BELL, R. and MEISELMAN, H. L. (1995). The role of the eating environment in determining food choice. In D. W. Marshall, *Food Choice and the Consumer*. London: Blackie Academic.

BELL, R. and UELAND, Ø. (2005). Defining complexity is not so simple. Presented at the 6th Pangborn Sensory Symposium, Harrogate, UK, August 7–11, 2005.

BLISARD, N. (2001). *Income and Food Expenditures Decomposed by Cohort, Age, and Time Effects,* US Department of Agriculture, Economic Research Service, TB-1896.

BLOCK, G., DRESSER, C. M., HARTMAN, A. M. and CARROLL, M. D. (2004). Nutrient sources in the American diet: quantitative data from the NHANES II survey. II. Macronutrients and fats. *American Journal of Epidemiology*, 122(1), 27–40.

BLOCK, N. (1994). *Qualia.* In S. Guttenplan (ed.) *A Companion to Philosophy of Mind*, Oxford: Blackwell.

BOTER, J. and WEDEL, M. (1999). Segmentation of hedonic consumption: an application of latent class analysis to consumer transaction databases. *Journal of Market-Focused Management*, 3 (3–4), 295–311.

CARDELLO, A. V. (1994). Consumer expectations and their role in food acceptance. In H. J. H MacFie and D. M. H. Thomson (eds), *Measurement of Food Preferences*, Glasgow: Blackie Academic, pp. 253–297.

CARDELLO, A. V. (1996). The role of the human senses in food acceptance. In H. L. Meiselman and H. J. H. MacFie (eds), *Food Choice, Acceptance and Consumption*, London: Blackie Academic.

CONNER, M. T. (1994). An individualized psychological approach to measuring influences on consumer preferences. In H. J. H MacFie and D. M. H. Thomson (eds), *Measurement of Food Preferences*, Glasgow: Blackie Academic, pp. 167–201.

DAY, J. C. (1996). Population projections of the United States by age, sex, race, and Hispanic origin: 1995 to 2050. Current Population Reports, Series P-25, No. 1130. Washington, DC: US Bureau of the Census.

DREWNOWSKI, A. (1997). Taste preferences and food intake. *Annual Review of Nutrition*, 17, 237–253.

EDWARDS, J. S. A., MEISELMAN, H. L., EDWARDS, A. and LESHER, L. (2003). The influence of eating location on the acceptability of identically prepared foods. *Food Quality and Preference*, 14, 647–652.

FRIDDLE, C., MANGARAJ, S. and KINSEY, J. (2001). 'The Food Service Industry: Trends and Changing Structure in the New Millenium,' Working Paper #01-02, The Retail Food Industry Center, University of Minnesota.

GAINS, N. (1994). The repertory grid approach. In H. J. H MacFie and D. M. H. Thomson (eds), *Measurement of Food Preferences*, Glasgow: Blackie Academic, pp. 51–76.

GARDYN, R. (2002). Food for thought – trends in food spending – brief article, *American Demographics*, March 1.

GREENHOFF, K. and MacFIE, H. J. H. (1994). Preference mapping in practice. In H. J. H MacFie and D. M. H. Thomson (eds), *Measurement of Food Preferences*, Glasgow: Blackie Academic, pp. 137–166.

HALEY, R.I. (1985). *Developing Effective Communication Strategy: A Benefit Segmentation Approach.* New York: John Wiley & Sons.

HIRSCH, E. S. and KRAMER, F. M. (2002). Hedonic ratings as food choice predictor. Presented at the Xth Food Choice Conference, Wageningen, The Netherlands, July 1, 2002.

HOLLMAN, F. W., MULDER, T. J. and KALLEN, J. E. (2000). Methodology and assumptions for the population projections of the United States: 1999 to 2100. Population Division Working Paper, No. 38. Washington, DC: US Bureau of the Census.

JEKANOWSKI, M., BINKLEY, J. and EALES, J. (2001). 'Convenience, accessibility, and the demand for fast food,' *Journal of Agricultural and Resource Economics*, 26: 58–74.

JOHNSON, D., LEWIS, L. and BRAGG, L. (1996). *1996 National Farmers' Market Directory.* USDA Marketing Service, Transportation and Marketing Division, Wholesale and Alternative Markets, Washington, DC.

KING, S. C., MEISELMAN, H. L., HOTTENSTEIN, A. W., WORK, T. M. and CRONK, V. (2006). The Effect of contextual variables on food acceptability: A confirmatory study. *Food Quality and Preference*, In press.

KRIEGER, N. (2000). Refiguring 'race': epidemiology, racialized biology, and biological expressions of race relations. *International Journal of Health Services*, 30(1), 211–226.

KRIEGER, N., ROWLEY, D. L., HERMAN, A. A., AVERY, B. and PHILLIPS, M. T. (1993). Racism, sexism, and social class: implications for studies of health, disease, and well-being. *American Journal of Preventive Medicine*, 9, 82–122.

LEONARD, R.E. (1982). Nutrition profiles: Diet in the '80s. *Community Nutrition*, 1(5), 12–17.

LIU, S., STAMPFER, M. J., HU, F. B., GIOVANNUCCI, E., RIMM, E., MANSON, J. E., HENNEKENS, C. H. and WILLETT, W. C. (1999). Whole-grain consumption and risk of coronary heart disease: results from the Nurses' Health Study. *American Journal of Clinical Nutrition*, 70(3), 412–419.

LOGUE, A. W. and SMITH, M. E. (1986). Predictors of food preferences in adult human. *Appetite*, 7(2), 109–125.

LOGUE, A. W., LOGUE, C. M., UZZO, R. G., McCARTY, M. J. and SMITH, M. E. (1988). Food preferences in families. *Appetite*, 10(3), 169–180.

MANCHESTER, A. and CLAUSON, A. (1996). 'Spending for food up slightly in 1995.' *Food Review*, USDA Economic Research Service, Washington, DC.

MARSHALL, D. W. and BELL, R. (1996). The Influence of Meal Occasions and Physical Situations on Food Choice in Australia and Britain. *Food Quality & Preference*, 7(3/4), 325–326.

MARSHALL, D. W. and BELL, R. (2003). Meal construction: exploring the relationship between eating occasion and location. *Food Quality & Preference,* 14(1), 53–64.

MARSHALL, D. W. and BELL, R. (2004). Relating the food involvement scale to demographic variables, food choice and other constructs. *Food Quality & Preference*, 15 (7–8), 871–879.

MITCHELL, A. (1983). *The Nine American Lifestyles: Who we are and where we are going.* New York: Macmillan Publishing Co.

MOSKOWITZ, H. R. (1985). *New Directions for Product Testing and Sensory Analysis of Foods.* Westport, CT: Food and Nutrition Press.

MOSKOWITZ, H. R. and RABINO, S. (1994). Sensory segmentation: an organizing principle for international product concept generation, *Journal of Global Marketing*, 8, 73–93.

MYERS, J. H. (1996). *Segmentation and Positioning for Strategic Marketing Decisions.* Chicago, IL: American Marketing Association.

PALMER, S. and HUMPHREY, J. A. (1990) *Deviant Behaviour: Patterns, Sources and Control.* New York: Plenum Press.

PERYAM, D. and PILGRIM, R. (1957). Hedonic scale method of measuring food preferences. *Food Technology*, 11, 9.

PLINER, P. and HOBDEN, K. (1992). Development of a scale to measure the trait of food neophobia in humans. *Appetite*, 19, 105–120.

PLINER, P. L., BELL, R., MEISELMAN, H. L., KINCHLA, M. and MARTINS, Y. (2004). A layperson's perspective on meals. *Food Quality & Preference*, 15 (7–8), 902–903.

PUNJ, G. N. and STEWART, D. W. (1983). Cluster Analysis in Marketing Research: Review and Suggestions for Application. *Journal of Marketing Research*, 20, 134–148.

PUTNAM, J. J. and ALLSHOUSE, J. (1997). *Food Consumption, Prices, and Expenditures 1970–95.* USDA Economic Research Service, Washington, DC, Statistical Bulletin No. 939.

RAPPOPORT, L., PETERS, G. R., DOWNEY, R., MCCANN, T. and HUFF-CORZINE, L. (1993). Gender and age differences in food cognition. *Appetite*, 20(1), 33–52.

Restaurant Industry Forecast 2003, National Restaurant Association, Washington, DC.

ROKEACH, M. (1973). *The Nature of Human Values*. New York: Free Press.

ROLLS, B. J., ROLLS, E. T., ROWE, E. A. and SWEENEY, K. (1981). Sensory specific satiety in man. *Physiology & Behavior*, 27(1), 137–142.

ROZIN, P. (1979). Preference and affect in food selection. In J. H. A. Kroeze (ed.), *Preference Behavior and Chemoreception*, London, pp. 289–302.

ROZIN, P. (1990). Acquisition of stable food preference. Marabou Symposium: Factor influencing food intake in man. Sundbyberg, Sweden, June 16–18, pp. 75–82.

SENAUER, B., ASP, E. and KINSEY, J. (1991). *Food Trends and the Changing Consumer*, St. Paul, Minn.: Eagen Press.

SHEPHERD, R. and SPARKS, P. (1994). Modelling food choice. In H. J. H MacFie and D. M. H. Thomson (eds), *Measurement of Food Preferences*, Glasgow: Blackie Academic, pp. 203–226.

SPREITZER, G. M. and SONENSHEIN, S. (2004). Toward the construct definition of positive deviance. *American Behavioral Scientist*, 47, 828–847.

STAYMAN, D. M. and DESHPANDE, R. (1989). Situational ethnicity and consumer behavior. *Journal of Consumer Research*, 16, 361–371.

STEWART, H., BLISARD, N., BHUYAN, S. and NAYGA, R. M. (2004). The demand for food away from home: Full-service or fast food? USDA Economic Research Services, Agricultural Economic Report Number 829.

STUNKARD, A. J. and MESSICK, S. (1985). The three-factor eating questionnaire to measure dietary restraint, disinhibition and hunger. *Journal of Psychosomic Research*, 29(1), 71–83.

THE HARTMAN REPORT (1996). *Food and the Environment: A Consumer's Perspective*, Bellevue, Washington, DC.

TURRELL, G. (1998). Socioeconomic differences in food preference and their influence on healthy food purchasing choices, *Journal of Human Nutrition & Dietetics*, 11(2), pp. 135–149.

UNITED STATES, CENTERS FOR DISEASE CONTROL AND PREVENTION (2001). First Marriage Dissolution, Divorce, and Remarriage: Advanced Data, Number 323.

US DEPARTMENT OF COMMERCE, BUREAU OF THE CENSUS (1984). Projections of the Population of the U.S. by age, sex, and race: 1983 to 2020. Current Population Reporting Service. P-25, No. 952, Washington, DC.

US DEPARTMENT OF COMMERCE, BUREAU OF THE CENSUS (1998). Current Population Reports, Series P20-514, 'Marital Status and Living Arrangements: March 1998 (Update)'.

WEDELL, M. and DESARBO, W. S. (2002). Market segment derivation and profiling via a finite mixture model framework. *Marketing Letters*, 13 (1) 17–25.

WILLIAMS, D. R. (1997). Race and health: basic questions, emerging directions. *Annals of Epidemiology*, 7, 322–33.

WILLIAMS, D. R. (1999). Race, socioeconomic status and health. The added effect of racism and discrimination. *Annals of the New York Academy of Science 896* (Socioeconomic Status and Health in Industrial Nations: Social, Psychological and Biological Pathways), 173–188.

WILLIAMS, D. R. and COLLINS, C. (1995). US socioeconomic and racial differences in health: patterns and explanations. *Annual Review of Sociology*, 21, 349–86.

13

Cross-cultural dimensions in food choice: Europe

E. Risvik, M. Rødbotten and N. Veflen Olsen, Matforsk, Norway

13.1 Introduction: the importance of understanding cross-cultural dimensions in food choice

Obvious and important hindrances for food choice are dislikes and aversions. Both are to some extent culturally learned, therefore the hard fact hits hard the first time a European drinks Root beer from the US. The American, on the other hand, may react in a similar way when he/she meets liquorice chewing gum in Scandinavia. That Asians actually eat the fruit Durian creates maybe the greatest disbelief of all among both Europeans and Americans. The examples are many, and not necessarily as blatant as these. Understanding cross-cultural dimensions both positive and negative in food choice is therefore critical, as the global village develops, with national borders being boundaries of cultural belonging, rather than physical hindrances for exchange of foods and information. With increasing migration of people, information and foods (especially to urban areas) we often see a bigger difference within a country, compared to what we used to find between countries, as city areas are universally more alike than city and rural areas are becoming. To develop foods for changing consumer demands and markets is already a cultural challenge of proportions. Determinants of importance for understanding these differences are many, in this chapter and context we have focused on relevance for healthiness, and with a strong relationship to taste, as these are both critical for successful product development.

Europe has gone through dramatic changes during the last 50 years. The first bridges across the cultural divide were first built with the post-war Rome treaty (early EU developments), and since then the incentives have flourished, with the inclusion of the new membership countries in the EU in 2004 as one of the biggest late changes. The internal market is now close to 500 million people, probably the third biggest internal market in the world. Still, the cultural differences persist and will for a long time represent a challenge for product developers of foods.

13.2   Cross-cultural dimensions of healthiness and food choice

When the Second World War was over, Europe experienced an influx of many 'exotic' food products for the first time. In remote European countries tomatoes had not been marketed before the war, certainly not to the working class, so before being tried they were admired through shop windows and perceived through their appearance. Frequent stories exist of how many people imagined them to be sweet as strawberries, juicy, aromatic and fruity. Huge was the disappointment when the first bite revealed what was perceived to be a slimy core, and a watery, slightly tarty flavour. These kinds of experiences seem odd, as tomatoes are such an integrated part of our daily diet today, but only 50 years ago tomatoes were perceived as repulsive by many Europeans. Expectations have dramatically been modified since, through repeated exposure, and most disappointments have long since changed in favour of a passion for sun-ripe juicy tomatoes. Still, many countries in Europe have considerable differences in preference for food. When Europe united again in 1989, many Germans threw their favourite beer out for large international brands, before they discovered after a few months that 50 years of separation had actually done something to their preference for beer. They missed the old brands and wanted them back. This shows that preference has some permanency to it also, and abrupt changes may not be preferred, even when the attitudes are in favour of change.

In Europe of today differences are just as big, maybe bigger between Paris and rural France than between Paris and Reykjavik. This comes as a consequence of an increasingly open European food market. One implication of this for the food industry is that the market for any given product does not necessarily lie within the boundaries of one country as it used to do. Opportunities lie also in the marketing of traditional foods, from one country to a similar preference segment of consumers in another. Migrating traditional foods have also attained quite considerable markets for products like Parma Ham, Parmesan cheese, Aceto Balsamico and Pesto to mention a few from northern Italy. Marketing of traditional foods cater to a consumer group where price is not the most important feature of the food, but rather the intrinsic sensory attributes and the myth built around the product and its origin. In this aspect the Latin sphere of Europe has had longer traditions than the rest. With common legal protection of foods based on tradition, origin and quality, throughout Europe, this gives a completely new situation for migration and marketing of these products.

Healthy food eating represents a strong political motivator in favour of food habit changes. It has become apparent that the costs related to disease, as a consequence of lifestyle are strongly connected with eating habits, and the costs are becoming too high for society to maintain for long without serious negative effects on living standards. If food habits remain unchanged, they represent a significant threat to our welfare system. Changes are required, but must be favoured by the individual as the economic and market situation for most Europeans allows for freedom of choice when it comes to food. Society still has an ambition to remain democratic, and this limits the selection of means available to moderate consumer behaviour. In this situation, governments, legal bodies, the food industry and professionals in medicine, nutrition, food science and all social sciences need to work together, to provide understanding underpinning the good, healthy food behaviour of alternatives preferred and chosen by the European food consumer.

European unification works strongly towards a homogenisation of Europe, but does this also apply to food, food habits and eating? The question is interesting as it raises issues related to whether there are strong needs for geographically defined and homogeneous behaviours in order to be unified, or whether subculture development across national borders may just as well replace the need for identity; identity till now provided by national states.

A definition of 'cross-cultural' is necessary at this point, as the term is often confused with 'national differences' as expressed through the existence of European national states. The true meaning of cross-cultural slightly diverts from these restrictions as it contains 'the complete results of a group's spiritual and material activities, at a certain time' (*Norsk fremmedordbok*, (Norwegian dictionary of foreign words/phrases) Gyldendal) or 'collective programming of the mind' (Hofstede, 2001). For many practical purposes there is a strong correlation to nation or state, but strictly considered, any grouping within the given boundaries of the definition can be described by the term. For this chapter we have chosen examples to illustrate both national differences, and cross-cultural consumer segments not connected with traditional national borders.

13.2.1   Diets and traditions

The Mediterranean diet, the Balkan diet, the Polar diet, the Atkins diet and the Japanese diet are all examples of accumulated choice of diets with a healthy connotation attached to them. The question still remains what a healthy food diet is, as individual variation will hardly favour the same choice for all individuals at all times. The answer is still not obvious, although there is no lack of advice, as the media are full of this every day. Still, few studies exist where cultural dimensions are significantly related to healthiness of food. Conclusive evidence therefore cannot be given at present on this point, although it might exist.

Understanding of how different cultural segments (in this case exemplified by national states as cultures) vary in their relationship to aspects with relevance for perceived healthiness is not easily found in literature. National statistics for consumption patterns with health connotations are not harmonised and therefore not really comparable yet, although several EU Framework 6 programmes aim to do so. Still it is possible to assume that cross-cultural studies on issues like trust in food (Poppe and Kjaernes, 2003; Ferretti and Magaudda, 2003) and Pan-

EU surveys on consumer attitudes to physical activity, body-weight and health (EC, 1999) give indications with strong relevance and correlation to food choice. When Bergeaud-Blackler (2004) defines five dimensions with importance for trust to be safety, nutrition, quality, value and ethics, these cannot be without effect on choice. It will be interesting to see how comparative studies (both across national cultures and subcultures) will link these perceptions to behavioural patterns in the development of these programmes.

Studies with closer relationships between attitudes and choice can be found for specifically aimed studies such as Genetically Modified Foods (GMO) (Springer et al., 2005), where socio-demographic and subjective belief reasons are recorded in a large number of EU countries. The attitudes to GMO could not be explained only as a function of socio-demographic variables, but to a larger extent on subjective beliefs. When the socio-demograpics were corrected for, it was still not possible to explain national differences, which implies that subjective belief segments are not necessarily country specific. Unique value priorities may still exist and play a vital role in attitude formation, but so far no country-specific attributes are identified.

Looking at generic confidence in food bought for one's own household, on the other hand, shows strong north-south correlation, where the south shows less trust in food being safe, but greater confidence in what they choose to bring home (Ferretti and Magaudda, 2003; Poppe and Kjaernes, 2003). It is not difficult to assume that this difference will also be manifested in behavioural differences.

Cultural boundaries exploring Europe in regions are found in very few reports (Askegaard and Madsen, 1998). In the local and the global approach to traits they are looking for patterns in consumption with a root in statements of different food dimensions describing factors like style, trends, preferences, habits and dieting behaviour. The patterns are an empirical description of food-related behaviour and attitudes, and show greater similarities across language boundaries than national borders. This clusters the British Isles, the Netherlands, Flanders, France and the French-speaking part of Switzerland, while peripheral countries like the Scandinavian tend to form individual clusters. In total the study describes 12 clusters where 7 are nation-states.

Several models have roots in theories based in cognitive paradigms of consumer behaviour (Brunsø et al., 1996; Hofstede, 2001; Schwartz and Bilsky, 1990). Several commercial methods have also been developed to define attitude or value maps for lifestyle segments (RISC, CCA and VALS), but as the theoretical base for these methods is not readily available, the results are difficult to validate in research. Among the documented methods Food-Related Lifestyle (Brunsø et al., 1996) come close to explaining a mental construct which explains, but is not identical to actual behaviour. The construct assumes that behaviour to some degree can be explained by a cognitive paradigm where the person recognises the self as mirrored in the values associated with a product and therefore seeks behaviour directed by this linkage. In a European study across four nations (Denmark, Great Britain, France and Germany) (Brunsø et *al.*, 1996), four clusters with great commonalties across countries could be defined, but also country-specific characteristics such as Great Britain having, to a greater extent, both Careless and Uninvolved consumers than the other countries. The Eco-moderate Danish consumers were best grouped with the Adventurous consumers of the other countries, while the French Hedonistic consumers were best clustered with the Careless consumers from the other four countries. These models have been evaluated for cross-cultural validity within Western Europe (Scholderer *et al.*, 2004).

Other cognitive rooted approaches with background in clinical psychology are found in Hofstede's analyses, and categorisation of value-based 'mental programmes' defining cultural differences (Hofstede, 2001). This assumes that the concept of dimensions of culture makes sense, and that the sub-dimensions are possible to interpret as fundamental problems of societies. The empirical works of Hofstede describe five independent dimensions; power distance, uncertainty avoidance, individualism/collectivism, femininity/masculinity and long-term orientation. Nations are characterised based on these dimensions. European commonalties are difficult to see in Hofstede's maps as both the north-south differences and also the east-west differences become apparent. The north-south differences are best illustrated with the north being more feministic (consensus) oriented and with a shorter power distance than in the south. In the east, collectivism is more pronounced, while uncertainty avoidance remains high. These are major empirical differences across Europe, with much greater detail than explained here, and it shows the need for work where these dimensions are related to behavioural patterns. It is not difficult to understand that factors such as power distance have an effect on shopping attitudes and eventually also on choice, but these studies are not yet performed on a pan-European selection of consumers.

Schwartz and Bilsky (1990) developed a theory of the psychology and structure of human values. This value structure has been tested in several large studies, and in a comparison of Australia, Finland, Hong Kong, Spain and the United States, only Hong Kong fell outside of a model which revealed seven distinct motivational types: achievement, enjoyment, maturity, pro-social, restrictive conformity, security, and self-direction. In Hong Kong social power also emerged as an important value. The study confirmed the structure revealed in a previous study from Germany and Israel. Value structures in these studies suggest that the motivational dynamics underlying value priorities are quite similar across societies in their dimensionality, while weights on each dimension will vary between cultures and for individuals within cultures.

The patterns in models developed by Brunsø *et al.* (1996), Hofstede (2001) and Schwartz and Bilsky (1990) contain simplified structures, both efficient for reduction of dimensionality in complex analyses and with interpretable structures that make common sense. These and similar models are in this way good candidates for tools in cross-cultural studies. Because the dimensionality needed for interpretation often is as low as two dimensions (or may be portrayed in two dimensions), this makes mapping exercises and predictive modelling in conjunction with other data very favourable for these models.

We have chosen an example to illustrate these perspectives, where Schwartz's values have been used to characterise two distinct segments of organic consumers inside one country, which have earlier been identified as prototype consumers from two different countries. This illustrates that segments within a country may just as well exist, even when comparisons of countries show the same characteristics to be typical for distinguishing between two countries.

13.2.2   Organic bread as an example of value-based models

Organic, biodynamic, ecological or sustainable grown foods have very different status in European countries, including market share and media coverage. Within the Scandinavian countries (perceived by many to be a fairly homogenous market), sales of organic produce vary by a factor of ten, where Norway has 2% market share, Denmark more than 20% and Sweden in the middle around 12% (2002 data). A study of consumer attitudes (A C Nielsen, 1997) revealed striking differences between the country 'attitude prototypes'. Norway, with the smallest market share, was dominated by consumer motivations interpreted to be 'like in the old good days'. This is supported by independent studies where Norwegian consumers confirm beliefs about Norwegian food being 'almost organic'. The prototype Swede was motivated by health as in 'good for you and your family', and the Dane had a motivation admixed with political overtones 'local in a global perspective'. These cliché segments also represent an age profile where the Norwegian consumers had the highest average age, the Swedes were in the middle, and the Danes were the youngest consumers. These findings led to a follow-up study in Sweden (Kihlberg and Risvik, 2006) where the question was to see if the 'Swedish' as well as the 'Danish' prototypes were both to be found in the Swedish national market. The Norwegian market segment was defined as not very interesting, both because of its size and high average age of the consumers, and also because the attitudes were considered to be 'old fashioned'. The momentum in attitude changes represented by the segments could be interpreted as a movement starting with the Norwegian 'as it used to be', to the Swedish 'good for you' and to the young politically aware Danish 'local in a global perspective'. This can also be understood as a movement starting from the remote, protected Norwegian market towards the more exposed, open and international markets in Sweden and Denmark. It was therefore also hypothesised that the Danish segment would be more demanding and less accepting of poor quality (both demanding market situation and demanding consumers). Product perception, in part related to preference, would then be expected to become more important for these younger consumers. To understand how the three segments related themselves to perception of sensory attributes of products, it became central to the study to see if preference patterns and attitude characteristics varied between the groups.

Bread from organically produced wheat was used as a vehicle for this study and presented to 400 consumers, selected among regular consumers of organic foods. The younger consumer groups in this sample indeed confirmed both attitudes similar to Danish young consumers, and they also scored higher (both verbally and through preference scores for actual samples) for samples with strong 'fresh' bread characteristics.

In particular, women showed positive attitudes towards organic food, and this also influenced their preference in a positive way, while men did not show such an effect. For women, preference improved for all samples except the highest preferred ones, when they were said to be organic, and especially the least preferred sample was given a significant lift in preference. For consumers younger than 30, the attitudes coincided with what was found in Denmark for younger consumers, while the older consumers (30+) responses were condensed with all attitudes from the other two segments in one place together (Kihlberg and Risvik, 2006).

The study supports the impressions of younger organic consumers being more demanding and dedicated, while female (older) consumers are more accepting. This was also supported through add-on questions to the survey, where they were asked to rate importance of attributes (sensory and others) of organic foods. The implications are obvious, as information about growth systems is no longer enough to make young consumers choose organic; the product quality must also deliver, no less than an alternative product of conventional origin. For marketing, the consequence must be that organic consumers can no longer be interpreted to be a homogenous segment where the ethical aspects of environmental issues dominate communication, food quality is equally important. Availability, the place of exposure, degree of processing, packaging and context, it seems, are critical factors for successful marketing of organic produce. This has been given little systematic attention in scientific literature, and would be of interest for defining the cross-cultural segment to be targeted through successful product development.

Similar differences to what has been seen in Scandinavia would obviously be found throughout Europe, but few studies have been found to document this.

13.3   Cross-cultural dimensions of food choice

Health as a market segment may also represent a window of opportunity, as a 'healthy choice' does not necessarily represent the same selection of foods across segments and countries. For Europe, to cope with the social consequences of unhealthy food choices will become too expensive very soon, with the increasing average weight. Unhealthy food choice then becomes a threat to our lifestyle and welfare development. This motivator favours strongly increased cross-cultural understanding of food choice. In this context it is possibly more correct to talk about cultural differences as differences between sub-groups with common cultural denominators, rather than national differences. This is not the main aim of this chapter, but strategies and theories would apply in similar ways.

13.3.1   Basics

The most basic question to be asked related to taste is whether people in one culture physically perceive food differently from people in another culture. The Japanese have a large vocabulary for food texture and also distinct preference for textures. The natural question to ask is whether this is because of a unique gift for perception of texture (genetically) or is this because their culture pays more attention to texture? There is little evidence to support the hypothesis of physiological differences. In colour perception it is possible to define objective references for comparisons of visual perceptions, this shows, with very strong indices (as causal proof is not possible) that colours are seen in the same way for all of us (NCS, 2000). Detection of deviations in colour sight can easily be seen as atypical evaluations for individuals, and this gives opportunities for testing with fairly high reliability. For texture and odour perception no similar references can be made, but individuals show little sign of variation related to genetically different make-up. The only logical assumption is that the observed differences in preference have a strong cultural origin, caused by time and evolution in food culture and tradition, rather than a biological origin.

Assuming all normal persons perceive about the same from a given stimulus, biology can be down-played as the most influential factor causing cross-cultural variation in taste preference and food choice. Left to consider as important is a mixture of food-related internal and human-related external determinants, such as image and taste (and taste meaning: taste, smell, texture and appearance). As Lutheran restrictive attitudes ('eat your food regardless of whether you like it') give way to more Latin attitudes (pleasure and image) even in the most northern and cold parts of Europe, all determinants are expected to increase in importance. The economic situations for most Europeans allow freedom of choice; opportunities for choice are abundant and are expected to continue to be the norm. External factors including image will be thoroughly discussed in other sections, this chapter will therefore continue with a discussion of how taste segmentation can be used to understand preference segments in a market, using coffee as an example.

Before we get to coffee, there is a need to ask one more fundamental question. Do national taste and flavour preferences exist, and are these of a permanent nature or changeable? This is necessary to answer before we discuss segmentation of consumer preferences as a function of national states or whether what is perceived as predominantly national segment, in reality is one of many segments, but weighted differently in countries.

Relevant questions can be posed related to unique national or cross-cultural taste preference for a country. Do Swedes have preference unique to Sweden and Swedish products? It is possible to believe that liking may be strongly affected by positive repeated exposure, such as is the case for sweetness in a sugar producing country. Norway and Sweden have large commonalties in food culture, dark full grain bread being one of them. But traditions are distinctly different as Swedish bread contains sugar, while the Norwegian does not. Swedish sugar production has made sugar a cheap and convenient ingredient in recipes in the baking industry, while in Norway it has for long been considered an expensive ingredient. In preference studies of full grain bread in those countries, consumers seemed to have a higher preference for their own tradition. This is not at all surprising, but few studies have documented these situations. Available resources and climatic advantages have caused national differences in many characteristic products, although there are strong indices that these differences are consciously being reduced over time, as a consequence of multinational companies wanting to rationalise production across national borders. Multinational companies hold detailed information about product formulations in different markets, and use this for internal and competitive adaptive changes of products towards one another. Changes occur in imperceptible increments to harmonise markets, cut costs and rationalise marketing strategies. For consumers, this makes familiar products available in several markets, although variation and choice is being reduced.

13.3.2   Coffee as an example

Coffee habits (style, volume, strength and context) and preference vary across Europe, although differences here are diminishing also. What used to be national differences, are becoming segments across countries, meaning that 'cross-cultural' is truly not synonymous with 'national borders' for coffee, still there are national tendencies, stronger in one country than another.

In a study performed by the European Sensory Network (1996) 16 coffees were selected to span the variation in 'the world of coffee'. Differences in variety, degree of roast, country of origin and use of production technologies made a sample set where the most important variations were covered. Trained taste panels in eight countries described differences between coffees, and later consumers were asked for preference for a subset of eight samples. These coffees were chosen because they represented unique variation in the overall set, and therefore described a concept through which potential new coffee samples could be understood.

The first analysis was performed to evaluate trained panel performance as an indicator for differences in taste perception. The descriptive trained panels in each country were given a large degree of freedom to develop their own vocabulary, and between 8 and 54 variables were used by these panels for this description. Despite the large variation, it was still not surprising to see a large degree of consensus in the interpretation of data describing the most important features of coffee perception (see Fig. 13.1). The first two dimensions describe the degree of maturity and roasting of coffee beans, not very surprising as this also comprises the greatest variation in the material.

Based on the global structure evaluated by the trained panels, eight coffees were selected for consumers, asking for preferences. Based on a global analysis of all consumers, five patterns emerged as a suggested definition of sub-segments with distinct different preference patterns. These segments were found to some extent in all the seven participating countries. The size of each segment

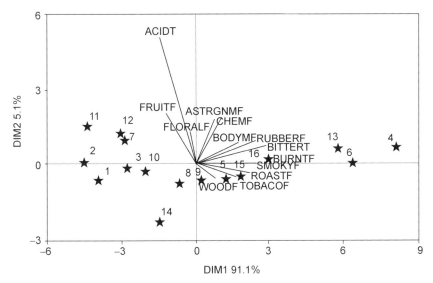

Fig. 13.1 Sample and attribute biplot derived from covariance PCA on common European vocabulary for eight countries, samples shown as 1–16.

varied a lot for each country. Based on these results it was possible to suggest a 'dominant' country for almost all segments (see Fig. 13.2).

Each of the clusters was dominated by one or a few samples significantly different from the others and also of a significantly different position in the market, seen from a marketing angle. It was therefore no surprise to see that, for instance, the cluster dominated by flavours characteristic of instant coffee also had the greatest representation with preference from UK consumers. For some of the other countries where the consumption of instant coffee is very low, like Germany and Norway, the preference segment was correspondingly very small. The same countries had the largest segment related to medium roasts, while France and Poland had highest contributions to the dark roast cluster.

Compared to sales figures for coffee roasts and blends, there was no great surprise where preference was high. A bit more surprising was that there was preference for all samples in all countries, and that some of these segments were similar in size to the 'best selling' coffees. For marketing, this opened opportunities of selling more than one type of coffee to consumers in all these countries and it also guided towards a launch strategy for the countries according to size of the market segment. For communication of coffee as a segmented product category, language and perception of product quality became important, something we have seen in market strategies during the first years of the 21st century.

A similar study to this repeated now, after a wide variety of coffees have been on the market for a few years, would most likely show a harmonisation of preference for coffee throughout Europe, as a result of market strategy and exposure.

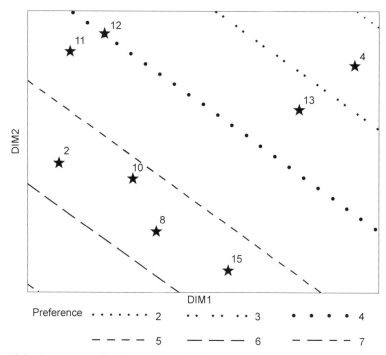

Fig. 13.2   Iso-contour plot for vector model showing one segment (dominated by UK consumers) as high in preference for decaffeinated and light/medium blended coffee, selected samples shown as numbers between 1 and 16, and preference (from 1 = low to 10 = high) shown through line intensity for preferences 2 to 7.

13.4   Cross-cultural dimensions of other factors in food choice

13.4.1   Language-related cross-cultural segments?

Disgust represents a cross-cultural segment not necessarily connected with national borders, and for the consumer this may be negatively correlated to health perception and choice, through 'I do not like the food, so it is not good for me'. Rozin (Rozin and Fallon, 1987; Rozin et al., 2000; Rozin, 2003) has developed a theory on relevance of negative emotional responses to foods and compared this across national borders in several studies. With relevance for Europe he found national differences between French and American consumer attitudes (Rozin et al., 2002), where the French gave much more significance to sensory perspectives to food, while the Americans' attitudes were more based on health beliefs. This is similar to findings by Cervellon and Dubé (2005), where the French (in Canada) also came out with more affective components in their attitudes to food as compared to Chinese, which were equally influenced by both affective and cognitive factors. In a third study with the root in sociology by Eertmans et al. (2006) the Food Choice Questionnaire (FCQ) developed by Steptoe et al. (1995) was evaluated on urban populations of Canadians, Belgians and Italians. Again the French-speaking Belgians came out as not very interested in health, while sensory appeal was scored the highest. In a fourth study on meat Grunert (1997) developed models for quality perception, and again the French came out as different from the British, Spanish and German. A uni-dimensional and common model for purchasing motives and quality perception was found for Germany, Spain and the UK, while France needed a multidimensional model to explain their motives.

In all these four studies, French-speaking consumers came out with attached motives and perceptions, which distinguish them from others. Whether this is distinctly French or appears as a consequence of the French language is not possible to resolve at this point as these effects are found in all studies.

13.4.2   Meat as an example of a sub-cultural segment related to disgust

The examples of odd, strange or disgusting food habits in 'other' cultures are many. Escargots, snake, horsemeat and haggis are obvious candidates. Although aversive differences do not have to be so obviously negative or blunt in order to be significant. Sushi, olives and avocado represent more subtle differences between European consumer groups based both in tradition and speed of adaptation to the migrating food trends. Common cultural food products include Coca-Cola, Champagne and French fries and increasingly foods like olive oil, espresso coffee and Pata Negra hams from Spain. These products have achieved common acceptance in larger segments, but originally they also represented small and local consumer markets. This shows that food preference is a strongly culturally linked entity, where exposure and cultural status are important factors for migration.

Disgust, on the other hand, plays a major role in food avoidance behaviour. Aversions are also central to Chapter 1 in this book, so they will not be discussed in detail, only used as a representative of cross-cultural dimensions where national borders are less important to define the segment than, for example, a combination of attitudes, sex and age and other sub-cultural phenomena.

It has been observed in France, that certain young consumer groups are consuming consistently less red meat, and it seems that the colour of the meats is inversely related to consumption frequency. This trend has been observed since the late 1960s (Gregory, 1997) for young females. In contrast, the average consumption of red meats has increased in countries like Norway, where the price of red meat has been reduced over the same period. In a market situation with increased polarisation between segments, it is still possible that some segments are decreasing their consumption while the average in the whole population goes up, as the reduction is more than compensated for by other and larger increasing segments. This was indeed the case when the decrease among young Norwegian females was investigated for their attitudes, motivation and liking for red meat. The main driver in focus groups was said to be a lack of liking, and this was interpreted as related to potential aversions for red meat. A quantitative study on drivers for reduced consumption is presented by Kubberød *et al.* (2006).

In a laddering exercise (Grunert *et al.*, 1995), performed during focus group interviews, red meat associated values were found to be linked to the feeling of being full, being fat, of poverty and of having a miserable life. This explains some of the higher level motivations for reduced red meat consumption, with strong leads for product exposure and communication.

In the quantitative conceptual model, several factors were evaluated for their contribution. In a questionnaire filled in by 866 young females (Kubberød *et al.*, 2006) several factors gave significant contribution to the understanding. On an ideational level, moral concerns related to animal handling and carnivorous consumption contributed strongly to significance. In the sensory and affective domain, perceived texture was the most important, and with an ideational component related to blood in the meat. The anticipated consequences were related to satiety from meat, and individual traits were related to negative self-esteem from eating red meat. The ideational and sensory affective components came out as the most important factors predicting red meat avoidance behaviour. This is not surprising as female ideals are portrayed as, amongst others, physically very thin being very attractive in most media exposure. To achieve this, 'hunger' is an emotion often associated with positive consequences for this group of consumers. For those who can manage this emotional manipulation, food may come as far as being a threat, while others give in to craving and show excessive food behaviour in the opposite direction.

Image is everything, or so it seems. Food exposure may prevent consumers from choosing, what from a nutritional point of view would have been optimal. Health as food image must also coincide with other values and perceptions conveyed through the food. Food is no longer sold by its nutritive and taste attributes alone. To make a consumer choose a food, the concept as a whole must convey the same message without sending conflicting messages to the consumer. To give an example of a message that most likely would not go down well, we have constructed this objectively correct commercial message: 'Try this young girls' nutritional "slim-snack to go" made from fresh cows' liver and blood'. It is much easier to go wrong than this. Success demands that nothing goes wrong in the formulation, nor in the message accompanying the product and that these fit together perfectly.

13.5   Understanding cross-cultural dimensions in food choice for food product development

Defining consumer segments and understanding cross-cultural dimensions in food choice is a key issue in order to achieve successful product development. In the early part of the last century it was a problem to provide enough food for the European population. In the second half of the last century the market was still production driven, that is getting the food out to people, establishing distribution systems and providing raw materials of good quality and establishing a food industry. In the market-driven food market of the 21st century, focus has changed. There is no shortage of food, and food has impeccable quality, while food costs only take up a small fraction of available private economy. Food choice behaviour is a challenge to understand, because of the increased complexity, and this is therefore a major driver for success in product development. For product developers it is a major challenge to understand factors underlying choice in consumer segments. Behind this lie trends at different levels, where the slow changing trends reflect a combination of cross-cultural phenomena and major world trends like 'food and health'. Modulated on top is fashion, with faster changes and finally fads coming and going without very predictable intervals.

The cross-cultural dimensions are important features in this, as they give guidelines for definitions of consumer segments and through this also for product development. Changes in cross-cultural segments over time will provide learning for understanding of cultural development.

For the increasingly important political dimensions of health perspectives, the unwanted and not anticipated effects of food becoming increasingly cheaper is the effect of over exposure to carbohydrate and especially sugar-rich foods. Biologically, humans are inclined to favour sweet foods, and with societal values like 'freedom of choice' and 'reward yourself', resistance becomes unequally distributed among economically different segments. This is easily seen in cultures like the American where weight problems increasingly are associated with poverty and lack of education. As a cross-cultural phenomenon this is currently among the most interesting to be understood. While some wear, inhabit or drive their rewards, other cultural strata with less economic muscular power eat their rewards. Food as a reward is still among the cheaper rewards, this leaves heavy responsibilities on the food industry for developing 'rewarding' foods with less caloric density, with increased satiety and higher fibre content for reduced energy adsorption. Problems with these foods are many, as fibre and increased protein content often is associated with dry, less palatable foods. For sugar-reduced foods, sweetness changes and it is not clear if 'sugar hunger' actually is reduced through consumption. Many studies exist where it is shown that consumers correct for the reduced caloric intake already within the same meal. Fibre is associated with constipation, and for segments with low physical activity, this is a problem. Research in this area now relies heavily on good communication and collaboration between medical research, nutrition, psychology, market research, but also on emerging knowledge from genomics (including nutrigenomics, metabolomics and all other -omics). The interplay between biology and psyche opens up a whole new understanding and consequently a whole new product spectrum on the arena of food and health.

13.5.1   Two alternative strategies for product development

Two competing strategies are quite common for product development, evolved for handling of knowledge about cross-cultural differences in food liking and preferences.

For markets with similar, but not identical products, it is possible to see an evolution towards stronger similarities and joint/identical branding of products. For multinational companies there is a great advantage in moving products closer to one another in such small steps that the market not will notice the changes. This will eventually cut production costs and rationalise the product portfolio, when the same prototype can cover several varieties in smaller markets. In a strongly competitive market this is the only option available for producers of commodity products in large volumes. For consumers, the product becomes available in several markets, price can be kept low, but there is also a danger in loss of variety and characteristic product quality. For the producer, the danger lies in the product slowly losing its characteristics to such an extent that the consumer segment loses interest and disintegrates.

The alternate strategy to this will be the introduction of characteristic food products with the protection of origin, production method, raw materials and taste characteristics. The philosophy here is radically different as the aim of the process is not to change the product, but to teach the market to learn new food habits and get new preferences. Traditional foods, which these often are, encompass increasing variety, but often a smaller and stratified market segment, in the higher price ranges. Marketing to and defining the segment for these products is often challenging and often relies on myth building and story telling to create an air of sophistication and of being a connoisseur.

In between the two strategies are lots of intermediate strategies, but only the two extremes are exemplified to illustrate a major phenomenon of importance for choice of strategy for product development.

Context creates additional complexity to the understanding of segments as individuals do not necessarily belong to the same cross-cultural segments over time. Increasingly consumers also tend to 'shop' for their behavioural patterns according to who they are shopping with, and through this give a varied signal to the environment. 'I shop therefore I am', is increasingly a slogan to be used to characterise modern consumers as identity increasingly is associated with shopping behaviour.

13.6   Future trends

Healthy food eating is more than a trend; it is a growing awareness that our welfare is at stake if we continue the eating habits from the last part of the 20th century. The political will to enforce change is growing, but the motivation also has to come from the individual consumer to achieve the wanted effects. Freedom of choice is, and will be a value of great importance to the bulk of European consumers and is also built into the European political system. Enforcement of changes through legal measures will be seen as an effort towards change, but must not be expected to have the foreseen effect as long as economic growth in private economy disfavours political and legally enforced action through external means on a societal level. Understanding of consumer motivation, attitudes and food choice behaviour is an alternative route with a better chance of achieving the anticipated effects. Results can only be expected through collaboration between food producers, food distributors, food providers, teachers, legal authorities and NGOs in a joint effort for a common goal. All actors in the food value chain must be expected to join forces for an improved diet, healthier choice, better alternatives, and at the same time be perceived to be better for the consumer through a complex set of criteria. First of all, the product must be liked better, then it must be more convenient, be available at consumption points at an acceptable price, and finally provide documented health benefits of value for the consumer. This means the food must contain improved information for choice, processes must be optimised in order for products to deliver as much as possible of taste and health benefits. Consumers must be exposed to and learn about products through systematic early exposure in the school system, through work and public exposure, such as catering for elderly and in hospitals. This must be a result of a conscious process, underpinned by correct information.

The new complexity describes a situation where eating out of home still increases; and this leaves increased responsibilities on food providers, private and public. Providers of food therefore share an increased need for guidance towards improved alternatives, specialised needs and options for substitutes.

13.7   Sources of further information

The area covered in this chapter is truly multidisciplinary, which implies that literature of relevance is spread in journals from very different sectors and with several mental paradigms represented in thinking and representation of theories and data. For searchers of information it is therefore necessary to use several channels, from www.Cordis.lu where the information from EU projects such as the Trust in Food project is found, to journals from social sciences, market research, sensory science and psychology, on the other hand. For more fundamental knowledge, the three very different approaches by Brunsø et al. (1996), Hofstede (2001) and Schwartz and Bilsky (1990) may give a very complementary introduction to the diversity of thinking within this field.

Cross-cultural factors affecting food choice are many, some directly related to perception and focus of attention, while others are related to language and how language directs focus of attention, while others have social, or cultural origin related to group or sub-culture. Culture is also no longer synonymous with nationality or geographic belonging. Sub-culture development gains in importance and will for food choice situations play a greater role, the problem is often to define criteria for recognition and measurement of sub-cultures as these will also change over time.

13.8   References

A C NIELSEN AIM (1997), Scandinavian value card MINERVA, http://www.acnielsen.dk/produkter/livsstil/minerva/minerva.asp.

ASKEGAARD S and MADSEN T K (1998), 'The local and the global: exploring traits of homogeneity and heterogeneity in European food cultures', *International Business Review*, 7, 549–568.

BERGEAUD-BLACKLER F (2004), *Consumer trust in food – A European study of the social and institutional conditions for the production of trust*, (EC, 5th Framework Programme, QLK1-CT-2001-00291), CRIC, University of Manchester, UK.

BRUNSØ K, GRUNERT K G and BREDAHL L (1996), 'Surveying the European food consumer. The importance of consumer market surveillance', in *An analysis of national and cross-national consumer segments using the food-related lifestyle instrument in Denmark, France, Germany and Great Britain*, MAPP Working paper no. 35, Arhus, Denmark.

CERVELLON M-C, DUBÉ L (2005), 'Cultural influences in the origins of food likings and dislikes', *Food Quality and Preference*, 16, 455–460.

EC (1999), *A Pan-EU survey on consumer attitudes to physical activity, body-weight and health*, The Institute of European Food Studies, Trinity College, Dublin, Ireland.

EERTMANS A, VICTOIR A, NOTELAERS G, VANSANT G and VAN DEN BERGH O (2006), 'The Food Choice Questionnaire: Factorial invariant over Western urban populations?' *Food Quality and Preference*, 17, 344–353.

EUROPEAN SENSORY NETWORK (ESN) (1996) *A European sensory and consumer study. A case study on coffee*. Campden & Chorleywood Food RA, UK.

FERRETTI P and MAGAUDDA P (2003), 'Consumer trust in food. Italy: Between local traditions and global aspirations', UNBO, University of Bologna, Italy. (EC, 5th Framework Programme: Trust in Food, QLK1-CT-2001-00291).

GREGORY N G (1997), 'Meat, meat eating and vegetarianism. A review of the facts', 43rd *ICoMST, Vitality of meat*, Auckland, New Zealand.

GRUNERT K G (1997), 'What's in a steak? A cross-cultural study on the quality perception of beef', *Food Quality and Preference*, 8 (3), 157–174.

GRUNERT K G, GRUNERT S C and SØRENSEN E (1995), 'Means-end chains and laddering: An inventory of problems and an agenda for research. *MAPP Working Paper No. 34*.

HOFSTEDE G (2001), *Culture's consequences*, 2nd edn, Thousand Oaks, CA: Sage Publications.

KIHLBERG I and RISVIK E (2006), 'Consumers of organic foods – value segments and liking of bread', *Food Quality and Preference* (in press).

KUBBERØD E, UELAND Ø, LEA P, MEVIK B-H and RISVIK E (2006), The mediating role of disgust in the prediction of red meat consumption among young women. *J Consumer Behaviour*, 5 (4), 281–291.

NCS – NATURAL COLOR SYSTEM (2000), Scandinavian Color Institute B, Stockholm, Sweden.

POPPE C and KJÆRNES U (2003), 'Trust in food in Europe. A comparative analysis', (EC, 5th Framework Programme: Trust in Food, QLK1-CT-2001-00291).

ROZIN P (2003), 'Five potential principles for understanding cultural differences in relation to individual differences', *J Research in Personality*, 37, 273–283.

ROZIN P and FALLON A E (1987), 'A perspective on disgust', *Psychological Review*, 94, 23–41.

ROZIN P, HAIDT J and McCAULEY C (2000), 'Disgust', in Lewis M and Haviland-Jones J M, *Handbook of Emotions*, 2nd edn, New York, The Guilford Press, 637–653.

ROZIN P, KURZER N and COHEN A B (2002), 'Free associations to "food": the effects of gender, generation, and culture', *J Research in Personality*, 36, 419–441.

SCHOLDERER J, BRUNSØ K, BREDAHL L and GRUNERT K G (2004), 'Cross-cultural validity of the food-related lifestyles instrument (FRL) within Western Europe', *Appetite*, 42, 197–211.

SCHWARTZ S H and BILSKY W (1990), 'Personality processes and individual differences. Towards a theory of the universal content and structure of values: extensions and cross-cultural replications', *J Personality and Social Psychology*, 58 (5), 878–891.

SPRINGER A, PAPASTEFANOU G, TSIOUMANIS A and MATTAS K (2005), 'Sociodemographic and subjective belief reasons for inter-EU differences of attitudes towards genetically modified food', *ZUMA-Nachrichten*, 56 (29), 78–93.

STEPTOE A, POLLARD T M and WARDLE J (1995), 'Development of a measure of the motives underlying the selection of food: the Food Choice Questionnaire', *Appetite*, 25, 267–284.

14

Gender differences in food choice

Ø. Ueland, Matforsk, Norway

14.1   Introduction

The first question asked on the birth of a child is whether it is a boy or a girl, and the first thing people notice when meeting a stranger is whether the person is male or female. All our subsequent attitudes and behaviour toward the individual will be coloured by this fact although an individual's *awareness* of the influence of gender on their subsequent reactions and responses towards the individual might be very low. Socially and culturally, the very obvious biological differences between men and women have led to differential cultural roles, expectations, and behaviour being associated with gender. As a consequence, the differences observed between the genders can be of both genetic and societal origin, or the result of an interaction between the two, although it is often difficult to establish where the influence of genetics ends and the influence of the environment begins (Maccoby and Jacklin, 1974). Independent of the causation of gender differences, it is important to understand to what extent these differences might be a factor in food choices of men and women, as there are implications for healthy eating as well as new product development.

Literal interpretation of the expression 'you are what you eat' implies that if an individual consumes a lot of fat and calories in the diet, obesity may result as a consequence of poor dietary choices. However, it also means that an individual will give other people an impression of themselves based on the foods chosen. The expression applies to both men and women, but what the different genders wish to signal to others may be very different (Bourdieu, 1984). People eat because they need nourishment, because eating has a social function, and because, for most people, eating produces a hedonistic sensory response. Differences between men and women regarding food choices can apply across all three situations.

In Western affluent societies of the present time, food choice plays an increasingly important role. The marketplace abounds with a large variety of foods, both regarding the amount of food product categories, and in the number of product alternatives within these categories. Furthermore, the increase in various contexts in which food is consumed has resulted in a diversification of food product alternatives and combinations (Risvik *et al.*, 2003). The possibilities for the food industry to tailor foods to any occasion are immense, providing they know what factors drive consumer food choices. In this chapter some of the differences between men and women in their attitudes and behaviour toward food will be considered. Further, possible implications of these differences with respect to development and marketing of food products will be discussed.

14.2   Food choice and consumer health

Today's food consumption in Western, affluent societies has little to do with getting enough food to survive, although food security still represents a concern in many parts of the world. In general, food is plentiful, and although variations in food consumption are large between social classes and populations, products other than food account for the greatest financial expenditure of most households (Kristiansen, 2001). Thus, most households have the resources to make healthy food choices. This section will mainly focus on the choices made by women and men relative to the healthiness of foods with respect to optimal nutrition for bodily image and prevention of illness. It has been well established that women are more health conscious regarding their food choices than are men (Turrell, 1997; Wardle *et al.*, 2004). Women count calories, they are concerned with their weight, they eat a lot of fruit and vegetables, and they consume less food than they probably would like to do. In a study by Rozin *et al.* (2003), 78% of a sample of younger American women were concerned with how food intake might influence their appearance. The meaning of health with respect to food choice and, consequently, the strategies used to achieve a healthy diet may, however, vary with age. Health may be considered synonymous with a fit and slim body in young persons, while health in older people means the absence of sickness and bodily frailties. The image of a slim and fit body is of particular concern among young women. Interestingly, this has been shown not to be just a figment of women's imagination but may be a real threat to women's quality of life. A study conducted at the New York University by Conley and Glauber (2005) showed that there was a direct negative correlation between level of overweight and expected income, ability to acquire a job or having a family life. The same was not found for men. As women are more concerned with their appearance than men, overweight women may have lower self-esteem and, thus, be less active or display lower ambitions with respect to career and income planning.

It is still hard to explain the impact of health on differences in food choice between genders. Health beliefs as measured by the International Health

Behaviour Survey (Steptoe and Wardle, 1996) were found to explain up to 40% of the differences in food choices between women and men (Wardle *et al.*, 2004). One recurring argument for women choosing healthier foods has been attributed to their role as the main provider in the household; it is their duty to ensure the healthiness of the family (Lupton, 1996; Fagerli and Wandel, 1999; Bäckström *et al.*, 2004). However, women also eat differently from men even if they live in single households (Berg, 2005). They consume more fresh fruit, vegetables, and fish whereas men consume more bread, cereals, and meat.

Wardle *et al.* (2004) found that men did not really mean to eat unhealthily or have unhealthy eating attitudes. Their food choices were the result of lower levels of concern about the health aspect of their dietary choices than women. It naturally follows that if you are not really very interested in a topic, there is no good reason to spend a lot of energy to find out more about it. This may explain why men generally have less knowledge about health aspects of food and why they display a simpler cognitive structure when selecting healthy or unhealthy foods (Rappoport *et al.*, 1993). In general, women are more interested than men in acquiring health information and they more often comply with dietary guidelines in their food choice and are more often considering health aspects of foods when they make a choice (Hunt *et al.*, 1997; Turrell, 1997; Schafer *et al.*, 1999).

Attitudes to foods produced by emerging technologies such as genetic modification are often associated with health arguments (Frewer *et al.*, 1996). As a consequence, women, who display higher health interest than men, are often more negative towards genetically modified foods (Cox *et al.*, 2004). However, in Europe attitudes towards genetic modification of foods are generally negative and there are studies that show little apparent difference between genders (Frewer *et al.*, 1997; Lähteenmäki *et al.*, 2002). Other studies have shown that genetically modified foods are associated with feelings of uncertainty concerning risks and benefits (Grünert *et al.*, 2001) and these concerns are particularly characteristic of women (Bäckström *et al.*, 2004). A natural consequence of this is that women display more positive attitudes and are more committed towards consumption of foods perceived as natural than are men, and consequently, are more frequent users of organic foods (Lockie *et al.*, 2004; Bäckström *et al.*, 2004).

Level of education appears to have some effect on men's knowledge about, and attitudes toward, healthy food choices while this was not equally apparent among women (Fagerli and Wandel, 1999). Men's views on what are considered healthy or what they can achieve with their diet also tend to differ from the views of women (Fagerli and Wandel, 1999). For example, men choose more carbohydrate-rich foods and meats than do women (Fagerli and Wandel, 1999; Berg, 2005). Building muscle mass both for performance and appearance requires some protein and although consumption of raw eggs and large steaks is associated with a certain type of movies, the positive masculine perception regarding body building may be underlying some of the food choices men make. Women will say that red meat is unhealthy while quite the opposite is true for men (Kubberød *et al.*, 2002a,b). Both genders, however, regard fish as a healthy food. In a study from Norway (Fagerli and Wandel, 1999), an equal proportion of men and women reported to have changed their diet in a more healthy direction. However, while men changed from regular fat alternatives to low-fat alternatives of the same foods, the dietary changes reported by women consisted of increasing the consumption of healthy foods already present in their diets. There were some indications that the changes seen among males were just delayed changes in the diet that had already been implemented among females.

14.2.1   Food choice in a social context

Following what we know about health-related food choice attitudes we would expect that food choice behaviour in a social context would be different for men and women. No one would be surprised to see a woman choosing a main course from a salad buffet and, likewise, no one would find it strange that a man chooses a steak when he eats out. This stereotypical behaviour fits well with the health attitudes displayed by the two genders. Does this stereotypical behaviour of men and women apply in a social context?

Eating at home

Historically, women have had the responsibility of providing and preparing food for the family. In many families and social contexts, men were served the choicest and most energy-dense foods. Men were usually responsible for the family's income and their work was hard labour, which required a high calorific intake, while the rest of the family had to make do with what was left. If meat were available, it would be served to the man first, because meat was a good source of energy but also because of meat's high status as a food. Although meat's status has changed somewhat, the connotations of meat as a prestigious and nourishing food still persists (Fagerli and Wandel, 1999; Kubberød *et al.*, 2002b). The lingering importance of meat as a man's food, and the notion that men need a lot of food to keep fit, support food choice behaviours that are in line with our stereotypical ideas regarding gender differences in nutritional requirements (Bourdieu, 1984).

In today's society the focus of the family meal has changed to revolve around children and their needs, likes, and dislikes. The meals served will thus often be a compromise between several desirable ends. The meals served to the family will provide the proper nutritional balance, they will be acceptable by all family members; and it should be possible to prepare the food in the context of time and economic constraints. In this setting it is most common to find that the mother, who still is the main food provider, frequently puts the needs or preferences of other family members before her own (Lupton, 1996; Fagerli and Wandel, 1999). Women in single households will, for example, consume considerably less meat than women living with a partner. Furthermore, a woman who becomes single after cohabiting with a partner or husband will change her diet back to what she used to eat prior to cohabitating, even if the marriage or partnership has lasted for many decades (Fagerli and Wandel, 1999). However, this study also showed that frequency of consumption of meat did not vary too much between men and women. The authors hypothesize that women, in addition to catering to their partners preferences, may influence meat consumption by creatively providing many additional non-meat dishes.

As we have seen, health issues play a large part in how gendered food habits evolve. The assumptions consumers make regarding the appropriateness of foods related to gender would therefore be the result of the interaction of health and traditions. Furthermore, new social trends will exert influence over the appropriateness of food choice behaviour. While food historically has been used as a means of showing off wealth and status, type of food served is now used as a means to show other people what kind of person an individual consumer is and to what social group they belong. In this setting gender differences are very visible. It is said that men are best described by what they eat, while women are verified by what they are not eating (Fürst, 1994; Fagerli and Wandel, 1999). The expectations following these observations are that foods that are considered to be heavy or rich, tough to chew or digest, are associated with men, and conversely, light and easily consumed foods are associated with women. Men would be expected to eat red meat and have large helpings, while women would choose lighter meat or salads and eat considerably less (Bourdieu, 1984; Lupton, 1996). Bourdieu (1984) argues that the eating habits of men are a way to show power. According to Bourdieu it is therefore fitting that men eat more than women, drink more than women, and consume stronger drinks.

Eating out
Different rules in food choice may apply when eating out. Woods (1992) argued that patterns revealed in the consumption of food at home were confirmed in the eating out context. Martens (1997), however, found that eating out behavior of men and women in some instances deviated from expectations. For some women the eating out experience is a time to relax. When women were asked what they would choose if they were invited to dine out, it turns out that quite a few would actually like to have a steak (Martens, 1997). The arguments were that it was a rare treat and they really liked it or it was ordinarily too expensive and they used this opportunity to eat it (Kubberød *et al.*, 2002a). It also turns out that the eating-out experience may be a time-out experience from the daily calorie counting so frequently found among women (Martens, 1997). Several women said that they would choose both steaks and rich, creamy sauces. Despite this observation, there is an overwhelming overrepresentation of red meat choices among men and, conversely, of white meat among women. With regard to fish, however, the results are varied. Bourdieu (1984) very clearly stated that fish was perceived to be an appropriate food choice for women and therefore also more frequently eaten by them because it has to be eaten with restraint, it is light and with insubstantial texture. Some studies confirm these results (Martens, 1997; Warde and Martens, 2000; Verbeke and Vackier, 2005), while other studies do not find differences between genders in fish consumption (Døving, 1997; Fagerli

and Wandel, 1999). Cultural and socio-economic differences between and within countries may account for these divergent findings. However, the healthy aspect of fish, as well as its light characteristics makes it a typically 'feminine' food according to the societal attitudes and stereotypes described earlier.

Usually, portion size is a factor for food choice when eating out. Many women would settle for only the starter rather than the main course, if the starters were just a little bit bigger. But choice of courses has also been found to differ between men and women. Martens (1997) found that if men and women chose more than one course, men were more likely than women to have a starter while women would select a dessert. The observation that women are more likely to order desserts than starters may also relate to differences in hedonic preferences between women and men. This will be discussed in the next section.

14.2.2   Hedonic aspects of food choice

Sweet taste

Men and women have different preferences for foods with different sensory characteristics. A preference for sweetness is a characteristic that is particularly associated with women (Lupton, 1996; Grogan et al., 1997; Kähkönen and Tuorila, 1999). Considering the current societal focus on healthy foods and slim bodies, a craving for sweet stuff must be an additional problem for women.

Children have an inborn preference for sweetness. This preference abates as children grow up and their tastes become more sophisticated. However, women seem to continue their preferences for sweet foods into adulthood to a much larger extent than do men, and they express higher pleasantness ratings for sweet foods (Drewnowski et al., 1992; Grogan et al., 1997; Wansink et al., 2003). The female preference for sweet foods has long been acknowledged by industry and is reflected in how sweet foods are marketed and used in the society. Chocolate and ice cream are typically advertised with feminine values and using female models and the typical Mother's Day gift is a box of chocolates (Lupton, 1996).

Food product preferences

Foods with a strong and rich taste, high colour intensity, and chewier texture tend to be preferred more by men than by women. Again, red meat such as beef or pork is a prime example of a food product that fulfils all of these characteristics (Bourdieu, 1984; Kähkönen and Tuorila, 1999; Wansink et al., 2003). Women are more likely to choose pale, light foods such as white meat, fish, and also crispy salads that have no troublesome textural properties. In addition to the purely sensory characteristics, women will show higher prefer-ence for foods that are associated with healthiness (Lupton, 1996; Kähkönen and Tuorila, 1999). In this setting, the combination of high perceptions of healthi-ness and sweet taste that we find in fresh fruits could well explain women's higher consumption of fresh fruits (Berg, 2005). However, it is important to note that while men have higher liking for red meats and high-fat products than women, men also profess high liking for white meat, fish and vegetables (Martens, 1997; Fagerli and Wandel, 1999; Kubberød *et al.*, 2002b).

Many studies have shown that women tend to exhibit more negative attitudes towards foods than men (Rozin *et al.*, 2003) and are more likely to actually reject certain foods (Santos and Booth, 1996; Worsley and Skrzpiec, 1997; Kubberød *et al.*, 2002a,b; Nordin *et al.*, 2004). The rejection of foods may be for health reasons, but often the reasons for rejection are based on sensory cues. Women, in particular, express disgust when confronted with certain types of foods, especially meats, and the reasons given for the rejections are related to textural properties, smell, taste, and visual characteristics. One of the food items most commonly rejected is red meat, and the characteristics cited are that meat is hard and tough to chew, that raw meat smells disgusting, that meat gives a fatty feeling in the mouth, and it looks bloody.

Portion size
Women are more comfortable than men with eating small portions, bites, and items (Bourdieu, 1984; Lupton, 1996). Typically, sweet foods are presented in smaller portions as compared to non-sweet foods. Chocolates are often bite-sized, chocolate bars are easy to divide and share, and sweet cakes and confectionery are not served in huge portions, except in comic strips. This conforms to the feminine image of not eating large food portions.

Women and men do not differ when it comes to how many meals they eat or how many times they snack, but they differ in the size of the helpings and in the types of food they choose (Martens, 1997; Wansink *et al.*, 2003). The foods normally associated with substantial meals often consist of items that are served in larger pieces, such as steak, baked potato, and apple pie. The manner in which these types of foods are served makes it more difficult to have small helpings, which may partly account for many women choosing other types of foods for their meals. Thus, food choice and meal composition may not only be due to preferences but also to convenience and the possibility of choosing the size of the product.

14.3   Methodological considerations for gender differences in food choice

In almost all research involving animals or humans gender has an unavoidable and large impact on the results. Gender has been found to complicate findings of food intake and models of behaviour (Silverman *et al.*, 2002; Lien *et al.*, 2002) and is a confounding factor which must be accounted for or the findings may be misinterpreted (Stevens, 1996). Gender, as such, may be seen as a general descriptor to characterize a number of factors that influence food choice (Stevens, 1996). According to Stevens (1996), '... gender defines differences in perceived expectancies, environments, opportunities, income level, interactions with children, and experience in food selection and preparation, and many other variables in addition to genetic, hormonal and anatomical differences' (p. 305).

It is important to bear in mind if and how gender can affect the results when studies are designed. Depending on the intended outcome of the study, a theoretical approach that considers gender aspects should always be included at some time during the planning stage.

14.3.1   Investigating gender and the influence of health on food choices

Health attitudes and behaviour have been studied with respect to gender differences in particular. The findings indicate that although health attitudes and behaviour can explain 30–40% of variance in food choices between genders, there is still a large percentage not accounted for by psychological factors (Grogan *et al.*, 1997; Turrell, 1997; Wardle *et al.*, 2004). However, in all studies where one might expect health concerns to influence the results, great attention should be paid to gender given that health may be perceived differently by men and women.

14.3.2   Investigating the influence of other factors on gender and food choices

Even in research where health issues are not thought to have an impact, gender should be considered as a variable, which should be analysed and controlled. For example, even if a product is solely targeted towards men, the producers should also consider how women might be involved with the product. In some cases women would be the most likely purchasers of the product and the producers should make sure that women also perceive the product as appealing or appropriate for the intended use.

Women are frequently targeted in food preference studies because they are considered knowledgeable about food products and food preparation, and often have responsibility for household food purchases. However, it is more common for women than men to have more than one agenda when they make food choices, and it is important to consider this when designing studies. One might find that women will express preferences for a product they will not buy in consumer tests, as the product is not preferred by other family members. As a woman once said when she was asked about her buying behaviour for a preferred bread: 'Why should I buy this bread when I am the only one in the family who would eat it?' and she would go on to explain that, as she didn't eat much bread personally, it would go stale, and it would be a waste of money. She concluded by saying she would continue to buy the dull bread they usually had because her children liked it, but that it was nice to know that there was good bread to be had (Ueland, 2000). In this case, it was important to understand the respondent's food choices by applying more than one question. Hedonic measures by themselves were not sufficient to determine possible buying behaviour.

14.4 Understanding gender differences in food choice for food product development

We have seen that women respond to different cues to food than men when they make food product comparisons. For food developers this represents a challenge when products will be targeted for a consumer segment that consists of both men and women.

14.4.1 Health considerations

First of all, a food product will have a higher chance of being selected by women if it signals health benefits associated with its consumption. The same product may be equally acceptable to men, but they may not pay as much attention to the health aspects, and, as a consequence, health profiling would not necessarily be a success factor regarding men's food choice. Health messages should take into consideration that calorie counting is important with a larger proportion of women than men and that fitness and strength are important for men (Oakes and Slottenback, 2001a,b).

Healthy eating is no problem for women, but men may need the support of others or better facilitation to change eating habits in a healthy direction to a greater extent (Øygard and Klepp, 1996). An example of how this may be facilitated is to provide healthy food alternatives in conspicuous locations where one would normally buy typical fast food such as hamburgers or pizzas. Men are more prone to eat a hamburger or pizza as a fast food than women are. Thus, giving men an opportunity to select a healthier, but easily consumed and convenient product is a challenge which will become increasingly important as food choice related problems such as obesity have a greater impact on population health levels.

In light of emerging food technologies and novel food alternatives, the perceived health benefits of products will be of particular importance with respect to women. Especially as women are more sceptical than men are towards products that are not perceived as natural.

14.4.2 Sensory considerations

The main differences in food choices between men and women concern the usage of meats in general and red meat in particular. Considering the problems many women experience when having to prepare and eat red meat, special attention should be placed on reducing the effect of the sensory cues that women find offensive (Kubberød, 2005; Kubberød et al., 2006b). For example, the less blood-like the product appears the better, and as red meat is considered fattening by women, visible fat should definitely be removed (although it should be noted that visible fat is also a negative issue for men in many societies). Reduced red meat consumption has been observed for women of all ages, but is particularly prevalent among young women. For younger women, handling of meat is also a problem, and meat that is ready to prepare is more acceptable than whole meat

(Kubberød *et al.*, 2006a). Furthermore, processed meats may be less repulsive than meat with a minimal degree of processing. In some instances, young women will describe themselves as vegetarians while they still profess liking for pizza and hamburgers (Kubberød *et al.*, 2002a). The vegetarianism they display is selectively concerned with avoidance of identifiable red meat. It is not clear at the present time whether these preferences will be maintained by younger women as they get older, or whether such food choices are attributable to youthfulness. Information of this type will become increasingly important given ageing Western populations if the food industry is to align product development and production with consumer demands in the future, and merits further investigation.

Another consideration in product development is the change in dietary preferences that occur among women when they change status from married or cohabiting to living alone (Fagerli and Wandel, 1999). The observed dietary change can have health overtones but it is also anchored in sensory preferences. Thus, meat is among the food products that are most sensitive to changes in living situation. Developing and presenting meat products with less meat-like properties is one possibility. Another strategy may be to incorporate meat with other food products, for example in a meal context, where its importance is on a less prominent level.

Another characteristic of food that influences food choice between genders is the size of the product. Large portions or items are more appealing to men than to women. Men are more concerned with achieving fullness than women (Zylan, 1996) and big bite sizes give more satisfaction to men, while products that enable nibbling and small bite sizes are more preferable to women (Bourdieu, 1984). Bourdieu (1984) states that it is not considered appropriate for women to open their mouths wide and consequently they will refrain from selecting products that requires big bites. Chocolate is a product that, although it is well liked by men, is typically associated with women and that is often made in small pieces. Developing chocolate bars for young male consumers would typically include a size dimension where it would be very difficult to take small bites of the product.

14.5   Future trends

This chapter has briefly mentioned age in connection with gender issues. Age seems to moderate the effect of gender somewhat to the extent that older women are more pragmatic than younger women. The avoidance of red meat, for instance, is not so apparent in older women, but this may be a cohort effect. Longitudinal studies are needed to follow attitudes towards meat at different stages of life. It has been speculated that older women are better at camouflaging their food choices so that they appear the same as men's. In the younger generations of women, however, one might expect a more independent approach and that they uphold their preferred diet as they age. Increased age reflects experience and the impact of societal and family traditions, which again are influenced by gender. Thus, in designing studies, the age effect on gender should also be considered.

Differences in food choices between genders that are apparent today may not be so in the future. The types and varieties of food products that are available increase steadily, and following this comes the opportunity to prepare and serve meals that consist of many items and choices. In many instances this means that it is possible to cater to most people's preferences within a meal. Being able to choose and combine ingredients in a way that is optimal for the individual is desirable to women and also appreciated by men. And even if selection of items may vary between genders, differences may not be so obvious.

Men, especially those with higher education, are to a larger extent than previously focusing on the health aspects of food which again may reduce food choice differences between genders in this segment. Although socio-economic factors influence food choice between different levels, economic constraints affect both genders and this may to some degree restrict differences in food choice between genders (Darmon *et al.*, 2003). In line with this, health-related problems due to dietary practices are becoming a large societal challenge, and this applies to both men and women. This again may affect the attitudes towards healthy eating of both men and women to a point where they converge.

14.6   References

BÄCKSTRÖM A, PIRTTILÄ-BACKMAN AM, TUORILA H (2004), 'Dimensions of novelty: a social representation approach to new foods', *Appetite*, 40 (3): 299–307.

BERG H (2005), 'Forbruksundersøkelsen, aleneboende kvinner og menn, 2001–2003.' [The study on consumption, women and men living alone, 2001–2003]. Oslo: Statistics Norway.

BOURDIEU P (1984), *Distinction; A Social Critique of the Judgement of Taste.* London: Routledge and Kegan Paul Ltd.

CONLEY D, GLAUBER R (2005), 'Gender, Body Mass and Economic Status.' NBER working papers by David Conley: w11343, May 2005.09.09.

COX DN, KOSTER A, RUSSELL CG (2004), 'Predicting intentions to consume functional foods and supplements to offset memory loss using an adaptation of protection motivation theory', *Appetite*, 43 (1): 55–64.

DARMON N, FERGUSON E, BRIEND A (2003), 'Do economic constraints encourage the selection of energy dense diets?', *Appetite*, 41 (3): 315–322.

DØVING R (1997), 'Fisk – En studie av holdninger, vurderinger og forbruk av fisk i Norge.' [Fish – A study of mechanism behind the fish consumption pattern in Norway.] SIFO-report no 12, Lysaker, Norway: The National Institute of Consumer Research.

DREWNOWSKI A, KURTH C, HOLDEN-WILTSE J, SAARI J (1992), 'Food preferences in human obesity: Carbohydrates versus fats', *Appetite,* 18 (3): 207–221.

FAGERLI R, WANDEL M (1999), 'Gender differences in opinions and practices with regard to a healthy diet', *Appetite,* 32: 171–190.

FREWER LJ, HOWARD C, SHEPHERD R (1996), 'The influence of realistic product exposure on attitudes towards genetic engineering of food', *Food Quality and Preference*, 7: 61–67.

FREWER LJ, HOWARD C, HEDDERLEY D, SHEPHERD R (1997), 'Consumer attitudes towards different food-processing technologies used in cheese production – the influence of consumer benefit', *Food Quality and Preference*, 8: 271–280.

FÜRST EI (1994), 'Kvinnelig og mannlig mat. Mat – et annet språk. En studie av rasjonalitet, kropp og kvinnelighet belyst med litterære tekster.' [Feminine and masculine food. Food – a different language. A study in rationality, body and femininity by literary references.] ISO report no. 7. Oslo, Norway: Department of Sociology, University of Oslo.

GROGAN SC, BELL R, CONNER M (1997), 'Eating sweet snacks: Gender differences in attitudes and behaviour', *Appetite*, 28: 19–31.

GRÜNERT KG, LÄHTEENMÄKI L, NIELSEN, NA, POULSEN, JB, UELAND, Ø. AND ÅSTRÖM A (2001), 'Consumer perceptions of food products involving genetic modification – results from a qualitative study in four Nordic countries', *Food Quality and Preference*, 12: 527–542.

HUNT MK, STODDARD AM, GLANZ K, HEBERT JR, PROBART C, SORENSEN G, THOMSON S, HIXSON ML, LINNAN L, PALOMBO R (1997), 'Measures of food choice behavior related to intervention messages in worksite health promotion', *Journal of Nutrition Education,* 29 (1): 89–103.

KÄHKÖNEN P, TUORILA H (1999), 'Consumer responses to reduced and regular fat content in different products: effects of gender, involvement and health concern', *Food Quality and Preference*, 10 (2): 83–91.

KRISTIANSEN JE (2001), 'Dette er Norge.' [This is Norway]. Oslo, Norway: Statistics Norway.

KUBBERØD E (2005), 'Not just a matter of taste – disgust in the food domain.' Thesis. Sandvika, Norway: Norwegian School of Business Management.

KUBBERØD E, UELAND Ø, TRONSTAD Å, RISVIK E (2002a), 'Attitudes towards meat and meat-eating among adolescents in Norway: a qualitative study', *Appetite*, 8: 53–62.

KUBBERØD E, UELAND Ø, RØDBOTTEN M, WESTAD F, RISVIK E (2002b), 'Gender specific preferences and attitudes towards meat', *Food Quality and Preference*, 13: 285–294.

KUBBERØD E, DINGSTAD GI, UELAND Ø, RISVIK E (2006a), 'The effect of animality in disgust response at the prospect of meat preparation – an experimental approach from Norway', *Food Quality and Preference*, (In press).

KUBBERØD E, UELAND Ø, RISVIK E, HENJESAND IJ (2006b), 'A study on the mediating role of disgust with meat in the prediction of red meat consumption among young females', *Journal of Consumer Behaviour,* (In press).

LÄHTEENMÄKI L, GRUNERT K, UELAND Ø, ÅSTRÖM A, ARVOLA A, BECH-LARSEN T (2002), 'Acceptability of genetically modified cheese presented as real product alternative', *Food Quality and Preference*, 13 (7–8): 523–533.

LIEN N, LYLE LA, KOMRO KA (2002), 'Applying theory of planned behavior to fruit and vegetable consumption of young adolescents', *American Journal of Health Promotion*, 22: 29–59.

LOCKIE S, LYONS K, LAWRENCE G, GRICE J (2004), 'Choosing organics: a path analysis of factors underlying the selection of organic food among Australian consumers', *Appetite*, 43 (2): 135–146.

LUPTON D (1996), *Food, the Body and the Self,* London: Sage Publications.

MACCOBY EE, JACKLIN CN (1974), *The Psychology of Sex Differences*, Stanford: Stanford University Press.

MARTENS L (1997), 'Gender and the eating out experience', *British Food Journal*, 99 (1): 20–26.

NORDIN S, BROMAN DA, GARVILL J, NROOS M (2004), 'Gender differences in factors affecting rejection of food in healthy young Swedish adults', *Appetite*, 43: 295–301.

OAKES ME, SLOTTERBACK CS (2001a), 'What's in a name? A comparison of men's and women's judgements about food names and their nutrient contents', *Appetite*, 36: 29–40.

OAKES ME, SLOTTERBACK CS (2001b), 'Judgements of food healthfulness: food name stereotypes in adults over age 25', *Appetite*, 37: 1–8.

ØYGARD L, KLEPP KI (1996), 'Influences of social groups on eating patterns: A study among young adults', *Journal of Behavioral Medicine*, 19 (1): 1–15.

RAPPOPORT L, PETERS GR, DOWNEY R, McCANN T, HUFF-CORZINE L (1993), 'Gender and age difference in food cognition', *Appetite*, 20: 33–52.

RISVIK E, UELAND Ø, WESTAD F (2003), 'Segmentation strategy for a generic food product' (Abstract). The fifth Pangborn Sensory Symposium – A sensory revolution.

ROZIN P, BAUER R, CATANESE D (2003), 'Food and life, pleasure and worry, among American college students: Gender differences and regional similarities', *Journal of Personality and Social Psychology*, 85 (1): 132–141.

SANTOS MLS, BOOTH D (1996), 'Influences on meat avoidance among British students', *Appetite*, 27: 197–205.

SCHAFER E, SCHAFER RB, KEITH PM, BOSE J (1999), 'Self-esteem and fruit and vegetable intake in women and men', *Journal of Nutrition Education*, 41 (2): 104–111.

SILVERMAN P, HECHT L, McMILLIN JD (2002), 'Social support and dietary change among older adults', *Ageing and Society*, 34 (4): 184–193.

STEPTOE A, WARDLE J (1996), 'The European Health and Behaviour Survey: The development of an international study in health psychology', *Psychology and Health*, 11: 49–73.

STEVENS DA (1996), 'Individual differences in taste perception', *Food Chemistry*, 56 (3): 303–311.

TURRELL G (1997), 'Determinants of gender differences in dietary behavior', *Nutrition Research*, 17 (7): 1105–1120.

UELAND Ø (2000), 'Forbrukertest av nye brødtyper' [Consumer test of new bread varieties] Ås: Matforsk. Report 2000/1.

VERBEKE W, VACKIER I (2005), 'Individual determinants of fish consumption: application of the theory of planned behaviour', *Appetite*, 44 (1): 67–82.

WANSINK B, CHENEY MM, CHAN N (2003), 'Exploring comfort food preferences across age and gender', *Physiology & Behavior*, 79: 739–747.

WARDE A, MARTENS L (2000), *Eating Out: social differentiation, consumption and pleasure*. Cambridge: Cambridge University Press.

WARDLE J, HAASE AM, STEPTOE A, NILLAPUN M, JONWUTIWES K, BELLISLE F (2004), 'Gender differences in food choice: the contribution of health beliefs and dieting', *Annals of Behavioral Medicine*, 15 (4): 361–374.

WOODS RC (1992), 'Dining out in an urban context', *British Food Journal*, 94 (9): 3–5.

WORSLEY A, SKRZYPIEC G (1997), 'Teenage vegetarianism: beauty or the beast?', *Nutrition Research*, 17: 391–404.

ZYLAN KD (1996), 'Gender differences in the reasons given for meal termination', *Appetite*, 26: 37–44.

15

Children and food choice

S. Nicklaus and S. Issanchou, UMR FLAVIC INRA-ENESAD, France

15.1 Introduction: importance of understanding children's food choices

The current increase of children's obesity prevalence in all industrialised countries (Lobstein and Frelut, 2003; Ogden *et al.*, 2004) has put the question of children's food choices forward. Recent studies indicate that weight status, as early as the first year of life, is likely to track further on into childhood (Danielzik *et al.*, 2004), which in turn will track into adulthood (Deshmukh-Taskar *et al.*, 2005). In this context, it seems particularly relevant to understand how food choices and preferences are shaped in early infancy and track into later in life. A couple of studies have started to show the precocity of food preference development (Skinner *et al.*, 2002a; Nicklaus *et al.*, 2004, 2005a).

In this chapter we will describe how food choices and preferences are acquired through childhood, i.e. the period from birth to pre-adolescence (~12 years). It is somehow difficult to refer to children's food choice since on many occasions, in particular during meals, children are not given the complete choice of the food they eat but are rather offered few alternatives which they might eat or not. This is why most of the research concerning children has focused on preference or liking. Rozin (1979) clarifies the distinction between preference and liking by defining preference as the choice of one food over another one, and liking as the affective response to a food. The same distinction will be made when specific studies are reported. For more general matters, preference and liking might be used indistinctively, in particular because in infants, it is not always possible to separate the two notions.

A general conceptual overview of potential determinants of food choices might include physiological, genetic, sensory, experiential, affective, psychological, social, extrinsic and economic factors in an attempt to cover the range of parameters a consumer trades off in order to make decision about food consumption. We propose that in children, not all factors are likely to play a role. Indeed, cognitive abilities develop from birth to preadolescence (Piaget and Inhelder, 2003; see also Roedder John 1999). From 0 to 2 years, infants are in a 'sensorimotor' stage mostly characterised by the importance of perception; they do not develop symbolic thoughts. Between ~2 and ~8 years, they are in a perceptual stage: symbolic thoughts develop but reasoning abilities are still limited. From ~8 to ~12 years, children are in a reflective stage, reasoning abilities develop but are mostly limited to concrete objects. Therefore, because young children might not have full reasoning abilities to process information concerning food, such as nutritional information, extrinsic characteristics are less likely to be involved in their food choice decision-making. Also, economic constraints might not play a role in children's decisions, either because they have little money to spend on food or because they are less likely to be involved in the decision-making for the full range of food products consumed in their household. However, one must recognise that children could orientate many of their parents' decisions related to food purchase, but this aspect will not be developed here. Furthermore, compared to adults, children might be more sensitive to the sensory properties of food and to the physiological consequences of their consumption.

Figure 15.1 proposes a framework of potential determinants of children's food choices. It outlines physiological consequences of ingestion, sensory and affective determinants, genetic inheritance, experiential influences, social factors and some extrinsic factors. In infancy, physiological, genetic and sensory/affective factors might account for most of the determinants of food intake. With age, of course, experiential factors will play a role and we will describe how the earliest experiences are likely to imprint food liking. Social or extrinsic factors are more likely to be involved in the shaping of older children's food choices.

Although every human is born consuming a single food, milk, by the age of one, infants already exhibit a variety of preferences which will lead them eventually to like and consume the 'adult' foods from their culture, which vary greatly from one culture to another. This suggests that plasticity in food preferences at the beginning of life is tremendous, which makes it possible for learning about the 'how's, 'what's and 'when's of food consumption to be achieved rapidly.

To describe the acquisition of food preference, we will start by showing the role of physiological factors, e.g. chemosensory acuity or chewing abilities, then we will explain in particular how genetics can help to understand variations in sensory preferences. The role of experience in the acquisition of food preference is central and therefore will be developed next. Some mechanisms of learning will first be exposed, to frame how experience might act. Then, we will define stages in the development of food preferences and show how food experience during one stage is likely to model preferences and to track into later stages. We will explain how social context might interact with experience by depicting the role of parents and

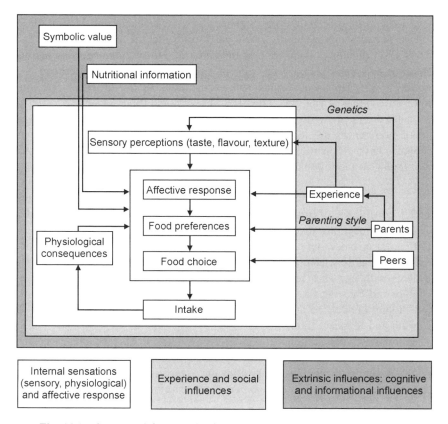

Fig. 15.1   Conceptual framework of determinants of children's food choices.

peers. Then cognitive influences on food preference and choices will be outlined. To conclude, some methodological issues concerning the assessment of children's food preference for food development will be discussed.

15.2   Physiological influences on food choice

Children's food choices are assumed to be more influenced by sensory perceptions than those of adults, which are more likely to take on cognitive, economic or even practical influences (Drewnowski, 1997). For instance, when given the choice to select items among a variety of foods, 2–3-year-old children avoid vegetables, probably because of their tough texture or their strong flavour (Nicklaus *et al.*, 2005b), but adults find vegetables, even the bitter ones, more likely to be consumed, maybe because they believe that they are beneficial to health (Drewnowski, 1997).

Sensory qualities are finely discriminated at birth. Neonate babies exhibit consistent preference and liking for sweet taste and rejection of bitter taste and to a lesser extent of sour taste (Steiner, 1979; Beauchamp and Moran, 1982, 1984). The olfactory system is functional at birth (Schaal, 1988) and capable of finely discriminating odours such as the odour of human milk and formula milk (Marlier and Schaal, 2005) or even of two lactating mothers (Cernoch and Porter, 1985). However, despite some indications of less negative reactions toward the odour of vanillin compared to butyric acid (Soussignan *et al.*, 1997), neonates do not show strong odour preference at birth which suggests that most odour preferences are learned through experience (Engen, 1986). So human neonates have efficient chemosensory senses at birth, which are, however, still developing in the following months (Ganchrow and Mennella, 2003).

15.2.1 Influence of oral abilities on preference for texture in childhood

Textural perceptions strongly influence food intake in childhood. Szczesniak (1972), for instance, has clearly shown that children's liking for foods of various textures is tightly related to their ability to deal with each specific texture, which evolves with age as a result of the development of oral parameters such as denture and chewing forces. Table 15.1 summarises how chewing abilities and food liking are related and co-evolve during development (Szczesniak, 1972; Carruth and Skinner, 2002).

In infancy, oral abilities also develop as a result of exposure to certain textures. For instance, introduction of solid foods delayed beyond the first year might result in difficulties in swallowing (Northstone *et al.*, 2001), which suggests the existence of a critical period for learning to cope with a variety of textures.

Children's attention to texture could be more important than adults'. Preference mapping in 6- to 11-year-old children showed that liking of apple varieties are influenced by their texture (Thybo *et al.*, 2004), in particular by the skin roughness, which is a criteria for rejection (Kühn and Thybo, 2001). In children aged 8 to 10 years, the image of vegetables is influenced by their texture: cauliflower, raw carrots, turnips and cabbage are described as 'crunchy' (Baxter *et al.*, 2000).

15.2.2 The case of taste preference: influence of perception and of genotype

The relationship between taste perception and preferences was particularly studied in the case of bitter taste. Researchers have long noticed that individual levels of perception of two bitter compounds of close chemical structure, phenylthiocarbamide (PTC) and 6-n-propylthiouracyl (PROP) vary greatly in the population (Bufe *et al.*, 2005). These variations in perception were tentatively related to variations in liking for bitter foods. In children (aged 5–7 or 4–5), the relationship between dislike for bitter foods and PROP bitterness perception was demonstrated for some foods such as cheese, spinach, broccoli (Anliker *et al.*, 1991; Turnbull and Matisoo-Smith, 2002; Keller *et al.*, 2002) but

Table 15.1 Development of oral mechanical functions and of preferred texture (Adapted from Szczesniak, 1972 and Carruth and Skinner, 2002)

Age range	Oral mechanical function development	Accepted/rejected textures
Birth	Only a rudimentary outline of the temporomandibular articulation; poor coordination of tongue and lip movements	Liquid
4 months	Greater mobility of tongue and lips; tongue moves gently back and forth as food enters mouth	Semi solid
7 months	Teeth eruption begins with mandibular central incisors	
8–9 months		Eats food with tiny lumps without gagging
9–10 months	Eruption of maxillary central incisors	Chews softer foods, keeps most in mouth
10–11 months	Beginning of lateral chewing movements; eruption of maxillary lateral incisors	Chews firmer foods, keeps most in mouth
12–14 months	Eruption of mandibular lateral incisors	Chews and swallows firmer foods without choking
15–16 months	Eruption of maxillary and mandibular first primary molars	Chews foods that produce juice
17–18 months	Eruption of canines	
21 months	Jaw and tongue movements are well developed	
2 years	Eruption of second primary molars	Tendency to greater acceptance of chewy foods
2–4 years	Good development of lateral chewing movements	Preference for moist/soft or crisp/crunchy
4 years	5/6 mature width of the palatal arch	
5 years	Permanent premolars begin to move towards eruption	Dislike of soft, mushy
6 years	Eruption of first permanent teeth, beginning of rapid growth of muscles of mastication	Refusal of lumpy, stingy; preference for raw vegetables.
8–9 years	Eruption of permanent maxillary incisors	Dislike of greasy, slippery
10 years	Eruption of permanent maxillary first premolars	Preference for firm/chewy/rough; aversion to soft/smooth/slippery

not for all bitter foods studied, suggesting that the influence of PROP sensitivity on children's food liking, if any, is moderate.

A recent study showed more directly the link between allelic forms of the TAR2R38 gene which codes for the taste receptor for PTC and PROP sensitivity (Mennella *et al.*, 2005). In general, the AA allelic form is associated with low sensitivity to PROP, the AP allelic form with medium sensitivity to PROP and the PP allelic form with high sensitivity to PROP. Noticeably, AP children are more sensitive to PROP than AP mothers. Moreover, different allelic forms are also associated with different sugar preference in children but not in adults: AP and PP children prefer higher concentrations of sucrose than AA children; PP children consume cereal with higher sugar content than AA or AP children; and PP children consume beverages with higher sugar content that AA children. However, such discrepancies are marked in black children but not in white children. These results suggest that genotype related to taste perception might indeed directly affect taste perception and also taste preference phenotype in children, but this might be modulated by experience, either across ethnic groups (white and black children might have experienced different foods) or across age groups (adults' phenotype is less related to their genotype than that of children).

Relationships between the perception of other tastes and food preferences were studied without much success (Olson and Gemmill, 1981; Fischer *et al.*, 1961). However, children find sucrose solutions less sweet and prefer them compared to adolescents and adults (Zandstra and de Graaf, 1998). Children's preference for very sweet solutions was put forward to explain their high consumption of sweet foods and beverages (Drewnowski, 1989). Some children (about one third) show higher preference for sour foods than adults and than children of the same age (Liem and Mennella, 2003). This higher preference is not related to a difference in sensitivity but might be related to previous experience: children who liked extremely sour foods and children who did not like them differed in their previous experience with sour candies and their mothers' reports of going through a 'sour food eating phase'.

The same type of study performed to relate children's sensitivity to odour compounds and preference for food containing such compounds was attempted in the case of sensitivity to trimethylamine and rejection of fish; however no direct relationship was shown (Solbu *et al.*, 1990). Due to the large numbers of the olfactory compounds which constitute food flavour and of the human olfactory receptors, it is less likely that a direct relationship between allelic forms, sensitivity and preference will be discovered in the case of olfaction as it was in the case TAS2R38 variants and PTC/PROP perception.

Early in life, food liking strongly depends on sensory perception, which implies that it might be influenced by sensory capacities; however, experience can modulate perceptions either in their functioning (such as in the case of exposure to a variety of texture and oral abilities), or in their affective dimension and thus alter food liking.

15.3   Mechanisms involved in the acquisition of food preference

Beyond the physiological impact of perception on affective response which is especially influential in children, food likes are mostly acquired with experience. Several theoretical frames might help to understand the mechanisms involved in the acquisition of liking with experience. We will expose some of them here but one might also gain further insight on psychological mechanisms involved in learning (notably in adults) by consulting Chapter 4 in this volume.

15.3.1   The effect of mere exposure

Developmental psychologists generally acknowledge the role of mere exposure to a stimulus in the acquisition of liking for this object, in the domain of vision or audition, for instance. This theory might be transferred in the case of the acquisition of food liking, and it was indeed shown that following repeated exposure, children acquired a preference for a food (Birch and Marlin, 1982). However, one important difference between exposure to a sound or an image and exposure to a food is the fact that the food is ingested and therefore has physiological consequences, at several levels. Moreover, food is generally offered in a social context which is also likely to reinforce affective orientation. So the exposure to a food is rarely if ever mere. We will therefore develop further how the frame of associative conditioning helps to understand how experience induces preference.

15.3.2   Learned safety and diminution of neophobia

Having learned that the consumption of a food is safe can help to increase its liking (Birch et al., 1998). This effect is psychological: following the consumption of the (safe) food, the neophobia or fear of novelty is diminished (Zajonc, 1968). This evolutionary perspective might help to interpret neophobia as an adaptive response of the organism to novel substances in an environment where novel foods are actually unknown, but might not be as relevant to account for neophobia in industrialised countries where most foods are well known and selected, often industrially processed and very carefully controlled for safety. Children are, however, not aware of the 'regulated' safety of the foods they are offered and might still display neophobia. Nevertheless, neophobia might help to prevent consumption of bad tasting foods (Pliner et al., 1993). A new food, especially if its appearance is unfamiliar, is likely to trigger neophobia, which could be dissipated if it tastes good. This is confirmed by experimental study showing that neophobia was indeed dissipated after consumption of good tasting novel food in 10- to 12-year-old but not in 7- to 9-year-old children (Loewen and Pliner, 1999).

15.3.3   Conditioning with physiological consequences of the ingestion

One important learning mechanism is the association of a flavour to the physio-logical consequences of the ingestion of the food. Thus, it has been clearly demonstrated that pairing a flavour with calories increases the preference for this particular flavour (Birch, 1990; Booth *et al.*, 1982). The flavours that 2–5-year-old children learn to prefer more quickly are indeed those of high-caloric foods, even when the flavour is later presented in a different, low-caloric food (Birch and Deysher, 1985; Birch *et al.*, 1987b, 1990). This conditioning for energetic foods could explain in particular the development of preference for fat foods (Birch, 1992; Johnson *et al.*, 1991; Kern *et al.*, 1993).

Beyond the context of experimental studies, in everyday situations, the relationship between caloric content and preference seems to be maintained: children's liking for fruits and vegetables (Gibson and Wardle, 2003) as well as rates of choice of a variety of foods by toddlers in a self-service setting (Nicklaus *et al.*, 2005b) are positively correlated to their energy density, as illustrated in Fig. 15.2. Going one step further, some authors proposed that children's and adolescents' liking for energy-dense foods was related to their high energy need, in relation to growth and development (Drewnowski, 1998, 2000; de Graaf and Zandstra, 1999). In a society of plenty, in particular where many palatable, energy-dense foods are available, this efficient conditioning mechanism could be one of the numerous factors involved in the development of children's obesity epidemic (Birch, 1992).

15.3.4   Social learning
According to Rozin (1990) social factors may be the most potent means of enhancing liking in humans. This affirmation is in accordance with results from Ton Nu *et al.* (1996) who indicated that social motivation was one of the two main reasons given by subjects to explain why they had begun to like a food. Several types of mechanisms can occur.

Social modelling
Various experimental studies have shown that preference for initially neutral or disliked foods might increase if a young child can observe a peer eating the same food (Zellner, 1991; Hobden and Pliner, 1995; Birch, 1980; Birch *et al.*, 1980). This effect is more important if the model is the same age rather than older (Hendy and Raudenbush, 2000), or if the food is initially neutral or liked, or energy-dense. However, such a modelling effect might not last more than one month if the model is not present anymore (Hendy, 2002). In everyday situations where children remain in contact with a model, the model's influence might last longer.

Social conditioning
Several types of social conditioning have been described in literature.

Social evaluative conditioning
According to Rozin (1990), social evaluative conditioning may be viewed as a form of Pavlovian conditioning in which the positive effect shown by another

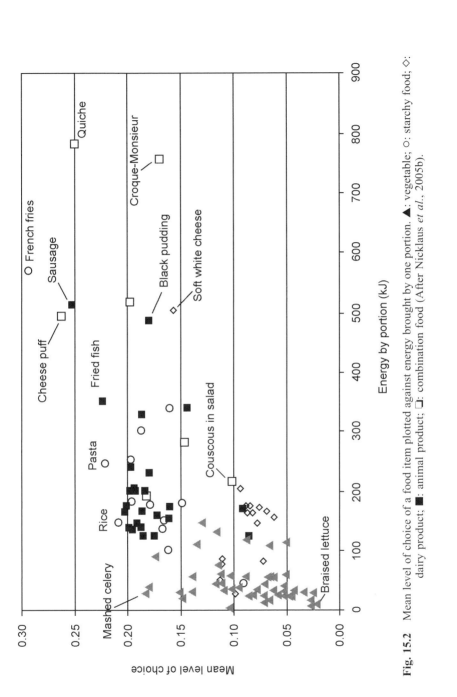

Fig. 15.2 Mean level of choice of a food item plotted against energy brought by one portion. ▲: vegetable; ○: starchy food; ◇: dairy product; ■: animal product; □: combination food (After Nicklaus *et al.*, 2005b).

person is the unconditioned stimulus. If an adult demonstrates pleasure when consuming a food, children note this positive response which can influence their own behaviour and later their own affective response (Rozin, 1990).

Social affective context

Food preferences might also be acquired in situations by association with either positive or negative reinforcers. Very often, social interactions, in particular with parents, provide opportunities for such a mechanism to take place.

Parents often attempt to deliberately influence the food choices of their offspring. One strategy consists in offering a reward in exchange for the consumption of a food (eat your vegetables and you'll be able to watch TV). Experimental evaluations of such a situation showed that when the consumption of an initially neutral food gives children access to a reward, their preference for this food decreases (Birch *et al.*, 1982, 1984b). The subjective value of the reward for the child might alter the resulting preference: if the reward value is high (watching TV), a strong negative reinforcement might happen whereas in the case of a low reward value (obtaining an image), a slight increase in preference might take place. This was observed experimentally (Wardle *et al.*, 2003b): when associated with repeated exposures, a reward of low subjective value led to an increase in liking for a new vegetable. However, exposure alone led to a higher increase in liking than exposure associated with reward, as shown in Fig. 15.3.

To alter their children's preference, parents also offer a food as a reward or in a positive social context. To evaluate this effect, Birch and her colleagues offered neutral foods to 2- to 5-year-old children as a reward after good behaviour or in association with an adult's positive attention: after a couple of presentations, preference for this initially neutral food increased (Birch *et al.*, 1980; Birch, 1981).

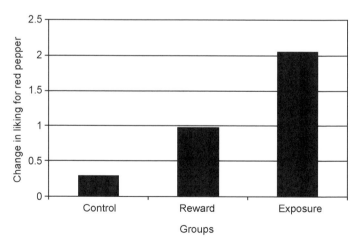

Fig. 15.3   Evolution of liking for a new vegetable in 5–7-year-old children after 'mere' exposure (Exposure), after exposure associated with reward (Reward) or after no exposure (Control) (After Wardle *et al.*, 2003b).

Another common strategy consists of restricting access to a much liked food. However, in 3–6 year olds restriction leads to an increase in preference for the restricted food (Fisher and Birch, 1999a) and to an increase in its consumption when the restriction ends (Fisher and Birch, 1999b). Conversely, forcing a child to eat a food leads to the development of a rejection of this food, which can be long lasting (Batsell *et al.*, 2002; Carper *et al.*, 2000).

Initiation rite
Stable liking can be acquired during childhood when products, previously forbidden to children, are introduced with a certain initiation rite (Köster, 1991). Köster gives the example of a child who is offered to have a sip of its father's glass of beer and indicates that the child 'will shout "wonderful" notwithstanding the horribly bitter sensation he experiences'. As pointed out by Ton Nu *et al.* (1996), children's desire to enter the adult world contributes to make an initially unpalatable sensation acceptable. 'Being an adult and doing what members of one's society do' would be perceived as a reward and would explain the acquisition of liking for chili pepper, a similar case of an innately unpalatable sensory characteristic which later becomes pleasant (Rozin and Schiller, 1980).

15.3.5   Flavour-flavour associations
New flavours are experienced within a food context where other tastes or aromas are present. Thus flavour-flavour conditioning can occur. For example, an initially neutral flavour associated to a negative taste such as a bitter taste will induce a negative hedonic response when presented in the absence of the bitter taste. On the contrary, if a neutral flavour is associated to a positive taste such as sweetness, an enhancement of liking of this flavour is observed (Zellner *et al.*, 1983). However, beyond experimental situations, when children experience sweetness, this pleasant taste generally also brings calories. So, a flavour-flavour association mechanism can occur at the same time as a positive association with post-ingestive effects. Moreover, adding sugar in a product can also have a direct effect as it modifies the sensory perception and, for example, can diminish the perception of bitterness.

15.4   Stages of acquisition of food liking in children

In this section we will describe acquisition of liking with experience (through the different underlying mechanisms described previously) according to various stages during childhood. We propose that these stages are delimited to best describe major transitions in the development of food behaviour. The tracking of food preference from one stage to the other will be outlined when possible. First the impact of prenatal experience will be described, and then the influence of neo- and postnatal experiences, mostly acquired through liquid feeding (from birth to the end of exclusive milk feeding, generally around 4 to 6 months) will be developed. We will then focus on the transition phase to a solid diet (~6 months–2 years) before exploring food preferences during the 'neophobic phase' (~2–8 years) and briefly during preadolescence (~8–12 years).

15.4.1   Prenatal experiences

The acquisition of food preference might start very early in life. The odour compounds from foods (such as garlic, for instance) consumed by the pregnant mother can pass through the placenta barrier and impregnate the amniotic fluid (e.g., Mennella *et al.*, 1995). Those flavours are accessible to the foetus, which olfactory system is functional by the last trimester of pregnancy. This early exposure to specific flavours from the mother's diet in utero modifies the preference for the same odour in neonates (Schaal *et al.*, 2000) and also alters the preference for a food similarly flavoured in 5- to 6-month-old babies (Mennella *et al.*, 2001). Moreover, the events occurring during pregnancy might impact much later in life: the attraction to salt is higher in neonates whose mothers had suffered morning sickness (Crystal and Berstein, 1998), and it is maintained until adulthood and is accompanied by a higher intake of salty foods (Crystal and Berstein, 1995).

15.4.2   Neo and postnatal experiences, experiences through milk feeding (0 to 4/6 months)

Maternal milk, such as amniotic fluid, carries some flavours of the foods previously ingested by the nursing mother (e.g., Mennella and Beauchamp, 1991). The exposure to those flavours also modifies the children's behaviour. In 5- to 6-month-old babies, a higher intake of cereal prepared with carrot juice is observed if the mother had drunk carrot juice during the first two months of nursing (Mennella *et al.*, 2001). Moreover, the variety of preceding flavour experiences might preset the infant to accept more easily new food flavours during weaning: 4- to 6-month-old babies better accept an unknown food when they had previously been breastfed than bottle fed (Sullivan and Birch, 1990), see Fig. 15.4. So breastfeeding incidentally provides children with a variety of flavours from the culture of their future adult diet, which might help them to bridge prenatal experiences (with the same flavours experienced in utero) with postnatal experiences.

Moreover, it must be noted that breastfeeding is not evenly distributed across social classes (in France, for instance, the percentage of breastfeeding mothers is more significant in low and high social classes as opposed to middle class, see Gojard, 2000). These variations in early imprinting through breastfeeding according to social classes might provide a first general frame for further acquisitions of food liking. This will be further developed.

Direct experiences with foods are also likely to influence the child's prefer-ences during the first months of life, when plasticity is high. An infant who receives sweetened water during the first six months of life will maintain the preference for sugar exhibited at birth, whereas this preference is not maintained

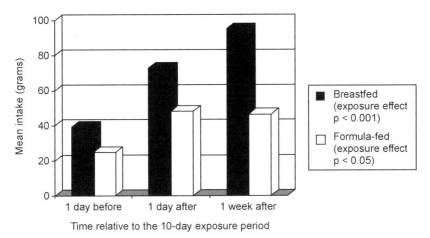

Fig. 15.4   Intake of a novel vegetable before and after 10 exposures, in previously breastfed ($n = 19$) or formula-fed infants ($n = 17$) (After Sullivan and Birch, 1994).

in an infant not exposed to sweetened water. Such an effect is still apparent at the age of 24 months (Beauchamp and Moran, 1982, 1984). Formula milk is generally recognised as 'bland' compared to breast milk, however there are sensory variations across the different types of formula milk (e.g., regular, soy-based or hydrolysed protein-based). These variations have an impact on the preferences of children previously fed those formulas: at the age of 4 to 5 years, their preferences for the odours of those formulas and for a variety of taste solutions differ (Mennella and Beauchamp, 2002; Liem and Mennella, 2002). Experiences during this early phase of development are likely to imprint later preference. For instance, adults who were fed a formula flavoured with vanilla in infancy, exhibit higher preference for a vanilla-flavoured ketchup compared to breastfed adults (Haller *et al.*, 1999).

15.4.3   The transition from milk to baby food and from baby food to table food (4/6 months to 2 years)

The transition from an exclusive milk diet to a diet including semi-solid, then solid foods is a very important step for the acquisition of food liking. At this stage, the decision to expose the infant to solid foods will partly depend on the mother's evaluation of her infant's mouthing and chewing abilities. Conversely, the development of chewing abilities also partly depends on the type of food stimulations the infant is exposed to (Szczesniak, 1972; Northstone *et al.*, 2001), as was detailed in Section 15.2. So eating abilities, food preferences and experiences are tightly related to one another during this period of early childhood.

Nine-month-old infants, weaned by switching from exclusive breastfeeding to exclusive solid feeding, accepted surprisingly well the foods they were offered in general (though some children oriented their choices toward certain foods and ignored others), even though some foods (raw eggs, raw offal) might have repelled older children (Davis, 1928). The type of food a child is likely to accept shortly after this transition phase from milk to a mixed diet might be conditioned by the timing of introduction: introduction to solid foods before six months leads to fewer food rejections at the age of 15 months than introduction delayed after 10 months (Northstone *et al.*, 2001).

Around the age of 4 to 6 months (the age of weaning for most children in most developed societies), repeated exposure to a novel food increases its acceptance (Sullivan and Birch, 1994), see Fig. 15.4. Moreover, this increased acceptance of a new food is enhanced by exposure to a variety of foods, rather than to the single new food (Gerrish and Mennella, 2001). That exposure to a variety of foods produces different preference than exposure to a single food implies that sensory qualities of baby foods are discriminated by infants. Furthermore, the frequency of exposure needed to increase acceptance might well depend on the food and/or on its flavour: a vanilla-flavoured dessert seems to be accepted as early as the first exposure (Melcer and Murphy, 1997), intake of some vegetables increases as early as the second exposure (Birch *et al.*, 1998), whereas acceptance of peas does not increase after three exposures (Gerrish and Mennella, 2001). Learned preference for a vegetable or a fruit purée can generalise to the same vegetable or fruit from another brand, i.e. for a sensorily close food, but the extension of preference for other vegetables was not always observed if there was sensory disruption, e.g. between peaches and pears (Birch *et al.*, 1998).

The period of transition from baby food to adult foods has not been much researched, in part probably because children of this age are difficult to study. Observational studies showed that dislike of vegetables became more frequent during the second year of life (Skinner *et al.*, 1997a,b, 1999), despite the fact that vegetables were well accepted (after repeated exposure) during the transition from exclusive milk feeding to semi-solid diet (Birch *et al.*, 1998; Gerrish and Mennella, 2001). Exposure to fruits before the age of 2 seems to be positively related to the variety of fruit consumption at 8 years old, but the same result was not found for vegetable consumption (Skinner *et al.*, 2002a), which highlights again the resistance of children to eating vegetables.

Infants present an innate ability to learn to like foods, which is facilitated by prior exposure to a variety of flavours either in utero or in breast milk. Learned likes for one food are probably not generalisable between or within food groups, and mostly depend on exposure to each specific food. Though the period of exposure to the first solid foods is probably very important in the development of eating habits and food preferences, few data are available to document its long-term impact on later preferences. The transition to table foods also represents an important step because it often implies a switch to the foods liked by the parents (Skinner *et al.*, 1997b, 1999).

15.4.4   The neophobic phase (2 years to 8/9 years)
The concept of food neophobia was proposed to describe the human dilemma of eating a diet both varied enough to achieve good nutritional quality and

restricted enough to guarantee consumption of foods recognised for their safety (Rozin, 1976). The initial reaction to novel foods would be of mistrust, to protect against any harmful substance. According to this concept, neophobic reactions are very likely to show in young children for whom most foods are novel. It was also proposed that an individual's level of neophobia could vary within a population, and an instrument was designed to evaluate this dimension of adults' and children's temperament (Pliner and Hobden, 1992; Pliner, 1994). Although a high level of neophobia might characterise some adults, most children go through a specific neophobia phase, which levels off during mid-childhood (8–9 years). In young children, the development of neophobia might not be specific to food and might correspond to the 'no' phase when the child's own personality develops, partly by standing against adults' attitudes. Moreover, in this 'no phase', children tend to refuse not only novel foods but also foods they previously consumed. Most of them become 'picky eaters' (i.e. children who resist eating many familiar foods).

Different investigations make it possible to frame this neophobic phase in childhood. Only 19% of 4–6-month-old infants are perceived as 'picky eaters' by their mothers, as opposed to 50% of 19–24-month-old infants (Carruth et al., 2004). Other parental reports confirmed the increase of neophobic or refusal behaviour in childhood (Pelchat and Pliner, 1986), notably during the third year (Cashdan, 1994; Hanse, 1994). Food intake logically increases during early childhood, however food preferences emerge during the second year of life: food refusals appear (Davis, 1939), the variety of free food choices decreases between the age of 2 and 3 years (Nicklaus et al., 2005c) and the diversity of intake lowers between the age of 2 and 5 years (Cox et al., 1997). Experimental studies also showed that food neophobia is stronger at 2 than at 3 or 5 years (Birch et al., 1987a) and that, after the age of 3, a child will prefer familiar foods to unfamiliar foods (Birch, 1979). Between 3½ and 7 years old, neophobia does not evolve clearly (Carruth and Skinner, 2000), but it levels off after the age of 8–9 years (Hanse, 1994). This neophobic phase has consequences on the child's food choices. Picky children have a less varied diet (Carruth et al., 1998; Falciglia et al., 2000). Between the ages of 2 and 9 years, neophobic children eat fewer fruits and especially fewer vegetables (Jacobi et al., 2003; Cooke et al., 2003; Galloway et al., 2003, 2005).

Despite the characteristic food neophobia of children aged 2–8/9 years, it remains possible to induce preference for a novel food with repeated exposure during this period. However, in children aged 2–5½, visual or gustatory experience to novel fruits increase respectively visual or gustatory preferences but a visual exposure is not sufficient to increase gustatory preference (Birch et al., 1987a). In 2–5-year-old children, 8 to 15 exposures are needed to reduce the initial neophobia (Birch and Marlin, 1982; Birch et al., 1987a; Sullivan and Birch, 1990). The speed of learning depends on the food: at the age of 2 to 3 years, preference increases after at least ten exposures for cheese, but five exposures are sufficient for fruits (Birch and Marlin, 1982), see Fig. 15.5. One might think that cheese, an energy-dense food, might be preferred more readily than fruit due to associative conditioning

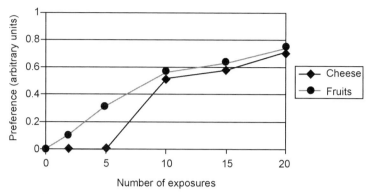

Fig. 15.5   Preference for cheese and fruits as a function of number of exposures (After Birch and Marlin, 1982).

with caloric content; however, in this case, it seems that sensory properties override caloric content: the sweetness of the fruit might facilitate the acquisition of preference compared to the bitterness or the unpleasant flavours of cheese. In children aged 3–7, ten exposures to a novel vegetable are necessary to increase its liking, but intake of this vegetable starts to increase before the tenth exposure (Wardle *et al.*, 2003a). In children aged 5–7, eight exposures to red pepper lead to an increase in liking and in intake; but an increase in intake, not in liking, is observed after only one exposure (Wardle *et al.*, 2003b). This suggests that the affective reaction (liking) is more difficult to modify than the behavioural reaction (intake). It must be noted that, as during infancy, learned preferences are food-specific: 4–5-year-old children, after 15 exposures to tofu, either salty or sweetened, have learned to prefer the version they had been exposed to but not the other version, even though the sweetened version was initially preferred to the salty version (Sullivan and Birch, 1990). The acquired preference for tofu did not generalise to another similar food, ricotta, even when it was offered with the same flavour (salty or sweetened). The specific combination (flavour + taste + texture) was learned as a whole and any variant should be learned *de novo*.

Even if the size of the food repertoire increases after the 'neophobic period', the number of foods liked by a child at the age of 3 and 7 years are correlated (Skinner *et al.*, 2002b). Fruit intake at school age is related to fruit intake during the first two years of life, however the same result was not shown for vegetables (Skinner *et al.*, 2002a). By initiating observations at the age of 2 years and by following up subjects until early adult life, we showed that, for most food categories (except for starchy foods), liking in childhood, adolescence and early adult life is related to food choices and consumption between the age of 2 and 3 years (Nicklaus *et al.*, 2004). Food variety for all food categories in childhood, adolescence and early adult life is also related to food variety in early childhood (Nicklaus *et al.*, 2005a). Also, not surprisingly, children eat more fruit and vegetables when more fruits and vegetables are available, at school or at home (Hearn *et al.*, 1998, in Hill, 2002).

Foods have more chances to be liked before the appearance of the period of neophobia than during this period, so offering children a variety of foods should be encouraged from a very young age.

15.4.5   The preadolescence

Food behaviour acquired at the beginning of the preadolescence period is likely to track within this period. For instance, children who consume few fruit and vegetables at the age of 9 years are also the ones who consume few fruit and vegetables at the age of 11 years (Resnicow *et al.*, 1998). Nevertheless pre-adolescence and adolescence are periods where changes in food preferences are likely to be observed (Nicklaus *et al.*, 2004; Ton Nu *et al.*, 1996). In a survey of 222 French adolescents of 10 to 20 years old (Ton Nu *et al.*, 1996), one of the reported reasons for tasting foods that were previously disliked was social influence and more precisely initiation by an adult within the family, or outside the family for the oldest ones. The other reason was a personal decision to taste foods they previously disliked such as vegetables, condiments, coffee and tea.

To conclude this section, one might wonder if there are certain periods during which food exposure is more likely to impact strongly and for a longer period of time. In early infancy, there seems to be such a favourable window: introduction of hydrolysate formula milk before the age of 4 months is more successful than after this age (Mennella and Beauchamp, 1996; Mennella *et al.*, 2004) and has an impact on later preferences (Mennella and Beauchamp, 2002). However, food liking, preferences and choices remain flexible later in life, even in adulthood (Stein *et al.*, 2003). Nevertheless, we proposed that food preferences and food habits acquired during the first four years of life are likely to be deeply imprinted and to influence food choices later on (Nicklaus *et al.* 2004, 2005a).

15.5   Parental and other social influences on food choice

Children generally consume foods in social circumstances which are likely to influence their acceptability. Foods also carry a cultural and even sometimes an individual identity 'I am what I eat'. The child, then the teenager, will progressively develop a higher sensitivity to cognitive aspects and will develop his own representation of the social and individual value of a food, which might largely influence its acceptance and its consumption.

15.5.1   Parental influence

Social factors might act according to three pathways (Zellner, 1991). The family or cultural group might give access to certain foods, favouring the development of a preference for those foods. The preference might also result from the imitation of another's person food consumption (modelling effect described in Section 15.3). Finally, through different conditioning mechanisms described in

Section 15.3, the preference might increase or decrease depending on the social context in which the food is presented.

Very early, parents have a strong influence on their children's food preferences. For instance, as was developed above, prenatal mother's food choices are likely to impact on their child's further preference (Mennella *et al.*, 2001). Then, the decision to breastfeed will modify the flavour world of the infant (Mennella, 1995) and in addition it is very often associated with different maternal attitudes towards child's feeding, which are in turn likely to impact on the child's later food behaviour (Fisher *et al.*, 2000; Taveras *et al.*, 2004). Moreover, the prevalence of breastfeeding differs across socio-cultural contexts (Gojard, 1998, 2000; Dubois and Girard, 2003; Celi *et al.*, 2005; Forste *et al.*, 2001). So as early as the first months of life, the child is bathed in a different sensory, attitudinal and socio-economic context if he is breast or bottle fed. Beyond early infancy, parents have an implicit influence by selecting the foods they will offer their child, which correspond to their cultural, regional and family standards (Lewin, 1943, in Koivisto Hursti, 1999; Fischler, 1990).

Moreover, parents are probably the most important model for a young child, which will also impact on preference. Between 14 and 48 months, a child is more likely to taste a novel food if he sees an adult eating it, and more so if the adult is one of his parents rather than a stranger (Harper and Sanders, 1975). The modelling role of parents could be more efficient for energy-dense foods (Jansen and Tenney, 2001). This greater efficiency could be linked to the co-occurrence of social modelling and conditioning with post-ingestive consequences.

We have described extensively in Section 15.3 how several types of social reinforcement are spontaneously used by parents in an attempt to influence their child's food choices: they might offer a reward in exchange for the consumption of a food, offer a food as a reward in exchange for certain behaviour, restrict the access to a much liked food or force eating of a disliked food. It is interesting to note that parents generally use such contingencies in a different way according to the food: sweets and cakes are instrumentalised in a positive context (used as reward or for celebrations or birthdays), whereas vegetables are used in a negative context to give access to a pleasant activity. Restrictions are more often applied to preferred foods whereas disliked foods are the object of episodes of forced feeding. Overall, parents' intervention generally produces effects which are contrary to intentions and often reinforces pre-established affective orienta-tions. The most efficient strategy to enhance preference seems to repeatedly provide children with a variety of foods, trying to decontextualise their con-sumption from any external contingency.

Despite a rich web of parental influences on children's food preference and choices, and despite the fact that food intake in children is linked to that of parents (Fisher *et al.*, 2002; Wardle *et al.*, 2005), neither preference (Borah-Giddens and Falciglia, 1993) nor neophobia (Koivisto Hursti and Sjöden, 1997) are correlated between parents and children. This paradox of family (Rozin, 1991) underlines the idiosyncrasy of food preferences and choices, probably as the result of different physiological equipment (at sensory, digestive and cognitive levels) and might result from counterproductive strategies to influence children's food preference. Different parenting styles (e.g., authoritarian, authoritative, indulgent and uninvolved, Hughes *et al.*, 2005) are indeed associated with difference in children's intake in particular of fruit and vegetables (Fisher *et al.*, 2002; Wardle *et al.*, 2005). This highlights the importance of characterising the type of parenting attitude to better understand the development of food preferences in children. Moreover, parents, in particular, mothers, who are traditionally more frequently in charge of feeding, might exert control on their child's eating habits in a different way for each gender. Daughters appear to be subject to more control than sons (Tiggemann and Lowes, 2002). This fact led some researchers to specifically study the mothers' influence on girls' feeding (Carper *et al.*, 2000). At teenage, when the sexual identity and concerns about body image are reinforced, the role of parents in feeding might evolve differently according to gender, and the pressure on teenage girls might increase.

15.5.2   Influence of peers

The influence of parents on their offspring's food preference and choices might be more important during childhood than during teenage (Hill, 2002). As children grow into teenage, preponderant social influences might shift from parents to peers (Shepherd and Dennison, 1996; Worsley and Skrzypiec, 1998; Rolls, 1988). However, even very early, a child's food preferences are likely to be modelled by other children's preferences, by simple imitation or to conform socially as described in Section 15.3.4.

15.6   Extrinsic influences on food choice

15.6.1   Psychological influences

Acquisition of the symbolic value of food

Each culture develops rules about what is edible and what is not, depending on local resources. Sociocultural rules also define what time is appropriate for eating a given food: a child implicitly learns them from his daily experience and such rules seem to be acquired at 3–5-years-old (Birch *et al.*, 1984a). In infants aged 16 to 29 months, a variety of objects appear to be edible but during the third year the number of edible objects decreases (Rozin *et al.*, 1986). Different reasons might explain this avoidance: distaste, based on sensory cues; danger, based on anticipated consequences of the food's consumption; disgust, based on the symbolic and psychological value of the food and appropriateness (Fallon and Rozin, 1983; Fallon *et al.*, 1984). These categories for rejection do not all appear at the same age: distaste might show up very early in life but danger and disgust might not appear before Piaget's pre-operational developmental stage, around the age of 6 to 7 years, when the child is actually able to develop more elaborate reasoning (Contento, 1981). When growing older, children might better understand the symbolic value of a food (Fallon *et al.*, 1984), which they might in turn use to establish their own identity by selecting to eat some foods they judge appropriate for themselves. For instance, teenagers might refuse their mothers' food and prefer to eat junk food not only because they think it tastes better, but to demonstrate their autonomy in food choices (Hill, 2002). In adolescence, increased consciousness of ideational values might explain the rejection of meat products (especially red meat) in girls (Santos and Booth, 1996; Kubberød *et al.*, 2002; Martins *et al.*, 1999).

Attitude toward dieting
Concern with body shape, which eventually leads to dieting, alters food choices, especially in girls at puberty (Contento *et al.*, 1995). Recent studies demonstrated that such a concern appears earlier and earlier in girls: it is sometimes expressed at the age of 6 years (Tiggemann and Lowes, 2002), and certainly acts at the age of 9 years (Shunk and Birch, 2004).

15.6.2   Role of external information on children's preference and choice
As introduced earlier in this chapter, cognitive abilities evolve greatly during childhood and they have an important influence on how children understand external information, whether it is objective or derived from advertisement (Roedder John, 1999).

Role of information about the food
The role of information might increase with developing cognitive abilities. For the young children, sensory qualities (which are implicitly associated to post-ingestive effects) are probably the most influential characteristics. Even in 9 year olds, dislikes have more influence on fruit and vegetable consumption than expected benefits (Resnicow *et al.*, 1997). Since children's abstraction abilities are limited, they might interpret an expected benefit claim such as, 'It's good for you' (i.e., good for your health) as 'It tastes good' (Contento, 1981). However the claim, 'it's good for you' is very often used in an attempt to increase the acceptance of disliked foods, so a child might quickly learn to find such a claim suspicious. A child aged 9–11 years exposed to an initially neutral food will like it less if it is repeatedly associated with claims concerning its health benefits (Wardle and Huon, 2000). At the age of 2 to 6 years, exposing a child to a food leads to an increase in preference, but exposing parents to nutritional information about the food, which they are supposed to communicate to the child, has a very slight impact on preference (Wardle *et al.*, 2003a).

   The effect of information is likely to be more important in preadolescents, at least from a declarative point of view but the receptivity to information might differ according to each individual's health concern (Engell *et al.*, 1998). The sensory properties and acceptability of standard and reduced-fat cookies are judged the same way with or without information, but when the information is provided the reduced-fat cookies are preferred, especially by the highly-concerned preadolescents, see Fig. 15.6 (Engell *et al.*, 1998).

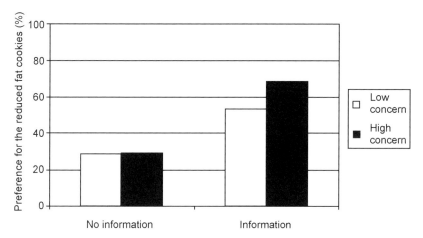

Fig. 15.6 Effect of information about fat content on preference for reduced fat cookies (versus a standard version) in preadolescents (after Engell *et al.*, 1998).

Information is likely to alter preference but also willingness to try a novel food. Claiming that a novel food 'tastes good' increases the number of children willing to try it, especially if they are 7 rather than 4 years old (Pelchat and Pliner, 1995). Despite this apparent resistance of children to information, different campaigns tried to modify food habits by providing information about the benefits or about the risks of consuming certain foods (Gribble *et al.*, 2003; Räsänen *et al.*, 2003). The objective is often to decrease the consumption of fatty foods and to promote that of fruits and vegetables (Nicklas *et al.*, 2001). The effect of such intervention on preferences and behaviour is not systematically assessed; when it is, results are generally disappointing (Nicklas *et al.*, 2001; Basdevant *et al.*, 1999). This is not completely surprising: nutritional information reinforces nutritional knowledge but might not change behaviour or consumption (Shepherd and Dennison, 1996).

Role of publicity and advertising
Some authors consider that the role of advertising about food on children's preferences is not proved (Young, 1997, in Hill, 2002). If this was true, one could wonder about the time devoted to children-targeted foods on TV advertisements!

When children are in the perceptual stage (from 3 to 7 years), they develop the ability to distinguish ads from other TV programmes but they think that information provided by ads is objective; they globally have positive attitudes towards ads (Roedder John, 1999). When growing older and reaching the analytical stage (7–11 years), they understand that ads contained biased information and are intended to influence their choice, but they are still unable to use cognitive defence against ads. Only in the reflective stage (11 to 16 years) do they develop sceptical attitudes against ads.

Advertisements might influence a child's short-term choices, by enhancing certain of the food's features such as the pleasure derived from its consumption (Kapferer, 1985). Such short-term choices are more likely to be modified at the age of 8 than 13 years: at 13, beliefs are more stabilised and less likely to be influenced by external factors. One concern about advertising is that it generally promotes industrial, 'unhealthy' foods. So it creates a hiatus in children between beliefs derived from TV ('eat (unhealthy) foods such as chocolate, cereals, cakes, sodas') and information given by parents or institutions ('you should eat fruits and vegetables'): this discrepancy is likely to alter the child's value system.

Children's exposure to advertising might be viewed as a stage in the consumer socialisation process, as proposed by Roedder John (1999). It is often the first step into the world of consumption, exposing them to products and brand names. Other steps in consumer socialisation include transaction knowledge, shopping knowledge and skills, decision-making abilities and acquisition of consumption motives and skills. They all evolve according to cognitive ability stages of the child (Roedder John, 1999).

15.7   Understanding children and food choice for food product development

As underlined several times in this chapter, children perceive food differently than adults because physiologically and psychologically their development is not achieved. Their choices are generally bound by the range of foods made available by their parents. They tend to be highly influenced by sensory and affective determinants especially when they are young. Their food preferences might be very strongly driven by experience, and even very early experience might play a role. This clearly suggests that the process of developing food targeted at children should incorporate children in the testing phase, to best take into account their specific features.

15.8   Sources of further information and advice

Research on food choice and more likely on food preference and food liking in children might be reported in journals such as *Appetite, Journal of the American Dietetic Association, Journal of Nutrition Education and Behaviour, Child Development, Pediatrics, Chemical Senses, Food Quality and Preference*. Reports and recommendations issued by organisations such as the American Academy of Pediatrics might give interesting insights on this topic.

15.9   References

ANLIKER, J. A., BARTOSHUK, L., FERRIS, A. M. and HOOKS, L. D. (1991), Children's food preferences and genetic sensitivity to the bitter taste of 6-*n*-propylthiouracil (PROP), *Am. J. Clin. Nutr.*, 54, 316–20.

BASDEVANT, A., BOUTE, D. and BORYS, J. M. (1999), Who should be educated? Education strategies: could children educate their parents?, *Int. J. Obes. Relat. Metab. Disord.*, 23, S10–2.

BATSELL, J., W.R. , BROWN, A. S., ANSFIELD, M. E. and PASCHALL, G. Y. (2002), 'You will eat all of that!': A retrospective analysis of forced consumption episodes, *Appetite*, 38, 211–19.

BAXTER, I. A., SCHÖDER, M. J. A. and BOWER, J. A. (2000), Children's perceptions of and preferences for vegetables in the west of Scotland: the role of demographic factors, *J. Sens. Stud.*, 15, 361–81.

BEAUCHAMP, G. K. and MORAN, M. (1982), Dietary experience and sweet taste preferences in human infants, *Appetite*, 3, 139–52.

BEAUCHAMP, G. K. and MORAN, M. (1984), Acceptance of sweet and salty tastes in 2-year-old children, *Appetite*, 5, 291–305.

BIRCH, L. L. (1979), Dimensions of preschool children's food preferences, *J. Nutr. Educ.*, 11, 77–80.

BIRCH, L. L. (1980), Effects of peer model's food choices and eating behaviors on preschoolers' food preferences, *Child Dev.*, 51, 489–96.

BIRCH, L. L. (1981), Generalization of a modified food preference, *Child Dev.*, 52, 755–8.

BIRCH, L. L. (1990), The control of food intake by young children: the role of learning, In Capaldi, E. D. and Powley, T. L., *Taste, Experience & Feeding*, American Psychological Association, Washington, USA, pp. 116–35.

BIRCH, L. L. (1992), Children's preferences for high-fat foods, *Nutr. Rev.*, 50, 249–55.

BIRCH, L. L. and MARLIN, D. W. (1982), I don't like it; I never tried it: effects of exposure on two-year-old children's food preferences, *Appetite*, 3, 353–60.

BIRCH, L. L. and DEYSHER, M. (1985), Conditioned and unconditioned caloric compensation: evidence for self regulation of food intake in young children, *Learn. Motiv.*, 16, 341–55.

BIRCH, L. L., ZIMMERMAN, S. I. and HIND, H. (1980), The influence of social-affective context on the formation of children's food preferences, *Child Dev.*, 51, 856–61.

BIRCH, L. L., BIRCH, D., MARLIN, D. W. and KRAMER, L. (1982), Effects of instrumental consumption on children's food preference, *Appetite*, 3, 125–34.

BIRCH, L. L., BILLMAN, J. and RICHARDS, S. S. (1984a), Time of day influences food acceptability, *Appetite*, 5, 109–16.

BIRCH, L. L., MARLIN, D. W. and ROTTER, J. (1984b), Eating as the 'Means' activity in a contingency: Effects on young children's food preference, *Child Dev.*, 55, 431–9.

BIRCH, L. L., McPHEE, L., SHOBA, B. C., PIROK, E. and STEINBERG, L. (1987a), What kind of exposure reduces children's food neophobia? Looking vs. Tasting, *Appetite*, 9, 171–8.

BIRCH, L. L., McPHEE, L., SHOBA, B. C., STEINBERG, L. and KREHBIEL, R. (1987b), 'Clean up your plate': effects of child feeding practices on the conditioning of meal size, *Learn. Motiv.*, 18, 301–7.

BIRCH, L. L., McPHEE, L., STEINBERG, L. and SULLIVAN, S. (1990), Conditioned flavor preferences in young children, *Physiol. & Behav.*, 47, 501–5.

BIRCH, L. L., GUNDER, L., GRIMM-THOMAS, K. and LAING, D. G. (1998), Infants' consumption of a new food enhance acceptance of similar foods, *Appetite*, 30, 283–95.

BOOTH, D. A., MATHER, P. and FULLER, J. (1982), Starch content of ordinary foods associatively conditions human appetite and satiation, indexed by intake and eating pleasantness of starch-paired flavours, *Appetite*, 3, 163–84.

BORAH-GIDDENS, J. and FALCIGLIA, G. A. (1993), A meta-analysis of the relationship in food preference between parents and children, *J. Nutr. Educ.*, 25, 102–7.

BUFE, B., BRESLIN, P. A. S., KUHN, C., REED, D. R., THARP, C. D., SLACK, J. P., KIM, U. K., DRAYNA, D. and MEYERHOF, W. (2005), The molecular basis of individual differences in phenyl-thiocarbamide and propylthiouracil bitterness perception, *Curr. Biol.*, 15, 322–7.

CARPER, J. L., ORLET FISCHER, J. and BIRCH, L. L. (2000), Young girls' emerging dietary restraint and disinhibition are related to parental control in child feeding, *Appetite*, 35, 121–9.

CARRUTH, B. R. and SKINNER, J. D. (2000), Revisiting the picky eater phenomenon: neophobic behaviors of young children, *J. Am. Coll. Nutr*, 19, 771–80.

CARRUTH, B. R. and SKINNER, J. D. (2002), Feeding behaviors and other motor development in healthy children (2–24 months), *J. Am. Coll. Nutr*, 21, 88–96.

CARRUTH, B. R., SKINNER, J. D., HOUCK, K., MORAN III, J., COLETTA, F. and OTT, D. (1998), The phenomenon of 'Picky Eater': a behavioral marker in eating patterns of toddlers, *J. Am. Coll. Nutr*, 17, 180–6.

CARRUTH, B. R., ZIEGLER, P. J., GORDON, A. and BARR, S. I. (2004), Prevalence of picky eaters among infants and toddlers and their caregivers' decisions about offering a new food, *J. Am. Diet. Assoc.*, 104, S57–S64.

CASHDAN, E. (1994), A sensitive period for learning about food, *Hum. Nat.*, 5, 279–91.

CELI, A. C., RICH-EDWARDS, J. W., RICHARDSON, M. K., KLEINMAN, K. P. and GILLMAN, M. W. (2005), Immigration, race/ethnicity, and social and economic factors as predictors of breastfeeding initiation, *Arch. Pediatr. Adolesc. Med.*, 159, 255–60.

CERNOCH, J.-N. and PORTER, R. H. (1985), Recognition of maternal axillary odors by infants, *Child Dev.*, 56, 1593–8.

CONTENTO, I. (1981), Children's thinking about food and eating – A Piagetian-based study, *J. Nutr. Educ.*, 13, S86–S90.

CONTENTO, I. R., MICHELA, J. L. and WILLIAMS, S. S. (1995), Adolescent food choice criteria: role of weight and dieting status, *Appetite*, 25, 51–76.

COOKE, L., WARDLE, J. and GIBSON, E. L. (2003), Relationship between parental report of food neophobia and everyday food consumption in 2–6-year-old children, *Appetite*, 41, 205–6.

COX, D. R., SKINNER, J. D., CARRUTH, B. R., MORAN III, J. and HOUCK, K. S. (1997), A food variety index for toddlers (VIT): development and application, *J. Am. Diet. Assoc.*, 97, 1382–6.

CRYSTAL, S. R. and BERSTEIN, I. L. (1995), Morning sickness: impact of offspring salt preference, *Appetite*, 25, 231–40.

CRYSTAL, S. R. and BERSTEIN, I. L. (1998), Infant salt preference and mother's morning sickness, *Appetite*, 30, 297–307.

DANIELZIK, S., CZERWINSKI-MAST, M., LANGNASE, K., DILBA, B. and MULLER, M. J. (2004), Parental overweight, socioeconomic status and high birth weight are the major determinants of overweight and obesity in 5–7-y-old children: baseline data of the Kiel Obesity Prevention Study (KOPS), *Int. J. Obes.*, 28, 1494–502.

DAVIS, C. M. (1928), Self selection of diet by newly weaned infants. An experimental study, *Am. J. Dis. Child.*, 36, 651–79.

DAVIS, C. M. (1939), Results of the self-selection of diets by young children, *Can. Med. Assoc. J.*, Sept., 257–61.

DE GRAAF, C. and ZANDSTRA, E. H. (1999), Sweetness intensity and pleasantness in children, adolescents, and adults, *Physiol. Behav.*, 67, 513–20.

DESHMUKH-TASKAR, P., NICKLAS, T., MORALES, M., YANG, S. J., ZAKERI, I. and BERENSON, G. S. (2005), Tracking of overweight status from childhood to young adulthood: the Bogalusa Heart Study, *Eur. J. Clin. Nutr.*, doi:10.1038/sj.ejcn.1602267.

DREWNOWSKI, A. (1989), Sensory preferences for fat and sugar in adolescence and adult life, In Murphy, C., Cain, W. S. and Hegsted, D. M., *Nutrition and the chemical senses in aging: Recent advances and current research needs*, The New York Academy of Sciences, New York, pp. 243–50.

DREWNOWSKI, A. (1997), Taste preference and food intake, *Annu. Rev. Nutr.*, 17, 237–53.

DREWNOWSKI, A. (1998), Energy density, palatability, and satiety: implications for weight control, *Nutr. Rev.*, 56, 347–53.

DREWNOWSKI, A. (2000), Sensory control of energy density at different life stages, *Proc. Nutr. Soc.*, 59, 239–44.

DUBOIS, L. and GIRARD, M. (2003), Social determinants of initiation, duration and exclusivity of breastfeeding at the population level: the results of the Longitudinal Study of Child Development in Quebec (ELDEQ 1998–2002). *Can. J. Public Health*, 94, 300–5.

ENGELL, D., BORDI, P., BORJA, M., LAMBERT, C. and ROLLS, B. (1998), Effects of information about fat content on food preferences in pre-adolescent children., *Appetite*, 30, 269–82.

ENGEN, T. (1986), Children's sense of smell, In Meiselman, M. L. and Rivlin, R. S., *Clinical measurement of taste and smell*, Macmillan, New York, pp. 316–25.

FALCIGLIA, G. A., COUCH, S. C., GRIBBLE, L. S., PABSTA, S. M. and FRANK, R. (2000), Food neophobia in childhood affects dietary variety, *J. Am. Diet. Assoc.*, 100, 1474–81.

FALLON, A. E. and ROZIN, P. (1983), The psychological bases of food rejections by humans, *Ecol. Food Nutr.*, 13, 15–26.

FALLON, A. E., ROZIN, P. and PLINER, P. (1984), The child's conception of food: the development of food rejections with special reference to disgust and contamination sensitivity, *Child Dev.*, 55, 566–75.

FISCHER, R., GRIFFIN, F., ENGLAND, S. and GARN, S. M. (1961), Taste thresholds and food dislikes, *Nature*, 191, 1328.

FISCHLER, C. (1990), *L'Homnivore*, Odile Jacob, Paris.

FISHER, J. O. and BIRCH, L. L. (1999a), Restricting access to foods and children's eating, *Appetite*, 32, 405–19.

FISHER, J. O. and BIRCH, L. L. (1999b), Restricting access to palatable foods affects children's behavioral response, food selection, and intake, *Am. J. Clin. Nutr.*, 69, 1264–72.

FISHER, J. O., BIRCH, L. L., SMICIKLAS-WRIGHT, H. and PICCIANO, M. F. (2000), Breast-feeding through the first year predicts maternal control in feeding and subsequent toddler energy intakes, *J. Am. Diet. Assoc.*, 100, 641–6.

FISHER, J. O., MITCHELL, D. C., SMICIKLAS-WRIGHT, H. and BIRCH, L. L. (2002), Parental influences on young girls' fruit and vegetable, micronutrient, and fat intakes, *J. Am. Diet. Assoc.*, 102, 58–64.

FORSTE, R., WEISS, J. and LIPPINCOTT, E. (2001), The decision to breastfeed in the United States: does race matter?, *Pediatrics*, 108, 291–6.

GALLOWAY, A. T., LEE, Y. and BIRCH, L. L. (2003), Predictors and consequences of food neophobia and pickiness in young girls, *J. Am. Diet. Assoc.*, 103, 692–8.

GALLOWAY, A. T., FIORITO, L., LEE, Y. and BIRCH, L. L. (2005), Parental pressure, dietary patterns, and weight status among girls who are 'picky eaters', *J. Am. Diet. Assoc.*, 105, 541–8.

GANCHROW, J. R. and MENNELLA, J. A. (2003), The ontogeny of human flavor perception, In Doty, R. L., *Handbook of olfaction and gustation*, Dekker, M., New York.

GERRISH, C. J. and MENNELLA, J. A. (2001), Flavor variety enhances food acceptance in formula-fed infants, *Am. J. Clin. Nutr.*, 73, 1080–5.

GIBSON, E. L. and WARDLE, J. (2003), Energy density predicts preferences for fruit and vegetables in 4-year-old children, *Appetite*, 41, 97–8.

GOJARD, S. (1998), L'allaitement: une pratique socialement différenciée, *Rech. Prévis.*, 53, 23–34.

GOJARD, S. (2000), L'alimentation dans la prime de l'enfance. Diffusion et réception des normes de puériculture, *Rev. Fr. Sociol.*, 41, 475–512.

GRIBBLE, L. S., FALCIGLIA, G., DAVIS, A. M. and COUCH, S. C. (2003), A curriculum based on social learning theory emphasizing fruit exposure and positive parent child-feeding strategies: A pilot study, *J. Am. Diet. Assoc.*, 103, 100–3.

HALLER, R., RUMMEL, C., HENNEBERG, S., POLLMER, U. and KÖSTER, E. P. (1999), The influence of early experience with vanillin on food preference later in life, *Chem. Senses*, 24, 465–7.

HANSE, L. (1994), La néophobie alimentaire chez l'enfant, Thèse de Doctorat, Université Paris X – Nanterre.

HARPER, L. V. and SANDERS, K. M. (1975), The effect of adults' eating on young children's acceptance of unfamiliar foods, *J. Exp. Child Psychol.*, 20, 206–14.

HENDY, H. M. (2002), Effectiveness of trained peer models to encourage food acceptance in preschool children, *Appetite*, 39, 217–25.

HENDY, H. M. and RAUDENBUSH, B. (2000), Effectiveness of teacher modeling to encourage food acceptance in preschool children, *Appetite*, 34, 61–76.

HILL, A. J. (2002), Developmental issues in attitudes to food and diet, *Proc. Nutr. Soc.*, 61, 259–66.

HOBDEN, K. and PLINER, P. (1995), Effects of a model on food neophobia in humans, *Appetite*, 25, 101–14.

HUGHES, S. O., POWER, T. G., ORLET FISHER, J., MUELLER, S. and NICKLAS, T. A. (2005), Revisiting a neglected construct: parenting styles in a child-feeding context, *Appetite*, 44, 83–92.

JACOBI, C., AGRAS, W. S., BRYSON, S. and HAMMER, L. D. (2003), Behavioral validation, precursors, and concomitants of picky eating in childhood, *J. Am. Acad. Child Adolesc. Psychiatr.*, 42, 76–84.

JANSEN, A. and TENNEY, N. (2001), Seeing mum drinking a light product: is social learning a stronger determinant of taste preference acquisition than caloric conditioning?, *Eur. J. Clin. Nutr.*, 55, 418–22.

JOHNSON, S. L., McPHEE, L. and BIRCH, L. L. (1991), Conditioned preferences: young children prefer flavors associated with high dietary fat, *Physiol. Behav.*, 50, 1245–51.

KAPFERER, J.-N. (1985), Les effets de la publicité sur le comportement nutritionnel des enfants, *Cah. Nutr. Diét.*, XX, 269–73.

KELLER, K. L., STEINMANN, L., NURSE, R. J. and TEPPER, B. J. (2002), Genetic taste sensitivity to 6-n-propylthiouracil influences food preference and reported intake in preschool children, *Appetite*, 38, 3–12.

KERN, D. L., McPHEE, L., FISHER, J., JOHNSON, S. and BIRCH, L. L. (1993), The postingestive consequences of fat condition preferences for flavors associated with high dietary fat, *Physiol. & Behav.*, 54, 71–6.

KOIVISTO HURSTI, U.-K. (1999), Factors influencing children's food choice, *Ann. Med.*, 31, 26–32.

KOIVISTO HURSTI, U.-K. and SJÖDEN, P.-O. (1997), Food and general neophobia and their relationship with self-reported food choices: familial resemblance in Swedish families with children of ages 7–17 years, *Appetite*, 29, 89–103.

KÖSTER, E. P. (1991) In *Food Ingredients Asia Conference Proceedings 1991*, Expoconsult,

Maarssen, The Netherlands, pp. 223–7.

KUBBERØD, E., UELAND, Ø., RØDBOTTEN, M., WESTAD, F. and RISVIK, E. (2002), Gender specific preferences and attitudes towards meat, *Food Qual. Pref.*, 13, 285–94.

KÜHN, B. F. and THYBO, A. K. (2001), The influence of sensory and physiochemical quality on Danish children's preferences for apples, *Food Qual. Pref.*, 12, 543–50.

LIEM, D. G. and MENNELLA, J. A. (2002), Sweet and sour preferences during childhood: role of early experiences, *Dev. Psychobiol.*, 41, 388–95.

LIEM, D. G. and MENNELLA, J. A. (2003), Heightened sour preferences during childhood, *Chem. Senses*, 28, 173–80.

LOBSTEIN, T. and FRELUT, M.-L. (2003), Prevalence of overweight among children in Europe, *Obes. Rev.*, 4, 195–200.

LOEWEN, R. and PLINER, P. (1999), Effects of prior exposure to palatable and unpalatable novel foods on children's willingness to taste other novel foods, *Appetite*, 32, 351–66.

MARLIER, L. and SCHAAL, B. (2005), Human newborns prefer human milk: Conspecific milk odor is attractive without postnatal exposure, *Child Dev.*, 76, 155–68.

MARTINS, Y., PLINER, P. and O'CONNOR, R. (1999), Restrained eating among vegetarians: does a vegetarian eating style mask concerns about weight, *Appetite*, 32, 145–54.

MELCER, T. and MURPHY, C. (1997), The human infant's ingestive and hedonic responses during spoon feeding following repeated exposure to novel vanilla flavoring in bananas, *Chem. Senses*, 22, 745.

MENNELLA, J. A. (1995), Mother's milk: a medium for early flavor experiences, *J. Hum. Lact.*, 11, 39–45.

MENNELLA, J. A. and BEAUCHAMP, G. K. (1991), Maternal diet alters the sensory qualities of human milk and the nursling's behavior, *Pediatrics*, 88, 737–44.

MENNELLA, J. A. and BEAUCHAMP, G. K. (1996), Developmental changes in the acceptance of protein hydrolysate formula, *J. Dev. Behav. Pediatr.*, 17, 386–91.

MENNELLA, J. A. and BEAUCHAMP, G. K. (2002), Flavor experiences during formula feeding are related to preferences during childhood, *Early Hum. Dev.*, 68, 71–82.

MENNELLA, J. A., JOHNSON, A. and BEAUCHAMP, G. K. (1995), Garlic ingestion by pregnant women alters the odor of amniotic fluid, *Chem. Senses*, 20, 207–9.

MENNELLA, J. A., JAGNOW, C. P. and BEAUCHAMP, G. K. (2001), Prenatal and postnatal flavor learning by human infants, *Pediatrics*, 107, 1–6.

MENNELLA, J. A., GRIFFIN, C. E. and BEAUCHAMP, G. K. (2004), Flavor programming during infancy, *Pediatrics*, 113, 840–5.

MENNELLA, J. A., PEPINO, M. Y. and REED, D. R. (2005), Genetic and environmental determinants of bitter perception and sweet preferences, *Pediatrics*, 115, e216–22.

NICKLAS, T. A., BARANOWSKI, T., BARANOWSKI, J. C., CULLEN, K., RITTENBERRY, L. and OLVERA, N. (2001), Family and child-care provider influences on preschool children's fruit, juice, and vegetable consumption, *Nutr. Rev.*, 59, 224–35.

NICKLAUS, S., BOGGIO, V., CHABANET, C. and ISSANCHOU, S. (2004), A prospective study of food preferences in childhood, *Food Qual. Pref.*, 15, 805–18.

NICKLAUS, S., BOGGIO, V., CHABANET, C. and ISSANCHOU, S. (2005a), A prospective study of food variety seeking in childhood, adolescence and early adult life, *Appetite*, 44, 289–97.

NICKLAUS, S., BOGGIO, V. and ISSANCHOU, S. (2005b), Food choices at lunch during the third year of life: high selection of animal and starchy foods but avoidance of vegetables, *Acta Pædiatr.*, 94, 943–51.

NICKLAUS, S., CHABANET, C., BOGGIO, V. and ISSANCHOU, S. (2005c), Food choice at lunch during the third year of life: increase in energy intake but decrease in variety, *Acta Pædiatr.*, 94, 1023–9.

NORTHSTONE, K., EMMETT, P. and NETHERSOLE, F. (2001), The effect of age of introduction to lumpy solids on foods eaten and reported feeding difficulties at 6 and 15 months, *J. Hum. Nutr. Diet.*, 14, 43–54.

OGDEN, C. L., FRYAR, C. D., CARROLL, M. D. and FLEGAL, K. M. (2004), Mean body weight, height, and body mass index, United States 1960–2002, *Adv. Data*, 347, 1–17.

OLSON, C. M. and GEMMILL, K. P. (1981), Association of sweet preference and food selection among four to five year old children, *Ecol. Food Nutr.*, 11, 145–50.

PIAGET, J. and INHELDER, B. (2003), *La psychologie de l'enfant*, PUF, Paris.

PELCHAT, M. L. and PLINER, P. (1986), Antecedents and correlates of feeding problems in young children, *J. Nutr. Educ.*, 18, 23–9.

PELCHAT, M. L. and PLINER, P. (1995), 'Try it. You'll like it.' Effects of information on willingness to try novel foods, *Appetite*, 24, 153–66.

PLINER, P. (1994), Development of measures of food neophobia in children, *Appetite*, 23, 147–63.

PLINER, P. and HOBDEN, K. (1992), Development of a scale to measure the trait of food neophobia in humans, *Appetite*, 19, 105–20.

PLINER, P., PELCHAT, M. and GRABSKI, M. (1993), Reduction of neophobia in humans by exposure to novel foods, *Appetite*, 20, 111–23.

RÄSÄNEN, M., NIINIKOSKI, H., KESKINEN, S., HELENIUS, H., TALVIA, S., RONNEMAA, T., VIIKARI, J. and SIMELL, O. (2003), Parental nutrition knowledge and nutrient intake in an atherosclerosis prevention project: the impact of child-targeted nutrition counselling, *Appetite*, 41, 69–77.

RESNICOW, K., DAVIS-HEARN, M., SMITH, M., BARANOWSKI, T., LIN, L. S., BARANOWSKI, J., DOYLE, C. and WANG, D. T. (1997), Social-cognitive predictors of fruit and vegetable intake in children, *Health Psychol.*, 16, 272–6.

RESNICOW, K., SMITH, M., BARANOWSKI, T., BARANOWSKI, J., VAUGHAN, R. and DAVIS, M. (1998), 2-year tracking of children's fruit and vegetable intake, *J. Am. Diet. Assoc.*, 98, 785–9.

ROEDDER JOHN, D. (1999), Consumer socialization of children: A retrospective look at twenty-five years of research, *J. Consumer Res.*, 26, 183–213.

ROLLS, B. J. (1988), Food beliefs and food choices in adolescents, *Med. J. Aust.*, 148, 59–73.

ROZIN, P. (1976), The selection of foods by rats, humans and other animals, In Rosenblat, J. S., Hinde, R. A., Shaw, E. and Beer, C., *Advances in the study of behaviour*, Academic Press, New York, pp. 21–71.

ROZIN, P. (1979), Preference and affect in food selection, In Kroeze, J. H. A., *Preference Behaviour and Chemoreception*, IRL, London, pp. 289–302.

ROZIN, P. (1990), Acquisition of stable food preferences, *Nutr. Rev.*, 48, 106–13.

ROZIN, P. (1991), Family resemblance in food and other domains: The family paradox and the role of parental congruence, *Appetite*, 16, 93–102.

ROZIN, P. and SCHILLER, D. (1980), The nature and acquisition of a preference for Chili pepper by humans, *Motiv. Emot.*, 4, 77–101.

ROZIN, P., HAMMER, L., OSTER, H., HOROWITZ, T. and MARMORA, V. (1986), The child's conception of food: differentiation of categories of rejected substances in the 16 months to 5 year age range, *Appetite*, 7, 141–51.

SANTOS, L. and BOOTH, D. (1996), Influences on meat avoidance among British students, *Appetite*, 27, 197–205.

SCHAAL, B. (1988), Olfaction in infants and children: developmental and functional perspectives, *Chem. Senses*, 13, 145–90.

SCHAAL, B., MARLIER, L. and SOUSSIGNAN, R. (2000), Human foetuses learn odours from their pregnant mother's diet, *Chem. Senses*, 25, 729–37.

SHEPHERD, R. and DENNISON, C. M. (1996), Influences an adolescent food choice, *Proc. Nutr. Soc.*, 55, 345–57.

SHUNK, J. A. and BIRCH, L. L. (2004), Validity of dietary restraint among 5- to 9-year-old girls, *Appetite*, 42, 241–7.

SKINNER, J. D., CARRUTH, B. R., HOUCK, K., MORAN III, J., COLETTA, F., COTTER, R., OTT, D. and McLEOD, M. (1997a), Transitions in infant feeding during the first year of life, *J. Am. Coll. Nutr*, 16, 209–15.

SKINNER, J. D., CARRUTH, B. R., HOUCK, K. S., COLETTA, F., COTTER, R., OTT, D. and McLEOD, M. (1997b), Longitudinal study of nutrient and food intakes of infants aged 2 to 24 months, *J. Am. Diet. Assoc.*, 97, 496–504.

SKINNER, J. D., CARRUTH, B. R., HOUCK, K. S., BOUNDS, W., MORRIS, M., COX, D. R., MORAN III, J. and COLETTA, F. (1999), Longitudinal study of nutrient and food intakes of white preschool children aged 24 to 60 months, *J. Am. Diet. Assoc.*, 99, 1514–21.

SKINNER, J. D., CARRUTH, B. R., BOUNDS, W., ZIEGLER, P. and REIDY, K. (2002a), Do food-related experiences in the first 2 years of life predict dietary variety in school-aged children?, *J. Nutr. Educ. Behav.*, 34, 310–15.

SKINNER, J. D., CARRUTH, B. R., BOUNDS, W. and ZIEGLER, P. J. (2002b), Children's food preferences: A longitudinal analysis, *J. Am. Diet. Assoc.*, 102, 1638–47.

SOLBU, E. H., JELLESTAD, F. K. and STRAETKVERN, K. O. (1990), Children's sensitivity to odor of trimethylamine, *J. Chem. Ecol.*, 16, 1829–40.

SOUSSIGNAN, R., SCHAAL, B., MARLIER, L. and JIANG, T. (1997), Facial and authomatic responses to biological and artificial olfactory stimuli in human neonates: Re-examining earlt hedonic discrimination of odors. *Physiol. & Behav.*, 62, 745–758.

STEIN, L. J., NAGAI, H., NAKAGAWA, M. and BEAUCHAMP, G. K. (2003), Effects of repeated exposure and health-related information on hedonic evaluation and acceptance of a bitter beverage, *Appetite*, 40, 119–29.

STEINER, J. E. (1979), Human facial expressions in response to taste and smell stimulation, *Adv. Child Dev. Behav.*, 13, 257–95.

SULLIVAN, S. A. and BIRCH, L. L. (1990), Pass the sugar, pass the salt: Experience dictates preference, *Dev. Psychol.*, 26, 546–51.

SULLIVAN, S. A. and BIRCH, L. L. (1994), Infant dietary experience and acceptance of solid foods, *Pediatrics*, 93, 271–7.

SZCZESNIAK, A. S. (1972), Consumer awareness of and attitudes to food textures. 2: Children and teenagers, *J. Texture Stud.*, 3, 206–17.

TAVERAS, E. M., SCANLON, K. S., BIRCH, L. L., RIFAS-SHIMAN, S. L., RICH-EDWARDS, J. W. and GILLMAN, M. W. (2004), Association of breastfeeding with maternal control of infant feeding at age 1 year, *Pediatrics*, 114.

THYBO, A. K., KUHN, B. F. and MARTENS, H. (2004), Explaining Danish children's preferences for apples using instrumental, sensory and demographic/behavioural data, *Food Qual. Pref.*, 15, 53–63.

TIGGEMANN, M. and LOWES, J. (2002), Predictors of maternal control over children's eating behaviour, *Appetite*, 39, 1–7.

TON NU, C., MACLEOD, P. and BARTHÉLÉMY, J. (1996), Effects of age and gender on adolescents' food habits and preferences, *Food Qual. Pref.*, 7, 251–62.

TURNBULL, B. and MATISOO-SMITH, E. (2002), Taste sensitivity to 6-*n*-propylthiouracil predicts acceptance of bitter-tasting spinach in 3–6-y-old children, *Am. J. Clin. Nutr.*, 76, 1101–5.

WARDLE, J. and HUON, G. (2000), An experimental investigation of the influence of health information on children's taste preferences, *Health Educ. Res.*, 15, 39–44.

WARDLE, J., COOKE, L. J., GIBSON, E. L., SAPOCHNIK, M., SHEIHAM, A. and LAWSON, M. (2003a), Increasing children's acceptance of vegetables; a randomized trial of parent-led exposure, *Appetite*, 40, 155–62.

WARDLE, J., HERRERA, M. L., COOKE, L. and GIBSON, E. L. (2003b), Modifying children's food preferences: the effects of exposure and reward on acceptance of an unfamiliar vegetable, *Eur. J. Clin. Nutr.*, 57, 341–8.

WARDLE, J., CARNELL, S. and COOKE, L. (2005), Parental control over feeding and children's fruit and vegetable intake: How are they related?, *J. Am. Diet. Assoc.*, 105, 227–32.

WORSLEY, A. and SKRZYPIEC, G. (1998), Teenage vegetarianism: prevalence, social and cognitive contexts, *Appetite*, 30, 151–70.

ZAJONC, R. B. (1968), Attitudinal effects of mere exposure, *J. Pers. Soc. Psych.*, Monograph Supplement. 9, 1–27.

ZANDSTRA, E. H. and DE GRAAF, C. (1998), Sensory perception and pleasantness of orange beverages from childhood to old age, *Food Qual. & Pref.*, 9, 5–12.

ZELLNER, D. A. (1991), How foods get to be liked: some general mechanisms and some special cases, In Bolles, R. C., *The Hedonics of Taste*, Lawrence Erlbaum Associates, Hillsdale, NJ, pp. 199–217.

ZELLNER, D. A., ROZIN, P., ARON, M. and KULISH, C. (1983), Conditioned enhancement of human's liking for flavors by pairing with sweetness, *Learn. Motiv.*, 14, 338–50.

16

Understanding Asian consumers of food products

D. N. Cox, Commonwealth Scientific and Industrial Research Organisation, Australia

16.1   Introduction

This chapter will first outline the geographical scope; next, advise caution on 'units of analysis'; emphasise the importance of understanding culture (particularly values and beliefs) and provide a theoretical framework and some tools to do this. Diet-health beliefs are explored with reference to traditional Chinese foods and recent Japanese 'functional foods'. Cultural flavour princicals are briefly described, the influence of genetics explored and a detailed review of cross-cultural sensory perception and preference provided. Metric equivalence across Asian and Western cultures for theoretically based models of behaviour and response scales are explored. Some practical advice is provided on how to undertake Asian studies. Finally two brief case studies relate, firstly, the success of Australian wheat for noodle manufacture and, secondly, the future of genetically modified foods in Asia.

16.1.1   Scope

To understand the food consumption behaviour of 50% of the world's population (Schutte and Ciarlante, 1998) is a challenge that can only be touched upon in this chapter. Furthermore Asia is the fastest growing continent in terms of population and economy (Newman, 1999) adding a dynamic aspect to any understanding. For example, it has been argued that for consumers in the People's Republic of China (PRC) 'choice' is a relatively new concept (Schutte and Ciarlante, 1998). Fast growing economies provide marketing opportunities and it is no accident that much of what is known about Asian consumers focuses upon developed Asia (principally Japan, Hong Kong and Taiwan); rapidly growing economies (some ASEAN countries[1]) and the huge but only recently 'accessible' PRC. Consequently this chapter reviews what is published about consumers from these essentially Eastern and South-East Asian states.

The recent two systems political economy (powerful state and 'free market capitalism') in PRC has led to a scramble of various Western commercial interests for the huge Chinese market. However, income distribution appears to be polarising and distribution difficulties combine to create, at present, a limited and difficult market. In many respects understanding consumers is possibly helped by the Chinese diaspora (Douglas and Craig, 1997) and some of our understanding is derived from migrant Chinese (especially in the USA) however care must be exercised in extrapolating from consumers that may be acculturated to Western thinking and behaviour (Wills and Wooton, 1997). An understanding (Cox and Anderson, 2004; Suinn *et al.*, 1992) of acculturation is likely to be valuable if migrant groups of consumers are used as surrogates to test product acceptance (further explored below).

16.1.2   The importance of understanding Asian consumers

Various national government trade departments provide valuable overviews of the potential markets for foods in various Asian countries (for example, Supermarket to Asia/Austrade, 1998) and the increasing demand for basic commodities and valued added products is well known, although the suitability of value added food products for particular Asian consumers is not always well understood. Equally well documented (Supermarket to Asia/Austrade, 1998; Schutte and Ciarlante, 1998) is the recent economic crisis that was prevalent throughout East Asia in the late 1990s. Hangovers of that crisis still remain, particularly in Thailand and Indonesia; however, there is general consensus that many markets have recovered or are recovering and that the East Asian market provides a huge opportunity for the marketing of new food products.

16.1.3   Asian diversity and units of analysis

It is both inaccurate and dangerous to make generalisations about geographic, national or cultural 'units of analysis' (Douglas and Craig, 1997). Asia is highly differentiated, for example, nationally encompassing some of the poorest nations in the world (e.g., Laos) and the wealthiest (e.g., Japan) with variation in income distribution (and other socio-economic status attributes) and changing values and beliefs within countries (Shutte and Ciarlante, 1998).

---

1. The Association of Southeast Asian Nations (ASEAN) comprises Indonesia, Malaysia, Philippines, Singapore, Thailand, Brunei Darussalam, Vietnam, Laos, Myanmar and Cambodia. The ASEAN region has a population of about 500 million, a total area of 4.5 million square kilometers, a combined gross domestic product of US$737 billion, and a total trade of US$720 billion. http://www.aseansec.org/64.htm

Malaysia is, for example, multi-ethnic with Malay, Chinese and Indian communities with differing religious beliefs and values (see below). Cross-cultural studies (Bech-Larsen and Grunert, 2003) comparing Western cultures (USA, Denmark and Finland) have found nationality to be problematic when seeking to characterise cultural differences in consumers attitudes towards functional foods. Douglas and Craig (1997) discuss appropriate units of analysis and care should be taken to account for segmentation and/or define homogenous groups of consumers.

16.1.4   Food industry interests and globalisation

Multinational food companies have a strong presence in Asia (particularly ASEAN countries and India) and similarly Western supermarket chains (e.g., Carrefour) have a growing interest in the ASEAN countries (Supermarket to Asia/Austrade, 1998). Whilst globalisation is an increasingly dominant phenomenon, global products are often the result of producer 'push' for greater efficiency not necessarily based upon consumer demand. Furthermore, specific cultural values are considered most strong in respect of non-durable consumer goods such as foods (Schutte and Ciarlante, 1998).

16.1.5   Eurocentric approaches to understanding consumers' food choice

The complexity of food choice has been described elsewhere almost exclusively in terms of Western (European origin) cultures (for example, Cox and Anderson, 2004; Meisleman and MacFie, 1996; MacFie and Thomson, 1999; Frewer *et al.*, 2001; Marshall, 1995) yet remains a relatively young science with a huge need for more information. The importance of cultural variation in food choice (also described as 'foodways') is well recognised (Rozin, 1990; E. Rozin 1982, 2000; Fieldhouse, 1995; Cox and Anderson, 2004).

16.2   Principles of cross-cultural data collection

16.2.1   Social psychology

A solid foundation for an understanding of (Asian) consumers is social psychology; however, the cross-cultural psychologist Harry Triandis stated that 'Almost all the theories and data of contemporary psychology come from western populations (e.g., Europeans, North Americans, Australians, etc.)' (Triandis, 1996, p. 407). It has been noted that 50% of the world's population is Asian and yet, according to Triandis, data on Asian behaviour and attitudes are rarely included in this body of knowledge. For example, only relatively recently has psychology been explored from Asian national perspectives (for example, Harris Bond, 1996; Yoon and Choi, 1994). Without a universal psychology any attempt to understand Asian consumers must acknowledge that theories and tools may not be culturally appropriate and that each theory and tool needs to be tested and validated within a specific culture. For a detailed description of cross-cultural research methodology from a psychological and statistical perspective readers are referred to van de Vijver and Leung's (1997) *Methods and data analysis for cross-cultural research* (see references).

In practice, cultural issues of bias and equivalence are applicable to established theoretical behavioural models, for example the Theory of Reasoned Action or Planned Behaviour (Ajzen and Fishbein, 1980, Ajzen, 1985; Lee, 1990; Lee and Green, 1991) and to methodology, for example Asian consumers use of hedonic scales (Yao *et al.*, 2003; Cox *et al.*, 2001; Yeh *et al.*, 1998). These examples are further explored below.

Triandis (1994) and Douglas and Craig (1997) discuss a particularly useful starting point that can be used for undertaking cross-cultural work, namely recognising the difference between universal behaviour and culturally specific behaviour. Borrowing from linguistics (phon*etics* meaning universal sounds and phon*emics* meaning sounds unique to one language) *etic* describes universal behaviours and *emic* culturally specific. Triandis asserts that science seeks to understand generalisations, for example common elements in food choice (*etics*), however these may manifest as culturally specific *emic* elements. For example, it is thought that *etic* universal values (Schwartz, 1992) can drive consumer behaviour (Reynolds and Olson, 2001) and that such an approach has been found to be useful in the study of food choice (for example, Flight *et al.*, 2003b) however differing *emic* values may be important in different cultures (Grunert, 1997). Perhaps a classic example of *emic* is Japanese perception of the fifth basic 'taste' *umami*, the word itself lacking any satisfactory English translation but often translated as 'savoury' (O'Mahony and Ishii, 1986). In another example, Prescott (1998) reports that Koreans identify an important flavour construct pertaining to noodles as '*kusu*' (a composite flavour also lacking English translation). Importantly, Prescott suggests that whilst *kusu* may be important to one food, it may also be important to one culture and hence, using Triandis schema *emic*. Prescott indicates that qualitative work (e.g., focus groups) may be necessary to identify which particular issues or constructs are salient to a particular culture, hence avoiding '*pseudo-etic*' approaches that mistakenly assume universality (Triandis, 1994). Such an approach would certainly be more efficient than Berry's (1989) schema (imposed *etic*, comparison across cultures and identification or failure to identify *etics*; and if the latter, *emic* studies) for testing whether one culture's methods are universal (Douglas and Craig, 1997).

16.2.2   Cultural or cross-cultural studies?

Given the many problems of comparison between cultures, including the potential inappropriateness of theory and tools, i.e. a lack of *etic* (Douglas and Craig, 1997) within-culture studies as opposed to cross-cultural studies, depending upon the research questions, may be more appropriate.

16.3   The importance of culture

It is thought that culture is a major determinant of human food choice. There is evidence that traditions, beliefs, and values are amongst the main factors influencing preference, mode of food preparation, serving and nutritional status. Indeed Rozin (1990) stated that, 'If one wanted to know about an individual's food preference, the best question to ask would be, "What is your culture or ethnic group?"'

Hence understanding a given Asian culture is an essential first step in understanding Asian food consumers. The disciplines of sociology; psychology; food science and marketing have all sought to assist in this respect; for example, respectively, Fieldhouse (1995); Triandis (1994, 1996); Hofstede (1984); Ang *et al.* (1999); Schutte and Ciarlante (1998).

16.3.1   Definitions of culture

Triandis (1994, 1996) reviewed many different definitions of culture and concluded that most agreed that culture was reflected in 'shared cognitions, standard operating procedures and unexamined assumptions'. A common theme in most definitions is that of *sharing* and it is not difficult to apply elements of the definition to food choice.

Triandis (1996) further describes culture in terms of seven constructs:

- *Tightness* – cultural and domain variation in the strength of social norms, rules and regulations. Japanese culture is thought to be 'tighter' (than the USA) generally across mainly domains.
- *Cultural complexity* – the multiplicity of religious, economic, political, educational, social and aesthetic standards. Population density (a common feature of East and South Asia) is thought to be a driver of this complexity, for example, recognisable in the multiplicity and strength of religions in certain Asian states.
- *Active – passive*
- *Honour*
- *Collectivism* – in many Asian cultures 'the self' is defined as an aspect of the collective (e.g., family). Personal behaviour is subordinated to the collective. This is a particularly important construct within many Asian societies (Yoon and Choi, 1994; Leung in Harris Bond, 1996).
- *Individualism* – The self is defined as independent and autonomous from collectives. Personal goals are primary and social behaviour is shaped by attitudes, hedonics and judgements of utility. There is considerable evidence that this construct dominates Western behaviour in contrast to Asia.
- *Vertical and horizontal relationships* – hierarchy contrasts with egalitarian social behaviour. Many Asian societies are characterised by vertical or hierarchical relationships.

Triandis concedes that each construct can be inter-related. Nevertheless, constructs found to be important in Asian societies may provide useful ways of understanding Asian consumer behaviour. Collectivism and individualism receive much attention when comparing Asian and Western cultures. Triandis suggests that four attributes characterise the constructs: self (unit of social space is the individual or group); goal structures (individualists give priority to individual goals); behaviour as a function of norms and attitudes (individualists are driven by attitudes, collectivists more by social norms); focus on the needs of the in-group (collectivists) in contrast to social exchange or contractual relationships (individualists). In addition as many as 60 other attributes of collectivist culture (relative to individualists) have been identified, including some that may be relevant to food choice: lower self-esteem; lower self-efficacy; lower use of affect in decision making and group decision making. Additionally others have sought to go beyond collectivist – individualist characterisations by understanding values in more depth (Schwartz, 1994), see below. However these attitudes and behaviours have only rarely been applied to issues of Asian food choice. Readers may also wish to refer to some of the early explorations of cultural values (applied to organisational psychology) by referring to Hofstede's work (Hofstede, 1984; Shutte and Ciarlante, 1998).

16.3.2   Values as characteristics of cultures and the potential influence on consumer behaviour

Underpinning such cultural characteristics (Triandis' schema) are notions of motivating beliefs or values (Schwartz, 1992; Rokeach, 1973) which have been used both in food conjoint studies (Bech-Larsen and Grunert, 2003) and means-end-chain (laddering) studies (Leppard *et al.*, 2004) cross-culturally (albeit the former only in a Western context and with limited success).

Schwartz's (1992) study of 20 countries included samples (n ~ 200/sample) from the PRC (5 samples); Hong Kong (2); Japan (3) and Taiwan (1). Indicative of differences in cultures, Schwartz (1992) found that most of the Chinese samples (Shanghai teachers, students and factory workers) deviated from the theoretical universal structure of values the researchers were seeking. Notably 'societal harmony'; 'virtuous interpersonal behaviour' and 'personal and interpersonal harmony' were found to be the most important motivational goals (values) and were interpreted as pertaining to, respectively, Taoism; Confucianism and Buddhism. Such constructs are similar to the observations made by Triandis (1996) above. In contrast, data from other Asian samples, Hong Kong, Japan, Taiwan and Guangzhou (PRC) did fit the universal structure of values.

In making a comparison between hypothesised 'communal' (Taiwan) and 'contractual' (New Zealand) societies, again reflected in Triandis' (1996) schema (above), cultural differences were found. Specifically, differences were found in terms of the importance ratings by the Taiwanese sample to values (and their antecedents/dimensions) labelled 'power' (authority, wealth, public image, social recognition); 'conformity' (obedience, honouring parents and elders, politeness, detachment) and 'security' (social order, family security, sense of belonging, reciprocation of favour, national security, clean, healthy). These data reflect consistent differences in the relative importance of values between Asian and Western societies and a suggestion of unique values pertaining to particular Asian cultures. The importance of these values remains largely untested within Asian food choice; however, if it is accepted values drive choices (Reynolds and Olson, 2001) then these cultural differences need to be tested in the context of understanding the food choices of Asian consumers. Furthermore these values may influence attitudes and beliefs (Le Page *et al.*, 2005) that, in turn, can influence food choice (Shepherd and Raats, 1996). For example, social norms, attitudes, self-esteem, self-identity, self-efficacy have all been found to be influential in food choice to a greater or lesser degree (Armitage and Conner, 2001). In Western cross-cultural studies (Bech-Larsen and Grunert, 2003) a contrast between the ratings of 'mastery' and 'harmony' values (respectively antecedents of 'openness to change' and 'conservatism') were used to help understand attitudes towards genetically modified (GM) foods. However, the study was limited to comparisons between the USA and two European sites, whereas the Asian samples in Schwartz's original data (for example, China, Malaysia and Hong Kong) scored highest for 'mastery' (Schwartz, 1994) tantalisingly suggestive of further study.

16.4   Eastern philosophy and religion

Underpinning values are the philosophy of Confucius and a multiplicity of religions that pervade East Asia. Additionally, religion has been recognised as a direct influence upon food choice (Fieldhouse, 1995; Cox and Anderson, 2004).

16.4.1   Confucianism

The philosophy of Confucius (*circa* 500 BC) is said to dominate East Asian culture (Shutte and Ciarlante, 1998) with its focus upon stability achieved by the management of interpersonal relationships (particularly the importance of family) and the relationship between individuals and society. Self-regulation of desires, relationships, politeness, respect, hierarchy, and social harmony dominate guidelines for a way of earthly life that often has diverse separate religious beliefs (e.g., Shintoism, Buddhism) overlaying the philosophy. Confucianism is thought to remain strong throughout Chinese cultures (despite the Cultural Revolution, 1966–69, in the PRC which sought to negate these values). Also Confucianism is thought to be strong in Japan, especially strong in Korea, whilst influencing other SE Asian countries where Chinese communities exist. There appear to be many similarities between the central constructs of Confucianism and the empirical assessments of values (Schwartz, 1992). There is evidence that Confucian values underpin the importance of social norms which in turn can be associated with product choice (see below).

16.4.2   Buddhism

Whilst religious avoidance of the flesh of animals (but not fish) is advocated (Cwietrtka, 1998), in practice in East Asia meat is consumed frequently albeit in small quantities or as flavouring in nominally Buddhist countries (e.g., Thailand, Japan, and Vietnam).

16.4.3   Shintoism

It is thought that most Japanese simultaneously practise animistic Shintoism and Buddhism (Shutte and Ciarlante, 1998, p. 20). Apart from ritual feasts, little is known on how Shintoism influences food choice.

16.4.4   Christianity (Catholicism)

It should be acknowledged that the Philippines is a predominantly Catholic Christian country with few proscribed foods and some rituals influencing food choice (Cox and Anderson, 2004). See Asian consumers and genetic modification (below).

16.4.5   Islam

South-East Asia represents the largest concentration of people following Islam (~250m) in the world with Indonesia and Brunei almost wholly Islamic, over half the population of Malaysia and sizable minorities in Singapore, the Philippines and China. Whilst Islamic law is important, governments tend to be strongly secular, in contrast to the Middle-East, and whilst more fundamentalist groups exist (particularly in parts of rural Malaysia and Indonesia), interpretations of Islamic law tend to be relatively liberal and co-exist with rapid free market capitalist development (although fundamentalist elements may be reactions to secular liberalism and Western capitalism). For example, alcohol is tolerated in Indonesia and Malaysia despite religious proscription. However, there is strict avoidance of pig meat (pork) in all Islamic communities and halal ('what is permitted') is generally rigorously followed in all East Asian Islamic communities as is daylight fasting during Ramadan. In recent years a huge industry has developed in Malaysia supplying halal food products throughout the world.

16.5   Diet – health beliefs (food as medicine)

16.5.1   Chinese food's medicinal or functional qualities

The Chinese have appreciated food for its medicinal (preventative and therapeutic) qualities in addition to its sensory and nutritional properties (Huang and Huang, 1999) for at least 4000 years. The Chinese used foods for their 'functional' properties long before the concept 'functional food' (Kwak and

Jukes, 2001) was used in the West or indeed in Japan (see below). Chinese medicine, with its emphasis on prevention and health promotion, traditionally emphasises the maintenance of health by prioritising the use of diet above herbal medicines, above pharmaceuticals.

Principles and properties
The balance of *Yin* (negative, cooling, feminine) and *Yang* (positive, male, heating) underpins Chinese food as medicine. Furthermore the basic dominant 'taste' of foods (sweet, sour, salt, bitter, pungent[2]) and five 'energies' or 'properties' (hot, warm, neutral, cool and cold) are considered to be interrelated and foods are classified and used by taste and their energy/property (Huang and Huang, 1999) to 'balance' health. Whether there is evidence to support such diet-health relationships remains questionable; however, there is evidence that such beliefs do influence Chinese individuals' food choices even after migration (Wheeler and Tan, 1983). There have been recent attempts to regulate both traditional and new 'items classified as food and drugs' in the light of the recent popularity of 'functional foods' (Huang and Huang, 1999).

Marketeers' assertions (Shutte and Ciarlante, 1998) that many Asian cultures have strong beliefs in 'fate' (perception that external forces may determine or guide various behaviours or situations), and that these are attributable to an external locus of control (Wallston *et al.*, 1978) may well be true of certain behaviour; however, it remains untested in respect of diet-health behaviour. Such generalisations ignore that locus of control is domain specific (Norman, 1995) and contradicts the use of diet to protect health (amongst Chinese) which may be a manifestation of internal locus of control of health.

16.5.2   Functional foods in Japan
Whilst having origins in addressing malnutrition post-Second World War, the Japanese were the first to formalise the concept of functional foods (1984) and by 1991 the government had set up legal formalised approval systems (Arai, 1996, 2002; Arai *et al.*, 2001). Functional foods were originally described as 'foods for specified health uses' (FoSHU) and are, uniquely, regulated within this framework (Richardson, 1996). However, the term functional foods is now commonly used (Kwak and Jukes, 2001). Notwithstanding the rigorous approval system, the potential for greater use of functional foods in Japan, and other parts of Asia, including China and Korea (Tee *et al.*, 2004) has been noted as being enormous (Kojima, 1996; Milo Ohr, 2003). For example, in Japan the market has been estimated to be worth US$16 billion. The Japanese spent US$126 per person per year on functional foods compared with US$67.9 per person per year in the US, US$51.2 for Europeans and US$3.20 (estimated) for other Asians. The Japanese spent more than twice as much on functional foods than people in

---

2. Pungency is considered distinct from taste within Western science as, physiologically, pungency is perceived through the trigeminal nerve system not taste receptors.

the US on a per capita basis and almost four times more than Europeans. Six percent of Japanese food expenditure is on functional foods; that is higher than any other country (http://www.functionalfoodsjapan.com/pages/3/).

The market has potential because of traditional beliefs about food-health relationships but also the relative maturity of the market. Amongst Japanese consumers, however, there is evidence that knowledge amongst consumers is not extensive. In a review (Tee *et al.*, 2004) of three Japanese market research surveys (1999, 2001 and 2003) awareness of FoSHU foods was 25% of those sampled in 1999. In 2001, 38% of housewives and only 21.5% of male respondents were aware of the FoSHU label, however, importantly, only 11% of males and 27% of females reported consuming FoSHU products. In the 2003 survey of fermented milk products (the largest product segment, Arai, 2002) most respondents (80%) consumed such products for health reasons, however, the importance of the FoSHU label was relatively minor and considerably more important amongst older consumers than younger consumers. Nevertheless the number of products on the market (particularly probiotics) continues to grow with an estimate of ~400 approved products and ~1600 unapproved at last count (http://www.functionalfoodsjapan.com/pages/3/). There would appear to be some discrepancy between consumer surveys and market intelligence reports and the suggestion that unapproved (non-FoSHU) products may have more commercial value.

16.6   Cultural flavour principles

Elizabeth Rozin (2000; 1982) describes how 'flavour principles' characterise particular cultures' food preferences. For example, soy sauce, rice wine and ginger root are characteristic of Chinese cuisine. Chinese regional differences are characterised by the additional flavours of fermented black beans and garlic (Canton, Southern); soy bean paste, garlic and sesame oil (Beijing, Northern) and pepper or chillies, bean paste and oil (Szechuan, Southern). Some claim additional regional cuisines across China (Newman, 1999). Rozin's 'flavour principles' appear to be strongest in regions where agriculture was first developed including South East Asia and India.

Rozin (1982) additionally structured cuisine as:

* basic foodstuffs;
* manipulative techniques;
* cultural flavour principles.

Basic foodstuffs can vary culturally and regionally with different preferences, for example types of rice by variety, grain size and shape, glutinous versus non-glutinous types (Luh in Ang *et al.*, 1999, pp. 5–42). Furthermore there are a myriad of manipulation, particulation, incorporation, separation or extraction, marinating, fermentation, and the various applications of heat (see Ang *et al.*, 1999) within Asian food preparation techniques. Flavour is achieved by primarily using oils, liquid sauces and aromatics/pungency. Rozin further suggests that Chinese consumers value the 'natural' (unadulterated) intrinsic subtle flavours of foods as an ideal but concede that additional 'cheap' flavours are required because such natural flavours are often prohibitively expensive. However, such values remain untested.

Rozin (2000) gives the example of how a basic foodstuff can be the same in European and Asian cultures (e.g., grilled chunks of lamb) but flavourings differ, so for Indonesians *satay* is seasoned with a sweet and spicy blend of soy sauce, coconut, chillies and ground peanuts. In contrast, Greeks flavour grilled lamb (*souvlaki*) with lemon and oregano.

The historical perspective given to these 'shared' processes and cultural favours introduces the concept of *exposure* and *familiarity*, which has been demonstrated to be a significant predictor of food acceptance (see below). However, traditional culinary traditions are not without change and there is recognition, for example, that Japanese food choices, whilst having origins in Chinese cuisines (Otsuka, 2000), later developed unique foods but have, since the 1950s, been influenced by the West (Cwiertka, 1998).

16.7   Factors affecting food choice in Asian consumers

16.7.1   A brief summary of a complex history of sophisticated food processing and food cultures

For a food science perspective of Asian foods, readers are referred to Ang *et al.*'s (1999) review which provides an in depth description of the sophistication of Asian foods beyond the ubiquitous cookery books and restaurant menus. Secondly, Newman (1999) provides an overview of South and East Asian foodways and thirdly Newman (2000) and Otsuka (2000) provide in-depth descriptions of Chinese and Japanese meals.

Certain Asian cuisines (particularly Chinese cuisines) can be found all over the world (although the foods are sometimes 'acculturated' to the host nation) and are partly a reflection of the strength of that cuisine which in turn demonstrates the integrity of those Asian cultures. It has been observed that different cultural groups within a given country (for example, Malaysia) retain cultural integrity and therefore cultural food choices differentially. It has been generally observed that Chinese consumers retain their beliefs about food long after migration to Western countries (Murray *et al.*, 2001) and that Chinese culture is one of the most resilient cultures in the world (Trey Denton and Kaixuan, 1995). For example, in Malaysia, the Chinese community is thought to retain its cultural food choice (Cox *et al.*, 2001) in contrast to Malay Malaysians who are thought to be more open to change, specifically, openness to Western-style foods (Meudic and Cox, 2001). Consistent with theories of exposure and age (largely derived from the acculturation literature, Cox and Anderson, 2004) younger people appear to be more open to different food choices. Hence a study of perceptions of 'Western' breakfast cereals (Meudic and Cox, 2001) focused upon young Malays. Identifying segments of populations open to novel foods requires both an understanding of cultural integrity and market intelligence (for example, Jolly and Breddin, 1995; Cheeseman *et al.*, 1995; Trey Denton and Kaixuan, 1995).

16.8   Genetics

16.8.1   Genetic markers

It would seem reasonable that genetically moderated differences in physiology may account for individual differences in sensory perception (Drewnowski and Rock, 1995). The most well used marker for genetic difference, 6-*n*-propyl-thiouracil (PROP), in taste sensitivity has identified that certain Asian people are more likely to be more sensitive to such a bitter chemical. For example, Malaysian subjects were found to rate PROP more bitter than European-Australians did (Holt *et al.*, 2000); however, that difference was unrelated to perceptions of sweetness intensity or liking for sweetness. However, there is a paucity of data on whether such sensitivity determines avoidance of, for example, bitter foods.

Despite some early work on low preferences for brassica vegetables by European-American 'supertasters' of PROP, the evidence for the specific effects upon actual food choice are weak and poorly studied. This may be explained by the lack of associations between PROP sensitivity and other bitter compounds, for example a vegetable may contain hundreds of potentially interactive bitter compounds that may or may not be perceived. PROP sensitivity, which is more prevalent among some Asian people, is associated with capsaicin sensitivity, however, chili is an important part of many Asian cuisines (Cox and Anderson, 2004; Newman, 1999).

16.8.2   The genetic inability to metabolise alcohol

Despite the ubiquitous consumption of alcoholic beverages across almost all cultures including Asian, there is evidence that approximately 50% of Asians lack the appropriate enzyme aldehyde-dehydrogenase 2 isozyme (ALDH2) required to convert alcohol's toxic acetaldehyde to acetate (Morimoto *et al.*, 1994). Consequently such consumers should not consume alcohol. Acute symptoms of alcohol consumption on an ALDH2 deficient individual include flushing, dizziness, perspiration, palpitations and nausea. Long-term effects are chromosome damage. This does not necessarily stop ALDH2 deficient people consuming alcohol, for example, in China red faces are perceived as a source of amusement when binge drinking at weddings (Chen *et al.*, 1999). Similarly a large range of domestic alcoholic beverages are common throughout Asia, for example, China (Chen *et al.*, 1999, pp. 383–408) although mostly used for cooking. There may be marketing opportunities for low or zero alcohol beverages (e.g., non-alcoholic wines) which remain largely unexploited.

The available evidence, reviewed below, suggests that social and cognitive factors and learning processes override individual physiological differences in respect of food and beverage choice.

16.9   Sensory perception and preference

Cross-cultural studies have mainly focused on comparisons of sensory perception (threshold, sensitivity or discrimination).

16.9.1   Taste

Most studies looking at perception of tastes in solutions have found no differences between cultures. Druz and Baldwin (1982) recorded no difference in the four basic taste thresholds between Nigerian, Korean and American subjects; American and Japanese thresholds for sweetness, saltiness and umami did not significantly differ in a study by Yamaguchi *et al.* (1988). Perception of small differences in taste intensity (Laing *et al.*, 1993) and intensity ratings for various basic tastes levels in foods or drinks showed no differences in discrimination either between Caucasian Australians and Japanese (Prescott and Bell, 1993), or between Caucasian Australians and Malaysians (Holt *et al.*, 2000).

There is less research studying cultural differences in perception of sensory properties other than tastes (odour, flavour, texture).

16.9.2   Odour

In a comparative study (Distel *et al.*, 1999) of Mexican and Japanese perceptions of odorants, perceptual similarity was found. Familiarity and liking were found to be the most important perceptual factors. Other data (Ayabe-Kanamura *et al.*, 1998) comparing Japanese and Germans also found that ability to provide meaningful descriptors was based upon cultural experience and that perception of intensity may even be driven by experience. These observations were particularly true for food odours. A more recent study of Americans, French and Vietnamese (Chrea *et al.*, 2004) found any differences in perceptions of odours were largely attributable to 'cultural differences in food and household habits'. Furthermore, whilst American and French subjects could discriminate between two broad categories of odorants ('fruit' and 'flower'), the Vietnamese could not and that lack of discrimination was attributed to a lack of familiarity.

16.9.3   Colour

A recent multi-cultural study (Kay and Regier, 2003) resolved the question of colour naming universals. Sampling 110 cultures with an emphasis upon pre-industrial societies, including a few Asian cultures, the authors concluded that all cultures perceive colours similarly. Colours do, however, have different meanings for different cultures; for example, red is associated with happiness and good luck (particularly in Chinese societies) whereas white is strongly associated with death throughout Asia (Shutte and Ciarlante, 1998, p. 64).

The available evidence seems to suggest that there is some universality of cognitive perceptions of sensory stimuli, however, there is a need for more cross-cultural studies.

16.9.4   Flavour preference

Prescott *et al.* (2001) sought to understand whether flavour characteristics of pasture-fed sheep meat were leading to poor acceptability of such meat in Asia, particularly Japan. Volatile branched chain fatty acids (BCFAs) and skatole were implicated in reported negative flavours. Indeed, *soo*, in Chinese, is a hedonically negative word to describe sweaty or sour flavour characteristics of the cooking odour of sheepmeat. A model food system using BCFAs and skatole added to beef (originally containing neither of these flavours and acceptable and familiar to Japanese consumers) was used as stimuli. New Zealand trained sensory panellists found BCFAs to be associated with negative flavours. Japanese consumers reported significant decreases in preference for meat with added BCFAs at both low and high concentrations whereas New Zealand consumers reported significantly lower preference for the meat only with high concentrations of BCFAs. Skatole was a significant but lesser factor in variation in liking. The authors acknowledge several limitations of their study, particularly the use of a model system, however they assert that evidence of the role of BCFAs was found and that very lean meat, grain-fed finished meat and flavouring could be strategies to reduce (perception or content of) BCFAs and hence improve acceptability of New Zealand lamb amongst certain Asian consumers.

Further work (Prescott *et al.*, 2004) on a similar theme investigated one of those strategies, the use of flavouring in a study of ethnic Chinese Singaporean and European origin New Zealand consumers. Surprisingly both groups preferred the unflavoured control as opposed to culturally flavoured versions of lamb although the Chinese-Singaporeans had a significantly higher preference rating for the 'Chinese' flavoured lamb in comparison to New Zealanders. Neither labelling nor health beliefs had any effects.

Clearly these studies represent an important start in understanding flavour preferences for certain 'Western' foods amongst certain Asia consumers, however, there may be a need to improve methodologies particularly in terms of choosing more appropriate food stimulus.

16.9.5   Familiarity and preference

It is well accepted that there is evidence of variation between cultures in preference for food (Abdullah, 1995; Ward *et al.*, 1998). Past research has therefore looked at cultural patterns for liking of specific sensory properties.

Taste intensity liking was found to vary between cultures depending on the context or food studied, with no consistency in direction or magnitude across products (Druz and Baldwin, 1982; Bertino et al., 1983; Prescott et al., 1997, 1998; Holt et al., 2000). These product-dependent differences appeared to be related to consumer familiarity and exposure (Laing et al., 1994; Prescott and Bell, 1995; Prescott, 1998). In a cross-cultural study of Chinese-origin Australian versus European-origin Australians (Murray et al., 2001) there were no differences for preferences of textures of novel extruded cereal snack products. Importantly, because the snacks were novel to both cultural groups, the artefact of experience was controlled. In contrast to culture, age was found to be a factor in discriminating between preferences for textures.

The evidence available to date suggests that chemosensory abilities appear generally similar between cultures. Therefore, as for individual differences (Rozin, 1990), very little of the cross-cultural variation in food preferences appears to be genetically based, seeming rather to arise from experience, dietary habits, and attitude to food (Rozin et al., 1999).

16.9.6   Motives for food choice across cultures (beyond sensory attributes)
Assuming we have the appropriate tools, there is a need to expand research to other aspects than just chemosensory sensitivity or liking for a specific sensory property when attempting to assess Asian food preferences. In this respect reference to the general food choice literature (which is largely Western-based) provides a starting point for exploring our lack of understanding of Asian consumers. Understanding factors that influence Asian food choice may be useful in understanding how best to market food products.

A study by Rozin et al. (1999) compared consumers from Belgium, France, the USA and Japan on beliefs about diet-health link, worry about food, the degree of consumption of food modified to be healthier, the importance of food as a positive force in life, the tendency to associate foods with nutritional vs. culinary contexts, and satisfaction with the healthiness of one's own diet. The authors concluded that in all domains, except beliefs about the importance of diet for health, there were substantial country (and usually gender) differences. Interestingly, the Japanese sample tended to be the most diet-health conscious but least anxious about their diet. Furthermore, they also rated culinary associations with food most highly.

Steptoe et al.'s (1995) Food Choice Questionnaire (FCQ) has been applied across European cultures and found to be useful in describing motivations for food choice based upon nine factors: health; mood; convenience; sensory appeal; natural content; price; weight control; familiarity and ethical concern. Prescott et al. (2002) elicited responses to this questionnaire from convenience samples of Japanese, Taiwanese, Chinese-Malaysian and New Zealand female consumers, seeking to understand if there were cultural differences. The authors acknowledge the FCQ factors are UK/European and may not reflect the most important factors in Asian cultures, furthermore, they did not reanalyse the items to see if the factors were similar across cultures nor did they report internal consistency of the factors. However, within these limitations some consistent significant differences between cultures were found when these data were analysed by three methods (scores, ranking and discriminant analysis). Notably the two Chinese groups (Malaysian and Taiwanese) were found to similarly rate and rank (and be discriminated by) health, natural content, weight control and convenience highest and the authors interpreted the first three as possibly being related to Chinese beliefs about food as medicine. In discriminate analysis New Zealanders uniquely placed emphasis on sensory appeal whereas the Japanese were clearly separated from the other countries; with the factor rankings suggesting greater importance placed upon price and ethical concerns. Curiously, familiarity was found to be the least important factor for all cultures and weakly correlated with neophobia (Pliner and Hobden, 1992) suggesting that there is a need for better measures of (Asian) neophilia required, particularly in respect of marketing novel and/or 'Western' foods. Indeed, Murray *et al.* (2001) found that whilst more Australian-resident-Chinese were classified as neophobic (compared to their European-origin counterparts) those that were classified as neophobic scored preferences for novel foods higher. The authors questioned the appropriateness of a scale that is Anglo-centric for Asians (several items contain the word 'ethnic', meaning 'non-English speaking') and difficulties in translation. For use in multicultural Australia, Flight *et al.* (2003a) changed the term 'ethnic' to 'cultures other than your own', however, such language should be validated and tested cross-culturally.

Prescott *et al.*'s (2002) study is a useful first step but raises many questions that can be generalised to the principals of cross-cultural studies (van de Vijver and Leung, 1997) of Asian consumers:

- Were the items, derived from Europeans, appropriate for Asians?
- Are the items interpreted in a similar way?
- Can factors of European origin be used with Asians?
- Are the scales used in a similar way? The authors report that Chinese consumers tended to use midpoints of the scale and generally rate most items highly. This is consistent with avoidance of negative ends of a scale by Chinese (see below).
- Whilst culture does discriminate, there is likely to be segmentation within cultures that could transcend cultural differences (Bech-Larsen and Grunert, 2003; van de Vijver and Leung, 1997).

There is clearly a need for fundamental research addressing these and other issues of cross-cultural validity.

16.9.7   Health and the role of product information

Given the evidence that particular Asian consumers may be health conscious, Mialon *et al.* (2002) asked whether health information attributed to food products influences Asian consumers in a similar way to Western consumers with particularly reference to Western foods, such as bakery products. It is well established (Aaron *et al.*, 1994) that attitudes, beliefs and label information influence perceptions and liking for health-associated foods. The effect of information about dietary fibre content was investigated (Mialon *et al.*, 2002) for Chinese Malaysian and Australian consumer perceptions of bread and 'English muffins' for (a) white low-fibre, (b) white fibre enriched and (c) wholemeal/ grain versions of the products.

Acceptance measures, perception of sensory intensities and health- and nutrition-related attributes were rated before and after information about fibre content was presented. Information strongly affected the perceived healthiness and nutrition value of the breads and muffins for both cultural groups, and Australians were more receptive to information in their ratings for fibre content. Perceived sensory intensities were also influenced by information, and a cultural difference in direction of change was observed for the muffins. For both cultures, bread liking and muffin likelihood of consumption were enhanced for the white product labelled as high in fibre, while no changes were noted for the wholemeal/grain ratings. Whilst limited to one study and Asian surrogate consumers, there is a suggestion that particular Asian consumers behave in a similar way to their European-origin counterparts.

16.10   Predicting consumer behaviour

16.10.1   Culturally neutral tools and metric equivalence

Acknowledgement that behavioural theories originate in North America or Europe and may not be appropriate in Asian settings (Douglas and Craig, 1997) determines the need to test the 'metric equivalence' or dimensionality and internal consistency (Durvasula *et al.*, 1993). Surprisingly few studies have tested metric equivalence.

Attitude expectancy models

The Theory of Planned Behaviour, (TPB, Ajzen, 1985) or its predecessor the Theory of Reasoned Action (TRA, Ajzen and Fishbein, 1980) have been the most widely used attitude-expectancy models for predicting drivers of behaviour (Shutte and Ciarlante, 1998; Armitage and Conner, 2001) nevertheless questions remain over whether the model is appropriate for use amongst Asian consumers (Schutte and Ciarlante, 1998).

Lee (1990) first modified TRA for a Korean 'Confucian' culture by including *emic* constructs, 'face-saving' and 'group conformity pressure', as TRA's 'social (subjective) norms' and regressing attitudes upon those constructs (as opposed to the conventional regressing of subjective norms directly on intention to behave).

However, Lee and Green (1991) tested cross-culturally the conventional design of TRA, using structural equation modelling (LISREL) for 'goodness of fit'. The model was found to fit the Korean culture and, consistent with cultural differences in values (see above), purchase intentions were predicted by subjective norm (others influence) for Koreans ($\beta = 0.87$) and by attitudes in the USA ($\beta = 0.90$). The authors concluded the TRA was a valid model in a Korean setting albeit the percentage variance explained was less amongst the Korean sample than the USA sample.

Using a non-specific model of attitudes Durvasula *et al.* (1993) undertook a similar exercise looking at attitudes to television advertisements using a comparative fit index (LISREL) and found that New Zealanders, Americans (USA) and Indians were metrically similar.

The suggestion is that TPB and Western models of attitudes may be appropriate for understanding some Asian consumers, however more work is required.

Construct theory and repertory grid methodology

Kelly's (1955) construct theory and its application as repertory grid methodology (RGM) for measuring perceptions of foods (Gains, 1994) and understanding consumers' perceptions of new product development (van Kleef *et al.*, 2005) is well established. As an idiographic technique it is particularly suitable in cross-cultural (McCoy, 1983; Scriven and Mak, 1991) and cultural work with Asian consumers (Meudic and Cox, 2001) being free of researcher or semantic bias. In a comparative study of Hong Kong Chinese (studying in Australia) and European-origin Australian students, Scriven and Mak (1991) used RGM (and also conventional descriptive techniques) to look at perceptions of a range of meat products. They found little difference between the two methods but concluded that RGM successfully elicited similar perceptions of the various meat products regardless of culture.

Nantachai *et al.* (1990/91) introduced context to a RGM study of Thai (resident in Australia for less than 6 months) and European-Australian consumers' perceptions of Thai and Australian meat products in an attempt to gain a better understanding of export potential. RGM used in conjunction with principal components analysis and Generalised Procrustes analysis (GPA) successfully found unique (meal use and flavouring *emic* constructs) and common (social function *etic* constructs) contexts in which meat products were used.

In a cultural study of Malay Malaysians, Meudic and Cox (2001) found RGM to be useful in eliciting responses from consumers who were considered to be early adopters of novel Western foods, general and discriminating attributes of breakfast cereals.[3] These young Malay Malaysians were found to be health conscious but to prefer cereals with medium (not, as expected, high) sweetness.

RGM remains an exploratory (not predictive) technique (van Kleef *et al.*, 2005); however, it is theory based, quantifiable (using GPA), reasonably easy to interpret and, most importantly, appears to be culture neutral and therefore a useful tool for understanding Asian consumers.

---

3. Generally, age has not been found to be a consistent predictor of innovation adoption (Rogers, 1971, pp. 185–186) and readers should refer to Rogers' (1971) book that explores predictors and other aspects of communicating innovation.

16.10.2   Asian use of hedonic response scales

Serious questions have been raised about appropriate response scales across a range of domains. The use of the labelled category hedonic scale has been popular (Lawless and Heymann, 1998), in particular the labelled nine-point category scale (Peryam and Pilgrim, 1957), however, computerisation of sensory responses offers a large choice of response scale options. The labelled-category-scale is considered advantageous in pairing the (semantic) label with a number allowing for ease of use by (untrained) consumers (Lawless and Malone, 1986), ease of interpretation and, furthermore, greater and relatively simple analysis (data can be treated as both continuous and categorical data). Within 'Western' cultures there has been little evidence to suggest major differences in use between the category scale and other selected scales (Lawless and Malone, 1986; Mattes and Lawless, 1985). However, it is well documented (reviewed by Cardello, 1996) that there are problems inherent in category scales, particularly with regard to the fact the category labels do not constitute equal intervals, the neutral category reduces its efficiency, and the avoidance of end-categories. In contrast, it has been suggested that unstructured line scales may reduce bias because of behaviours attached to semantic labels. Specifically, for example, unstructured line scales may minimise Asian 'politeness' or 'positive bias' (Christopher, 1983; Triandis, 1994; Shutte and Ciarlante, 1998) in avoiding negative responses (as found on a labelled category scale). It should be noted, however, little empirical evidence for such cultural bias exists and that such assertions are often based upon understandings of cultural values (see above). Moreover, some researchers (Prescott, 1998) have found that Japanese respondents were quite capable of expressing dislike of bitter stimuli. Should an unstructured scale be used there may be disadvantages for the untrained consumer who has no reference point to assist in translating his or her response into a position on the scale (Lawless and Malone, 1986).

Yeh *et al.* (1998) reported that, in comparison to US consumers, Korean, Chinese and Thai consumers, regardless of residency (USA or country of origin), systematically used a narrower range and avoided 'dislike' categories of the nine-point labelled category scale, suggesting a possible cultural bias. Other work (Prescott, 1998), on comparisons of Japanese and Australian consumers' sensory preferences, whilst claiming to find no evidence of non-equivalence between labelled and line scales, did not specifically test for equivalence. However, Prescott (1998) recommended that, because of the lack of equivalence tests for spacing of labelled categories across cultures, Asian consumers should use line scales.

Because of uncertainty we undertook a study (Cox *et al.*, 2001) testing cultural, scale and gender interactions between European-origin Australian and Chinese Malaysian consumers' hedonic responses to food and drink stimuli. A between-groups design, one group using a labelled nine-point category scale and the other an unstructured-anchored line scale, both using computerised responses, found no systematic cultural bias. The anonymity facilitated by the computerised administration may have overcome cultural 'politeness' bias. In addition, a task in which semantic labels (from the labelled hedonic category scale) were assigned to a line scale found no statistically significant cultural differences in the scores attributed to the labels.

These data indicated that, in the context of computerised responses, young Malaysians used both scales in a similar way to their Australian counterparts. However, further non-parametric analysis suggested that the unstructured line scale encouraged greater use of a range of possible responses and therefore line scales may be a preferred option for use in this population. This study was limited in its use of the English language and surrogate Asian consumers.

Subsequent studies (Yao *et al.*, 2003) have tested the issue using the respondents own languages, comparing structured, unstructured and labelled hedonic scales, this time across US, Japanese and Korean cultures, on dental (not food) products. The unstructured scale elicited a wider range of scores for US and Japanese respondents; however, Koreans gave very narrow responses regardless of the scale used, which may have been partially due to semantics. The authors concluded that there was only partial support for the 'politeness' hypothesis and that many unanswered questions remain. Secondly there is a definite need for 'culture-free' scales, tested across a range of cultures and thirdly, that ranking might be an alternative to scoring (O'Mahony *et al.*, 2004).

In summary, the available evidence suggests that scales that minimise semantic labelling may be more appropriate for (some) Asian consumers, however, there is need for more research both in terms of cross-cultural studies and alternative response scales.

16.11   Guide to undertaking Asian consumer research

16.11.1   Use of surrogates (Asians resident overseas)

Pragmatism often dictates eliciting the food choice opinions and behaviour of Asians in residence in the Western researchers' country. This is not necessarily best practice and the use of such surrogates will depend upon the research questions being asked. Within the context of measuring hedonic responses, some are positive but cautious (Yeh *et al.*, 1998) in terms of scale, use that was found to be similar regardless of residency in USA or country of origin (Taiwan, Korea and Thailand). Careful examination of the ex-patriate group is recommended particularly in terms of 'cultural integrity' (i.e., has that group resisted acculturation). For example, Murray *et al.* (2001) justified the use of Australian resident Chinese on the basis that 71% spoke Chinese as a first language and all associated with a Chinese institution or community group (the study's recruitment framework). Indeed Chinese culture seems to be particularly resistant to acculturation in many Western countries particularly in regard to food choice.

Others have justified use of Chinese Malaysian surrogates by a relative short length of residence (visiting students) and maintaining cultural integrity by living together in university halls of residence (Cox *et al.*, 2001). Wills and Wooton (1997), in testing Koreans' preferences for wheat noodles, found that length of residency (> 7 years) in Australia influenced Korean surrogates to report preferences more similar to European-Australians than Koreans resident in Korea. Moreover, whilst Koreans resident less than two years were more similar to Koreans resident in Korea, there was considerable within-group variability attributed to degrees of acculturation.

Some researchers have deliberately sought cultural groups that may represent early adopters of novel Western foods (Meudic and Cox, 2001), for example Malay Malaysian students resident (less than 2 years) in Australia. Careful consideration should be paid to the acculturation literature (Cox and Anderson, 2004) if surrogates are used.

16.11.2   Administration

Given the information on cultural differences examined above, the administration of studies in Asian countries is fraught with challenges. Clearly, matching resources to the research questions is paramount and the expense of working overseas in a different culture is considerable. Accessing either appropriate purposeful samples or representative samples of consumers is never easy and becomes increasingly difficult as people appear to become more time poor. It is our laboratory's experience that using an accredited market research company registered, for example, with the European Society for Opinion and Marketing Research (http://www.esomar.org/), now a worldwide organisation, has proven useful but not entirely sufficient as a recruitment resource. In addition it is often necessary to consult local collaborators with specialist knowledge of the field (e.g., university nutrition, and/or food science departments and/or local representatives of the client) with the time to explain culturally specific sensitivities that may affect the study.

Triandis (1994) and Schutte and Ciarlante (1998) provide considerable anecdotal guidelines on the dangers of culturally specific factors that influence both administration and responses. The use of mail questionnaires in countries with unreliable postal services or poor literacy are some of the most obvious ones that are considered. There is general preference to use personal interviews in many South East Asian countries particularly in countries where it is culturally important to establish personal relationships and/or receive hospitality before making enquiries about people's attitudes or preferences. Projective techniques ('why do *people* buy X products?' rather than 'why do *you* buy X products?') have also been used to minimise the interrogative nature of questioning thought to be culturally inappropriate in some circumstances although such techniques appear to be losing popularity (Cooper *et al.*, 1998).

However, no amount of secondary guidance can compensate for a lack of in-country knowledge, personal experience and investigators in-country. Proximity to Asia, a large and varied Asian-origin population segment and a large Asian student population, as found in Australia, have proven useful in respect of pilot testing and checking appropriate administration.

Translation

The principles of translation and back translation are well established in cross-cultural work (Brislin, 1970; Triandis, 1994) and numerous examples exist in the published literature. What is rarely reported is the detail of the process and the acknowledgement that translation is an art as much as it is a science. Our experience of employing 'accredited' translators (for example, in Australia, accredited with the National Accreditation Authority for Translators and Interpreters, NAATI) has proven to be necessary but, again, insufficient to ensure correct translation. Engagement with nationals of the culture of interest (often, in our experience, bi-lingual Asian students working in our laboratory) has often indicated errors in translation. For example, the appropriate use formal and non-formal variations of Bahasa Malay was only revealed though discussions with Malaysian nationals and reliance solely upon an accredited translator would have resulted in errors of translation (unpublished data). A further detailed example (Small *et al.*, 1999a,b) explains the process of testing translations of instruments used in cross-cultural studies (within multi-cultural Australia), illustrating a need for numerous iterative rounds of pilot testing with bi-lingual researchers and respondents.

Questions or statements (items) of interest do not always survive translation (Small *et al.*, 1999a,b) particularly when items are idiomatic in the original English versions. The huge vocabulary of the English language is rarely matched by other languages and subtleties (adjectives) in semantic scale labels are sometimes lost. In some cases negative labels do not exist in some Asian languages. For example, our experience (unpublished) in the use of age appropriate hedonic scales for children (Kroll, 1990) amongst Indonesian children resulted in the negative side of the scale (degrees of 'bad') being translated to 'not good' in Bahasa Indonesian.

16.12   Case studies

16.12.1   Case study 1: Australian wheat for noodle manufacture

Whilst China is the largest producer of wheat in the world (Corke and Battachatya, 1999), it is also the largest importer, furthermore, other South East Asian nations cannot produce wheat for climatic reasons but have a demand for imported wheat, particularly for noodles and more recently bread and bakery products (Corke and Battachatya, 1999; Supermarket to Asia/Austrade, 1998). Furthermore, the demand is growing (Miskelly, 1993)

The matching of imported wheat to diverse types of noodles and consumer preferences has been a success story that has utilised agronomy, food and sensory science (Corke and Battachatya, 1999) and, in the case of Australia, the benefit of geographical proximity (www.graingrowers.com.au/_data/page/43/what_the_world_wants.pdf). Challenges remain in new markets, adding value and convenience to noodle products (Miskelly, 1993).

16.12.2   Case study 2: Genetically modified (GM) foods in Asia
Given the size of the Asian market and agricultural potential some have asked
whether the fate of GM foods will be decided in Asia (Mackenzie, 2002)
particularly since China, India and Indonesia (amounting to 2.5 billion people)
have already invested in GM (for example, *bt* cotton).

There is a wealth of market research type information on numerous Asian
countries' reactions to GM (most of it 'grey' literature) and summarised by
MacKenzie (2002). Furthermore, a conference of the International Consortium
on Agricultural Biotechnology Research (ICABR, 2004) recently reported on
cross-cultural consumer attitudes, including PCR (Lin *et al.*, 2004) and Taiwan
(Chiang, 2004). A brief summary is made in Table 16.1 (MacKenzie, 2002,
drawing upon various un-refereed publications unless indicated otherwise).

Table 16.1   Asian consumers and genetically modified (GM) foods (MacKenzie, 2002,
drawing upon various un-refereed publications unless indicated otherwise)

Country	Summary
Hong Kong	Active anti-GM lobby groups; GM identified in many foods; consumer demand for mandatory labelling; almost half surveyed report they would be willing to pay more for non-GM food.
Indonesia	Active anti-GM lobby groups; High public awareness but low knowledge of GM.
Japan	Majority perceive GM negatively. Macer and Chen Ng (2000) reported high awareness and high understanding; 31% (32% of scientists) support GM food with a declining trend; Scientists sample less favourable towards specific GM foods. Likely segmentation; low trust in national institutions and government regulation.
Korea	Mandatory labelling following lobbying. High awareness of GM but poor acceptance. Realisation that many soy products contained GM led to sourcing non-GM soy.
Malaysia	High public awareness but low knowledge of GM. Active lobbyists.
Philippines	High public awareness but low knowledge of GM. High profile anti-GM campaigns including the Catholic Church.
Peoples Republic of China	Very low (Lin *et al.*, 2004) or superficial awareness and low knowledge of GM. However generally favourable or neutral attitudes towards 'Biotech foods' with suggestions that recent knowledge increases acceptance (Lin *et al.*, 2004).
Singapore	High SES (n = 417) attendees prior to a public lecture on GM, expressed high knowledge, over half were concerned about GM and the majority wanted labelling (Subrahmanyan and Cheng, 2000). Another small qualitative study found low knowledge.
Taiwan	2002 survey (n = 257) indicated moderate knowledge; 40% concerned about GM and health, majority demanded labelling (now mandatory) and only 28% willing to purchase (Chiang, 2004).
Thailand	Low awareness; low concern; active anti-GM lobbyists.

Numerous cases of GM ingredients found in supposedly non-GM food have been catalogued in addition to other food scares, for example an outbreak of bovine spongiform encephalitis (BSE) in Japan in 2001. As in Europe these have created a climate of fear and distrust towards food technology in certain Asian cultures (MacKenzie, 2002).

With such variation in quality of information (most superficial and not peer reviewed) it is difficult to draw firm conclusions; however, tentatively there would appear to be variation in knowledge, attitudes and willingness to buy, suggesting, not surprisingly that Asia is highly differentiated. There is a suggestion that Asia generally falls somewhere in between Europe and the USA on the fear of GM continuum. The PCR stands out as being a special case in being generally unaware of GM but providing positive responses, and contrary to the European and Australian literature (Scholderer and Frewer, 2003; Wilson *et al.*, 2004) knowledge was thought to increase acceptance. However, the evidence is only preliminary and more data are required. It is possible that the PCR is special in many respects including political economy, value and belief systems, being relatively new to 'choice' and even in the use of response scales, all of which may contribute to such reported attitudes towards GM.

16.13   Conclusions

Understanding Asian consumers remains a huge task given the size, dynamics and diversity of the populations. This chapter has reviewed some of the knowledge of South East and East Asian consumers but there are clearly many gaps and a paucity of published empirical studies. There is evidence that differences in physiology and genetics are unlikely to be major contributors towards differences in perceptions or preferences for foods. Culture as manifested in belief systems, values and experiences is likely to be central to consumer preferences, yet whilst there is a wealth of information on these aspects, there are few published studies that link culture and consumer behaviour. Some questions remain over the universality of behavioural theories on which consumer understanding is based, although some culturally neutral approaches and tools have been identified. Asia has led the thinking and behaviour concerning the relationship between diet and health, which has only been a relatively recent phenomenon in European-origin cultures. The Asian market for functional foods is established and continues to grow. It is possible that any changes in the adoption (or consumer resistance) of genetically modified foods may well be decided in Asia.

16.14   References

AARON J, MELA D and EVANS R (1994) The influence of attitudes beliefs and label information on perceptions of reduced-fat spread. *Appetite* **22**, 25–37.

ABDULLAH A (1995) Consumer preferences in fruit juice formulation. In Merican and Yeoh (eds), *5th ASEAN Food Conference Plenary Papers Conference Proceedings*, pp. 165–170.

AJZEN I (1985) From intentions to actions: a theory of planned behaviour. In Kuhl J and Beckermann (eds), *Action Control from Cognition to Behaviour* (pp 11–39) Berlin: Springer.

AJZEN I and FISHBEIN M (1980) *Understanding Attitudes and Predicting Social Behaviour.* Englewood Cliffs: Prentice Hall.

ANG CYW, LIU K and HUANG YW (1999) *Asian Foods: Science and Technology*. Lancaster: Technomic Publishing Co Inc.

ARAI S (1996) Studies on functional foods in Japan – State of the art. *Bioscience, Biotechnology, Biochemistry* **60**, 9–15.

ARAI S (2002) Global view on Functional foods: Asian perspectives. *British Journal of Nutrition* **88** Suppl, S139–S143.

ARAI S, ASAWA T, OHIGASHI H, YOSHIKAWA M, KAMINOGAWA S, WATANABE M, OGAWA, T, OKUBO K, WATANABE S, NISHINO H, SHINOHARA K, ESASHI T and HIRAHARA T (2001) A mainstay of functional food science in Japan – history, present status, and future outlook. *Bioscience, Biotechnology, Biochemistry* **65**, 1–13.

ARMITAGE CJ and CONNER M (2001) Efficacy of the theory of planned behaviour: a meta analytic review. *British Journal of Social Psychology* **40**, 471–499.

AYABE-KANAMURA S, SCHICKER I, LASKA M, HUDSON R, DISTEL H, KOBAYAKAWA T and SAITO S (1998) Differences in perceptions of everyday odors: a Japanese-German cross-cultural study. *Chemical Senses* **23**, 31–38.

BECH-LARSEN T and GRUNERT K (2003) The perceived healthiness of functional foods. A conjoint study of Danish, Finnish and American consumers' perception of functional foods. *Appetite* **40**, 9–14.

BERRY JW (1989) Imposed etics – emics – derived etics: the operationalization of a compelling idea. *International Journal of Psychology* **24**, 721–735.

BERTINO M, BEAUCHAMP G and JEN K (1983) Rated taste perception in two cultural groups. *Chemical Senses* **8**, 3–15.

BRISLIN RW (1970) Back-translation for cross-cultural research. *Journal of Cross Cultural Psychology* **1**, 185–216.

CARDELLO AV (1996) The role of the human senses in food acceptance. In Meisleman HL and MacFie HJH (eds) *Food Choice Acceptance and Consumption*, London: Blackie Academic & Professional, pp. 1–82.

CHEESEMAN N, JOLLY H, SMITH K, WILKINSON M and BREDDIN R (1995) Food retailing in East Asia and Australia, Agribusiness Marketing Services Information Series Q195031. Brisbane: Department of Primary Industries.

CHEN TC, TAO M and CHENG G (1999) Perspectives on alcoholic beverages in China. In Ang CYW, Liu K and Huang Y-W (eds) *Asian Foods Science and Technology*. Lancaster and Basel: Technomic Publishing Co Inc., pp. 383–408.

CHIANG F-S (2004) An analysis of consumer perception and acceptance of genetically modified foods in Taiwan. 8th International conference on agricultural biotechnology: trade and domestic production. International Consortium on agricultural biotechnology research: Ravello (Italy) July 2004.

CHREA C, VALENTIN D, SULMONT-ROSSE C, LY MAI H, HOANG NGUGEN D and ABDI H (2004) Culture and odor categorization: agreement between cultures depends upon the odors. *Food Quality and Preference* **15**, 669–679.

CHRISTOPHER RC (1983) *The Japanese Mind*, New York: Fawcett Columbine.

COOPER HR, HOLWAY A and ARSAN M (1998) Cross cultural research – should stimuli be psychologically pure or culturally relevant? *Marketing and Research Today* **February** 67–72.

CORKE HD and BATTACHATYA M (1999) Wheat products: 1. Noodles. In Ang CYW, Liu K and Huang Y-W (eds) *Asian Foods Science and Technology*. Lancaster and Basel: Technomic Publishing Co Inc., pp. 43–70.

COX DN and ANDERSON AS (2004) *Food Choice* in Gibney MJ, Margetts BM, Kearney JM and Arab L (eds) *Public Health Nutrition*. London: The Nutrition Society/Blackwell.

COX DN, CLARK MR and MIALON VS (2001) Cross-cultural methodological study of the uses of two common hedonic response scales. *Food Quality and Preference* **12**, 119–131.

CWIERTKA K (1998) A note on the making of culinary tradition – an example of modern Japan. *Appetite* **30**, 117–128.

DISTEL H, AYABE-KANAMURA S, MARTINEZ-GOMEZ M, SCHICKER I, KOBAYAKAWA T, SAITO S and HUDSON R (1999) Perception of everyday odors – correlation between intensity, familiarity and strength of hedonic judgement. *Chemical Senses* **24**, 191–199.

DOUGLAS SP and CRAIG CS (1997) The changing dynamic of consumer behavior: implications for cross-cultural research. *International Journal of Research in Marketing* **14** 379–395.

DREWNOWSKI A and ROCK CL (1995) The influence of genetic taste markers on food acceptance. *American Journal of Clinical Nutrition* **62**, 506–511.

DRUZ L and BALDWIN R (1982) Taste thresholds and hedonic responses of panels representing three nationalities. *Journal of Food Science* **47**, 561–569.

DURVASULA S, ANDREWS JC, LYSONSKI S and NETEMEYER RG (1993) Assessing the cross-national applicability of consumer behaviour models: a model of attitude toward advertising in general. *Journal of Consumer Research* **19**, 626–636.

FIELDHOUSE P (1995) *Food and Nutrition Customs and Culture*. London: Chapman & Hall.

FLIGHT I, LEPPARD P and COX DN (2003a) Food neophobia and associations with cultural diversity and socio-economic status amongst rural and urban Australian adolescents. *Appetite* **41**, 41–59.

FLIGHT I, RUSSELL CG, BLOßFELD I and COX DN (2003b) From sensory attributes to marketing hooks: using laddering to understand consumer perceptions of red meat. *Food Australia* **55**, 418–424.

FREWER L, RISVIK E and SCHIFFERSTEIN H (eds) (2001) *Food, People and Society*. Berlin: Springer-Verlag.

GAINS N (1994) The repertory grid approach. In MacFie H and Thomson D (eds), *Measurement of Food Preference*. Gaithersburg: Aspen (a Chapman Hall Food Science Book), pp. 51–73.

GRUNERT K (1997) What's in a steak? A cross-cultural study on the quality perception of beef. *Food Quality and Preference* **8**, 157–174.

HARRIS BOND M (ed.) (1996) *The Handbook of Chinese Psychology*. Hong Kong: Oxford University Press.

HOFSTEDE G (1984) *Culture's Consequences: International Differences in Work Related Values*, Beverley Hills, CA: Sage.

HOLT SHA, COBIAC L, BEAUMONT-SMITH N, EASTON K and BEST DJ (2000) Dietary habits and the perception and liking of sweetness by Australian and Malaysian students: a cross-cultural study. *Food Quality and Preference* **11**, 299–312.

HUANG AND HUANG (1999) Traditional Chinese functional foods. In Ang CYW, Liu K and Huang Y-W (eds), *Asian Foods Science and Technology*. Lancaster and Basel: Technomic Publishing Co Inc., pp. 409–452.

ICABR (2004) Papers from 8th International conference on agricultural biotechnology: trade and domestic production. International Consortium on agricultural biotechnology research: Ravello (Italy) July 2004.

JOLLY H and BREDDIN R (1995) An overview of food retailing in East Asia, Agribusiness Marketing Services Information Series Q195025. Department of Primary Industries: Brisbane.

KAY P and REGIER T (2003) Resolving the question of color naming universals *Proceedings of the National Academies of Sciences* **15**, 9085–9089.

KELLY G (1955) *The Psychology of Personal Constructs*. New York: Norton.

KOJIMA K (1996) The Eastern consumer viewpoint: the experience of Japan. *Nutrition Reviews* **54**, S186–S188.

KROLL B (1990) Evaluating rating scales for sensory testing with children. *Food Technology* **44**, 78–86.

KWAK N-S and JUKES D (2001) Functional foods part 1: the development of a regulatory concept. *Food Control* **12**, 99–107.

LAING D, PRESCOTT J, BELL G, GILLMORE R, JAMES C, BEST D, ALLEN S, YOSHIDA M and YAMAZAKI K (1993) A cross-cultural study of taste discrimination with Australians and Japanese. *Chemical Senses* **18**, 161–168.

LAING D, PRESCOTT J, BELL G, GILLMORE R, ALLEN S and BEST D (1994) Responses of Japanese and Australians to sweetness in the context of different foods. *Journal of Sensory Studies* **9**, 131–155.

LAWLESS H and HEYMANN H (1998) *Sensory Evaluation of Food*. New York: Chapman & Hall.

LAWLESS H and MALONE G (1986) A comparison of rating scales: sensitivity, replicates and relative measurement. *Journal of Sensory Studies* **1**, 155–174.

LE PAGE A, COX DN, RUSSELL CG and LEPPARD PI (2005) Assessing the predictive value of means-end-chain theory: an application to meat product choice by Australian middle-aged women. *Appetite* **44**, 151–162.

LEE C (1990) Modifying an American consumer behaviour model for consumers in Confucian culture; the case of the Fishbein behavioural intention model. *Journal of International Consumer Marketing* **3**, 27–50.

LEE C and GREEN RT (1991) Cross-cultural examination of the Fishbein behavioral intentions model. *Journal of International Business Studies* **22**, 289–305.

LEPPARD P, RUSSELL CG and COX DN (2004) Improving means-end-chain studies by a ranking method to determine hierarchical value maps. *Food Quality & Preference* **15**, 489–497.

LIN W, SOMWARU A, TUAN F, HUANG J and BAI J (2004) Consumer attitudes towards biotech foods in China. 8th International conference on agricultural biotechnology: trade and domestic production. International consortium on agricultural biotechnology research: Ravello (Italy) July 2004.

LUH BS (1999) Rice products. In Ang CYW, Liu K and Huang Y-W (eds) *Asian Foods Science and Technology*. Lancaster and Basel: Technomic Publishing Co Inc.

MACER D and CHEN NG MA (2000) Changing attitudes to biotechnology in Japan *Nature Biotechnology* **18**, 945–947,

MacFIE HJH and THOMSON DMH (1999) *Measurement of Food Preferences*. Gaithersburg ML: Aspen/Chapman & Hall.

MacKENZIE V (2002) GM foods will their fate be decided in Asia? Hong Kong: ASRIA/ Cazenove Asia.

MARSHALL (1995) *Food Choice and the Consumer*. Glasgow: Blackie Academic & Professional.

MATTES R and LAWLESS H (1985) An adjustment error in optimization of taste intensity. *Appetite* **6**, 103–114.

McCOY M (1983) Personal construct theory and methodology in intercultural research. In Adams-Webber J and Manusco JC (eds) *Applications of Personal Construct Theory*. Ontario: Academic Press Canada, pp. 173–186.

MEISLEMAN HL and MacFIE HJH (1996) *Food Choice, Acceptance and Consumption*. Glasgow: Blackie Academic & Professional.

MEUDIC B and COX DN (2001) Understanding Malaysian consumers' perceptions of breakfast cereals using free choice profiling. *Food Australia* **53**, 303–307.

MIALON VS, CLARK MR, LEPPARD PI and COX DN (2002) The effect of dietary fibre information on consumer responses to breads and 'English' muffins: a cross-cultural study. *Food Quality and Preference* **13**, 1–2.

MILO OHR L (2003) Neutraceuticals and functional foods. *Food Technology* **57**, 59–73.

MISKELLY DM (1993) Noodles, a new look at an old friend. *Food Australia* **45**, 496–501.

MORIMOTO K, TAKESHITA T, MIURA K, MURE K and INOUE C (1994) Does the genetic deficiency in ALDH2 determine the alcohol-drinking behaviour and the induction of chromosome alterations in peripheral lymphocytes by alcohol? In Obe G, Natarajan AT (eds) *Chromosomal Alterations*. Berlin and Heidelberg: Springer Verlag, pp. 293–306.

MURRAY JM, EASTON K and BEST DJ (2001) A study of Chinese-origin and European-origin Australian consumers' texture preferences using a novel extruded product. *Journal of Sensory Studies* **16**, 485–504.

NANTACHAI K, PETTY MF and SCRIVEN FM (1990/91) An application of contextual evaluation to allow simultaneous food product development for domestic and export markets. *Food Quality and Preference* **3**, 13–22.

NEWMAN JM (1999) Cultural aspects of Asian dietary habits. In Ang CYW, Liu K and Huang Y-W (eds) *Asian Foods Science and Technology*. Lancaster and Basel: Technomic Publishing Co. Inc., pp. 455–459.

NEWMAN JM (2000) Chinese meals. In Meisleman HL (ed.) *Dimensions of the Meal*. Gaithersburg, MD: Aspen, pp. 163–175.

NORMAN P (1995) Health locus of control and health behaviour: an investigation into the role of health value and behaviour-specific efficacy beliefs. *Personality and Individual Differences* **18**, 213–218.

O'MAHONY M and ISHII R (1986) A comparison of English and Japanese languages: taste descriptive methodology, codability and the umami taste. *British Journal of Psychology* **77**, 161–174.

O'MAHONY M, PARK H, PARK JY and KIM K-O (2004) Comparison of the statistical analysis of hedonic data using analysis of variance and multiple comparisons versus an R-index analysis of the ranked data. *Journal of Sensory Studies* **19**, 519–529.

OTSUKA S (2000) Japanese meals. In Meisleman HL (ed.) *Dimensions of the Meal*. Gaithersburg, MD: Aspen, pp. 178–190.

PERYAM D and PILGRIM F (1957) Hedonic scale method of measuring food preferences. *Food Technology* **11**, 9–14.

PLINER P and HOBDEN K (1992) Development of a scale to measure the trait of food neophobia in humans. *Appetite* **19**, 105–120.

PRESCOTT J (1998) Comparisons of taste perceptions and preferences of Japanese and Australian consumers: overview and implications for cross-cultural sensory research. *Food Quality and Preference* **9**, 393–402.

PRESCOTT J and BELL G (1993) Tailoring food exports to Japanese taste preferences. *Search* **24**, 161–162, 167–168.

PRESCOTT J and BELL G (1995) Cross-cultural determinants of food acceptability: Recent research on sensory perceptions and preferences. *Trends in Food Science and Technology* **6**, 201–205.

PRESCOTT J, BELL G, GILLMORE R, YOSHIDA M, KORAC S, ALLEN S and YAMAZAKI K (1997) Cross-cultural comparisons of Japanese and Australian responses to manipulations of sweetness in foods. *Food Quality and Preference* **8**, 45–55.

PRESCOTT J, BELL G, GILLMORE R, YOSHIDA M, O'SULLIVAN M, KORAC S, ALLEN S and YAMAZAKI K (1998) Cross-cultural comparisons of Japanese and Australian responses to manipulations of sourness, saltiness and bitterness in foods. *Food Quality and Preference* **9**, 53–66.

PRESCOTT J, YOUNG O and O'NEIL L (2001) The impact of variations in flavour compounds on meat acceptability: a comparison of Japanese and New Zealand consumers. *Food Quality and Preference* **12**, 257–264.

PRESCOTT J, YOUNG O, O'NEILL L, YAU NJN and STEVENS R (2002) Motives for food choice: a comparison of consumers from Japan, Taiwan, Malaysia and New Zealand *Food Quality and Preference* **13**, 489–495.

PRESCOTT J, YOUNG O, ZHANG S and CUMMINGS T (2004) Effects of added 'flavour principles' on liking and familiarity of a sheepmeat product: a comparison of Singaporean and New Zealand consumers. *Food Quality and Preference* **15**, 187–194.

REYNOLDS TJ and OLSON JC (2001) *Understanding Consumer Decision Making*. Mahwah, NJ: Lawrence Erlbaum Associates.

RICHARDSON DP (1996) Functional foods – shades of grey: an industry perspective. *Nutrition Reviews* **54**, S174–S185.

ROGERS EM (1971) *Communication of Innovations*. New York: The Free Press/Macmillan.

ROKEACH M (1973) The nature of human values and value systems. In *The Nature of Human Values*. New York: The Free Press, pp. 3–52.

ROZIN E (1982) The structure of cuisine. In Barker LM (ed.) *The Psychobiology of Human Food Selection*. Westport, CT: AVI Publishing Company Inc., pp. 189–203.

ROZIN E (2000) The role of flavor in the meal and the culture. In Meiselman HL (ed.) *Dimensions of the Meal*. Gaithersburg, MD: Aspen, pp. 134–142.

ROZIN P (1990) Acquisition of stable food preferences. *Nutrition Reviews* **48**, 106–111.

ROZIN P, FISHLER C, IMADA S, SARUBIN A and WRZESNIEWSKI A (1999) Attitudes to food and the role of food in life in the USA, Japan, Flemish Belgium, and France: possible implication for the diet-health debate. *Appetite* **33**, 163–180.

SCHOLDERER J and FREWER LJ (2003) The biotechnology communication paradox: Experimental evidence and the need for a new strategy. *Journal of Consumer Policy* **26**, 125–157.

SCHUTTE H and CIARLANTE D (1998) *Consumer Behaviour in Asia*. Basingstoke and London: Macmillan Press Ltd.

SCHWARTZ S (1992) Universals in the content and structure of values: theoretical advances and empirical tests in 20 countries. *Advances in Experimental Social Psychology* **25**, 1–65.

SCHWARTZ S (1994) Beyond individualism and collectivism: new cultural dimensions of values. In Kim U, Triandis HC, Kagitcibasi C, Choi S-C and Yoon (eds)

Individualism and Collectivism: Theory, Method and Applications. Newbury Park, CA: Sage.

SCRIVEN F and MAK Y (1991) Usage behavior of meat products by Australians and Hong Kong Chinese: A comparison of free choice and consensus profiling. *Journal of Sensory Studies* **6**, 25–36.

SHEPHERD R and RAATS MM (1996) Attitudes and beliefs in food habits. In Meisleman HL and MacFie HJH (eds) *Food Choice, Acceptance and Consumption.* Glasgow: Blackie Academic & Professional, pp. 346–362.

SMALL R, YELLAND J, LUMLEY J and RICE PL (1999a) Cross-cultural research: trying to do it better 1. Issues in study design. *Australian and New Zealand Journal of Public Health* **23**, 385–389.

SMALL R, YELLAND J, LUMLEY J, RICE PL, COTRONEI V and WARREN R (1999b) Cross-cultural research: trying to do it better 2. Enhancing data quality. *Australian and New Zealand Journal of Public Health* 23, 390-395

STEPTOE A, POLLARD TM and WARDLE J (1995) Development of a measure of the motives underlying the selection of food: the food choice questionnaire. *Appetite* **25**, 267–284.

SUBRAHMANYAN S and CHENG PS (2000) Perceptions and attitudes of Singaporeans toward genetically modified food. *The Journal of Consumer Affairs* **34**, 269–289.

SUINN R, AHUNA C and KHOO G (1992) The Suinn-Lew Asian self identity acculturation scale: concurrent and factorial validation. *Educational and Psychological Measurement* **52**, 1041–1046.

SUPERMARKET TO ASIA/AUSTRADE (1998) *12 Asian Food Markets.* Barton ACT: Australian Trade Commission.

TEE E-S, CHEN J and ONG C-N (2004) *Functional Foods in Asia Current Status and Issues.* Singapore: International Life Sciences Institute South East Asia Region, monograph series.

TREY DENTON L and KAIXUAN X (1995) Food selection and consumption in Chinese markets. *Journal of International Food Agribusiness Marketing* **7**, 55–77.

TRIANDIS HC (1994) *Culture and social behaviour.* New York: McGraw-Hill.

TRIANDIS HC (1996) The psychological measurement of cultural syndromes. *American Psychologist* **51**, 407–415.

VAN DE VIJVER FJR and LEUNG K (1997) *Methods and Data Analysis for Cross-cultural Research.* Thousand Oaks, CA: Sage Publications.

VAN KLEEF EL, VAN TRIJP, HCM and LUNING P (2005) Consumer research in the early stages of new product development: a critical review of methods and techniques. *Food Quality and Preference* **16**, 181–201.

WALLSTON K, WALLSTON B and DEVELLES R (1978) Development of the multi-dimensional health locus of control scale. *Health Education Monographs* **6**, 160–169.

WARD C, RESURRECTION A and McWATTERS K (1998) Comparison of acceptance of snack chips containing cornmeal, wheatflour and cornpea meals by US and West-African consumers. *Food Quality and Preference* **9**, 327–332.

WHEELER E and TAN SP (1983) Food for equilibrium: The dietary principals and practices of Chinese families in London. In Murcott A (ed.) *The Sociology of Food and Eating.* Aldershot: Gower, pp. 84–89.

WILLS RBH and WOOTON M (1997) Sensory perceptions by Koreans of dry-salted wheat noodles. *Journal of the Science of Food and Agriculture* **74**, 156–160.

WILSON C, EVANS G, LEPPARD P and SYRETTE J (2004) Reaction to genetically modified food crops: How does the perception of risk and benefits impact on subsequent information gathering? *Risk Analysis* **24**, 1311–1321.

YAMAGUCHI S, KIMURA M and ISHII R (1988) Comparison of Japanese and American taste thresholds. 22nd Japanese symposium on taste and smell **19**, 73–76. Abstract.

YAO E, LIM J, TAMAKI K, ISHII R, KIM K-O and O'MAHONEY M (2003) Structured and unstructured 9-point hedonic scales: a cross cultural study with American, Japanese and Korean consumers. *Journal of Sensory Studies* **18**, 115–139.

YEH L, KIM K, CHOMPREEDA H, RIMKEEREE H, YAU N and LUNDAHL D (1998) Comparison in use of the 9-point hedonic scale between Americans, Chinese, Koreans and Thai. *Food Quality and Preference* **9**, 413–419.

YOON G and CHOI S-C (eds) (1994) *Psychology of the Korean People*. Seoul: Korean Psychological Society/Dong-A Publishing.

Part IV

Consumers, food and health

17

Liking, wanting and eating: drivers of food choice and intake in obesity[1]

D. J. Mela, Unilever Food and Health Research Institute, The Netherlands

17.1 Introduction

In science and anecdote, it is often assumed that obesity is in part driven by a heightened hedonic response to specific foods or a greater pleasure from eating in general. For at least 50 years, initially focusing on sweet taste perception (e.g., Pangborn and Simone, 1958), a large volume of research on food choice and intake has been driven by this intuitively-appealing idea (Mela and Rogers, 1998; Nasser 2001; de Graaf, 2005). However, much of this literature has blurred together aspects of perception, physiology and behaviour. Recent developments highlight important distinctions to be made between 'liking' (pleasure derived from oro-sensory stimulation of food) and 'wanting' (incentive salience, the motivation to engage in eating), and how these might relate to human food consumption behaviours (Mela, 2001; Blundell and Finlayson, 2004; Berthoud, 2004). The theoretical and neurophysiological underpinning for these concepts have been particularly articulated by Berridge and colleagues (Berridge, 1996, 2004; Berridge and Robinson, 1995; Winkielman and Berridge, 2003). Although the operationalization and measurement of these is still relatively undeveloped in eating behaviour research with humans, recent work has begun to reveal scenarios where the liking vs wanting distinction can be observed and characterized (Saelens and Epstein, 1996; Epstein *et al.*, 2003, 2004; Finlayson *et al.*, 2005). This chapter highlights why this distinction may

---

1. This chapter is an extended version of Mela DJ (2006), 'Eating for pleasure or just wanting to eat? Reconsidering sensory hedonic responses as a driver of obesity', *Appetite*, 47(1), 10–17. Extensive use is made of the original text and figures, with permission of the publishers.

be important for how we model human eating motivations in general, and its relevance to weight control and obesity.

17.1.1   Obesity and 'homeostatic' control of appetite

It is now apparent that variation in obligate energy metabolism plays at most a minor role in the aetiology of obesity (Weinsier *et al.*, 1993, 2003; Mela and Rogers, 1998; Hill and Melanson, 1999; Ravussin and Bogardus, 2000). Thus, food intake and physical activity behaviours have (belatedly) become major foci for weight control research and intervention. With regard to food intake and choice, the preponderance of research has been directed towards understanding the hormonal and neuronal machinery that presumably evolved to maintain immediate and longer-term energy needs and metabolic health. This knowledge may be exploited in understanding certain pathologies and developing (pharmaceutical) treatments. However, there is frankly little evidence for 'defects' in these homeostatic systems as a cause of obesity except in relatively unusual cases (Hellström *et al.*, 2004). Even while some investigators may speak of 'insensitivity' of hormonal systems that signal suppression of feeding, the evidence actually points to systems that are intact and functioning at the limits of their biological potency, but apparently overridden by other sources of positive stimulation to eat (Berthoud, 2004).

17.1.2   Obesity and 'non-homeostatic' influences on appetite

Studies of endogenous physiological mediators of eating have, arguably, highlighted potential therapeutic targets for obesity, but told us little or nothing about its aetiology or prevention. There is a growing consensus that observed variation (across time, place and people) in energy intake relative to obesity largely reflects variation in the influence of (and susceptibility to) factors which stimulate non-homeostatic eating (eating not driven by biological need state or signals), and their interactions with or overriding of the internal homeostatic systems (Schlundt *et al.*, 1990; Yeomans *et al.*, 2004; Blundell and Findlayson, 2004; Berthoud, 2004). This view is prompting the development of hypotheses and research approaches to identify physiological correlates that may explain the variance in predisposition toward non-homeostatic eating, and its contribution to human obesity. Note that common obesity in humans is not necessarily the result of higher energy intakes, as opposed to lower activity energy expenditure, relative to the non-obese population. The degree of 'overconsumption' in obesity may be even less if account is made of the larger obligate metabolic costs associated with a greater body mass. Thus, the terms 'high', 'excessive', or 'over-' -eating and -consumption simply refer to a sustained excess of energy intake relative to the level required by an individual to maintain a stable body weight in the desired range.

There is a large literature on eating motivations and also characterization of eating behaviours, though the mechanistic links between these remain a subject of considerable conjecture. Academic and industrial research on food choice frequently makes an underlying assumption that taste (here used in the generic sense, although much of the work focuses on gustation alone) is *the* driver of food selection and purchase. In other words, there is a latent or stated presumption that food choices are a close or even direct reflection of sensory hedonic responsiveness. This continues to be a justification for research on orosensory perception and liking, despite frankly minimal and inconsistent evidence that these explain meaningful variation in energy intakes or nutritional status.

17.2   Terminology: liking, desire and preference

In order to explore this further, finer and more consistent distinctions need to be made in the terminology used to refer to different aspects of food acceptance, such as the one which will be used here (Mela, 2000).

- *Liking*   The immediate experience or anticipation of pleasure from the orosensory stimulation of eating a food; hedonic value or 'palatability' (but see later text).
- *Desire*   Wanting; the intrinsic motivation to engage in eating a food, now or in the (near) future.
- *Preference*   Selection of a food over relevant alternatives at the point of choice, including intrinsic and extrinsic factors (for humans, this would include evaluation of liking and desire, plus consideration of health, brand, cost, convenience, etc.).

Particular confusion surrounds the term 'preference', which is variably used to mean liking (e.g., 'sensory preference tests'), or choice in a limited situation (e.g., product comparison tests in marketing; two-bottle 'preference' tests with rats), or actual purchase decisions or food intake ('consumer preference' in sales data, or measures of nutrient consumption) (Mela, 2000). I will restrict myself insofar as possible to the latter uses of the term, by which 'preference' is largely an outcome rather than a cause.

There is clearly an implied hierarchy here when related to everyday conscious experience (Fig. 17.1), where liking is seen as an essential component of desire, and desire is a major contributor to preference. However, the disconnections amongst these may actually be more interesting to explore, from the research as well as public health and commercial perspectives. As noted, several authors have begun to draw particular attention to the distinction between 'liking' and 'wanting', which is less readily amenable to conscious introspection (Berridge, 2004), and will be considered further in this text.

A preference (selection) of less desired alternatives is more immediately recognized from common experience. Less 'desired' (in terms of intrinsic motivations) or less liked foods may be chosen where these positive drivers for choice are outweighed by, for example, physical/economic or cognitive/attitudinal constraints. This is an obvious but certainly not trivial issue: A goal

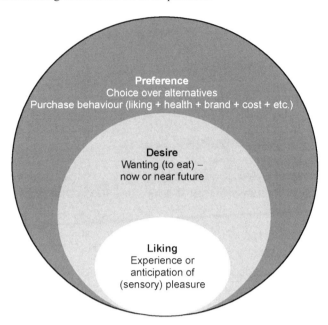

Fig. 17.1 Hierarchical view of determinants of everyday food choice in humans, where liking (hedonic value, pleasure) is seen as part of desire (state of wanting, intrinsic motivation), which is one of many factors influencing preference (actual food choice). In reality, an individual may prefer something (e.g., for reasons of health or cost) which is not very greatly desired, though less likely to experience desire for a food that is not also liked.

of many public health campaigns is to try to shift consumer choice toward foods that are (by the definitions above) initially *less liked and less desired.* This fact alone positions the understanding and ability to guide food likes and wants as a central challenge to academic and industrial nutritionists and consumer research.

17.3   Obesity, food liking and choice

It is worth exploring these concepts further, in order to understand potential sources of variance in human eating behaviour related to obesity. One can start by asking a simple question: What is the evidence that obese individuals experience greater food 'liking'?

17.3.1   Obesity, sensory function and liking in the laboratory

Previous research on this issue (see Mela and Rogers, 1998; de Graaf, 2005) leads to the general conclusion that, compared to normal weight individuals, obese subjects exhibit normal chemosensory function (e.g., detection, recognition), and no consistent differences in liking for specific tastes (notably including sweetness, which has been the subject of particular interest) or aromas (though this has not been well-studied). The question of texture perception and acceptance remains rather less clear, particularly in relation to the role of fats in foods. This partly reflects the relative difficulty of objectively characterizing and controlling the physical properties of (test) foods, but also the wide and non-specific range of terminology available to describe textural sensations (e.g., Cox et al., 1998).

In different studies, population groups at elevated risk of obesity (already obese, reduced-weight former-obese, non-obese overweight, children of obese parents) have been found to report greater liking for higher fat items in laboratory sensory tests, or freely choose to consume foods higher in fat or energy density (reviewed by Mela and Rogers, 1998; de Graaf, 2005; Rissanen et al., 2002; Nasser, 2001). However, the literature and data on this are far from consistent (e.g., Salbe et al., 2004), and may suffer from a number of difficulties in design (e.g., choice, control and presentation of food stimuli) and interpretation. Paradoxically, these methodological issues may be most problematic when real examples of everyday foods are used in testing. This is compounded by the fact that obese subjects will often carry experiences and thoughts of food that differ from lean subjects (Herman et al., 2005). Particularly relevant to this discussion is the possible conflation of affective 'liking' ratings with subjects' attitudes and cognitions, and expectations or interpretation of the rating task (e.g., pleasure/liking vs. wanting/desire to eat the test items). In line with this view, implicit rating procedures may give a somewhat different picture (Roefs and Jansen, 2002; Roefs et al., 2005).

17.3.2   Obesity, food pleasure and choice in the real world

Direct extrapolation from laboratory tests may lead to unwarranted conclusions regarding the role of hedonic responses to specific foods in obesity-related eating. Indeed, 'liking' ratings elicited using either lists of foods (Fig. 17.2, Cox et al., 1998) or during consumption of freely-selected foods (Fig. 17.3, Cox et al., 1999) fail to support the notion that obese individuals select or judge foods overall as being more pleasant, with limited evidence for differences in ratings of specific food types. Furthermore, cross-sectional and longitudinal studies find little consistency in the apparent food intake patterns associated with body mass index (BMI) or obesity (Togo et al., 2001; Halkjær et al., 2004; Drapeau et al., 2004; Mela and Rogers, 1998; de Graaf, 2005). Nevertheless, obese individuals may disproportionately consume higher energy content (density or volume) foods or meals (Westerterp-Plantenga et al., 1996; Cox et al., 1999), though this may partly depend on the method of dietary analysis (Cox and Mela, 2000). Those observations are also consistent with current views of the effect of energy density on weight control.

These data suggest that obesity may be associated with greater motivation for food consumption, possibly directed at energy-dense foods, without any clear difference in the pleasure derived from the orosensory experience of eating.

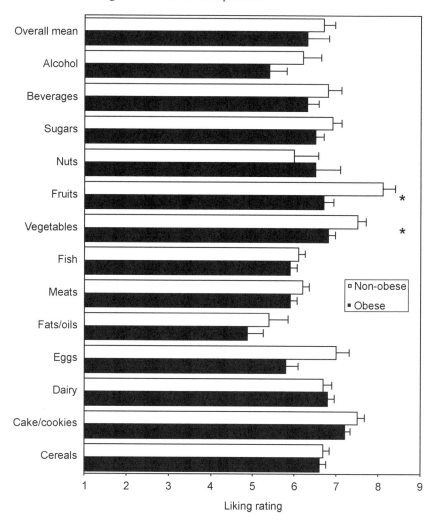

Fig. 17.2   Liking ratings (questionnaire, 1 = 'dislike extremely', 9 = 'like extremely') for listed foods comprising 14 food groups, by non-obese (BMI < 25, n = 20) and obese (BMI > 30, n = 23) subjects. *p < 0.05 between groups. Data from Cox *et al.* (1998).

17.4   'Palatability' and food intake

17.4.1   What do we mean by 'palatability'?

Research on the role of food pleasure and hedonics in intake commonly refers to the role of 'palatability', although the literature is often unclear and inconsistent in its use of this term. Fifteen years ago, Ramirez (1990) noted that 'palatability' may be used to refer to any or all of '(1) a simple observation that some foods stimulate more intake than others, (2) an innate response to the taste of foods

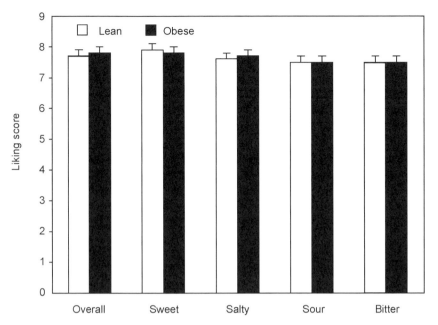

Fig. 17.3   Mean liking ratings (questionnaire, 1 = 'dislike extremely', 9 = 'like extremely') of foods freely chosen and eaten by lean (BMI < 25, $n = 41$) and obese (BMI > 30, $n = 35$) subjects. Foods were also categorized by predominant taste characteristic as judged by subjects at the time of eating. No significant differences between groups. Data from Cox *et al.* (1999).

that alters appetite, (3) a correlate of food intake that does not itself affect intake, and (4) a link in a causal chain involving prior associations between foods and their postingestive consequences.'

Research with animals has often used intake as the indicator of presumed 'palatability', leading to circular arguments about influences of palatability on intake (Yeomans, 1998). In other words, the conclusion that a more 'palatable' food stimulates intake may be derived from a prior position that if a food stimulates intake it must be more 'palatable'. More recently, the term has even been used as a shorthand for a certain nutritional composition presumed to drive excessive intakes: 'palatable food, i.e., food rich in fat and sugar' (Erlanson-Albertsson, 2005, p. 65). Yet, when energy density is controlled, the effects of macronutrient composition on energy intake in both rats (Ramirez and Friedman, 1990) and humans (Rolls and Bell, 1999; Stubbs *et al.*, 2000) is greatly attenuated, suggesting that it is energy density rather than fat or sugar that makes a diet 'palatable', if defined by energy intakes. On the other hand, if defined by liking, then reduced-energy foods (using fat replacers and intense sweeteners) may match the 'palatability' of their more energy-dense counterparts, yet lead to reduced energy intake (e.g., DellaValle *et al.*, 2005). In rats, lowering the energy density of dry feeds and purified macronutrient sources by addition of water can also make them more 'palatable' and stimulate a prolonged increase in consumption (Ramirez, 1987; Sclafani, 2004). And there are even examples where an apparently less palatable (less preferred in choice test) food may nevertheless lead to greater consumption and weight gain (Ramirez, 1988). Thus, experimental manipulations of energy density or oro-sensory quality can achieve almost any directional effect on intake, at least in the short-term.

Nutrient- and intake-based definitions and assumptions of 'palatability' are clearly both unhelpful and unreliable. Most researchers in human food intake and acceptance (including this author) prefer to use the term in its colloquial sense, referring to a (positive) hedonic evaluation (pleasantness, liking) under a given set of conditions (e.g., Yeomans *et al.*, 2004; de Castro and Plunkett, 2001; Blundell and Rogers, 1991). This makes no *a priori* assumptions about the nutritional composition or other aspects that make the food palatable, nor its effects on intake. In the best case, those factors are either controlled or their nature objectively explored. Furthermore, unlike nutritional composition, palatability is not seen as a fixed property of a food, but a momentary evaluation of it (Blundell and Rogers, 1991), which can change with experience. Often though, the term is not defined at all, and various meanings may be implied. The bottom line is that confounders and unclarities present in research on both animals and humans may unduly influence discussions of 'palatability' and its presumed role in stimulating (over-)eating.

17.4.2 'Palatability', liking and intake in humans

Everyday experience and controlled research confirm the superficially trivial fact that people often eat more of better-liked foods when offered an *ad libitum* choice, and that the experience or anticipation of highly-liked food can stimulate its consumption in the absence of an energy deficit or perceived state of hunger (Sørensen *et al.*, 2003). Given that most food likes are acquired through past experience indicating the safety and nutritional value of the liked food, this is evolutionarily adaptive behaviour. Furthermore, there is a body of evidence that obese (or, at least, weight-concerned) individuals may have a heightened appetitive response to the availability of well-liked foods (de Graaf, 2005; Herman *et al.*, 2005; Schacter, 1971). But this does not mean that changes in food 'palatability' are necessarily causally implicated in trends toward increased energy intake and obesity. Even in rats, there may be threshold effects, where adding a preferred flavour does not enhance the response to an already 'palatable' feed (Sclafani, 2004). In traditional and modern undeveloped human societies, access to highly palatable foods may once have been limited. However, where humans have the economic and physical opportunity to select a highly palatable diet, they rarely select foods that they do not like. Thus, in wealthy modern societies, we observe that liking for what people select is uniformly very high (Cox *et al.*, 1999; Fig. 17.3), and variation in food 'palatability' may explain only a very small fraction of variation in intake (de Castro and Plunkett, 2001).

17.4.3   Palatability and obesity trends

The preceding section is not intended to trivialize the role of liking as a general driver of intake, but to question how and where the immediate pleasure derived from eating is a useful focus for understanding or intervening in obesity. Are foods today generally more pleasant to eat than the foods of 25 years ago? Are today's manufactured foods more 'palatable' than the home-cooked food from a generation ago? These unanswerable questions reflect anecdotal opinions heard by the author, although there is no objective evidence at all for a generalized increase in food 'palatability' paralleling the global and socio-economic trends in obesity over the past 25 years. Furthermore, the view that 'food is more palatable, and that is driving people to eat too much of it' does not lead to any credible course of action (make worse-tasting food?). Economic and availability arguments are far more objective and compelling (Blaylock *et al.*, 1999; Cutler *et al.*, 2003; Wansink, 2004; Drewnowski and Darmon, 2005), and can be linked to hedonic arguments if we consider the increased presence and accessibility of food and also food composition (Lowe and Levine, 2005). It is therefore important to draw a distinction between, on the one hand, variance in the palatability of the foods eaten (which is small), and on the other hand, (1) variance in the accessibility and composition of those foods (addressed largely by reference to the food economics and marketing literature), and (2) variance in individual responsiveness to environmental cues that prompt a desire to eat, which may include the presence and availability of those foods (addressed below).

17.5   Food 'wanting' and intake

Commenting on Ramirez (1990), Rogers (1990) proposed that 'a distinction should be made between the pleasantness of the taste of food (influenced by palatability) and the pleasantness of ingesting that food'. This usefully breaks down the issue into two separate but related questions: 1) What are the factors that make a food 'palatable'?, and 2) What are the factors (including and beyond 'palatability') that motivate desire to (over)-ingest a given food? The first of these has been the subject of a large volume of research on origins and acquisition of food likes (Brunstrom, 2005; Mela, 2000; Sclafani, 1995, 2004) and will not be discussed further here, while the second is a wider issue that is now of renewed interest.

17.5.1   'Non-homeostatic' eating: animal models

There has been growing interest in the use of various novel paradigms to stimulate non-homeostatic overeating, 'bingeing', 'cravings' and nutrient-associated 'dependencies' in rodents (Corwin and Buda-Levin, 2004; Corwin and Hajnal, 2005; Avena *et al.*, 2005; Colantuoni *et al.*, 2002). The relevance and value of these models in understanding and ultimately treating human obesity is not yet clear, and will only be given limited consideration here. These models typically depend on use of particular types of food stimuli and access schedules to produce the relevant behavioural responses (e.g., Corwin and Buda-Levin, 2004; Corwin, 2004). However, there are still many potential confounders in the interpretation of these data, not limited to issues raised already in relation to 'palatability'. When viewed from a nutritional perspective, most of these models use extreme dietary conditions (purified sources of fat or carbohydrate). Indeed, effects may be diminished or lost when more nutritionally balanced materials are used (Simpson and Raubenheimer, 1997, 2005; Melhorn et al., 2005). Thus, it becomes unclear whether some responses reflect an inappropriate (in human terms) overconsumption, or perhaps an appropriate bio-behavioural strategy in response to a contrived nutritional challenge. This alternative interpretation may still have relevance to humans (e.g., nutrient balancing hypotheses such as that proposed by Simpson and Raubenheimer, 2005). However, there are undoubtedly *caveats* to consider in assigning causality to the animal data, perhaps analogous to the issues of variety vs composition vs 'palatability' associated with the 'cafeteria feeding' rodent model of obesity 20 years ago (Naim et al., 1985; Sclafani, 2004).

Experimental animal research has been crucial to identifying common substrates and pathways associated with (homeostatic) eating, but there will always be doubts as to whether animal models can faithfully represent the natural history and phenomenology of non-homeostatic eating behaviours in humans (Herman, 1996). Nevertheless, there are precedents where this has led to valuable insights and hypothesis generation (Schacter, 1971; Schacter and Rodin, 1974; Berridge, 1996).

17.5.2 Food pleasure and reward in obese humans

There has been a recent resurgence of interest in the potential role of pleasure and related aspects of food wanting and 'craving' in promoting energy consumption and obesity (Yeomans et al., 2004; Erlanson-Albertsson, 2005; Berthoud, 2004). This has stimulated a wave of empirical studies and speculation related to central neuroanatomical responses and reward systems in (obese) humans (e.g., Di Chiara, 2005; Wang et al., 2001, 2004a, 2004b; Volkow et al., 2002; Volkow and Wise, 2005; Del Parigi et al., 2003). Much of this literature makes analogies between food and drugs. It is not surprising that these share common reward pathways, which presumably originated to reinforce appropriate food choices and intake, and it is possible to draw qualitative parallels between selected behavioural and neurobiological aspects of obesity and drug abuse (Volkow and Wise, 2005). These papers typically focus on possible obesity-related deficits in dopaminergic systems underlying a reward deficiency syndrome (Blum et al., 1996), whereby additional compensatory food stimulation is needed to adequately satisfy desires. This notion derived empirical support from the work of Wang et al. (2001), who reported low striatal dopamine D_2 receptor availability in association with high body mass index in a group of extremely obese subjects. Those authors suggest that their obese subjects might manifest blunted reward responses to food, and that their presumed overeating could be a way of compensating for this deficit.

Further work is needed to confirm this view and its broader relevance and possible causality. A reduction in dopamine receptor activity in obesity could also develop secondary to repeated stimulation of dopamine release from chronic overeating, or as a response to other obesity-related neuroendocrine changes, rather than being a primary cause of the condition. Or it might have little to do with actual body weight, and be better related to past behaviours and cognitions associated with weight concern and dieting (Herman *et al.*, 2005; Lowe and Levine, 2005), which have repeatedly been shown to influence food perceptions and appetitive responses to food exposure.

There are clearly limitations in the degree to which to analogies between obesity and drug (ab)use can be made. This in part reflects differences in the characteristics of these conditions, especially quantitative intensity, and also gaps in knowledge, especially – and importantly – cause-and-effect (Di Chiara, 2005; Del Parigi *et al.*, 2003). Others (Lowe and Levine, 2005; Myslobodsky, 2003) argue that the innate dependence on food, and motivational aspects of energy deprivation, set food apart from, for example, drugs, alcohol and nicotine. Nevertheless, the research on dopaminergic systems and 'non-hedonic' aspects of food motivation in humans (Volkow *et al.*, 2002; Wang *et al.*, 2002) serves to underscore the potential relevance of the 'liking' vs 'wanting' differentiation in relation to food intake and obesity.

17.6   Behavioural discrimination of food liking and wanting in (obese) humans

What is the additional behavioural evidence for this distinction in humans, and its relevance to obesity? Most of the behavioural research addressing food acceptance as a driver of intake in obese humans has focused either on food 'preferences' (in the broad sense, including both hedonic responses and choice) or the related area of hunger and satiety. There are considerably fewer empirical studies that focus on incentive salience and non-hedonic reward aspects of food and eating, or possible discrimination between explicit liking and the desire to eat a food. The subtlety of this distinction in relation to everyday, conscious experience means that it probably cannot be consistently captured using traditional, explicit line or category scales of 'liking', 'desire to eat', etc. Berridge (2004, p. 196) very effectively articulates how the unconscious 'incentive salience' aspects of wanting differ from the cognitive, conscious, experienced sense of wanting.

Implicit tests or other sorts of behavioural tasks or physiological correlates are probably needed to isolate and characterize the liking vs. wanting discrimination for food in humans. For example, Johnson (1974) linked obesity to increased willingness to work for food rewards in the presence of food cues.

Blood flow to the right temporal and parietal cortices in response to food cues has also been shown to be significantly elevated in obese relative to non-obese women, in spite of similar self-reports of liking and desire to eat the foods (Karhunen *et al.*, 1997). Most convincingly, Epstein and colleagues (Saelens and Epstein, 1996; Epstein *et al.*, 2003, 2004) have successfully developed and applied behavioural tests that reveal this distinction between hedonic and reinforcement value of foods. Saelens and Epstein (1996) used a test of reinforcement value of food and non-food rewards with equivalent perceived hedonic value, based on subjects' willingness to 'work' for the rewards in computer tasks. In this model, the food reward had much greater reinforcement value for obese vs lean women. This type of test has also been used to differentiate food reinforcement and liking in smokers, with reinforcing value found to be more strongly related than liking to energy intake (Epstein *et al.*, 2004). Lastly, Epstein *et al.* (2003) have shown that food deprivation selectively influences the reinforcing value of food, relative to its hedonic value, a result which may help resolve past discrepancies in the literature regarding the effect of hunger and satiety on perceived pleasantness of foods (Yeomans *et al.*, 2004). Finlayson *et al.* (2005) have also reported progress in developing computerized tests that could be used to experimentally dissect liking and wanting in humans.

The recent data linking obesity to sensitivity to food reinforcement recall finds further support in observations from the original studies of 'external' eating begun in the 1960s (Lowe and Levine, 2005; de Graaf, 2005; Herman *et al.*, 2005). As described by Schacter (1971; Schacter and Rodin, 1974) and subsequent experiments, a heightened appetitive response to food cues including palatability was frequently observed in obese human subjects. This research suffered from a number of well-known problems in interpretation, including assignment of cause-effect and the role of hunger, weight concern and dieting. Nevertheless, the general notions and original findings fit a (re-)emerging pattern of thinking which is not so different in its essence, though which now benefits from a greater sophistication of both cognitive and neurophysiological conceptual frameworks (Herman *et al.*, 2005; Lowe and Levine, 2005, Berthoud, 2004; Volkow and Wise, 2005; Wang *et al.*, 2004).

17.7   Conclusions

17.7.1   Summary

There is a growing consensus that overeating in obesity reflects responsiveness to non-homeostatic stimuli, rather than a defect or failure of endogenous homeostatic systems involved in energy balance. Food liking and pleasure are viewed as elements that can stimulate non-homeostatic eating, and therefore are also viewed as potential contributors to obesity. However, variation in obesity is not reliably associated with variation in liking of foods or pleasure of eating. Greater understanding of this apparent discrepancy may come from the recognition that 'liking' (hedonic response) may be usefully differentiated from 'wanting' (incentive salience or motivation to eat). This is supported by behavioural and neurophysiological data on responsiveness to food-related cues indicating that obesity may be associated with increased motivation for food consumption, without necessarily any greater explicit pleasure being derived from the orosensory experience of eating. This view and the emerging research related to it should have important and broader general implications for research in food choice and intake, and for its application in commercial and public health approaches to modifying energy intakes.

17.7.2 Relevance: a model of food desire and eating in humans

Does this have any real-world relevance? Figure 17.4 attempts to illustrate this using concepts discussed here, in terms of understanding 'why we want to eat what we want to eat'. This schema recognizes that at any given moment a conscious feeling of the desire to eat a particular food is the outcome of a balance of: (1) physiological state and cues (especially, but not limited to hunger and thirst); (2) anticipated pleasure of eating, (largely acquired from learned ingestive and post-ingestive associations); and (3) external associations and cues (also largely learned, with cognitive components but also elements that may be unrecognized and unconscious). For any given internal or environmental condition, there may be more or less stimulation (or suppression) of desire due to the relative strength of signals from these different drivers (Mela, 2000).

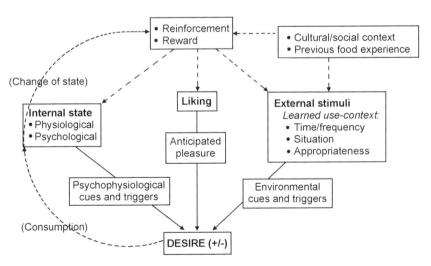

Fig. 17.4 Simplified operational schematic combining the influences of liking (pleasure), internal state (psychophysiology), and external stimuli (learned cues) in the acquisition and activation of desire for foods in everyday situations. Solid lines reflect proximate drivers, dashed lines are underlying processes. (Modified from Mela (2000)). Copyright Society of Chemical Industry. Reproduced with permission granted by John Wiley & Sons Ltd on behalf of the Society of Chemical Industry.

17.7.3 Implications

For research in food intake and appetite regulation
This view places renewed emphasis on understanding the nature of externally-cued eating behaviour, and especially the extent and cause of variations in sensitivity to this (including cognitive restraint and dieting). We have moved beyond the stage where we can look at obesity as a 'defect' in homeostatic systems, or believe that the study of these in isolation can be justified outside of a pharmaceutical approach. For research with humans and especially animals, there is also a need to take heed of past research, and be extremely cautious in design and interpretation in behavioural studies that combine manipulations of both nutritional composition and orosensory stimulation.

For research in sensory food science and food acceptance
In order to understand why certain food stimuli are liked and also have a high and sustained desired frequency of consumption, this field should escape the bounds of its traditionally narrow focus on immediate orosensory responses. Instead, this body of knowledge should be integrated with consideration of the dynamics of the acquisition and changes in liking, and perhaps the *activation of desires*.

For guidance in prevention and treatment of obesity
This seems to give renewed justification for approaches that emphasize environmental control, including structuring and limiting food stimulation. This is already a longstanding feature of behavioural strategies; however, their long-term success remains disappointing. Improved understanding and ability to assess the nature of the individual responsiveness to environmental cues could improve the personalization of such programmes. It may also highlight the need to find ways to fit opportunities to eat highly-liked 'palatable' foods into a controlled environment and structured, balanced eating pattern.

For commercial food developers
It is increasingly clear that limited frequency or changes in food preferences (purchases) over time often do not reflect a poor or loss of orosensory quality, but a sort of product boredom (a change in 'wanting'?) that can be distinguished from liking (Mela, 2000; Stubenitsky *et al.*, 1999; Zandstra *et al.*, 2004). The challenge in relation to weight control remains to offer lower energy foods that are not just 'liked', but also 'wanted'.

For commercial food marketers
This stream of research clearly points to the role of environment cues, including but not limited to food itself as a stimulus to eat. This will place increased pressure on marketers to ensure that they are not unduly adding to the environmental stimulation to eat inappropriately. International food trade bodies have developed guidelines for responsible marketing to adults and children in light of the obesity epidemic (e.g., CIAA, 2004). Advertising and communica-

tion material should be scrutinized to ensure such guidance is being followed in spirit as well as letter.

17.8 A final word

Lastly, I have observed a subjective element of Calvinism colouring even professional discussions of food pleasure, an unstated expression of the view that 'If it tastes nice, it must be bad for you'. From the perspective of evolutionary biology this is clearly a modern and unnatural view of the relationship between mechanisms of food selection and nutritional needs. This also positions pleasure as a foe, rather than a potential ally of healthy eating behaviours. The corollary, 'In order to be good for you, it cannot taste nice' will assuredly get us nowhere in making progress toward attracting consumers to a healthy, balanced food choice and intake. The challenge is to understand the drivers of variance in eating behaviour, and to apply this knowledge to food development, marketing, and public health guidance in ways that make healthy, appropriate eating something that is liked, desired, and preferred.

17.9 References

AVENA NM, LONG KA, HOEBEL BG (2005), 'Sugar-dependent rats show enhanced responding for sugar after abstinence: evidence of a sugar deprivation effect', *Physiol Behav*, 84(3), 359–362.

BERRIDGE KC (1996), 'Food reward: Brain substrates of wanting and liking', *Neurosci BioBehav, Rev*, 20, 1–25.

BERRIDGE KC (2004), 'Motivation concepts in behavioral neuroscience', *Physiol Behav*, 81 (2), 179–209.

BERRIDGE KC, ROBINSON TE (1995), 'The mind of an addicted brain: Neural sensitization of "wanting" versus "liking"', *Curr Dir Psycholog Sci*, 4, 71–76.

BERTHOUD HR (2004), 'Neural control of appetite: cross-talk between homeostatic and non-homeostatic systems', *Appetite*, 43(3), 315–317.

BLAYLOCK J, SMALLWOOD D, KASSEL K, VARIYAM J, ALDRICH L (1999), 'Economics, food choices, and nutrition', *Food Policy*, 24 (2), 269–286.

BLUM K, CULL JG, BRAVERMAN ER, COMINGS DE (1996), 'Reward deficiency syndrome', *Am Sci*, 84, 132–145.

BLUNDELL JE, FINLAYSON G (2004), 'Is susceptibility to weight gain characterized by homeostatic or hedonic risk factors for overconsumption?', *Physiol Behav*, 82(1), 21–25.

BLUNDELL JE, ROGERS PJ (1991), 'Hunger, hedonics, and the control of satiation and satiety', In Friedman M, Kare M, eds, *Chemical Senses Volume 4: Appetite and nutrition*, New York, Marcel Dekker, 127–148.

BRUNSTROM JM (2005), 'Dietary learning in humans: directions for future research', *Physiol Behav*, 85(1), 57–65.

DE CASTRO J, PLUNKETT SS (2001), 'How genes control real world intake: palatability-intake relationships', *Nutrition*, 17(3), 266–268.

CIAA [CONFEDERATION DES INDUSTRIES AGRO-ALIMENTAIRES DE I'UE – CONFEDERATION OF EU FOOD AND DRINK INDUSTRIES] (2004), 'Principles of Food and Beverage Product Advertising', *http://www.gwa.de/fileadmin/download/Kommbranche/ CIAA_Principles.pdf.* Accessed 6 March 2006.

COLANTUONI C, RADA P, MCCARTHY J, PATTEN C, AVENA NM, CHADEAYNE A, HOEBEL BG (2002), 'Evidence that intermittent, excessive sugar intake causes endogenous opioid dependence', *Obesity Res*, 10(6), 478–488.

CORWIN RL (2004), 'Binge-type eating induced by limited access in rats does not require energy restriction on the previous day', *Appetite*, 42, 139–142.

CORWIN RL, BUDA-LEVIN A (2004), 'Behavioral models of binge-type eating', *Physiol Behav, 82*, 123–130.

CORWIN RL, HAJNAL A (2005), 'Too much of a good thing: Neurobiology of non-homeostatic eating and drug abuse', *Physiol Behav*, 86, 5–8.

COX DN, MELA DJ (2000), 'Determination of energy density of freely selected diets: methodological issues and implications', *Int J Obesity*, 24(1), 49–54.

COX DN, VAN GALEN M, HEDDERLEY D, PERRY L, MOORE P, MELA DJ (1998), 'Sensory and hedonic judgments of common foods by lean consumers and consumers with obesity', *Obesity Res*, 6, 438–447.

COX DN, PERRY L, MOORE PB, VALLIS L, MELA DJ (1999), 'Sensory and hedonic associations with macronutrient and energy intakes of lean and obese consumers', *Int J Obesity*, 23(4), 403–410.

CUTLER DM, GLAESER EL, SHAPIRO JM (2003), 'Why have Americans become more obese?', *J Econ Perspect*, 17, 93–118.

DEL PARIGI A, CHEN K, SALBE AD, REIMAN EM, TATARANNI PA (2003), 'Are we addicted to food?' *Obesity Res*, 11(4), 493–495.

DELLAVALLE DM, ROE LS, ROLLS BJ (2005), 'Does the consumption of caloric and non-caloric beverages with a meal affect energy intake?', *Appetite*, 44(2), 187–193.

DI CHIARA G (2005), 'Dopamine in disturbances of food and drug motivated behavior: a case of homology?' *Physiol Behav*, 86(1–2), 9–10.

DRAPEAU V, DESPRES JP, BOUCHARD C, ALLARD L, FOURNIER G, LEBLANC C, TREMBLAY A (2004), 'Modifications in food-group consumption are related to long-term body-weight changes', *Am J Clin Nutr*, 80(1), 29–37.

DREWNOWSKI A, DARMON N (2005), 'The economics of obesity: dietary energy density and energy cost', *Am J Clin Nutr*, 82(1 Suppl), 265S–273S.

EPSTEIN LH, TRUESDALE R, WOJCIK A, PALUCH RA, RAYNOR HA (2003), 'Effects of deprivation on hedonics and reinforcing value of food', *Physiol Behav*, 78(2), 221–227.

EPSTEIN LH, WRIGHT SM, PALUCH RA, LEDDY J, HAWK LW JR, JARONI JL, SAAD FG, CRYSTAL-MANSOUR S, LERMAN C (2004), 'Food hedonics and reinforcement as determinants of laboratory food intake in smokers', *Physiol Behav*, 81(3), 511–517.

ERLANSON-ALBERTSSON C (2005), 'How palatable food disrupts appetite regulation', *Basic Clin Pharmacol Toxicol*, 97(2), 61–73.

FINLAYSON G, KING N, BLUNDELL J (2005), 'Experimental dissociation of 'liking' and 'wanting' in humans', *Appetite*, 44(3), 360 (abstract).

DE GRAAF C (2005), 'Sensory responses, food intake and obesity', In Mela D, ed., *Food, Diet and Obesity*, Cambridge UK, Woodhead Publishing Limited, 137–159.

HALKJÆR J, SØRENSEN TIA, TJØNNELAND A, TOGO P, HOLST C, HEITMANN BL (2004), 'Food and drinking patterns as predictors of 6-year BMI-adjusted changes in waist circumference', *Br J Nutr*, 92 (4), 735–748.

HELLSTRÖM PM, GELIEBTER A, NÄSLUND E, SCHMIDT PT, YAHAV EK, HASHIM SA, YEOMANS MR (2004), 'Peripheral and central signals in the control of eating in normal, obese and binge-eating human subjects', *Br J Nutr*, 92 (Suppl. 1), S47–S57.

HERMAN CP (1996), 'Human eating: diagnosis and prognosis', *Neurosci Biobehav Rev*, 20(1), 107–111.

HERMAN CP, POLIVY J, LEONE T (2005), 'The psychology of overeating', In Mela D, ed., *Food, Diet and Obesity*, Cambridge UK, Woodhead Publishing Limited, 114–136.

HILL JO, MELANSON EL (1999), 'Overview of the determinants of overweight and obesity: current evidence and research issues', *Med Sci Sports Exercise*, 31(11, Suppl 1), S515–S521.

JOHNSON WG (1974) 'Effect of cue prominence and subject weight on human food-directed performance', *J Pers Social Psychol*, 29, 843–848.

KARHUNEN LJ, LAPPALAINEN RI, VANNINEN EJ, KUIKKA JT, UUSITUPA MIJ (1997), 'Regional blood flow during food exposure in obese and normal-weight women', *Brain*, 120, 1675–1684.

LOWE MR, LEVINE AS (2005), 'Eating motives and the controversy over dieting: Eating less than needed versus less than wanted', *Obesity Res*, 13, 797–806.

MELA DJ (2000), 'Why do we like what we like?', *J Sci Food Agric*, 81, 10–16.

MELA DJ (2001), 'Determinants of food choice: Relationships with obesity and weight control', *Obesity Res*, 9, S249–S255.

MELA DJ, ROGERS PJ (1998), *Food, Eating & Obesity: The psychobiological basis of appetite and weight control*, London UK, Chapman & Hall.

MELHORN SJ, HEIMAN JU, STRADER AD, CLEGG DJ, BENOIT SC (2005), 'Comparison of binge-like behavior in rats with limited access to Crisco or nutritionally complete high fat', *Appetite*, 44(3), 367 (abstract).

MYSLOBODSKY M (2003), 'Gourmand savants and environmental determinants of obesity', *Obesity Rev*, 4(2), 121–128.

NAIM M, BRAND JG, KARE MR, CARPENTER RG (1985), 'Energy intake, weight gain and fat deposition in rats fed flavored, nutritionally controlled diets in a multichoice ("cafeteria") design', *J Nutr*, 115(11), 1447–1458.

NASSER J (2001), 'Taste, food intake and obesity', *Obesity Rev*, 2(4), 213–218.

PANGBORN RM, SIMONE M (1958), 'Body size and sweetness preference', *J Am Dioet Assoc*, 34, 924–928.

RAMIREZ I (1987), 'Feeding a liquid diet increases energy intake, weight gain and body fat in rats', *J Nutr*, 117(12), 2127–2134.

RAMIREZ I (1988), 'Overeating, overweight and obesity induced by an unpreferred diet', *Physiol Behav*, 43(4), 501–506.

RAMIREZ I (1990), 'What do we mean when we say "palatable food"?', *Appetite*, 14(3), 159–161.

RAMIREZ I, FRIEDMAN MI (1990), 'Dietary hyperphagia in rats: role of fat, carbohydrate, and energy content', *Physiol Behav*, 47(6), 1157–1163.

RAVUSSIN E, BOGARDUS C (2000), 'Energy balance and weight regulation: genetics versus environment', *Br J Nutr*, 83(Suppl s1), 17–20.

RISSANEN A, HAKALA P, LISSNER L, MATTLAR C-E, KOSKENVUO M, RÖNNEMAA T (2002), 'Acquired preference especially for dietary fat and obesity: a study of weight-discordant monozygotic twin pairs', *Int J Obesity*, 26(7), 973–977.

ROEFS A, JANSEN A (2002), 'Implicit and explicit attitudes toward high-fat foods in obesity', *J Abnormal Psychol*, 111(3), 517–521.

ROEFS A, HERMAN CP, MACLEOD CM, SMULDERS FT, JANSEN A (2005) 'At first sight: how do restrained eaters evaluate high-fat palatable foods?', *Appetite*, 44(1), 103–114.

ROGERS PJ (1990), 'Why a palatability construct is needed', *Appetite*, 14(3), 167–170.

ROLLS BJ, BELL EA (1999), 'Intake of fat and carbohydrate: role of energy', *Eur J Clinl Nutr*, 53(Suppl 1), S166–S173.

SAELENS BE, EPSTEIN LH (1996), 'Reinforcing value of food in obese and non-obese women', *Appetite*, 27, 41–50.

SALBE AD, DEL PARIGI A, PRATLEY RE, DREWNOWSKI A, TATARANNI PA (2004), Taste preferences and body weight changes in an obesity-prone population', *Am J Clin Nutr*, 79(3), 372–378.

SCHACHTER S (1971), 'Some extraordinary facts about obese humans and rats', *Am Psychol*, 26, 129–144.

SCHACHTER S, RODIN J (1974), *Obese Humans and Rats*, Washington DC, Erlbaum/Halsted.

SCHLUNDT DG, HILL JO, SBROCCO T, POPE-CORDLE J, KASSER T (1990), 'Obesity: a biogenetic or biobehavioral problem', *Int J Obesity*, 14(9), 815–828.

SCLAFANI A (1995), 'How food preferences are learned: laboratory animal models', *Proc Nutr Soc*, 54, 419–427.

SCLAFANI A (2004), 'Oral and postoral determinants of food reward', *Physiol Behav*, 81(5), 773–779.

SIMPSON SJ, RAUBENHEIMER D (1997), 'The geometric analysis of feeding and nutrition in the rat', *Appetite*, 28, 201–213.

SIMPSON SJ, RAUBENHEIMER D (2005), 'Obesity: the protein leverage hypothesis', *Obesity Rev*, 6(2), 133–142.

SØRENSEN LB, MØLLER P, FLINT A, MARTENS M, RABEN A (2003), 'Effect of sensory perception of foods on appetite and food intake: a review of studies on humans', *Int J Obesity*, 27(10), 1152–1166.

STUBBS J, FERRES S, HORGAN G (2000), 'Energy density of foods: effects on energy intake', *Crit Rev Food Sci Nutr*, 40(6), 481–515.

STUBENITSKY K, AARON JI, CATT SL, MELA DJ (1999), 'Effect of information and extended use on the acceptance of reduced-fat products', *Food Qual Pref*, 10, 367–376.

TOGO P, OSLER M, SØRENSEN TIA, HEITMANN BL (2001), 'Food intake patterns and body mass index in observational studies', *Int J Obesity*, 25(12), 1741–1751.

VOLKOW ND, WISE RA (2005), 'How can drug addiction help us understand obesity?', *Nature Neurosci*, 8(5), 555–560.

VOLKOW ND, WANG GJ, FOWLER JS, LOGAN J, JAYNE M, FRANCESCHI D, WONG C, GATLEY SJ, GIFFORD AN, DING YS, PAPPAS N (2002), '"Nonhedonic" food motivation in humans involves dopamine in the dorsal striatum and methylphenidate amplifies this effect', *Synapse*, 44(3), 175–180.

WANG GJ, VOLKOW ND, LOGAN J, PAPPAS NR, WONG CT, ZHU W, NETUSIL N, FOWLER JS (2001), 'Brain dopamine and obesity', *Lancet*, 357, 354–357.

WANG GJ, VOLKOW ND, FOWLER JS (2002), 'The role of dopamine in motivation for food in humans: implications for obesity', *Expert Opin Ther Targets*, 6(5), 601–609.

WANG GJ, VOLKOW ND, TELANG F, JAYNE M, MA J, RAO M, ZHU W, WONG CT, PAPPAS NR, GELIEBTER A, FOWLER JS (2004a), 'Exposure to appetitive food stimuli markedly activates the human brain', *Neuroimage*, 21(4), 1790–1797.

WANG GJ, VOLKOW ND, THANOS PK, FOWLER JS (2004b), 'Similarity between obesity and drug addiction as assessed by neurofunctional imaging: a concept review,' *J Addict Dis*, 23(3), 39–53.

WANSINK B (2004), 'Environmental factors that increase the food intake and consumption volume of unknowing consumers', *Ann Rev Nutr*, 24, 455–479.

WEINSIER RL, BRACCO D, SCHUTZ Y (1993), 'Predicted effects of small decreases in energy expenditure on weight gain in adult women', *Int J Obesity*, 17(12), 693–700.

WEINSIER RL, HUNTER GR, ZUCKERMAN PA, DARNELL BE (2003), 'Low resting and sleeping energy expenditure and fat use do not contribute to obesity in women', *Obesity Res*, 11(8), 937–944.

WESTERTERP-PLANTENGA MS, PASMAN WJ, YEDEMA MJ, WIJCKMANS-DUIJSENS NE (1996), 'Energy intake adaptation of food intake to extreme energy densities of food by obese and non-obese women', *Eur J Clin Nutr*, 50(6), 401–407.

WINKIELMAN P, BERRIDGE K (2003), 'Irrational wanting and subrational liking: How rudimentary motivational and affective processes shape preferences and choices', *Political Psychology*, 24 (4), 657–680.

YEOMANS MR (1998), 'Taste, palatability and the control of appetite', *Proc Nutr Soc*, 57, 609–615.

YEOMANS MR, BLUNDELL JE, LESHEM M (2004), 'Palatability: response to nutritional need or need-free stimulation of appetite?', *Br J Nutr*, 92 (Suppl. 1), S3–S14.

ZANDSTRA EH, WEEGELS MF, VAN SPRONSEN AA, KLERK M (2004), 'Scoring or boring? Predicting boredom through repeated in-home consumption', *Food Qual Pref*, 15, 549–557.

18

Consumer attitudes towards functional foods

L. Lähteenmäki, M. Lyly and N. Urala, VTT Technological
Research Centre of Finland

18.1   Introduction

Functional foods are a fast growing market segment in Europe, USA and Japan.
The growth of the functional food market is predicted to continue in the near
future as well (Menrad, 2003), although estimates of the market size vary hugely
depending on the source (Verbeke, 2005). Functional foods have a small market
share, but they can be marketed with premium price and thus are appealing to
the producers. New options are launched rapidly and according to some views in
Europe, improving the gut health is the fastest growing application, with dairy
products as the main provider of probiotics and prebiotics to improve the gut
health (Hilliam, 2002; Stanton *et al.*, 2001).

   As the wide range of market share estimates indicate, functional foods are a
difficult term to define as there is no commonly accepted official definition.
Therefore a diverse variety of products have been included under this umbrella
(Menrad, 2003). In USA the figures describing market shares for functional
foods often include products that contain added vitamins, minerals and other
nutrients. In Section 18.2 some definitions related to so-called functional foods
are described and their implications for potential consumers of functional foods
are discussed.

   Basically, functional foods are products that are marketed with health-related
claims. The acceptability of the product is thus influenced, not only by the usual
interaction between product and consumer, but also by appropriateness and
appeal of the claim. Factors that affect the consumer perception of functional
claims and carrier products are described in Section 18.3. These are the aspects

of novelty, the impact on product quality and price, and whether the product becomes more like a medicine than a food.

Although functional foods are regarded as one of the fastest growing market sections for the future, their success will depend on how consumers accept them as part of their daily diet. The factors that promote or hinder consumer acceptance of functional products will be examined in more detail in Section 18.4. These include the socio-demographic factors, content and type of claim attached to the product and motivational factors.

Finally, based on the existing literature, some implications for the future development of functional products are gathered together in Section 18.5. Although functional foods are often presented as a single abstract food category, the actual functional products belong to several different product categories, have varying raw materials with varying health claims. Owing to lack of clear legislation in this area the scientific substantiation behind the claims varies widely. Some products typically have relatively well established clinical studies to back their claims, whereas others base their claims on assumptions that have not been verified. Therefore functional foods are a category of products that are widely discussed as unity, but in reality the essence of the term can differ between contexts.

18.2   Functional foods and their role in diet

18.2.1   What are functional foods?

Functional food is a commonly used term to describe products that promise a health-related benefit. However, there is no officially approved definition of functional foods in Europe and this has been regarded as an obstacle for their acceptance among consumers. Japan has its own legislation for 'food for specified health use' called FOSHU, which are clearly regarded as food products that are eaten as part of an ordinary diet. Many of the foods marketed with health-related claims have been processed through the system and thus gained the official status of FOSHU-foods (Arai, 2002). In Europe, according to a widely used definition functional foods are 'satisfactorily demonstrated to affect beneficially one or more target functions in the body, beyond adequate nutritional effects in a way that is relevant to an improved state of health and well-being and/or reduction of risk of disease' (Diplock et al., 1999). Criteria for satisfactory scientific demonstration have been suggested by an EU-supported PASSCLAIM – project (Aggett et al., 2005). Several European countries have self-administered systems for endorsing health-related claims in products. Although there is no unanimous definition or conformity on what functional foods are, these definitions and approval systems widely agree in two major aspects: functional products have special scientifically substantiated health-related claims and they should be foods eaten as part of a normal diet.

According to this strict demand for scientific evidence, very few products can be categorised as true functional foods. Yet, an increasing variety of food products are marketed with health-related arguments. From the consumers' point of view it is hard to make a difference between the products that qualify as true functional foods and those that do not, since both deliver consumers the same kind of message that is different from conventional nutritional messages. Therefore in this text, the functional food term can be broadly interpreted as daily foods marketed with health-related arguments without taking a view on the sufficiency of scientific evidence behind the claims. Similarly, foods with modified nutrient content such as low-fat or low-salt products are excluded since they can be regarded as products that support nutritional recommendations, but do not promise specific health effects.

18.2.2   Health claims

The health-related claims can vary in their nature. The claims can be broadly categorised into 'enhanced function' and 'reduction of risk of disease' – claims depending on the effect promised (Council of Europe's ..., 2001). A majority of products marketed with health arguments belong to the enhanced function claim group. Many of these products have been enriched with vitamins, minerals or trace elements. Their physiological activity has been proved in nutritional literature that provides causal explanations for effects. The effects of these products do not typically go beyond their nutritional impact.

Enhanced function claims promise that food produces certain physiological responses, such as lowering of blood cholesterol. These claims leave open the link between the function and its actual health effect, e.g. reduced level of cholesterol lowers the risk of heart disease. Thus, consumers need to be informed about the link from sources other than the product itself. Therefore marketing functional food products is a challenging task, as consumers have to gather the product information from various sources. This requires skilful marketing strategies which allow consumers to build up a coherent jigsaw from pieces of information that support each other.

In reduced risk claims the product or its component with the promised function are directly linked with the possible risk of disease or improvement of health. The reduced risk claims require more complex evidence, as they are based on probabilistic evidence between risk factors of diseases and likelihood of developing the illness or disorder.

Consumers did not seem to make much difference between different types of claims when assessing the possible benefits of products with health claims (Urala *et al.*, 2003). If the component providing the health benefit was familiar, such as calcium or probiotics, the perceived advantage was as good with just the presence of the component as with a promise to reduce the risk of disease or even prevent a disease (Urala *et al.*, 2003). The last claim is illegal, but the results suggest that for consumers, making a difference between reduced risk and prevention of a disease may be difficult. Furthermore, just knowing the link between the component and disease is sufficient for creating the positive reaction among consumers.

Another distinction in claims is whether they are generic in their nature or product specific. In generic claims the connection is made between the active component or ingredient in the product and its possible consequence, whether the claim is an enhanced function or a risk reduction claim. The claim does not cover the product itself and can be used in all products that contain sufficient amount of the component. Most of the allowed health-related claims are generic in their nature and several countries have a list of accepted claims. For example, in Sweden the voluntary code 'Health claims in the labelling and marketing of food products' (2004) in its revised version defines nine generic claims, which link certain components with a physiological consequence. In UK five generic claims have been accepted and three are considered in the 'Joint Health Claims Initiative' (2005). In UK generic claims are allowed for reduced saturated fat, soya protein and oats with the aim of reducing blood cholesterol, and for omega-3-fatty acids and whole grain with the aim of enhancing heart health.

18.2.3   The role of functional foods in promoting health

Health is mostly a credence character in food products as in most cases the possible health effects cannot be directly observed or experienced (Grunert *et al.*, 2000). Functional foods offer a new kind of positive health message to the consumers. Instead of having to avoid certain types of foods, these food products can be favoured in order to gain positive physiological effects on bodily functions or even reduced risk levels of diseases through eating a single product. The messages in functional foods promising a positive health effect can be more appealing to the consumers than the avoidance messages that are typical with nutritional recommendations.

Conventionally, 'healthy' foods are products that help to comply with nutritional guidelines. These products should be low in fat, high in fibre, contain a moderate amount of salt and energy. The promised reward of following the nutritionally good diet is a higher probability of maintaining a good health status and lowering the risks of lifestyle-related diseases. In the nutritional messages the role of whole diet is emphasised, not single products. For ordinary consumers this message may be tough to comprehend as the reward typically is unsure and far away, and no direct benefits can be observed or sensed in most cases while pursuing the goal. Lambert *et al.* (2002) have pointed out that even the five-a-day campaign promoting the use of fruits and vegetables in the UK and USA is hard to comprehend for consumers, partly because they cannot define what can be counted as vegetables and fruits or what constitutes a portion.

Functional foods, on the other hand, put emphasis on single products and their effects in the body. The promised outcome, whether lowering cholesterol or improving gut health, is well defined and can be achieved in a relatively short time. In most cases the result can be measured, even if it cannot be directly sensed. The level of cholesterol or blood pressure are easily available measurements. Sometimes consumers need to rely purely on the promise of the effect. Avoidance of symptoms (stomach upset) or improved well being may be difficult to verify as there are no valid and distinct markers of these functions. Consumers' own reported well being cannot be used as an objective measure of any physiological effects, as many psychological studies have shown that the mere effect of paying attention to a phenomenon always affects subjective responses.

One of the worries brought forward by consumer organisations and health professionals has been that health-related claims may confound consumers by offering easy options to promote health, so that people assume that healthy diet is not essential anymore. Instead of making disciplined choices, health effects can be gained through purchasing and consuming particular products. Therefore one basic condition for allowing a health claim is that they should only be combined with foods that are part of a balanced diet and have nutritional quality equal to their conventional counterparts (Health claims in the …, 2004; Joint Health Claim Initiative, 2005). When Quaker Oats were testing how consumers understand health-related claims, those receiving an oatmeal or a fibre claim did not more often believe that one does not need to pay attention to the rest of the diet than those who received no claim in the package (Paul *et al.*, 1999). However, a Finnish study that measured indirectly consumers' impressions of buyers of functional foods with a shopping list method, found that users of functional foods were perceived as more innovative and more disciplined than buyers of similar conventional products (Saher *et al.*, 2003). The latter was true only when the other food choices in the shopping list contained neutral items. Those shoppers who had a basic list of products with a high health image were rated as more disciplined, regardless of their functional choices. Choosing functional foods thus required some discipline but less than buying conventionally healthy foods. These not consciously recognised impressions may have a great impact on consumers' decision-making process. According to Cox *et al.* (2004) self-efficacy was a good predictor for functional food choices. These results indicate that functional foods present to people a new way of implementing healthiness. Functional foods are regarded as a convenient way to acquire the required nutrients and other beneficial compounds (Poulsen, 1999), which may not be required if the diet were otherwise more varied and balanced (Bogue and Ryan, 2000). The functional food may be used to some extent to mend the possible flaws in the diet.

18.3   Functional foods – a new category of products or new alternatives within existing product categories

Functional foods are marketed with health-related arguments and therefore consumer reception of claims is a crucial fact for their acceptability. However, most of the functional products are typical daily foods that have a certain role and place in people's diets. The reasons for choosing functional products depend on the product categories they belong to (Urala and Lähteenmäki, 2003) and different factors explain the behavioural responses to individual products (de

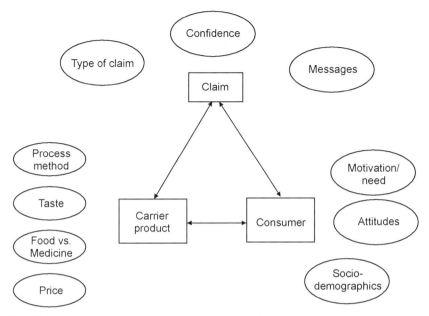

Fig. 18.1 Consumer responses to functional foods depend on the relationship between consumer, carrier product and the claim attached to the product. The factors affecting the relationships in this triangle are difficult to define as they tend to be interactions between these components. In this chapter those issues that have special relevance for functional foods are considered in more detail. This includes the changes in product characteristics, issues related to claims and their messages, and consumers' background and motivation to use functional foods.

Jong *et al.*, 2003; Urala and Lähteenmäki, 2007). The health claim, carrier product and consumer characteristics all influence how the functional food is perceived (Fig. 18.1). Furthermore, functional foods are a relatively novel category of foods that may require novel production or processing technologies and consequently, the concerns consumers have for these technologies need to be taken account (Frewer *et al.*, 2003). Although functional foods are clearly defined as food products, they also promise benefits that are similar to those normally obtained from medicines.

18.3.1 Novelty and production method
Functionality also brings a novel component to the food. Novelty in food presents a challenge for consumers: on one hand it brings variety into our food choices, but at the same time it also contains a possible risk of containing something harmful or at least giving an unpleasant sensation (Pliner and Hobden, 1992). The novelty in functional foods, however, is related to new ingredient components and their physiological effects although the food itself, in most cases, looks very much the same as its conventional alternative.

Technology required to produce functional foods may also raise doubts in consumers' minds. One technology that consumers are suspicious about is genetic modification. The responses have also been very negative when consumers have been asked if functional foods with beneficial health claims had been produced using gene technology. In Italy ($n = 120$) 31% of the respondents felt strongly that if the functions have been achieved by genetic modification, the product is not acceptable (Saba and Rosati, 2002). In an Australian study ($n = 290$) (Cox et al., 2004) gaining functionality by genetic modification was the least acceptable choice.

Claimed health benefits can be regarded as artificially added properties that do not naturally belong to food products. Whether functional foods are perceived as natural or not natural varies widely (Bäckström et al., 2004; de Jong et al., 2003). Almost half of the Dutch representative sample ($n = 1183$) rated yoghurt with lactic acid bacteria as natural and less than one third regarded it as not natural. Lemonade or sweets with added vitamins were regarded as not natural by almost half of the population, but with cholesterol-lowering spread only about half could make up their mind about the naturalness and even they were evenly split for and against (de Jong et al., 2003). Willingness to try modified dairy products including functional yoghurt, functional ice cream and calcium-fortified milk was positively associated with both adherence to technology and naturalness and negatively to suspicion of new foods in a study applying social representations in measuring responses to novel foods in Finland ($n = 743$) (Bäckström et al., 2004). This suggests that although these foods were regarded with suspicion, there is no obvious contradiction between naturalness and technology in the acceptability of functional foods.

Consumer perception of health-related claims depends on suitability to the product in question. In Denmark consumers were more positive about functional effects that were created through adding components that are naturally occurring ingredients or components in that food product (Poulsen, 1999; Bech-Larsen et al., 2001). Adding calcium into milk products may be more acceptable in consumers' minds than adding other minerals that do not originally belong to milk in any significant quantities. Among some Americans, living bacteria are considered as repellant and the idea of beneficial probiotic yoghurt with living bacteria was deemed as an unacceptable product (Bruhn et al., 2002). The bacteria, in the mind of these consumers, were linked with harmful pathogens. The strong reaction is to some extent irrational since yoghurts as such already contain bacteria, whether probiotic or not.

18.3.2   Taste vs. health

Consumers regard health benefits as positive factors, but only as long as they do not have to make any compromises with taste characteristics (Tepper and Trail, 1998). Consumers are not ready to compromise taste for functionality (Tuorila and Cardello, 2002; Verbeke, 2005). In Belgium ($n = 245$) half of the respondents were ready to accept the concept of functional foods if they tasted good, whereas only 9% accepted the concept if the products tasted worse than their conventional counterparts (Verbeke, 2005). In a three-country study, health claims increased the willingness to use beverages and ready-to-eat frozen soups slightly, but acceptability was mostly determined by tasted pleasantness (Lyly *et al.*, in press).

In an Irish sample ($n = 425$) over 80% of consumers believed that the juice they preferred was also the one with special health effects (Luckow and Delahunty, 2004). The food products with special effects are not necessarily regarded as inferior in taste; however, choosing the preferred version as the healthy option may reflect the likely positive health image of the carrier product, namely blackcurrant juice.

18.3.3   Foods or medicine
Most functional foods are seen more as food products than as medicines (Consumers' Association, 2000; de Jong *et al.*, 2003). The emphasis in all definitions and prerequisite for acceptance of claims is that they are attached to products that are eaten as part of normal daily diet. The food and meal systems are culturally determined and the role of a carrier product in this system is usually well defined. Therefore the products and their health-related claims need to fit into this food system in order to be accepted. In some cases the functional food products have bypassed the food system by offering an additional product that is eaten or drunk separately in small quantity, in the same way as medicines are taken apart from dietary rules. An example of these products are the small bottles of probiotic drinks originally introduced by Yakult, which can be taken almost as a medicine or preventive action against possible stress. More of these kinds of products with new claims have appeared in recent years. The acceptability of these products can be built in a different way as they do not require a role in the food system.

18.3.4   Price
Functional foods with health-related claims tend to be more expensive than their conventional alternatives. One of the questions raised is whether this higher price will be an obstacle for the success of functional foods. Within such a new product category, consumers' willingness to pay is hard to estimate, since consumers may lack a reference point for making reliable judgments. There seems to be a group of consumers who reported that they are ready to pay a premium for functional foods (Poulsen, 1999; Bogue and Ryan, 2000) or the price had very little impact on intention to buy (Bech-Larsen *et al.*, 2001). Health claims did not increase the willingness to buy chocolate bars among Finnish students ($n = 79$), and increase in price lowered the ratings similarly, regardless of the claim (di Monaco *et al.*, 2005). Price has been regarded as a barrier in adopting functional foods as part of the diet (Wilkinson *et al.*, 2004). In a study where two types of products were tasted and tried, in three countries

consumers ($n = 1157$) were not ready to pay extra for the products that contained a health claim (Lyly *et al.*, in press).

18.4   Acceptability of functional foods

18.4.1   Socio-demographic variables

The results on the role of gender, age and education have been contradictory from one study to another. According to some studies the most positive group towards functional foods have been women (Bogue and Ryan, 2000; Poulsen, 1999; Urala *et al.*, 2003) and the middle-aged or elderly consumers (Bogue and Ryan, 2000; Poulsen, 1999). In some studies there have been no differences (Verbeke, 2005) or the differences have been product-specific (de Jong *et al.*, 2003, Urala *et al.*, 2003).

Motivational factors related to promised functions influence the findings. Women were more responsive to products that are associated with breast cancer (Urala *et al.*, 2003) and men to products that lower the risk of prostate cancer (Hilliam, 2002). In The Netherlands ($n = 1183$) women were more likely to use foods containing extra calcium, but the difference between genders was small (de Jong *et al.*, 2003). In the same study, the oldest age group with respondents age 65 and over was the least willing to use probiotic yoghurt but this may reflect the lower overall acceptance of yoghurt in that age group.

Although women are, in general, more health conscious (Rozin *et al.*, 1999), socio-demographic variables do not appear such strong factors in explaining consumer responses to functional foods as a concept, but their role in some products can be crucial and needs to be verified on a product-by-product basis. Instead, belief in health benefits and an ill family member increased acceptance (Verbeke, 2005). When the motivational basis is taken into account the socio-demographic differences may exist, but they tend to be rather small. In a study with a population of those who had been diagnosed with high or elevated cholesterol levels in blood ($n = 2950$), 9% of men and 12 % of women reported use of cholesterol-lowering spread, and the use became more frequent with age: only 7% of those under 45 used cholesterol-lowering spread, but over 11% of those who were 45 or older used it (de Jong *et al.*, 2004).

18.4.2   Acceptance of different types of claims

Consumers have started to become aware of foods that have health benefits attached to them (Bogue and Ryan, 2000; Bech-Larsen *et al.*, 2001). In a UK study, consumers ($n = 100$) were asked opinions about three example products with health-related claims (Consumers' Association, 2000). The respondents were most aware of cholesterol-lowering spread (72%) and probiotic drink (58%), but the orange juice with bone benefits was only recognised by few respondents (7%). To be able to choose functional foods consumers have to be aware of the health benefits offered by them, but knowledge in itself is not a sufficient condition for willingness to use these products. The crucial factor in noticing and adopting messages related to the health effects of foods is the motivation of the respondent. The acceptability of new foods depends on the benefits these new foods provide for consumers (Frewer et al., 1997). Wrick (1995) divided the potential consumers of functional foods into those who recognise themselves at risk for a disease and those who are health conscious. The motivational expectations of these two groups vary: the 'at risk' group wants to get measurable results whereas for the 'health conscious' group the motivation is nutritional insurance and ability to take care of oneself.

In a cross-cultural study carried out in Denmark, Finland and USA (Bech-Larsen et al., 2001), the most attractive influences of functional foods were heart health or cardiovascular diseases mentioned by 54–59%, prevention of stomach cancer mentioned by 34–48% depending on the country, and enhancing immune system mentioned by 36–39% of the respondents. In an Irish study, the first two were the same, but maintaining the health of teeth and bones rose as the third most appealing influence (Bogue and Ryan, 2000).

Some studies have looked into whether the type of the claim influences consumer responses. In the UK claims that promised to improve health received higher approval than claims that promised to prevent diseases or disorders (Consumers' Association, 2000). Allowing respondents to have a closer examination of the product had an impact on opinion. The benefits of yoghurt and spread became more agreeable, whereas fewer respondents agreed with the preventive power of orange juice after studying the package. The product-dependent differences were large, so that 82% of the interviewees agreed that cholesterol-lowering spread would improve specific health problems, whereas the agreement figure was 66% for the probiotic drink and 49% for orange juice.

Contrary to earlier findings by Urala et al. (2003), van Kleef et al. (2005) observed that consumers ($n = 50$) seem to find claims that promise to reduce risks of diseases more attractive than those claims that enhance physiological functions. However, there seemed to be an interaction between the benefit and its framing. With cardiovascular disease the reduced risk claim increased willingness to buy the products in comparison to enhanced heart function claim, whereas for energy level claims the results were the opposite. In addition to just framing the messages, the content of the message should also be considered: heart-related functions can easily be connected with disease, whereas impact on energy/activity level is a function that improves overall well being. Ability to imagine consequences may influence the effects of framing the messages. Broemer (2004) found that negatively framed messages are more effective when symptoms are easy to imagine, whereas positively framed messages are more effective when picturing outcomes is difficult.

Responses to physiological enhancement and reduction of risk of disease claims, both applied to three types of products, namely juice, yoghurt and spread, were studied in three countries: Denmark, Finland and USA ($n = 1533$) (Bech-Larsen and Grunert, 2003). Adding the claim in the products increased the perceived wholesomeness of the products, regardless of the type of the claim. The effect was product-dependent, so that in juice and yoghurt the enriched products were rated as less wholesome, whereas in spread the situation was the other way around.

The results are somewhat contradictory and require further research to improve our understanding of the personal motivation and information-processing styles that may have an impact on our perceptions of claims. Furthermore, the responses to claims also depend on how they are worded and whether the actual product can deliver the benefits that have been pledged. In a study carried out with focus groups on probiotic cultures ($n = 100$) in USA the claims that promised to prevent or reduce the incidence of a disease were perceived less positively than claims that made less definite promises saying that they help to reduce the disease or may reduce it. Overstating the effect was deemed as inappropriate and unacceptable (Bruhn et al., 2002).

18.4.3   Motivational factors

Consumers attitudes towards functional foods vary among countries (Bech-Larsen and Grunert, 2003). In Finland the attitudes were more positive than in USA, and Denmark had the lowest attitude ratings. In Finland consumer attitudes and factors influencing willingness to choose functional foods have been studied in more detail. Consumers' attitudes towards functional foods contained four dimensions in a study carried out in Spring 2004 (Urala and Lähteenmäki, 2007). The strongest predictor in willingness to use functional foods was perceived personal reward of taking care of oneself by choosing the functional options. This personally felt reward was the best predictor for almost all functional foods used in the study. Whether functional foods were perceived as necessary, in general, predicted the willingness to use some products, and confidence in their promised claims seems to be important in those products whose effects are hard to verify by measuring any specific outcomes, such as probiotic products or calcium added beverages.

The attitudes towards functional foods in Finland had changed from year 2001 so that the seven dimensions (reward, necessity, confidence, safety, medicine vs. food, taste vs. functionality, wholesome diet vs. functionality) appearing in 2001 had combined into three dimensions in 2002 (reward, necessity and confidence), and then in 2004 the last dimension had divided into two, namely confidence and safety (Urala and Lähteenmäki, 2004; Urala and Lähteenmäki, 2007). The dimensions inter-correlated moderately (around $r = 0.5–0.6$). The varying factor structure indicates that the attitudes towards functional foods have not yet been fully established and thus may change further.

Confidence and perceived need for the promised health effect are important motivators for paying attention to functional foods. Person's own belief in the effectiveness of the product together with self efficacy was the strongest predictor for intentions to choose functional products that combat memory loss (Cox et al., 2004). The belief in health benefit increased the acceptance of the concept of functional foods (Verbeke, 2005). In addition to own motivation, a need in the close environment such as an ill family member improved the acceptance of the functional food concept (Verbeke, 2005).

If a consumer feels a need to lower blood cholesterol level or reduce level of stress then the products promising these effects are appealing. Messages tailored according to the receivers' own food habits were more effective in adding fibre consumption than messages that gave general advice on fibre-containing foods (Brinberg et al., 2000). The tailored messages appeal to a certain group of people whereas more general messages can be significant to a larger group although their weight is not as strong. Producers of functional foods have to balance between messages and claims that are highly relevant to a small group of consumers, and messages that aim to produce general well being to most consumers. With the targeted approach one can reach a small number of highly motivated consumers, but with the latter approach the interested group is larger, although the motivation may be lower.

18.5 Implications for developing and marketing functional foods

The first requirement for successful functional foods is product quality that can compete with conventional alternatives. Functional foods are perceived as members of a product category and they need to compete with other alternatives within these product categories (Urala and Lähteenmäki, 2003). The health claim showing the functionality is just one product characteristic among others and its importance varies among consumers. The studies carried out with functional products demonstrate that consumers are not ready to make sacrifices in hedonic pleasure (Lyly et al., in press; Tuorila and Cardello, 2002).

In a study with focus groups ($n = 35$) in the UK, participants suggested different strategies for developing functional foods (Wilkinson et al., 2004). The first strategy suggested that foods causing health problems could be modified to remove or eliminate the hazards. The second one proposed that health effects could be added as part of normal everyday foods. The functional products should be easily accessible, suitable for everyone and provide the benefits they promise. The first strategy may be hard for consumers to accept since eliminating possible negative aspects of 'bad' foods reduces the need for making 'healthy' choices and following a wholesome diet.

From the manufacturer's point of view product-related claims can be more appealing as they link not only the compound or ingredient, but also the specific product with the promised benefit in the claim. The product-specific claim offers better protection against competitors and gives a chance to use the claim effectively in marketing. The claims have to be based on adequate and sound scientific evidence, which means that effect has to be proven in independently conducted clinical studies, which are expensive to complete. The studies should give sufficient support for the evidence that products provide the promised benefits with portions that are eaten as a part of normal diet.

Substantiating the health effects and ensuring their presence in the final product requires scientifically sound research (Kwak and Jukes, 2001). The true challenge in marketing functional foods is how to make the claims understandable and appealing to consumers. Vast amounts of research effort have been targeted at finding functional components (Lucas, 2002). Similarly serious thought has also been put into substantiating the effectiveness of these compounds and finding rules for what can be considered sound scientific evidence (see Contor and Asp, 2004; Aggett et al., 2005). However, although the problem of conveying the information to consumers has been recognised, much less research effort has been devoted to understanding how consumers receive possible health claims and what expectations ordinary people have for these products at different stages of life. One problem in translating science-based information to consumer language is the different rules of thinking that are followed in solving everyday problems and scientific questions.

The science tries to reach the best knowledge on the phenomenon at the moment. Scientific knowledge increases all the time and new information challenges and replaces existing truths. Furthermore, the scientific evidence on health-related risks is based on probabilities. Consumers, on the other hand, base their food-related beliefs in approximation and rather crude rules of reasoning. Rather than using probabilities, consumers tend to make clear-cut assumptions on links between two statements or phenomena. If there is a reduced risk of a disease, for consumers this may mean that the risk has been removed. The equation where the product is effective in five cases out of ten is hard to comprehend; for the consumer the product should either be effective or not. An additional problem on probabilistic information is that the result cannot be promised to everyone. Although something is effective in most cases, the possibility of not personally gaining the benefit should be made understandable to the consumers. Consumers tend to be wary of linking food with reduction of disease, as this implies associating food that is commonly considered as a source of pleasure with something unpleasant, even if the promised effect would be positive. This conforms with another typical feature in everyday thinking which tends to associate any two things that appear at the same time regardless whether they are causally connected or not. Food and health, in general, are both very sensitive topics and this adds to the challenge of communicating possible health-related claims to consumers in comprehensible, usable and non-alarming manner.

As approximation is one of the typical features in consumers' way of handling information, the health-related messages in products have to be simple and clear enough to be understood, but at the same time they need to be distinguishable from other messages. The targeted functions also require messages targeted to consumers with varying age and cultural background. The issue of communicating health benefits to consumers effectively remains the ultimate challenge in the future success of functional foods. Only products that will be accepted and consumed can have a role in improving human health.

Products with specified health claims have entered the market providing a new way to express healthiness in food choices. The best known functional foods at the moment are cholesterol-lowering products and products containing fibre or beneficial bacteria. Although these foods have a new message, basically they respond to needs that have been recognised in medical and nutritional sciences for a long time. Consumers are becoming more familiar with these products and at the same time, while losing their novelty they seem to gain acceptance from those who wish to follow nutritional recommendations and are interested in health in food choices in general (Urala and Lähteenmäki, 2007). The question for the future is then what is the role of foods with specific health claims?: are they offering an additional benefit to consumers who try to avoid medication?, are they just one more issue to consider for those who wish to eat in a healthy way in general?, or do they become a standard that most people want to choose in order to enhance their well being and lower the risks of diseases just as a precautionary measure?

18.6   References

AGGETT PJ, ANTOINE JM, ASP NG, BELLISLE F, CONTOR L, CUMMINGS JH, HOWLETT J, MÜLLER DJG, PERSIN C, PIJLS TJ, RECHKEMMER G, TUIJTELAARS S, VERHAGEN H (2005), 'PASSCLAIM Consensus on criteria', *Eur J Nutr*, 44, S1, I/5–30.

ARAI S (2002), 'Global view on functional foods: Asian perspectives', *Br J Nutr*, 88 Suppl. 2, S139–S143.

BÄCKSTRÖM A, PIRTTILÄ-BACKMAN AM, TUORILA H (2004), 'Willingness to try new foods as predicted by social representations and attitude and trait scales', *Appetite*, 43, 75–83.

BECH-LARSEN T, GRUNERT KG (2003), 'The perceived healthiness of functional foods. A conjoint study of Danish, Finnish and American consumers' perception of functional foods', *Appetite*, 40, 9–14.

BECH-LARSEN T, GRUNERT K G, POULSEN J B (2001), *The Acceptance of functional foods in Denmark, Finland and the United States*, Working paper no 73, MAPP, Aarhus, Denmark.

BOGUE J, RYAN M (2000), *Market-oriented New Product-Development: Functional Foods and the Irish Consumer*, Agribusiness Discussion Paper No. 27, University College Cork.

BRINBERG D, AXELSON M L, PRICE S (2000), 'Changing food knowledge, food choice, and dietary fiber consumption by using tailored messages', *Appetite*, 35, 35–43.

BROEMER P (2004), 'Ease of imagination moderates reactions to differently framed messages', *Eur J Soc Psychol*, 34, 103–119.

BRUHN CM, BRUHN CJ, COTTER A, GARRETT M, KLENK M, POWELL C, STANFORD G, STEINBRING Y, WEST E (2002), 'Consumer attitudes toward use of probiotic cultures', *J Food Sci*, 67, 1969–1972.

CONSUMERS' ASSOCIATION (2000), *Functional food – health or hype?*, London, Consumers' Association.

CONTOR L, ASP NG (2004), 'Process for the assessment of scientific support for claims on foods (PASSCLAIM) Phase Two: Moving forward', *Eur J Nutr*, 43, Suppl.2, II3–II6.

COUNCIL OF EUROPE'S POLICY STATEMENTS CONCERNING NUTRITION, FOOD SAFETY AND

CONSUMER HEALTH (2001), *Guidelines concerning scientific substantiation of health-related claims for functional food*, Technical Document, 2 July, 2001, http://www.coe.int/soc-sp

COX DN, KOSTER A, RUSSELL CG (2004), 'Predicting intentions to consume functional foods and supplements to offset memory loss using an adaptation of protection memory theory', *Appetite*, 43, 55–64.

DE JONG N, OCKÉ MC, BRANDERHORST HAC, FRIELE R (2003), 'Demographic and lifestyle characteristics of functional food consumers and dietary supplement users', *Br J Nutr*, 89, 273–281.

DE JONG N, SIMOJOKI M, LAATIKAINEN T, TAPANAINEN H, VALSTA L, LAHTI-KOSKI M, UUTELA A, VARTIAINEN E (2004), 'The combined use of cholesterol-lowering drugs and cholesterol-lowering bread spreads; health behavior data from Finland', *Prev Med*, 39, 849–855.

DI MONACO R, OLLILA S, TUORILA H (2005), 'Effect of price on pleasantness ratings and use intentions for a chocolate bar in the presence and absence of a health claim', *J Sens Stud*, 20, 1–16.

DIPLOCK A T, AGGETT P J, ASHWELL M, BORNET F, FERN EB, ROBERTFROID MB (1999), 'Scientific concepts of functional foods in Europe: Consensus Document', *Brit J Nutr*, 81 (4), S1–S27.

FREWER LJ, HOWARD C, HEDDERLEY D, SHEPHERD R (1997), 'Consumer attitudes towards diferent food-processing technologies used in cheese production – the influence of consumer benefit', *Food Qual Pref*, 8, 271–280.

FREWER L, SCHOLDERER J, LAMBERT N (2003), 'Consumer acceptance of functional foods: issues for the future', *Br Food J*, 105, 714–731.

GRUNERT K G, BECH-LARSEN T, BREDAHL L (2000), 'Three issues in consumer quality perception and acceptance of dairy products', *Int Dairy J*, 10, 575–584.

'Health claims in the labelling and marketing of food products' Revised version September, 2004. Downloaded from *www.snf.ideon.se*, 28 June, 2005.

HILLIAM M (2002), 'Functional Food update', *The World of Food Ingredients*, April/May, 52–53.

'Joint health claims initiative.' Downloaded from *www.jhci.org.uk*, 28 June, 2005.

KWAK N S, JUKES D (2001), 'Issues in the substantiation process of health claims', *Critical Rev Food Sci Nutr*, 41, 465–479.

LAMBERT N, DIBSDALL LA, FREWER LJ (2002), 'Poor diet and smoking: the big killers. Comparing health education in two hazard domains', *Br Food J*, 64–75.

LUCAS J (2002), 'EU-funded research on functional foods', *Br J Nutr*, 2, S131–132.

LUCKOW T, DELAHUNTY C (2004), 'Which juice is "healthier"? A consumer study of probiotic non-dairy juice drinks', *Food Qual Pref*, 15, 751–759.

LYLY M, ROININEN K, HONKAPÄÄ K, LÄHTEENMÄKI L (in press), 'Factors influencing consumers' willingness to use beverages and ready-to-eat frozen soups containing oat β-glucan in Finland, France and Sweden', *Food Qual Pref*.

MENRAD K (2003), 'Market and marketing functional food in Europe', *J Food Eng*, 56, 181–188.

PAUL G L, INK S L, GEIGER C J (1999), 'The Quaker Oats health claim: A case study', *Journal of Nutraceuticals, Funct & Med Foods*, 1(4): 5–32.

PLINER P, HOBDEN K (1992), 'Development of a scale to measure the trait of food neophobia in humans', *Appetite*, 19, 105–120.

POULSEN J B (1999), *Danish consumers' attitudes towards functional foods*, Working paper no 62, MAPP, Aarhus, Denmark.

ROZIN P, FISCHLER C, IMADA S, SARUBIN A, WRZESNIEWSKI A (1999), 'Atttiudes to food and the role of food in life in the USA, Japan, Flemish Belgium and France: possible implications to diet-health debate', *Appetite*, 33, 163–180.

SABA A, ROSATI S (2002), 'Understanding consumer perceptions of fermented yoghurt products using conjoint and generalised procrustes analysis', *Ital J Food Sci*, 14, 339–350.

SAHER M, ARVOLA A, LINDEMAN M, LÄHTEENMÄKI L (2003), 'Impressions of functional food consumers', *Appetite*, 42, 79–89.

STANTON C, GARDINER G, MEEHAN H, COLLINS K, FITZGERALD G, LYNCH P B, ROSS P (2001), 'Market potential for probiotics', *Am J Clin Nutr*, 73, (Suppl): 476S–483S.

TEPPER BJ, TRAIL AC (1998), 'Taste or health: a study on consumer acceptance of corn chips', *Food Qual Pref*, 9, 267–272.

TUORILA H, CARDELLO AV (2002), 'Consumer responses to off-flavour in juice in the presence of specific health claims', *Food Qual Pref*, 13, 561–569.

URALA N, LÄHTEENMÄKI L (2003), 'Reasons behind consumers' functional food choices', *Nutr Food Sci*, 33, 148–158.

URALA N, LÄHTENEMÄKI L (2004), 'Attitudes behind consumers' willingness to use functional foods', *Food Qual Pref*, 15, 793–803.

URALA N, ARVOLA A, LÄHTEENMÄKI L (2003), 'Strerngth of health-related claims', *Int J Food Sci Tech*, 38, 1–12.

URALA N, LÄHTEENMÄKI L (2007), 'Consumers changing attitudes towards functional foods', *Food Qual Pref*, 18, 1–12.

VAN KLEEF E, VAN TRIJP HCM, LUNING P (2005), 'Functional foods: health claim – food product compatibility and the impact of health claim framing on consumer evaluation', *Appetite*, 44, 299–308.

VERBEKE W (2005), 'Consumer acceptance of functional foods: socio-demographic, cognitive and attitudinal determinants', *Food Qual Pref*, 16, 45–57.

WILKINSON SBT, PIDGEON N, LEE J, PATTISON C, LAMBERT N (2004), 'Exploring consumer attitudes towards functional foods: a qualitative study', *J Nutraceut*, 4(3/4), 5–28.

WRICK KL (1995), 'Consumer issues and expectations for functional foods', *Critical Rev Food Sci*, 35, 167–173.

19

The priorities of health and wellness shoppers around the globe

R. Vaidya and M. Mogelonsky, HealthFocus International, USA

19.1   Introduction

Shoppers around the world are becoming increasingly aware of the need for healthful foods in their diets. The level of awareness and the impetus for embracing behaviours that promote healthy lifestyles vary from country to country. Using up-to-date research, HealthFocus International is able to present a profile of 'Health Active' shoppers in twelve Western European countries (Denmark, Finland, France, Germany, Greece, Italy, Netherlands, Norway, Portugal, Spain, Sweden, United Kingdom); five Central European and Middle Eastern countries (Poland, Russia, Saudi Arabia, Turkey, Ukraine); four Latin American countries (Argentina, Brazil, Mexico, Venezuela); eight countries in the Asia/Pacific region (Australia, China, India, Indonesia, Japan, Malaysia, Philippines, Thailand), and the USA. This profile examines who they are and what drives them to make health and nutrition choices. This survey has been carried out in the USA since 1990; data for other countries has been collected for 2003 and in some cases, 2000/2001 as well.

19.2   Study background

The HealthFocus Global Survey of Health and Nutrition attitudes and behaviours is conducted in 31 countries around the world. Based on the longest running consumer health and nutrition survey in the USA, the survey's key objectives include:

- identifying current issues in consumer health and nutrition behaviour and attitudes

- assessing the trends in consumer priorities regarding nutrition issues such as fat, calories, vitamins, etc.
- understanding where consumers are headed in their behaviour towards their health and diet
- showing what nutritional issues will be important over the near horizon
- providing detailed data on regional and global differences/commonalities, and track trends worldwide.

Designed to meet these key objectives, the survey instrument collects in-depth information on the full spectrum of associated topics:

- actions toward food, food choices
- attitudes toward food, nutrition and personal health
- health interests, concerns and problems
- food and food processing concerns
- labelling
- about organics, biotechnology
- shopping patterns and brand influences
- children's health issues and concerns
- most important benefits of foods/beverages
- meal habits
- products used, increasing or decreasing
- foods used for disease prevention
- information awareness, nutrition interests, information sources
- dieting and weight management
- exercise habits
- making changes for health
- demographics.

The data in most of the countries was collected using face-to-face interviews. In the USA it was based on a mail survey, supplemented by online data collection.

This chapter addresses only a few selected elements of the information collected. A particular focus is on the motivations for healthy food choices and on how consumers segment on their food choice behaviours. The HealthFocus segments are derived on the basis of a linear additive combination of scores on an extensive range of questions associated with each of the segments. Respondents are assigned to the segment on which they score the highest.

Note that the ratings provided by consumers in different countries are subject to cultural influences that can limit comparisons. In some countries, respondents may be eager to please or not offend the interviewer and will give responses they consider more positive in nature. That said, it is possible to make inferences about the relative importance of measures across countries. For example, we can draw conclusions about differences between countries on the most important motivation for selecting healthful food and beverage products.

19.3   Why shoppers choose the foods they eat

While shoppers in many countries select foods and beverages for their own personal health, both currently and in the future, in some countries, shoppers base their choices on the needs of their families. Following is a regional analysis of the primary motivators that drive shoppers to choose healthy food and beverage products.

19.3.1   'To ensure my future good health'

For many shoppers around the world, the primary reason for making healthy choices is to ensure their own future good health. According to HealthFocus' global surveys, the percentage of shoppers in the different regions of the world that cite this as their primary reason for choosing healthy foods includes 30% in the USA, 23% each in Western Europe and Asia-Pacific, 20% in Central Europe and Middle East, and 18% in Latin America.

Within Western Europe, shoppers reporting such forward-looking motivations range from a high of 37% in the Netherlands to a low of 7% in Italy (Table 19.1). In the Asia-Pacific region, Indian (29%) shoppers are the most likely to feel this way while Indonesian (12%) shoppers are the least likely to do so (Table 19.2). As shown in Table 19.3, in Central Europe and the Middle East, shoppers in Turkey (37%) are significantly more likely to cite future good health as their primary motivation for healthy food choices than are those in the Ukraine (11%). Latin American shoppers in general are less likely to cite their future good health as a primary motivator for making healthy food choices. Within the Latin American region, shoppers in Brazil (22%) are most likely to mention this as their primary reason for choosing healthy foods while those in Argentina (9%) are the least likely to do so (Table 19.4). As Table 19.5 shows, 30% of American shoppers choose healthy food and beverages to ensure their future good health, and this is the primary motivator for US shoppers.

19.3.2   'To enhance my daily health'

Also important to shoppers is the enhancement of their daily health. According to HealthFocus' global surveys, US shoppers are the most likely to cite this as their primary reason for choosing healthy food or beverages (20%), followed by those from Latin America (19%), Western Europe (18%), the Asia Pacific region (18%), and Central Europe and the Middle East (15%).

In Western Europe, Spanish shoppers are the most likely to cite this as their major motivation for making such food and beverage choices (40%), while Swedish shoppers are the least likely to do so (7%; see Table 19.1). In the Asia-Pacific region, shoppers in Thailand are the most likely to select this response (25%) while those in the Philippines are the least likely to do so (11%; see Table 19.2).

Among Central Europe and Middle Eastern shoppers, those in Poland (21%) and Saudi Arabia (21%) are more likely cite enhancing daily health as a primary motivator than are shoppers in Turkey (6%; see Table 19.3). In Latin America,

Table 19.1 The one primary reason for choosing health foods or beverages, Western European respondents, 2003

	Total Western Europe	Denmark	Finland	France	Germany	Greece	Italy	Nether-lands	Norway	Portugal	Spain	Sweden	U.K.
To ensure my future good health	23	25	27	15	35	33	7	37	27	20	16	13	24
To feel good	21	23	12	22	23	5	28	30	8	34	9	43	11
To enhance my daily health	18	12	9	18	12	17	19	9	33	13	40	7	22
To meet the health needs of family members	14	20	12	22	11	14	33	4	6	12	10	9	14
To treat or control an existing health problem	7	4	12	4	6	6	2	6	5	14	9	6	6
To lose weight	5	6	11	4	2	5	3	4	2	2	1	7	10
To provide extra day-to-day stamina – an energy boost	5	5	10	9	4	8	4	5	6	1	4	5	3
To improve my appearance	2	1	2	1	1	3	2	1	1	2	1	1	2
None of the above	6	4	6	4	6	8	1	3	12	2	9	7	8

Source: HealthFocus International, Western European survey, 2003

Table 19.2 The one primary reason for choosing health foods or beverages, Asia-Pacific respondents, 2003

	Total Asia Pacific	Australia	China	India	Indonesia	Japan	Malaysia	Philippines	Thailand
To meet the health needs of family members	27	17	33	26	39	53	60	0	18
To ensure my future good health	23	28	28	29	12	21	22	19	23
To enhance my daily health	18	23	23	14	21	13	14	11	25
To provide extra day-to-day stamina – an energy boost	12	11	5	15	14	5	16	12	20
To feel good	10	8	6	5	6	2	4	44	3
To treat or control an existing health problem	6	7	3	4	5	3	12	6	10
To lose weight	2	5	1	4	1	2	1	4	1
To improve my appearance	1	1	2	1	1	1	1	1	0
None of the above	1	1	0	2	1	1	0	1	0

Source: HealthFocus International, Asia-Pacific survey, 2003

Table 19.3   The one primary reason for choosing health foods or beverages, Central European and Middle Eastern respondents, 2003

	Total Central Europe and Middle East	Poland	Russia	Saudi Arabia	Turkey	Ukraine
To meet the health needs of family members	23	13	29	16	24	34
To ensure my future good health	20	12	12	30	37	11
To enhance my daily health	15	21	10	21	6	14
To provide extra day-to-day stamina – an energy boost	13	9	16	8	12	20
To feel good	8	8	16	6	3	9
To treat or control an existing health problem	7	8	8	3	13	5
To improve my appearance	5	7	3	9	1	2
To lose weight	3	6	1	5	2	0
None of the above	5	14	5	2	1	3

Source: HealthFocus International, Central European and Middle Eastern survey, 2003

shoppers in Mexico (25%) are the most likely to cite this reason while shoppers in Argentina (15%) are the least likely to do so (see Table 19.4). In the United States, 20% of shoppers cite this as their primary reason for choosing healthy food or beverages (see Table 19.5).

19.3.3   'To meet the health needs of family members'
Meeting the health needs of family members is cited by 14% of Western European shoppers. Italian shoppers are the most likely to choose healthy foods and beverages for the wellness of their family members (33%), while those in the Netherlands are the least likely to do so (4%), as shown in Table 19.1.

In the Asia-Pacific region, 27% of shoppers choose healthy foods or beverages to meet the health needs of their family members. Shoppers in Malaysia (60%) are the most likely to give this response while those in the Philippines do not cite this reason at all, as shown in Table 19.2.

Twenty-three percent of Central European and Middle Eastern shoppers cite this response, with shoppers from the Ukraine (34%) the most likely to do so and those from Poland are the least likely (13%; see Table 19.3). As well, 29% of Latin American shoppers cite this response with Mexicans (15%) the least likely to do so and Argentinean shoppers (50%) the most likely, as shown in Table 19.4. American shoppers (15%) are only somewhat more likely than those in

Table 19.4   The one primary reason for choosing health foods or beverages, Latin American respondents, 2003

	Total Latin America	Argentina	Brazil	Mexico	Venezuela
To meet the health needs of family members	29	50	29	15	32
To enhance my daily health	19	15	20	25	17
To ensure my future good health	18	9	22	19	21
To provide extra day-to-day stamina – an energy boost	10	6	16	11	8
To feel good	10	8	11	12	9
To treat or control an existing health problem	8	8	6	9	8
To improve my appearance	2	0	3	2	2
To lose weight	2	1	3	2	2
None of the above	5	3	1	6	1

Source: HealthFocus International, Latin American survey, 2003

Western Europe to cite the health needs of family members as their prime reason for making healthy food and beverage choices (Table 19.5).

19.3.4   'To feel good'

Choosing healthy foods and beverages simply to feel good, is the primary reason given by 21% of Western European shoppers, as well as 10% of Asian-Pacific and Latin American shoppers, and 8% of those from Central Europe/Middle East, and 7% of those from the USA.

In Western Europe, 43% of Swedish shoppers choose this response, compared with only 5% of Greek shoppers (see Table 19.1). In the Asian-

Table 19.5   The one primary reason for choosing healthy foods or beverages, United States respondents, 2004

	United States
To ensure my future good health	30
To enhance my daily health	20
To meet the health needs of family members	15
To treat or control an existing health problem	10
To lose weight	8
To feel good	7
To provide extra day-to-day stamina – an energy boost	3
To improve my appearance	2
None of the above	5

Source: HealthFocus International, United States survey, 2004

Pacific region, response rates are lowest in Japan (2%), Thailand (3%), and Malaysia (4%) and highest in the Philippines (44%), as shown in Table 19.2. Among Central European and Middle Eastern shoppers, fewer than one in ten Polish, Saudi Arabian, Turkish, and Ukrainian shoppers choose this as their primary motivation for healthy food choices, while nearly one in five Russian shoppers choose this reason, as shown in Table 19.3.

19.3.5 'To treat or control an existing health problem'

Only 7% of Western European shoppers cite this as their primary reason for healthy food and beverage choices (see Table 19.1). Response rates were low in the Asia-Pacific region as well – where only 6% of shoppers cite health issues as their primary motivation for making food choices (Table 19.2); while in Central Europe and the Middle East, only 7% of shoppers select this as their top response (see Table 19.3) and just 8% of shoppers in Latin America provide this response (see Table 19.4). In the USA, 10% of shoppers cite control or treatment of existing health conditions as their primary motivator for making food and beverage choices (see Table 19.5).

19.3.6 'To lose weight'

While just 5% of Western European shoppers choose healthy food and beverages as a way of losing weight, only 2% of shoppers in the Asia-Pacific region do so (see Tables 19.1 and 19.2). Only 3% of shoppers in the Central European/Middle Eastern region make healthy food and beverage choices to lose weight, and just 2% in Latin America do so (see Tables 19.3 and 19.4). In the USA, weight loss is somewhat more of a motivator, but it is still selected by only 8% of shoppers (see Table 19.5).

19.3.7 'To provide extra day-to-day stamina – an energy boost'

Among Western European shoppers, 5% make their primary food and beverage choices based on the need for an energy boost (Table 19.1); but in the Asia-Pacific region, 12% of shoppers make this their primary reason for specific food choices, as shown in Table 19.2. In Central Europe and the Middle East, 13% of shoppers cite the desire for extra stamina or an energy boost as their primary reason for choosing healthy food and beverages (see Table 19.3). Among Latin American shoppers, 10% make their food and beverage choices primarily based on their need for an energy boost (see Table 19.4) but in the USA, this response is selected by just 3% of shoppers (Table 19.5).

19.3.8 'To improve my appearance'

Few shoppers in any region look to healthy foods and beverages primarily as a way of improving their appearance. In Western Europe, just 2% of shoppers select this as their top reason for choosing such foods, while in the Asia-Pacific region, only 1% do so (Tables 19.1 and 19.2). Five percent of shoppers in Central Europe and the Middle East choose foods primarily to enhance their appearance (Table 19.3); while 2% in Latin America and the USA (Tables 19.4 and 19.5).

19.4   The 'Health Active' shopper

'Health Active' shoppers are defined as those shoppers who select food and beverages for reasons related to health at least some of the time. Those who only rarely or never make food and beverage choices for healthful reasons are considered 'Unmotivated' shoppers. Table 19.6 shows the percentage of shoppers in each country whose behaviours fit this definition. At least eight in ten shoppers in all markets studied by HealthFocus are 'Health Active;' those in Western Europe (84%) are the least likely to fit this definition while those in the Asia-Pacific region are the most likely to do so (97%).

Within Western Europe, there is a wide variation in commitment to a 'Health Active' lifestyle. Only 66% of Danish shoppers, for example, are considered to

Table 19.6   Health Active shoppers around the world, 2003*

	Health Active %		Health Active %
Asia-Pacific	97	**Western Europe**	84
Australia	97	Denmark	66
China	95	Finalnd	83
India	99	France	89
Indonesia	94	German	85
Japan	95	Greece	85
Malaysia	94	Italy	96
Philippines	99	Netherlands	81
Thailand	100	Norway	93
		Portugal	85
Central Europe/Middle East	91	Spain	88
Poland	83	Sweden	74
Russia	92	UK	87
Saudi Arabia	91		
Turkey	94	**United States**	95
Ukraine	94		
Latin America	86		
Argentina	78		
Brazil	90		
Mexico	90		
Venezuela	87		

*United States data are from 2004
Source: HealthFocus International

be Health Active shoppers. Danish shoppers take a different approach to healthful eating: they are considerably more likely than Western European shoppers to always or usually balance healthy foods with less healthy foods that they enjoy more (66%, compared with 40% of Western European shoppers overall). Italian shoppers are the most likely to be Health Active (96%). They are more likely than Western European shoppers to always or usually avoid some favourite foods in order to eat more healthfully (32%, vs. 29% of all Western European shoppers).

While shoppers in the Asia-Pacific and Central European/Middle Eastern regions are highly invested in healthful eating, those in Latin America are somewhat less likely to be concerned about healthful eating. Overall, 86% of Latin American shoppers are Health Active, but in Argentina, only 78% may be classified as Health Active.

19.5   Segmenting Health Active shoppers

HealthFocus has segmented Health Active shoppers based on their motivations for making healthy choices and the sense of control they have over their own personal health and the health of their families as a result of these choices (see Table 19.7).

Each shopper has a different set of motivations and a different sense of control over his or her own health and wellness. On the more proactive side, HealthFocus has identified three segments: Disciples, Managers, and Investors. Two other segments, Healers and Strugglers, tend to be more reactive, only making changes in their lifestyles after they have encountered some setbacks or difficulties. The final HealthFocus segment is the Unmotivated shopper, who is the least likely to be motivated to make healthy choices of any sort.

Disciples, Managers and Investors make healthy food choices with a strong sense of control over their health. Strugglers and Healers are more likely to feel that they have less control over their health from the point of view of diet and lifestyles. Unmotivated shoppers – as their name implies – have little motivation to make any connection between diet and health.

19.5.1   Managers

Managers are an optimistic group. They make healthy choices from a proactive position of control over their health and well being. Their focus is daily health, looking good, and feeling good. Their proactive approach leads them to believe that life will continue to be even better in the future, especially as a result of the steps they are taking in the present.

Around 40% of shoppers globally are Managers, as shown in Table 19.7. The Asia-Pacific region has the greatest percentage of this type of Health Active shopper – 44% overall and 66% of shoppers in the Philippines. Western Europe has the lowest concentration of Managers – 35% overall and as few as 24% of

Table 19.7   Health Active segments around the world, 2003*

	Health Active %	Managers %	Investors %	Healers %	Strugglers %	Disciples %
Asia-Pacific	**97**	**44**	**26**	**8**	**11**	**7**
Australia	97	50	22	11	11	3
China	95	38	32	7	9	10
India	99	55	21	3	4	15
Indonesia	94	47	27	5	12	2
Japan	95	25	35	6	23	6
Malaysia	94	35	25	13	13	9
Philippines	99	66	15	8	4	6
Thailand	100	38	33	8	12	8
Central Europe/ Middle East	**91**	**40**	**23**	**5**	**20**	**3**
Poland	83	31	26	6	18	2
Russia	92	39	22	4	24	3
Saudi Arabia	91	47	22	3	13	6
Turkey	94	50	23	5	13	3
Ukraine	94	31	21	7	32	2
Latin America	**86**	**39**	**26**	**9**	**9**	**4**
Argentina	78	40	20	8	8	3
Brazil	90	33	38	9	6	3
Mexico	90	44	22	8	12	3
Venezuela	87	38	22	10	9	7
Western Europe	**84**	**35**	**25**	**9**	**13**	**3**
Denmark	66	34	28	3	3	1
Finland	83	24	27	8	9	5
France	89	35	34	9	18	2
Germany	85	36	32	5	9	4
Greece	85	37	17	16	12	3
Italy	96	25	23	10	21	5
Netherlands	81	25	34	7	14	2
Norway	93	41	28	7	16	1
Portugal	85	35	20	13	14	2
Spain	88	33	23	11	18	3
Sweden	74	57	11	1	4	2
UK	87	32	20	10	21	3
United States*	**95**	**42**	**22**	**13**	**16**	**2**

*United States data are from 2004
Source: HealthFocus International

Finnish shoppers. But 57% of Swedish shoppers are managers, showing the wide spread of shopper behaviours in that region.

19.5.2   Investors
Investors are motivated to make healthy food choices by a concern for their future good health. As their name suggests, they invest by eating well now to make sure that they can reap the benefits as they get older. Investors also have some concern for ensuring the health needs of others around them. They are more likely than average to cite 'to meet the health needs of family members' as their primary reason for choosing healthy foods or beverages, for example.

About a quarter of shoppers worldwide are Investors, as shown in Table 19.7. The highest percentage of Investors is found in Brazil (38% of shoppers), Japan (35%), the Netherlands (34%), and France (34%).

19.5.3   Healers
Healers feel compelled to follow a healthy diet plan because of current health problems or because of the risk of health problems in the future. They may be under the care of a medical professional and they often look to foods for thera-peutic solutions. This segment is the most likely to trade taste and convenience for health benefits. Healers base their health and nutrition choices on their current health problems or on the fact that they run a strong risk of health problems in the future.

While 13% of American shoppers exhibit these characteristics (see Table 19.7), the average is lower in other regions. Only 5% of Central European/ Middle Eastern shoppers are Healers, and on a country-by-country basis, just 1% of Swedish shoppers belong to this segment.

19.5.4   Strugglers
Strugglers find it difficult to make healthy choices although they acknowledge that there is a connection between diet and health. They tend to look for the 'quick fix' for health and diet problems. If they do not find an instant solution they are likely to say that staying healthy is a matter of luck, rather than something they can control.

The percentage of Strugglers is highest in the Central European/Middle East region (20%) and lowest in Latin America (9%). Fully 32% of Ukrainian shoppers are Strugglers, as are 24% of Russian shoppers. On the other hand, only 4% of Indian, Filipino, and Swedish shoppers and just 3% of Danish shoppers exhibit the characteristics of this segment.

19.5.5   Disciples
Disciples are committed to healthy food and lifestyle choices. Whether moti-vated by moral, ethical or philosophical principles, they tend to be knowledge-able about the most current health and nutrition information and are more likely

than any Health Active segment to use that information to improve their lives. Disciples are early adopters of emerging health and nutrition trends and products, and they are committed to long-term health and wellness.

Disciples comprise a small segment of Health Active shoppers in every market studied. In Western Europe and Central Europe/Middle East, they make up only 3% of shoppers while in Latin America they comprise 4% of the total of shoppers; and in the United States, only 2% of shoppers are Disciples. There is a higher percentage of Disciples in the Asia-Pacific region than anywhere else: fully 7% of shoppers in this area fall into this group. Table 19.7 shows the percentage of Disciples by region and country.

The Asia-Pacific region boasts the greatest percentage of Disciples (7%), with India (15%) and China (10%) leading the way. Denmark and Norway have the smallest percentage of Disciples – only 1% of shoppers in each country have this profile.

19.6   Trends in the USA

HealthFocus' analysis of global shopper habits with regard to health and wellness is a 'snapshot in time' that benchmarks the current attitudes and habits of shoppers around the world. To see how this type of information can be used to chart future trends, HealthFocus herewith presents some of the data that has been collected for the American Health Active shoppers since 1990.

Table 19.8 illustrates US Health Active segments between 1990 and 2004, the latest data year available. A combination of factors has influenced shoppers' attitudes towards health and nutrition since 1990. These factors have changed over time, which explains the shifts in size of specific Health Active segments.

Some ten to fifteen years ago, shoppers equated health with the avoidance of disease – to be healthy was to be 'not ill.' Today, however, there is a much greater emphasis placed on health as 'wellness' – an ongoing lifestyle, not a response to an existing or potential condition. Health is also about 'feeling good' – for the long term, in a proactive sense, not as a response to 'feeling bad.'

Table 19.8   US Health Active segments, 1990–2004

Segment	1990 %	1992 %	1994 %	1996 %	1998 %	2000 %	2002 %	2004 %	Change % point
Managers	33	45	47	48	41	44	40	42	+9
Investors	41	25	17	25	24	22	23	22	−19
Healers	8	10	13	8	14	6	15	13	+5
Strugglers	6	9	14	13	15	19	14	16	+10
Disciples	3	2	2	1	1	2	2	2	−1
Unmotivated	10	10	7	5	6	6	6	5	−5

Source: HealthFocus US Trend Report

The percentage of shoppers who feel confident that they can manage their long term health has declined since 1990 and shoppers are less likely to feel they can control the future, especially when it comes to such major health issues as cancer.

There have been only minor changes in the distribution of shoppers among Health Active segments between 2002 and 2004; the biggest attitudinal shifts took place in the mid-1990s. Managers eclipsed Investors in 1992 and have stayed the largest segment since then. The percentage of Strugglers has steadily increased – not surprising in an aging population for which health issues are becoming more relevant. The steadiest segment is the Disciples – this is also the smallest segment, one that is likely to remain in the minority.

As HealthFocus continues its study of shoppers around the world, this data base will be enriched. For now, however, global findings serve as a benchmark against which future behaviours can be measured.

19.7 Future trends

As the American trends show, a number of concerns have influenced the behaviour of shoppers. The most important one is the aging of the Baby Boom population. In the USA, the Baby Boom generation denotes people born between 1946 and 1964 and is made up of some 76 million adults, a large percentage of whom are over the age of 50 and about to turn 60. This age change leads to a number of changes in health, in diet, and in attitude.

It is not clear the extent to which similar patterns will be visible in other areas of the globe as only the United States, Canada, Australia, and New Zealand exhibit this population pattern. In countries in which World War II was actually fought, post-war reconstruction and economic factors did not lend themselves to as dramatic a population shift. The patterns of aging, therefore, vary from country to country.

Attitudes about health, diet, nutrition, and exercise also vary and it is beyond the scope of this article to chart patterns on a country by country basis. It is hoped, however, that continued monitoring and surveying of populations in the countries included in this study will further enrich our knowledge base in the years to come.

19.8 Meta-analytic postscript

Meta analyses of the data recently conducted by HealthFocus International elicited underlying attitudinal and behavioural dimensions and investigated how countries grouped on these dimensions.

Two key dimensions identified by this analysis were: (1) the extent of control over food choices respondents felt they have; and (2) their level of individual engagement in food choice decisions.

The resulting country groupings reveal that in some countries with strong prevailing dietary norms, respondents exhibit low individual engagement in food choice decisions but a stronger sense of control over their food choices. This includes countries such as China, India, Japan, Malaysia, Philippines, and Saudi Arabia. On the other hand, in some other countries where culturally driven dietary norms may not be as deep-seated, respondents exhibit high individual levels of engagement in food choice decisions but a weaker sense of control over their food choices. This includes Western countries such as United States, Australia, Germany, United Kingdom, Denmark, Netherlands, Norway, and Sweden.

Thus it appears that in countries with strong culturally driven dietary norms, consumers may feel less need to be personally engaged in food choice decisions, perhaps because of confidence in their inherited traditions. On the other hand, in countries where a normative dietary framework may not be as strong, consumers feel compelled to be more engaged in their food choice decisions.

20

Consumers, communication and food allergy

M. C. van Putten, M. F. Schenk, B. Gremmen and L. J. Frewer, Wageningen University, The Netherlands

20.1   Introduction

Foods and food consumption can have various and diverse impacts on daily life beyond basic nutrition, including sensory enjoyment of foods, or the social meanings that eating with other people may have. The daily life of consumers, and their immediate families, may be influenced in many other ways. Functional foods and nutrigenomic products offer the possibility of disease prevention over and above that offered by basic nutrition. Some food products offer opportunities for fitting the needs of changing consumer lifestyles, for example the need for quickly and conveniently prepared meals. Other issues related to food consumption and food choice focus on the problems encountered by specific groups of consumers. Food allergic consumers face specific problems in the area of food choice, and this chapter will describe these potential problems, and discuss some potential solutions to these problems.

Food allergy is a condition in which the body's immune system responds to substances which do no harm to most people, and, in terms of human health, are potentially harmless. There is some (equivocal) evidence that the prevalence of food allergy is increasing in various parts of the world, particularly in developed countries (Helm and Burks, 2000; Jackson, 2003). Current estimates of the incidence of food allergy suggest that approximately 6% of children are affected in Western countries, and, although many of those who suffer from food allergy find that the condition disappears in adult life, around 1–2% of the adult population are afflicted by the condition (Sampson, 2001). Food allergy can have a profoundly negative impact on quality of life, extending beyond the immediate clinical effects of the individuals' allergic condition to have a negative impact on quality of life and economic function of individuals and households (Fernández Rivas and Miles, 2004). Indeed, food allergy is occasionally fatal (as in the case of anaphylaxis), and individuals suffering from severe food allergy need to be ever vigilant. More frequently, dietary restrictions may compromise social activities such as dining outside of the home, or attending social functions (Knibb et al., 2000). Food allergy has been found to exert a significant impact on the perception of general health of affected children, as well as an emotional impact on the parent and limitations regarding family activities (Sicherer et al., 2001). Children may also experience learning impairment, problems with peer group socialisation, anxiety and family dysfunction (Meltzer, 2001). For example, Primeau et al. (2000) found that families of peanut allergic children experienced significantly more disruption in their familial and social interactions than families of a child with chronic rheumatologic disease. They suggest that this finding may be due to the constant risk of sudden death from anaphylactic reaction in the peanut allergy group, leading to parental restriction of activities. Affected adults may also experience negative impacts on quality of life, including restricted leisure activities, loss of working time or days, impaired social functioning and psychological distress (Knibb et al., 1999).

Taken together, this implies that food allergy must be managed in some systematic way to limit the impact it has on daily lives. Since there is currently no cure available for food allergies, prevention of exposure to allergens is the only way to prevent the occurrence of food allergy. This chapter will consider two potential mitigation strategies to prevent the exposure to allergens, avoidance and replacement. For example, consumers themselves can adopt an avoidance strategy only under circumstances where they can identify what ingredients are contained in food products, for example through effective information provision on food labels. This is only possible if effective traceability systems are implemented throughout the food chain, as manufacturers cannot signal the presence of potentially problematic ingredients if they do not know whether foods contain them or not.

The alternative to avoidance, the adoption of replacement strategies for problematic ingredients, will of course benefit allergic patients in that the foods and ingredients which may provoke an allergic reaction will be removed and replaced by non-allergenic equivalents. This may necessitate the development and market introduction of *novel foods*. For the purpose of this chapter, novel foods are defined as food or food ingredients that have no history of safe use in the European Union. This not only includes foods that have been genetically modified or that have been produced using genetically modified organisms, but also exotic foods imported from non-EU Member States. Some novel foods may be of benefit to food allergic consumers. However, the introduction of novel foods into local food chains may also create new problems related to food allergy. In other words, novel foods can increase the risk of food allergies by introducing novel allergens into the food chain and the human diet. Against this, novel foods can be developed which remove allergy risks through elimination of allergenic proteins. In both examples, the issue is further complicated by societal concerns about novel foods, particularly those produced by potentially controversial processing technologies, such as genetically modified foods.

The introduction of novel foods may have a profound impact on the lives of food allergic consumers. Therefore, due consideration must be given to the communication needs of different stakeholders, including consumers, regarding food allergy, as well as the broader societal issues which may influence the acceptability of different preventative approaches by allergic consumers and consumers more generally.

The communication needs of different stakeholders, including food allergic consumers, will be discussed. Two mitigation strategies will be described, and contextualised by discussion of related consumer concerns, both those relevant to food allergic consumers and the broader population, and illustrated with a case study focusing on the application of genetic modification for allergy prevention. Ethical perspectives relevant to food allergy mitigation and potential ethical questions pertinent to the debate about food allergy will also be discussed. Finally, the conclusions will focus on the identification of future research needs regarding consumers and food allergy.

20.2   Communication needs

Different stakeholders in the area of food allergy can be described as members of society who are affected by food allergy in some direct or indirect way. As a consequence of different stakeholders having different interests and requirements relating to food allergy information, it is intuitive that this information must be targeted towards these needs and preferences (Mills *et al.*, 2004; van Putten *et al.*, 2006).

Empirical analysis of the information requirements of consumers has not been extensive. There is some evidence that the *method* of communication (for example, the different media used to transmit the information), as well as the *source* of the information (for example, medical sources versus the news media), can have an effect on the psychological distress of parents with food allergic children, who often get their general knowledge about food allergy from media sources (such as television, radio, newspapers, magazines and the Internet) or from local community contacts. Many parents self-diagnose and self-treat allergy in their children, which means that erroneous diagnosis may result in unnecessary dietary restrictions for the food allergic child (indeed, the whole family may also be affected by such reduced food choice) (Eggesbo *et al.*, 2001; Young *et al.*, 1994). Effective communication with the parents of food allergic children is thus an important priority for those with responsibility for consumer protection (for example, regulators, the food industry and the medical profession).

Adolescents and young adults are the groups most at risk from severe food allergy reactions (Bock, 1987). Hourihane (2001) has commented that there are additional difficulties in developing communication about food avoidance for these individuals who are at the stage of their lives where they are developing independent lifestyles. On reaching adulthood, further communication needs may be identified. For example, the communication needs of a person who has suffered from allergy all of his/her life may be very different from someone who has just recently been diagnosed as suffering from the condition.

20.3   Mitigation strategies

20.3.1   Food allergen avoidance

Avoidance of allergens in foods can be achieved by applying various methods of communication and end-point labelling to facilitate consumer choice in the retail environment. Labels on the food product are one way this can be done (Robinson, 1998). Food allergic consumers rely on the information on food labels to ensure that they do not consume products containing potentially allergenic proteins. All ingredients in a food need to be declared on the label in a clear and under-standable format (Mills *et al.*, 2004). There is consensus among patient groups that ingredients and substances recognised as causing allergies must be listed, labelled and updated according to new and possibly emerging scientific evidence about food allergies and allergens (Mills *et al.*, 2004). These practices should apply to ingredients as well as whole foods, and restrictions should be placed on the flexibility with which ingredients can be replaced in different food products to optimise safe food choices for allergic consumers. Information delivery in the retail environment may need to become more sophisticated in order to tailor the information to the needs of different customers. For example, there may be limitations to the amount of information that can be included on a label, particularly if this information needs to cover other issues as well as allergy matters (for example, production method, place of production, quality indicators, and nutritional information). New approaches to delivering information need to be developed (for example, developing new information and communication technologies, e.g. ICT approaches) to target information to the needs of specific consumers. For example, a peanut allergic consumer could be alerted as to whether a particular product contained peanuts at the supermarket checkout following presentation of a personalised smart card to checkout personnel, which could be scanned and the information used to check for peanut inclusion in each product being purchased as it to is being scanned at the checkout.

Of course, the introduction of innovative labelling systems is dependent on effective traceability systems for allergic ingredients being introduced throughout the food chain. For example, North American research has indicated that, in 1999, around 40% of undeclared allergens in foods resulted from unintended contact of the food with a substance containing the allergens. Five percent of undeclared allergens resulted from errors being made by manufac-turers within the supply chain (Vierk *et al.*, 2002). Similarly 31% of chocolate bars from western Europe and 62% of those from eastern Europe were found to contain undeclared peanut proteins (Vadas and Perelman, 2003) although it was not clear whether the levels of these undeclared allergens in products were at a threshold level where an allergic reaction might be triggered.

Is the regulatory framework evolving in such a way as to facilitate avoidance of potentially problematic foods by allergic consumers? On one hand, one might argue that society has an obligation to protect vulnerable consumers from ingesting potentially dangerous products. On the other hand, the increase of 'litigation culture' means that the food industry is increasingly liable for unsafe products should a consumer experience adverse effects resulting from consuming them. As a consequence, the food industry's risk management strategy will often focus on informing the consumer of the presence of the allergen, particularly if consumer exposure is likely to be over the threshold required to trigger an allergic reaction (Crevel, 2001). The industry must also conform to food labelling legislation which has been enacted in order to protect allergic consumers. Historically, the '25% rule' has been applied within the regulatory context. Consider a product made of different ingredients. If a particular ingredient constituted less than 25% of the *weight* of the food, only the additives themselves would have to be declared on the label, but not the ingredients. This legislation was insufficient and did not provide enough information to promote an effective consumer protection policy, as allergic consumers were still placed at risk. As a consequence, the 25% rule was replaced by a 5% rule and a list of major allergenic foods has been created for which labelling is always mandatory independent of whether the problematic food constitutes less than 5% of the final product weight or not (Mills *et al.*, 2004). This may have led to an increase in precautionary ('may contain') labelling practices (Gowland, 2001). Pre-cautionary labelling of this type is frequently used when a product contains a specific allergen when the content is so low as to pose an infinitesimal risk of reaction (Crevel, 2001).

To determine which allergen risks should be mentioned on a label and which should not, it is necessary to have information about the 'threshold for effect'. The problem with determining threshold levels is that, for the individual patient, the dose that elicits an allergic response may vary over time. If it is not possible to determine threshold levels for food allergens then risk assessment and management will be very difficult (Madsen, 2001). The implication would be that even an extremely low dose of allergen is not safe, since it is not possible to determine a level of allergens at which there is absolute certainty that no ill-effect will result to all food allergic consumers. As a consequence, the industry has no choice but to implement precautionary labelling. An alternative to labelling for allergens is replacing them with non-allergenic equivalents, such as foods genetically modified to be non-allergenic. However, societal acceptance of such approaches is not universal, and this will be discussed in the following section.

20.3.2   Consumer concerns about novel foods

Novel foods can be applied to the removal of allergy risks through elimination of allergenic proteins. They also can increase the risk of allergies by introducing novel allergens into the food chain. This issue is further complicated by societal concerns about novel foods, particularly those produced by the use of potentially problematic food processing technologies, where consumer perceptions of risk and other negative attributes associated with the processing technology may offset perceived benefit associated with the final product. Consumer protection regulation and risk assessment and management practices must take societal concerns as well as technical risk assessments into account, particularly in the context of societal introduction of controversial food technologies (Frewer *et al.*, 2004; De Jonge *et al.*, this volume (Chapter 5); Siegrist, this volume (Chapter 10)). To summarise, consumer concerns are likely to be higher if consumers *perceive that they have no personal control over exposure* to genetically modified foods and ingredients, if *effective and reliable traceability and labelling practices* are not implemented, if consumers perceive that consumer benefits associated with novel products do not offset potential risks (whether to consumers or to the environment), and if there is differential accruement of risk and benefit between consumers and producers, or between different groups of consumers across the population. Distrust in regulatory systems and institutions are also frequently cited as a potential problem regarding the acceptance of novel genetically modified products (for example, see De Jonge *et al.*, this volume).

Consumers may also have concerns about non-genetically modified novel foods. For example, there is a risk that proteins in 'exotic foods' cross-react with known allergens (Shewry *et al.*, 2000). Cross-reactions occur when the molecular structure of the allergen of the exotic food (for example, an imported fruit) is similar to the molecular structure of a known allergen. Antibodies that recognise the known allergen may also recognise the 'exotic' allergen and an allergic reaction may occur as a result. This may even be the case if the food allergic person has never been in contact with the exotic fruit before. Genetic differences may exist between consumers in the geographical region or consumer population where the product originated and in the new market where the product is subsequently introduced (Howlett *et al.*, 2003). In other words, absence of allergenic reactions in the region of *origin* of a particular product does not automatically imply that there is no risk of allergic responses in the *new region* into which the product is imported.

Food allergic consumers may need to increase the level of vigilance required to avoid accidental exposure to a particular food allergen following the introduction of novel foods with reduced allergenicity. For example, consider the case of a genetically modified food product with reduced allergenicity in comparison to its traditional counterpart. The traditional product would still be available in retail outlets, because consumers generally demand a choice between foods produced using traditional production methods and genetic modification used in processing. As a consequence, two variants of the product would be available for purchase; the traditional product which still includes allergenic proteins, and the new product which contains no allergenic proteins. For those consumers actively avoiding the products containing allergens, this will result in greater vigilance when shopping as they will have to pay attention to allergenic ingredients in the food product they intend to buy, and to ensure they purchase the non-allergenic variant of the product. Avoidance of potential problems by sensory selection is also made more difficult, as consumers may not be able to discriminate by smell or appearance which product variants are hazardous, and which are safe to consume (Gowland, 2001).

It may also be possible to develop products with reduced allergenicity using classical breeding techniques instead of genetic modification. It should be noted that, even if this solution is acceptable to consumers in general, the problems of increased vigilance on the part of food allergic consumers will not disappear following their introduction unless there is complete replacement of problematic foods throughout the food chain. Such replacement may be more pragmatic for some products compared to others.

20.4 Case study on the application of genetic modification for allergy prevention

The case study was designed to examine consumer attitudes towards the use of genetic modification applied to the prevention of allergy. Given the evidence for consumer negativity towards genetic modification of foods already discussed, two different applications of genetic modification (food *vs.* non-food) were considered. The attitudes of two consumer groups (allergic patients *vs.* non-patients) were compared, as it was expected that the attitudes of allergic patients would be more positive to genetic modification used as a mitigation strategy compared to non-allergic patients. In addition, we expected the attitude towards the food application to be less positive, given that food-related applications are generally rated by the public as being less acceptable than medical applications (Zechendorf, 1994). Allergic patients may directly benefit from products with a reduced allergenicity in terms of an increased dietary choice or reduced potential for an allergic reaction. On the other hand, these products may introduce increased dietary vigilance, increased shopping times, or an increased risk of inadvertent exposure resulting from a reduced ability to detect allergens through sensory evaluation. It is not known whether the perceived benefits of genetically modified low allergenic foods will be offset by these other factors. For consumer acceptance of a potentially controversial technology to occur, the perceived benefit must outweigh the perceived risk associated with a potentially hazardous behaviour or activity (Slovic 1987; Frewer *et al.*, 1997, 2004). The non-patient group has no direct benefits from the novel products and we therefore predicted that this group would be less willing to accept novel low allergenic products that are produced through genetic modification.

20.4.1 Methods
Survey design
The cases included in the survey focus on the application of genetic modification to develop novel products (apples, birch trees) with a reduced allergenicity.

Apple is a major allergenic food that causes relatively mild symptoms (Wensing *et al.*, 2002). Research towards the development of genetically modified low-allergenic apples has been undertaken in recent years (Gilissen *et al.*, 2005). Birch pollen is an important source of allergens that provokes hay fever in large parts of northern Europe. Moreover, the major birch allergen cross-reacts with several food allergens, such as in apple (Ferreira *et al.*, 2004). Recent research has focused on genetically modified non-flowering birch trees, which could mitigate allergic responses in patients (Lemmetyinen *et al.*, 2004).

The separation between patients and non-patients is case dependent. In the birch case, patients were defined as suffering from allergic rhinitis and in the apple case as suffering from food allergy. A more narrow definition of patients was also tested in which patients were defined as being allergic to pollen in the birch case and to fruits/nuts in the apple case. To enable the distinction between patients and non-patients, respondents were asked to indicate if they had suffered from allergic symptoms, separating between allergic rhinitis, allergic asthma, allergic eczema, food allergy, and other allergic disorders. If allergic, information was collected on allergy history, medication use, allergy diagnosis, timing of the symptoms, and the allergens to which respondents were allergic.

Standard demographic characteristics were recorded (age, gender, household composition, education, income, and employment). Items assessing environmental concern were also included, to determine whether consumers were concerned about the potential impact of genetically modified crops on the environment. The influence of the demographic characteristics on the acceptance of genetic modification strategies is analysed in Schenk *et al.* (in preparation), and will not be discussed further here. The importance of demographic variables on consumer risk perception is described extensively elsewhere (see, for example, Slovic, 1999; Siegrist, 2000; Titchener and Sapp, 2002).

Attitude items

The survey included items on respondent attitudes towards the application of genetic modification. These items were derived from previous research on genetic modification, which has focused on acceptance and rejection of specific applications of the technology (Frewer *et al.*, 1997). We added two items on health effects to the original set of 19 items, which was then reduced to 14 items during the pilot study. Two pilot studies were conducted (72 Dutch respondents were included in the first pilot study, and 60 Dutch respondents in the second). The first pilot was used to reduce the number of items through application of a principal component analysis (PCA), while the second was conducted to test the internal consistency of the items remaining after data reduction. Responses were collected on seven-point scales anchored by 'not at all' to 'to a large extent'.

To enable a comparison of the genetic modification strategy with more conventional strategies for allergy prevention, we collected data about consumer acceptance of a broader set of prevention strategies. One strategy was directed towards active involvement of the allergic patients themselves, and focused on *avoidance* of problematic allergens. The other three strategies focused on *replacement* of either the allergenic trees in the urban environment or apples in the human food chain. Allergenic trees are either replaced by *other tree species* or birch trees with a reduced allergenicity. Trees with a reduced allergenicity can either be produced by selection and breeding from existing varieties (*conventional breeding*) or application of *genetic modification*. Apple can be replaced by *other fruits* or apples with a reduced allergenicity, which can again be attained by *conventional breeding* or *genetic modification*. In addition, the acceptability of *maintaining the current situation* was included to compare consumer acceptance of the novel approaches to allergy mitigation with the current situation. For each strategy, respondents were asked to rate the desirability of implementing the strategy. Responses were collected on seven-point scales.

Survey implementation and study population
The main survey was carried out in the Netherlands by a professional social research company using quota sampling. The data set included 178 respondents for the birch case, 175 respondents for the apple case, and 179 respondents who filled questions on both cases (details on the survey design are described in Schenk *et al.*, in preparation).

The study population was characterised according to the allergy background of the respondents (Fig. 20.1a). Forty-six percent of the respondents claimed to suffer from allergic symptoms. This number is higher than indicated by research that uses objective markers, such as detection of elevated levels of specific IgE with skin-prick tests or blood testing, which indicates manifestations of allergic diseases in 35% of the general population (UCB Institute of Allergy, 1997). Questionnaire-based research generally finds higher allergy prevalence, due to an imperfect distinction between non-atopic and atopic symptoms. For example, rhinitis symptoms may either be 'true' rhinitis symptoms or allergic rhinitis symptoms, and food intolerance is often confused with food allergy by sufferers. It is therefore stressed that the distinction between patients and non-patients that is used throughout this study is based on *perceived* allergies. In terms of acceptance of novel products, perceived allergy is likely to influence consumer attitudes independent of whether a formal allergy diagnosis has been made. The majority of the allergic respondents indicated that they were suffering from an airway allergy (allergic asthma and/or allergic rhinitis) (Fig. 20.1a). The study population can also be divided according to the allergens to which respondents perceive an induced allergic reaction. The majority of the patient population (55%) was allergic to pollen allergens, while a large proportion (23%) didn't know to what they were allergic (Fig. 20.1b).

20.4.2   Results
Attitude of patients and non-patients towards application of genetic modification
A principal component analysis (PCA) was applied to investigate on how many dimensions the attitude items could be fitted. The PCA (rotated; varimax) was

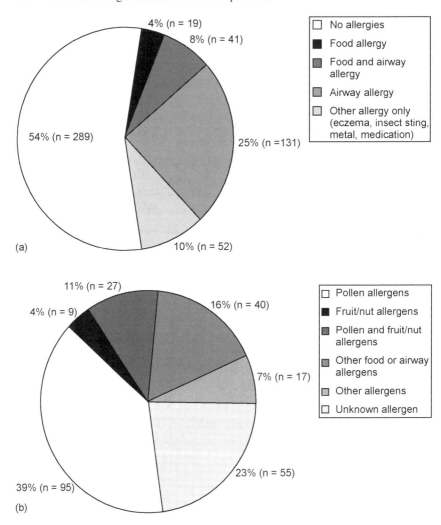

Fig. 20.1   Composition of the study population according to (a) the respondent's allergy history and (b) the allergens to which the patients are allergic.

first performed separately for both applications. No differences were observed between the factor structure of both applications. The same components were extracted in both cases and the items loaded in a similar fashion onto these components. The PCA for the combined cases indicated a two-component solution (rotated; varimax) accounting for 64% of the variance. The first component explained 42% of the variance and was labelled 'negative effects' (Table 20.1). The second component explained 22% of the variation and was labelled 'benefits' (Table 20.1). The internal consistency of the items loading on the first and second component was estimated using Cronbach's alpha. The reliability of the internal consistency was high, as revealed by an alpha of 0.93 for the first

Table 20.1   Loadings from the principal component analysis. Bold numbers indicate variables loading on a principal component

	Component	
	Negative effects	Benefits
Personal worries	**0.82**	−0.19
Risky	**0.81**	−0.18
Personal objections	**0.79**	−0.37
Damaging	**0.79**	−0.26
Unethical	**0.79**	−0.22
Tampering with nature	**0.78**	−0.30
Unnatural	**0.77**	−0.17
Negative health effects	**0.75**	−0.29
Long-term effects	0.45	0.37
Progressive	0.00	**0.58**
Positive health effects	−0.39	**0.67**
Beneficial	−0.42	**0.76**
Necessary	−0.30	**0.79**
Important	−0.36	**0.80**

and 0.85 for the second component. These subscales were used as dependent variables in further analysis.

The distinction between patients and non-patients was case-dependent, since the patient group as a whole is too diverse to test the hypothesis that respondents who directly benefit from the application of genetic modification have a more positive attitude. For example, a food allergic patient does not have a direct benefit from the use of genetically modified low allergenic birch trees. Therefore, the comparison between the two applications was performed separately from the comparison between patients and non-patients.

A paired-sample t-test on the respondents who filled in both cases was used to test for differences in attitude towards the apple and birch applications. The apple and birch cases were significantly different for 'negative effects' ($t = -5,307$; df 178; $p < 0.005$) and 'benefits' ($t = 5,329$; df 178; $p < 0.005$). Inspection of the means indicated that the birch case was perceived to have greater 'benefits', while the apple case was perceived to have greater 'negative effects' (Table 20.2).

Table 20.2   Mean item scores and standard deviation for the two GM application subscales of the apple and birch case

Case	Negative effects	Benefits
Apple	4.33 (1.66)[a]	3.43 (1.34)[a]
Birch	3.95 (1.55)[b]	3.87 (1.40)[b]

Means with different letters are significantly different at the $p < 0.05$ level using the paired-sample t-test

Subsequently, differences between patient and non-patient attitudes towards the application of genetic modification were examined for the attitude characteristics and for rating the set of prevention strategies. Two patient definitions were tested; the first definition depended on the general allergy type (food allergy, allergic rhinitis) and the second on the allergens to which respondents were allergic (fruit/nut allergens, pollen allergens). Using the broad patient definition, there was a significant main effect for self-reported history of allergic rhinitis in the birch case (Piliai's trace $F(2,354) = 5.489$, $p < 0.005$), while we did not find such an effect for food allergy in the apple case (Piliai's trace $F(2,351) = 0.417$, $p > 0.05$). The univariate test on the birch case was significant for the 'benefits'. Patients reporting allergic rhinitis perceived greater 'benefits' for the application of genetic modification than non-patients (not shown). As food allergy patients can be sensitised to a variety of foods and allergic rhinitis patients to pollen, house mite, and animal allergens, we applied a narrow definition of patients defining them as being allergic to respectively pollen or fruit/nuts. The results were similar to above. A significant multivariate effect was found for the birch case (Piliai's trace $F(2,354) = 8.871$, $p < 0.0005$), but not for the apple case (Piliai's trace $F(2,351)=1.404$, $p > 0.05$). The univariate test for birch was positive for the 'benefits'. Pollen allergic patients perceived greater 'benefits' for the application of genetic modification than non-patients (Table 20.3).

The independent sample t-test was used to examine differences between patients and non-patients regarding the desirability of the presented prevention strategies. The Mann-Whitney test was used to examine differences for the prevention strategy that involved the application of genetic modification, which had a bimodal distribution. Independent of whether food allergy or fruit/nuts allergens were used to define patients, no differences were found between patients and non-patients in the apple case regarding the application of any of the mitigation strategies. Differences were, however, found in the birch case (Table 20.3). Applying the patient definition based on allergic rhinitis complaints revealed differences regarding the desirability of implementing the strategies 'replacement with genetic modification' and 'maintaining the current situation'. Allergic rhinitis patients rated the application of genetic modification for allergy prevention as more desirable, and the maintaining of the current situation as less desirable, than non-patients (not shown). Using the patient definition based on pollen allergens gave similar findings with higher significance levels (Table 20.3). In addition, the strategy 'replacement with other trees' was rated as more desirable by the pollen allergic patient group.

Differences between allergy prevention strategies
A Friedman test was conducted to examine differences between the prevention strategies. Patients and non-patients were examined separately in the birch case, as significant attitude differences were identified between these groups. The desirability of implementing the prevention strategies differed for the apple case ($\chi^2(4, 354) = 228.986$, $p < 0.0005$) and for the birch case for patients ($\chi^2(4, 87)$

Table 20.3 Mean item scores and standard deviation of the attitude characteristics and strategy ratings of patients and non-patients. The apple and birch cases are tested separately because the patient group is case-dependent

	Apple		Birch	
	Non-patients	Patients[a]	Non-patients	Patients[b]
Attitude characteristics				
Negative effects	4.36 (1.59)	4.57 (1.80)[ns]	4.14 (1.49)	3.94 (1.65)[ns]
Benefits	3.38 (1.29)	3.65 (1.39)[ns]	3.62 (1.30)	4.27 (1.46)***
Desirability of implementation[c]				
Avoidance	4.66 (1.87)	4.68 (2.08)[ns]	4.81 (1.49)	5.16 (1.50)[ns]
Replacement with other products	4.86 (1.87)	4.47 (2.29)[ns]	4.84 (1.51)	5.31 (1.29)*
Replacement with classical breeding	4.77 (1.60)	4.63 (1.74)[ns]	4.85 (1.42)	5.05 (1.45)[ns]
Replacement with genetic modification	3.07 (1.70)	2.95 (2.01)[ns]	3.38 (1.73)	4.29 (1.82)***
Maintaining the current situation	4.07 (1.49)	4.21 (1.23)[ns]	3.67 (1.49)	3.23 (1.54)*

[ns] not significant; *** $p < 0.0005$; ** $p < 0.005$; * $p < 0.05$ using the t-test for the attitude characteristics and the prevention strategies, except the 'replacement with genetic modification' strategy, for which the Mann-Whitney is used
[a] respondents who are allergic to fruits and/or nuts
[b] respondents who are allergic to pollen
[c] see Section 20.4.1 (attribute items) for a detailed description of the strategies

$= 83.44$, p < 0.0005) and non-patients ($\chi^2(4, 270) = 183.24, p < 0.0005$). In the case of apple, the desirability was rated highest for the strategies 'replacement with other products' and 'replacement with conventional breeding' (Table 20.4). The 'replacement with other products' strategy scored significantly higher than 'avoidance', which was in turn rated significantly higher than 'maintaining the current situation'. 'Replacement with genetic modification' scored significantly lower than any other strategy. The main difference for non-patients in the birch case was that they rated 'maintaining the current situation' equal to 'replacement with genetic modification'. The ratings by the patients in the birch case were different in that they rate the 'replacement with genetic modification' significantly higher than 'maintaining the current situation'. In addition, 'replacement with other products' was rated higher than the conventional breeding strategy (Table 20.4).

20.4.3   The application of genetic modification for allergy prevention
The attitude towards the application of genetic modification for allergy prevention was measured on two subscales that were identified using a PCA. Both components that were found in the PCA are similar to the main variables in the proposed model of acceptance of gene technology by Siegrist (2000). This

Table 20.4   Comparison of the desirability of implementing different allergy mitigation strategies for apple and birch. Patients and non-patients are treated separately in the birch case as significant differences in preference for mitigation strategies were identified between the two groups

	Apple	Birch	
Desirability of implementation*		Non-patients	Patients[†]
Avoidance	4.66 (1.88)[b]	4.81 (1.49)[a]	5.16 (1.50)[ab]
Replacement with other products	4.84 (1.89)[a]	4.84 (1.51)[a]	5.31 (1.29)[a]
Replacement with conventional breeding	4.77 (1.60)[ab]	4.85 (1.42)[a]	5.05 (1.45)[b]
Replacement with genetic modification	3.06 (1.72)[d]	3.38 (1.73)[b]	4.29 (1.82)[c]
Maintaining current situation	4.08 (1.48)[c]	3.67 (1.49)[b]	3.23 (1.54)[d]

Means with different letters are significantly different at the $p<0.05$ level within each column using the Wilcoxon signed ranks test
* see Section 20.4.1 (attribute items) for a detailed description of the strategies
[†] respondents who are allergic to pollen

research indicated that the variables 'perceived benefit' and 'perceived risk' were influential factors on consumer acceptance. The attitude towards application of genetic modification differed between the two allergy prevention cases. The differences could be explained by the fact that birch can be viewed as a strictly medical application, while apple is both a medical and a food application. Medical applications of genetic modification are viewed somewhat more positively by consumers than food applications (Frewer *et al.*, 1997; Torgersen *et al.*, 2002; Zechendorf, 1994).

Differences between the attitude of patients and non-patients were found in the birch case. Allergic rhinitis patients perceived greater 'benefits' associated with the birch application compared to non-patients. There were no differences between the patient groups in terms of perceptions of 'negative effects'. If we focus on the patients that are allergic to pollen allergens, the level of perceived benefit was even higher. It is therefore interesting to note that there were no differences between food allergic patients and non-patients on the perceptions of genetically modified apples. This could be due to the fact that the food allergic group is too diverse (i.e., suffers from a wide range of allergies) to relate to the apple case, but even if we narrowed down on the fruit/nut allergic patients we observed no differences. The differences observed in the birch case relate to the perception of 'greater benefits', and not fewer 'negative effects', associated with genetic modification by patients, supporting the idea that consumer acceptance of genetically modified products is primarily a function of perceived personal benefit as opposed to personal or environmental risk perceptions *per se*. Further research is needed to clarify this issue.

The differential perception of the 'benefits' between patients and non-patients in the birch case was also reflected by the observation that patients rated the implementation of replacing birch trees with trees developed by genetic modification as more desirable than non-patients. The prevention strategy of replacement with other products was also rated as more desirable by patients, whilst maintaining the current situation was rated less desirable by these respondents. This effect was not observed in the apple case. Alternative explanations may relate to the difficulties experienced by patients avoiding birch pollen compared to apples. In addition, apple allergy is relatively low in severity, and even if a patient is inadvertently exposed to apples in the diet, the consequences are rather minor, and thus tolerable. A different response might be observed for consumers who experience severe allergic reactions (for example, anaphylaxis in response to peanut protein), particularly where the problematic product is used as an ingredient rather than a whole food, and thus more difficult to identify in food products.

The scores associated with genetic modification of apples are significantly lower in the apple case than the option of maintaining the current situation. Genetic modification of birch is rated equal to maintaining the current situation by non-patients and rated higher by pollen allergic patients. This is in agreement with the observation that the birch case is associated with higher 'benefits' than the apple case, and that patients in the birch case perceive higher 'benefits' than non-patients. However, the respondents were more positive about implementation of conventional breeding and about replacing apples in a food choice situation with other fruits, or replacing birch trees in the urban environment with other tree species. A preference for conventional breeding as opposed to genetic modification was also found by Miles *et al.* (2005), who also reported that some food allergic patients would actually purchase low allergen food produced by genetic modification (although, of course, this finding may be contingent on the severity of the allergy itself, as well as whether increased vigilance is needed to avoid the allergenic foodstuff).

One conclusion that may be drawn is that acceptance of mitigation strategies for allergy prevention may relate to whether the consumer is an allergic patient, the severity of the allergy, and the potential impact of avoidance of potentially problematic allergens on quality of life, and whether the putative mitigation strategy is perceived to be medical in nature or related to food production.

20.5   Ethical issues regarding different allergy mitigation strategies

Four potential allergy prevention strategies were described in the case study that compared acceptance of genetically modified apples and birch trees. In this section these mitigation strategies will be evaluated from a normative perspective. In other words, the rules about what counts as 'good' or as 'bad' in society need to be applied to understanding specific applications. The following ethical categories will be used in the discussion to assess each of the mitigation strategies; *responsibility and fairness from a deontological perspective* (Thompson, 2004), *benefits/costs from a utilitarian perspective* (LaFollete, 2002), and *biodiversity* and *sustainability from the perspective of societal values* (Mepham, 2005).

Table 20.4 summarised the attitudes of respondents towards expectations on the desirability of implementing a particular strategy. A distinction was made between the *avoidance strategy* and the *replacement strategies*. The avoidance strategy divides society into patients and non-patients. All mitigation efforts have to be conducted by the patients themselves. In other words, patients are made responsible for their own health, perhaps because the effects of apple- and birch-related allergies are not so severe (LaFollete, 2002). This mitigation strategy will only work when patients are able to avoid the allergens. In the case of apple allergens it is not costly or difficult to avoid them. In most culinary cultures whole apples are used, or the parts are recognisable in prepared food and so can be easily avoided. In addition, the consequences of ingestion are not severe. Birch pollen, on the other hand, can only be avoided by taking severe self-protection measures. Another aspect which must be considered, if the avoidance strategy is to be adopted, is the number of people affected in a given population. If many people are affected it is more appropriate to apply mitigation strategies based on replacement. The question that needs to be asked is whether it is fair to use the avoidance strategy when only a small group of patients are involved (Thompson, 2004).

The replacement strategy consists of three mitigation strategies: replacement by *other products*, replacement by products based on *classical plant breeding*, and replacement by products based on *genetic modification*. The extent to which these different strategies implicate societal values such as *biodiversity* and *sustainability* must be addressed (Mepham, 2005). We may also question whether it is fair to non-patients to allocate costly resources and societal effort to remove allergens, perhaps at the expense of some other societally beneficial activity. The mitigation strategy involving the replacement of apple and birch trees by other products is the most radical. It means cutting down all birch trees, while in the case of apple trees only the fruits are prevented from entering the human food chain. This is not only costly (LaFollete, 2002), but would also mean, from the perspective of societal values (Mepham, 2005), potentially unacceptable losses to biodiversity. In addition, the non-allergenic products that replace the allergenic products could cause new problems.

The replacement of apple and birch trees by non-allergenic trees based on classical breeding is generally considered to be very difficult (and especially time-consuming) in a scientific sense, as well as very expensive, but is nevertheless, as can be seen in Table 20.4, accepted by the respondents in the survey. However, the replacement of apple and birch trees based on genetic modification breeding is less acceptable to the respondents in this study, although respondents suffering from birch pollen allergy are more positive towards the modification of birch trees compared to respondents who do not suffer from birch pollen allergy. This comes as no surprise, since people are not enthusiastic about consuming foods that they associate with negative attributes (Fife-Schaw and Rowe, 2000; Gaskell *et al.*, 2002). All the concerns of the 1990s about genetic modification food in the case of apples, and in general in case of the birch trees, may be reasons to reconsider the potential impact of a genetic modification mitigation strategy.

Next to the four mitigation strategies, 'maintaining the current situation' was offered as a fifth option to the respondents in assessing the desirability of implementing the different strategies. The desirability of this strategy was rated as rather low by respondents. Only in the case of apple was the strategy to replace existing varieties with genetically modified varieties with a reduced allergenicity rated lower. The low rating may be explained by the fact that patients perceive no benefits resulting from maintenance of the current situation. However, it is possible that this strategy will be the actual strategy used within society. As we have seen, the other four mitigation strategies all have negative normative consequences: avoidance is unfair to the patients, genetic modification breeding evokes many concerns, the replacement of the products is a rather radical solution, and classical breeding is expensive and might even be impossible. However, changes in the dynamic of public opinion may result in changes in public perceptions regarding the implementation of these different strategies with time, and the situation must be closely monitored.

20.6   Conclusions

Food allergy appears to be an increasingly problematic issue in society, although further research is needed to clarify this issue regarding both the reported increase in incidence and the extent of the disease. In particular, the prevalence and causative factors of food allergy are not well understood at the present time, indicating the need to engage in an extensive epidemiological analysis of food allergy *per se*. In addition, the impact of food allergy on quality of life and economic impact also need to be systematically evaluated, in order to prioritise different mitigation strategies within society. Societal decision-making regarding the introduction and implementation of mitigation strategies to deal with food allergies is also likely to be dependent on developing systematic evaluations of patient quality of life and the economic impact of food allergy (both to the food allergic patient and to society in terms of lost working days and health costs). For example, replacement of problematic proteins may be acceptable if the socio-economic impact is high and many individuals are negatively affected. Research must also explore the effects of improved communication about food allergy, as well as improved diagnosis, on patient quality of life as this is not effectively understood at present. Some potential mitigation strategies (for example, the introduction of apples with low allergenicity) may not improve patient quality of life as there is a need for increased vigilance when making food choices, (particularly as at the present time it is unlikely that regulatory and industrial interests will permit complete replacement of the problematic foods throughout the food chain). Against this, specialist niche markets for low allergen products could be developed, and this may be particularly relevant for specific ingredients where the patients reaction may be severe should it occur.

In the debates of the 1990s a rather clear distinction in three fields can be found in the consumer acceptance of genetic modification: the 'green' agricultural genetic modification was rejected by the majority of the consumers, they were indifferent to the 'white' industrial genetic modification, and positive about the 'red' medical genetic modification. In agriculture, especially in research and development, there is a tendency to treat health problems by making changes to nutrition. Research into the reduction of allergens is one of the first examples of health claims about new cultivars. However, the genetic modification apple case shows that a combination of the medical and food-related applications of genetic modification does not necessarily increase consumer acceptance of the final food product. It is arguable that, as with other pharma applications, the benefits must be relatively high and desired by consumers.

Although some sectors of society may argue against this 'medicalisation' of food, the link between dietary choices and optimal health is becoming more obvious as new nutritional knowledge (for example, in the area of nutri-genomics) is developed. However, the problems encountered by food allergic patients must be addressed through either regulatory changes, changes in the supply chain, innovative new science strategies or a combination of these, if any real positive effects on patient quality of life are to occur.

20.7   References and further reading

BOCK, S. A. (1987). Prospective appraisal of complaints of adverse reactions to foods in children during the first 3 years of life. *Pediatrics* **79**, 683–688.

CREVEL, R. W. R. (2001). Risk assessment for food allergy – the industry viewpoint. *Allergy* **56**, S94–S97.

EGGESBO, M., BOTTEN, G. and STIGUM, H. (2001). Restricted diets in children with reactions to milk and egg perceived by their parents. *Journal of Pediatrics* **139**, 583–587.

FERNÁNDEZ-RIVAS, M. and MILES, S. (2004). Food allergies: Clinical and psychosocial perspectives. In: Mills, E. N. C. and Shewry, P.R. (eds). *Plant Food Allergies*. Blackwell Science Ltd., Oxford, UK, pp. 1–23.

FERREIRA, F., HAWRANEK, T., GRUBER, P., WOPFNER, N. and MARI, A. (2004). Allergic cross-reactivity: from gene to the clinic. *Allergy* **59**, 243–267.

FIFE-SCHAW, C. and ROWE, G. (2000). Extending the application of the psychometric approach for assessing public perceptions of food risk: Some methodological considerations. *Journal of Risk Research* **3**, 167–179.

FREWER, L., HOWARD, C. and SHEPHERD, R. (1997). Public concerns in the United Kingdom and specific applications of genetic engineering: risk, benefit, and ethics. *Science, Technology & Human Values* **22**, 98–124.

FREWER, L., LASSEN, J., KETTLITZ, B., SCHOLDERER, J., BEEKMAN, V. and BERDAL, K. G. (2004). Societal aspects of genetically modified foods. *Food and Chemical Toxicology* **42**, 1181–1193.

GASKELL, G., ALLUM, N., WAGNER, W., NIELSEN, T. H., JELSØE, E., KOHRING, M. and BAUER, M. (2002). In the public eye: representations of biotechnology in Europe. In: Gaskell, G. and Bauer, M. (eds). *Biotechnology 1996–2000 – The years of controversy*. Science Museum, London, UK, pp. 53–79.

GILISSEN, L., BOLHAAR, S. T. H., MATOS, C. I., ROUWENDAL, G. J. A., BOONE, M. J., KRENS, F. A., ZUIDMEER, L., VAN LEEUWEN, A., AKKERDAAS, J., HOFFMANN-SOMMERGRUBER, K.,

KNULST, A. C., BOSCH, D., VAN DE WEG, W. E. and VAN REE, R. (2005). Silencing the major apple allergen Mal d 1 by using the RNA interference approach. *Journal of Allergy and Clinical Immunology* **115**, 364–369.

GOWLAND, M. H. (2001). Food allergen avoidance – the patient's viewpoint. *Allergy* **56**, S117–S120.

HELM, R. M. and BURKS, A. W. (2000). Mechanisms of Food Allergy. *Current Opinion in Immunology* **12**, 647–653.

HOURIHANE, J. O. B. (2001). The threshold concept in food safety and its applicability to food allergy. *Allergy* **56**, S86–S90.

HOWLETT, J., EDWARDS, D. G., COCKBURN, A., HEPBURN, P., KLEINER, J., KNORR, D., KOZIANOWSKI, G., MULLER, D., PEIJNENBURG, A., PERRIN, I., POULSEN, M. and WALKER, R. (2003). The safety assessment of novel foods and concepts to determine their safety in use. *International Journal of Food Sciences and Nutrition* **54**, S1–S32.

JACKSON, W. F. (2003). *Food allergy*. ILSI Press, Brussels.

KNIBB, R. C., BOOTH, D. A., PLATTS, R., ARMSTRONG, A., BOOTH, I. W. and MACDONALD, A. (1999). Psychological characteristics of people with perceived food intolerance in a community sample. *Journal of Psychosomatic Research* **47**, 545–554.

KNIBB, R. C., BOOTH, D. A., PLATTS, R., ARMSTRONG, A., BOOTH, I. W. and MACDONALD, A. (2000). Consequences of perceived food intolerance for welfare, lifestyle and food choice practices, in a community sample. *Psychology, Health and Medicine* **5**, 419–430.

LAFOLLETE, H. (2002). *Ethics in practice: an anthology*, 2nd edn, Oxford: Blackwell.

LEMMETYINEN, J., KEINONEN, K. and SOPANEN, T. (2004). Prevention of the flowering of a tree, silver birch. *Molecular Breeding* **13**, 243–249.

MADSEN, C. (2001). Where are we in risk assessment of food allergens? The regulatory view. *Allergy* **56**, S91–S93.

MELTZER, E. O. (2001). Quality of life in adults and children with allergic rhinitis. *Journal of Allergy and Clinical Immunology* **108**, 545–553.

MEPHAM, B. (2005). *Bioethics: an introduction for the biosciences*, Oxford: Oxford University Press.

MILES, S., BOLHAAR, S., GONZÁLEZ-MANCEBO, E., HAFNER, C., HOFFMANN-SOMMERGRUBER, K., FERNÁNDEZ-RIAS, M. and KNULST, A. (2005). Attitudes towards low-allergen food in food allergic consumers. *Nutrition and Food Science* **35**, 220–228.

MILLS, E. N. C., VALOVIRTA E., MADSEN, C., TAYLOR, S. L., VIETHS, S., ANKLAM, E., BAUMGARTNER, S., KOCH, P., CREVEL R. W. R. and FREWER, L. (2004). Information provision for allergic consumers – where are we going with food allergen labelling? *Allergy* **59**, 1262–1268.

MOSELEY, B. E. B. (1999). The safety and social acceptance of novel foods. *International Journal of Food Microbiology* **50**, 25–31.

PRIMEAU, M., KAGAN, R., JOSEPH, L., LIM, H., DUFRESNE, C., DUFFY, C., PRHCAL, D. and CLARKE, A. (2000). The psychological burden of peanut allergy as perceived by adults with peanut allergy and the parents of peanut allergic children. *Clinical and Experimental Allergy* **30**, 1135–1143.

ROBINSON, C. (1998). Understanding the commercial and regulatory issues for genetically modified novel foods and food ingredients. *Trends in Food Science & Technology* **9**, 83–86.

SAMPSON, H. A. (2001). Immunological approaches to the treatment of food allergy. *Paediatric Allergy and Immunology* **12**, S91–S96.

SCHENK, M.F., FISHER, A. R. H., FREWER, L. J., GILISSEN, L. J. W. J. , JACOBSEN, E. and SMULDERS,

M. J. M. Acceptance of GM applications for hay fever and food allergy prevention by allergic patients and non-patients. In preparation.

SHEWRY, P. R., TATHAM, A. S. and HALFORD, N. G. (2000). Genetic modification and plant food allergens: risks and benefits. *Journal of Chromatography B* **756**, 327–335.

SICHERER, S. H,. NOONE, S. A. and MUNOZ-FURLONG, A. (2001) The impact of childhood food allergy on quality of life. *Annals of Asthma, Allergy and Immunology* **87**, 461–464.

SIEGRIST, M. (2000). The influence of trust and perceptions of risks and benefits on the acceptance of gene technology. *Risk Analysis* **20**, 195–203.

SLOVIC, P. (1987). Perception of risk. *Science* **236**, 280–285.

SLOVIC, P. (1999). Trust, emotion, sex, politics, and science: surveying the risk-assessment battlefield. *Risk analysis* **19**, 689–701.

THOMPSON, D.F. (2004). *Restoring responsibility: ethics in government, business and healthcare*, Cambridge: Cambridge University Press.

TITCHENER, G. D. and SAPP, S. G. (2002). A comparison of two approaches to understanding consumer opinions of biotechnology. *Social Behavior and Personality* **30**, 373–382.

TORGERSEN, H., HAMPEL, J., DURANT, J., EINSIEDEL, E., FJAESTAD, B., GASKELL, G., GRABNER, P., HIEBER, P., JELSOE, E., LASSEN, J., MAROUDA-CHATJOULIS, A., NIELSEN, T., RUSANEN, T., SAKELLARIS, G., SEIFERT, F., SMINK, C., TWARDOWSKI, T. and HOVMAND, M. W. (2002). Promise, problems and proxies: 25 Years of European biotechnology debate and regulation. In: Bauer, M. and Gaskell, G. (eds). *Biotechnology – the Making of a Global Controversy*. Cambridge University Press, Cambridge, UK, pp. 21–94.

UCB INSTITUTE OF ALLERGY (1997). European Allergy White Paper. Van Moerbeke, D. (ed.). Brussels.

VADAS, P. and PERELMAN, B. (2003). Presence of undeclared peanut protein in chocolate bars imported from Europe. *Journal of Food Protection* **66**, 1932–1934.

VAN PUTTEN, M.C., FREWER, L.J., GILISSEN, L.J.W.J., GREMMEN, B., PEIJNENBURG, A.A.C.M. and WICHERS, H.J. (2006). Novel foods and food allergies: a review of the issues. *Trends in Food Science and Technology* **17**, 6, 289–299.

VIERK, K., FALCI, K., WOLYNIAK, C. and KLONTZ, K. C. (2002). Recalls of foods containing undeclared allergens reported to the US Food and Drug Administration, fiscal year 1999. *Journal of Allergy Clinical and Immunology* **109**, 1022–1026.

WENSING, M., AKKERDAAS, J. H., VAN LEEUWEN, A., STAPEL, S. O., BRUIJNZEEL-KOOMEN, C. A. F. M., AALBERSE, R. C., BAST, B. J. E. G., KNULST, A. C. and VAN REE, R. (2002). IgE to Bet v 1 and profilin: Cross-reactivity patterns and clinical relevance. *Journal of Allergy and Clinical Immunology* **110**, 435–442.

YOUNG, E., STONEHAM, M. D., PETRUCKEVITCH, A., BARTON, J. and RONA, R. (1994). A population study of food intolerance. *The Lancet* **343**, 1127–1130.

ZECHENDORF, B. (1994). What the public think about biotechnology. *Bio Technology* **12**, 870–875.

21

Consumers of food products, domestic hygiene and public health

E. Redmond and C. Griffith, University of Wales Institute Cardiff, UK

21.1 Introduction

The health of the population is affected by many factors, and in reality a country gets the food poisoning it deserves, based on the foods eaten, where they are sourced and how they are processed and handled. For example, *Vibrio parahaemolyticus* is rare in the UK but is the leading cause of food poisoning in Japan, due to the greater consumption of specific, often raw, sea foods. Many foods that are bought into the home are frequently contaminated with naturally occurring pathogenic microorganisms (Ellard, 1999). Such pathogens cannot be seen, smelled, tasted or identified by touch (Roberts *et al.*, 1995) but when consumed can cause illnesses of varying severity, including death. Thus, food safety issues are of major importance to world health (WHO, 2000a). Safe food is a basic human right and in modern society prevention of disease and improvement of human health is of paramount importance, not only for governments and industries but also for consumers themselves.

21.1.1 Incidence of foodborne disease

Foodborne disease has emerged as an important and growing public health and economic problem in many countries in the last two decades (WHO, 2000a). It has been estimated that 130 million Europeans (WHO, 2000b), 2.4 million Great Britons from England and Wales (Adak *et al.*, 2002), 76 million Americans (Mead *et al.*, 1999) and 5.4 million Australians (Hall *et al.*, 2005) are annually affected by episodes of foodborne disease and food-related illnesses. However, the true incidence is difficult to obtain due to under-reporting (Lake *et al.*, 2000;

Robertson *et al.*, 2004). Although, foodborne illnesses can be severe and fatal, milder cases are often not detected through routine surveillance (Mead *et al.*, 1999). Given that most foodborne illnesses only cause discomfort for a short period of time, medical attention is frequently not sought (IID Executive Committee, 2000; Mead *et al.*, 1999; Rocourt *et al.*, 2003). Therefore, the small proportion of more severe food poisoning cases that are reported may only represent the '*tip of the iceberg*' (Maurice, 1995). Illustrating such under-reporting, the Food Standards Agency (FSA) in the UK found that 80% of people who suffered what they considered to be food poisoning failed to report it (FSA, 2001).

The significant incidence of foodborne disease results in substantial tangible (financial) and intangible (pain, suffering) costs to society – factors that highlight the need for effective methods of health education to reduce the incidence. In a recent strategy document, the UK FSA stated that the most significant reduction in the number of cases of foodborne disease over the next five years is likely to come from focusing attention on food preparation, particularly in the domestic setting (FSA, 2001).

21.1.2   The home as a location for foodborne disease
In England and Wales, 10–17% of general outbreaks of foodborne disease are reported to have originated in the home (Cowden *et al.*, 1995; Tirado and Schmidt, 2000; Rocourt *et al.*, 2003), and the home is also an important location of outbreaks in other countries (FAO/WHO, 2002). The estimated international incidence of foodborne disease attributable to the home can be seen in Table 21.1. However, the majority (>95%) of foodborne disease cases are sporadic (FSA, 2000) and less likely to be investigated by public health authorities. It has been suggested the actual proportion of cases that occur in the home is likely to be much larger than reported outbreak data suggests and may be the location for the majority of the sporadic cases (Redmond and Griffith, 2003a).

21.1.3   The microbiology of the domestic kitchen
Raw foods brought into the kitchen are invariably contaminated with microorganisms, some of which may be pathogenic. Poultry worldwide is known to be contaminated (Griffith and Redmond, 2005), e.g. in the UK at a frequency of 68% with *Campylobacter* at levels up to 7.4×10^4 (Harrison *et al.*, 2001). However, food is not the only route or vehicle by which microorganisms can enter the kitchen. The presence of pets and soiled laundry is not uncommon, although the domestic kitchen has also been found to be used for motor vehicle maintenance, gardening and even breeding chickens (Worsfold and Griffith, 1997), each bringing their own microbiological hazards. It is perhaps therefore not surprising that a wide range of pathogens have been isolated from the domestic kitchen (see Table 21.2). These data indicate the range of microorganisms present with other studies reporting the numbers isolated (Ojima *et*

Table 21.1 International incidence of foodborne disease attributable to the home (1982–2004)

Country	Years of data collection	Incidence
England and Wales, UK[1]	1992–1993	17% general foodborne outbreaks of infectious intestinal disease (IID) associated with food prepared in private house and served elsewhere.
England and Wales, UK[2]	1993–1998	12% general foodborne outbreaks of foodborne disease attributed to food consumed in a private house.
France[3]	1993–1997	40% foodborne disease outbreaks (microbiologically confirmed and suspected) associated to the private home (the place where food was eaten).
Spain[4]	1993–1998	49% foodborne disease outbreaks associated with the private home (the place where food was eaten or acquired).
Australia[5]	1999	Suggested between 20 and 40% of foodborne illness arises from private homes.
New Zealand[6]	1997	~50% cases of foodborne illness have been reported to be caused by poor handling techniques in the domestic kitchen.
USA[7]	1993–1997	20% reported bacterial foodborne disease outbreaks from place where food was eaten.
'The Americas'[8]	1998–2001	38.1% homes were implicated in foodborne outbreaks.
Canada[9]	1982	14% incidents (outbreaks and cases) caused by mishandling of foods in homes.

Sources: [1]Cowden *et al.*, 1995; [2,3,4]Tirado and Schmidt, 2000; [5]ANZFA, 1999; [6]Bloomfield and Neal, 1997; [7]Olsen *et al.*, 2000; [8]PAHO, 2004; [9]Todd, 1989.

al., 2002; Sharp and Walker 2003) with counts for some sites in excess of 10^8 cfu/ml (Hilton and Austin, 2000).

Problems with these types of studies, which may underestimate the presence of pathogens, include the random nature of the sampling, irrespective of the types of foods prepared and when. This may be compounded by relatively low numbers of pathogens in relation to non-pathogens, coupled with overgrowth of the latter. Other studies (Haysom and Sharp, 2005) have attempted to monitor trends in kitchen site microbial contamination over time and contamination was seen to peak after meal preparation, although other non-food preparation activities also contributed. Research studies (Redmond, *et al.*, 2004) starting with an uncontaminated kitchen, showed how contamination of specific sites with food pathogens was found to occur during food preparation. Given these types of data it is perhaps not surprising that contamination and recontamination of sites in the domestic kitchen is constantly changing. Coupled with often poor general design, construction, maintenance and cleaning compared to food

Table 21.2 Reported isolations of different potential pathogens from specific environmental sites within food preparation areas

Environmental site	*Campylobacter spp.*	*Salmonella spp.*	*Y. enterocolitica*	*S. aureus*	*E. coli*	*Bacillus spp.*	*B. cereus*	*L. monocytogenes*	*Listeria spp.*
Dish cloth	•			•	•	•		•	•
Cleaning cloth	•	•		•	•		•	•	•
Wash-up sponge	•	•		•	•				•
Wash-up brush						•		•	•
Wash cloth		•						•	
Floor mop				•		•			•
Tea/hand towel				•	•	•			
Sink		•	•	•	•			•	•
Taps				•	•			•	
Refrigerator/door	•			•	•			•	•
Waste/pedal bin	•			•	•		•		
Chopping boards	•			•	•				
Work surfaces	•				•		•		
Floors	•				•				

Adapted from Griffith (2000b)

processing plants, it is easy to envisage how the domestic kitchen could be a factor in domestic foodborne disease.

21.1.4 Consumer responsibilities

Consumers are the important final link in the food chain to assure safe food consumption and prevent foodborne illness (The Pennington Group, 1997; Zhang and Penner, 1999). Multiple food safety responsibilities are required by consumers because they not only purchase products but also process and provide foods for themselves and for others. Therefore consumers have responsibilities as purchasers, storers, providers and processors of food and need to be conscious of the nature and safety of food products (CDNANZ, 1997). Thus, food-handling practices employed by consumers in the domestic kitchen influence the risk of pathogen survival and multiplication, as well as cross contamination to other products (Roberts *et al.*, 1995). Given that 92% women and 61% men prepare meals (if not every day) at least once or twice a week (Nicolaas, 1995) it is extremely important that food is handled in a manner that does not increase the risk of foodborne disease. A great deal of research has been carried out on manufacturing, processing and distribution processes in the food industry, however, the consumer remains the least studied, yet crucial link in the food chain.

21.2   Consumer food safety research

Consumer food safety research is required to ascertain how consumers handle food in their homes, determine what consumers know about food safety and why some safe food-handling practices are implemented and others are not. The overall aim of international consumer food safety studies should be to provide information for the development of effective communication strategies to promote safe food handling practices, although this has not always been the case.

21.2.1   Mechanisms for assessing consumer food safety behaviour

Information relating to domestic food safety behaviour comes from two main sources, analysis of food poisoning outbreaks and consumer based research studies (Griffith and Worsfold, 1994). Outbreak investigations provide quantitative data regarding contributory food-handling malpractices that have resulted in foodborne disease. However, such retrospective analysis provides limited information about consumer food safety behaviour. Partly because sporadic cases are rarely investigated, but also because the accuracy and availability of data is limited due to difficulties of recalling exact food consumption details and handling practices that may have occurred some time before the illness was reported. Internationally, numerous consumer-based research studies have evaluated food safety practices of consumers. Different approaches have been adopted including questionnaire and interview surveys, focus group discussions and observational studies.

In the past 29 years 100 consumer food safety studies have been undertaken using different research methodologies (see Table 21.3). Eighty-five percent utilised survey techniques (questionnaires and interviews), 19% were based on

Table 21.3   Methods of data collection used for the assessment of consumer food safety behaviour (1977–2005) ($n = 100$)

Method of data collection		Frequency of use by specific methodology n (% of total studies)	Overall frequency of use n (% of total studies)
Self-completion questionnaires	Postal	16 (16%)	
	Self administered	14 (14%)	**32 (32%)**
	Online	2 (2%)	
Interviews	Telephone	22 (22%)	**53 (53%)**
	Face-to-face	31 (31%)	
Focus groups		–	**11 (11%)**
Observational studies		–	**19 (19%)**

Note: Figures may not add up to 100% due to some studies utilising more than one data collection method.

direct observation and 11% utilised focus groups. There are advantages and disadvantages associated with each of the different research methods used and these have previously been discussed (Redmond and Griffith, 2003b).

An analysis of studies has shown than survey responses (denoting knowledge, attitudes, intentions and self-reported practices) have provided a more optimistic portrayal of consumer food safety behaviour than data obtained from focus groups and direct observation. Although consumers have demonstrated knowledge, positive attitudes and intentions to implement safe practices, substantially larger proportions of consumers have been observed to frequently implement malpractices (Redmond and Griffith, 2003a).

21.3   Consumers' knowledge of food hygiene

Health-related behavioural research has suggested that individuals make rational decisions about behaviours when they are aware and have some knowledge of the associated health problems (McIntosh *et al.*, 1994). Although acquisition of knowledge alone does not change behaviour, knowledge makes it possible for the consumer to make more informed choices, for this reason, knowledge gaps may be barriers to establishing appropriate behaviours (Cody and Hogue, 2003). Therefore, it is considered that knowledge gain must precede behavioural change (Medeiros *et al.*, 2004) and knowledge of the consequences of unsafe food-handling practices can enhance consumer motivation to change behaviour (Bruhn, 1997). Assessment of consumer knowledge featured in 75% of consumer food safety surveys undertaken between 1970 and 2002 (Redmond and Griffith, 2003a). Research has shown that some consumers have exhibited adequate general knowledge of factors that affect foodborne illness, however, gaps in knowledge of specific practices have also been identified.

21.3.1   Food storage

Storage of food products above refrigeration temperature and below the recommended hot holding temperature of 63°C (DoH, 1995) encourages proliferation of bacterial cells, germination of spores and possible toxin production to potentially dangerous levels. Inadequate temperature control during storage is frequently implicated as a cause of foodborne illness (Knabel, 1995). In recent years inadequate temperature control during storage of foods has been associated with 28–46% of reported foodborne disease outbreaks in England and Wales (Tirado and Schmidt, 2000; Ryan *et al.* 1996; Cowden *et al.*, 1995). Thus, it is important that consumers correctly store foods in the home to reduce the risk of foodborne disease.

Determination of consumer knowledge of food storage at home has largely focused upon knowledge of refrigeration temperatures, and cumulatively many consumer food safety studies have found that knowledge to ensure safe storage of perishable foods is inadequate (O'Brien, 1996; Redmond and Griffith,

Table 21.4 Proportions of UK and US consumer where knowledge of key food safety practices is reported to be lacking (1995–2005)

Food safety issue	Country	% Lacking knowledge of food safety issue
Hand-washing/drying[1]	UK	20–33
	USA	14–21
Separation of raw and cooked meats during food preparation[2]	UK	0–36
	USA	20–22
Refrigeration temperatures[3]	UK	50–93
	USA	40–65
Correct heating temperature[4]	UK	80–85
	USA	80–93

Sources
[1] Mathias, 1999; Redmond et al., 2005; Altekruse et al., 1996; Albrecht, 1995.
[2] ADA Conagra, 1999; Altekruse et al., 1996; Griffith et al., 2001; Walker, 1996.
[3] Endres et al., 2001; Albrecht, 1995; Hudson and Hartwell, 2002; Wenrich et al., 2003
[4] Worsfold, 1994; AI, 1999; Griffith et al., 2001

2003a). Indeed, consumer food safety surveys have found that up to 93% of consumers were *unaware* of correct refrigeration temperatures (FSAI, 1998) (see Table 21.4) and similarly ~75% of consumers have been found to be unaware of the temperature of their own fridge (Marklinder et al. 2004). Data detailing measurements of consumers' refrigerator temperatures correspond with the apparent lack of knowledge. Results have shown that up to 70% of consumers' refrigerators exceeded recommended temperatures (Daniels, 2001; Johnson et al., 1998) therefore providing conditions that may encourage the proliferation of bacterial cells and increase the risk of foodborne disease.

Storage of leftover food is frequently undertaken in the domestic kitchen and it is of concern that knowledge of adequate cooling and subsequent storage practices has been found to be inadequate (Redmond and Griffith, 2003a) thus increasing the risk of foodborne disease. An American survey found that a large proportion of consumers (63%) were aware that leftovers should not be stored in the container they are cooked in, however 23% of consumers thought this was acceptable practice (Wenrich et al., 2003). Furthermore, research has shown that although 70–75% of consumers knew that bacteria responsible for causing foodborne illness grow at room temperature (Meer and Misner, 2000; Wenrich et al., 2003), a large number of consumers were unaware that storage of food at room temperature may cause food poisoning (Mathias, 1999).

Separation of raw and cooked foods during refrigerated storage is recommended to avoid risks of cross contamination (IFH, 1998). However, observations of inappropriate storage of foods in refrigerators by >80% Australian consumers (Mitakakis et al., 2004) indicated a substantial lack of knowledge, resulting in food storage practices that may increase the risk of foodborne disease. This finding concurs with previous research indicating 75% of respondents were unaware of potential risks associated with storing raw meat and poultry on upper shelves of refrigerators (Sammarco and Ripabelli, 1997).

21.3.2 Cooking

Inadequate heating is an important contributory factor in foodborne disease outbreaks (Olsen *et al.*, 2000; Evans et al., 1998). Indeed, cooking is considered to be an important control step in the food preparation process (Food Safety and Hygiene Working Group, 1997). The time and temperature association for heating should be such to ensure that heat penetration to the centre of the foodstuffs occurs and results in destruction of vegetative, non-sporeforming organisms (DoH, 1993).

Undercooking has been acknowledged by 89% of UK consumers as a risk factor associated with foodborne disease (Mathias, 1999) and data from USA has shown that 67–74% of consumers know that cooking meat well decreases the risk of food poisoning. However, it has also been reported that between 80 and 93% of consumers do not know what the temperature should be inside a piece of meat when it is considered to be safe to eat (see Table 21.4). Knowledge of how to determine heating adequacy has been evaluated by several studies and between 93 and 96% consumers have recognised that it is important to check the inside of the chicken to ensure that it is fully cooked (Bloomfield and Neal, 1997; Hodges, 1993). However, 88% of consumers thought that a subjective measurement was acceptable to assess end of cooking (Beddows, 1983). This is of concern as US research found that colour is not a reliable indicator that meat has reached a sufficiently high temperature to destroy pathogens such as *E.coli* O157:H7 (FSIS, USDA, 1998a). Thus, educational campaigns have been recommended that food thermometers should be used to determine end of cooking times of meat and meat products (FSIS, USDA, 1998b).

21.3.3 Cross contamination during food preparation

The microbiological risks associated with contamination of ready-to-eat (RTE) foods prepared using unclean utensils, previously used for preparation of raw meat and poultry are considered to be significant (DeBoer and Hahne, 1990; Humphrey *et al.*, 1994; Redmond *et al.*, 2004). Cross contamination as a risk factor has been implicated in 33% of reported, general outbreaks of foodborne disease in England and Wales (WHO, 2003). However, due to recall difficulties in retrospective epidemiological investigations, coupled with microbiological data on kitchen contamination, this percentage is likely to be an underestimation (Griffith, 2000a).

Lack of recognition of surface contamination levels, plus poor knowledge of appropriate food safety practices to prevent cross contamination could result in the transfer of pathogens to RTE foods and the potential for causing illness. Many international surveys assessed this aspect of safe food-handling and results are comparable. Whilst surveys have sometimes reported high percentages of people possessing knowledge, large numbers of consumers (albeit small percentages) also lack appropriate knowledge. Research has suggested that up to 36% of UK consumers and up to 22% of American consumers do not recognise the importance of using separate or adequately cleaned utensils for preparation of RTE foods after preparation of raw meat and poultry (see Table 21.4). Furthermore, 68–73% of US consumers considered themselves to be extremely likely/likely to experience food poisoning after consuming RTE foods placed on unwashed surfaces contaminated with raw meat (Cody and Hogue, 2003).

21.3.4   Hand decontamination

Effective hand-washing and hand drying is considered to be an important control measure for preventing the transmission of foodborne diseases in food-handling environments (Paulson *et al.*, 1999). Contamination of food via the hands may be through direct contact of the food with hands that are contaminated, or indirectly through poor practice such as handling and contaminating equipment that is subsequently used for food preparation (Taylor and Holah, 2000). Microbial transfer of bacteria from hands to other contact surfaces has been well documented (Chen *et al.*, 2001; De Boer and Hahne, 1990; Redmond *et al.*, 2001) and findings indicate that where hands are intermediary vectors, adequate hand decontamination behaviours are required to prevent microbial cross contamination.

Determination of knowledge of hand-washing practices has largely concentrated on the importance of hand-washing for decreasing foodborne disease or timing of hand-washing actions during food preparation. As expected, the majority (75–100%) of consumers recognise that hand-washing is a necessary food safety action (Redmond and Griffith, 2003a). Although data has suggested that consumers know the correct procedure for adequate hand-washing and drying (Griffith *et al.*, 2001), research has also indicated that nearly a fifth of US and UK consumers are unaware of desirable hand-washing and hand drying procedures (see Table 21.4). Limited consumer food safety research has evaluated knowledge of specific hand-washing and hand drying behaviours, however UK research found that 23% of consumers do not consider it necessary to use soap for hand-washing and 97% of the same consumers have indicated knowledge that rinsing hands is an important part of the hand-washing process (Redmond *et al.* 2005). Data from the same survey also indicated that a smaller proportion of UK consumers (54%) did not think it was essential to dry hands after hand-washing. This finding is a concern as the drying process is considered to be of critical importance to maximise reduction of transient and resident bacteria on hands (Michaels *et al.*, 2001).

21.4   Consumers' attitudes to food hygiene in the home

A key to the design of effective educational initiatives is an understanding of factors that influence an individual's behaviour (Middlestadt *et al.*, 1996).

Indeed, to raise awareness of food safety issues there is a need to determine baseline attitudes towards food safety behaviours. Furthermore, research has indicated that consumer attitudes towards food safety may be an important influence on performed behaviours (Saba and DiNatale, 1999; Westaway and Viljoen, 2000). Evaluation of consumer attitudes towards issues related to microbial food safety in the domestic kitchen and food safety education is required to inform the development of effective, consumer-orientated food safety communication strategies that aim to raise awareness of food safety issues and bring about behavioural improvement. Health-related behaviours, such as those associated with food safety, are considered to be influenced by a number of cultural, socio-economic and environmental factors, as well as psychological determinants, such as knowledge, attitudes, beliefs and values (WHO, 2000b). Such cognitive antecedents are considered as being important determinants for providing a rationale or motivation for implementation of behaviours (Connor and Norman, 1999; Levy, 2002) and the more that is known about factors underlying performance or non-performance of health-related practices, the more successful the design of an intervention can be (Strand, 1999).

21.4.1   General: specific attitudes

Overall, UK survey research has shown that cumulative consumer attitudes towards food safety in general have been found to be relatively positive (Redmond and Griffith, 2004b). This finding concurs with other studies where large proportions of consumers have expressed concern for food safety issues (Smith and Riethmuller, 2000; Redmond et al., 2005). However, despite this, consumers have expressed specific attitudes that were contrary to safe food preparation practices, for example 53–64% of consumers did not consider it essential to cool cooked food down quickly for subsequent storage (Redmond and Griffith, 2004b).

In the past decade, many UK food safety education efforts have provided generalised food safety information for the overall population. However, supporting previous qualitative research (Redmond et al., 2000; Redmond, 2002), quantitative findings indicate that older consumers are associated with an overall positive attitude towards food safety in the domestic kitchen and younger adults are associated with a more negative attitude. Furthermore, significant differences of cumulative food safety attitudinal responses have been identified between males and females (see Table 21.5). Thus, it is suggested that in the future, targeted food safety education strategies are required for different groups of consumers. For example, it may be necessary to improve attitudes towards specific behaviors for males and/or younger adults before attempting to change their actual food safety practices (Redmond, 2002). Whereas, females and/or older adults, who may possess a more positive attitude towards specific practices based on a food safety behaviour, may be more receptive towards interventions promoting a corresponding behavioural change.

Table 21.5 Attitudes towards food safety behaviours where statistically significant differences ($p < 0.05$) between male and female respondents have been identified

Attitude statement	Male % total no. of respondents $n = 23$	Female % total no. of respondents $n = 77$
It is essential for hot food to be cooled down quickly for storage		
Strongly agree/agree*	14	63
Neither	29	0
Strongly disagree/disagree	57	37
Cooked foods, once cooled should be refrigerated or frozen immediately		
Strongly agree/agree*	66	91
Neither	17	3
Strongly disagree/disagree	17	6
It is not all right to leave cooked rice in a bowl on a kitchen work surface overnight		
Strongly agree/agree*	50	81
Neither	25	11
Strongly disagree/disagree	25	8
Reheating food to a warm temperature is acceptable		
Strongly agree/agree	24	13
Neither	19	7
Strongly disagree/disagree*	57	80

* = positive response
Source: Redmond and Griffith, 2004b

21.4.2 Risk, control and responsibility

Concepts of risk, control and responsibility are present in many behavioural models used to aid health education processes (Bennett and Murphy, 1999). Risk perceptions are considered to form the basis of a heuristic framework that guides decisions about behaviour (Frewer et al., 1994) and perceptions of food safety risks may contribute to shape an individual's behaviour (Yeung and Morris, 2001). Controllability has been identified as an important determinant of the perceived risk associated with a hazard (Frewer et al., 1994) and perceptions of risk and control are important influences of health-related precautionary behaviours.

Research has suggested that consumer judgements of 'optimistic-bias' and the 'illusion of control' concerning food safety during domestic food preparation are prevalent (Redmond and Griffith, 2004a; Frewer et al., 1995). Identification of over-optimistic biases during evaluation of personal health risks are considered to be common (Bennett and Murphy, 1999) and have previously been associated with automobile accidents, crime and disease (Weinstein, 1980). Such judgements may contribute to continued implementation of unsafe food-handling behaviours associated with microbial risks during domestic food preparation (Redmond et al., 2004) and also hinder educational efforts to reduce risk-reducing behaviours (Miles et al., 1999).

It has been suggested that food poisoning incidence and frequency of serious consequences are underestimated by consumers (Bruhn, 1997). Research indicates that consumers associate a low personal risk of food poisoning from home-produced food (Redmond and Griffith, 2004a) and individuals believe themselves to be at less risk than 'other people' from food-related hazards (Frewer et al., 1995). Data presented in Table 21.6 illustrates such a concept whereby more UK consumers (91%) consider themselves to have a low risk of experiencing food poisoning from consuming food that they had prepared at

Table 21.6 Consumer perceptions of risk, control and responsibility

	% of UK consumers (n = 2014)	
	Very low risk /low risk	Very high risk/ high risk
Perceived *personal risk* of getting food poisoning after you have eaten food that you have prepared in your own home	91	1
Perceived *risk of other people* getting food poisoning after they have eaten food that they have prepared in their own homes	64	2
	Complete/ nearly complete responsibility	No/ very little responsibility
Perceived *personal responsibility* for ensuring that food prepared in your home is safe to eat	84	7
Perceived *responsibility that other people* have for ensuring that food prepared in their home is safe to eat	73	5
	Complete/ nearly complete control	No/ very little control
Perceived personal control over food hygiene in your kitchen	81	3
Perceived control that other people have over food hygiene in their kitchen	61	3

Source: Redmond et al., 2005

home compared to other people (64%). An underestimation of personal risk from food may prevent consumers from taking appropriate steps to reduce their exposure to food-related hazards (Frewer *et al.*, 1995).

Previous research has suggested there is no direct relationship between perception of risk and perception of control (Frewer *et al.*, 1994). However, McKenna (1993) has noted that an overly positive perception of events may stem from an illusionary perception of personal control. This may be particularly relevant for preventing cross contamination behaviours where the underlying principles may be more complex. Previous research has found that some consumers perceived themselves to have control of food safety during their own food preparation as well as a lower risk of experiencing food poisoning (Redmond, 2002). Such findings concur with research reported by Levy (2002) who indicated that consumers may not perceive a risk if they are confident that they are controlling the risks. Such a perception may be an obstacle for health educators when communicating information about food safety hazards and risks to consumers (Hoorens, 1994).

Recognition of personal responsibility for food safety is considered to be a prerequisite for implementation of appropriate food safety behaviours (Unklesbury *et al.*, 1998). Multiple food safety responsibilities are required by the consumer during domestic food preparation and failure to assume personal responsibility for food safety may result in increased potential for unsafe food-handling behaviours and consequential increased risks of food poisoning. Research evaluating consumer perceptions of responsibility for food safety has revealed inconsistent findings. Concurring with Hodges, (1993), the majority of consumers from a recent UK survey (see Table 21.6) perceived themselves to be responsible for their own food safety, however, other research has suggested that nearly three-quarters of consumers perceive food manufacturers to be ultimately responsible for the safety of their foods (Redmond and Griffith, 2004b). Epidemiological data suggests the home is an important point of origin for food poisoning incidence (Redmond and Griffith, 2003a) and therefore it is important for consumers to be responsible for safe food-handling in the home. Research suggests that consumers are beginning to recognise personal responsibility for food safety, but still consider external providers of food are also accountable to maintain levels of food safety. A notion of a shared responsibility has been suggested between industry and consumers (Griffith, 2000a) and this, along with increased consumer awareness of food safety responsibilities in the home needs to be communicated.

21.4.3 Perception of the home as a location for food poisoning

Consumers are reported to only think about safe food preparation behaviours when they perceive a food safety risk (Levy, 2002). However, research has shown that the majority of consumers perceive their own homes, and the homes' of friends and family to be the most unexpected locations to acquire food poisoning, indicating that risks of food poisoning associated with the home are underestimated. A recent survey showed that 85% of UK consumers perceived the home to be the location where they would *least* expect to acquire food poisoning (Redmond *et al.* 2005). Such a finding corroborates with international research indicating only 16–23% of North American consumers (CFIA, 1998; Fein *et al.*, 1995; Woodburn and Raab, 1997) perceived the home as a likely place to acquire food poisoning. Indeed, 70% of American consumers did not think that it is very common for people in the USA to become sick because of the way food is handled or prepared in homes (Cody and Hogue, 2003). Recent UK findings have indicated that mobile food outlets, takeaway outlets and fast-food outlets were perceived to be the most likely locations for acquiring food poisoning (Redmond *et al.*, 2005) and this is consistent with international con-sumer food safety research (Redmond and Griffith, 2003a). Despite increased media and educational attention, perception of the home as an unlikely location for getting food poisoning appears to be relatively unchanged over the past 15 years.

Failure to associate home food-handling practices with foodborne illnesses is considered to be a serious impediment to convincing consumers to change inappropriate food-handling behaviours (Fein *et al.*, 1995). The notion of optimistic-bias relating to the underestimation of risks and the occurrence of a negative events (such as experience of food poisoning from home prepared food) (Weinstein and Klein, 1996) needs to be addressed in efforts to reduce risk-reducing behaviours. It is recommended that future food safety com-munication initiatives specifically equate food poisoning incidence with the need for domestic food safety.

21.4.4   Perception of preferred sources and types of information

Behavioural scientists have stated that '*human beings are not empty vessels in which correct information can simply be poured which in turn will eliminate undesirable customs*' (Foster and Kaferstein, 1985). For communication to have the desired impact, a whole chain of responses needs to be elicited (McGuire, 1984). Therefore the development of community-based interventions for food safety initiatives is considered to be a complex process, owing to the need for provision of information for diverse target audiences in many different settings. Diverse strategies are required for many different groups of consumers, each having their own food preparation practices (Campbell *et al.*, 1998) and social and environmental influences. Of importance for food safety education is the message and the manner in which the message is communicated to and received by the public (Griffith *et al.*, 1994).

Channels and sources generally used for public communication of food safety issues include a variety of formats such as television, radio, posters, leaflets, newspapers, cookery books, magazines and reminder aids. However, TV, in particular, may not always transmit the right food safety message (Griffith, *et al.*, 1994) Although limited research has been conducted to evaluate the effec-tiveness of different intervention types, the potential effectiveness of different media is reported to vary considerably, despite having common characteristics (Tones and Tilford, 1996).

Preference for different sources of food safety information may impact upon source effectiveness. The population at large comprises individuals with different ages, sexes, social classes, family influences and educational backgrounds and not everyone has the same attitudes, perceptions or behavioural traits, nor do they have the same needs (Hastings and Haywood, 1991). Therefore, identification of preferable sources of interventions may aid the development of effective, targeted food safety education initiatives. UK research has found that the most preferred sources of consumer food safety information (see Table 21.7) include food packaging, leaflets and TV adverts (Redmond and Griffith, 2005a; Redmond et al., 2005). Television has been a channel of communication for health education issues and has been consistently shown to reinforce existing behaviour and raise awareness, however, it may have little or no effect on actual behavioural change (Bennett and Murphy, 1999). Placement of food hygiene information in FSA TV advert campaigns in the UK in recent years has been recalled by 61% of consumers (Redmond et al., 2005) and indeed, previous international research has indicated that between 50 and 67% consumers perceived television to be a common source of food safety information (Jay et al., 1999b; NCC, 1991; Meer and Misner, 2000).

The production and distribution of leaflets is considered to be a mainstay of health education and promotion activities (Fraser and Smith, 1997), however, the effectiveness of communication using written information in the form of leaflets has been widely debated. Although UK consumers indicated that leaflets were a preferred source of food safety information, the same survey also found that only 13% of the same consumers recalled previously seeing a leaflet detailing food safety advice (Redmond et al., 2005). Some workers have

Table 21.7 Preferable sources of information about food hygiene in the home

Source of information	Proportion of consumers who perceived source of information as 'preferable' % of UK consumers ($n = 2014$)
Television adverts	27
Leaflets	17
Food packaging	15
Television documentaries/news programmes	12
Television cooking programmes	10
Magazine articles	4
Kitchen aids (e.g., fridge magnets and T-towels)	3
Radio programmes	2
The Internet	2
Posters with food safety information	2
Recipes	1

Source: Redmond et al., 2005

reported that leaflets alone have resulted in an increase of knowledge for the short and long term (Paul and Redman, 1997), and other research has indicated that printed leaflets can bring about positive attitude change (Harvey et al., 2000). Other studies have reported the contrary; for example, after leaflets had been personally given to patients by GPs, recall of such leaflets was less than 50% (Tones and Tilford, 1996). Thus suggesting that the leaflet had little or no impact on a large proportion of persons that it was given to. Nevertheless, there is evidence to suggest the appropriate use of leaflets can be effective in *helping* people make changes (Fraser and Smith, 1997) yet no available data suggests that leaflets alone can bring about actual behavioural change. It is believed that the real value of leaflets lies in their combined use with other strategies, especially those concerning interpersonal support (Griffith et al., 1994; Tones and Tilford, 1996). Furthermore, it is considered that significant changes in recall, knowledge and behaviour are most likely when a leaflet comes from a reliable source and is used in conjunction with interpersonal communication and other educational resources in a familiar context (Bennett and Murphy, 1999).

Although food packaging has been noted as a preferable source of food safety information by consumers, previous research has questioned the effectiveness of placement of food safety advice on packaging of raw meat. In the USA research showed that such placement was not effective for improving food safety behaviours in the home (Yang et al., 2000). Observations of domestic food-handling behaviours showed that consumers frequently touch the inside of raw chicken packaging (Redmond et al., 2004). Concerns have been expressed about the external contamination of packaging (Burgess et al., 2005) and more than a third of internal raw meat packaging was found to be contaminated with pathogens such as *Campylobacter* and *Salmonella* (Harrison et al., 2001). The placement of instructions on the inside of raw meat packaging, as is often the case, could therefore increase the risk of microbial cross contamination during handling. Food safety advice on packaging should be easily visible before opening, simple and reading the instructions should require no additional handling actions.

Least preferable sources of food safety information found in Redmond and Griffith (2005a) included reminder aids such as fridge magnets and t-towels. This may reflect lack of use in the locality, rather than a general dislike for the type of educational aid. Perceived disfavour did not correspond with previous focus group findings whereby targeted sub-groups of consumers have responded positively to visual presentation of reminder aids such as magnets used for promoting food safety information (Redmond et al., 2000, 2001; Li-Cohen et al., 2002). Consumer responses in focus groups indicated that such reminder aids were novel and useful as a constant reminder to implement safe food preparation behaviours. Indeed, key interventions for the ongoing nationwide US Fight-Bac® food safety initiative are fridge magnets (Partnership for Food Safety Education, 2002) and previous research has shown that use of strategically placed reminders to aid hand-washing compliance in hospitals can help to improve behavioural compliance (Naikoba and Haywood, 2001).

Research has shown that use of multiple channels and sources of information may increase potential effectiveness of educational initiatives (Bruhn and Schultz, 1999). Research findings have shown that correlations between consumer perceptions of different sources of food safety information exist (Redmond and Griffith, 2005a) and such information should be used to aid effectual placement of interventions in strategies where multiple information sources are used.

One of the most important determinants of consumer reactions to food risk information is the extent to which the public trusts the source from which the information originates (Frewer *et al.*, 1995; Shepherd *et al.*, 1996). People are unlikely to change their attitudes or behaviour if they do not trust the source of information (Frewer *et al.* 1996), and information from a credible source is more likely to influence the public (FAO/WHO, 1998). For example, a scientist or other health care worker may seem the ideal source of public health information, however, a community activist or lay person affected by the disease may carry more credibility and have a greater public impact (Freimuth *et al.*, 2000). A source low in credibility may be discounted and have limited or no impact, whereas a highly credible source is likely to be more influential (Griffin *et al.*, 1991). In the UK, the individuals most trusted as deliverers of home food hygiene advice were Environmental Health Officers (EHOs), medical professionals and food scientists; The Chief Medical Officer, health educators and dieticians were also fairly well trusted (see Table 21.8). Concurring with previous research (Finn and Louviere, 1992; Redmond and Griffith, 2005a), politicians were considered to be the least trusted spokespersons, this may be due to perceptions of distortion of the facts, having a vested interest and concern with self-protection (Frewer *et al.*, 1996).

Table 21.8   Perceptions of spokespersons as deliverers of home food hygiene advice

Spokesperson	Proportion of consumers who perceived listed spokesperson as 'trusted' % of UK consumers ($n = 2014$)
Environmental health officer	76
Medical doctor, midwife, health visitor or nurse	74
Food/health scientist	70
Chief medical officer	69
Health educator or health promoter	64
Dietician	57
Television chef	40
Staff from a food store or supermarket	10
Farmer	10
News reader	14
Familiar TV personality	10
Politician	4

Source: Redmond *et al.*, 2005

The FSA has been determined as the most trusted organisation across the UK for provision of advice on home food hygiene (Redmond *et al.* 2005; Redmond and Griffith, 2005a), closely followed by Environmental Health departments and health-related organisations. Health organisations were trusted more than consumer organisations in each region across the UK, and, like individual politicians, the government as a whole was the least trusted organisation. Concurring with recent UK findings indicating Environmental Health departments being trusted providers of food safety information, other research has indicated that 81% consumers report that they would use Environmental Health departments to obtain information about food safety (Mathias, 1999). However, despite such positive perceptions of Environmental Health departments, such a location in practice is rarely approached for consumer food safety advice (Griffith *et al.* 1994; Mathias, 1999). Thus, it is suggested that Environmental Health departments and EHOs become more accessible to consumers and assume a more proactive role in future consumer orientated home food safety education strategies.

21.5   Consumer hygiene behaviour

Assessment of consumer behaviour can be based on self-reported practices or observations of food preparation. Self-reported practices are personal accounts of actions, which may or may not reflect actual behaviours. Data from self-report questions may provide valid information of awareness or indirect knowledge about 'correct' behaviours rather than 'actual' behaviours, so may not give an accurate representation of what a respondent's true behaviour actually is.

21.5.1   Self-report: actual behaviour

A comparison between self-reported practices and actual observed behaviour (Table 21.9) has found that substantially larger proportions of consumers reported to implement safe food-handling behaviours than actually performed them. For example, Anderson *et al.* (2000) found that although nearly all respondents (87%) reported to wash their hands before food preparation, observational findings showed that less than half (45%) actually did so. Similarly, 99% of consumers reported that they always discarded paper towels in the bin after single usage, however, observational findings showed that 30% consumers used paper towels for more than one task during food preparation (Redmond and Griffith, 2005b). Thus such findings suggest that self-reports of food safety practices may not be a reliable indication of actual behaviour.

21.5.2   Behavioural practices

Various methods of observation that have been applied to consumer food safety observational studies include personal direct observation or observation using

Table 21.9 A comparison of consumers' self-reported food safety practices and observed behaviours

	Self-reported food safety practice	Observed food safety behaviour
Hand-washing[1]	87% reported hand-washing all or most of the time before food preparation.	45% attempted to wash hands before beginning to prepare food.
Cooking[1]	30% reported to own a food thermometer.	5% used a food thermometer to determine doneness of their meat entrée.
Cross contamination	65–85% stated they wash or change cutting boards/plates for cutting up raw meat/poultry and RTE foods.[2]	52–75% failed to wash/dry c/board and/or knife for preparation of RC then salad ingredients.[3]
Dishcloth use[4]	92% consumers reported rinsing cloth-wipers immediately after use.	Cloth-wipers were rinsed on 11% occasions after use (23/204 occasions).
Paper towel use[4]	99% consumers reported discarding paper towels in the bin immediately after single use.	30% consumers used paper towels for more than one task during food preparation.

[1]Anderson et al., 2000; [2]Altekruse et al., 1996; Nunnery, 1997; Griffith et al., 2001; [3]Griffith et al., 1999; Griffith et al., 2001; [4]Redmond and Griffith, 2005b

video recordings. Nineteen observation studies of consumer food safety practices have been identified between 1977 and 2005. Data collected for the majority of the studies was collected by direct observation, where the observer openly watched participants' preparation of a meal in home kitchens and concurrently recorded preparation. Observations using video camera recording of consumer food-handling practice have been carried out in South Wales (UK), England (UK), Australia and USA. Research undertaken in Australia (Jay et al., 1999a) used time lapse video monitoring from a single mounted camera in home kitchens for periods of time lasting one or two weeks. The American study (Anderson et al., 2000) used portable video cameras to record the food preparation practices of one meal preparation in participant home kitchens. In South Wales, UK, a variety of observational studies have taken place in a model domestic kitchen using closed circuit television (CCTV) (Griffith et al., 1999; Redmond et al., 2001). An evaluation of repeatability and reproducibility of consumers' food safety behaviours found that specific food safety malpractices were consistent during repeated meal preparations and between preparation of different meals (Redmond et al., 2000; Redmond, 2002). Other studies have quantified food safety behaviours of a cross section of the population (Griffith et al., 1999; Redmond, 2002). A recent UK study has used the observational approach in conjunction with isolation techniques for *Campylobacter* and

Table 21.10   Observed consumer food safety behaviours

Food safety issue	Observed behaviour
Hand-washing/drying	~75–100% failed to wash and dry hands immediately and adequately after handling raw chicken (Griffith *et al.*, 1999; Griffith *et al.*, 2001; Jay *et al.*, 1999a; Worsfold, 1994; Redmond *et al.*, 2004).
Cross contamination	52–75% failed to wash/dry c/board and/or knife for preparation of RC then salad ingredients (Griffith *et al.*, 1999; Griffith *et al.*, 2001; Redmond *et al.*, 2004). 83–90% did not use separate areas of the kitchen for raw and RTE foods (Anderson *et al.*, 2000; Griffith *et al.*, 1999; Worsfold, 1994).
Heating efficacy	46–83% undercooked home-made burgers/meatloaf and chicken (Griffith *et al.*, 1999; Anderson *et al.*, 2000). 5% consumers used a food thermometer to evaluate the doneness of meat, poultry or seafood (Anderson *et al.*, 2004).
Cleaning	Direct contact between raw chicken and cloth-wipers was observed in 30% of meal preparations; 20% of consumers wiped unclean/contaminated hands on cloth wipers during meal preparations (Redmond and Griffith, 2005b).

Salmonella detection. This facilitated a detailed evaluation of the risk of cross contamination during food preparation and enabled identification of suspected exposure routes which linked naturally contaminated raw foods with important food-handling malpractices, contaminated contact surfaces and ready-to-eat foods (Redmond *et al.*, 2004).

Direct observations of consumer food safety behaviour in a model domestic environment and in consumer homes have indicated frequent food safety malpractices during food preparation. A summary of observational results from different studies can be found in Table 21.10, and these form a basis for a risk-based approach in the construction of educational initiatives.

21.6   Changing attitudes and behaviours

Development of effective communication strategies to raise awareness of hygiene issues and to bring about behavioral change needs to be based upon a consumer-orientated framework to maximise effectiveness (Redmond, 2002). Attitudes are considered to influence behaviours, differentiate between individuals and be open to change, thus they represent a route for influencing performance of health behaviours (Connor and Norman, 1999). It is therefore important to determine attitudes towards behaviour for the development of effective health education initiatives as strategies to address attitude change may be a requirement for behavioural change.

21.6.1   General food safety education

An improvement in consumer food safety behaviour is likely to reduce the risk and incidence of foodborne disease. A reduction of foodborne disease in the general population depends on positively altering the behaviour of food-handlers (Howes *et al.*, 1996). Food control authorities cannot intervene in every household (WHO, 2000c), therefore educational initiatives are required to reduce incidence of foodborne illness within the food safety continuum from 'farm to table' (Meer and Misner, 2000).

To effectively decrease food poisoning incidence, educational strategies are required to reduce prevalence of behaviours associated with foodborne illness, increase consumer awareness of risks, and motivate consumers to change unsafe behaviours (Yang *et al.*, 1998). It has been suggested that the use of information related to the food habits and beliefs of consumers is essential if the disease control messages are to effect behavioural change (Ehiri and Morris, 1996). To maximise the effectiveness of food safety educational initiatives, strategies should be based on knowledge of consumer attitudes towards food safety behaviours, actual food safety behaviours and an understanding of receptivity for advice and preference for sources and message types.

Traditional approaches to food safety education tend to have had a negative focus that addresses prevention rather than positive heath (Downie *et al.*, 1998). In addition, conventional approaches to food safety education have been mainly 'expert driven' and largely based on the provision of educational materials. A common fault of public health programmes is to rely solely on clinical and epidemiological research as the basis for message development. Thus, the 'facts' about a specific health behaviour may be presented upon the assumption that exposure to such 'facts' will lead to the desired behaviour (Sutton *et al.*, 1995). A problem common in food safety education is the assumption that food handlers are ignorant of hygiene principles (Ehiri and Morris, 1996). However, epidemiological evidence shows that most cases of foodborne disease result not only from ignorance of good practices, but also from a failure to apply learned techniques (Ehiri and Morris, 1994). On the whole, traditional food safety education interventions have aimed to provide knowledge and an increased awareness of food safety issues, on the assumption that consumers will make informed and correct decisions about their own food safety behaviours. Communication of these messages has mainly involved widespread distribution of knowledge-based information using the mass media directed at large numbers of people (Freimuth *et al.*, 2000). Although knowledge of the consequences of unsafe food-handling practices can enhance consumer motivation to change behaviour (Bruhn, 1997), a substantial amount of research has established that provision of knowledge does not necessarily translate into practice (Ackerley, 1994; Curtis *et al.*, 1993; Nichols *et al.*, 1988; Pinfold, 1999). The traditional approach to food safety education has had limited success and it is accepted that traditional methods have failed to meet the challenges of primary food safety problems (Ehiri and Morris, 1994).

21.6.2   Social marketing

A contemporary approach to structured behavioural change for health education initiatives has been the application of social marketing to a variety of public health-related disciplines (Andreason, 1995). The key feature of social marketing that distinguishes it from traditional public health approaches is the consumer orientation or 'audience centred thinking' (Bryant and Salazar, 1998) applied to all stages of initiatives. Social marketing is a social change strategy that focuses on voluntary behavioural change to benefit the individual and society, rather than coercing consumers to adopt healthy behaviours. At the centre of all stages of social marketing initiatives are the target audiences' needs, wants, attitudes and perceptions of aspects influencing the behavioural objective. Such variables need to be attended to and acted upon in social marketing programme planning, delivery, management and evaluation (Lefebvre et al., 1995).

It is considered that social marketing may be the most developed approach to public health communication (Maibach and Holtgrave, 1995) and use of social marketing has proven to be an immensely powerful tool for effecting massive behavioural change (Andreason and Kotler, 1991) particularly in developing countries (Ling et al., 1992). Examples of the numerous successful social marketing applications in developed countries include smoking cessation (Crowell, 1999), increased immunisations (Bryant and Salazar, 1998), nutritional supplementation (Hammerschmidt et al., 1999), cancer screening (McCormack-Brown et al., 1999), physical activity (Fridinger, 1999), adolescent drinking (Macintosh et al., 1997) and water fluoridation and dental anaesthesia (Hastings, 1999). Internationally, application of social marketing to food safety education has been more limited, and large scale food safety and social marketing initiatives have mainly been implemented in USA and developing countries. In the USA, Sutton et al. (1997) applied social marketing to food preparation behaviours at BBQs. Target audiences were defined as younger men and messages were targeted according to segmentation into 'low-germ concerns' and 'high-germ concerns' (Sutton et al., 1997). In the UK, an investigation to assess the potential application of food safety to social marketing was piloted in a small-scale study (Redmond et al., 2000). The pilot study showed that social marketing processes, principles and developmental techniques can be utilised for food safety education (Redmond et al., 2000). This was followed by the development and implementation of a consumer orientated, highly-focused community food safety education strategy with tailored intervention materials (Redmond, 2002). The effectiveness of the initiative, (often a forgotten element of food safety education) was evaluated using repeated observations of food-handling behaviours during meal preparations in a model domestic kitchen using CCTV. It was concluded that the social marketing initiative did result in immediate behavioural improvement, however sustained intervention is required to bring about long-term behavioural changes (Redmond, 2002). More recently the social marketing approach has been adopted to increase the impact of the nationwide Food Thermometer Education Campaign in the USA (FSES, FSIS, USDA, 2001).

21.6.3   Food hygiene initiatives

In the UK, to achieve the FSA target for foodborne disease reduction (Hilton, 2002) a national food hygiene campaign has been implemented. The campaign has been based upon increasing awareness and understanding of 'The 4 C's' (cleanliness, cooking, chilling and cross contamination) (Boville, 2002). To date, a variety of media-based interventions have been developed, for example a 'Preventing food poisoning' leaflet has been designed for all consumers and catering establishments (Boville, 2002). In the UK the FSA, The Food and Drink Federation (FDF) and the Food Safety Promotion Board (FSPB) (NI) are the largest providers of consumer food safety information. Although based in Southern Ireland the Food Safety Authority for Ireland are also significant providers of consumer food hygiene information in NI. UK local authorities (LAs), especially Environmental Health departments have been found to be the most significant disseminators of consumer food hygiene information in the UK. A large percentage (95%) of LAs claim they currently provide food safety advice (Redmond et al., 2005). In addition to national initiatives, a number of supermarkets, product specific and food industry organisations in the UK are known to provide food safety interventions for consumers. Furthermore, 37% of PCTs in England claim to distribute information on food safety to consumers (Redmond et al., 2005). It is also noted that a vast amount of consumer food safety advice for consumers is available on the Internet from smaller, less known organisations, local authorities, university research groups, extension services and government organisations.

In 1999, The Fight BAC! National Food Safety Initiative was set up to provide targeted information for consumers in USA. The Fight BAC! Campaign is a product of the Partnership for Food Safety Education which is a unique public-private partnership of government and consumer groups dedicated to increasing awareness of food safety and reducing the incidence of foodborne illness (Partnership for Food Safety Education, 2002). The Campaign is based on four food safety messages ('*Clean*' – wash hands and surfaces often, '*Separate*' – don't cross contaminate, '*Chill*' – refrigerate properly and '*Cook*' – cook to proper temperatures) (FSES, USDA, FSIS, 2001) and BAC! a big, green 'bacterium' character has served as the focal point to the campaign (Partnership for Food Safety Education, 2002). A recent addition to the Fight BAC! initiative has been the introduction of 'Thermy' a cartoon thermometer. Such a character has been used to support the Fight BAC! message of '*Cook*' based on studies that have indicated there is significant risks of foodborne illness when the colour is used to judge when a food has been cooked to a safe temperature (FSES, USDA, FSIS, 2001). Intervention materials have not only been targeted at specific food safety behaviours but also for specific groups of consumers and have included a wide range of media formats, some of which have been interactive.

21.7   Future trends

Consumer food safety research has, over the past 15 years, been increasingly recognised as both valid and needed. However, future random studies of consumers' knowledge are likely to have only limited value as existing international studies have shown similar patterns. Their greatest future use is likely to be as part of an evaluation strategy to assess the success of interventions. Future research should perhaps concentrate on linking microbiological assessment with observation of behaviour to further refine the risk ranking of the most frequently implemented malpractices. Qualitative studies on underlying attitudes would aid the development of interventions centred either around the highest risk behaviours or those that may not carry the highest risk, but could be most easily changed. Additionally both qualitative and quantitative studies are likely to be useful as part of the development of social marketing initiatives.

21.8   References

ACKERLEY, L. (1994) Consumer awareness of food hygiene and food poisoning. *Environmental Health*, March, 70–74.

ADAK, G.K., LONG, S.M. and O'BRIEN, S.J. (2002) Trends in indigenous foodborne disease and deaths, England and Wales: 1992 to 2000. *Gut*. 51, 832–841.

ALBRECHT, J.A. (1995) Food safety knowledge and practices of consumers in the USA. *Journal of Consumer Studies and Home Economics*. 19, 119–134.

ALTEKRUSE, S.F., STREET, D.A., FEIN, S.B. and LEVY, A. (1996) Consumer knowledge of foodborne microbial hazards and food handling practices. *Journal of Food Protection*. 59, 287–294.

AMERICAN DIETETIC ASSOCIATION (ADA) AND CONAGRA FOUNDATION. (1999) Home Food Safety Benchmark Survey. http://www.homefoodsafety.org/HomeFoodSafety/pr_key_find2.htm. (Accessed 23-09-00).

ANDERSON J.B., SHUSTER, T.A., GEE, E., HANSEN, K. and MENDENHALL, V.T. (2000) A Camera's View of Consumer Food Safety Practices. *Personal communication*. (04-02-02).

ANDERSON, J.B., SHUTSTER, T.A., HANSEN, K.E., LEVY, A.S. and VOLK, A. (2004) A Cameras View of Consumer Food Handling Behaviors. *Journal of the American Dietetic Association*. 104(2), 186–191.

ANDREASON, A. (1995) *Marketing Social Change*. Jossey-Bass. San Francisco.

ANDREASON, A. and KOTLER, P. (1991) *Strategic Marketing for Non-profit Organisations*. (4th edn). Prentice Hall.

AUDITS INTERNATIONAL. (1999) Audits International's Home Food Safety Survey. (Conducted 2nd Quarter of 1999). http://www.audits.com/Report.html. (Accessed 30-07-99).

AUSTRALIA NEW ZEALAND FOOD AUTHORITY (ANZFA). (1999) Food Standards Costs and Benefits: An Analysis of the Regulatory Impact of the Proposed National Food Safety Reforms. May. Internal Report.

BEDDOWS, C. (1983) Chicken research. *Home Economics*. April, 28–30.

BENNETT, P. and MURPHY, S. (1999) *Psychology and Health Promotion*. Open University Press. Buckingham.

BLOOMFIELD, A. and NEAL, G. (1997) Consumer Food Safety Knowledge in Auckland. Auckland Healthcare Public Health Protection. *Personal communication.* (September, 2000).

BOVILLE, C. (2002) Implementing the Agency's Foodborne Disease Strategy – Food Hygiene Campaign. Worshop on the Domestic Setting. 7th May. London.

BRUHN, C. (1997) Consumer concerns: motivating to action. *Emerging Infectious Diseases.* 3, (4), 511–515.

BRUHN, C.M. and SCHULTZ, H.G. (1999) Consumer food safety knowledge and practices. *Journal of Food Safety.* 19, 73–87.

BRYANT, C. and SALAZAR, B. (1998) Social Marketing – A Tool for Excellence. *Unpublished manuscript.*

BURGESS, F., LITTLE, C.L., ALLEN, G., WILLIAMSON, K. and MITCHEL, R. (2005) Prevalence of *Campylobacter*, *Salmonella*, and *Escherichia coli* on the external packaging of raw meat. *Journal of Food Protection.* 68, (3), 469–475.

CAMPBELL, M.E., GARDNER, C.E., DWYER, J.J., ISAACS, S.M., KRUEGAR, P.D. and YING, J.Y. (1998) Effectiveness of public health interventions in food safety: a systematic review. *Canadian Journal of Public Health.* May–June, 197–201.

CANADIAN FOOD INSPECTION AGENCY (CFIA). (1998) 1998 Safe Food Handling Study. A Report by Environics Research Group Ltd. PN4242 (June).

CHEN, Y., JACKSON, K.M., CHEA, F.P. and SCHAFFER, D.W. (2001) Quantification and variability analysis of bacterial criss contamination rates in common food service tasks. *Journal of Food Protection.* 64, (1), 72–80.

CODY, M.M. and HOGUE, M. (2003) Results of the Home Food Safety – It's in your hands 2002 Survey: Comparisons to the 1999 Benchmark Survey and Healthy People 2010 Food Safety Behaviours Objective. *Journal of the American Dietetic Association.* 103, (9), 1115–1125.

COMMUNICABLE DISEASES NETWORK, AUSTRALIA AND NEW ZEALAND WORKING PARTY (CDNANZ). (1997) Foodborne Disease: Towards Reducing Foodborne Illness in Australia. December. Technical Report series No. 2 Commonwealth of Australia.

CONNOR, M. and NORMAN, P. (1996) *Predicting Health Behaviour.* Open University Press. Buckingham.

CONNOR, M. and NORMAN, P. (1999) The role of social cognition in health behaviours. *In* Norman, P. and Connor, M. (eds.) *Predicting Health Behaviour.* Open University Press. Philadelphia.

COWDEN, J.M., WALL, P.G., LE BAIGUE, S., ROSS, D., ADAK, G.K. and EVANS, H. (1995) Outbreaks of foodborne infectious intestinal disease in England and Wales: 1992 and 1993. *Communicable Disease Report Weekly.* 5, (8), R109–R117.

CROWELL, T. (1999) Small budget, big value – including your audience in the fight against tobacco. Presented at The 9th Annual Conference of Social Marketing in Public Health. 23–26 June. Clearwater Beach, Florida.

CURTIS, V., COUSENS, S. MERTENS, T., TRAORE, T., KANKI, B. and DIALLO, I. (1993) Structured observations of hygiene behaviours in Burkina Faso: validity, variability and utility. *Bulletin of the World Health Organisation.* 71, 23–32.

DANIELS, R.W. (2001) Increasing food safety awareness. *Food Technology.* 55, 132.

DE BOER, E. and HAHNE, M. (1990) Cross contamination with *Campylobacter jejuni* and *Salmonella* spp. from raw chicken products during food preparation. *Journal of Food Protection.* 53, (12), 1067–1068.

DEPARTMENT OF HEALTH (DOH). (1993) Chilled and Frozen – Guidelines on Cook Chill and Cook Freeze Catering Systems. HMSO. London

DEPARTMENT OF HEALTH (DOH). (1995) Guidance on the Food Safety (Temperature Control) Regulations 1995. The Department of Health. September.

DOWNIE, R.S., TANNAHILL, A. and TANNAHILL, C. (1998) *Health Promotion: Models and Values*. Oxford University Press. New York.

EHIRI, J.E. and MORRIS, G.P. (1994) Food safety control strategies: a critical review of traditional approaches. *International Journal of Environmental Health Research*. 4, 254–263.

EHIRI, J.E. and MORRIS, G.P. (1996) Hygiene training and education of food handlers: does it work. *Ecology of Food and Nutrition*. 35, 243–251.

ELLARD, R. (1999) Consultation: Reforms of Food Controls, the First WHO Food and Nutrition Action Plan. 8–10 November. Malta

ENDRES, J. T., WELCH, T. and PERSELI, T. (2001) Use of a computerised kiosk in an assessment of food safety knowledge of high school students and science teachers. *Journal of Nutrition Education*. 33, 37–42.

EVANS, H.S., MADDEN, P., DOUGLAS, C., ADAK, G.K., O'BRIEN, S.J., DJURETIC, T., WALL, P.G. and STANWELL-SMITH, R. (1998) General outbreaks of infectious intestinal disease in England and Wales. *Communicable Disease and Public Health*. 1, (3), 165–171.

FEIN, S.B., JORDAN-LIN, C.T. and LEVY, A.S. (1995) Foodborne illness: perceptions, experiences and preventative behaviors in the United States. *Journal of Food Protection*. 58, (12), 1405–1411.

FINN, A. and LOUVIERE, J.J. (1992) Determining the appropriate response to evidence of public concern: the case of food safety. *Journal of Public Policy and Marketing*. 11, (1), 12–25.

FOOD AND AGRICULTURE ORGANISATION OF THE UNITED NATIONS (FAO)/WORLD HEALTH ORGANISATION (WHO). (1998) FAO/WHO Expert Consultation on the Application of Risk Communication to Food Standards and Food Safety Matters. 2–6 February. Rome, Italy. http://www.fao.org/ (Accessed 09/02).

FOOD AND AGRICULTURE ORGANISATION OF THE UNITED NATIONS (FAO)/WORLD HEALTH ORGANISATION (WHO). (2002) Statistical information on foodborne disease in Europe microbiological and chemical hazards. Conference Paper (Dec. 01/04. Agenda item 4b) presented at FAO/WHO Pan European Conference on food safety and quality. 25–28 February. Budapest, Hungary.

FOOD SAFETY AND HYGIENE WORKING GROUP. (1997) Industry Guide to Good Hygiene Practice: Catering Guide. Food Safety (General Food Hygiene) Regulations 1995. Chadwick House Group Ltd. London

FOOD SAFETY AUTHORITY OF IRELAND (FSAI). (1998) Public Knowledge and Attitudes to Food Safety in Ireland. Prepared by Research and Evaluation Services. Dublin. Ireland. (October).

FOOD SAFETY EDUCATION STAFF (FSES), FOOD SAFETY INSPECTION SERVICE (FSIS) AND UNITED STATES DEPARTMENT OF AGRICULTURE (USDA). (2001) Final Research Report – A Project to Apply the Theories of Social Marketing to the Challenges of Food Thermometer Education in the Unites States. Baldwin Group Inc. Washington DC.

FOOD SAFETY INSPECTION SERVICE (FSIS), UNITED STATES DEPARTMENT OF AGRICULTURE (USDA). (1998a) Thermometer use for cooking ground beef patties. Key facts. August. http://www.fsis.usda.gov/.

FOOD SAFETY INSPECTION SERVICE (FSIS), UNITED STATES DEPARTMENT OF AGRICULTURE (USDA). (1998b) Premature browning of cooked ground beef. http://www.fsis.usda.gov/OPHS/prebrown.htm (Accessed 09-17-98).

FOOD STANDARDS AGENCY (FSA). (2000), Foodborne disease: developing a strategy to deliver the agencies targets, Paper FSA 00/05/02, Agenda Item 4, 12 October.

FOOD STANDARDS AGENCY (FSA). (2001) Microbiological foodborne disease strategy. July. http://www.food.gov/

FOSTER, G.M. and KAFERSTEIN, F.K. (1985) Food safety and the behavioural sciences. *Social Science and Medicine*. 21, 1273–1277.

FRASER, J. and SMITH, F. (1997) Pretesting health promotion leaflets – a case study. *International Journal of Health Education*. 35, (3), 97–101.

FREIMUTH, V. LINNAN, H.W. and POTTER, P. (2000) Communicating the threat of emerging infections to the public. *Emerging Infectious Diseases*. 6, (4), 337–347.

FREWER L.J., SHEPHERD, R. and SPARKS, P. (1994) The interrelationship between perceived knowledge, control and risk associated with a range of food related hazards targeted at the individual, other people and society. *Journal of Food Safety*. 14, 19–40.

FREWER L.J., HOWARD, C. and SHEPHERD, R. (1995) Consumer perceptions of food risks. *Food Science and Technology Today*. 9, (4), 212–216.

FREWER, L.J., HOWARD, C., HEDDERLEY, D. and SHEPHERD, R. (1996) What determines trust in information about food-related risks? Underlying psychological constructs. *Risk Analysis*. 16, (4), 473–486.

FRIDINGER, F. (1999) Using market research to influence environmental policies to promote physical activity. 9th Annual Conference of Social Marketing in Public Health. Clearwater Beach, Florida.

GRIFFIN, M., BABIN, B.J. and ATTAWAY, J.S. (1991) An empirical investigation of the impact of negative public publicity on consumer attitudes and intentions. *Advances in Consumer Research*. 18, 334–341.

GRIFFITH, C.J. (2000a) Good hygiene practices for food handlers and consumers. *In* Blackburn, C.W. and McClure, P.J. (eds) *Foodborne Pathogens: Hazards, Risk and Control*. Woodhead Publishing Ltd. London.

GRIFFITH, C.J. (2000b) Food safety in catering establishments. *In* Farber, J.M. and Todd, E.C. (eds) *Safe Handling of Foods*. Marcel Dekker. New York.

GRIFFITH C.J. and REDMOND E. (2005) Handling poultry and eggs in the kitchen. *In* Mead, G.C. (ed.) *Food Safety Control in the Poultry Industry*. Woodhead Publishing Ltd, Cambridge, UK and CRC Press, USA. pp. 524–540.

GRIFFITH, C.J. and WORSFOLD, D. (1994) Application of HACCP to food preparation practices in domestic kitchens. *Food Control*. 5, 200–204.

GRIFFITH, C.J., MATHIAS, K.A. and PRICE, P.E. (1994) The mass media and food hygiene education. *British Food Journal*. 96 (9), 16–21.

GRIFFITH, C.J., PETERS, A.C., REDMOND, E.C. and PRICE, P. (1999) Food safety risk scores applied to consumer food preparation and the evaluation of hygiene interventions. Department of Health. London.

GRIFFITH, C. J., PRICE, P., PETERS, A. and CLAYTON, D. (2001). An evaluation of food handlers knowledge, belief and attitudes about food safety and its interpretation using social cognition models. FSA. London.

HALL, G., KIRK, M.D., BECKER, N., GREGORY, L., UNICOMB, L., MILLARD, G., STAFFORD, R., LALOR, K. and THE OZFOODNET WORKING GROUP. (2005) Estimating Foodborne Gastroenteritis, Australia. *Emerging Infectious Diseases*. 11, (8), 1257–1264.

HAMMERSCHMIDT, P., HIMEBAUCH, L., WRUBLE, C., SMYTH, P., HOLADAY, R.M. and COHEN, M. (1999) Eat Healthy. Your kids are watching – development and evaluation of a pilot nutrition education social marketing campaign. Presented at The 9th Annual Conference of Social Marketing in Public Health. 23–26 June. Clearwater Beach, Florida.

HARRISON, W.A., GRIFFITH, C.J., TENNANT, D. and PETERS, A.C. (2001) Incidence of *Campylobacter* and *Salmonella* isolated from retail chicken and associated packaging in South Wales. *Letters in Applied Microbiology*. 33, 450–454.

HARVEY, H.D., FLEMING, P., CREGAN, K. and LATIMER, E. (2000) The health promotion implications of the knowledge and attitude of employees in relation to health and safety leaflets. *International Journal of Environmental Health Research*. 10, 315–329.

HASTINGS, G. (1999) Whose behaviour is it anyway? The broader potential of social marketing. Presented at The 9th Annual Conference of Social Marketing in Public Health. 23–26 June. Clearwater Beach, Florida.

HASTINGS, G. and HAYWOOD, A. (1991) Social marketing and communication in health promotion. *Health Promotion International*. 6, (2), 135–145.

HAYSOM, I.W. and SHARP, A.K. (2005) Bacterial contamination of domestic kitchens over a 24 hour period. *British Food Journal*. 107(7), 441.

HILTON, A.C. and AUSTIN, E. (2000) The kitchen dishcloth as a source of and vehicle for foodborne pathogens in a domestic setting. *International Journal of Environmental Health Research*. 10, 257–261.

HILTON, J. (2002) Reducing foodborne disease: meeting the Food Standards Agency's targets. *Nutrition and Food Science*. 32, (2), 46–50.

HODGES, I. (1993) Raw to Cooked: Community Awareness of Safe Food Handling Practices. Internal Report for The Department of Health Te Tari Ora, Health Research and Analytical Service, Wellington.

HOORENS, V. (1994) Unrealistic optimism in health and safety risks. *In* Rutter, D. and Quine, L. (eds) *Changing Health Behaviour*. Open University Press, Buckingham.

HOWES, M., McEWEN, S., GRIFFITHS, M. and HARRIS, L. (1996) Food handler certification by home study: measuring changes in knowledge and behaviour. *Dairy, Food and Environmental Sanitation*.16, (11), 737–744.

HUDSON, P.K. and HARTWELL, H.J. (2002) Food safety awareness of older people at home: a pilot study. *The Journal of the Royal Society for the Promotion of Health*. 122, (3), 165–169.

HUMPHREY, T. J., MARTIN, K. and WHITEHEAD, A. (1994) Contamination of hands and work surfaces with *Salmonella enteritidis* PT4 during the preparation of egg dishes. *Epidemiology and Infection*. 113, 403–409.

INFECTIOUS INTESTINAL DISEASE EXECUTIVE COMMITTEE (IID). (2000) A Report of the Study of Infectious Intestinal Disease in England. The Stationery Office. London.

INTERNATIONAL SCIENTIFIC FORUM ON HOME HYGIENE (IFH). (1998) Recommendations For a Selection of Suitable Hygiene Procedures for Use in the Domestic Environment. Intramed Communications Ltd. Milano, Italy.

JAY, L.S., COMAR, D. and GOVENLOCK, L.D. (1999a) A video study of Australian food handlers and food handling practices. *Journal of Food Protection*. 62, (11), 1285–1296.

JAY, L.S., COMAR, D. and GOVENLOCK, L.D. (1999b) A national Australian food safety telephone survey. *Journal of Food Protection*. 62, (8), 921–928.

JOHNSON, A.E., DONKIN, A. J. M. MORGAN, K., LILLEY, J.M., NEALE, R.J., PAGE, R.M. and SILBURN, R. (1998) Food safety knowledge and practice among elderly people living at home. *Journal of Epidemiology and Community Health*. 52, 745–748.

KNABEL, S.J. (1995) Foodborne illnesses: role of home food handling practices. *Food Technology*. 49, (4), 119–131.

LAKE, R.J., BAKER, M.G., GARETT, N., SCOTT, W.G. and SCOTT, H.M. (2000) Estimated no. of cases of foodborne infections disease in New Zealand. *The New Zealand Journal.* 113, (1113), 278–281.

LEFEBVRE, R.C., LURIE, D., GOODMAN, L.S., WEINBERG, L. and LOUGHREY, K. (1995) Social marketing and nutrition education – inappropriate or misunderstood. *Journal of Nutrition Education.* 27, (3), 146–150.

LEVY, A. (2002) Cognitive antecedents of 'good' food safety practices. Presented at 'Thinking Globally – Working Locally' A Conference on Food Safety Education. September 18–20 September. Orlando, Florida.

LI-COHEN, A.E., KLENK, M., NICHOLSON, Y., HARWOOD, J. and BRUHN, C. (2002) Refining consumer safe handling educational materials through focus groups. *Dairy, Food and Environmental Sanitation.* 22, (7), 539–551.

LING, J.C., FRANKILIN, B.A.K., LINDESTEAAADT, J.F. and GEARON, S.A.N. (1992) Social marketing: its place in public health. *Annual Reviews in Public Health.* 13, 341–362.

MACINTOSH, A.M., HASTINGS, G., HUGHES, K., WHEELER, C., WATSON, J. and INGLIS, J. (1997) Adolescent drinking – the role of designer drinks. *Health Education.* 6, (November), 213–224.

MAIBACH, E. and HOLTGRAVE, D.R. (1995) Advances in public health communication. *Annual Review of Public Health.* 16, 219–238.

MARKLINDER, L.M., LINDBALD, M., ERIKSSON, L.M., FINNSON, A.M. and LINDQVIST, R. (2004) Home Storage Temperatures and Consumer Handling of Refrigerated Foods in Sweden. *Journal of Food Protection.* 67, 11, 2570–2577.

MATHIAS, K. (1999) The use of Consumer Knowledge, Beliefs and Attitudes in The Development of a Local Authority Strategy for Domestic Food Safety Education. Open University, Cardiff, UK. M.Phil. Thesis.

MAURICE, J. (1995) The rise and rise of food poisoning. *New Scientist.* (December), 28–33.

McCORMACK-BROWN, K., BRYANT, C., FORHOFER, M., PERRIN, K., QUINN, G. and FIGG, M. (1999) A social marketing plan to increase breast and cervical cancer screening in Florida. Unpublished material. In *PHC 6411 Introduction to Social Marketing.* Summer semester, 1998. University of South Florida, College of Public Health.

McGUIRE, W.J. (1984) Public communication as a strategy for inducing health-promoting behavioural change. *Preventative Medicine.* 13, 299–319.

McINTOSH, W.A., CHRISTENSEN, L. B. and ACUFF, G. R. (1994) Perceptions of risks of eating undercooked meat and willingness to change cooking practices. *Appetite.* 22, 83–96.

McKENNA, F.P. (1993) It won't happen to me: unrealistic optimism or illusion of control? *British Journal of Psychology.* 84, 39–50.

MEAD, P.S., SLUTSKER, L., DIETZ, V., McCAIG, L., BRESEE, J.S., SHAPIRO, C., GRIFFIN, P.M. and TAUXE, R.V. (1999) Food related illness and death in the United States. *Emerging Infectious Diseases.* 5, (5), 607–625.

MEDEIROS, L.C., HILLERS, V.N., CHEN, G., BERGMANN, V., KENDALL, P. and SCHROEDER, M. (2004), 'Design and development of food safety knowledge and attitude scales for consumer food safety education', *Journal of the American Dietetic Association,* 104, (11), 1671–1677.

MEER, R.R. and MISNER, S.L. (2000) Food safety knowledge and behaviour of expanded food and nutrition program participants in Arizona. *Journal of Food Protection.* 63, (12), 1725–1731.

MICHAELS, B., GANGAR, V., AYERS, T., MEYERS, E. and CURIALE, M.S. (2001) The significance of hand drying after hand-washing. *In* Edwards, J.S.A. and Hewedi, M.M. (eds) *Culinary Arts and Sciences III. Global and National Perspectives.* The Worshipful

Company of Cooks Centre for Culinary Research at Bournemouth University. Al-Karma, Egypt.

MIDDLESTADT, S.E., BHATTACHARYYA, K., ROSENBAUM, J., FISHBEIN, M. and SHEPHERD, M. (1996) The use of theory based semi-structured elicitation questionnaires: formative research for CDC's Prevention Marketing initiative. *Public Health Reports.* 111, (1), 18–27.

MILES, S., BRAXTON D.S. and FREWER, L.J. (1999) Public perceptions about microbiological hazards in food. *British Food Journal.* 101, (10), 744–762.

MITAKAKIS, T.Z., SINCLAIR, M.I., FAIRLEY, C.K., LIGHTBODY, K.P., LEDER, K. and HELLARD, M.E. (2004) Food Safety in Family Homes in Melbourne, Australia. *Journal of Food Protection.* 67, (4), 818–822.

NAIKOBA, S. and HAYWOOD, A. (2001) The effectiveness of interventions aimed at increasing hand-washing in healthcare workers – a systematic review. *Journal of Hospital Infection.* 47, 173–180.

NATIONAL CONSUMER COUNCIL (NCC). (1991) Time temperature indicators: research into consumer attitudes and behaviour. MAFF. London.

NICHOLS, S., WATERS, W., WOOLAWAY, M. and HAMILTON-SMITH, M. (1988) Evaluation of the effectiveness of a nutritional health education leaflet in changing public knowledge and attitudes about eating and health. *Journal of Human Nutrition and Dietetics.* 1, 233–238.

NICOLAAS, G. (1995) Cooking: Attitudes and Behaviour. OPCS Social Survey Division. Omnibus Survey Publications: Report 5. HMSO. London.

NUNNERY, P. (1997) Epidemiology of Foodborne Illness. In *Changing Strategies, Changing Behavior Conference.* Conference Proceedings. 13–14 June, 1997. Washington DC. United States Department of Agriculture (USDA), Food and Drink Administration (FDA), Centers for Disease Control and Prevention (CDC).

O'BRIEN, G.D. (1996) A pilot study to assess domestic refrigerator air temperatures and the public's awareness of refrigerator use. *Environmental Health Review.* Winter. 100–111.

OJIMA, M., TOSHIMA, Y., KAJA, E., ARA, K., KAURAN, S. and UEDA, N. (2002) Bacterial contamination of Japanese households and related concern about sanitation. *International Journal of Environmental Health Research.* 12, 41–52.

OLSEN, S.J., MACKINON, L.C., GOULDING, J.S., BEAN, N.H. and SLUTSKER, L. (2000) Surveillance for foodborne disease outbreaks United States 1993–1997. *Morbidity and Mortality Weekly Report Surveillance Summaries.* 49, (SS01), 1–51.

PARTNERSHIP FOR FOOD SAFETY EDUCATION. (2002) Fight Bac for Education. http://www.fightbac.org/main.cfm. (Accessed Sept. 02).

PAUL, C.L. and REDMAN, S. (1997) A review of the effectiveness of print material in changing health-related knowledge, attitudes and behaviour. *Health Promotion Journal of Australia.* 7, (2), 91–99.

PAULSON, D.S., RICCARD, C., BEAU-SOLEIL, C.M., FENDLER, E.J., DOLAN, M.J., DUNKERTON, L.V. and WILLIAMS, R.A. (1999) Efficacy evaluation of four hand cleaning regimes for food handlers. *Dairy Food and Environmental Sanitation.* 19, 680–684.

PINFOLD, J.V. (1999) Analysis of different communication channels for promoting hygiene behaviour. *Health Education Research.* 14, (5), 629–639.

REDMOND E.C. (2002), Food safety behaviour in the home: development, application and evaluation of a social marketing food safety education initiative, PhD Thesis, University of Wales, Cardiff, UK.

REDMOND, E.C. and GRIFFITH, C.J. (2003a) Consumer food-handling in the home: a review of food safety studies. *Journal of Food Protection*. 66, (1), 130–161.

REDMOND, E.C. and GRIFFITH, C.J. (2003b) A comparison and evaluation of research methods used in consumer food safety studies. *International Journal of Consumer Studies*. 27, (1), 17–33.

REDMOND, E.C. and GRIFFITH, C.J. (2004a) Consumer perceptions of food safety risk, control and responsibility. *Appetite*. 43, 309–319.

REDMOND, E.C. and GRIFFITH, C.J. (2004b) Consumer attitudes and perceptions towards microbial food safety in the domestic kitchen. *Journal of Food Safety*. 24, (3), 169–194.

REDMOND, E.C. and GRIFFITH, C.J. (2005a) Consumer perceptions of food safety education sources: implications for effective strategy development. *British Food Journal*. 107, (7), 467–483.

REDMOND, E.C. and GRIFFITH, C.J. (2005b) Consumer attitudes, self-reported and observed behaviors relating to cloth-wiper usage in the domestic kitchen. Accepted for presentation at the International Association for Food Protection Conference, 92nd Annual Meeting, August 14–17 Baltimore, Maryland. Poster presentation.

REDMOND, E.C., GRIFFITH, C.J. and PETERS, A.C. (2000) Use of social marketing in the prevention of specific cross contamination actions in the domestic environment. In *Proceedings of the 2nd NSF International Conference on Food Safety: Preventing Foodborne Illness through Science and Education*. 11–13 October, Savannah, Georgia.

REDMOND, E.C., GRIFFITH, C.J., SLADER, J. and HUMPHREY, T.J. (2001) The evaluation and application of information on consumer hazard and risk to food safety education. Food Standards Agency. London.

REDMOND, E.C., GRIFFITH, C.J., SLADER, J. and HUMPHREY, T.J. (2004) Microbiological and observational analysis of cross contamination risks during domestic food preparation. *British Food Journal*. 106, (8), 581–597.

REDMOND, E.C., GRIFFITH, C.J., KING, S. and DYBALL, M. (2005) Evaluation of consumer food safety education initiatives in the UK and determination of effective strategies for food safety risk communication (RRD-8). Food Standards Agency. London.

ROBERTS, T., AHL, A. and MCDOWELL, R. (1995) Risk assessment for microbial hazards. *In* Roberts, T., Jensen, H. and Unnovehr, L. (eds) *Tracking Foodborne Pathogens from Farm to Table*. Economic Research Service (ERS). Conference proceedings, Jan 9–10. Washington D.C. USDA, ERS. Miscellaneous Publication No. 1532.

ROBERTSON, A., TIRADO, C., LOBSTEIN, T., JERMINI, M., KNAI, JENSEN, J.H., FERRO-LUZZI, A. and JAMES, W.P.T (eds) (2004) Food and Health in Europe: a new basis for action. World Health Organisation Europe. WHO Regional Publications, European Series No 96.

ROCOURT, J., MOY, G., VIERK, K. AND SCHLUNDT, J. (2003) The present state of foodborne disease in OECD (Organisation for Economic Co-operation and Development) countries. Food Safety Department. WHO. Geneva.

RYAN, M.J., WALL, P.G., GILBERT, R.J., GRIFFIN, M. and ROWE, B. (1996) Risk factors for outbreaks of infectious intestinal disease linked to domestic catering. *Communicable Disease Report (Review)*. 6, (13), R179–R182.

SABA, A. and DiNATALE, R. (1999) A study on the mediating role of intention on the impact of habit and attitude on meat consumption. *Food, Quality and Preference*. 10, 69–77.

SAMMARCO, M., L. and RIPABELLI, G. (1997) Consumer attitude and awareness towards food related hygienic hazards. *Journal of Food Safety*. 17, 215–221.

SHARP, K. and WALKER, H. (2003) A microbiological survey of communal kitchens by undergraduate students. *International Journal of Consumer Studies*. 27, (1), 11–16.

SHEPHERD, R.. FREWER, L.J. and HOWARD, C. (1996) Trust and risk communication on food issues. *In:* Conference proceedings: Risk in a Modern Society: Lessons for Europe. 3–5 June. University of Surrey.

SMITH, D. and RIETHMULLER, P. (2000) Consumer concerns about food safety in Australia and Japan. *British Food Journal.* 102(11), 838–855.

STRAND, J. (1999) Summary of Change Theories. Training notes: The 9th Annual Conference Social Marketing in Public Health. 23–26 June. University of South Florida, College of Public Health. Clearwater Beach, Florida.

SUTTON, S.S., BALCH, G.L. and LEFEBVRE, R.C. (1995) Strategic questions for consumer-based health interventions. *Public Health Reports.* 110, (Nov/Dec), 725–733.

SUTTON, S., ANDREASON, A.R.., SMITH, W.A., MAIBACH, E. and LEFEBVRE, R.C. (1997) Social marketing and food safety education. In *Changing Strategies, Changing Behavior Conference.* Conference Proceedings. 13–14 June, 1997. Washington DC. United States Department of Agriculture (USDA), Food and Drink Administration (FDA), Centers for Disease Control and Prevention (CDC).

TAYLOR, J. and HOLAH, J.T. (2000) Hand Hygiene in the Food Industry: A review. (Review No. 18). Campden and Chorleywood Food Research Association Group.

THE PAN AMERICAN HEALTH ORGANISATION (2004) Protecting Food, Safeguarding the Publics Health. *In* Cooperating in Veterinary Public Health. Quadrennial Report of the Director, Centennial Edition. http://www.paho.org/English/AD/DPC/VP/ops98-02_ch04-vet.pdf (Accessed September, 2005).

THE PENNINGTON GROUP. (1997) Report on the circumstances leading to the 1996 outbreak of infection with *E.coli* O157 in Central Scotland, the implications for food safety and the lessons to be learned. The Stationery Office. Edinburgh.

TIRADO, C. and SCHMIDT, K. (eds) (2000) WHO Surveillance Programme for Control of Foodborne Infections and Intoxications in Europe. 7th Report, 1993–1998. BGVV-FAO/WHO Collaborating Centre for Research and Training in Food Hygiene and Zoonoses.

TODD, E.C.D. (1989) Preliminary estimates of foodborne disease in Canada and costs to reduce *salmonellosis. Journal of Food Protection.* 52, (8), 586–594.

TONES, B.K. and TILFORD, S. (1996) *Health Education; Effectiveness and Efficiency.* London, Chapman and Hall.

UNKLESBURY, N., SNEED, J. and TOMA, R. (1998) College students attitudes, practices and knowledge of food safety. *Journal of Food Protection.* 61, (9), 1175–1180.

WALKER, A. (1996) *Food Safety in the Home.* HMSO. London.

WEINSTEIN, N.D. (1980) Unrealistic optimism about future life events. *Journal of Personality and Social Psychology.* 39, (5), 806–820.

WEINSTEIN, N.D. and KLEIN, W.M. (1996) Unrealistic optimism: present and future. *Journal of Social and Clinical Psychology.* 15, (1), 1–8.

WENRICH, T., CASON, K., LV, N. and KASSAB, C. (2003) Food Safety Knowledge and Practices of Low Income Adults in Pennsylvania. *Food Protection Trends.* 23(4), 326–335.

WESTAWAY, M.S. and VILJOEN, E. (2000) Health and hygiene knowledge, attitudes and behaviour. *Health and Place.* 6, 25–32.

WOODBURN, M.J. and RAAB, C.A. (1997) Household preparers' food safety knowledge and practices following widely publicised outbreaks of foodborne illness. *Journal of Food Protection.* 60, (9), 1105–1109.

WORLD HEALTH ORGANISATION (WHO). (2000a) Address by the Director General to the 53rd World Health Assembly. Reference A53/3. 15 May. Geneva

WORLD HEALTH ORGANISATION (WHO). (2000b) Foodborne disease: a focus for health education. WHO. Geneva.

WORLD HEALTH ORGANISATION (WHO). (2000c) Foodborne disease: a focus for health education. WHO. Geneva.

WORLD HEALTH ORGANISATION (2003) WHO Surveillance Programme for the Control of Foodborne Infections and Ontoxications in Europe. 8th Report 1999–2000. Country Reports UK: England & Wales. *http://www.bfr.bund.de/* (accessed Sept 2005).

WORSFOLD, D. (1994) An evaluation of food hygiene and food preparation practices. Open University, UK. PhD Thesis.

WORSFOLD, D. and GRIFFITH, C.J. (1997). Food Safety Behaviour in the Home. *British Food Journal.* 99, 97–104.

YANG, S., LEFF, M.G., McTAGUE, D., HORVATH, K.A., THOMPSON, J., MURAYI, T. , BOESELAGER, G.K., MELRUK, T.A., GILDMASTER, M.C., RIDINGS, D.L., ALTEKRUSE, S.F. and ANGULO, F. J. (1998) Multi-state surveillance for food handling and preparation and consumption behaviours associated with foodborne diseases 1995 and 1996. *Morbidity Mortality Weekly Report.* 47, 33–54.

YANG, S., ANGULO, F.J. and ALTEKRUSE, S.F. (2000) Evaluation of safe food-handling instructions on raw meat and poultry products. *Journal of Food Protection.* 63, (10), 1321–1325.

YEUNG, R.M.W. and MORRIS, J. (2001) Food safety risk – consumer perception and purchase behaviour. *British Food Journal.* 103, (3), 170–186.

ZHANG, P. and PENNER, K. (1999) Prevalence of selected unsafe food consumption practices and their associated factors in Kansas. *Journal of Food Safety.* 19, 289–297.

22

Changing unhealthy food choices

A. S. Anderson, University of Dundee, UK

22.1 Introduction: importance of changing unhealthy consumer food choices

In 2003, The World Health Organisation reported non-communicable diseases (NCD), including obesity, diabetes, cardiovascular disease and cancer, accounted for almost 60% of the 56 million deaths annually and 47% of the global burden of disease. A small number of risk factors account for much of the observed mortality, including inadequate intake of fruits and vegetables, overweight and obesity, high blood pressure, hyperlipidaemias and physical inactivity. Other diseases related to diet including dental caries, osteoporosis and gut problems and account for further morbidity.

The disease burden of NCDs is greatest and continuing to grow in developing countries, and tends to affect younger people (compared to the developed world). Rapid changes in diet and activity are likely to cause chronic disease rates to rise and will be further exacerbated by tobacco use. Whilst there remains debate on the relative importance of energy expenditure and energy intake (gluttony or slothdom!) in the aetiology of obesity, it is clear that both sides of the energy balance equation must be tackled (Prentice and Jebb, 1995).

Of particular concern is the rise of obesity in children and younger adults with corresponding increases in the development of type 2 diabetes and associated co-morbidities. These findings have led to a recent emphasis on promoting healthy eating habits in childhood and attempts to decrease advertising and promotions for energy dense food to younger children. However, it is recognised that most of the medical and social burdens (and most of the costs) of obesity occur in adult life. Measures to decrease the incidence of obesity cannot ignore the increase in body fat across the whole population (Lean, 2005). Obesity plays a role in all three major causes of death (cardiovascular disease, stroke and cancer) independent to specific food and nutrient constituents of the diet (WCRF, 1997).

Whilst there is considerable investment within nutritional science to identify 'superfoods', protective nutrients or mechanisms by which individual components influence physiological systems, the bulk of disease undoubtedly relates to overall dietary patterns, e.g. high energy, high sugar and salt and low fruits and vegetables. The health merits of single nutrients have been exploited well beyond food sources to the marketing and sales of nutrient supplements changing emphasis from food to pharmacy. In the UK, the National Diet and Nutrition Survey undertaken in 2000/01 reported 40% of women and 29% of men reported taking dietary supplements which was an increase from 17% of women and 9% of men in 1986/87 (Henderson *et al.*, 2003). At the same time as this apparent interest in nutrition and health, the prevalence of obesity or overweight increased from 45% of men and 36% of women in 1986/87 to 66% of men and 53% of women in 2000/01.

For consumers, selecting foods to achieve a healthy balanced diet remains an area of tangled confusion with many opportunities for being seduced by misleading health suggestions and financial bargains. The challenge of assisting consumers towards making healthful dietary selections whilst maintaining freedom of choice and supporting the economic benefits of a thriving food industry are not easily resolved.

22.2   Factors inhibiting healthy food choices

It is recognised that any efforts to encourage healthful food choices need to be as effective as the food industry's excellence in selling their products.

Advertising and promotions of confectionery, chocolate, soft drinks and fast foods are well funded (Hastings *et al.*, 2003) and have successfully resulted in increased sales and consumption over a range of energy dense foods over the last decade. For example, sales of soft drinks have increased from 720 ml per adult per day in 1992 to 1284 ml in 2000 (DEFRA, 2000).

Affordable, value for money marketing approaches include free menu items when certain menu choices are made, larger portion sizes and two for price of one offers luring thrifty consumers to part with cash (often when financial resources are limited). These approaches are particularly noticeable in fast food restaurants for energy dense foods as opposed to fruits, vegetables and other healthier options.

Availability of energy dense snacks and drinks at every possible setting (e.g., worksites, leisure facilities, schools, hospitals and garages) at every possible time of day (e.g., vending) and in ever larger portion sizes, has undoubtedly contributed to the temptation to consume excess energy. Changes in US portion sizes between 1977 and 1998 have shown an increase of 93 kcals per portion of salty snacks, followed by a 49 kcal increase in soft drinks – and those might be consumed before the main course arrives (Neilson and Popkin, 2003)! Of particular concern is the increasing availability and consumption of soft drinks (liquid calories) and fast foods. In the US, the prevalence of soft drink intake among children aged 6 to 17 years increased from 37% in 1977/78 to 56% in 1994/98. Mean intake of soft drinks more than doubled, from 5 fl oz to 12 fl oz per day. Although the home environment was the main source of children's soft drink access, they were also obtained from restaurants and fast-food establishments (+53%), vending machines (+48%), and other sources (+37%) (French et al., 2003). The health risks of women consuming one or more sugar-sweetened soft drinks per day has been reported by Schulze (2004) who showed an increased relative risk (RR) of type 2 diabetes of 1.83 compared with those who consumed less than one of these beverages per month. In addition, another form of liquid calories (alcoholic drinks) has increased in women from a mean weekly consumption (g/alcohol/day) of 6.9 g in 1986/87 (Henderson et al., 2003) to 9.3 g in 2000/01.

The energy density of fast foods has been illustrated by Prentice and Jebb (2003) who showed that in typical fast-food outlets the average energy density of the entire menus was approximately 65% higher than the average British diet and more than twice the energy density of recommended healthy diets. Furthermore, Pereira et al. (2005) have demonstrated that fast-food consumption has strong positive associations with weight gain and insulin resistance, suggesting that fast food increases the risk of obesity and type 2 diabetes. Clearly the health risks of fast-food consumption are valued less by consumers who appreciate the apparent 'value for money' of supersize meals, meal deals, limited healthy options and limited signposting, marketing and availability of nutrient dense choices. Younger consumers with limited cognitive restraint over diet are efficiently lured by marketing strategies, including toys, familiar logos and 'familiar' icons.

Acceptability of calorie munching and liquid energy has resulted in relaxation of social norms around eating such that food is now consumed in all the places that smoking is banned, e.g. public transport, cinema, pubs, hospital waiting rooms.

22.3   Mechanisms to change unhealthy food choices

It is recognised that a range of approaches (NIH, 2001) should be employed to achieve dietary change and include strategies involving the following approaches:

- Intrapersonal (e.g., individual)
- Interpersonal (e.g., family)
- Institutional (e.g., school)
- Community (e.g., private, public, voluntary)
- Public policy (e.g., government policy).

Such strategies will move beyond the responsibility of the individual to make wise food selections and will necessitate engaging with the food industry for new product formulations (e.g., low fat) and retailing policies and be undertaken within an environment that facilitates healthy public policies.

Probably the best examples of effective dietary change have been in comprehensive community programmes (Anderson, 2004), such as The North Karelia Programme in Finland, which (over a 20-year period) resulted in tripling of vegetable intake, doubling of fruit consumption and clinically significant decreases in total fat and saturated fat. Importantly, these changes (and other changes in smoking, and activity) resulted in a major decrease in mortality from coronary heart disease followed by reduction in cancer (Puska *et al.*, 1993).

The North Karelia demonstration project focussed heavily on community organisation (e.g., NGOs, schools, health service) as a route to influencing social and health policy implementation. Puska (1999) has also highlighted wider aspects of public policy including intersectoral collaboration (e.g., agriculture and health policy), the role of a single agency in co-ordinating efforts, industry involvement and a range of food polices (including food-labelling and pricing policies). Communication approaches employing innovation – diffusion theory (through promotion of knowledge, persuasion, decision and confirmation) has also been considered an important part of the implementation of behavioural change.

In the last decade dietary intervention programmes have tended to focus on fruit and vegetable interventions rather than on total diet. The programme settings have varied widely from churches, schools and worksites, some have focussed on educational approaches only whilst others have emphasised wider changes so that local environments facilitate increased opportunities for purchase and consumption of healthy items. Such programmes have also varied in specific aims from the UK recommendation to increase intake to five a day (400 g fruits and vegetables) http://www.dh.gov.uk/PolicyAndGuidance/HealthAnd SocialCareTopics/FiveADay/fs/en, to the US where current recommendations are gender specific (seven servings for women and nine for men) and commend eating a variety of 'colourful' fruits and vegetables (http://www.5aday.gov/ homepage/index_content.html) to the Australian approach to aim for '2 fruit and 5 veg' (http://www.gofor2and5.com.au/benefits.asp) in an attempt to focus efforts to increase consumption of vegetables (includes potatoes) from 2.6 portions to 6 portions per day. Overall, the impact of these behaviourally focussed interventions shows significant but small increases in fruit and vegetable intake, with an average increase of 0.6 servings per day but with clear scope for greater change. Two intervention components were identified as particularly promising in modifying dietary behaviour – goal setting and small group approaches (Ammerman *et al.*, 2002). It is not clear what the long-term impact of these interventions is, especially when delivered in childhood. The recent English 'Fruit and Vegetables in Schools' scheme, where a free piece of fruit is given to all children aged 4 to 6 years, has been demonstrated to increase fruit consumption in the years fruit is provided but this does not appear to be habit forming and continued beyond the years of free provision (Wells and Nelson, 2005).

Puska *et al.* (1993) also reminds us that the food industry has an important role to play in influencing healthy food choices and that lifestyles and commercial products cross borders. The food industry can work in both directions; for example, in a positive way by formulating products with low fat content or producing prepared meals high in fruit and vegetable content, or in negative ways by effective promotion and retailing of excess consumption of energy dense foods.

Appropriate health claims, nutrition labelling and nutrition signposting can be a useful form of collaboration between industry and other sectors. It is likely that food product development will respond to labelling regulations that will conform to consumer information and mis-information. Thus health claims will be limited unless these can be substantiated by scientific evidence. Nutrient profiling and labelling in accordance with agreed nutrient standards is likely to act as an incentive to companies to re-formulate.

In the US, the concept of 'naturally nutrient-rich' is being promoted as a positive route to healthy living. This approach highlights nutrient density as a way for the population to 'get the most nutrition from their foods' and make their 'calories count more'. The first stage in this process is creating a nutrient density index as a tool to help consumers choose wisely within appropriate calorie levels.

A nutrient density index is used to identify the maximum nutrient content per calorie. The nutrients of interest include a wide range of micronutrients and will be influenced by water content as well as energy (Zelman and Kennedy, 2005). Nutrient density is clearly relevant and even the USDA is urging Americans to 'consume a variety of nutrient dense foods and beverages' (USDA, 2005). Creating an index or score is very similar to the approach being taken in the UK by the Food Standards Agency for nutrient profiling. Initially this system will be used to help guide the independent regulator and competition authority for the UK communications industries (OFCOM) in decisions over television advertising of foods marketed at children. However, it is possible that the 'score' may be used to develop a traffic lights system for front of package use in due course. The score would no doubt be used in the form of a nutrition signpost, which represents a range of positive attributes (e.g., nutrient density) and weighs up the balance with more negative nutrient characteristics.

This signposting approach is not new. The New Zealand 'tick the pick' system has already shown positive benefits. Food manufacturers whose products meet defined nutritional criteria are allowed to display the logo on labels. According to Young and Swinburn (2002), the symbol is used by 59% of shoppers in making healthy food choices and companies are encouraged to reformulate their products if they fail to meet the criteria. Between July 1988 and June 1999 this scheme influenced food companies to exclude approximately 33 tonnes of salt by reformulation. The largest reduction of sodium (salt) was found in breakfast cereals, with an average reduction of 61% sodium (378 mg sodium per 100 g product). Sodium in bread was reduced by an average of 26% (123 mg per 100 g product) and margarine by 11% (53 mg per 100 g product) without impacting on product taste or quality.

In the US, some companies have developed their own signposts and whilst these might indicate better choices (e.g., better than the worst), they might not indicate the best. The SMARTSPOT symbol (http://www.smartspot.com/) introduced by PepsiCo company (includes Tropicana and Quaker Oats) is one example.

Vested interests are never far away where food initiatives are concerned but such partnerships may offer the marketing that governments and health promotion agencies fail to provide. In the US, the naturally nutrient-rich coalition includes the National Dairy Council, Egg Nutrition Center, American Beef Producers, National Pork Board, California Kiwifruit, Wild Blueberry Association of North America, United States Potato Board, Wheat Foods Council and others. All of these agencies are involved in the marketing of basic commodities and working together must be applauded. A number of these companies will, however, also be involved in marketing foods with lower as well as high nutrient density but let us hope that commitment to naturally nutrient-rich might also mean minimally processed nutrient-rich.

Other approaches to modulating the impact of the food industry on eating habits may require fiscal measures such as taxation of foods high in sugar or fat, sponsorship, promotion and advertising restrictions. Models for government regulation on tobacco control indicate that raising the price of tobacco through taxation has been shown to be one of the most effective ways of reducing consumption (Sandford, 2003). There appears to be growing support for taxing soft drinks at a level, which will (at the very least) generate financial support for health promotion efforts (Jacobson and Brownell, 2000). This initiative can be justified on many grounds, given that most sweetened beverages supply little in the way of essential nutrients and need to be seen as a luxury rather than basic dietary item. All of these regulatory approaches remain to be assessed in terms of effectiveness for aiding consumers select a healthy balanced diet.

Policy and campaigning work by consumer groups (e.g., Centre for Science in the Public Interest http://www.cspinet.org/) working to promote healthy eating have identified the following areas as useful for facilitating dietary changes in the population:

- nutrition labelling on menus
- decrease marketing of low-nutrition foods to children
- improve school food
- promote fruit and vegetable intake
- increase resources for nutrition programmes (e.g., through soft drink taxes).

What has become clear in recent years is that the industry cannot ignore the power of consumer advocacy. For example, in England, school food has long been an area of emotive concern since nutrient standards were withdrawn in the 1980s. In 2005, television chef Jamie Oliver and his 'Feed me Better' campaign brought the topic on to centre stage for millions of British viewers, resulting in a government response to increase the budget available for school food and to re-introduce nutrient standards across England. Such action has lead to a view that government policy now seems to be more easily influenced by the media circus than evidence-based science (Crawley, 2005). There seems little doubt that 'Jamie Oliver has done more for the public health of our children than a corduroy army of health promotion workers or a £100m Saatchi & Saatchi campaign' (Spence, 2005). The power of advocacy has been declared!

In summary, the approaches to change unhealthy food consumption to some extent mirror the approaches taken in tobacco control with respect to: (a) legislative measures being used to foster a health promoting context where nutrient-dense foods are made widely available, e.g. school meals; (b) normative measures are used to re-enforce the social values behind food choices, e.g. through good social marketing campaigns; and (c) programmes are used to motivate and promote specific changes and actions, e.g. fruits and vegetables (Biedermann, 2004).

22.4   Implications for food product development

The implementation of nutrient standards in institutional catering environments has been shown to be an incentive for product re-formulation or development. For example, using current cultural food choices, limited cooking facilities, skilled staff and limited budgets, manufactured products (including frozen foods) provide a practical approach to food provision but all too often contribute to high levels of fat and sodium. The re-introduction of nutrient standards for school meals in Scotland (Scottish Executive, 2002) led to the removal of processed meat which did not meet the nutrient specifications and the addition of new formulated products which did meet nutrient specifications, e.g. reformulated turkey twizzlers. Other companies have recognised the economic opportunities of this approach and followed suit.

Meeting nutrient standards is only one aspect of product development and health. For example, many low calories drinks (e.g., diet drinks) fit with the scientific health limits but still contain a number of additives that many consumers suspect are unhealthy. The soft drinks area is interesting in that water and low fat milk provide the best examples of beverages for dietary promotion but these are given considerably less industry promotion and marketing in favour of artificially flavoured, coloured, nutrient-fortified concoctions. Developing *value-for-money* drinks with minimal additives and minimal calories that are fun and socially acceptable (and still manage to hydrate) remains a challenge which even bottled water companies have still to fully embrace.

In the food arena, there is increasing emphasis on calorie reduction, including the development of non-nutritive sweeteners and products such as olestra (a fat substitute used in foods and in processing, including frying and baking, which is not digested or absorbed by the body). The total cost of the development work on olestra has been estimated at \$200M and, after 25 years of research, approval was given by the US Food and Drug Administration in January 1996 for use of olestra as a partial replacer of fats in certain snack foods. At the time, it was a requirement that the label must also state that olestra may cause abdominal cramping and loose stools. However, in August 2003 the FDA concluded that the latter statement was no longer warranted. Fortification is necessary because some of the fat-soluble vitamins present in the gut at the same time (notably vitamins A and E) are preferentially dissolved in olestra and so partially lost to the body. Olestra is not yet approved in the UK and there is no application currently pending. The use of olestra has been criticised as being unnecessary and 'unnatural' (The Institute of Food Science & Technology, 2004). More recently, Olibra (currently used in Swedish yoghurt) has been developed from a fraction of palm and oat oils in a novel emulsion which reports to produce a 20–30% reduction in calorie intake in meals following its consumption.

Where it is desirable for intake of certain nutrients to be increased eating naturally rich sources of that nutrient are unavailable or unpopular (e.g., due to taste or cultural reasons), additional supplies can be made available through agricultural practices of food fortification or processing. Whilst there are a number of statutory fortification programmes in the UK (e.g., vitamin D is added to margarines), many food companies add nutrients to create nutritional enhancement of basic foods. The most well known example of this is the addition of vitamins and iron to breakfast cereals, and more recently the addition of calcium to certain fruit juices, and omega-3 fatty acids to orange drinks. It is important to note that these foods generally highlight their nutritional content for marketing, and further developments in nutritional enhancement of products may be dependent on whether these products will meet the criterion for government nutrition signposting guidelines.

22.5   Future trends

There is little doubt that future trends in the food and retailing industry will include the development and expansion of 'healthier' products. However, this is no guarantee that these products will provide the core foods of a healthy diet or that energy-dense food will diminish. Trends over the last two decades suggest that the food industry are good at selling their products in multiple formats in order to increase profits, e.g. increases in low-fat, semi-skimmed milk and increases in high-fat, luxury ice creams.

Our future is our children. Such sentiment is well recognised by current food industry trends. The developmental origins of health and disease point to the importance of fetal nutrient supply and, in turn, diet during pregnancy. Food marketing to pregnant women extends from self-prescribed vitamin supplements to special milk fortified with omega-3 fatty acids and other fatty acids, multi-fruit drink enriched with nutrients, and herbal teas. A new range of milk shakes and bars illustrates the future developments in this area. One company (which already has a track record in supplemental feeds) is marketing 'Homemade Vanilla' and 'Creamy Milk Chocolate' shakes that provide balanced nutrition, are high in calcium, and can also help with pregnancy cravings for sweet, dairy or chocolate flavours.

Media reporting of a study on the impacts of omega-3 supplementation on learning (reading and spelling) and concentration in children have lead to a range of new fortified products, plus traditional advertising and website backups. However, it is worth noting that Mintel (2005) is cited as reporting that one in three parents take little interest in their children's eating habits, suggesting that 'there is much talk but little action' about children's diet.

Beyond marketing health options to parents, food companies also excel at promoting their products to different child age groups, e.g. 'tweens' aged 8 to 11 years. This area is one that is likely to see innovation if regulatory authorities insist on limiting marketing to children.

The promotion of naturally-occurring (non-nutrient) bioactive food components with health benefits is likely to increase. For example, high melatonin levels in milk (sold as 'Nachtmilk') are marketed to address problems of sleeplessness. Probably the best example of this type of component is omega-3 fatty acids found in oil-rich fish which (as part of a healthy lifestyle) help maintain heart health. Probiotic foods (e.g., Yakult) remain an expanding market, sold in product sizes similar to pharmacy doses whilst vaguely resembling a mini milk bottle. The popularity of these types of food serves to remind health professionals that consumers can be more easily lured into the approach of 'natural pharmacy' than trying to achieve an overall healthy balanced diet. From a general health perspective, they may also lead to people taking considerable efforts over maintaining health in one physiological system (e.g., gut health) with less effort over other systems, e.g. cardiovascular health.

On the other hand, there is a considerable concentration of effort on developing functional foods to reduce cardiovascular disease, most recently in the form of blood-pressure-lowering drinks. Such products include Evolus (fermented milk which produces angiotensin-converting enzyme (ACE) inhibitors) with fruit juice added. Sales of this product reached 1.5 million litres a year in Finland two years after its launch (Anon, 2005).

With the threat of taxation or economic penalties a number of global companies are now seeking partnerships that enable a health image (or gloss!) to emerge. These include such actions as McDonald's' move to partner with the Produce for Better Health Foundation to promote five a day and a commitment to the provision of educational materials available in-restaurant: table tents, tray liners, brochures and packaging, new menu options that follow the '5 A Day for Better Health' programme guidelines, and foodservice research projects. In a similar way, PepsiCo (includes Pepsi, Tropicana and Quaker Oats) has partnered with a highly popular obesity prevention programme (America on the Move). As trends in eating out continue to rise, it is likely that more restaurants will start to use labelling or 'healthy eating' awards systems to promote their choices.

Time-limited consumers seek ready meals at home and it is likely that more convenience meals combined with functional foods may appear. However, recent trends suggest that the emphasis is on naturally and intrinsically healthy ingredients with less fat. In the US, Savvy Faire Lifestyle cuisine and in the UK the Get Real brand offer organic and meat-free pies and ready meals from the chiller cabinet. Apparently, each ready meal provides at least 2.5 vegetable portions, demonstrating that fast foods do not have to mean fatty foods (Mellentin and Foley, 2005).

Novel scientific theories to health have also led to the development of new product ranges. A number of nutritional and metabolic approaches have been postulated for improving health, especially in relation to body weight control and the development of diabetes and cardiovascular diseases. The Atkins diet is one example and Glycaemic Index (GI) another. Both of these concepts have moved from being of minority interest to generating considerable retail opportunity including merchandising involving supermarket labelling and popular science books. Whilst there is little doubt that GI holds considerable potential to improve glucose regulation (Brand Miller *et al.*, 2003), there is currently insufficient information to determine whether there is a relationship between glycaemic index or glycaemic load of diets and the development of diabetes (Sheard *et al.*, 2004). The evidence from randomised controlled trials showing that low-glycaemic-index diets reduce coronary heart disease and CHD risk factors is also weak (Kelly *et al.*, 2004). Thus the move from theoretical concept to practical application has skipped the need for a strong evidence base but (with strong marketing) still manages to entice consumers to purchase GI-labelled foods.

22.6   Conclusions

In conclusion, scientific evidence provides us with the fundamental basis for nutritional health. The broad knowledge about what constitutes a healthy diet to reduce disease risk is well described. The details about new emerging science (including what the genome tells us about nutritional requirements of individuals) provide industry with exciting opportunities to provide nutritional health in a short cut for busy consumers and families. However, promoting the uptake of a nutrient-dense diet remains a challenge that agriculture, health and industry need to work together to be able to address for mutual benefit.

The route by which 'joined-up' working can be made to happen is, however, far from clear. Finland and Norway have both had National Nutrition Councils (which act as government advisory bodies) concerning food supply, nutrition, physical activity and health for some years. Such councils have been described as 'a key structure for effective co-ordination of activities and for a clear division of responsibility for implementing the nutrition policy' (Roos *et al.*, 2002). More recently (2005) Scotland initiated a Food and Health council (http://www.scotland.gov.uk/Topics/Health/health/19133/17905) comprising the heads of key policy areas within the Scottish government, the FSA and appointed representatives (from National Farmers Union, food retailing, food processing and manufacturing as well as Public Health and health inequalities experts) to 'integrate cross-cutting elements of food and health policy and the strategies of the Food Standards Agency (FSA)'. It remains to be seen whether this grouping will be able to harness the political pressure and power to effect change at individual, commercial and legislative levels.

Ultimately, food choices are made by the individual and may be influenced to a greater or lesser extent by wider environmental factors. The success of initiatives to change food choices can be measured by changes in health outcomes (both morbidity and mortality) but also by changes in the food culture. We should perhaps eagerly await the day when menus provide a solitary item labelled 'unhealthy choice' just to ensure freedom of choice for those poor individuals who cannot cope with optimal nutrition and need to retain the last vestiges of slothdom!

22.7   Sources of further information and advice

Centre for Science in the Public Interest (http://www.cspinet.org/)
Food Standards Agency (http://www.food.gov.uk/)
New Nutrition Business (http://www.new-nutrition.com/indexDev.asp)
World Cancer Research Fund (http://www.wcrf-uk.org/)

22.8   References

AMMERMAN AS, LINDQUIST CH, LOHR KN, HERSEY J (2002), 'The efficacy of behavioral interventions to modify dietary fat and fruit and vegetable intake: a review of the evidence', *Prev Med*, 1, 25–41.
ANDERSON AS (2004), 'Evidence based dietary behaviour strategies to reduce cancer risk', in Sancho-Garnier H, Biedermaan A, Slama K, Anderson AS, Lynge E (eds), *Evidence-based Cancer Prevention Strategies for NGOs, A Handbook for Europe UICC*, Geneva, 97–111
ANON (2005), *Nutrition Business News*, 10 (10), August, 22.
BIEDERMANN A (2004), 'Recommendations for action', in Sancho-Garnier H, Biedermaan A, Slama K, Anderson AS, Lynge E (eds), *Evidence-based Cancer Prevention Strategies for NGOs, A Handbook for Europe UICC*, Geneva, 197–213.
BRAND-MILLER JC, THOMAS M, SWAN V, AHMAD ZI, PETOCZ P, COLAGIURI S (2003), 'Physiological validation of the concept of glycemic load in lean young adults', *J Nutr*, 133, 2728–2732.
CRAWLEY H (2005), 'Is school food finally getting the attention it deserves?', *J Hum Nutr Diet*, 18 (4), 239.
DEFRA (2000), *National Food Survey*, London, Office for National Statistics.
FRENCH SA, LIN BH, GUTHRIE JF (2003), 'National trends in soft drink consumption among children and adolescents age 6 to 17 years: prevalence, amounts, and sources, 1977/1978 to 1994/1998', *J Am Diet Assoc*, 103 (10), 1326–1331.

HASTINGS G, STEAD M, McDERMOTT L, FORSYTH A, MacKINTOSH AM, RAYNER M, GODFREY C, CARAHER M, ANGUS K (2003), 'Review of the research into the effects of food promotion to children'. Retrieved 26/07/06 from http://www.food.gov.uk/multimedia/pdfs/foodpromotiontochildren1.pdf.

HENDERSON L *et al.* (2003), 'The National Diet and Nutrition Survey: adults aged 19 to 64 years', London, The Stationery Office.

THE INSTITUTE OF FOOD SCIENCE & TECHNOLOGY (2004), 'IFST: current hot topics – olestra'. Retrieved 26/07/06 from http://www.ifst.org/hottop13.htm.

JACOBSON M AND BROWNELL KD (2000), 'Small taxes on soft drinks and snack foods to promote health' *Am J Public Health*, 90 (6), 854–857.

KELLY L *et al.* (2004), 'Low glycaemic index diets for coronary heart disease', *Cochrane Database of Systematic Reviews* 2004, Issue 4, Article CD004467.

LEAN MEJ (2005), 'Prognosis in obesity', *BMJ*, 330, 1339–1340.

MELLENTIN J AND FOLEY M (2005), 'Naturally wholesome and health makes for ready-meal success', *New Nutrition Business*, 10 (9), 28–29.

MINTEL (2005), 'Child Obesity', Kids Nutrition Report, July/August, 27.

NIELSEN SJ AND POPKIN BM (2003), 'Patterns and trends in food portion sizes 1977–1998', *JAMA*, 289 (4), 450–453.

NIH (2001), 'Theory at a glance'. Retrieved 26/07/06 from http://www.cancer.gov/cancerinformation/theory-at-a-glance

PEREIRA MA, KARTASHOV AI, EBBELING CB, VAN HORN L, SLATTERY ML, JACOBS DR JR, LUDWIG DS (2005), 'Fast-food habits, weight gain, and insulin resistance (the CARDIA study): 15-year prospective analysis', *Lancet*, 365 (9453), 36–42.

PRENTICE AM AND JEBB SA (1995), 'Obesity in Britain: gluttony or sloth?', *BMJ*, 311 (7002), 437–439.

PRENTICE AM AND JEBB SA (2003), 'Fast foods, energy density and obesity: a possible mechanistic link', *Obes Rev*, 4, 187–194.

PUSKA P (1999), 'The North Karelia Project: from community intervention to national activity in lowering cholesterol levels and CHD risk'. Retrieved 26/07/06 from http://www.cvhpinstitute.org/daniel/readings/northkareliacap.PDF

PUSKA P, KORHONENN HJ, TORPPA J, TUOMILEHTO J, VARTIAINEN E, PIETINEN P, NISSINEN A (1993), 'Does community-wide prevention of cardiovascular diseases influence cancer mortality?', *Eur J Cancer Prev*, 2 (6), 457–460.

ROOS G, ANDERSON A, LEAN M (2002) Dietary interventions in Finland, Norway and Sweden: nutrition policies and strategies *J Hum Nutr Diet*, 15 (2), 99–110.

SANDFORD A (2003), 'Government action to reduce smoking', *Respirology*, 8 (1), 7–16.

SCHULZE MB, MANSON JE, LUDWIG DS, COLDITZ GA, STAMPFER MJ, WILLETT WC, HU FB (2004), 'Sugar-sweetened beverages, weight gain, and incidence of type 2 diabetes in young and middle-aged women', *JAMA*, 292, 927–934.

SCOTTISH EXECUTIVE (2002), 'Hungry for Success: A whole school approach to school meals in Scotland', Edinburgh, The Stationery Office. Retrieved 26/07/06 from http://www.scotland.gov.uk/library5/education/hfs-00.asp.

SHEARD NF, CLARK NG, BRAND-MILLER JC, FRANZ MJ, PI-SUNYER FX, MAYER-DAVIS E, KULKARNI K, GEIL P (2004), 'Dietary carbohydrate (amount and type) in the prevention and management of diabetes: a statement by the American diabetes association', *Diabetes Care*, 9, 2266–2271.

SPENCE (2005), 'Jamies school dinners – review', *BMJ*, 330, 678.

USDA (2005), 'Dietary Guidelines for Americans 2005'. Retrieved 26/07/06 from http://www.healthierus.gov/dietaryguidelines

WELLS L AND NELSON M (2005), 'The National School Fruit Scheme produces short-term but not longer-term increases in fruit consumption in primary school children', *Br J Nutr*, 93 (4), 537–542.

WORLD CANCER RESEARCH FUND AND AMERICAN INSTITUTE FOR CANCER RESEARCH APPENDIX (1997), '*Food, Nutrition and the Prevention of cancer: A Global Perspective*', London, Washington DC, WCRF. Retrieved 26/07/06 from http://www.wcrf-uk.org/report/summary/lasso

WHO (2003), '*Diet, nutrition and the prevention of chronic diseases.* Report of a joint WHO/FAO expert consultation', Geneva, World Health Organisation. Retrieved 26/07/06 from http://www.who.int/hpr/nutrition/.

YOUNG L AND SWINBURN B (2002), 'Impact of the Pick the Tick food information programme on the salt content of food in New Zealand', *Health Promotion International*, 17 (1), 13–19.

ZELMAN K AND KENNEDY E (2005), 'Naturally nutrient rich … putting more power on Americans' plates', *Nutrition Today*, 40 (2), 60–68.

Part V

Consumer attitude, food policy and practice

23

Social factors and food choice: consumption as practice

U. Kjaernes and L. Holm, SIFO, Norway

23.1 Introduction

While the notion of the European citizen is still important in the discourse on food regulation, over recent years there have been increasing references to 'the consumer'. Policy papers are explicitly referring to consumer choice and consumers' own responsibility through 'informed choice' and labelling strategies (Commission of the European Communities, 2000; Reisch, 2004). The widespread usage of the term 'consumer' has coincided with neo-liberal precepts, thereby envisaging the consumer as an isolated self-interested individual. Indeed, this kind of consumer has been the model for neo-classical *Homo Oeconomicus*, an abstract and universal agent, conceived of as carrier of a black box of given preferences constrained by a given budgetary level and linked to an environment defined in terms of the goods available, the relative prices of these goods, and information made available about them. But how can we understand, from this model of consumption, the major variations that can be observed in consumption practices – between countries and between social groups? How can the stability and consistency in food choice that is often found within national and cultural contexts be understood? And why do large-scale shifts in food choice occur? This chapter will argue that a concept of food consumption understood as socially created sets of practices represents a viable approach to such questions.

Many contemporary theories of consumption suggest that the consumer is far from being a champion of individualistic forward-looking choices, based on deliberate calculation of self-interest. Food consumption can be fully appreciated as a form of *social* action only by leaving behind the idea that such action may be modelled exclusively as a conscious decision at the point of purchase. So as to avoid conceptualising consumption as a series of abstract and individualistic decisions, we have to consider that consumption practices happen within social institutions like the family, work – and the marketplace and that these practices are themselves institutionalised. By this we mean that there are predictable societal patterns of behaviour related to food provisioning and consumption, emerging from social structures, norms and conventions, as well as the particular contexts and situations within which consumption takes place. Food represents an intersection between public arenas and the private sphere, the collective and the individual. Meal structure and cuisine will affect how people do their food provisioning, but the character of various forms of supply will also form a significant context for people's expectations and actions.

In this chapter, we will concentrate on one element in food consumption, that of eating. Eating is an activity, which is based on physiological needs and closely linked to the organisation of society and social life. Understanding eating as embedded in social practices, introduces questions of not only what and how people eat, but also when they eat, where and with whom. National and regional cultures of eating, themselves historical and changing constructs, shape patterns of food consumption, and have been argued to be constitutive of national identity. A nation is (in part) what it eats (Mintz, 1996; Hogan, 1997; Appadurai, 1988). The ordering and significance of meals at different eating times, and the structure and content of the meals, are related to the social ordering of time, e.g. working hours, and the way they have historically evolved (Kjærnes, 2001). The place of the meal in the household as part of its social fabric is changing not only according to the internal dynamics of the household's social organisation, but also in relation to work, leisure activities, etc. The eating of food is a changing cultural and social institution that is structured and organised in the sphere of consumption partly beyond, but in interaction with, the food market and other forms of food provisioning (e.g. family and friends, work canteens, school meals, and meals on wheels). Eating is moreover subject to regulation and influence beyond these organisational aspects, through basic and very strong normative frames, as well as formal regulatory arrangements and extensive public discourses ranging from risk and health to political consumerism and fashion and trends in cooking.

We will start by briefly addressing the understanding of food consumption as a matter of individual choice. We then move on to point to elements in an alternative understanding of consumption as routine practices embedded in social institutions. We go into some detail on the particular character of eating and the role of meals, illustrated by empirical examples.[1] Emphasis is put on

1. Our main empirical reference will be a comparative survey of eating patterns that the authors carried out in Denmark, Finland, Norway, and Sweden, together with Jukka Gronow, Johanna Mäkelä and Marianne Pipping Ekström.. A telephone survey with representative samples (1200 in each country) was conducted in 1997. The main results are reported in Kjærnes (2001). The methodology has been described in (Mäkelä et al., 1999). The study was co-funded by the Nordic Research Council for Social Science.

national variations and social differentiation, first, by dealing with organisational aspects of eating and, second, by discussing normative elements, in particular the notion of 'proper meals'. We will do this in light of the sociological debate on individualisation as a characteristic of contemporary food consumption. We contend that while changes in the social organisation and norms of eating may have taken place in recent years, they seem overrated and, when they are observed, new patterns do not appear to be less socially determined. The concluding part will concentrate on some implications of a social as opposed to an individual understanding of food consumption, with some reflections also on implications on policy formulation.

23.2 Food consumption: from individual choice to social practices

In neoclassical economic theory consumption 'is the sole end and purpose of all production' (Adam Smith, The Wealth of Nations 1776, here quoted from Söderlind, 2001, p. 6) and the system of production is seen as responding as a servant to the needs and wishes of consumers (Fine and Leopold, 1993; Friedman and Friedman, 1981). These sovereign consumers are seen as rational choosers, driven by an individual utilitarian orientation, seeking to maximise personal benefits at the lowest possible cost. This conceptualisation implies a tendency to see consumers as independent and autonomous actors and their choices of consumer goods as driven solely by their individual needs and demands.

While economic theory takes needs and demand for granted, giving them little attention, consumer research has focused much more on preferences, i.e. on how consumer needs and demands are constituted. Much research on food consumption focuses on the formation of individual food choices. Several models have been suggested, which from different scientific disciplines and with different perspectives seek to conceptualise food choice and to organise how it should be studied. In an overview of factors influencing food choice, Shepherd reports on a series of models which summarise factors influencing food acceptance, food preferences, food selection, food choice and food intake (Shepherd, 1990). All models include factors related to the specific food product in question (brand attributes, food appearance, odour, temperature, flavour, etc.), and to the individual making the choice (personality, personal values and beliefs, attitudes and norms, socio-economic background, gender, biological factors, and cultural factors). Some models also include wider societal factors – sometimes named 'extrinsic factors' – such as environment, situation, advertising, culture, economy and society.

Even though the number, character and naming of the factors included in the models may vary, they all share a focus on individual choice as the key action in food consumption, which needs to be explained. The focus is usually on decisions understood as selecting between marketed goods.

Several models aim to quantify the relative importance of different types of factors. This endeavour is often unsuccessful, Shepherd says (*ibid*). Methods based on asking individuals about influences on their choice are problematic, as individuals may not be fully aware of the influences on their behaviour. Thus, other study designs have been developed, e.g. studies examining actual choices made and relating them to attitudes and beliefs of individuals.

One widely used model for this kind of study is the 'theory of planned behaviour' (TPB) proposed by Ajzen and Fishbein (Ajzen and Fishbein, 1980; Ajzen, 1991) which offers a systematic framework for empirical investigation. In this theory, behaviour (food choice) is seen as the result of behavioural intention. This again is determined by the individuals' attitudes (is the choice good, beneficial, etc.?) and the subjective norm (does the individual perceive any social pressure to make the choice?). While studies using this model are successful in terms of producing high correlations between intentions and actual food choice, and between attitudes and/or norms and intentions – they do typically deal with a very narrow set of variables. Consequently, in many studies, extra variables are added to the model in order to strengthen its predictive power (e.g., habits, knowledge, perceived control; Shepherd, 1990). Studies based on Ajzen and Fishbein's theory often focus on choice of specific types of food, and results are presented in a manner which profiles food items by assigning specific configurations of influences of choice to them (e.g., drinking milk is influenced strongly by subjective norms – others think it is good to drink milk) whereas eating ice-cream and chocolate are not – instead these are first of all influenced by individual attitude (I eat it because I like it – not because others think it is good to eat) (Lähteenmäki and Van Trijp, 1995; Tuorila and Pangborn, 1988).

Studies like these are based on a notion of reasoned action and planned behaviour, which assume that choices are reflected and deliberate. The theory of planned behaviour is only one of several social psychological models used in studies of food consumption. Currently, a large number of studies, for example, direct attention towards how to modify eating through various types of intervention. Experiments and intervention programmes aim at promoting healthy eating, understood as individual behaviour, by changing factors in the individual's environment, including peer groups, information in shops, campaigns, etc. (Glanz *et al.*, 1997; Neumark-Sztainer, 1999; Richter *et al.*, 2000).

The TPB model is brought out here, because it clearly illustrates how the focus is on the individual making choices. Socio-demographic variables may be included in the studies, thus differentiating results and presenting systematic differences in attitudes, social norms and intensions, e.g. related to age, gender and social class (Shepherd and Dennison, 1996; Shepherd and Stockley, 1985). Further, perceptions of risk and danger, credibility of and trust in public authorities and other actors in the food chain (see de Jonge *et al.*, Chapter 5) may be included too. This type of research appears to see food choice as a reflection of aspects of specific types of food as perceived by different types of individuals, often in the light of specific public debates or issues. It follows from this cognitive framing of choice that reflection precedes practice. The possibility of an opposite causal direction is not considered. Tacit routines and the situational character of many individual choices have no place in this theory; neither does the wider societal and political framing of choices. Norms and values are a matter of individual priority, not socially structured. Consequently, the rationale for these kinds of studies is to inform actors – whether public policy makers or private commercial actors – who seek to alter the choices people make through persuading them to change their intentions. We will in the following argue that food consumption does not emerge directly from individual intentions and that reflection does not always precede practice.

23.3   Food consumption as sets of practices

Theories of practice are manifold and represent a wide and versatile sociological field. Theories are – with few exceptions – generally not referring to food and eating. There is a need, though, to start discussing what implications the concept of practice may have for the study of food consumption. Alan Warde has presented a discussion of consumption as practice, which we see as a valuable starting point. His suggestion is that consumption be best understood as embedded in particular practices and not as a practice in itself (Warde, 2005).[2] Whereas much food consumption research constrains its interests to the choice situation, i.e. basically to market exchange – Warde's suggestion implies that food consumption must be understood as a much broader phenomenon which must be examined as an integral part of daily life. Consumption is thus a process whereby, 'agents engage in appropriation and appreciation, whether for utilitarian, expressive or contemplative purposes, of goods, services, performances, information or ambience, whether purchased or not, over which the agent has some degree of discretion' (Warde, 2005, p. 137). Consumption is understood as embedded in practice, and practices are constituted outside the individual. Consumption thus occurs as items are appropriated in the course of engaging in particular practices. This notion of practices can therefore account for both social order (emerging from the coordination and the norms that a practice represents) and individuality (differentiation and performance within a practice) (Schatzki, 1996).

In the context of food, the range of relevant practices may be very different in character and will vary according to historical circumstance. Eating is something that everybody does – usually every day. But practices that involve eating are highly variable. They may, for example, include the practices of making and consuming family meals, of maintaining health, strength and functionality as part of doing other things – work or leisure activities, as well as socialising with others, of pausing and resting, of celebrating, etc. (see also Gronow, 2004).

---

2. Warde's contribution builds partly on the conceptual framework developed by Pierre Bourdieu.

Unlike eating, food provisioning and preparation may or may not stand out as particular and significant parts of practices in which the individual is involved (because somebody else can take care of it). This introduces an important issue of division of labour and responsibility. This approach implies a shift in focus from individual perception of particular foods to the logic of situations where food is purchased, cooked, served, and eaten. It also shifts focus from attitudes and gender, age or social class, seen as individual properties, to institutionalised practices, and how they shape food and eating. The focus is more on how activity generates wants, rather than *vice versa*.

This understanding does not exclude the influences of socio-demographic or cultural background or of personal skills and interests. The point is that this influence takes place within a practice which is already established before the individual enters the scene. Before individuals want milk instead of chocolate for breakfast, there are already institutionalised sets of practices which define milk as a healthy drink, a drink to be taken with ordinary everyday breakfast, and a drink to be readily available in settings such as schools due to established welfare policy programmes – and there are other institutionalised sets of practices defining chocolate as not part of ordinary meals, but a snack readily available across a wide range of outlets to be found almost everywhere in urban life. This is, of course, not deterministic or static. People may and will often challenge or alter practices through their actions, but they will do so with reference to the established practices and to the structures that form these practices.

Attention is directed towards how groups of people understand a practice, the values to which they aspire, and the procedures they adopt within practices. 'The patterns of similarity and difference in possessions and use within and between groups of people, often demonstrated by studies of consumption, may thus be seen as the corollary of the way the practice is organized, rather than as the outcome of personal choice, whether unconstrained or bounded. The conventions and the standards of the practice steer behaviour' (Warde, 2005, p. 137). Social categories will then be socially conditioned rather than representing individual properties. Gender will imply socialisation into gender roles and the gendered labour and responsibility within practices. Age may, on the one hand, indicate a generation with particular experiences and expectations related to sets of practices and, on the other hand, a phase in life, which places the individual within specific sets of practice, such as being under education, having small children, being a divorced single person, or retirement. Social differentiation and inequality are then not a matter of randomly choosing a lifestyle, but a structurally contingent disposition influenced by economic resources, upbringing, social networks, and cultural codes.

Practices have a trajectory or path of development, a history. The substantive forms that practices take will always be conditional upon the institutional arrangements characteristic of time, space and social context. The practice requires that competent practitioners avail themselves of the requisite services, possess and command the capability to manipulate the appropriate tools, and devote a suitable level of attention to the conduct of the practice. This is, of course, in addition to exhibiting common understanding, know-how, and commitment to the value of the practice. As a consequence, the focus is directed towards the routine, ordinary, collective, conventional nature of much consumption, but also towards the fact that practices are internally differentiated. Focus is on use of goods rather than (only) on acquisition. In the area of food, the focus is then more often on menu planning, food preparation, and conduct of meals in different social contexts, which also form an important background for the selection of specific products in a shopping situation. Purchases do, however, form a connection between the commercial systems of provisioning and what we do outside the market sphere. It is one act in a chain of decisions on both the consumer and the supply sides. We can only purchase what is available in the shops. Purchases are, of course, not completely predetermined, as commercial actors do what they can to make people select their products instead of those from their competitors. However, social consumption practices are not to be reduced to acquisition and most certainly not to the choice of products or to commercial consumption, which is only one way in which goods and services are obtained for the purposes of practice – here first of all eating (Harvey *et al.*, 2001, p. 45). Practices represent an appropriation of the purchased goods that, in turn, will influence what we want to buy.

Everyday food-related practices represent a typical example of mundane, routinised consumption practices, closely associated with how we carry out our daily activities. Ordinary shopping, breakfast eating and school lunches are examples of such organised practices. Overall, societal organisation and institutional structures will have impacts on the social conventions and standards that regulate food consumption and they will form rather concrete frames for how practices are organised.

This section has pointed to a number of aspects that characterise food consumption activities as social, conceptually framed as practices, rather than representing individual choices. In the following, we will discuss how food consumption can be analysed as a matter of socially contingent practices. First, we will explore a bit more what this perspective means in an analysis of variations in eating practices. In the next two sections, we then move on to characterise contemporary eating practices in terms of social coordination and normative regulation, respectively.

23.4   Eating as a practice

How can observations about eating fit into this understanding of consumption as part of socially formed practices rather than being framed as individualised actions? What does this perspective imply in terms of understanding variations in food consumption? The conceptualisation of eating as part of socially contingent practices means that we must expect to find distinct patterns of similarity and difference according to location, time, and social and institutional context. These patterns can be described in terms of situated events (who, what, when, where, and with whom). Patterns emerge in interplay between practical coordination, institutional restrictions and opportunities, and normative regulation. We are therefore interested in patterns of eating within certain contexts, for example a country, basic characteristics of such patterns, how patterns can be differentiated, as well as how various types of eating practices are modified and changed.

Considering the wide selection of food items offered to us in supermarkets, and the speed by which the supply changes through the success and failures in the sale of a growing range of new products, it might be anticipated that food choices are very varied, unstable and quite unpredictable. There is no doubt that food consumption is highly diverse and also dynamic. But, at the same time, social scientific studies reveal food-related practices as strongly habitual and predictable. This can be observed both in individual routines and in national consumption patterns. People do more or less the same things every day and they do so in rather coordinated ways. Within a given context, for example in a country at a certain point in time, major decisive features of consumption patterns may not be easily detected because what people do when they shop, prepare and eat food is so normalised, trivial and taken for granted. Comparisons across social and cultural contexts or over time reveal that such 'normalities' may vary considerably and that they change over time. Every person involved in the marketing of food across national borders – and every tourist too – will know that meals served in Italy are not the same as meals served in Norway or in Hungary. Nor are the food items available in the shops – or the shops themselves – identical. These differences may represent a trivial observation, but we argue that they are significant for understanding food consumption embedded in social practice. But we need to be more precise and look into the specific features of eating practices.

It is evident that eating is influenced by our social background and experiences as well as by the practical considerations of everyday life. The particular social contexts of eating out can illustrate this (Warde and Martens, 2000). While a purchased lunch is common in many social groups in Britain, young people will typically purchase their lunch in fast food outlets, while those doing so in restaurants are predominantly from higher social classes. Different contexts, therefore, are dominated by different sorts of people. Warde contends that: 'If eating is the sum effect of many situated events, the sociologically appropriate question is whether there is a social logic to the situations in which people find themselves' (Warde, 1997). As a consequence, neither items nor situations should be observed separately. Rather, the focus should be on the sequence of situations and bundles of items. The organisation of eating depends on factors like the job and family situation, as well as social forces associated with events, such as the frequency of attendance at celebrations or the habit of eating Sunday lunch.

Combining the overall institutional perspective and the understanding of eating as situated events means that we must expect to find variations in how eating practices are organised. Lunch eating in the Nordic countries is, for example, varied and may also be different from what is found in Britain. While a cold lunch dominates in Norway and Denmark, a hot lunch is much more common in the neighbouring countries Sweden and Finland. The two types of lunch are both dependent on specific forms of historical social organisation – and both are likely, in their specific national context, to be seen as the most healthy solution (Gronow and Jääskeläinen, 2001; Holm, 2001b). In Sweden and Finland, lunch at work and in school is a cooked meal served by the institution or bought in a café. For a majority of the population, the lunch away from home typically consists of fish or meat as a main ingredient, potatoes or another staple, a salad, and bread on the side. The provision of these cooked lunches represents a major public or workplace welfare service and an important part of eating out. In Denmark and Norway, lunch remains a matter of household provisioning even when eaten away from home. People usually bring a packed lunch, consisting of open sandwiches with a topping. This packed lunch is no less social, most often eaten together with colleagues and schoolmates (Holm, 2001b). In these two countries, the only cooked meal of the day is eaten in the evening. Turning to another country, Germany, lunch for employed people is usually cooked and served in a work canteen, while schoolchildren go home for their cooked midday meal (to be prepared by their – working or non-working – mothers). So even within these quite similar and geographically adjacent societies, eating practices differ considerably, and they depend upon and influence significant differences not only with respect to what is eaten, but also with respect to the division of labour between men and women, and between private households, public institutions, and commercial services.

Practices such as workday lunches are typically socially coordinated and strongly linked to shared norms and expectations within specific social contexts. Such 'normal' practices describe not only how things are usually done, but often also how things should be done. But as a practice, there may be internal differentiation based on competence and commitment, for example by women adding more vegetables to these lunches than men. And evidently, there is individual variation. Even though acting in direct contrast to social norms may raise practical difficulties as well as social sanctions, it will also form a rich ground for, for example, teenager opposition to parental authority and dominant culture (Andersson, 1983; Shepherd and Dennison, 1996).

A practice is neither static nor necessarily consensual. Mismatches and tensions may occur if some of the social conditions for practices change. For example, the two open sandwiches eaten for lunch in Norway may be too little when the organisation of work and leisure activities pushes dinner time to later hours in the day, and the Danish lunch-pack may be threatened by the fact that many schoolchildren refuse to eat it and prefer to buy cheap junk-food in shops which are increasingly established near schools. So, here we see, on the one hand, clear patterns of what is ordinary – the lunch-pack – but, on the other hand, that not everybody is equally committed and engaged in the same practices. We also see that practices must be expected to change over time, both due to internal tension, changes in the frames of everyday life, and the character of the supply.

23.5   De-structuration of contemporary eating practices?

One of the decisive features of eating as a social practice is, as noted by many authors, the close relation between the eating of meals and social interaction. Meals organise social groups and is important for their cohesion (DeVault, 1991; Otnes, 1991). The commensality of eating implies that we try to coordinate our actions. As indicated by Simmel already in 1910, the importance of meals taken together will inevitably lead to temporal regularity (Simmel, 1957). To Simmel, the sociability of eating is related to the refinement of social forms of interaction (Gronow, 1997) and Rotenberg argues that a meal is essentially a social affair, 'a planned social interaction centred on food' (Rotenberg, 1981, p. 26). Correspondingly, (Mäkelä, 1995) found that for working mothers in Southern Finland, the idea of sharing food with others proved to be essential for the concept of a meal.

The organisation and social context of eating has changed considerably over time. A classical study by Rotenberg shows how eating patterns in Vienna shifted from the early 1900s, throughout the interwar years, and up until the 1980s (Rotenberg, 1981). These changes were closely associated with the shift from traditional to industrial society in Austria. Around 1900, the first meal of the day was eaten at home, the second at work or in a cafe, the third, main meal at home, the fourth again in a cafe, and the fifth in the evening at home. Meals at home with the family alternated with meals with colleagues or friends away from home throughout the day. The condition for this pattern was that the workplace was located quite close to home. The men did not have to travel far, and the women, who often contributed by working in their husbands' work-shops, could also tend to cooking during morning hours. By the 1930s, however, industrialisation had produced a three-meal pattern, where the men did not go home for the midday meal (women and children thus eating at home without the male breadwinner), and where socialising with friends took place mainly during weekends. The family meal was relocated to the end of the working day. Similar shifts in meal patterns were identified in a Finnish study, where urbanisation and industrialisation were found to be accompanied by a move towards a three-meal pattern and fewer hot meals (Prättälä and Helminen, 1990).

These studies serve as illustrations of how eating patterns change and how those changes are influenced, even driven, by major societal shifts. So, what is happening in contemporary society? Modernisation theorists often point to individualisation as a key characteristic of contemporary societies. Individuali-sation means the dissolution of tradition and the disruption of strong social regulation. It is often presented as implying more perceived flexibility and freedom for the individual to choose according to his or her personal tastes and preferences (Bauman, 2001). Many have also emphasised negative aspects when each act of choice is to be reflexively considered (Giddens, 1994; Sulkunen, 1997). A high degree of unpredictability and dissolution is extremely imprac-tical within an everyday context, and for the individual it may create a certain basic uncertainty, even anxiety. More perceived individual freedom is assumed to be accompanied by increased pressure on subjectivity (Ziehe, 1989), as more responsibility is pushed over to private individuals. This leads to considerable tension within individuals, because these shifting responsibilities are not matched by similar shifts in the distribution of power and resources (Beck and Beck-Gernsheim, 2002).

This perspective on modern everyday life has been important in debates on contemporary trends in eating. They have often come down to a discussion of the status of meals in contemporary society. Meals are said to be a practice of declining importance, as other kinds of eating are emerging. A key issue has been whether ongoing changes in the social organisation and structuring of eating imply a de-structuring,[3] with less social regularity in what people eat, where and when, associated first of all with processes of individualisation. This has been conceptualised as for example 'grazing' and 'vagabond feeding'. Picking up on Durkheim's concept 'anomie', Fischler suggested that modernisation is associated with 'gastro-anomie' (Fischler, 1988). This concept indicates high degrees of deregulation and de-structuring of food selection and eating, accompanied by individual anxiety and uncertainty. However, unlike many of the pessimistic accounts of modernisation theorists, he sees our dealings with food as basically social and foresees a re-identification of foods and a process of re-structuration. With the increasing complexity of modern societies we need both flexibility and daily routines, and conventionality and individuality are not necessarily social opposites (Gronow and Warde, 2001). This is supported by Campbell (1996, p. 149), who contends that life in modern societies has become de-traditionalised (less dependent on what people used to do in earlier times), but at the same time also more habitual and predictable from one day to the next. From this, we can take that eating patterns may be changing, perhaps even towards more individual flexibility. In a practice perspective this is not a question of whether eating is socially contingent or not, but rather that existing practices may be challenged and changing due to societal change. Meals are situated events, but so are also other forms of eating. So we can ask whether the social organisation of eating is changing.

Theories about modernity and modern eating have been followed by a number of empirical studies. These studies indicate, on the one hand, considerable stability in European eating patterns, but some change, variation and flexibility on the other. Still, this does not add up to the disappearance of social regularities and influences (Aymard et al., 1996; Kjærnes, 2001; Poulain, 2002; Mestdag, 2005). Larger parts of the populations' habits fit into what might be characterised as 'ordinary' or 'conventional' eating patterns. The order and rhythm of meals, the meal patterns, represent intersections between the public sphere of production and the private sphere of reproduction, of family and recreation. Eating contributes to ordering our days into segments: morning, midday, afternoon, evening. Attention is thereby directed towards the organisation of schedules, the

---

3. By the 'structuring' of eating we here refer to organisational and normative frames that shape the patterns of eating, making people adhere to a socially shared and predictable practice.

particular modes in which food preparation and meals interchange with work and other activities, as part of cyclical calendars as well as throughout the day. Aymard *et al.* (1996) have stated that: 'food practices make the greatest contribution to the structuring of social time and ... these practices are in turn strongly influenced by the place reserved for them in daily routines as well as by the role they play in the organization of the latter'. A detailed study of changes in Flemish meals 1988–1999 revealed remarkable stability in when people eat, where and with whom (Mestdag, 2005). Findings from the Nordic study already mentioned add to this picture (Gronow and Jääskeläinen, 2001; Holm, 2001b). However, we should be careful about making statements that are too bold. Searching for social regularities in what people do in a complex field like food may lead to an overemphasis on the stability and coherence of daily schedules. Ideas of 'normal' meal patterns are so strong and taken for granted that emerging disruptions or new patterns are not easily recognised and conceptualised by either researchers or respondents.

None the less, we can observe relatively clear and consistent meal patterns across Europe, including an early morning meal, a midday meal, perhaps an afternoon snack, and then an evening meal. One or two of these meals are usually cooked. These patterns show that eating practices are socially coordinated and regulated. However, even if it is not difficult to identify the regularities of eating in terms of national, regional and socially stratified meal patterns, such practices are flexible. At all hours and in all situations, the Nordic study found events that are not in accordance with the general code of practice (Gronow and Jääskeläinen, 2001). On the other hand, it was difficult to identify social groups with particularly 'de-structured' food habits in term of contents of meals, rhythm and/or social context in this study (Kjærnes 2001). Groups or individuals that display particularly 'gastro-anomic' ways of eating were few and far between. Given the institutionalised character of everyday life, this is not surprising. Life phase and household structure matter, and eating patterns depend on the structure of the private and public food provisioning system and the labour market. These structures vary considerably between countries, and form an important reason why eating patterns differ cross-culturally.

But even though eating practices appear to be relatively stable, and there seem to be few who provide empirical indications of a development towards an 'anomic' situation, this does not mean that changes are not taking place. Everyday life has changed considerably over the past two or three decades. In particular, the family system and family life have been transformed (Beck-Gernsheim, 1998). In countries like Britain and the Nordic countries, most married women are now working outside the home either part-time or full-time. The number of divorces has multiplied, and more people live alone. This has not happened to the same extent in a country like Italy, where the 'single' category is much smaller and includes mostly elderly widows and widowers, and where the gender division of labour is quite different from Nordic contexts (Barbagli and Saraceno, 1997; Jensen, 2000). It may seem a trivial observation that people who live alone more often eat alone, and people who are employed eat their lunch with colleagues. But it is important to recognise in order to understand how structural change may influence eating practices. Growing numbers of people living alone are clearly influencing demand (smaller portions and packages, for example), but it is not clear whether the increase in single person households will have other consequences on the practices of eating (Poulain, 2002, p. 104).

One important element in driving these changes seems to be the division of labour between the household and other institutions. As discussed previously, this is influenced by interactions between changes in family life and changes in the food market. Changes can be observed at two levels, partly in terms of the amount of time and labour put into provisioning and preparation within the household, partly in terms of meals served outside the home or in the household. The empirical studies already referred to indicate that the midday meal – lunch – is the meal that is most clearly undergoing change. Lunch is today the main meal eaten away from home (Gronow and Jääskeläinen, 2001; Poulain, 2002). This is first of all due to more people spending their time at work and in schools, a major factor being the proportion of female employment. This will mean not only that more women will, as a consequence, tend to eat away from home, their absence from home will also mean that other members of the family may not go home to be served a cooked meal during the day. Still, the strongly habitual and normative basis of eating practices are demonstrated when employed women spend their lunch breaks by hurrying home to cook for their school children (and/or husband), and then return to work.

In spite of all this, eating all over Europe appears dominantly to be taking place in people's homes (Holm, 2001b; Warde and Martens, 2000; Poulain, 2002; Mestdag, 2005). But even if most eating events take place within the context of the home, the division of labour between provisioning systems and households, between public and private spheres, varies across countries. While the framing of the practices in terms of timing and degree of social coordination has not changed as much as is commonly assumed, the character of food preparation and what is served have been considerably altered. In many countries, much of the domestic production of food has been displaced by market provision, as elements of food preparation have been transferred from the household kitchen to the food-processing industry. Instead of buying a relatively fixed set of raw ingredients, perhaps to be subject to elaborate cooking, we increasingly buy a wide range of pre-packaged, partially processed, complex foods, as well as ready-made meals. What we see is first of all a process of commodification, where the household expenditure of time on cooking has been replaced by spending money, a process paralleled by and made possible through women increasingly being wage earners. On the one hand, our timing of eating is strongly influenced by the organisation of the working day, while, on the other hand, we have more income to spend on purchasing processed foods or meals. Appadurai calls this a 'commodification of time' (Appadurai, 1997, p. 38). These changes appear not only to have altered and redirected food preparation processes, and the appearance of dishes on the dinner table. They have also influenced the interrelations between consumers and other actors in the food chain in terms of control and power, responsibility and competence.

Food and eating are embedded in close network relations, and these relations serve to modify and obstruct processes like commodification (Fine *et al.*, 1996). We also find striking variations in the degrees of displacement of household food production with industrial processing across Europe (Kjærnes *et al.*, 2005). At the extremes of this range, the Portuguese still rely heavily on unpackaged, raw ingredients bought from butchers and fruit and vegetable shops, to be prepared within the household, while British practices are characterised by a significant proportion of very varied ready-made dishes bought from large, integrated supermarket chains, taken home and heated in microwave ovens.

In order to further explore the make-up of eating practices, this section has discussed ongoing trends in the organisation of everyday eating. Changes do take place, but empirical research suggests that worry about the 'individualisation' of eating is overrated and, we contend, changes do not make eating less socially embedded.

Food habits and eating practices are, of course, not merely a matter of organisation. We have, earlier in this chapter, described food habits and eating practices as being influenced by social distinction, by competence and engagement, and by norms and expectations. Through the concept of 'a proper meal', the next section will take a closer look at the strong normative foundations that regulate food habits.

23.6   Norms and expectations: the notion of a proper meal

So far we have focused on practical aspects of eating practices in terms of food preparation and coordination of eating. But practices are, as discussed earlier, not only about how things are usually done, but also about how they should be done. Norms and expectations are defined by their social character, emerging over time and through mutual recognition. Normative regulation can take place through application of strict norms, accompanied by sanctions, but also through more flexible conventions and expectations that frame what is generally considered right and wrong, good and bad, edible and non-edible, appropriate and non-appropriate. Through such processes of classification, items are transformed from being a plant, a dead animal, or a manufactured good into becoming a 'food item' with a particular role and position in our diet. Such norms and conventions are not only related to cuisines in a narrow sense, i.e. defining proper dishes and meal elements for specific social contexts (O'Doherty Jensen, 2003), preparation techniques etc. They include, as indicated by Murcott in the British research programme: 'The Nation's Diet', an array of 'cultural conventions governing the micro-politics of social relationships in the food-sharing group' (Murcott, 1998, p. 150). Food, and what we do to and with it, is proclaimed to lie at the very core of sociality: it signifies 'togetherness' (Murcott *et al.*, 1992, p. 115). Special events and meals for particular groups as well as everyday, routine meals of home, school or workplace, reveal the symbolic significance of eating in everyday life. In spite of social change and conflicts at different levels, important elements of meal conventions seem to be altered only very slowly (Caplan *et al.*, 1998). One example is the division of labour between the genders, a division that has proved to be deeply embedded in the conception of eating as such, and – therefore – seems to be very persistent despite rather comprehensive changes in gender roles in society in general (O'Doherty Jensen and Holm, 1999).

Several researchers suggest that, even though alterations in food preferences and cooking may take place, neither normative regulations nor the role of commensality will necessarily disappear (Murcott, 1998). Our meal patterns are continuously being transformed, but these changes may, as we have seen, not be as dramatic and unambiguous as common notions suggest. The normative regulation of eating seems to be one very important stabilising factor. For example, in Norway, Bugge and Døving contend that the meal as a social, moral and family institution is very strong, perhaps even stronger than before (Bugge and Døving, 2000). However, a widespread protective attitude towards the family meal institution may indicate that while important, it is also challenged and in need of special arguments and actions of support (Swidler, 1986). In order to understand the debate on eating practices and the fate of meals, we will therefore have to take a closer look at the normative framing of meals. We will do this by focusing on 'proper meals', a notion encompassing a number of aspects that define meals as practices.

'Proper meals' have been an important concept in popular discussions about the status of meals. The notion appears to be an integral part of the everyday conceptualisation and classification of eating; representing a more or less explicit expression of what is acceptable as a meal. In Norwegian, expressions like '*ordentlige måltider*' and '*skikkelig mat*' – meaning 'square meals' and 'proper food' – frequently appear in qualitative interviews (Bugge, 1995; Guzman *et al.*, 2000), in Denmark '*rigtige måltider*' (Iversen and Holm, 1999), and in Finland – '*kunnon ateria*' (Mäkelä, 2000). These everyday notions were introduced as an analytical concept by Murcott, based on the British understanding of 'proper food' (Murcott, 1982, 1983). The British proper meal consisted (at the time of Murcott's study, i.e. the late 1960s) of roasted meat, potatoes, (boiled) vegetables, and gravy. A proper meal was to be served by the wife for her husband, symbolising the woman as a caregiver and the husband as a breadwinner. Nordic studies indicate that similar ideas are recognised within these countries in even more recent times (Jansson, 1988; Holm, 1996a; Ekström, 1991; Mäkelä, 1995; Bugge and Døving, 2000).

The British descriptions of a proper meal concentrate on hot, cooked meals. Similarly, in Swedish and Finnish discourses, only hot meals conforming to a certain format are seen as proper meals, and eating a sandwich is regarded as a snack and a poor substitute. This is expressed in Swedish fears of 'breakfastisa-tion' ('*frukostisering*') where increased eating of cold meals is seen as signifying a process of cultural dissolution. But conventions and routines may be very strong even in cases where the food eaten is uncooked, as in the case of the Danish and Norwegian '*mat/d-pakke*', i.e., open-faced sandwiches brought from home for lunch (Kjærnes, 2001). These meals are certainly also 'proper' in the popular notion of the concept. The preparations for serving and eating of cooked, hot meals are usually more elaborate, thus allowing for more multi-dimensional expectations to the practices related to preparing and eating the meal and to a more detailed discourse on norms and cultural dissolution.

Whether limiting the concept of 'the proper meal' to cooked meals, or whether to include cold meals requiring less immediate preparation in the concept, the 'proper meal' designates a certain division of labour, based on clear roles and competencies, and a clear classification of which types of food to be included in which types of meals (and other eating events). A number of distinct dimensions can be singled out (Kjærnes, 2001).

As indicated by Murcott, there are rules regarding which foods and dishes should be included in a 'proper meal'. We can call this the *food dimension*. The 'proper meal' indicates, first of all, a certain format, a structure. It is this format that suggests which food items and dishes are to be included (Douglas and Nicod, 1974; Mäkelä, 1991). The North European meal format described in many studies includes a 'centre' with meat or fish, a 'staple' (often potatoes), one or two vegetables on the side, and a sauce or condiment. This meal format includes (at least on certain occasions) a starter course and a dessert. In Southern European versions, the 'plateful' can be divided into a series of dishes, thus making the whole meal structure quite different. In earlier times, and in more distant places, we have seen yet other dominant meal formats (Mäkelä, 2000). Within a certain geographical and historical setting, however, the proper meal format seems to be a stable way of classifying what should be included in meals for significant situations even though the proper meal format may also be subject to social distinction (O'Doherty Jensen, 2003).

A 'proper meal' is supposed to be eaten together with the family, and serves as a kind of material confirmation of the existence of a (proper) family – thus there is a *family dimension*. It has been suggested that the couple is the essential element (Charles and Kerr, 1988; Murcott, 1983), while others emphasise the presence of children (Wandel *et al.*, 1995).

A 'proper meal' is defined by how the food has been prepared and by whom. A *cooking dimension* points to the role of preparing the food 'properly' – i.e., generally at home from raw and fresh ingredients – the preparation activities frequently expected to be carried out by the mother of the family (Ekström, 1990; Murcott, 1983). The division of labour – between the genders and between the household and other institutions – is not only a practical matter, but is associated with strong norms and expectations.

Then there is a *health dimension*. The food items included and the preparation of a 'proper meal' are also supposed to imply that proper meals are healthier than other meals (Charles and Kerr, 1988). Healthiness as a special dimension may then not easily be disentangled from other dimensions. When there is no family, there may be no 'proper meals' and hence no healthy meals either

(Guzman *et al.*, 2000). Turned the other way around, health education may face difficulties when arguing that dishes included in 'proper meals' can be unhealthy. In public discourse, it is often snacking (especially when eaten away from home) that is seen as unhealthy.

Notably, there does not seem to be a significant and explicit *taste dimension* in the North European accounts of proper meals. If we had studied French or Italian expectations of meals, that dimension might have been more significant. However, as taste is developed within a social context (Gronow, 1997), enjoyable taste more or less follow from the norms for proper meals and taste does appear indirectly, e.g. in reports about meal practices in families which often include accounts of difficulties with bringing together individual taste preferences of family members. There may also be important distinctions in taste expectations towards everyday meals and meals served on special occasions.

The expectations that emerge from what constitutes a proper meal clearly illustrate the social character of eating as sets of practices existing outside the individual. From qualitative studies we know that 'proper meals' are important as a convention and an ideal. But as representing a practice, we must expect that people may be more or less committed, more or less competent, with more or less resources and possibilities to take part in the practice. The various dimensions bring up a number of conditions and influences, which suggest that in everyday life there will be tensions and tradeoffs. While the concept of 'a proper meal' closely combines the various dimensions of food, cooking, and context, this is not as evident in everyday habits. Eating pizza while sitting on the sofa watching TV does not live up to most concepts regarding the structure or preparation of 'proper meals', but it may take place together with family members and at dinnertime, thus fulfilling some idea of sociality within the family. De-structuring of meals is often associated with people eating alone. But people living alone have meals with a 'proper' format almost as often as those living in a family situation (Holm, 2001a). 'Proper meals' clearly exist in the Nordic food cultures, but in its fully-fledged form not as a predominant everyday practice (Mäkelä, 2001). Still, the norms are strong and people try to adhere to at least part of them. While everyday meal formats are often simple, families do eat together at home and mothers do prepare the meals.

Practices vary in different living conditions. Variations in eating patterns are found between social groups, representing different living situations and life phases, as indicated by age and household type. As within families, single people often eat meals at home. In the Nordic countries, singles' eating patterns are generally quite similar to those of bigger households (Holm, 2001b). But this is also a matter of life phase. A Norwegian qualitative study among young people who had moved out from their parents' house, but who had not yet started a family, showed that this 'un-established' situation made them feel free from the need to adhere to the norms of having proper meals. Cooking and eating was not particularly experimental or varied, but their habits departed considerably from conventions regarding a proper meal format. However, these flexible habits were linked explicitly to this specific phase in life; living alone, with irregular hours, and with low income. As soon as the young people became part of couples, more traditional norms about meals, often passed on from their own upbringing, became relevant (Guzman *et al.*, 2000).

In this section we have discussed dominant norms about eating. But within a particular society, people may not understand norms in the same manner. Norms about eating may be shared in general terms as a 'good cause'. People may have learnt that you should eat breakfast and dinner, and that vegetables should accompany a cooked meal. However, such normative rules may not be transformed into practical expectations in the same ways within all social groups. Disparities between peoples' statements about eating, and observations of what they do, are perhaps not surprising. Importantly, however, such disparities seem to vary between social groups (Poulain, 2002, 98). One can expect class differentiations not only in the degree to which such normative expectations are put into practice, but as much in how proper meals are defined in concrete terms. Studies have, for example, indicated that middle-class people (especially those with higher education) put stronger emphasis on health considerations, of fat and sugar reduction and 'lighter' components, such as vegetables and fish. Working-class people more often seem to concentrate on substantial, calorie rich components and meat (Holm, 1996b; Roos and Prättälä, 1999). We must be careful of stereotyping regarding social differentiation. But we must also be careful of extrapolating from dominant social norms into the priorities and expectations that people may have in different life situations, places and periods of time.

23.7   Conclusion

In this chapter, we have argued against individualistic approaches to how social factors influence food choice. Instead, a social understanding is needed in order to capture how major social processes influence our eating. 'Practice' is suggested as a concept that builds on a social understanding of consumption, encompassing institutional, organisational, and normative elements. It follows that consumption is seen not only as a matter of choice in purchase situations, but rather as a series of activities involving both the acquisition, appropriation, and use of goods. It is also implied that individual choice is seen to be shaped by institutionalised structures, conventions and normative rules. Individual choice is not ruled out, since the ways in which individuals engage in practices are not uniform. But individual choice is not the focus. Rather it is seen as embedded in the events and social contexts in which practices are organised. This means that when searching for how eating patterns change, we must direct attention towards the social processes that influence these practices, rather than individual decision-making. This understanding of food consumption has implications for the study of food habits as well as for policy in the food area.

In Section 23.1, the potential policy implications of a social approach to food choice and eating were mentioned. Attempts to promote healthy eating often concentrate on information and education. There seem to be three important considerations following from our line of argument.

First, conceptualising eating practices as driven by social factors means that the logic of how changes take place is not derived from individual value hierarchies and preferences. Focusing on eating as practice, as routine, and as situated events means that we must bring in the structures that frame these events, as well as the strong, socially developed norms and conventions that form people's expectations towards them. Understanding peoples' expectations of what constitutes a proper meal is important when developing policies encouraging healthy eating. Healthy eating is interpreted and handled within this context, usually not as explicit reflections, but as part of the 'muddling through' that characterises mundane practices such as everyday eating (Holm and Kildevang, 1996).

Second, influences on eating cannot be sought only internally within practices. We have, in this chapter, presented several examples of how eating is influenced by the institutional context, of changes in family structures and the labour market, as well as in public and private food provisioning systems. Tensions and mismatches may occur between such changes, on the one hand, and the habits and norms of eating, on the other. We think that policy-making should pay particular attention to such tensions and mismatches.

Third, following from our understanding of food consumption, the presently contentious issues of risk and trust should also be treated as social phenomena, not primarily as a matter of individuals' perceptions and the trustworthiness of information sources. Trust and distrust does not seem to be determined mainly by individual strategies and information input. Varying levels of trust appear in different national contexts, which are characterised by different types of interrelations between consumers and consumption practices, on the one hand, and the organisation and performance of food provisioning systems and regulatory institutions, on the other (Kjærnes et al., 2006). Tensions, mismatches and failing expectations between these institutions and arenas seem to be a major source of distrust. That means, among other things, that trust in public authorities seems to depend on their overall organisation and performance, in which communication constitutes only one element (Halkier and Holm, 2003).

As a final point, the emphasis on the social embeddedness of eating, as opposed to individual choice, does not mean that people are seen as not having agency. Practices are differentiated and flexible regarding how people carry them out, and people may also act in direct opposition to the 'normality' of the practices. Importantly, also, institutional contexts constitute strong and powerful frames for practices, but they are not static or impossible to influence. It is, for example, of importance whether or not people as consumers respond to changes in the food supply by boycotting products and demanding organic food, or whether or not they as citizens resist GM food through collective action. The choices people make are formed by social structures, but these are in turn changed through the choices people make.

23.8   References

AJZEN, I. (1991). The theory of planned behaviour, *Organisational Behaviour and Human Decision Processes*, **50**(2), 179–211.

AJZEN, I. and FISHBEIN, M. (1980). *Understanding Attitudes and Predicting Social Behaviour*. Englewood Cliffs, NJ: Prentice-Hall International.

ANDERSSON, S. (1983). *Matens roller. Sosiologisk gastronomi [Role of Food. Sociological Gastronomy]*. Oslo: Universitetsforlaget.

APPADURAI, A. (1988). How to make a national cuisine: cookbooks in contemporary India, *Comparative Studies in Society and History*, **30**, 3–24.

APPADURAI, A. (1997). Consumption, duration, and history. In: D. Palumbo-Liu and H. U. Gumbrecht (eds.), *Streams of Cultural Capital. Transnational Cultural Studies*, pp. 23–45. Stanford: Stanford University Press.

AYMARD, M., GRIGNON, C. and SABBAN, F. (1996). Food allocation of time and social rhythms. Introduction. *Food and Foodways*, **63**(3–4), 161–185.

BARBAGLI, M. and SARACENO, C. (1997). *Lo stato delle famiglie in Italia*. Bologna: Il Mulino.

BAUMAN, Z. (2001). Consuming life, *Journal of Consumer Culture*, **1**(1), 9–29.

BECK, U. and BECK-GERNSHEIM, E. (2002). *Individualization. Institutionalized Individualism and its Social and Political Consequences*. London: Sage.

BECK-GERNSHEIM, E. (1998). On the way to a post-familial family. From a community of need to elective affinities, *Theory, Culture and Society*, **15**(3–4), 53–70.

BUGGE, A. (1995). *Mat til begjær og besvær. Forbrukernes vurderinger og kunnskaper om helse, miljø og etiske aspekter ved mat [Food as Desire and Bother. Consumer Evaluations and Knowledge on Health, Environmental and Ethical Food Issues]*. Report No. 6. Lysaker: The National Institute for Consumer Research.

BUGGE, A. and DØVING, R. (2000). *Det norske måltidsmønsteret – ideal og praksis [The Norwegian Meal Pattern – Ideal and Practice]*. Report No. 2. Lysaker: The National Institute for Consumer Research.

CAMPBELL, C. (1996). Detraditionalization, character and the limits of agency. In: P. Heelas, S. Lash and P. Morris (eds), *Detraditionalization. Critical Reflections on Authority and Identity*, pp. 149–169. Cambridge, Mass.: Blackwell.

CAPLAN, P., KEANE, A., WILLETTS, A. and WILLIAMS, J. (1998). Studying food choice in its social and cultural contexts: approaches from a social anthropological perspective. In: A. Murcott (ed.), *The Nation's Diet. The Social Science of Food Choice*, pp. 168–182. London, New York: Longman.

CHARLES, N. and KERR, M. (1988). *Women, Food and Families*. Manchester: Manchester University Press.

COMMISSION OF THE EUROPEAN COMMUNITIES (ed.) (2000). *White Paper on Food Safety. COM (1999) 719 final*. Brussels.

DEVAULT, M. L. (1991). *Feeding the Family. The Social Organization of Caring as Gendered Work*. Chicago, London: The University of Chicago Press.

DOUGLAS, M. and NICOD, M. (1974). Taking the biscuit. The structure of British meals, *New Society*, **30**, 744–747.

EKSTRÖM, M. (1990). *Kost, klass och kön [Food, Class and Gender]*. Umeå: University of Umeå, Department of Sociology.

EKSTRÖM, M. (1991). Class and gender in the kitchen. In: E. L. Fürst, R. Prättälä, M. Ekström, L. Holm and U. Kjærnes (eds), *Palatable Worlds. Sociocultural Food Studies*, pp. 145–160. Oslo: Solum forlag.

FINE, B. and LEOPOLD, E. (1993). *The World of Consumption*. London, New York: Routledge.

FINE, B., HEASMAN, M. and WRIGHT, J. (1996). *Consumption in the Age of Affluence. The World of Food.* London: Routledge.

FISCHLER, C. (1988). Food, self and identity, *Social Science Information*, **27**(2), 275–292.

FRIEDMAN, M. and FRIEDMAN, R. (1981). *Free to Choose. A Personal Statement.* New York: Avon.

GIDDENS, A. (1994). Replies and critiques. Risk, trust, reflexivity. In: U. Beck, A. Giddens and S. Lash (eds), *Reflexive Modernization. Politics, Tradition and Aestethics in the Modern Social Order*, pp. 184–197. Cambridge: Polity Press.

GLANZ, K., LEWIS, F. M. and RIMER, B. K. (1997). *Health Behavior and Health Education. Theory, Research and Practice.* San Francisco: Josey-Bass Inc.

GRONOW, J. (1997). *The Sociology of Taste.* London, New York: Routledge.

GRONOW, J. (2004). Standards of taste and varieties of goodness: the (un)predictability of modern consumption. In: M. Harvey, A. McMeekin and A. Warde (eds), *Qualities of Food*, pp. 38–60. Manchester: Manchester University Press.

GRONOW, J. and JÄÄSKELÄINEN, A. (2001). The daily rhythm of eating. In: E. Kjærnes (ed.), *Eating Patterns. A Day in the Lives of Nordic Peoples.* Report No.7, pp. 91–124. Oslo: The National Institute for Consumer Research.

GRONOW, J. and WARDE, A. (2001). *Ordinary Consumption.* London and New York: Routledge.

GUZMAN, M., BJØRKUM, E. and KJÆRNES, U. (2000). *Ungdommers måltider. En studie av livssituasjon, mat og kjønn [Meals among Young People].* SIFO Report No 6. Lysaker: The National Institute for Consumer Research.

HALKIER, B. and HOLM, L. (2003). Tillid til mad – forbrug mellem dagligdag og politisering [Trust in food – the politics of food consumption], *Dansk sociologi*, **15**(3), 9–28.

HARVEY, M., McMEEKIN, A., RANDLES, S., SOUTHERTON, D., TETHER, B. and WARDE, A. (2001). *Between Demand and Consumption: A Framework for Research. CRIC Discussion Paper No 40.* Manchester: Centre for Research on Innovation and Competition, University of Manchester and UMIST.

HOGAN, D. G. (1997). *Selling 'em by the Sack. White Castle and the Creation of American Food.* New York: New York University Press.

HOLM, L. (1996a). Identity and dietary change, *Scandinavian Journal of Nutrition*, **40**, 595–598.

HOLM, L. (1996b). Food and identity among families in Copenhagen – A review of an interview study. In: H. J. Teuteberg, G. Neumann and A. Wierlacher (eds), *Essen und kulturelle Identität – Europäische Perspektiven*, pp. 356–371. Bonn: Akademie Verlag.

HOLM, L. (2001a). Family meals. In: U. Kjærnes (ed.), *Eating Patterns. A Day in the Lives of Nordic Peoples.* Report No. 7, pp. 199–212. Oslo: The National Institute for Consumer Research.

HOLM, L. (2001b). The social context of eating. In: U. Kjærnes (ed.), *Eating Patterns. A Day in the Lives of Nordic Peoples.* Report No.7, pp. 159–198. Oslo: The National Institute for Consumer Research.

HOLM, L. and KILDEVANG, H. (1996). Consumer's views on food quality. A qualitative interview study, *Appetite*, **27**, 1–14.

IVERSEN, T. and HOLM, L. (1999). Måltider som familieskabelse og frisættelse [Meals for the Making of Family and Individualisation], *Tidsskriftet Antropologi*, **39**, 53–64.

JANSSON, S. (1988). Maten och myterna, *Vår Föda*, **40**(suppl.2), 1–203.

JENSEN, P. H. (2000). Kontekstuelle og tværnationale komparative analyser [Contextual and Cross-National Comparative Analyses], *Dansk sociologi*, **11**(3), 29–48.

KJÆRNES, U. (ed) (2001). *Eating Patterns. A Day in the Lives of Nordic Peoples.* SIFO Report No.7. Oslo: The National Institute for Consumer Research.

KJÆRNES, U., POPPE, C. and LAVIK, R. (2005). *Trust, Distrust and Food Consumption. A Study in Six European Countries.* Project Report No 15. Oslo: The National Institute for Consumer Research.

KJÆRNES, U., HARVEY, M. and WARDE, A. (2006) *Trust in Food in Europe.* Houndmills: Palgrave Macmillan.

LÄHTEENMÄKI, L. and VAN TRIJP, H. C. (1995). Hedonic responses, variety-seeking tendency and expressed variety in sandwich choices, *Appetite*, **24**, 139–152.

MÄKELÄ, J. (1991). Defining a meal. In: E. L. Fürst, R. Prättälä, M. Ekström, L. Holm and U. Kjærnes (eds), *Palatable Worlds. Sociocultural Food Studies*, pp. 87–96. Oslo: Solum forlag.

MÄKELÄ, J. (1995). The structure of Finnish meals. In: E. Feichtinger and B. M. Köhler (eds), *Current research into eating practices. contributions of social sciences. Proceedings of the European Interdisciplinary Meeting, 1–16 Oct 1993, Potsdam,* pp. 112–114. Frankfurt am Main: Umschau Zeitschriftenverlag.

MÄKELÄ, J. (2000). Cultural definitions of the meal. In: H. L. Meiselman (ed.), *Dimensions of the Meal. The Science, Culture, Business, and Art of Eating*, pp. 7–18. Gaithersburg: Aspen Publication.

MÄKELÄ, J. (2001). The meal format. In: U. Kjærnes (ed.), *Eating Patterns. A Day in the Lives of Nordic Peoples.* Report no. 7, pp. 125–158. Oslo: The National Insitute for Consumer Research.

MÄKELÄ, J., KJÆRNES, U., EKSTRÖM, M. P., FÜRST, E. L., GRONOW, J. and HOLM, L. (1999). Nordic meals. Methodological notes on a comparative survey, *Appetite*, **32**, 73–79.

MESTDAG, I. (2005). Disappearance of the traditional meal: Temporal, social and spatial destructuration, *Appetite*, **45**, 62–74.

MINTZ, S. W. (1996). *Tasting Food, Tasting Freedom. Excursions into Eating, Culture, and the Past.* Boston: Beacon Press.

MURCOTT, A. (1982). On the social significance of the 'cooked dinner' in South Wales, *Social Science Information*, **21**, 677–696.

MURCOTT, A. (1983). 'It's a pleasure to cook for him': Food, mealtimes and gender in some South Wales households. In: E. Gamarnikow (ed.), *The Public and the Private*, pp. 78–90. Heinemann Educational Books.

MURCOTT, A. (1998). *The Nation's Diet. The Social Science of Food Choice.* London/New York: Longman.

MURCOTT, A., MENNELL, S. and VAN OTTERLOO, A. (1992). *The Sociology of Food. Eating, Diet and Culture.* Sage.

NEUMARK-SZTAINER, D. (1999). The social environment of adolescents: Associations between socioenvironmental factors and health behaviors during adolescence, *Adolesc Med State Art Rev*, **10**, 41–55.

O'DOHERTY JENSEN, K. (2003). Hvad er 'rigtig mad'? [What is proper food?]. In: L. Holm (ed.), *Mad, Mennesker og Måltider [Food, People, and Meals]*, pp. 65–80. Copenhagen: Munksgaard.

O'DOHERTY JENSEN, K. and HOLM, L. (1999). Preferences, quantities and concerns: Sociocultural perspectives on the gendered consumption of foods, *European Journal of Clinical Nutrition*, **53**, 351–359.

OTNES, P. (1991). What do meals do. In: E. L. Fürst, M. Ekström, L. Holm, U. Kjærnes and R. Prättälä (eds.), *Palatable Worlds. Sociocultural Food Studies*, pp. 97–110. Oslo: Solum Forlag.

POULAIN, J. P. (2002). *Manger Aujourd'hui. Attitudes, Normes et Pratiques.* Toulouse: Éditions Privat.

PRÄTTÄLÄ, R. and HELMINEN, P. (1990). Finnish meal patterns. In: J. C. Somogyi and E. H. Koskinen (eds), *Nutritional adaptation to new life-styles. Bibl. Nutritio et Dieata* **45**, pp. 80–92. Basel: Karger.

REISCH, L. A. (2004). Principles and visions of a new consumer policy, *Journal of Consumer Policy*, **27**, 1–27.

RICHTER, K. P., HARRIS, K. J., PAINE-ANDREWS, A., FAWCETT, A., SCHMID, T. L., LANKENAU, B. H. and JOHNSTON, J. (2000). Measuring the health environment for physical activity and nutrition among the youth: A review of the literature and applications for community initiatives, *Preventive Medicine*, **31**, S98–111.

ROOS, G. and PRÄTTÄLÄ, R. (1999). *Disparities in Food Habits. Review of Research in 15 European Countries.* Report B24. Helsinki: The National Public Health Institute.

ROTENBERG, R. (1981). The impact of industrialization on meal patterns in Vienna, Austria, *Ecology of Food and Nutrition*, **11**(1), 25–35.

SCHATZKI, T. (1996). *Social Practices: A Wittgensteinian Approach to Human Activity and the Social.* Cambridge: Cambridge University Press.

SHEPHERD, R. (1990). Overview of factors influencing food choice. In: M. Ashwell (ed.), *Why we eat what we eat. Suppl.* London: Nutrition Bulletin/British Nutrition Foundation.

SHEPHERD, R. and DENNISON, C. M. (1996). Influences on adolescent food choice, *Proc Nutr Soc*, **55**, 345–357.

SHEPHERD, R. and STOCKLEY, L. (1985). Fat consumption and attitudes towards food with a high fat content, *Hum Nutr: Appl Nutr*, **39A**, 431–442.

SIMMEL, G. (1957). Soziologie der Mahlzeit. In: M. Landmann (ed.), *Brücke und Tür. Essays des Philosophen zur Geschichte, Religion, Kunst und Gesellschaft*, pp. 243–250. Stuttgart: K.F. Koehlers Verlag.

SÖDERLIND, S. D. (2001). *Consumer Economics. A Practical Overview.* M.E. Sharpe, Armonk, New York.

SULKUNEN, P. (1997). Introduction: the new consumer society – Rethinking the social bond. In: P. Sulkunen, J. Holmwood, H. Radner and G. Schulze (eds.), *Constructing the New Consumer Society*, pp. 1–20. Houndmills, Basingstoke: Macmillan Press.

SWIDLER, A. (1986). Culture in action: symbols and strategies, *American Sociological Review*, **51**(2), 273–286.

TUORILA, H. and PANGBORN, R. M. (1988). Behavioural models in the prediction of consumption of selected sweet, salty and fatty foods. In: D. M. H. Thomson (ed.), *Food Acceptability*, pp. 267–279. London, New York: Elsevier Science Publishers Ltd.

WANDEL, M., BUGGE, A. and RAMM, J. S. (1995). *Matvaner i endring og stabilitet [Change and Stability of Food Habits].* Report no. 4. Lysaker: The National Institute for Consumer Research.

WARDE, A. (1997). *Consumption, Food and Taste.* London: Sage.

WARDE, A. (2005). Consumption and theories of practice, *Journal of Consumer Culture*, **5**(2), 131–153.

WARDE, A. AND MARTENS, L. (2000). *Eating Out. Social Differentiation, Consumption and Pleasure.* Cambridge: Cambridge University Press.

ZIEHE, T. (1989). *Kulturanalyser. Ungdom, utbildning, modernitet. [Cultural Analysis. Youth, Education, Modernity].* Stockholm/Stehag: Symposion Bokförlag.

24

Developing a coherent European food safety policy: the challenge of value-based conflicts to EU food safety governance

M. Dreyer and O. Renn, DIALOGIK gGmbH, Germany

24.1 Introduction

This chapter reviews recent regulatory and institutional changes in EU food safety governance which have been designed to rebuild public confidence in the safety of food and restore regulatory credibility and authority of the EU food safety managing institutions. The authors argue that, although in the right direction, the reforms are insufficient by themselves to address the concerns underpinning especially challenging food risk controversies. The underlying assumption is that controversies over technological innovations in the production of food taking place in the public domain are likely to be driven foremost by different worldviews and corresponding divergent perspectives on nature and the meaning and importance of 'naturalness'. It is these cognitive 'deep structures' rather than concerns about the adequate balancing and equal distribution of substantive risks and benefits of technological innovations which fuel public controversies. They essentially shape conflicts between regulators, economic actors, and civil society actors and the attitudes of the wider public towards new and emerging food production technologies. This is particularly the case if the associated risks are perceived as persistent, uncertain and undetectable (as with BSE and genetically modified food). The EU-level reforms can be interpreted as responding (indirectly) to the challenging features of conflicts over food production and food safety driven by deeply held value-orientations, especially through the endorsement of the precautionary principle, and by making provision for comprehensive labelling schemes. However, they do not provide arrangements which enable the risk analysis and regulatory process to address the important value dimension of food safety conflicts in a proactive, systematic, and direct manner. The authors claim, however, that institutional responsiveness which places special attention on value-based conflicts at the level of public policy is a necessary prerequisite to facilitate the handling of food risks in a more effective and socially legitimate manner.

This chapter is organised in five sections. The section following the introduction explains our conceptual approach using a three-level analysis of risk debates. The second section draws on this analytical perspective and introduces a case study on the British conflict over genetically modified food (GM food) to illustrate the relevance of divergent notions of nature as drivers of food debates. The third section proposes to interpret the reforms at EU-level as a response, however indirect, to value conflicts of this kind. The fourth section will make some suggestions for responding directly to value conflicts at the level of public policy. The last section will draw main conclusions and comment on likely future trends in food safety governance.

24.2   The three levels of risk debates

24.2.1   Understanding the structure of risk debates

In order to analyse the structure and dynamics of risk controversies, it is useful to distinguish different styles of argumentation which characterise different levels of risk debates (Renn 2004: 299; Renn and Klinke 2001: 247; Hampel and Renn 2000; van den Daele 1990). Although topics vary from risk source to risk source, most risk controversies reflect three major issues:

- factual evidence, probabilities, and uncertainties
- institutional performance, expertise, and experience
- conflicts about worldviews and value systems.

These three levels are illustrated in Fig. 24.1. The higher the level, the higher the probability that the conflict will gain in intensity. The degree of complexity may be high in levels 1 and 3 but less so on level 2.

The *first level* involves factual arguments about risk probabilities and the extent of potential damage. One of the main problems with respect to the first level of risk debates is the issue of framing. Depending on the issue or the wording of the problem (for example: stating probabilities in terms of losses or gains), people will change their preference order for decision options with identical outcomes (Tversky and Kahneman 1981; Fischhoff 1985; Levin *et al.* 1998; see also Chapter 5 by de Jonge *et al.*, this volume). The effects of framing occur firstly after the initial conceptualisation of the issue (for example as a food safety issue or a nutritional problem or a lifestyle aspect) and later when the factual information is compared with the values of the respondents. (Do the 'facts' relate to the concerns of respondents?) To avoid confusion about the effects of framing, it is essential that the assumptions that underlie specific

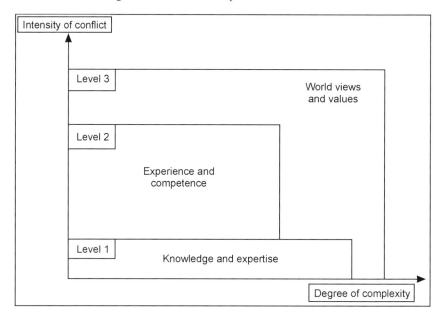

Fig. 24.1   The three levels of concern in risk debate.

frames about the problem need to be communicated along with the factual information.

The *second, more intense level* concerns institutional competence to deal with risks. At this level the focus of the debate is on the distribution of risks and benefits, and the trustworthiness of the risk management institutions. This type of debate does not rely on technical expertise, although reducing scientific uncertainty may help. Second level debates focus on the question of whether risk managers have met their official mandate and whether their performances match public expectations. In a complex and multifaceted society, evidence for trustworthiness is difficult to provide. Gaining trustworthiness requires a continuous dialogue between risk managers, stakeholders, and representatives of the public.

At the *third level* the conflict is defined along different social values, cultural lifestyles, visions of the future and their impact on risk management. In this case, neither technical expertise nor institutional competence and openness are adequate conditions for meeting public concerns. Decision making and policy design rely on a wider societal discourse about the frames and philosophies underlying the present risk management approaches.

24.2.2   Reframing third level debates to lower level issues

As long as fundamental value issues remain unresolved, even the best technical expertise and the most profound competence will not suffice to produce mutual agreements. Furthermore, knowledge, values, and worldviews are not independent from each other. Many groups have constructed a coherent body of beliefs that integrate cognitive, evaluative and normative claims about the world. These belief systems can form epistemic communities, which offer a complete, often holistic view of the world and define the legitimate realm of rules for evaluating claims of evidence (Jasanoff 2004). Once such a belief system is established, it is almost immune to any type of counterclaims. The only path to agreement will be through the creation of mutual gains for all parties (win-win-situation) or the generation of overarching values that are evoked or generated through dialogue-based sessions (Renn 2004). Both resolution strategies require that the value issues are taken as the starting point of discourse and not the level of factual knowledge. This strategy does not guarantee a resolution of conflict. Many value conflicts that arise on the third level of conflict cannot be resolved at all. In such a case, collectively binding decisions rely on compromises or majority votes rather than consensus.

There is a strong tendency for risk management agencies to re-frame higher level conflicts into lower level ones: third level conflicts are presented as first or second level conflicts, and second level conflicts as first level. This is an attempt to focus the discussion on technical evidence, in which the risk management agency is fluent. Stakeholders who participate in the discourse are thus forced to use first level (factual) arguments to rationalise their value concerns. As we will demonstrate later, much of the institutional reform of the food safety agencies was driven by the motivation to address third level concerns by changing institutional structures that incorporate reforms of the first and second level.

24.3   Divergent notions of nature as primary drivers of public food risk controversies

24.3.1   Three divergent notions of nature and corresponding worldviews

While there is wide agreement at the theoretical and abstract level on this multi-level character of risk controversies (Funtowicz and Ravetz 1985; Rayner and Cantor 1987), a relative paucity of empirical research has been conducted examining the third level systematically. One instructive study which has taken on this task deals with worldviews in conflicts over technology and the environment and analyses their role on the basis of two case studies of which one is the conflict over genetically modified (GM) food in Great Britain (Gill 2003;[1] see also chapters by Rowe (27), de Jonge *et al.* (5), Korthals (29), and Siegrist (10), all this volume). This case study confirms the overall study's main hypothesis: Conflicts over technology and the environment which take the form of public controversies, focus on uncertain consequences, and are long-lived are primarily due to divergent representations of nature. Different images of what 'natural' means unfold entirely different concepts of what are to be considered good or bad consequences of the product or process under discussion. Disputes

---

1. The other case study deals with the conflict over the application of modern biomedicine to humans.

on the relevance, quality, and sufficiency of knowledge and divergent interests concerning material risks and benefits usually also fuel such conflicts but do not act as the prime driver.

The interpretation provided in this chapter regarding the British GM food conflict is based on a heuristic which distinguishes three (ideal type) concepts of nature and corresponding worldviews, termed conservative identity-orientation, utilitarian progress-orientation, and romantic *Alterität*-orientation.[2] This heuristic is applied to the results of a discourse-analytical study of the historical context of the GM food conflict. The result of this analysis is that nutritional preferences since the 1970s have been changing. Over the last three decades, diet rich in proteins, fat and sugar has lost its predominance. Instead there is a growing demand for fast-food (for those who are comfort-loving and/or short of time) and designer food (for those with a neo-utilitarian sensibility for health and slimness). And even more importantly, there is a growing quest by consumers for natural foods and food of foreign origin which come to present the *ideal of cultivated food* in terms of being sophisticated in food choice as a consumer (Gill 2003, p. 245). This quest corresponds to the concept of nature in the *Alterität*-oriented worldview which is based on the principle of 'longing' (*Sehnsucht*).[3] In this worldview, nature is one of the embodiments of the 'different' (*das Andere*) – opposite to society, technology, industry, capitalism, etc. It is a contemplative refuge and a place for experiences and adventures in which the 'very different' is imagined.

24.3.2   Hidden motives: the search for the 'very different'

According to Gill's analysis, opposition to GM foods was pushed and continues to be fuelled by this concept of nature. GM foods are the products of plant biotechnology and derived from organisms in which the genetic material has been modified in a way that does not occur naturally by fertilisation or natural recombination. In an *Alterität*-oriented perspective such foods are identified as 'artificial' and rejected by many consumers because they perceive that the purity of nature needs to be protected as a genuine counter-world to industrial society. Accordingly, the inevitable pollen drift of transgenic crops across to non-GM crops is determined as genetic contamination and harm.[4] In this conceptualisa-tion nature is not understood as a threat to be brought under control, or as a condition which is in need of optimisation (as in the utility-oriented worldview)

---

2. In the cultural sciences, *Alterität* is sometimes used as a term in contrast with identity which denotes the 'different' or 'foreign' (*das Andere*) from which the 'own' (*das Eigene*) is dissociated. Gill uses the term in a slightly different way by relating it to the 'different' which raises curiosity, promises variety and relief, and is regarded as a challenge to revisit identity; thus it is basically positively connoted (Gill 2003: 51).

3. The utility-oriented worldview is based on the principle of benefit, the identity-oriented worldview on the principle of origin.

4. Also the identity-oriented concept of nature contributes to opposition but it is of minor importance. In this perspective, the use of genetic engineering in food production is perceived as an interference with the natural or creational order, especially gene transfer across species is considered unacceptable.

but is perceived as a representation of something 'ideal' and 'authentic'. Interventions which harm or substantially change nature are understood as violations of a pre-given harmony and *uncertainties* about consequences and the threat of the vague and infinite increasingly become matters of concern (Gill 2003: 277).[5]

Genuine expressions of the *Alterität*-oriented motive in the GM foods debate, as Gill underlines, are not widely found. When they can be identified, they are most likely to occur in social movement arenas and discourses. There are two main reasons for this: firstly, the level of abstraction of the ideal types used for interpretation is very high. Normally, they would not be used for self-interpretation and self-expression. Instead, they exist and take effect as cognitive 'deep structures'. Secondly, in the arenas of official decision-making the utility-oriented discourse is predominant. Resistance to novel foods based on motives relating to the value of 'naturalness' may be translated into arguments linked to human health risks, such as increased allergenicity, toxicity, antibiotic resistance, or unintended effects, such as unexpected compositional changes in foods (Gill 2003, p. 246). The actual motives of opposition and criticism towards new and emerging technologies in the food production field may be masked, in order to improve chances to be heard.[6] This shift from the third to the second and first level is therefore not only pursued by the risk management agencies but also echoed by the representatives of social groups that are motivated by third level concerns yet express them in surrogate lower level terms.

24.3.3   Food as a symbol of nature, tradition and the exotic

Many sociological (and indeed ethical, social psychological, and marketing) studies on food have shown that food is more than a collection of partly beneficial and partly harmful substances, as the approach of natural science appears to suggest. They have pointed out that food is a symbol of cultural identity, of origin, tradition and community (cf. Caplan 1997). Gill's study has highlighted another symbolic meaning of food which has gained in importance in the last three decades mainly in the urban middle and upper classes which take a particular interest in exotic and 'natural', genuine foods.[7] For these post-

---

5. The analysis by Zwick (1999) based on guided interviews regarding attitudes towards genetic engineering also provides rich empirical material on the role of this concept of nature in the evaluation of GM food by consumers.

6. Gill claims empirical plausibility for the effectiveness of 'deep structures' by arguing that only these can make sense of many surface phenomena (Gill, 2003, p. 248).

7. While inspired by the structural approach of cultural theory (as developed by social anthropologist Mary Douglas and political scientist Aaron Wildavsky in the often cited book *Risk and Culture*, 1982), Gill's concept is clearly distinguished from it. In his approach, different worldviews and concepts of nature do not only generate quantitative differences in commitment, scepticism or aversion as regards anticipated benefits and risks but entirely different ideas about what should count as benefit or harm. Empirically, the concepts of nature are not embodied in social structure (as supposed by cultural theory) but in discourses and collective action contexts between which the individuals move in their daily activities. The concepts of nature vary less on the individual level than on the level of practices so that the same person may oppose transgenic food but choose transgenic drugs.

industrial urban milieus food is a search for a contrasting experience of country life, tradition, originality, and authenticity. The objective is not reassurance of oneself (i.e., identity-oriented) but border crossing towards what is different, wild, free and original and thus a counter-world to the manifestations of modern civilisation, which is perceived as monotone and technologically imbued.

While intuition might assume that 'romantic' orientations are specific to Germany, it was the explicit objective of the British case study to show that also in Great Britain, often typified as the stronghold of pragmatism and rationality, such orientations are living elements of contemporary thinking and acting. The pivotal BSE event and its manifold consequences cannot be fully accounted for without considering the high symbolic power of feeding carnivores to 'vegetarian' bovine animals (cp. Grove-White *et al.* 1997; Jasanoff 1997).[8] The British case study indicates that *Alterität*-oriented motives are not cultural idiosyncracies but an inherent element and consequence of wider processes of modernisation which can be traced throughout Europe and probably most of the OECD countries (Gill 2003: 164).[9]

24.3.4   The 'irresponsibility' of food risk regulating institutions

Food debates pushed and fuelled by deeply held value orientations such as those related to nature and naturalness stay persistently in the public arena. This is because of the lack of institutional provisions and expertise to handle such value conflicts (Renn and Klinke 2001: 248).[10] In the arenas of regulatory decision making, scientific discourse and scienticised debates over risks dominate the communication among and between the major actors. While science is understood as a universal principle, ideas of nature appear subjective and arbitrary to most regulators. Institutional routines are designed to mitigate risks and balance divergent interests on the basis of factual evidence and utility-based arguments. In the following section we will argue that the recent EU-level reforms do not deviate from this approach, but attempt to meet the value-based concerns of food risk conflicts through mechanisms aimed at improving the scientific foundation of risk management (first level) and providing structural changes for enhancing transparency of decision making (second level). Although not directly linked to third level concerns, the changes have profound *indirect* implications for the ongoing third level debate.

---

8. Jasanoff points out that even in the 'dry, restrained text' of the 1989 British Southwood expert advisory report BSE is traced back to 'unnatural feeding practices as found in modern agriculture' (quoted in Jasanoff 1997: 226).
9. Gill understands the three worldviews as deeply anchored in European culture and also determined in our contemporary societies as latent background assumptions thinking and acting in daily discourses and practices.
10. In this perspective, it is the conflict of divergent worldviews which overstrains the responsible institutions and not the lack of insurability, as argued by Ulrich Beck in his 'classic' book on the *risk society* (1986).

24.4   Regulatory and institutional reforms at EU-level: indirect responses to third level concerns

During the past 15 years there have been major regulatory and institutional reforms of food safety governance at the level of the European Union and within several of its Member States. These reforms have been provoked foremost by a series of food-related scares (with 'mad cow disease' or BSE being the most prominent) and disputes (most notably the persistent GM food debate). As several scholars have observed, these incidents have raised the public's awareness about food safety issues, rendered the accountability of regulatory authorities and the reliability of scientific experts into matters of high political salience, and put food safety regimes under considerable reform pressure (Vos 2004, 2000; Millstone and van Zwanenberg 2002; Millstone 2000; Skogstad 'regulatory policy styles'). The 2000 report of the European Parliament's Scientific and Technological Options Assessment unit (STOA) on 'European Policy on Food Safety' states that, 'not a single Member State reported that its food safety policy-making system is experiencing a period of organisational and/ or institutional stability' (Trichopoulou et al. 2000: 68).[11] Since the 1990s, food policy regimes have been in a 'state of flux' (Millstone and van Zwanenberg 2002: 607).

24.4.1   Responding to the challenge of 'perceived' risks

Within the European Union the main reforms have been the General Food Law, adopted in January 2002 (CEC 2002), which sets forth principles and requirements for all future food laws in Europe, and the establishment of the European Food Safety Authority (EFSA) provided for by the Law. The General Food Law and EFSA are the main pillars of an emerging more generic governance system for food safety, which supersedes the former ad-hoc and committee-based approach to food regulation (Vos 2000).

These legislative and institutional reforms do not expressly address world-view conflicts and do not provide for mechanisms of regulatory governing for an explicit and systematic handling of value conflicts such as those nurtured by the quest for the experience of 'naturalness' in food and the refusal of 'unnatural' agricultural methods. Nevertheless, our argument goes, they should be read, at least in part, as a response to these challenges of conventional risk management routines, however an indirect one. The 'philosophy' underlying the reforms subsumes the reactions of individuals, groups, and societies when faced with risk situations, which, as the former European Commissioner for Health and Consumer Protection, David Byrne, puts it, 'can often be difficult to predict and indeed may appear irrational' (Byrne 2004), under the general heading of diffuse

---

11. The report cites the OECD which also observes '... many countries [have] re-examined their institutional and regulatory frameworks governing food safety' (OECD 2000, quoted in Trichopoulou et al. 2000: 79).

societal concerns over 'perceived' risks. These concerns apparently are immune to science-based messages and defy the utilitarian logic of conventional risk regulation.[12] Therefore they can hardly be accommodated by improving 'network governance', i.e. the mediation between and best possible integration of organised stakeholder interests (cf. Skogstad 2003).[13] They can no longer be ignored, however, as they increasingly put the regulatory and economic system under considerable pressure and impede upholding the internal market, which is an obligation under the EU treaties. The remedy in line with the general overhaul of EU legislation is fourfold and designed to restore public confidence in the competence and credibility of the authorities formally charged with food risk analysis (Skogstad 2003).[14]

Independent scientific advice generated in transparent procedures
One way in which the problem of confidence is addressed is through institutional and procedural reforms designed to strengthen procedural legitimacy by guaranteeing independent scientific advice generated in procedures which are open and transparent (cf. Millstone and van Zwanenberg 2002: 604). The European Food Safety Authority was created.[15] One goal of the agency was to provide independent scientific expertise underpinning food risk assessment. This could provide the basis for policy decisions in respect of risk management. EFSA is committed to adhere to the principles of openness and transparency in the exercise of its tasks (Vos *et al.* 2005: 9). The main activities which the newly established authority undertakes in order to develop and maintain transparency include public meetings of the Management Board, and publications on the EFSA website of the agendas and minutes of meetings of the Scientific Committee and the Scientific Panels; the opinions adopted including minority options;[16] the results of scientific studies; its final accounts; the annual report of activities; and the annual declarations of interest made by the members of the

12. According to the utilitarian logic, regulatory measures need to be based on evidence about short-term or middle-term risks to constitutionally protected collective goods such as human health and life (and increasingly also the environment).
13. According to Skogstad, network governance in terms of 'a pattern of policy-making in which state and non-state actors together make binding decisions in the EU' (2003: 322) presents a mechanism of input legitimacy on which the EU increasingly resorts in order to promote policy implementation.
14. Points 1 to 3 are informed by the EU-level report compiled by *Vos et al.* (2005) for subproject 5 ('Investigation of the Institutional Challenges and Solutions to Systemic Risk Management') of the EC Framework Programme 6 Integrated Project on SAFE FOODS which is coordinated by Harry Kuiper and Hans Marvin of the RIKILT-Institute of Food Safety, Wageningen, in the Netherlands. One key task of subproject 5, which is coordinated by the authors of this chapter through the DIALOGIK gGmbH, is a review of institutional arrangements of food safety regulation in Europe. The review takes a comparative perspective and includes five country studies – food safety regulation in Hungary, Sweden, United Kingdom, France, and Germany – and the EU-level study. For the results of the review see Vos and Wendler (2006).
15. EFSA commenced its operations on 1 January 2002.
16. As Vos *et al.* (2005: 56) underline there is a caveat to publication responsibilities in that European Union law stipulates that commercially sensitive information be kept confidential (CEC 2001a).

Management Board, the Advisory Forum and the Scientific Panels. One goal of the Authority is to ensure that the public is given objective and easily accessible information on existing and emerging food risks. Public risk communication through an independent scientific advisory authority is considered the best means to forestall public controversy such as that over GM foods which former European Commissioner David Byrne has depicted as 'only one example of how Europe's consumers seek verifiable and trustworthy information about food' (Byrne 2004; but see Chapters 10 and 5 by Siegrist and de Jonge *et al.*, respectively, this volume).

Incorporation of political accountability

In accordance with the legal structure in the EU, and in line with Codex Alimentarius risk analysis guidelines (CAC 2004: 101–107), risk management responsibilities were not delegated to EFSA, which operates in an advisory capacity. Instead they will remain with the Commission, the Council of Ministers, and the European Parliament as appropriate. Allocation of the task of deciding which risks are acceptable and how they should be managed, and accountable and identifiable political officials and institutions is another strategy to promote procedural but also substantive legitimacy. The underlying assumption is that risk management decisions necessitate political rather than purely scientific judgements – all the more so if these judgements are charged with scientific uncertainty. According to this concept, risk management is an inherently subjective undertaking which reflects the values of society and its choices. The idea behind this division of labour is that 'hard facts' come from selected scientists whose task is to counsel democratically legitimised politicians in such a way that they, when determining the desirable, keep in mind the cognitive basis of the scientific investigation of the likely consequences. Policy makers are asked to evaluate and weight the results obtained in independent and 'objective' assessment against costs, benefits, feasibility, and desirability of different management options. The trust-building quality of the division of responsibility between risk assessment and risk management is assumed to lie in the fundamental acknowledgement by food safety regulators that there are differences in the substantive concerns and values that determine evaluations of the acceptability of food risks.

Endorsement of the precautionary principle

Reflecting wider developments in different sectors and jurisdictions, the General Food Law holds that food safety regulation should be subject to principles both of risk analysis and precaution (Stirling 2005). Article 7 of the General Food Law establishes and governs the use of the precautionary principle in the food arena. It places an obligation on food policy makers to take appropriate (provisional) risk management measures when the assessment of available information demonstrates the possibility of harmful effects but scientific uncertainty persists. These measures were designed to ensure the high level of health protection chosen in the Community. The incorporation of the precautionary principle in

European food law suggests that the knowledge of scientific experts is limited and shifts the burden to business operators to demonstrate that their food products and production processes are safe. Endorsement of this principle is an expression of growing and deep consumer concerns that uncertain and long-term consequences of technical interventions into natural products and processes, characteristic of an *Alterität*-oriented view of nature, are being increasingly diffused into legislative and regulatory arenas. They compel the use of 'technologies of humility' (Jasanoff 2003)[17] for systematically exploring the limits of knowledge and the scope of ignorance and thereby (re)gaining public trust in regulatory competencies to cope with the uncertain and the unpredictable.[18]

Incorporation of popular accountability[19]
The former European Commissioner David Byrne stated, slightly apologetically, at an international conference on food safety: 'I acknowledge that friends in the US find the European public's attitude to GM difficult to understand. However, our consumers demanded clear labelling and traceability as essential prerequisites' (Byrne 2004). A key regulatory instrument to restore public confidence in food safety management, in the view of the European Commission, consists of mandatory and comprehensive labelling to identify products containing or derived from GMOs and mandatory traceability systems which enable food to be traced back to its origin (CEC 2003). The latter systems constitute a general requirement under the General Food Law. The Law adopts a comprehensive, integrated 'From the farm to the fork' approach[20] which requires that the traceability of food must be established at all stages of production, processing and distribution. To this end, business operators are required to apply appropriate systems and procedures. Traceability is a necessary prerequisite for comprehensive labelling as it is mandatory for GM food products. The introduction and extension of GM food labelling meets citizen demands by enabling them to exercise choice over whether or not they choose to buy these products. Consumers are provided with the possibility to translate their 'diffuse' concerns over 'perceived' risks into purchase decisions. As Skogstad puts it, mandatory labelling schemes and traceability systems are measures to bestow legitimacy of regulatory performance through 'popular accountability' (2003: 333). Similar regimes of traceability and labelling are presently implemented within the fish industry. Consumers should be reassured that the fish they consume is not deploring the biodiversity of marine wildlife.

17. Jasanoff argues that there is an increasing need for methods which acknowledge the limits of prediction and control in the management of technology and address the normative implications of these limits. These methods referred to as 'technologies of humility' should be developed around the focal points of framing ('what is the purpose?'), vulnerability ('who will be hurt?'), distribution ('who benefits?'), and learning ('how can we know?'). On all these points wider public engagement is identified as a means to improve the capacity for analysis and reflection (2003: 240).
18. For this argument in the context of GM crop and food regulation see Gill (2003: 228–235).
19. The term 'popular accountability' is borrowed from Skogstad's analysis of GMO regulation in the EU (2003: 333).
20. By the 'from farm to fork' approach, food safety regulation has to address all safety issues along the whole food and feed chain.

The four features of the emerging food safety policy regime represent efforts to accommodate citizen demands into food policy in response to information gained through public opinion polls and also other methods such as modelling procedures and citizen panels. These concerns, however, are still poorly understood by the risk regulating authorities and appear irrational from a purely utilitarian perspective. At the same time they are experienced as challenges to political legitimacy and economic stability. The way in which such consumer concerns are addressed is through establishing a food safety governance system which amalgamates the authority of scientific expertise, restricted through the endorsement of the precautionary principle, with the authority of democratic criteria including transparency, political accountability and popular account-ability.[21] Emphasis on democratic criteria in food safety governance corres-ponds to the more general tendency of the transformation of European governance structures directed towards reducing the 'democratic deficit' of EU institutions – 'their deficiencies in representation, representativeness, account-ability and transparency' (Skogstad 2003: 321). The rationale of, for instance, enhancing the policy role of the European Parliament or involving stakeholders in policy formulation, is that the strengthening of democratic principles may improve the effectiveness and efficiency of policy making and thereby increase the legitimacy of EU policies (cf. Abels 2002: 2).[22]

24.5   Some suggestions for making value conflicts operational for food risk management

In the last section we made the argument that recent reforms of the EU food safety regulatory system are in part a response, however indirect, to nature-related value orientations underlying intense public food risk controversies in Europe. This corresponds to Echols earlier observation that 'generally, the EC's system of food regulation more readily supports traditional food practices. Regulators and consumers believe that these practices and the foods they produce are safe, and, importantly for most European consumers, close to nature and naturalness. . . . Links to the land, provenance and limited processing remain important to many consumers' (1998: 528).[23] Major contributing factors to regulatory responsiveness are the increasing importance which the EU's decision-making framework assigns to stakeholder consultation processes making it more sensitive to wider concerns of the European populations (cf. CEC 2001b), and the over-riding objective of an internal common market.

---

21. Skogstad concludes this way of meeting consumer demands from her analysis of the EU GMO regulatory framework (2003: 333).
22. In the Commission's White Paper on European Governance it says: 'The goal is to open up policy-making to make it more inclusive and accountable. A better use of powers should connect the EU more closely to its citizens and lead to more effective policies' (CEC 2001b: 8).
23. Echols traces differences in food safety regulation in the EU and the United States which have resulted in trade conflicts or tensions back to cultural influences.

The responses of the emerging food safety governance system to non-scientific, non-technical, and non-economic issues are promising but still insufficient instruments for improving mitigation or resolution of conflicts based on deeply held value orientations. Examples include those related to cultural, religious and philosophical beliefs, concerns regarding the welfare of animals used in food production, or the desire for more 'natural' and fewer 'techno-logical' production methods. They need to be supplemented, our argument goes, by arrangements which enable the risk analysis and regulatory process to address the potential importance of such orientations to the risk issue in question in a proactive, systematic, and direct manner.

The main problem with the present institutional response to the food crisis is the reliance on the traditional *decisionistic model* of decision making. The decisionistic model represents a connecting link between science-oriented information input and political selection of desirable regulatory options. The experts are responsible for inputs regarding specialist knowledge. Politically legitimised decision makers then carry out the decisions based on this knowledge transfer and their political preferences (Hajer and Wagenaar 2003; Roqueplo 1995; Renn 1995: 149). Underpinning this process is the assumption that knowledge and interests can be separated in the course of analysis and deliberation. However, such a separation presents conceptual and practical problems. Firstly, expert knowledge must always refer to those dimensions which are of central significance to the political decision makers as the basis for the knowledge-oriented formation of their own preferences (relating again to the issue of framing). Secondly, there are always conceivable or even probable consequences for which only an insufficient knowledge base or none at all is available (the incorporation of non-knowledge). Thirdly, expert knowledge is not independent of biased basic assumptions; simultaneously the influencing preferences are not independent of the associated knowledge bases. Fourthly, this model also does not indicate how decision-makers should deal with experts who disagree about information. Moreover, it is the characteristic of many social conflicts that the preferences of decision-makers (within administrative or political institutions) do not coincide with the preferences and concerns of citizens. The decisionistic model cannot solve these structural problems (Shrader-Frechette 1995).

So what can be done to improve the situation and address in more adequate terms the concerns of citizens and social groups in terms of the development of food safety?

24.5.1   Inclusion of concern assessment in threat appraisal[24]

Constitutional and case law, as well as the 1994 Agreement on Sanitary and Phytosanitary Measures (SPS), limit the influence of value-based concerns regarding food safety measures in the EU. Accordingly, the authorities at both

---

24. The term threat appraisal has sometimes been used in the risk governance literature to include all knowledge elements necessary for risk evaluation and risk management (Stirling 1998, 2003).

national and EU levels with responsibility for generating the knowledge on which management decisions will be based largely restrict their assessment work to scientific analyses of potential harm to human health and life which are protected by the law. There is, however, some degree of discretion by national regulators to respond to cultural attitudes in the decisions on the most appropriate risk management measures to take. Millstone and van Zwanenberg (2000) point this out in the context of the SPS Agreement on the cases of beef hormones and recombinant bovine somatotropin (rBST). The Agreement permits the authorities to take measures which are more stringent than international standards in response to consumer concerns, when risk assessments are carried out along the lines requested by the WTO system and when these measures are applied consistently and non-discriminatorily.[25] Risk management decisions need not be based exclusively on risk assessment, but may also take into account wider socio-economic issues and the relevance of scientific uncertainties (Millstone and van Zwanenberg 2000: 118).

In order to deal with highly ambiguous risks,[26] risk managers need data which help to understand the wider, sometimes concealed, concerns of stakeholders, as well as the wider public, as an information input. When dealing with such risks, the results of scientific risk assessment are a necessary but not a sufficient basis for making decisions. What is needed is the provision of data on risk perception and social responses (Renn 2005: 19). The collection and analysis of results from risk perception studies in the threat appraisal process can improve the understanding of deeper lying motives of politically salient public risk controversies such as those related to nature and naturalness. The results of this 'concern assessment' are needed if a risk management process is to be applied which is sensitive to culture and (potentially divergent and contradicting) value preferences and able to produce socially legitimate and robust decisions. In relation to the recent reforms at EU-level the concrete proposal would be that EFSA's risk assessment activities carried out by natural scientists are complemented by the analysis of results of studies on risk perception and public risk controversies carried out by social scientists. In order to understand

25. EU-food safety regulation is embedded and operating in a new global framework for the setting, arbitrating and harmonising of world food standards. The global framework is made up of WTO dispute procedures, the SPS Agreement which is part of the GATT Final Act and stipulates global harmonisation of standards for which the Codex Alimentarius Commission plays the pivotal role, and the Technical Barriers to Trade agreement. The SPS Agreement essentially requires Members to justify the food safety regulations that they apply and demonstrate that any trade distorting effects are proportionate. Justification may be done by adoption of international standards and/or through risk assessments which must satisfy certain requirements (cf. Poli 2003; Henson and Caswell 1999).

26. Ambiguity is the result of divergent or contested perspectives on the justification, severity or wider 'meanings' associated with a given threat (Stirling 2003; Klinke and Renn 2002). It can be divided into *interpretative ambiguity* (different interpretations of an identical assessment result) and *normative ambiguity* (different concepts of what can be regarded as tolerable) referring, e.g. to ethics, quality of life parameters, distribution of risks and benefits, etc. (Renn 2005: 11–12). Klinke and Renn (2002) determine ambiguity as one of three key generic challenges (besides complexity and uncertainty) of risk handling.

the concerns of the various stakeholders and public groups, information about risk perceptions is needed and should be collected by EFSA and other risk management agencies. In addition, other aspects of the potentially hazardous activity that are relevant for the evaluation and the selection of risk management options should be collated and analysed. Based on such a wide range of information, risk managers can make more informed judgements and design the appropriate risk management options (Clark 2001). One way in which social scientists could appraise food risks is by using a 'mobilisation index' which includes the following four sub-criteria reflecting many factors that have been proven to influence risk perception (Renn 2005: 20):[27]

- inequity and injustice associated with the distribution of risks and benefits over time, space and social status;
- psychological stress and discomfort associated with the risk or the risk source (as measured by psychometric scales);
- potential for social conflict and mobilisation (degree of political or public pressure on risk regulatory agencies);
- spill-over effects that are likely to be expected when highly symbolic losses (including perceived losses in relation to deeply held values such as 'wild, pure nature' and 'naturalness') have repercussions on other fields such as financial markets or loss of credibility associated with management institutions.

These criteria were designed for the appraisal of environmental harm and might require adaptation to suit the appraisal of food threats. In principle, depending on the risk under investigation, additional criteria might need to be included or proposed criteria neglected.[28] The inclusion of non-scientific concerns helps risk managers to identify second and third level concerns and raise their awareness for associations and connotations that are not linked to the results of the scientific assessments of consequences or probabilities. In addition, this screening exercise for concerns may help risk assessors to address new research questions and to find more appropriate ways to characterise uncertainties.

24.5.2   Participatory discourses in risk evaluation/management

If the results of risk appraisal confirm that a food risk is charged with major ambiguities, risk managers are well advised to organise a systematic feedback from society. In these cases it is not enough to demonstrate that risk regulators are open to public concerns and address the issues that many people wish them to take care of. To make underlying value conflicts or differences in vision

27. This index was suggested by experts and risk managers after they had reviewed the proposal by the German Council for Global Environmental Change (WBGU) for a set of risk characterisation criteria beyond the established assessment criteria of which 'potential for mobilisation' was one of eight criteria (WBGU 2000).
28. A similar decomposition has been proposed by the UK-government (Environment Agency 1998; Pollard *et al.* 2000).

operational for risk management, the processes of risk evaluation and management option assessment need to be open to public input and new forms of
deliberation for conflict resolution and joint vision building. The aim of
deliberative procedures is to provide more reliable information on the question
of proper framing[29] (Is the issue really a risk problem or is it in fact an issue of
lifestyle and future vision?), and on the dimensions of ambiguity that need to be
addressed in balancing the pros and cons of the risk generating source concerned
(Renn 2005).

While the EU-institutions increasingly give attention to involving stakeholders in food safety regulation, deliberation-based involvement of the general
public as a nongovernmental actor is neglected. Indeed, at the European level
there is even a lack of discussion about adequate models for and the experience
of involving citizens in a decision-making process that spans national boundaries. This chapter does not provide the scope to spell out the possibilities and
limits of trans-boundary citizen deliberation.[30] We would like to suggest, however, that such procedures should be organised at the *national* level (to avoid demotivating language and logistical problems) and designed according to the
issue concerned and country-specific socio-cultural and political factors (Renn
and Klinke 2001: 259).[31] The result of each process could be fed back into EU-
decision making in form of consultancy reports (Renn and Klinke 2001: 255).
Even if the individual processes would not produce any tangible results in form
of consensus positions, the facilitators' analyses in terms of value violations and
the suggested modifications for risk management – e.g., in terms of introducing
labelling or monitoring schemes – would be valuable input to making a better
and probably more acceptable decision by the EU risk managers.

The degree of public participation can vary depending on the characteristics
of the risk and the three levels of the risk debate. For orientation purposes it
might be useful to distinguish *four stages of involvement* (Renn 2004; 2005).
The first stage refers to routine risks (Stage 1) which require hardly any changes
to the traditional decision-making framework. For assessing, evaluating and
managing risks a discourse among agency staff and enforcement personnel is
sufficient *(instrumental discourse)*. After the risk is taken, monitoring the risk
situation is important as a reinsurance that no unexpected consequences may
occur.

If the risk problems are large in scope or high in complexity, the conventional
methods of risk assessment and risk management do not suffice (Stage 2 and
corresponds to level 1 of the risk debate). Data collection and interpretation are
less obvious than in the routine case and demand more sophisticated methods.

---

29. Framing encompasses the selection and interpretation of phenomena as relevant risk topics
    (Renn 2005: 8).
30. A broad discussion of this issue can be found in Linnerooth-Bayer *et al.* (2001).
31. A substantial proposal of how to organise trans-boundary participation involving the general
    public and stakeholder groups is provided in Renn and Klinke (2001). The authors suggest how a
    three-stage 'cooperative discourse' procedure might be adapted to public participation in a trans-
    boundary setting (see also Renn *et al.* 1993).

Simple statistical data is either not available or insufficient to calculate the risks for humans or the environment. This is the place for participatory procedures for data collection and interpretation such as the Delphi process as a means to get the best expertise and experience represented in characterising causal chains from the initiating event to the final damage. This exercise of collecting and interpreting the causal relationships can be provided by an *epistemological discourse* aimed at finding the best estimates for characterising and evaluating the risks under consideration. This phase includes the screening of social concerns and perceptions as they may relate to different interpretations of the observed or measured cause-effect relationships. Once the risks are characterised and assessed risk managers can proceed in a similar way as they have done in the routine case.

If uncertainty plays a large role, in particular ignorance, the risk-based approach becomes counter-productive (Stage 3 corresponding to level 2 of risk debates). Judging the relative severity of risks on the basis of uncertain parameters, leads to unsatisfactory results. Under these circumstances, management strategies belonging to the precautionary approach are required. Precaution in this context means to avoid irreversible commitments to one option or activity. Risk managers are well advised to search for risk management options that promise to enhance resilience and decrease vulnerability. At the same time, however, risk managers face a serious dilemma: How can one judge the severity of a situation when the potential damage and its probability are unknown or highly uncertain? Facing this dilemma, it is appropriate to include the main stakeholders in the assessment process and ask them to find a consensus for determining the extra margin of safety that they would be willing to invest in exchange for avoiding potentially catastrophic consequences. This type of deliberation called *'reflective discourse'* rests on a collective reflection about balancing the possibilities for over- and under-protection based on uncertain data and ignorance.

In this context, it is important to distinguish clearly between uncertainty and ambiguity (Step 4 corresponding to level 3 of risk debates). Uncertainty refers to a situation of being unclear about factual statements; ambiguity to a situation of contested views about the implications (interpretative ambiguity) and tolerability (normative ambiguity) of a given hazard. Uncertainty can be resolved in principle by more cognitive advances (with the exception of indeterminacy and ignorance), ambiguity only by discourse. Discursive procedures include legal deliberations as well as novel participatory approaches. If ambiguities are associated with a risk problem, the process of risk evaluation itself needs to be open to public input and new forms of deliberation. The type of discourse required here is called *participative discourse*. There are sets of deliberative processes available that are, at least in principle, capable of resolving ambiguities in risk debates. Those processes include citizen panels, consensus conferences, ombudspersons and other participatory instruments.

Figure 24.2 illustrates the four stages of risk management specifying the different needs for public participation and stakeholder involvement.

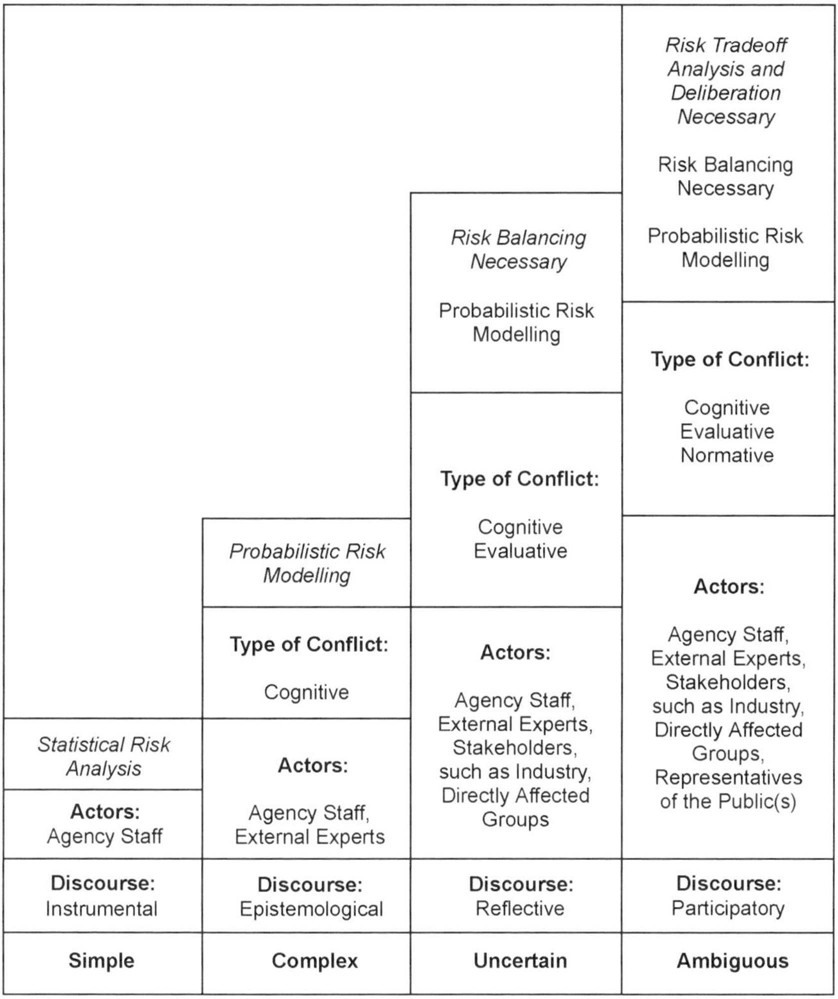

Simple	Complex	Uncertain	Ambiguous
			Risk Tradeoff Analysis and Deliberation Necessary
			Risk Balancing Necessary
		Risk Balancing Necessary	Probabilistic Risk Modelling
		Probabilistic Risk Modelling	**Type of Conflict:**
			Cognitive Evaluative Normative
	Probabilistic Risk Modelling	**Type of Conflict:**	
		Cognitive Evaluative	**Actors:**
	Type of Conflict:	**Actors:**	Agency Staff, External Experts, Stakeholders, such as Industry, Directly Affected Groups, Representatives of the Public(s)
	Cognitive	Agency Staff, External Experts, Stakeholders, such as Industry, Directly Affected Groups	
Statistical Risk Analysis	**Actors:**		
Actors: Agency Staff	Agency Staff, External Experts		
Discourse: Instrumental	**Discourse:** Epistemological	**Discourse:** Reflective	**Discourse:** Participatory
Simple	**Complex**	**Uncertain**	**Ambiguous**

Fig. 24.2   The risk management escalator (from simple via complex and uncertain to ambiguous phenomena).

24.6   Conclusions

The main thesis of this chapter has been to point out that recent reforms in EU food safety legislation – the General Food Law and also issue-specific regulation on GMOs – form an institutional response to the challenging features of food risk controversies fuelled by deeply held value-orientations. These controversies, for example the debates over GM food and mad cow disease, are persistent, remain in the public realm, and exert considerable pressure on the responsible policy-making institutions to demonstrate adequacy of their decision-making procedures and results. The EU-level response consists of a revised food safety governance system which interlinks the authority of scientific expertise, restricted through the endorsement of the precautionary principle, with the authority of democratic criteria including transparency, political accountability and popular accountability. According to official statements, the overarching objective of these innovations is trust-building. Trust-building appears as the 'general cure' for public pressures resulting from perceptions of food threats which diverge from the science-based expert perspective.

The emphasis on independent scientific review and trust-building measures can be interpreted in the light of the analytic concept of risk debates that we have used as an analytic tool to investigate the present risk management situation in Europe with respect to food safety. The three levels refer to: (i) factual evidence, probabilities, and uncertainties; (ii) institutional performance, expertise, and experience; (iii) conflicts about worldviews and value systems. The new regulatory European regime seems to address issues of levels 1 and 2, but may, however, fail to address the third level, i.e. the level of values and visions, although the intention of the EU has been to respond to the concerns raised on the third level.

We have hence argued, that the new features of EU food safety governance are going in the right direction but are insufficient for improving mitigation or resolution of conflicts based on value orientations such as those related to cultural, religious and philosophical beliefs, concern for the welfare of animals, or the desire for more 'natural' and less 'technological' production methods. They need to be supplemented by arrangements which enable the risk analysis and regulatory process to address the potential importance of such orientations to the respective risk issue in a proactive, systematic, and direct manner.

We have made two suggestions for rendering underlying value conflicts or differences in vision operational for risk management: If pre-assessment recognises potential food risks as charged with major ambiguities, risk appraisal should systematically compile results from risk *and* concern assessment so that risk managers can consider wider societal concerns in designing and evaluating risk handling options. Second, if the results of risk appraisal confirm the ambiguity challenge, risk managers should organise a public participation procedure designed to provide more reliable information on the question of framing of the issue concerned and on the dimensions of ambiguity that need to be addressed in the management decision-making process. We introduced a stepwise process of including experts, stakeholders and representatives of the general public for addressing the three levels of risk debates in a more effective and efficient manner. While these arrangements cannot resolve conflicts based on divergent notions of nature and underlying worldviews, they can guide risk managers in the choice of management instruments. With labelling, these include a powerful (however, also trade sensitive) tool for letting citizens and consumers make their own value judgements about food issues and incorporating socio-cultural perceptions and preferences across Member States and lifestyles into EU food safety regulation.

24.7   Future trends

If Gill (2003) is right, that tradition and naturalness will continue to gain in importance across Europe as motives of consumer behaviour, the development and application of new food production, processing, and packaging technologies are likely to produce further public controversies, only at the surface concerned with risks and benefits. If, in response, regulatory attention to and transparency of scientific uncertainty will continue to grow, more culturally informed strategies to handle food risks could be expected in EU food safety governance (Boholm 2003: 168; Beck and Grande 2004: 306). These are, as stated above, limited by an international framework which encourages reliance on a standardised, international, and science-based approach to regulation. In this approach, risk assessment is a technical, non-political, value-free exercise. Accordingly, cultural factors may enter the stage of choosing risk management measures, but not the stage of determining whether there is a food safety risk (Skogstad 2001: 302).

Institutional responses to wider concerns, however, must not be restricted to legislative and regulatory reforms. As the GM food conflict shows, the corporate sector may much more easily react to such concerns than governmental or supranational authorities. GM food was first boycotted by British supermarket chains before retailers across Europe followed the example (Gill 2003: 235–241; Dreyer and Gill 2000: 223). The responsiveness of the food industry and the retail sector – for instance, by increasing differentiation of products, production methods, and channels of distribution – will depend, amongst others, on the extent of mobilisation by consumer and environmental organisations. At any rate, social analysis should not be restricted to how public actors deal with the challenge of value-based food conflicts but give equal importance to the roles of the corporate sector and civil society. All of these actors contribute to EU-*governance* of food safety.[32] So in essence, we are convinced that the incorporation of concerns into the assessment phase, the stepwise inclusion of experts, stakeholders and the general public in designing risk management strategies and the framing of risk issues involving civil society, economic actors and regulators are the three major drivers for institutional change in the years to come.

24.8   References

ABELS, G. (2002). Experts, citizens, and eurocrats towards a policy shift in the governance of biopolitics in the EU. *European Integration online Papers* (EIoP), **6**, 19, published on 3 December 2002. Available at: http://eiop.or.at/eiop/texte/2002-019a.htm.

32. In the last decade, the term 'governance' has experienced tremendous popularity in the literature on risk research. Governing choices in modern societies is seen as an interplay between governmental institutions, economic forces and civil society actors (such as NGOs).

BECK, U. (1986). *Risikogesellschaft. Auf dem Weg in eine andere Moderne*. Frankfurt/M.: Suhrkamp.

BECK, U. and GRANDE, E. (2004). *Das kosmopolitische Europa*. Frankfurt/M.: Suhrkamp.

BOHOLM, A. (2003). The cultural nature of risk: can there be an anthropology of uncertainty? *Ethnos*, **68**, 2 (June 2003), 159–178.

BYRNE, D. (2004). The Regulation of Food Safety and the Use of Traceability/Tracing in the EU and USA: Convergence or Divergence? Speech at the Food Safety Conference in Washington DC, 19 March 2004. Accessed on: http://www.foodlaw.rdg.ac.uk/eu/doc-56.htm

CAPLAN, P. (ed.) (1997). *Food, Health and Identity*. London: Routledge.

CEC, COMMISSION OF THE EUROPEAN COMMUNITIES (2001a). Regulation (EC) No 1049/2001 of the European Parliament and of the Council of 30 May 2001 regarding public access to European Parliament, Council and Commission documents. *Official Journal of the European Communities*, 31.05.2001, L145/43.

CEC, COMMISSION OF THE EUROPEAN COMMUNITIES (2001b). *European Governance. A White Paper*, COM (2001) 428 final, 25 July 2001. Brussels.

CEC, COMMISSION OF THE EUROPEAN COMMUNITIES (2002). Regulation (EC) No. 178/2002 of the European Parliament and of the Council of 28 January 2002 laying down the general principles and requirements of food law, establishing the European Food Safety Authority and laying down procedures in matters of food safety. *Official Journal of the European Communities*, 01.02.2002, L31/1.

CEC, COMMISSION OF THE EUROPEAN COMMUNITIES (2003). Regulation (EC) No 1830/2003 of the European Parliament and of the Council of 22 September 2003 concerning the traceability and labelling of genetically modified organisms and the traceability of food and feed products produced from genetically modified organisms and amending Directive 2001/18/EC. *Official Journal of the European Union*, 18.10.2003, L268/24.

CLARK, W. (2001). Research systems for a transition toward sustainability, *GAIA*, **10**, 44, 264–266.

CODEX ALIMENTARIUS COMMISSION (CAC) (2004). *Procedural Manual* (14th edn). Joint FAO/WHO Food Standards Programme. Rome: World Health Organization/Food and Agriculture Organization of the United Nations.

VAN DEN DAELE, W. (1990). Risiko-Kommunikation: Gentechnologie. In: H. Jungermann, B. Rohrmann and P. M. Wiedemann (eds), *Risiko-Konzepte, Risiko-Konflikte, Risiko-Kommunikation* (pp. 11–58). Jülich: Forschungszentrum Jülich.

DOUGLAS, M. and WILDAVSKY, A. (1982). *Risk and Culture*. Berkeley: University of California Press.

DREYER, M. and GILL, B. (2000). Die Vermarktung transgener Lebensmittel in der EU – die Wiederkehr der Politik aufgrund regulativer und ökonomischer Blockaden. In: A. Spök, K. Hartmann, B. Wieser, A. Loinig, C. Wagner (eds), *Genug gestritten?! Gentechnik zwischen Risikodiskussion und gesellschaftlicher Herausforderung* (pp. 125–148). Graz: Leykam.

ECHOLS, M. A. (1998). Food safety regulation in the European Union and the United States: different cultures, different laws. *Columbia Journal of European Law*, **4**, 525–543.

ENVIRONMENT AGENCY (1998). *Strategic Risk Assessment. Further Developments and Trials*. RandD Report E70. London: Environment Agency.

FISCHHOFF, B. (1985). Managing risk perceptions. *Issues in Science and Technology*, **2**, 1, 83–96.

FUNTOWICZ, S.O. and RAVETZ, J.R. (1985). Three types of risk assessment: Methodological analysis. In: V.T. Covello, J.L. Mumpower, P.J.M. Stallen, and V.R.R. Uppuluri (eds), *Environmental Impact Assesment, Technology Assessment, and Risk Analysis* (pp. 831–848). Springer: New York.

GILL, B. (2003). *Streitfall Natur. Weltbilder in Technik- und Umweltkonflikten.* Wiesbaden: Westdeutscher Verlag.

GROVE-WHITE, R., MACNAGHTEN, P., MAYER, S. and WYNNE, B. (1997). *Uncertain World. Genetically Modified Organisms, Food and Public Attitudes in Britain.* A report by the Centre for the Study of Environmental Change in association with Unilever, and with help from the Green Alliance and a variety of other environmental and consumer non-governmental organisations (NGOs), University of Lancaster.

HAJER, M. and WAGENAAR, H. (2003). *Deliberative Policy Analysis: Understanding Governance in the Network Society.* Cambridge: University Press.

HAMPEL, J. and RENN, O. (eds) (2000). *Gentechnik in der Öffentlichkeit. Wahrnehmung und Bewertung einer umstrittenen Technologie.* 2nd edn. Frankfurt/M.: Campus.

HENSON, S. and CASWELL, J. (1999). Food safety regulation: an overview of contemporary issues. *Food Policy*, **24**, 589–603.

JASANOFF, S. (1997). Civilization and madness: the great BSE scare of 1996. *Public Understanding of Science*, **6**, 221–232.

JASANOFF, S. (2003). Technologies of humility: citizen participation in governing science. *Minerva*, **41**, 223–244.

JASANOFF, S. (2004). Ordering Knowledge, Ordering Society, in: S. Jasanoff (ed.): *States of Knowledge: The Co-Production of Science and Social Order* (pp. 31–54), London: Routledge.

KLINKE, A. and RENN, O. (2002). A new approach to risk evaluation and management: risk-based, precaution-based and discourse-based management. *Risk Analysis*, **22**, 6 (December 2002), 1071–1094.

LEVIN, I.P., SCHNEIDNER, S.L. and GAETH, G.J. (1998). All frames are not created equal: A typology and critical analysis of framing effects. *Organizational behavior and human decision processes*, **76**, 2, 149–188.

LINNEROOTH-BAYER, J., LÖFSTEDT, R.E. and SJÖSTEDT, G. (eds) (2001). *Transboundary Risk Management.* London and Sterling, VA: Earthscan.

MILLSTONE, E. (2000). Recent developments in EU food policy: Institutional adjustments or fundamental reforms? *Zeitschrift für Lebensmittelrecht*, **27**, 6, 815–829.

MILLSTONE, E. and VAN ZWANENBERG, P. (2000). Food safety and consumer protection in a globalized economy. *Swiss Political Science Review*, **6**, 3, 109–118.

MILLSTONE, E. and VAN ZWANENBERG, P. (2002). The evolution of food safety policy-making institutions in the UK, EU and Codex Alimentarius. *Social Policy and Administration*, **36**, 6, December, 593–609.

ORGANISATION FOR ECONOMIC CO-OPERATION AND DEVELOPMENT (OECD) (June 2000). *Agricultural Policies in OECD Countries 2000: Monitoring and Evaluation.* Paris: OECD.

POLI, S. (2003). Setting out international food standards: Euro-American conflicts within the Codex Alimentarius Commission (pp. 125–147). In: G. Majone (ed.), *Risk Regulation in the European Union: Between Enlargement and Internationalization.* Florence: European University Institute, Robert Schuman Centre for Advanced Studies.

POLLARD, S.J.T., DUARTE DAVIDSON, R., YEARSLEY, R., TWIGGER-ROSS, C., FISHER, J., WILLOWS, R. and IRWIN, J. (2000). *A Strategic Approach to the Consideration of 'Environmental Harm'.* Bristol: The Environment Agency 2000.

RAYNER, S. and CANTOR, R. (1987). How fair is safe enough? The cultural approach to societal technology choice. *Risk Analysis*, **7**, 1, 3–13.

RENN, O. (1995). Style of using scientific expertise: a comparative framework. In: *Science and Public Policy*, **22**, 147–156.

RENN, O. (2004). The challenge of integrating deliberation and expertise: participation and discourse in risk management. In: T. McDaniels and M.J. Small (eds), *Risk Analysis and Society. An Interdisciplinary Characterization of the Field* (pp. 289–366). Cambridge: Cambridge University Press.

RENN, O. (2005). *White Paper on Risk Governance. Towards a Harmonised Framework*. Document Version: Draft for Peer Review (Version 6), May 2005. Geneva: International Risk Governance Council.

RENN, O. and KLINKE A. (2001). Public participation across borders. In: J. Linnerooth-Bayer, R.E. Löfstedt, and G. Sjöstedt (eds), *Transboundary Risk Management* (pp. 245–278). London and Sterling, VA: Earthscan.

RENN, O., WEBLER, T., RAKEL, H., DIENEL, P.C. and JOHNSON, B. (1993). Public participation in decision making: A three-step procedure. *Policy Sciences*, **26**, 189–214.

ROQUEPLO, P. (1995). Scientific expertise among political powers, administrators and public opinion. *Science and Public Policy*, **22**, 3, 175–182.

SHRADER-FRECHETTE, K. (1995). Evaluating the expertise of experts. *Risk: Health, Safety and Environment*, **115**, 115–126.

SKOGSTAD, G. (2001). Internationalization, democracy, and food safety measures: The (il)legitimacy of consumer preferences? *Global Governance*, **7**, 293–316.

SKOGSTAD, G. (2003). Legitimacy and/or policy effectiveness?: Network Governance and GMO regulation in the European Union. *Journal of European Public Policy*, **10**, 3, June, 321–338.

SKOGSTAD, G. ('Regulatory policy styles'; Internet publication without specification of date). Regulating Food Safety Risks in the European Union and North America: Distinctive Regulatory Policy Styles. First Draft for: C. Ansell and D. Vogel (eds), *European Food Safety Regulation: The Politics of Contested Governance*. University of California, Berkeley. Accessed on: http://www.polisci.berkeley.edu/faculty/bio/permanent/ansell,c/foodsafety/Sk.pdf.

STIRLING, A. (1998). Risk at a turning point? *Journal of Risk Research*, **1**, 2, 97–109.

STIRLING, A. (2003). Risk, uncertainty and precaution: some instrumental implications from the social sciences. In: F. Berkhout, M. Leach and I. Scoones (eds), *Negotiating Change* (pp. 33–76). Edward Elgar.

STIRLING, A. (January 2005). *A General Framework for the Integration of 'Risk Assessment' and 'Precaution' in Appraisal for the Governance of Food Safety*. Draft paper for the EC Framework Programme 6 Integrated Project on SAFE FOODS. SPRU – Science and Technology Policy Research, University of Sussex, Brighton.

TRICHOPOULOU, A., MILLSTONE, E., LANG, T., EAMES, M., BARLING, D., NASKA, A., VAN ZWANENBERG, P. and CHAMBERS, G. (September 2000). *European Policy on Food Safety*. Report to the European Parliament's Scientific and Technological Options Assessment Programme (STOA), PE number: 292.026/Fin.St. Accessed on: http://www.europarl.eu.int/dg4/stoa/en/publi/default.htm.

TVERSKY, A. and KAHNEMAN, D. (1981). The framing of decisions and the psychology of choice. *Science*, **211**, 30, 453–458.

VOS, E. (2000). EU Food Safety Regulation in the Aftermath of the BSE Crisis. *Journal of Consumer Policy*, **23**, 227–255.

VOS, E. (2004). *Overcoming the Crisis of Confidence: Risk Regulation in an Enlarged*

European Union. Inaugural lecture of 23 January 2004. University of Maastricht.

VOS, E. and WENDLER, F. (2006). *Food Safety Regulation in Europe: A Comparative Institutional Analysis* (Serier las Commune). Antwerp: Intersentia Publishing.

VOS, E., NÍ GHIOLLARNÁTH, C. and WENDLER, F. (2005). *EU-Level Report* (July 2005). For the EC Framework Programme 6 Integrated Project on SAFE FOODS, contribution to workpackage 5 on the 'Investigation of the Institutional Challenges and Solutions to Systemic Risk Management'. University of Maastricht, Faculty of Law.

WBGU, WISSENSCHAFTLICHER BEIRAT DER BUNDESREGIERUNG GLOBALE UMWELT-VERÄNDERUNGEN (2000). *World in Transition: Strategies for Managing Global Environmental Risks.* Berlin: Springer.

ZWICK, M.M. (1999). Gentechnik im Verständnis der Öffentlichkeit – Intimus oder Mysterium? In: J. Hampel and O. Renn (eds), *Gentechnik in der Öffentlichkeit. Wahrnehmung und Bewertung einer umstrittenen Technologie* (pp. 98–132). Frankfurt/M. and New York: Campus.

25

Science, society and food policy

D. Coles, Enhance International Limited, The Netherlands

25.1 Introduction

Over the last 20 years dramatic advances in science and technology have taken place in many fields. Huge steps in the development of computing power coupled with its increasingly ubiquitous use by scientists in all areas of their work, has of course been a key driving force. This in turn has facilitated new breakthroughs across many science disciplines resulting in new applications which have themselves also found important applications in other fields, so creating synergies which generate even more rapid advances in new technological applications.

   Although the focus of this book is on understanding consumers' attitudes to food, the impact of the new technologies on all aspects of the quality of life has been profound (Frewer *et al.*, 2004; Frewer and Salter, 2002). For the most part such changes have been positive for those in a position to benefit from them. However, many of these advances have had, or are having significant impacts on society as a whole and in many cases the jury is still out on whether the impact is beneficial or harmful in respect of our quality of life or whether they make a positive contribution to fairness, equity and justice for humankind as a whole (Mehta, 2004). Developments at the boundaries of science and technology, in particular, give rise to consumer concerns especially where the consumer is unable to identify either a personal or societal benefit (Frewer, 2003). One outcome of this is a rise in consumer concern about where new developments in science and, in particular, food technology may lead, and what unforeseen implications some of the cutting edge innovations may have for our society and that of future generations.

It is hardly necessary to point out that issues such as BSE, GM crops and other such high profile crises have adequately demonstrated the necessity for food policy and food governance structures to take account of public perceptions and concerns (Frewer and Salter, 2002; OECD, 2000, Phillips, 2000). Not only do such concerns need to be identified, the underlying rationale that drives them also needs to be understood, if society is not to simply stagger from one food crisis to another (see also de Jonge *et al.*, Chapter 5, this volume).

This chapter will look at some of the indicators of the way consumers view the increasing impact of science and technology on their lives, discuss some of the reasons that lie behind this and point to some of the policy actions that need to be considered.

25.2   Human 'well-being' or human progress?

It is generally well established that in developing and industrial societies, scientific advances and new technologies are seen as important factors in improving the quality of life. However, in post-industrial societies there is evidence (World Values Surveys 1999–2001), that those who perceive themselves to be more highly emancipated are more sceptical about the positive impacts of science upon their quality of life (Welzel *et al.*, 2003). It is argued that economic development changes the value priorities in advanced industrial societies. 'As survival becomes increasingly secure, a "materialist" emphasis on economic and physical security diminishes, and people increasingly emphasise post materialist goals such as freedom, self-expression and the quality of life' (Ingelhart and Abramson, 1999).

Results from a recent Eurobarometer study (Special Eurobarometer 225, Social Values, Science and Technology, European Commission, March 2005), appear to confirm this hypothesis. In response to a question about what values it is important to teach children, values such as tolerance and respect for others, sense of responsibility and independence score much more highly than do the more 'traditional' values of obedience, thrift and hard work.

The Eurobarometer results also suggest that many such communities have an ambivalent attitude to many types of research and new technological developments, particularly in relation to their implications for ethical issues such as our responsibility toward the environment, use of animals, respect for human life, personal choice and freedom and the stability of society.

The Eurobarometer survey mentioned above indicates that almost 90% of Europeans believe we have a responsibility to protect the environment, even if this means limiting human progress and over 80% believe we have a duty to protect animals, whatever the cost. However, at the same time, around 43% say that we have the right to exploit nature for the benefit of human well-being and 51% believe that exploitation of the environment is inevitable if humankind is to progress. There are also large variations in responses in different Member States. The survey concludes that this ambivalence is, in part, explained by a distinction in people's minds between 'human progress' (developments in science, technology or innovation) and 'human well-being' (the state of being happy and healthy). In other words while most people would accept the importance and in the case of some, the right of humans to have happy and healthy lives, many no longer assume this as necessarily being linked to progress in science and technology. As a result these people have become much more cautious about new scientific advance *for its own sake* and wish to subject it to a much more critical and value-laden appraisal as to whether, in fact, it is likely to be of benefit or harm to society as we know it or as they would wish it to be.

25.3   How to approach societal change?

It may be worth a little philosophical reflection on how societal change might be approached. This chapter proposes three extreme models that might be considered.

The first, could be called the 'steady state model', and is represented by the view that today's society provides a good quality of life, a sense of well-being and a good set of values by which to live. They accept that scientific progress might produce further improvements in the quality of life but essentially this group wish to preserve the current set of societal norms and values. They do recognise that there are others who have a lesser quality of life and are unable to enjoy all the benefits of this society, and so they also wish to see progress towards a universal ideal in which the societal benefits and values they enjoy might be shared with all humankind. They acknowledge that science and technology has an important role to play in extending the compass of this social model to all, but that the pursuit of science must operate within the constraints and values of established society and should be governed and where necessary, constrained, to ensure that it does not follow paths that might challenge or jeopardise this 'steady state' of currently held values.

The second approach, which might well be favoured by those with a particular ideology in view, either secular or religious, could be described as the 'social engineering model'. While in some ways similar to the steady state model, it does not consider present-day society but some future idealised social model as the objective to be achieved. Again, as with the steady state model, science and technology are seen as having a subservient role in facilitating the attainment of this ideal social model, with any avenues of exploration that appear to threaten or challenge this ideal model or its underlying values, being curtailed. Both these first two approaches implicitly grant moral superiority to society and its values, with science having an important role within that society but subservient to it.

The third approach is that of the 'evolutionary model'. Here the desire of humankind to pursue knowledge for its own sake and seek to apply what it learns to facilitate progress, has a much more dominant role, with science and technology having free reign to pursue whatever avenues of exploration present themselves. At its extreme, this approach sees the pursuit of science as being separated from (and superior to), moral values and society because the intellectual patterns of science are of a higher evolutionary order than the older biological and social patterns (Pirsig, 1991). Hence it is better (and therefore more morally acceptable) for an idea to destroy a society or its values than for a society to destroy an idea (Pirsig, 1991). The implication of this approach is that scientific intellectualism provides a better moral guardian of humankind and its progress than any particular society or set of values. Being both reductionist and relativist it perceives science as being 'value-free' and so envisages humankind together with its values and social models continually evolving and adapting to accommodate new knowledge and its application.

Are any of these models the right one?

From a historical perspective, it is clear that societies that do not progress eventually stagnate and die or are overtaken by more developed societies and that 'social engineering' proposed above also always seems destined to fail. However it is also evident that while the pursuit of new knowledge has greatly benefited the well-being of humankind, the exploration and application of science and technology, contrary or without reference to social and ethical values or without philosophical reflection, has unleashed some extremely destructive forces upon humanity and our environment and has resulted in indescribable suffering for many.

> Philosophy is not science, it is not made of scientific facts, it is not even based on such facts. It is quite the opposite: Philosophy is the foundation of science, giving it aim, direction, and background. Wisdom is not science. Science has structure, clear concepts, logic, and rules. Wisdom has none of these, because wisdom is about simply being, and therefore knowing. But being and calmness is slowly being substituted by doing and business as the world develops its economical wealth, and likewise is wisdom slowly being substituted by science. To keep wisdom in our human culture in the future, we must try to make a science that contains it, preserves it, and – if possible – even develops it (Ventegodt et al., 2003).

It may be true that ideally, science should be considered to be value-free. However, those who practice science, together with those who apply it and the technologies and tools it produces, are themselves never value-free. Also science is after all dependent for its existence on the society that supports it and utilises its outputs. To claim exemption from the values framework of that society is itself morally questionable. However, this does not mean that science should not be able to challenge accepted societal norms – in fact it should be encouraged to do so. But science is not alone in this. Every member of a society, in whatever role they function, should be free to explore their environment, whether it be physically, emotionally, socially or intellectually, and to challenge the accepted values of that society. However, no activity, be it science, politics, economics, religion or individual enterprise, can operate without restraint. It is the role of a healthy and dynamic society to utilise all aspects of human endeavour to move forward with balance within as safe as possible a framework, seeking to achieve an ever-increasing state of well-being for its members and its environment.

Therefore, as for any other endeavour emerging from a given society, both scientific research and the application of its results must take place within some system of governance and ethics where certain avenues of research or particular applications may be closed off because they present either too great a risk or are contrary to, or offend, the prevailing ethical framework or morality of that society. This is clearly evident from the fact that certain research is, quite rightly, universally unacceptable, for example research on people against their will or where it could be foreseen that it might result in great pain, suffering or even death to research subjects (WMA Declaration of Helsinki, 2000; Charter of Fundamental Rights of the European Union, 2000). This would be the case even if there were potentially significant benefits to society as a whole. Similarly scientific research involving highly dangerous or toxic substances and organisms is only permitted under strict regulation where the risks can be minimised (EU Directive 2001/20/EC; EU Directive 2001/18).

Thus, while it is important to recognise and respect the right of scientists to explore new frontiers, it must also be recognised that they do so at the frontiers of the society of which they are part. However, the new, and possibly greatest challenge that we now face is that our society is becoming increasingly global, in which it is necessary to accommodate much greater diversity and inequality in human well-being. How can this best be achieved?

25.4   The challenge to policy makers

The results of the Eurobarometer survey described earlier indicate a much greater reluctance on the part of European society to exploit nature for scientific advance *per se*. This has important implications for scientists, industry and policy makers. A key underlying political driver for more scientific research and greater and more rapid innovation is, of course, economic competitiveness and economic growth. This is based on the premise that as the economy grows this translates into an improvement in the quality of life.

At its Lisbon Summit in 2000 the European Union announced its Lisbon Strategy: 'to become the most competitive and dynamic knowledge-based economy in the world, capable of sustainable economic growth with more and better jobs and greater social cohesion.' R&D in biotechnology was recognised to be a key driver in achieving this agenda with the European Commission declaring that: 'Life sciences and biotechnology are widely recognised to be, after information technology, the next wave of the knowledge-based economy, creating new opportunities for our societies and economies' (COM (2002) 27).

However, for much of Europe there are significant and vocal publics who consciously or subconsciously perceive themselves as being emancipated from significant economic need and for whom, therefore, economic growth is not their highest priority. Increasingly therefore, those wishing to promote the pursuit of new research or implement new technologies, especially in areas which are particularly novel or ethically sensitive, must keep in mind the need to present their case to those who both fund scientific research and consume its products, in terms of its contribution to the well-being of humankind.

It is no longer acceptable for scientists to assume they can tell people what is good for them, or to consider themselves above the constraints of the society to which they are a part. In an increasingly global environment it is becoming more and more important for science to operate within a broadly accepted framework of governance and code of ethical conduct. In most countries, for example, scientific research, particularly that which involves human or animal subjects is subject to a local ethical review. It is becoming less and less acceptable for researchers to try to sidestep this process by seeking to carry out research in countries that have little or no system of ethical review. Increasingly there are calls for the definition of and adherence to, basic underlying principles of ethics and governance, even thought both the interpretation and application of those principles may differ between cultures (Dal-Ré *et al.*, 2005; Maschke and Murray, 2004).

> Even though the rapid advances taking place in science and technology mean that such principles have to be subject to ongoing review and may have to adapt to rapid changes in both society and technology, they are an essential factor in enabling humankind to develop in a stable and coherent manner. Indeed the overwhelming majority of scientists themselves recognise and support committees that have institutional oversight of their work (Orlans, 1987).

25.4.1 EU Ethical Review

One example of how this process can operate successfully is the ethical review procedure adopted by the European Union for its 6th Framework Programme for research and technological development (FP6).

Prior to FP6, the European Commission operated a limited ethical review process for certain areas of research that it funded under previous framework programmes; mainly research in the fields of medicine and food, including red and green biotechnology. However, under FP6 the EU decided that all research that was to be funded by the European Community research framework programme should be subject to ethical review.

What was the reason behind this decision? Was it simply to provide a means of allaying public fears of the EU funding 'unethical' research or to deflect criticism of the 6th Framework Programme agenda? Whatever the original intention, the FP6 ethical review system has developed into a well-respected ethical review process, not only ensuring that the EU does not fund unethical research but also raising awareness amongst researchers and complementing the local ethical review process in validating the research itself.

Because the European Union is a rich multinational and multicultural tapestry, it was recognised that different societies may well have different cultural values and that there has to be room for this diversity within any framework of ethical review. Although the establishment of an agreed set of 'ethical rules' for FP6 was not without its difficulties, particularly in relation to highly sensitive areas of research such as the use of human embryonic stem cells, where fierce debate about the ethical issues almost delayed the start of the 6th Framework Programme, the European Commission managed to establish on behalf of the EU an ethical review process that retains both adherence to a set of fundamental ethical principles and a considerable degree of flexibility which takes into account not only regulatory and value differences within different Member States but also broader changes in public attitudes and values, reflected in national, EU and international regulations.

Research proposals for funding by FP6 that are identified as giving rise to ethical issues, are reviewed at EU level by an independent and multidisciplinary panel of experts whose task is to ensure that the researchers understand the ethical issues involved in their proposed work and that the research is in line with the fundamental ethical rules established by the European Commission and adopted by the European Parliament and European Council. This includes making sure that it complies with relevant EU directives and international regulations on research such as the Helsinki Declaration (*ibid.*) and the Council of Europe's Convention on Human Rights and Biomedicine (1997). This EU level ethical review does not, however, replace local or national ethical review in the individual countries where the research will take place; which is especially important for multinational co-operative research projects. Researchers also have to make sure that they comply with the regulations in the particular countries in which they carry out their research and the EU makes clear that it will not fund any research in a country where that research is not permitted, nor will it fund in any third country, research that is forbidden in *all* Member States. Therefore even in those countries that have few or no procedures for ethical review, the EU will only fund research that meets its own fundamental ethical standards. So, for example, research involving humans would only be funded if the European Commission was satisfied that each individual subject were in a position to give informed consent to their involvement.

There have been demands by some for the EU to adopt a more closely defined and specific ethical code for its research programmes which specifies in detail exactly what research is and is not allowed. However, not only would this be extremely cumbersome, it would also place a much more restrictive burden on the research community and could as a consequence also seek to impose the specific values of one community on others.

Much of the strength and stability of the European Community lies in its diversity and its ability to find positions of compromise, accommodation and flexibility that enable it to embrace both different and changing cultural norms.

25.5   The challenge of changing perceptions of acceptable risk

In an environment where consumers and other publics have an increased desire for involvement in decision-making that affects not only themselves but society as a whole, the process of ethical review described above is one example of how policy-makers must ensure that they put in place systems of governance that are not only open and transparent but which also reflect the underlying values of the society itself. They must also demonstrate that they have measures in place to protect both society and individuals against any risks that may arise from the new technologies. Because of the fast pace of change in science and technology, it is essential that such governance systems have the capacity for ongoing engagement with all stakeholders including scientists and the publics in order to adapt quickly to new developments and change in societal attitudes. Without such openness, engagement and assessment of potential risks, policy-makers expose themselves to the very real danger of important new technologies not being accepted because they have failed to secure public confidence in their development and use.

25.5.1   'Blame and claim' culture

Better informed and more critical publics have resulted in consequent growth of a culture in which there is a greater tendency to identify fault as a result of some realised risk or disaster. Depending on the nature of the disaster, the fault may be considered to lay with a manufacturer, regulator, public or private service provider or policy makers. Even unforeseen 'natural' disasters such as volcanoes, tsunamis, earthquakes, epidemics or pandemics may result in governments, local authorities and others being subject to charges of lack of preparedness

One significant impact of this is that it motivates governments, local authorities, policy makers, regulators, industry and others in positions of responsibility to take measures to ensure that they protect themselves from public accusations of blame. Some of these measures are both practical and helpful such as clear labelling of components on food products and chemicals. Sell-by or 'best before' dates on food also have a very practical application in improving public health and raising awareness of health risks.

However, many other initiatives may be more dubious in their benefit to the public, being much more intrusive and impinging on personal freedoms and choice. These so-called 'nanny state' initiatives often include public health issues such as obesity, healthy eating, vaccination programmes and/or compulsory vaccination, smoking bans, etc. Other measures to reduce public 'risk' are increasingly being introduced under the label of 'security measures' following '9/11' and other terrorism attacks. However, they increasingly go far beyond anti-terrorist measures and include, for example collection of biometric data including DNA collection and testing for large proportions of populations (Nelkin and Andrews 1999; Btihaj Ajana, 2005), collection and use of other personal data (medical, food, lifestyle, information on personal movements, telephone, email, banking, CCTV and other surveillance data, etc.) (Lyon,

2006). Even where this data is stored so-called 'anonymously', it may well be identified by postcode for health or crime mapping, for example (Cummins *et al.*, 2005), which may restrict it to 1–5 households and so make individual identification relatively simple.

In many cases the argument put forward is that this information is extremely useful in identifying and apprehending both terrorists and criminals and that 'if you have done nothing wrong you have nothing to fear'. On the whole this argument is accepted by a significant proportion of those societies which have benign government – but is an issue of far greater concern for those with more repressive regimes.

Technological advance means that this data is much more accurate, comprehensive and specific as well as being more readily interrogated and cross-referenced, so increasing the interrogative power by orders of magnitude. As a result, on a global level we are rapidly approaching a point where either further advances in this direction may need to be seriously curbed or it will become necessary to completely revise traditional and almost universally held concepts of personal rights to confidentiality and privacy.

It would be interesting to investigate the extent to which the sensitivity of a society to risk, whether real or perceived, affects its willingness to accept intrusive monitoring, regulation and governance.

25.5.2   The role of information

New technologies have enabled people to be better informed. This information access operates on many levels. Two of the most significant levels are superficial or 'soundbite' awareness and what I shall call 'interest group' awareness. The first of these can be considered as the perception of those whose knowledge comes almost exclusively from the mass media and social contacts. Here the influence of the media and associates is high and while the impact of any given individual is usually very small, the sheer weight of numbers can generate a powerful momentum of generally unfocused opinion which, nevertheless, can force significant changes in or even abandonment of a particular public policy.

Interest group awareness is very different. For the most part those operating at this level obtain their information from much more accurate and information-rich sources. Here also developments in information technology have played a crucial role. The impact here is much more focused, and is usually characterised by demands for much greater direct engagement with policy-makers in a process of advising or negotiating on policy (Bishop and Davis, 2002). As a result the impact is usually in the area of amending and refining policy to improve its public acceptability or of contributing to the formulation of new policy (see also Chapters 5 and 27 by de Jonge *et al.* and Rowe, respectively, in this volume)

25.5.3   More sensitive measures

Another important effect of technology on people's perception of risk has been the development of ever more sensitive methods for identifying hazardous substances. As a result, where a substance is known to present a hazard (such as toxicity or mutagenicity), society increasingly demands that even trace amounts of it should be eliminated even where 'safe' levels are well-established. This is particularly evident in the cases of the environment, consumer products and, of course, food (Daft, 1991; Daughton, 2004).

Food products are exceptionally sensitive and vulnerable. Public perception of a possible health risk from or contamination of a food product can result in significant and extremely rapid changes in public behaviour and purchasing patterns even when there is no scientific basis for this. For example, the recent high level of awareness in the UK of the potential for an avian flu pandemic resulted in a rapid fall in the consumption of eggs and poultry.

Others have described in some detail elsewhere in this publication, examples of food scares that have arisen from both real and perceived health risks, and highlighted the importance of regulators and policy-makers taking into account not only scientifically demonstrable risks but also public perceptions of risk.

25.6   Citizens input to the decision-making process

Food and health risks continue to generate high levels of publicity, with even low-level risks being highlighted, which lead to even more demands by society for an increasingly risk-free environment. There is much greater direct public access to sources of information (both good and bad sources) and as a result people are much more reluctant to accept being told what is good for them or what is safe. They want not only more opportunity to exercise personal choice but also to have a much greater say in decision-making processes. For example, 75% of Europeans believe that they should be more involved in political affairs although over 80% think that they have too little influence on policy-making (Special Eurobarometer 225, 2005).

Despite differing perceptions of particular risks or avenues of scientific exploration and concerns about challenges to sincerely held value systems, citizens as a whole have a pretty balanced and common-sense approach to the governance and management of scientific research and the challenges and risks that it may generate. Even though two-thirds of Europeans still think that policy decisions about science and technology should be based primarily on the advice of experts rather than the public (EB 225, 2005), in a post-industrial society, particularly, ready access to information and greater perceived emancipation leads to more demands to have a meaningful contribution to the decision-making process.

It is therefore important for policy-makers, industry and scientists to be seen to take public views into account. There is pressure to ensure that policy decisions take on board not only scientific, technical and economic (and, of course, political) perspectives, but that values are also part of the decision-making process. The danger here, of course, lies in how and to what extent the views of the public are actually taken into account and have a genuine role in contributing to policy. Public debate or consultation exercises which are little more than 'window-dressing' for predetermined decisions by policy-makers, in the medium to long term only diminish public confidence in those responsible for decision-making. There has to be sufficient transparency in the process for today's publics to be able to appreciate how their input contributes to the policy decision. The examples of public GM debates in the UK, Australia and New Zealand all demonstrate the problems of badly thought through consultative processes and the difficulties of linking regulatory concerns with the values of significant segments of society (Walls *et al.*, 2005).

There is always a difficulty for policy makers when they perceive the views or values of the public to be at odds with commercial interests or their own long-term strategies for economic growth. As an example we might consider the EU Biotechnology Strategy referred to earlier in this chapter. In 2001 the European Council in Stockholm invited the European Commission:

> to examine measures required to utilise the full potential of biotechnology and strengthen the European biotechnology sector's competitiveness in order to match leading competitors while ensuring that those developments occur in a manner which is healthy and safe for consumers and the environment, and consistent with common fundamental values and ethical principles.

Note the emphasis in this statement on the need for biotechnology to develop with the health, safety, values and ethical principles of consumers very much in mind.

This approach was reaffirmed in the 2002 biotechnology strategy document itself which emphasised that it was essential to maintain broad public support and to address ethical and societal implications and concerns so that it could deliver effective, credible and responsible policies on biotechnology which enjoyed the confidence and support of its citizens (COM (2002) 27 - *ibid.*). This document contains over 40 references to ethics and values and 18 references to consumer choice.

Each year the Commission produces a report on the Biotechnology Strategy. However, as it became clear that the original objectives of the Lisbon Strategy were not on track to be met by 2010, it is interesting to note that the emphasis in these reports on bringing products to market has increased markedly while reference to ethics, values and consumer choice has significantly declined. The 2005 report on the Biotechnology Strategy, which followed the 'refocusing' of the Lisbon Strategy (Barosso, 2005), has no reference at all to consumer choice and only 11 references to ethics and values. References to stakeholder consulta-tion appear to focus on industry, regulators and the professions rather than any consumer consultation. The refocused Lisbon Strategy does not mention these aspects at all.

It is true that ethics is a politically sensitive issue at the EU level and some ethical issues (in particular those related to the use of human embryos) have threatened to seriously disrupt the EU research agenda. Consumer choice is also politically sensitive, particularly at the international level, where, with GMOs, for example, consumer reluctance to accept even low levels of GM products has brought the EU into conflict with the United States and the WTO. Even within Europe this results in tensions because the failure of Member States to reach an agreement on whether or not to allow a particular GM food product onto the European market means that the Commission, through its comitology rules can implement the licensing unilaterally. This, of course, removes the choice from consumers (even when represented by their own national governments) and risks the possibility of some serious backlash in the future.

There is always a risk either that economic imperatives might seek to override ethical and societal concerns or consumer choice or that such concerns might retard progress and make the economy uncompetitive on the international stage.

It is therefore understandable that policy-makers may sometimes be nervous of engaging the public in any governance or decision-making process, particularly where they have concerns that societal perceptions may restrict development or impede the market. However, policy-makers ignore the consumer at their peril. It is essential for the future of both science and society, that policy-makers maintain the right balance between the economic imperatives of a developing knowledge society, societal cohesion and the concerns and priorities of the consumer, by ensuring wide scientific, philosophical and consumer involvement in any debate.

25.7   References

BAROSSO, Robert Schuman Lecture for the Lisbon Council, 2005.

BISHOP, P. and DAVIS, G. (2002), Community consultation symposium, mapping public participation in policy choices, *Australian Journal of Public Administration*, **61**, 14.

BTIHAJ AJANA (2005), Surveillance and biopolitics, *Electronic Journal of Sociology*.

Charter of Fundamental Rights of The European Union (2000/C 364/01).

COM (2002) 27 Life sciences and biotechnology – A Strategy for Europe.

COUNCIL OF EUROPE (1997), Convention on Human Rights and Biomedicine, Oviedo, Spain, April 4.

CUMMINS, S., MACINTYRE, S., DAVIDSON, S. and ELLAWAY, A. (2005), Measuring Neighbourhood Social and Material Context: Generation and Interpretation of Ecological Data from Routine and Non-routine Sources, *Health and Place*, Elsevier.

DAFT, J.L. (1991), Fumigants and related chemicals in foods: review of residue findings, contamination sources, and analytical methods, *The Science of The Total Environment*, March, **100**, Spec No. 501-18, Pub-Med.

DAL-RÉ, R., ORTEGA, R. and MOREJÓN, E. (2005), Multicentre trials review process by research ethics committees in Spain: where do they stand before implementing the new European regulation?, *J. Med Ethics*, **31**. 344–350.

DAUGHTON, C.G. (2004), Groundwater recharge and chemical contaminants: challenges in communicating the connections and collisions of two disparate worlds, *Ground Water Monitoring & Remediation*, Spring, **24**, 2, 127–138.

EU Directive 2001/20/EC of The European Parliament and of The Council of 4 April 2001 on the approximation of the laws, regulations and administrative provisions of the Member States relating to the implementation of good clinical practice in the conduct of clinical trials on medicinal products for human use.

EU Directive 2001/18 on the deliberate release into the environment of genetically modified organisms.

FREWER, L. (2003), Societal issues and public attitudes towards genetically modified foods, *Trends in Food Science and Technology*, **14**, 5–8, 319–332.

FREWER, L. and SALTER, B. (2002), Public attitudes, scientific advice and the politics of regulatory policy: the case of BSE, *Science and Public Policy*, **29**, 2, 137–145.

FREWER, L., LASSEN, J., KETTLITZ, B., SCHOLDERER, J., BEEKMAN, V. and BERDAL, K.G. (2004), Societal aspects of genetically modified foods, *Food and Chemical Toxicology*, **42**, 11811193.

INGLEHART, R. and ABRAMSON, P.R. (1999), Measuring postmaterialism, *American Political Science Review*, **93**, 3, September.

LYON, D. (2006), 'Surveillance, power and everyday life', in *Oxford Handbook of Information and Communication Technologies*, Oxford: Oxford University Press.

MASCHKE, K.J. and MURRAY, T.H. (2004), Ethical issues in tissue banking for research: the prospects and pitfalls of setting international standards, *Theoretical Medicine and Bioethics*, **25**, 2, 143–155.

MEHTA, M.D. (2004), From biotechnology to nanotechnology: what can we learn from earlier technologies?, *Bulletin of Science, Technology & Society*, **24**, 1, 34–39.

NELKIN, D. and ANDREWS, L. (1999), DNA identification and surveillance creep, *Sociology of Health and Illness*, **21**, 5, 689–706.

OECD (2000), Genetically Modified Food, Widening the Debate on Health and Safety, The OECD Edinburgh Conference on the Scientific and Health Aspects of Genetically Modified Foods.

ORLANS, F.B. (1987), Scientists' attitudes toward animal care and use committees, *Laboratory Animal Science*, **37**, Spec No. 162-6. Pub Med.

PHILLIPS, L. (2000), *The Phillips Report on the BSE* Crisis, London: HMSO, Crown Copyright.

PIRSIG, R. (1991), *Lila: An Inquiry into Morals*, Chap. 29. Bantam.

Special Eurobarometer 225, Social Values, Science and Technology, European Commission, March 2005.

VENTEGODT, S., ANDERSEN, N.J. and MERRICK, J. (2003), Quality of life philosophy: when life sparkles or can we make wisdom a science?, *The Scientific World JOURNAL*, 3, 1160–1163.

WALLS, J., ROGERS-HAYDEN, T., MOHR, A. and O'RIORDAN, T. (2005), Seeking citizens views on GM crops, *Environment*, **47**, 7, 22–36.

WELZEL, C., INGLEHART, R. and KLIGEMANN, H.D. (2003), The theory of human development, a cross-cultural analysis, *European Journal of Political Research*.

World Medical Association Declaration of Helsinki, latest revision, Edinburgh 2000.

World Values Surveys 1999–2001.

26

Planned promotion of healthy eating to improve population health

J. Brug and B. Wammes, Erasmus University Medical Centre Rotterdam, The Netherlands

26.1 Introduction: a simple model for planned promotion of population health

Ecological studies, migrant studies and case-control studies in the 1980s suggested that a high total fat intake was associated with higher breast cancer incidence (e.g. ref. 1). This association was one of the reasons for fat reduction campaigns in many countries. The preliminary evidence that fat intake increased cancer risk was subsequently communicated to the general public in order to increase risk perceptions to motivate dietary change. Mass-media approaches were often applied to spread the fat-and-cancer-risk message (e.g. refs 2 and 3).

However, later prospective cohort studies, i.e. studies with a much stronger design to investigate possible dietary risk factors for cancer, failed to confirm that dietary fat intake was an independent risk factor for breast cancer (e.g. ref. 4). Furthermore, results of health psychology research have provided strong evidence that fear appeals often fail to motivate people to change their risk behaviours[5,6] and health education studies have indicated that mass-media interventions alone are not well-suited to initiate behaviour changes.[7,8] In short: a doubtful behaviour change goal was pursued, by targeting a doubtful determinant of behaviour change, with a doubtful intervention strategy.

A healthy existence is to a large extent dependent on health behaviour,[9–11] and next to smoking and lack of physical activity, unhealthy eating has been identified as one of the important behavioural risk factors for important burdens of disease worldwide[9,11] and the main means to promote healthy dietary habits are health education and health promotion. To avoid non-optimal use of the sparse resources available for healthy diet promotion, careful evidence-based

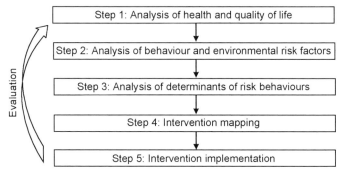

Fig. 26.1   A model for planned health education and health promotion.

planning of such interventions should be standard procedure in order to increase the likelihood that the right behaviour change goals are pursued by targeting the right behavioural determinants with the right intervention strategy.

This careful planning of health promotion has especially been advocated since the publication of Green and Kreuter's Precede and Precede-Proceed models.[12] These and other planning models[13,14] show great similarities. Comparison and integration of the available health promotion planning models identified five important steps or phases in health promotion planning[15] (Fig. 26.1).

The first two steps cover the epidemiological analysis. When applied to public health nutrition, these steps should identify the most important health and quality of life issues and their nutritional risk factors. These first two steps thus result in setting priorities for dietary change interventions and identification of behaviour change goals.

The third step is to investigate the mediators or determinants of these dietary risk factors. Since diet is subject to free choice for most people, especially in affluent societies where the variety in possible food choices is overabundant (with exceptions for specific population groups such as small children and some institutionalised people), eating behaviours cannot be influenced directly. In healthy diet promotion we need to be able to influence people's choices. What, when and how much people choose to eat is influenced by a complex, inter-related set of so-called behavioural 'mediators' or 'determinants' and successful dietary behaviour change interventions are dependent on the identification of the important and changeable determinants. This step of determinants analysis thus helps to identify more proximal, intermediary intervention goals, but also to identify specific target groups for interventions.

In step 4 of the planning process, intervention strategies, methods and materials need to be selected or developed that are tailored to the target popula-tions and the most important and best modifiable determinants of behaviour change. In step 5 the intervention should be implemented and disseminated so that the target-population is reached and exposed to the nutrition education messages. Each step should preferably be evidence-based and evaluation of each step is necessary.

In brief, the above-explained planning model states that we need to understand the food consumers' major health problems and related dietary behaviours, as well as what drives the consumer's food choice in order to be able to come up with effective healthy diet promotion interventions. In the remainder of this chapter we will use this planning model to first briefly describe some of the main dietary factors related to population health, after which we will elaborate on determinants of nutrition behaviours and different healthy diet intervention strategies for promotion of population health.

26.2   The epidemiological analysis: a selection of issues in diet, nutrition and population health

Recent reviews of the most important determinants of the major burdens of disease indicate that diet is important.[9,11] Dietary factors – energy, fat, and fruit and vegetable intake in particular – influence risk for the main health threats of today, such as cardiovascular disease (CVD), cancer, type 2 diabetes mellitus, and obesity. There are many other dietary factors that contribute to health, but for brevity this section will be restricted to a short overview of the presumed health effects of excess energy, fat, and fruit and vegetables, since these are arguably the main dietary factors for population health promotion efforts, at least in affluent countries.

Obesity is expected to become the most important determinant of preventable disease, surpassing smoking in this respect within the next few years[16] (although to date underweight is still responsible for a larger proportion of avoidable loss of life years than overweight[9]). Obesity is an independent risk factor for a wide range of diseases including CVD, arthritis, breast cancer, and type 2 diabetes.[17] Weight gain, which may eventually lead to being grossly overweight or obese, is caused by a long-lasting positive energy balance: energy intake (diet) is larger than energy output (basal metabolic rate + physical activities). Reducing energy intake is therefore one means of preventing weight gain. However, people do not eat energy as such but choose a variety of foods that are combined and prepared in different ways. Swinburn and colleagues[18] reviewed the evidence on specific dietary risk behaviours for weight gain. They concluded that there was convincing evidence that a high consumption of energy-dense foods, normally containing large amounts of fat (fat is the most energy-dense macronutrient) and/or sugar, is associated with higher risk for weight gain, overweight and obesity. The evidence for a protective effect of a high intake of fibre-rich foods was also classified as convincing. A high intake of sugar-containing drinks was identified as a probable risk factor for weight gain.

Total fat intake has a long history of smear. The only remaining real suspicion related to total fat intake is related to fat's high energy density; high-fat diets may induce weight gain. However, high-fat low-carbohydrate weight loss diets appear to be somewhat successful, and at least not less effective than low-fat high-carbohydrate diets.[19] Total fat in an energy-balanced diet is thus not regarded as a public health issue anymore, but there is convincing evidence that the quality of dietary fat is of importance, since different kinds of fatty acids contribute to either promotion or to prevention of CVD-risk. A high intake of saturated fatty acids, especially present in foods of animal origin (meat, dairy), is associated with elevated serum LDL cholesterol levels, which is a risk factor for CVD. Poly and mono-unsaturated fat (oils from nuts and seeds and fatty fish, and olive oil respectively, are important dietary sources) on the other hand, are associated with lower LDL and higher HDL cholesterol levels, predicting lower CVD-risk.

Fruit and vegetable promotion has received enormous attention in the past decades because of evidence, mostly from observational studies, that a high fruit and vegetable intake is associated with lower risk for different cancers, especially of the digestive and respiratory tracts, CVD, chronic obstructive pulmonary disease, and obesity.[20] Fruit and vegetables have high nutrient densities and low energy densities. The protective effect of fruit and vegetables has been ascribed to their high contents of antioxidants and fibre. However, recent research suggests that the protective effect of fruit and vegetable for cancer and CVD may be less strong than earlier studies indicated.[21]

Studies investigating possible dietary risk and protective factors are overabundant. In an ongoing review of the World Cancer Research Centre on the dietary factors related to cancer risk numerous original studies were identified, of which a majority was published since the completion of the review that resulted in the much cited World Cancer Research Fund's report on Diet, Nutrition and Cancer, published in 1997.[22,23] A library of papers of similar magnitude has been published on investigations of dietary risk factors for CVD.[24] Although there is still controversy on which specific dietary factors, ingredients, nutrients and foods are most important for health, there is some consensus that diets low in saturated fats, high in fruits and vegetables, and that contribute to a neutral energy-balance should be promoted. There is, however, much less evidence available and thus also much less certainty about why many people do not choose such healthy diets, and how healthy diets can successfully be promoted.

26.3   Determinants of healthy food choice

26.3.1   Motivation, ability and opportunity

People in general choose what, when and how much they eat. To induce dietary change, one needs to change people's food choices. To be able to do that, insight into why people choose to eat what they eat is necessary.

Studies on such determinants of food choice and eating behaviours have used learning theory, social cognition models and ecological models of determinants of human behaviour.[25,26] A framework proposed by Rothschild[27] provides a simple and easy to understand model to categorise the large and diverse number of potential determinants recognised in the different more specific behaviour theories. Rothschild identifies three categories of determinants: motivation, ability and opportunity.

26.3.2   Motivation: do you want to eat a healthy diet?

In nutrition education and other health education research, determinants of behaviour have been studied mostly from a social psychology perspective. Within social psychology, different theories and models have been proposed to study nutrition behaviours. These theories and models include the Health Belief Model, Protection Motivation Theory, Social Learning Theory and the Theory of Planned Behaviour, and these theories share a common feature in that they recognise behavioural decision, motivation or intention as the primary determinant of behaviour. Each theory proposes different but similar determinants of intentions. Based on an integration of insights from the aforementioned theories, four groups of determinants that predict intention can be recognised: attitudes, perceived social influences, self-efficacy and self-representation.

Attitudes are based on a subjective weighing of expected positive and negative consequences or outcomes of the behaviour. Closely related constructs are decisional balance, outcome expectations, and perceived threat. But which expected outcomes are important for most people in making a balanced decision on what to eat? In general, expectations about short-term outcomes are more important than longer-term outcomes. Taste, satiety and pleasure are of major importance for most people. People will eat what they like, and disliked foods will not be chosen.[28] Certain taste preferences are innate, such as a liking for sweet, a dislike for bitter. However, taste preferences can be learned and unlearned.[29] Satiety is a strong reinforcer and we therefore quickly learn to like and appreciate energy-dense foods,[28,29] but the fact that many people like the taste of coffee and beer illustrates that we can even unlearn our innate dislike of bitter tastes. (Learning to like and dislike certain tastes are basic classical and operant conditioning processes.) Some specific types of learning strategies have been identified related to food and eating. Taste-nutrient learning is based on the aforementioned reinforcing character of satiety. Taste-nutrient learning means that people easily learn to like tastes of foods that lead to the pleasant feeling of satiety, and is an example of operant conditioning. Evolution psychologists claim that this makes much sense given the fact that learning to like such energy-dense foods improved chances for survival in the long history of evolution in which times of energy shortage were much more likely than times of abundance. The present day obesity epidemic has, however, been attributed to this innate tendency to learn to prefer energy-dense foods in combination with an 'obesogenic' environment[30] (see Section 26.3.4). In the last decades, a period that is not more than a blink of the eye in the history of mankind, in which an over-abundance of palatable foods have become available and accessible for many people, we still tend to choose foods as if we expect meagre years. Since most fruits and vegetables have low-energy densities, preferences for these foods are not so easily learned.

Two other food preference-learning strategies are examples of classical conditioning and are referred to as taste-taste learning and taste-environment learning. If a new, unfamiliar, taste is combined with a taste for which a preference already exists, people will more easily learn to like the new taste. Almost all lovers of black coffee and tea have learned to like the taste by starting out with sugared drinks. (However, learning to like coffee or tea is also a result of operant conditioning: the caffeine stimulant works as a behaviour reinforcer.) Similarly, tastes that people are exposed to in pleasant physical and/or social environments are also more easily learned to like. Foods first encountered during a pleasant holiday, may become favourite foods this way.

Health-related outcome expectations or beliefs are also important in food choice; 'health' usually comes second after 'taste', if people are asked about what they find important in their diet and food choice,[31,32] especially in women.[33] Nevertheless, 40% of Americans and 57% of Europeans indicated rarely or never to compromise on taste to improve the healthfulness of their diets.[34] Furthermore, in practice, health expectations may only significantly influence food choices for most people when the health consequences are expected to be soon, severe and easy to recognise. People may therefore very quickly develop negative attitudes toward foods for which they are allergic or intolerant, i.e. foods that literally make you sick.[28] But remember that energy-dense foods provide a comfortable feeling of satiety. The potential negative consequences, like obesity, type 2 diabetes and heart disease will present only to some and probably only decades later. Convenience is a third important factor in decisional balance (e.g. ref. 35). In Europe 42% of consumers indicate to rarely or never give up convenience for good health compared to 24% in US and Australia.[34]

Perceived social influence is the second category of determinants of intention, and includes subjective norms and descriptive norms. Subjective norms are expectations about what 'important others' want us to do. If, for example, someone expects that her partner and children want her to eat a diet high in fruit and vegetables, this person will be more motivated to do so. Descriptive norms are based on observed behaviour of important others. If a person's partner and children eat diets high in fruit and vegetables, she will be more likely to be motivated to do so herself.

Self-efficacy, or *perceived behavioural control*, is the third determinant category, and refers to the perception of, or confidence in, one's abilities and skills to engage in certain behaviour. A person who is confident that he can cut back on saturated fat intake will be more motivated to do so. Perceived control is behaviour and context specific. A person can, for example, have high confidence to be able to eat less fat, but not to increase vegetable intake; and confidence to cut back on fat may be high for regular meals prepared at home, but not for eating out. Perceived control is strongly related to abilities and skills, for which we refer to Section 26.3.3.

Finally, *self-representations* or self-identity reflect what a person thinks of as important and stable characteristics of the self. Such representations can importantly influence food choice if related to one's personal moral values and norms. People may see themselves, for example, as health conscious, environment conscious, or animal friendly. Such personal norms may induce specific dietary habits such as healthy eating, choosing organically grown foods or adopting a vegetarian diet.[26]

Additionally, it has been suggested that the aforementioned rationale and conscious decision-making factors can only predict eating behaviour to a limited extent because many eating behaviours are habitual. Different eating behaviours are indeed repeated often and may therefore become habitual. Thus, a conscious decision-making process (as proposed in models like TPB and ASE) may be less likely to occur. *Habitual behaviour* is considered to be 'automatic', triggered by environmental cues instead of conscious evaluations of possible outcomes, the opinion of other people, and confidence about being able to perform the behaviour.[36] Studies show that inclusion of an assessment of past behaviour, in addition to attitudes, norms and PBC, has demonstrated higher explained variance and non-significant associations of attitudes, norms and PBC with behaviour.[37] Such findings support the habit hypothesis.[38] However, tracking of past behaviour to the present is not the same as habitual behaviour. Further, even if past behaviour is a *strong* determinant of present dietary practices, past behaviour is not *changeable*. In contrast, habit strength, a concept that is more than just past behaviour, may be changeable. More comprehensive tools to measure habit strength have been tested successfully and used in previous research.[39] Such measures include assessments of repetition as well as 'automaticity' of eating behaviours. A series of studies that we conducted recently in which we applied such habit strength measures shows that habit strength is indeed a strong predictor and correlate of a range of dietary behaviours (e.g., fat, fruit, soft drink intake), in study populations of adults, adolescents as well as children[40,41] and that habit strength may modify the association between attitudes and intentions as well as intention–behaviour associations.[42]

26.3.3 Ability: what enables people to eat a healthy diet?

In Section 26.3.2 we indicated that people with high confidence in their skills and abilities to make healthy dietary choices, will be more motivated to do so. If such confidence is based on true personal abilities and skills, people can translate their motivation into action. Skills and abilities are to some extent dependent on practical knowledge. For example, knowledge of recommended intake levels and healthy alternatives for unhealthy choices helps to enable voluntary dietary change. To make dietary changes for better bodyweight maintenance, knowledge is necessary about which dietary changes will be most effective. Some knowledge about which foods are high in calories, and which preparation techniques help to avoid caloric enrichment of foods, is helpful to be able to avoid high calorie foods and for self-monitoring of caloric intake. Nevertheless, earlier research has shown (e.g. ref. 43) that knowledge is often not a direct determinant of eating behaviours.

The complexity of energy balance behaviours has been mentioned before. Since caloric intake and expenditure are determined by such complex collections of different specific acts, from choosing foods, portion sizes and preparation methods, to transportation, work and leisure time physical activities, it takes a great deal of food knowledge and good arithmetic skills to monitor one's calorie intake or day-to-day energy balance. If the opportunities for objective self-assessment are lacking, people tend to search for other comparison possibilities, and often social-comparisons are used: people compare their own actions and performance with what they perceive that others do and accomplish. Such social-comparisons tend to be liable to a so-called optimistic bias, especially when people perceive a high personal control. Different studies have shown that many people think of themselves as complying with recommendations for complex behaviours such as low fat intake, fruit and vegetable consumption (e.g. ref. 44, 45), as well as physical activity,[46] while their actual behavioural patterns are not in line with the recommendations. This may not be surprising given the fact that behaviours such as fat intake or energy intake are in fact the result of a series of interrelated specific actions, such as buying, preparing, combining and eating specific foods in different serving sizes. To calculate one's total calorie or fat intake requires intensive self-monitoring as well as advanced arithmetic skills. In studies conducted in the Netherlands we found that up to 10% of the population thought that their diets were too high in fat, while more objective food consumption research showed that about 80% of the population had high fat diets.[47] Similar results were found for fruit and vegetable intakes.[44,45] If people think that they already comply with dietary recommendations, they will not be motivated to change.[45,47] Awareness of personal intake levels is thus an additional predictor of motivation, as well as a moderator of the importance of the other aforementioned determinants of motivation. Studies have shown that awareness of unhealthy eating habits was a strong positive correlate of intentions to make dietary changes and that social cognitions such as attitudes, perceived control and subjective norms were only significantly associated with intentions to change among respondents who were aware of their intake levels.[44,45,47] These findings indicate that improving awareness of personal intake levels is an important first step in improving motivation to change.

26.3.4   Opportunity: availability and accessibility of healthy choices

In promotion of health behaviours in recent decades, most attention has been given to health education as the primary tool to encourage the general public to adopt healthy lifestyles. Health education has been defined as 'planned learning experiences to facilitate voluntary change in behaviour'.[12] Health education, including nutrition education, thus strongly focuses on conscious behaviour change and on improving individuals' motivations and skills to increase likelihood of adopting healthy diets. It is therefore not surprising that TPB, its predecessor the Theory of Reasoned Action, and the Health Belief Model were among the theories most often applied to shape health education interventions.[14] However, people's opportunities to make health behaviour changes may be strongly dependent on the environments they live in. The health promotion movement recognised this ecological focus; health promotion has been defined as 'the combination of educational and environmental supports for actions and conditions of living conducive to health'.[12] This health promotion approach has resulted in a stronger attention for environmental barriers and opportunities for health behaviours, which has resulted in a large number of studies on the associations between environmental characteristics and health behaviours, as well as on the effectiveness of environmental change interventions.

The environment can be defined as everything and anything outside the person. Environmental factors are often believed to influence health behaviour via the personal determinants (motivation and abilities). Environments that offer appealing and tasty opportunities for healthy eating may improve motivation to do so, in an environment that offers easy opportunities, a person may need less motivation and fewer skills to engage in healthy eating, and people who have strong motivation and self-efficacy will be more likely to pursue healthy eating despite environmental barriers. Social Cognitive Theory[48] makes this interaction between person and environment explicit in predicting health behaviour. Just as personal factors have been further subdivided in more specific determinant constructs and proposed pathways of mediation, so can and should the environment be further defined by means of distinguishing various environmental factors.

Different classifications of possible environmental determinants of health behaviours have been proposed,[14,49–51] and these classifications show great overlap and similarities. So-called ecological models of health behaviour arguably put most emphasis on the environmental factors in shaping health behaviours.[52] In early ecological models of health behaviour, five levels of influence were distinguished, intra-personal factors, interpersonal processes, institutional factors, community factors and public policy. Story et al.[51] recognise social environmental influences (interpersonal influences), physical environmental influences (influences within community settings), and macro-systems influences (influences at the societal level). Flay and Petraitis[50] distinguish between the social environment and the cultural environment as important categories of environmental determinants of health behaviour, and within these categories they make a further distinction between ultimate, distal and proximal factors. Based on the distinctions within the environment combined with the proximity of the factors within these broad categories, a matrix or grid could be designed with six cells that represent different classes of environmental influences.

Such a grid structure is explicitly proposed in the ANGELO framework (Table 26.1).[53] This framework was specifically developed to conceptualise health behaviour environments, and enables the identification of potential intervention settings and strategies. ANGELO was developed for investigation and classification of so-called obesogenic environments, i.e. environments that

Table 26.1   The ANGELO grid (based on ref. 53)

	Micro-environment	Macro-environment
Physical environment		
Economic environment		
Political environment		
Socio-cultural environment		

promote excess energy intake and lack of physical activity, but the categorisation of environmental factors seems also applicable for other nutrition behaviours. The ANGELO framework is a grid with two axes.

On the first axis two 'sizes' of environment (micro and macro) are distinguished. Micro-environments are defined as environmental settings where groups of people meet and gather. Such settings are often geographically distinct and there is often room for direct mutual influence between individuals and the environment. Examples of micro-environments are homes, schools, work places, supermarkets, bars and restaurants, other recreational facilities, and also include neighbourhoods. Macro-environments, on the other hand, include the broader, more anonymous infrastructure that may support or hinder health behaviours. Examples of macro-environments are how food products are marketed, taxed and distributed; the media are also included in the macro-environment.

On the second axis four 'types' of environments, are distinguished: physical, economic, political, and socio-cultural. The physical environment refers to availability of opportunities for healthy and unhealthy choices, such as points-of-purchase for fruit and vegetables, soft drink vending machines, availability of low saturated fat spreads in worksite cafeterias, etc. The economic environment refers to the costs related to healthy and unhealthy behaviours, such as the costs of soft drinks, fruit and vegetables or energy-dense snacks. The political environment refers to the rules and regulations that may influence food choice and eating behaviour. Bans on soft drink vending machines in schools, rules on what treats can and cannot be brought to school, nutrition policies in worksites and institutions, but also family food rules are examples of political environmental factors. The socio-cultural environment refers to the social and cultural subjective and descriptive norms and other social influences such as social support for adoption of health behaviour, social pressure to engage in unhealthy habits, and thus show overlap with some of the motivational factors.

The aforementioned social cognition models, such as TPB, assume that the environmental influences are mediated by cognitions: the environment is observed and perceived by the person and these perceptions of the environment will influence attitudes, subjective norms or perceived control. TPB thus, for example, assumes that an environment with little opportunities to eat fruits and vegetables will result in lower perceived behavioural control related to fruit and vegetable consumption and therefore in weaker intentions to eat fruits and vegetables. However, there is some evidence that physical and social environmental factors are significantly associated with dietary behaviours after TPB variables have been accounted for (e.g. ref. 54).

We recently conducted two systematic reviews to evaluate the evidence for environmental influences on dietary behaviours in children and adults.[55] For children, especially, family environmental influences, such as parental support, parenting styles, and parental modelling, as well as availability and accessibility of foods seem important. For adults, availability and accessibility of health foods, portion sizes, family income and possibly labelling of healthy choices are important. Studies in which the interrelations and possible mediation between personal and environmental determinants of food choice are studied are especially needed.

26.4   Interventions to promote healthy eating

Interventions to promote healthy eating should address the most important and changeable determinants of healthy eating. To promote healthy eating, people should be motivated to do so, should be confident about their abilities, and should preferably be exposed to environments that offer them easy opportunities.

Different intervention strategies have been applied to encourage healthy eating. In this chapter we make a distinction between mass media nutrition education interventions, personalised nutrition education interventions and environmental change interventions as strategies that differ in their potential to target the different categories of important determinants of dietary behaviours.

26.4.1   Agenda setting and motivation: mass media interventions

Since the major dietary risk factors are present in large proportions of populations worldwide, interventions are needed that reach many people. Mass media approaches are therefore often chosen to communicate dietary change messages. Effective use of mass communication via mass media may indeed be essential to communicate health information to large audiences.[56–58]

Mass communications has been defined as any form of communication with the public, which does not depend on person-to-person contact, and as: 'purposive attempts to inform, persuade, and motivate a population (or sub-group of a population) using organized communication activities through specific channels, with or without other supportive community activities'.[59] There are currently many examples of using mass media channels for health promotion including broadcast (TV, radio), print (newspapers and magazines, billboards and leaflets) and, more recently, electronic media and Internet (see also Section 26.4.2).

Mass media campaigns to promote healthy eating and discourage unhealthy eating have also often been used in healthy diet promotion. In the past, mass media campaigns were often aimed at behavioural change, but with limited success.[60,61] Such mass media approaches aiming to change behaviours were often based on the assumption that if people knew the facts, i.e. that their diet put them at risk, they would act accordingly: change their eating habits to reduce their health risks. Unfortunately, few of the mass media health promotion campaigns that have been evaluated have demonstrated successful behaviour change, especially in the longer term.[62–64] Despite the fact that mass media interventions may often not lead to behaviour changes, mass media communication campaigns can serve a very useful purpose in health education when their inherent limitations are recognised.[61,65] The limitations of the mass media are that they are less effective in conveying complex information, in teaching skills, in shifting attitudes and beliefs, and in changing behaviour, especially in the absence of other forms of communication or environmental changes. Mass media campaigns are, however, well-suited for public and community agenda-setting, and for influencing potential early mediators of motivation and behaviour change such as awareness, health beliefs and risk perceptions, but more tailored or interpersonal communication is necessary to establish behaviour change.[56,66,67] Furthermore, a mass media campaign that is successful in raising public awareness and agenda setting can lead to social changes, changes in social norms, especially if the messages raise a public discussion. Such a public discussion and social changes may encourage politicians and other decision-makers to make policy changes to promote healthier diets.[57,60,68] For example, during the Fat Watch campaign, a mostly mass-media campaign to reduce fat intake levels in the Netherlands, a public-private collaboration between government and the food industry resulted in production and marketing of lower fat foods.[2]

A more recent example is the 'Maak je niet dik!' campaign of the Netherlands Nutrition Centre, a five-year initiative to promote weight gain preventive actions among Dutch adults 25–35 years of age. A study was conducted to evaluate the first campaign phase aimed at placing the issue of weight gain prevention on the public agenda, by creating awareness of a need to act to prevent weight gain, and to induce more positive attitudes and intentions towards prevention of weight gain. The first campaign phase reached a large proportion of the population and initiated some positive changes in attitudes but did not achieve significant improvements in other determinants of the prevention of weight gain such as awareness of personal bodyweight status, overweight related risk perceptions and motivation to prevent weight gain. Despite the limited results on determinants of prevention of weight gain the campaign created a lot of free publicity. The fact that during and soon after the campaign, several television and radio programmes and national newspaper and journal articles on the issue were published, may indicate that the campaign was successful in placing the issue of prevention of weight gain on the public agenda.[8]

In conclusion, the strength of the mass media lies in helping to put issues on the public agenda, in reinforcing local efforts, in raising consciousness about health issues and in conveying simple information. The mass media may not tell people what to think, but they may help to tell people what to think about. In addition, they can play a strong supportive role in drawing attention to programmes and strategies, in disseminating information and in setting the agenda for future health promoting initiatives.[58,68]

26.4.2 Motivation and abilities: personalised interventions

To go beyond agenda setting and the first stages of motivating people to make dietary changes, more intensive and interactive intervention strategies are needed. Based on the most comprehensive review of nutrition education to date,[69] three important conditions for likelihood of effect were identified. Nutrition education should:

- be tailored to personally relevant motivators and reinforcers;
- apply personalised self-assessment and self-evaluation techniques;
- enable and encourage active participation in the intervention.

Personal nutrition counselling may offer the best opportunities to meet these conditions. A personal counselling technique that shows promise and can be applied in person-to-person nutrition education is motivational interviewing (MI).[70,71] The basic framework of MI reflects stages of change theory[72] and self-determination theory;[70] motivation is regarded as a modifiable state of readiness to change, and not as a stable trait. MI tries to facilitate patients resolving their ambivalence about changing their behaviour and avoids taking a confrontational approach that may lead to arguments between patient and counsellor. MI encourages the patient to do most of the talking during consultations and encourages the counsellor to facilitate the patient to express what she or he thinks and feels, to explore ambivalence about behaviour change, so that the patient reaches a decisional balance and eventually chooses what to change and decides on a change plan and strategy, assisted by the counsellor. An empathic counselling style is essential to MI. The counsellor should expect ambivalence in patients and help to explore resistance to change; this is a first step towards change. Self-efficacy is another key-issue in MI: the patient is responsible for change but the counsellor should help increase self-confidence, by helping to strengthen the patient's abilities to change. Counselling techniques, such as reflective listening and expressing acceptance, are part of MI, affirming the patient's own responsibility and freedom of choice. Advice is provided on patient's request without judgement, so that patients are indeed encouraged to make their own choices. MI has been applied in nutrition education, but good trials testing such applications are few.[71] Our centre recently conducted a trial testing the impact of MI for dietary change in diabetes patients. Dieticians were randomly allocated to receive MI-training or not and the results showed that patients of MI-dieticians had significantly lower fat intake levels at post-test compared to patients of controls. No differential effects on blood glucose levels (HbA$_{1c}$), body mass index (BMI) and waist circumference were found but patients in both study groups showed significant improvements in mean fat intake, HbA_{1c}, BMI and waist circumference from baseline to follow-up.[73] Interpersonal counselling techniques like MI are, however, not suitable to reach the majorities of populations that have unfavourable diets; MI takes time, needs trained counsellors, and is thus expensive.

In recent decades several potentially important new channels for health communication have emerged, such as interactive computer programmes, mobile technologies like mobile phones with text messaging and hand-held computers, interactive television, and maybe most importantly, the Internet with its World Wide Web (WWW) and email applications.[74–76] Such interactive technologies can be used to tailor nutrition education for larger groups of people[77] and especially the WWW shows promise since it is a preferred source of health information for many consumers.[78,79]

Computer-tailored nutrition education has been identified as one of the more promising nutrition education techniques.[77] Computer-tailored interventions mimic personal nutrition counselling to a certain extent. Computer-tailored nutrition education provides people with information that is based on personal characteristics, such as personal dietary intake data, personal motivation, attitudes, knowledge, self-efficacy, and abilities for dietary change. Earlier reviews of the literature suggest that computer-tailored interventions are more effective than generic nutrition education.[80,81] In a recent systematic review 26 studies on computer-tailored nutrition education were identified of which 20 studies found significant effects in favour of the tailored interventions. The evidence was most consistent for tailored interventions on fat reduction.[82]

The process of computer-tailoring is similar to personal counselling: people are surveyed or interviewed and the results are used to develop individualised feedback and advice. The pooled expertise of nutrition counsellors is documented in a computer-program, making 'mass-customisation' possible: provision of individualised feedback and advice to large groups of people. In most computer-tailored interventions evaluated to date, the surveys were written self-administered questionnaires or occasionally administered by telephone and the survey results were keyed or automatically scanned into a data file.[77] The tailoring expert system analyses these data and links them with a feedback library-file that contains feedback and advice messages tailored to each survey response. The tailored feedback in such interventions are mostly print computer-tailored personal feedback letters or newsletters, and the aforementioned evidence for the effectiveness of tailored nutrition education is largely based on such 'first generation' computer-tailoring.[77] However, several limitations of printed tailored feedback have been noted.[83] Computer-tailored print materials only utilise part of the potential of computer-tailoring, since interaction and immediate feedback is not possible. First generation computer-tailored interventions are also more expensive than generic nutrition education, since it requires at least some handling of the survey questionnaires and the feedback letters.[77]

Using interactive technology in computer-tailoring may offer better opportunities to tackle these issues: lower costs, better interaction, less time between screening and feedback, and opportunities for combining computer-tailored feedback with Internet-based social support, for example via email, forum, MSN or chat applications. Interactive technology allows participants to enter the answers to the survey questions directly into the interactive system by means of, for example, mouse clicks, keyboard, voice recording or touch screen video. Feedback is then given almost immediately on the computer screen.[83] Furthermore, such systems allow much better interaction; one or few survey questions followed directly by feedback on the answers given, followed again by a small set of questions, answers and feedback, until the entire tailored-advice system has been completed. It also better allows so-called iterative feedback, since 'diagnosis' and initial feedback can be saved and retrieved to inform follow-up feedback for respondents who repeatedly use the tailored system.

It has, however, also been argued that personalised advice may not be enough because dietary habits are often not volitional, or personally determined, since food is often bought or prepared by others, and dietary choices may be largely dependent on what is available and other environmental determinants of food choice, i.e. the opportunities for healthy eating.

26.4.3   Opportunities: environmental change interventions

If the environment offers good opportunities for health behaviour, people can more easily turn their motivation into action and may need fewer skills to do so. There is a strong belief that the changes in our eating environments, from an environment with a high likelihood of shortage of food toward an environment that offers and encourages plenty of opportunities to eat palatable energy-dense foods almost always and everywhere, contributed to the present-day obesity epidemic.[18,84–86] Is it also possible to make planned changes in the environment to encourage healthy eating?

The case of the reductions in number of smokers in industrialised countries offers a good example that deliberately changing the physical, social, political and financial environment may contribute to positive health behaviour changes. To promote smoking prevention and cessation health education has been backed up by smoking bans in public buildings, taxation to increase the price of smoking, and increasingly more negative social norms toward smoking.[87] Fewer examples are available for environmental interventions to promote healthy eating. In a recent review of the literature[55] increasing availability and accessibility of health foods and decreasing availability and accessibility of less healthy choices proved to be successful strategies to promote healthier diets.

Since children and adolescents are less autonomous in their dietary choices, environmental interventions to improve availability and accessibility of healthy foods in the home and school environment may be especially relevant.[88] The Pro Children project is a cross European study to develop and test effective strategies to promote adequate consumption levels of fruits and vegetables among schoolchildren.[89] This project is an interesting example of an intervention study that attempted to promote healthy eating, i.e. fruit and vegetable consumption, especially by improving availability and accessibility of fruit and vegetables at school[90] with provision of fruit and vegetable in the schools as the main element, either as free in-school distribution, a subscription programme, or as part of school meals. The Pro Children intervention programme has been tested in a group-randomised trial design where schools in three of the participating countries (i.e., the Bilbao region in Spain; in Rotterdam, the Netherlands; and in Buskerud county of Norway) have been randomly allocated to an intervention arm and a delayed intervention arm (comparison group). Surveys among all participating children and their parents were conducted prior to the initiation of the intervention, immediately after the end of a one school year intervention period, and at the end of the subsequent school year.[89]

Preliminary analyses of the effects after the one-year implementation, shows that fruit and vegetable intake in children in the Intervention group was approximately 20% higher than in children in the control group.[91]

26.5   Conclusions and future trends

Promotion of healthy eating should be based on thorough analyses of the epidemiology of healthy eating and mediators of such eating behaviours. Such mediators can be individual or environmental and can be categorised in three groups: motivational factors, abilities and opportunities. In healthy diet promotion efforts most emphasis has been put on improving motivation and abilities to eat more healthily, but more recently more attention has been given to the environmental opportunities that may encourage or hinder healthy eating.

Mass media interventions are important in promotion of healthy eating to raise attention and awareness, but mass media interventions alone are unlikely to lead to behaviour changes. Behaviour change may be more likely when interventions are more personalised and tailored and when environmental changes that improve the availability and accessibility of healthy choices are incorporated.

26.6   References

1. WYNDER EL, ROSE DP, COHEN LA. Diet and breast cancer in causation and therapy. *Cancer* 1986; **58**:1804–1813.
2. VAN DER FEEN DE LILLE JCJF, RIEDSTRA M, HARDEMAN W, WEDEL M, BRUG J, PRUYN JFA, LÖWIK MRH. Fat Watch: a three year nationwide campaign in the Netherlands to reduce fat intake. Process evaluation. *Nutrition and Health* 1998; **12**:107–117.
3. VAN ASSEMA P, STEENBAKKERS M, KOK GJ, ERIKSEN M, DE VRIES H. Results of the Dutch community project 'Healthy Bergeyk' (short report). *Prev Med* 1994; **23**:394–401.
4. WILLETT WC, HUNTER DJ, STAMPFER MJ, COLDITZ G, MANSON JE, SPIEGELMAN DR, B., HENNEKENS CH, SPEIZER FE. Dietary fat and fiber in relation to risk of breast cancer. An 8-year follow-up. *JAMA* 1992; **268**:2037–2044.
5. WITTE K, ALLEN M. A meta-analysis of fear appeals: implications for effective public health campaigns. *Health Educ Behav* 2000; **27**:591–615.

6. RUITER RAC, ABRAHAM C, KOK G. Scary warnings and rational precautions: a review of the psychology of fear appeals. *Psychol Health* 2001; **16**:613–630.

7. RANDOLPH W, VISWANATH K. Lessons learned from public health mass media campaigns: marketing health in a crowded media world. *Annu Rev Public Health* 2004; **25**:419–437.

8. WAMMES B, BREEDVELD B, LOOMAN C, BRUG J. The impact of a national mass media campaign in the Netherlands on the prevention of weight gain. *Public Health Nutr* 2005; **18**:1250–1257.

9. EZZATI M, LOPEZ AD, RODGERS A, VANDER HOORN S, MURRAY CJ. Selected major risk factors and global and regional burden of disease. *Lancet* 2002; **360**:1347–1360.

10. EZZATI M, VANDER HOORN S, RODGERS A, LOPEZ AD, MATHERS CD, MURRAY CJL. Estimates of global and regional potential health gains from reducing multiple major risk factors. *Lancet* 2003; **362**:271–280.

11. MOKDAD AH, MARKS JS, STROUP DF, GERBERDING JL. Actual causes of death in the United States, 2000. *JAMA* 2004; **291**:1238–1245.

12. GREEN LW, KREUTER MW. *Health promotion planning: An educational and ecological approach*, 3rd edn. Mountain View, CA: Mayfield, 1999.

13. MCKENZIE JF, SMELTZER JL. *Planning, implementing, and evaluating health promotion programs: A primer*, 3rd edn. Boston: Allyn and Bacon, 2001.

14. GLANZ K, LEWIS FM, RIMER BK. *Health behavior and health education: theory, research and practice*, 2nd edn. San Francisco: Jossey-Bass, 1997:496.

15. BRUG J, OENEMA A, FERREIRA I. Theory, evidence and intervention mapping to improve behavioral nutrition and physical activity interventions. *Int J Behav Nutr Phys Act* 2005; **2**:2.

16. PEETERS A, BARENDREGT JJ, WILLEKENS F, MACKENBACH JP, AL MAMUN A, BONNEUX L. Obesity in adulthood and its consequences for life expectancy: a life-table analysis. *Ann Intern Med* 2003; **138**:24–32.

17. CATERSON ID, HUBBARD V, BRAY GA, GRUNSTEIN R, HANSEN BC, HONG Y, LABARTHE D, SEIDELL JC, SMITH SC, JR. Prevention Conference VII: Obesity, a worldwide epidemic related to heart disease and stroke: Group III: worldwide comorbidities of obesity. *Circulation* 2004; **110**:e476–483.

18. SWINBURN BA, CATERSON I, SEIDELL JC, JAMES WP. Diet, nutrition and the prevention of excess weight gain and obesity. *Public Health Nutr* 2004; **7**:123–146.

19. DANSINGER ML, GLEASON JA, GRIFFITH JL, SELKER HP, SCHAEFER EJ. Comparison of the Atkins, Ornish, Weight Watchers, and Zone diets for weight loss and heart disease risk reduction: a randomized trial. *Jama* 2005; **293**:43–53.

20. WHO. Promoting fruit and vegetable consumption around the world: http://www.who.int/dietphysicalactivity/fruit/en/, 2005.

21. HUNG HC, JOSHIPURA KJ, JIANG R, HU FB, HUNTER D, SMITH-WARNER SA, COLDITZ GA, ROSNER B, SPIEGELMAN D, WILLETT WC. Fruit and vegetable intake and risk of major chronic disease. *J Natl Cancer Inst* 2004; **96**:1577–1584.

22. WCRF. Food, Nutrition and the Prevention of Cancer: a Global Perspective, 1997.

23. WISEMAN M. Food, nutrition, physical activity and the prevention of cancer: the second World Cancer Research Fund report (Abstract). *Fourth Annual Conference of the International Society of Behavioral Nutrition and Physical Activity (ISBNPA), Amsterdam* 2005.

24. WATSON RR, PREEDY VR. *Nutrition and Heart Disease: Causation and Prevention*. Boca Raton, FL: CRC Press, 2004.

25. BARANOWSKI T, CULLEN KW, NICKLAS T, THOMPSON D, BARANOWSKI J. Are current

health behavioral change models helpful in guiding prevention of weight gain efforts? *Obes Res* 2003; **11 Suppl**:23S–43S.

26. CONNER M, ARMITAGE CJ. *The social psychology of food*. Buckingham: Open University Press, 2002.

27. ROTHSCHILD ML. Carrots, sticks, and promises: a conceptual framework for the management of public health and the social issue behaviors. *J Mark* 1999; **63**:24–37.

28. CAPALDI EDE. *Why we eat what we eat: The psychology of eating*. Washington, DC: American Psychological Association, 1996:339.

29. BIRCH LL. Development of food preferences. *Annu Rev Nutr* 1999; **19**:41–62.

30. BRAY GA. The epidemic of obesity and changes in food intake: the Fluoride Hypothesis. *Physiol Behav* 2004; **82**:115–121.

31. WARDLE J. Food choices and health evaluation. *Psychol Health* 1993; **8**:65–75.

32. STEPTOE A, POLLARD TM, WARDLE J. Development of a measure of the motives underlying the selection of food: the food choice questionnaire. *Appetite* 1995; **25**:267–284.

33. LENNERNAS M, FJELLSTROM C, BECKER W, GIACHETTI I, SCHMITT A, REMAUT DE WINTER A, KEARNEY M. Influences on food choice perceived to be important by nationally-representative samples of adults in the European Union. *Eur J Clin Nutr* 1997; **51 Suppl 2**:S8–15.

34. HEALTH-FOCUS. HealthFocus: Study of Public Attitudes and Actions Toward Shopping and Eating. Health Focus International. St. Petersburg, 2005.

35. GLANZ K, BASIL M, MAIBACH E, GOLDBERG J, SNYDER D. Why Americans eat what they do: taste, nutrition, cost, convenience, and weight control concerns as influences on food consumption. *J Am Diet Assoc* 1998; **98**:1118–1126.

36. AARTS H, PAULUSSEN T, SCHAALMA H. Physical exercise habit: On the conceptualization and formation of habitual health behaviors. *Health Educ Res* 1997; **12**:363–374.

37. AJZEN I. The theory of planned behavior. *Organ Behav Hum Decis Process* 1991; **50**:179–211.

38. VERPLANKEN B, AARTS H. Habit, attitude, and planned behaviour: Is habit an empty construct or an interesting case of goal-directed automaticity? *European Review of Social Psychology* 1999; **10**:101–134.

39. VERPLANKEN B, ORBELL S. Self-reported habit: A self-report index of habit strength. *J Appl Soc Psychol* 2003; **33**:1313–1330.

40. BRUG J, DE VET E, DE NOOIJER J, VERPLANKEN B. Predicting fruit consumption: cognitions, intention and habits. *J Nutr Educ Behav* 2006; **38**:73–81.

41. BRUG J, KROEZE W, WIND M, VAN DER HORST K, FERREIRA I. The importance of habit strength in dietary behaviors (Abstract). *Fourth Annual Conference of the International Society of Behavioral Nutrition and Physical Activity (ISBNPA), Amsterdam* 2005.

42. KREMERS SPJ. Habit Strength of energy balance-related behaviors among children and adolescents (Abstract). *Fourth Annual Conference of the International Society of Behavioral Nutrition and Physical Activity (ISBNPA), Amsterdam* 2005.

43. PATTERSON RE, KRISTAL AR, WHITE E. Do beliefs, knowledge, and perceived norms about diet and cancer predict dietary change? *Am J Public Health* 1996; **86**:1394–1400.

44. LECHNER L, BRUG J, DE VRIES H. Misconception of fruit and vegetable consumption: differences between objective and subjective estimation of intake. *J Nutr Educ* 1997; **29**:313–320.

45. BOGERS RP, BRUG J, VAN ASSEMA P, DAGNELIE PC. Explaining fruit and vegetable

consumption: the theory of planned behaviour and misconception of personal intake levels. *Appetite* 2004; **42**:157–166.

46. RONDA G, VAN ASSEMA P, BRUG J. Stages of change, psychological factors and awareness of physical activity levels. *Health Promot Int* 2001; **16**:305–314.

47. BRUG J, VAN ASSEMA P, LENDERINK T, GLANZ K, KOK GJ. Self-rated dietary fat intake: association with objective assessment of fat, psychosocial factors and intention to change. *J Nutr Educ* 1994; **26**:218–223.

48. BANDURA A. *Social foundations of thought and action. A social cognitive theory.* New Jersey: Prentice Hall, 1986.

49. FRENCH SA, STORY M, JEFFERY RW. Environmental influences on eating and physical activity. *Annu Rev Public Health* 2001; **22**:309–335.

50. FLAY BR, PETRAITIS J. The theory of triadic influence: a new theory of health behavior with implications for preventive interventions. *Advances in Medical Sociology* 1994; **4**:4–19.

51. STORY M, NEUMARK-SZTAINER D, FRENCH S. Individual and environmental influences on adolescent eating behaviors. *J Am Diet Assoc* 2002; **102**:S40–51.

52. SALLIS JF, OWEN N. Ecological Models of Health Behavior. In: Glanz K, Rimer BK, Lewis FM, eds. *Health Behavior and Health Education*, 3rd edn, 2003:462–485.

53. SWINBURN B, EGGER G, RAZA F. Dissecting obesogenic environments: the development and application of a framework for identifying and prioritizing environmental interventions for obesity. *Prev Med* 1999; **29**:563–570.

54. DE BRUIJN GJ, KREMERS SPJ, SCHAALMA H, VAN MECHELEN W, BRUG J. Determinants of adolescent bicycle use for transportation and snacking behavior. *Prev Med* 2005; **40**:658–667.

55. BLANCHETTE L, BRUG J. Determinants of fruit and vegetable consumption among six to twelve-year-old children and effective interventions to increase consumption. *J Human Nutr Dietetics* 2005 **18**:431.

56. FLAY BR, BURTON D. Effective mass communication strategies for health campaigns. In: Atkin C, Wallack L, eds. *Mass communication and public health: complexities and conflicts*. London: Sage Publications, 1990:112–149.

57. CAVILL N, BAUMAN A. Changing the way people think about health-enhancing physical activity: do mass media campaigns have a role? *J Sports Sci* 2004; **22**:771–790.

58. MARSHALL AL, LESLIE ER, BAUMAN AE, MARCUS BH, OWEN N. Print versus website physical activity programs: a randomized trial. *Am J Prev Med* 2003; **25**:88–94.

59. RICE R, ATKIN C. *Public Communication Campaigns*. Thousand Oaks, CA: Sage, 2001.

60. REDMAN S, SPENCER A, SANSON-FISHER RW. The role of mass media in changing health-related behaviour: a critical appraisal of two models. *Health Promot Int* 1990; **5**:85–101.

61. REID D. How effective is health education via mass communications? *Health Educ J* 1996; **55**:332–344.

62. WARDLE J, RAPOPORT L, MILES A, AFUAPE T, DUMAN M. Mass education for obesity prevention: the penetration of the BBC's 'Fighting Fat, Fighting Fit' campaign. *Health Educ Res* 2001; **16**:343–355.

63. VAN WECHEM SN, BRUG J, VAN ASSEMA P, KISTEMAKER C, RIEDSTRA M, LÖWIK MRH. Fat Watch: a nationwide campaign in the Netherlands to reduce fat intake. Effect evaluation. *Nutrition and Health* 1998; **12**:119–130.

64. CAVILL N. National campaigns to promote physical activity: can they make a difference? *Int J Obes Relat Metab Disord* 1998; **22 Suppl 2**:S48–51.

65. WELLINGS K, MACDOWALL W. Evaluating mass media approaches to health promotion: a review of methods. *Health Education* 2000; **100**:23–32.
66. SCHOOLER C, FLORA J, FARQUHAR JW. Moving toward synergy: media supplementation in the Standford Five-City Project. *Communication Research* 1993; **20**:587–610.
67. MORTON TA, DUCK JM. Communication and Health Beliefs: Mass and Interpersonal Influences on Perceptions of Risk to Self and Others. *Communication Research* 2001; **28**:602–626.
68. FINNEGAN JR, JR., VISWANATH K, HERTOG J. Mass media, secular trends, and the future of cardiovascular disease health promotion: an interpretive analysis. *Prev Med* 1999; **29**:S50–S58.
69. CONTENTO I, BALCH GI, BRONNER YL, LYTLE LA, MALONEY SK, OLSON CM, SWADENER SS. The effectiveness of nutrition education and implications for nutrition education policy, programs and research: a review of research. *J Nutr Educ* 1995; **27**.
70. RESNICOW K, DILORIO C, SOET JE, BORRELLI B, HECHT J. Motivational Interviewing in Health Promotion: It sounds like something is changing. *Health Psychol* 2002; **21**:444–451.
71. DUNN C, DEROO L, RIVARA FP. The use of brief interventions adapted from motivational interviewing across behavioral domains: a systematic review. *Addiction* 2001; **96**:1725–1742.
72. PROCHASKA JO, VELICER WF. The transtheoretical model of health behavior change. *Am J Health Promot* 1997; **38**:38–48.
73. BRUG J, OENEMA A, KROEZE W, RAAT H. The Internet and nutrition education: challenges and opportunities. *Eur J Clin Nutr* 2005; **59**:S130–S139.
74. OENEMA A, TAN F, BRUG J. Short-term efficacy of a web-based tailored nutrition intervention; main effects and mediators. *Ann Behav Med* 2005; **29**:54–63.
75. OENEMA A, DE WEERD I, DIJKSTRA A, DE VRIES H, BRUG J. The impact of a web-based computer-tailored intervention to promote heart-healthy lifestyles. in preparation.
76. TATE DF, JACKVONY EH, WING RR. Effects of Internet behavioral counseling on weight loss in adults at risk for type 2 diabetes: a randomized trial. *JAMA* 2003; **289**:1833–1836.
77. KREUTER M, FARRELL D, OLEVITCH L, BRENNAN L. *Tailoring health messages: customizing communication with computer technology.* Mahway, NJ: Lawrence Erlbaum, 2000.
78. VAN DILLEN SM, HIDDINK GJ, KOELEN MA, DE GRAAF C, VAN WOERKUM CM. Understanding nutrition communication between health professionals and consumers: development of a model for nutrition awareness based on qualitative consumer research. *Am J Clin Nutr* 2003; **77(4 Suppl)**, 1065S–1072S.
79. WILLIAMS P, NICHOLAS D, HUNTINGTON P, McLEAN F. Surfing for health: user evaluation of a health information website. Part one: Background and literature review. *Health Info Libr J* 2002; **19**(2):98–108.
80. BRUG J, CAMPBELL M, VAN ASSEMA P. The application and impact of computer-generated personalized nutrition education: a review of the literature. *Patient Educ Couns* 1999; **36**:145–156.
81. SKINNER CS, CAMPBELL MK, RIMER BK, CURRY S, PROCHASKA JO. How effective is tailored print communication? *Ann Behav Med* 1999; **21**:290–298.
82. KROEZE W, WERKMAN AM, BRUG J. A systematic review of randomized trials on the effectiveness of computer-tailored education on physical activity and dietary behaviors. *Ann Behav Med* 2006; **31**:205–223.
83. OENEMA A, BRUG J, LECHNER L. Web-based tailored nutrition education: results of a

randomized controlled trial. *Health Educ Res* 2001; **16**:647–660.

84. WAMMES BM, BREEDVELD B, BRUG J. Possible determinants of willingness to prevent weight gain. International Society of Behavior Nutrition and Physical Activity. Québec City, Canada: ISBNPA, 2003:55.

85. MAYOR S. Government task force needed to tackle obesity. *BMJ* 2004; **328**:363.

86. SCHWARTZ MB, PUHL R. Childhood obesity: a societal problem to solve. *Obes Rev* 2003; **4**:57–71.

87. LANTZ PM, JACOBSON PD, WARNER KE, WASSERMAN J, POLLACK HA, BERSON J, AHLSTROM A. Investing in youth tobacco control: a review of smoking prevention and control strategies. *Tobacco Control* 2000; **9**:47–63.

88. SANDVIK C, DE BOURDEAUDHUIJ I, WIND M, BERE E, PEREZ RODRIGO C, WOLF A, DUE P, ELMADFA I, THORSDOTTIR I, DANIEL VAZ DE ALMEIDA M, YNGVE A, BRUG J. Personal, social and environmental factors regarding fruit and vegetable intake among schoolchildren in 9 European countries – The Pro Children Study. *Ann Nutr Met* 2005; **49**:255–266.

89. KLEPP KI, PEREZ RODRIGO C, DE BOURDEAUDHUIJ I, DUE P, ELMADFA I, HARALDSDOTTIR J, KÖNIG J, SJÖSTRÖM M, THORSDOTTIR I, DANIEL VAZ DE ALMEIDA M, YNGVE A, BRUG J. Promoting fruit and vegetable consumption among European schoolchildren: Rationale, conceptualization and design of the Pro Children Project. *Ann Nutr Met* 2005; **49**:212–220.

90. PEREZ RODRIGO C, WIND M, HILDONEN C, BJELLAND M, ARANCETA J, KLEPP KI, BRUG J. The Pro Children Intervention: Applying the Intervention Mapping Protocol to develop a school-based fruit and vegetable promotion programme. *Ann Nutr Met* 2005; **49**:267–277.

91. WIND M, KLEPP KI, PEREZ-RODRIGO C, BRUG J. Effect of Pro Children: an international multi-component school-based intervention to promote fruits and vegetables in school children (Abstract). Fourth Annual Conference of the International Society of Behavioral Nutrition and Physical Activity (ISBNPA), Amsterdam, 2005.

27

Public engagement in food policy

G. Rowe, Institute of Food Research, UK

27.1 Introduction: the issue of public engagement in food policy

What *should* people eat? What *should* they avoid? And *who* should decide this? The answers to these and similar questions are of high concern to contemporary society, touching upon issues such as personal choice, trust, knowledge, values, information dissemination, food safety, and ultimately, the way in which our food systems are managed. In this chapter I explore how relatively recent events have changed the way in which we view food policy, particularly with regards the management of food risks. Specifically, I discuss the increasing enthusiasm in contemporary democratic societies for public 'engagement' or 'participation' – a process through which the public is involved *in some limited manner* in the practices of policy-making bodies. This approach contrasts with the traditional manner in which public involvement in policy (not just regarding food, but more generally) is largely limited to voting at national elections, after which decisions are left to the informed agents of elected governments. This chapter traces the reasons for the rise in profile of public engagement practices; it discusses various ways in which public engagement is carried out; it provides practical examples of engagement in various food domains; and it discusses what such approaches have achieved. One particular emphasis of this chapter is on the issue of evaluation, and the necessity of conducting research into the effects (good and bad) of the different policy-making approaches. Finally, this chapter comments on likely future trends in this area and indicates sources of further information on the topics covered.

27.2   The concept of public involvement

27.2.1   Traditional policy making: before the crises

Governmental policies in most democratic societies (e.g., in Europe) are determined by elected representatives in government supported by a neutral administrative executive, frequently drawing upon advice from unelected experts – either individual advisors or expert committees (Jasanoff, 1990). In this system, public involvement is largely limited to voting at election times. Over recent years, however, the appropriateness of this system has been challenged as a result of various controversies, including those relating to food safety policy, one result of which has been a call for greater public involvement in policy making.

Perhaps the most notable challenge to the traditional policy-making model was thrown down by the Bovine Spongiform Encephalopathy (BSE) crisis of the 1990s, which was widely perceived to have been seriously mismanaged, particularly by the UK government. In the UK, the epicentre of the crisis, official pronouncements at one point declared the eating of beef to be 'safe' (in one now infamous incident, the then minister responsible was shown on television feeding a beef burger to his young daughter), only for a subsequent link to be found between BSE and Creutzfeldt Jakob Disease (vCJD), a fatal and incurable human illness that has been linked to more than 100 deaths in the UK to date. There have been numerous other high profile food 'crises' in the last decade or so, from the presence of *Salmonella* in eggs in the UK to contamination of meat with dioxins in Belgium (for discussions about some of these issues see, for example, Gregory, 2000; Verbeke, 2001). Other emerging risks lie on the horizon – such as avian influenza – the consequences of which (for human health and food policy) remain to be seen.

The net result of these various dramas and crises has been a reduction in trust in those responsible for managing food policy (in particular) (e.g., Frewer, 1999). This is important because the effective implementation of the traditional policy-making system relies in part on policy makers convincing the public of the appropriateness and accuracy of their decisions. In the arena of 'risk communication', the traditional policy-making approach is thus closely aligned to the so-called deficit model (Hilgartner, 1990), which assumes that discrepancies between experts' (i.e., policy makers') positions on risks and those of the public can be overcome by the effective *one-way* transfer of appropriate information *to* the public. Specifically, it is expected that information presented in the right (most persuasive) way will result in recipients aligning their views to those of the information source; hence, the key to effective communication is finding the magic presentation formula. Lack of trust in the source, however, leads to a lack of recipients' belief in the message, undermining potential alignment and leading to the recipients (the public) ignoring or disobeying pronouncements. In the food domain, this might result in people ignoring advice on what foods or products to avoid (as an example, foods with high saturated fat content), as well as advice on what products are safe and need not be avoided (such as foods with genetically modified ingredients). Of course, lack of trust is but one factor that might undermine message utility: successful communications are also liable to depend upon whether people perceive the messages to be relevant to them, whether they perceive they are able to perform suggested behaviours, and so on, and it is important not to attribute all failures of communication to the trust issue.

27.2.2 A new approach: the public involvement antidote to public distrust in government

If trust in current policy-making approaches is undermined, how might it be regained? Indeed, are there more effective ways of deciding upon policy? In the food domain, there have been arguably two responses to this problem, which are not unrelated. The first has been to revise, and indeed, replace the responsible agencies that have been seen to be undermined. For example, in the UK, the Ministry of Agriculture, Fisheries and Foods (MAFF) was dissolved following the BSE crisis and its responsibilities were shifted to other ministries, with the Food Standards Agency (FSA) (an independent food safety watchdog set up by an Act of Parliament in 2000) being formed to protect the public's health and consumer interests in relation to food. Other food safety institutions have arisen in Europe, for example in France (the *Agence Française de Sécurité Sanitaire des Aliments – AFSSA*) and Germany (the *Bundesministerium für Verbrauchers- schutz, Ernährung und Landwirtschaft – BMVEL*). Indeed, following the series of food scares in the 1990s, the European Union concluded that it needed to establish a new scientific body charged with providing independent and objective advice on food safety issues associated with the food chain. Its primary objective (as set out in a White Paper on Food Safety) was to: '. . . contribute to a high level of consumer health protection in the area of food safety, through which consumer confidence can be restored and maintained' (see EFSA website at: http://efsa.eu.int/). The result was the formation of the European Food Safety Authority (EFSA). What all of these bodies have in common is that they have been charged, either explicitly or implicitly, with restoring public confidence in the management of food safety, and their terms of reference are characterised by concepts such as 'openness', 'transparency', 'alertness', 'independence', 'integrity' and 'putting the consumer first' (Byrne, 2002; Food Standards Agency, 2000; Künast, 2001; Rowe *et al.*, 2001).

The second trend has been to increase public involvement in food policy – a process that has often been termed 'public engagement' or 'public participation'. In 'public participation' the public is involved *in some limited manner* in the practices of policy-making bodies. That is, information is not merely transmitted *to* the public, but is also elicited *from* it. One common assumption is that, by allowing people some input into decisions, 'trust' in the consulting policy makers will *somehow* be maintained or enhanced (though the exact mechanism for achieving this is not entirely clear). This approach thus reflects a transition from a position where information is seen as the key to resolving a knowledge deficit, and so resolving lay opposition, to one in which regaining trust in governments and regulators is seen as vital to solving a perceived legitimacy (or trust) deficit (Walls *et al.*, 2004).

Public participation is also assumed to have other benefits beyond that of enhancing trust. It has been argued by political theorists, for example, that increasing public involvement is essentially the 'right' thing to do, in the sense that this will enhance democracy, procedural justice, fairness and human rights (e.g., Fiorino, 1990; National Research Council, 1996). It is also increasingly recognised that in many policy contexts there exists a high degree of scientific uncertainty combined with a plurality of value-based perspectives. In such cases, decisions may be based to a significant extent upon the *values* of the involved experts/policy makers, which in themselves have no greater inherent validity than those held by lay publics (e.g., Renn, 1992). Furthermore, it has been suggested that there is also a need to contextualise scientific knowledge to take account of the specificity of the issue in question. In this regard, local lay knowledges may provide important insights in addressing a range of practical issues, and might arguably improve the *quality* of policy decisions (e.g., Funtowicz and Ravetz, 1992; Horlick-Jones, 1998, 2004; Wynne, 1991).

In this respect, the new food safety institutions in Europe have been concerned to involve the public more in their dealings than previous organisations. Thus, a recent decision by the Management Board of the European Food Safety Authority (EFSA) was to open up its work further to public scrutiny, most significantly, by involving consumers through a stakeholder forum, and by holding public hearings on significant scientific issues (EFSA, 2003). Commitment to the public involvement cause has also been backed in Europe through research funding decisions. For example, one project recently funded through the EU Framework VI Programme is entitled SAFEFOODS – Promoting Food Safety through a New Integrated Risk Analysis Approach for Foods (see: www.safefoods.nl). An important aim of this project is to consider how and when to involve the public and other stakeholders in the risk analysis process.

In the next section, the various manifestations of public involvement will be considered, and subsequently the issue of whether public participation generally achieves the aims often assumed will be addressed.

27.3   Ways of involving the public in policy formation

27.3.1   Types of public engagement

The public may be 'involved' in science and technology policy (in general) at a number of levels. One contrast often made, as discussed earlier, is between public communication and public participation – the former being typified by top-down communication and a one-way flow of information (e.g., from policy makers to public), and the latter by dialogue and two-way information exchange between the parties involved (e.g., Rowe and Frewer, 2000). However, Rowe and Frewer (2005) distinguish between *three* levels of involvement: in addition to public communication and public participation they identify public consultation, in which information essentially flows one-way from the public or appropriate stakeholders to sponsors (e.g., policy makers) without notable interaction. They refer to the three levels in combination as 'public engagement', and this is the terminology I will use henceforth in this chapter. Rowe and Frewer (2005) argue that the three different levels need to be distinguished because they involve three notably different processes and intentions, and hence need to be evaluated against different criteria. I will discuss the issue of evaluation in a subsequent section.

Traditionally, public engagement in the food domain has largely entailed public communication – for example, communicating the dangers of consuming too much salt and saturated fat, and providing information on how food should be properly stored and prepared – often carried out through media campaigns and information leaflets (these are strategies used, for example, by the UK FSA). Added to this, public consultation has also taken place to a limited degree, often via consultations (providing consultation documents on which interested parties – including members of the public – can comment), and through the use of surveys. Until the last few years, public participation has been much rarer in Europe, but it is now steadily growing in frequency.

27.3.2   Participation mechanisms

Participation is enacted though a variety of different mechanisms. In some cases, it is enacted through changing existing institutional mechanisms – such as through co-opting public members or stakeholders onto existing advisory committees. In many cases, however, participation is achieved through one-off events rather than continuous processes. Rowe and Frewer (2005) note over 100 different engagement (including participation) mechanisms, some of which are more formalised than others, and some of which have been used more frequently than others. Often, there are difficulties in classifying these mechanisms, since most participation approaches use essentially similar elements, differing only in subtleties of timescale, participant numbers, and ordering of processes. In order to add structure and clarity to this issue, Rowe and Frewer developed a typology of engagement mechanisms by identifying key structural components that theoretically might impact upon the effectiveness of engagement approaches (of communication, consultation, and participation types) and using these to define certain broad mechanism classes and to distinguish these from others. With respect to public participation mechanisms, they used several key factors to differentiate mechanism types (which invariably involve groups of participants since they rely upon human *interactions*), most importantly: whether participant selection is controlled or uncontrolled; whether group events are facilitated/moderated or not; and whether the outputs of the group processes are formally aggregated in some way or not. The four types identified are:

- *Type 1 mechanisms* are characterised by controlled selection of participants, facilitated face-to-face (FTF) group discussions, unconstrained participant responses, and flexible information input from the sponsors, often in the form of 'experts' who are available for questioning by the public participants over a number of days. The group output is not structured as such, and may be heavily influenced by social and psychological group factors (for example, dominance of the discussion and undue influence of dogmatic individuals). Examples of this type of mechanism include Action Planning Workshops, Citizens' Juries, and Consensus Conferences.
- *Type 2 mechanisms* are structurally similar to Type 1, but with a key difference being that there is no facilitation of the information elicitation process. In many ways, these are simple group processes, with no specific facilitation of input from group members, or aggregation of opinions. They rely upon small groups of participants (public representatives) to solve specific problems, with ready access to all pertinent information, and include Negotiated Rule-Making Committees and Task Forces.
- *Type 3 mechanisms* are also similar to Type 1, but with the essential difference that structured aggregation takes place. In the case of Deliberative Opinion Polling (an example), the selected participants are polled twice, before and after deliberation on the issue (and questioning of experts), and in this process structured aggregation of all participant opinions is attained. In the case of Planning Cells (a German mechanism), these tend to use various decision aids to ensure structured consideration and assessment, and hence aggregation, of opinions.
- *Type 4 mechanisms* differ from the other types on a number of dimensions. Importantly, selection is uncontrolled, and there is no facilitation of information elicitation, though aggregation *is* structured. The archetypal example is the town meeting (New England Model), in which voting (aggregation) takes place after debate between self-selected participants.

An important point to note is that there may appear to be great variability between mechanisms within these broad types, but Rowe and Frewer argue that these differences are largely superficial. For example, specific mechanisms might suggest a range of numbers of participants to be recruited, but by-and-large whether a mechanism has, for example, 12 participants or 20, the impact of this on the likely effectiveness of the respective approaches (in general) is liable to be relatively small in contrast to the differences due to the key differentiating variables. The essential similarity between all participation mechanisms is that they are group-based approaches in which public participants (or members of other relevant stakeholder groups) are provided with context-relevant information from the sponsors of the exercises (e.g., food policy makers), and required over a certain amount of time to debate this information between themselves and with the sponsors and to come to some sort of conclusion that might be used by the sponsors to somehow inform their decision/policy making (e.g., POST, 2001; Rowe and Frewer, 2000).

27.3.3   Participation examples from the food domain

A number of these participation methods have been used in the context of food policy issues in recent years. Perhaps the first instance of true participation in this context in the UK was the *National Consensus Conference on Plant Biotechnology*, held in 1994, which was funded by the forerunner to the Biotechnology and Biological Sciences Research Council (BBSRC). During this process (a Type 1 mechanism according to the described typology), sixteen lay people were selected to represent a typical cross-section of the UK public. These participants were given information packs on plant biotechnology and attended briefings on the subject given by a number of experts over a couple of weekends. Following this, the participants were involved in the selection of key questions they wished to have answered on the subject, as well as the selection of experts from whom they wished to hear at the conference proper. The main conference event lasted two days, in which presentations were made by a number of these experts, who were then cross-examined by the panel, which produced a report on the third day (with help from a facilitator). A full description of the process is provided by Ellahi (1995).

Interestingly, this conference raised a number of key concerns that, had they been addressed by relevant policy makers and producers nearer to that time, *might* have pre-empted some of the subsequent tensions that have emerged. For example, the panel concluded that there was a moral obligation for producers to label foods that incorporate animal genes to allow consumers to exercise the right to choose if they feel that this process is morally wrong (Ellahi, 1995). However, a statutory labelling solution has only emerged relatively recently, undoubtedly as a consequence of consumer concerns, which crystallised after the market introduction of GM foods and ingredients in the late 1990s. For example, new rules for GM labelling came into force in all EU Member States as of 18 April 2004 (The GM Food and Feed Regulation (EC) No. 1829/2003). These rules mean that if a food contains or consists of genetically modified organisms (GMOs), or contains ingredients produced from GMOs (regardless of the presence of any GM material in the final product), this must now be indicated on the label (previous regulations focused on detectable GM material only).

The Consensus Conference (Type 1 mechanism) has also been used in a number of other countries to address food-related issues. Einsiedel *et al.* (2001) described three Consensus Conferences – also held on the topic of food biotechnology – carried out in Denmark, Canada and Australia in 1999. Notably, the three conferences were funded and sponsored by different types of organisations: the Danish conference by the Danish Board of Technology (a publicly financed independent assessment unit set up by parliament in 1985); the Australian conference by the Australian Consumers' Association and Australian Science Museum (with funding from a range of government ministries), and the Canadian conference by a national research granting agency and a provincial government community grant. In the Danish case, this was not the first of such conferences: the Danish Board of Technology is widely credited with having developed the Consensus Conference in its present form, and has used it in a variety of other contexts, but also to consider gene technology in industry and agriculture (1987) and the issue of transgenic animals (1992). The importance of the 'institutionalisation' of this participation mechanism will be considered later with regards the effectiveness of participatory approaches.

The areas of concern raised by the three panels were 'striking in their similarity' (Einsiedel *et al.*, 2001, p. 90). For example, common concern was expressed about the use of antibiotic markers and the uncertainties posed by multi-gene modifications and the ability of the regulatory system to cope with this problem. There was also concern about the dominance (and potential monopoly) of a few players in the economic control of the food industry, especially regarding the power of patents and use of terminator seeds. The potential impact of this power on developing countries was another focus of concern (and other ethical issues were widely raised). As with the UK conference, a key issue was the quality, quantity and accessibility of information available to consumers, with various recommendations produced for more effective labelling policies. There were also a number of differences between the panels, for example, in terms of the extent of support for the technology (ranging from some to almost none), which might be explained by cultural differences between the panels.

A more recent example of participation in the food domain took place in the UK during 2003. The '*GM Nation?*' public debate was a wider engagement initiative that also addressed the topic of genetic modification (GM) of food and crops. This initiative was sponsored by the Agriculture and Environment Biotechnology Commission (AEBC) (a UK government strategic advisory body on biotechnology issues affecting agriculture and the environment), and was ultimately funded through central government money. The aim of the initiative was broadly to understand public framings of the GM issue, provide the public with information on the topic, and allow the public to engage in a debate through which they might reach their own informed judgements on this subject. The debate was overseen by an independent Steering Board drawn from members of the AEBC together with a number of co-opted individuals, and comprised stakeholders from across the spectrum of opinion on GM agriculture. Much of the day-to-day implementation of the debate was carried out by an 'arms length' agency of government, the Central Office of Information (COI), which acted as the main contractor to the Steering Board.

This initiative actually involved several phases. A preliminary consultation phase comprised a number of facilitated discussion groups that were convened in order to gain an understanding of natural public framings of the GM issue (i.e., identifying issues to be addressed in the debate proper, and informing the production of stimulus materials for use in subsequent stages of the debate). The main 'debate' process, conducted in the summer of 2003, entailed a series of open public meetings, which anybody could attend, organised into three levels or 'Tiers'. Tier 1 meetings (three in England and one each in Wales, Scotland and Northern Ireland) were conceived of as 'national' high profile events, and were professionally facilitated. These attracted approximately 1000 participants in total. Tier 2 meetings, of which there were about 40, were typically hosted by a local authority or other major organisation, often with the assistance of the main debate contractors. Tier 3 meetings, typically organised by local voluntary organisations, were highly variable in terms of their character and formality. The contractors *estimated* that over 600 of these events took place. In all some 20 000 individuals across the UK were estimated to have taken part in the various open meetings (PDSB, 2003). In these meetings, participants were exposed to a commissioned video in which people discussed the key GM issues, and a booklet that provided answers to key questions about GM (with statements highlighting both positive and negative aspects of the technology). Participants discussed the key issues in smaller groups, electing their own chairs, and then reported back the results of these discussions in plenary. A feedback questionnaire was also available to be completed at the end of the events.

There were two other components to the debate: first, there was a dedicated interactive debate website, which contained a range of debate materials and interactive resources, as well as the questionnaire through which visitors could record their attitudes to GM food and crops (this website recorded over 24 000 unique visitors during the course of the debate). Second, there was a series of ten closed discussions with ordinary members of the public, known as 'narrow-but-deep' groups, which were conceived as a 'control' on the character of the discussions produced by the self-selected participants in the open meetings. Here 'narrow' refers to the limited scope of representation (only 77 members of the public took part, albeit recruited to reflect a broad demographic cross-section of the UK population), and 'deep' refers to the anticipated extended level of engagement and deliberation in these groups, in comparison to that typically possible at the open meetings. These groups met twice, with a gap of two weeks between during which participants were invited to explore the GM issue individually, using official stimulus materials and any other information that they could access, and to keep diaries of their discoveries (including newspaper clippings, website downloads, etc.), thoughts, relevant conversations and so on (PDSB, 2003: para 194).

Interestingly, one intention of the sponsors of the debate was to conduct a 'novel mechanism', and in the sense that certain components do not fit into the typology of Rowe and Frewer (2005), they might be deemed to have achieved this. Thus, the main debate conferences, which we might refer to as Type 5 participation mechanisms, were characterised by *uncontrolled* participant selection, information facilitation, and structured aggregation of opinions (in the sense that opinions were collected via questionnaires that asked participants' opinions on GM food and crops, and these opinions were then aggregated in the report to government). Rowe and Frewer noted the possibility that other classes of mechanism might exist, though such mechanisms may not have emerged for functional reasons: in this case, the uncontrolled selection of participants completely undermined the process, as will be discussed in the section discussing the evaluation of participation exercises.

The Steering Board's final conclusions and report were largely based upon responses to the 13 standard attitude questions on the feedback questionnaires (given to people at the conference, replicated on the website, and given to the narrow-but-deep participants *twice*: once at the commencement of their involvement and again at the beginning of their second meeting), though it was also informed by rapporteurs' reports from meetings, analysis of the narrow-but-deep discussions, and open-ended feedback responses (letters and e-mails received). The feedback questionnaire proved pivotal, with a total of 36 553 responses. Results from these revealed significant public concern about GM technology, and though the participants in the 'narrow-but-deep' groups were not as negative initially as the self-selected participants, they appeared to become more negative between completing their two questionnaires. The government response to these results was essentially ambiguous, however, and the consequence of the debate, beyond the expenditure of a considerable sum of taxpayers' money, has been insignificant at best. The debate is discussed in detail in Horlick-Jones *et al.* (2006a,b).

Having discussed the nature of public participation in this section, and given a number of practical examples from the food domain, the next section considers what advantages, if any, participation confers on the management of food policy.

27.4   Advantages and disadvantages of public engagement

27.4.1   The rationale for evaluation

Does public participation confer any particular advantage in setting food policy over and above the more traditional policy management model? Is public participation somehow better than the other approaches in the public engagement spectrum (i.e. public communication and, to a lesser degree, public consultation), which are associated with that traditional expert/policy-maker-led model? To answer these questions requires evaluation of the different engagement approaches and comparison of 'effectiveness' across them. Indeed, the evaluation of engagement exercises is important for all parties involved as it addresses a range of functions: financial (to ensure the proper use of public and institutional money), practical (to learn from past mistakes to allow exercises to be run better in future), ethical/moral (to establish fair representation and ensure that those involved are not deceived as to the impact of their contribution) and research-related (to increase our understanding of human and organisational behaviour). As such, few would deny that evaluation *should* be done when possible.

Unfortunately, evaluations that involve the comparison of different approaches (e.g., communication versus participation) are essentially non-existent, undoubtedly in part due to the highly complex nature of engagement, which means that experimental comparisons are incredibly difficult to enact. In the absence of such evaluations, much of the commentary on *relative benefits* is highly subjective. At present – perhaps because of the lack of clear evidence of participation *failures* – there are more proponents of participation than of the other approaches, and hence their voices are currently the loudest and most easily heard.

Certainly, communication approaches *have* been evaluated in their own right; for example, the area of *risk communication* has been the focus of much research. Studies have been conducted looking at aspects such as the best way to *present* information (e.g., Golding *et al.*, 1992); the best *medium* for transmitting information to a 'target' audience (e.g., Chipman *et al.*, 1996); and the best people to impart such information (e.g., Frewer *et al.*, 1996). Although there is little evidence for the existence of any one magic formula for communication effectiveness (Bier, 2001), research has generally indicated the need to tailor one's communication approach to match the specifics of target audience and context, and is beginning to provide clues as to ways to do this. Considerable research has also been done on how to collect opinions (consultation), which methods lie at the heart of social science research, and such findings are detailed in many text books (e.g., Sudman and Bradburn, 1974). However, there is still an absence of evaluations of real-world consultative approaches, such as of consultation documents and referenda. Rowe *et al.* (2006) do provide one recent example of a study comparing two different data collection methods from the 'GM Nation?' debate – notably, comparing the nature and characteristics of the respondents and their responses from questionnaires presented in either paper or electronic format – but more 'real world' studies of this type are required.

In terms of participation approaches, however, there have been very few evaluations. Rowe and Frewer (2004) suggested that the main problem in conducting evaluations lies in defining what is actually meant by 'effectiveness'. Earlier, a number of aims linked to participation were noted, such as increasing trust, democracy, decision quality, and so on. But these ideals have rarely been operationalised and assessed. Instead, other evaluation criteria have been stipulated and used, which in certain cases do touch upon these concepts as well as implying other participation aims. One distinction often made is between process and outcome criteria (e.g., Chess and Purcell, 1999; Renn *et al.*, 1995; Rowe and Frewer, 2000), that is, criteria related to the efficient conduct and running of an exercise (process) and to the outputs at the end (outcome). In terms of conducting an evaluation, processes may be easier to measure and assess, since outcomes are often temporally removed from the end of an event or may be influenced by other external (e.g., political) factors out of the control or remit of the exercise (Chess and Purcell, 1999). As such, good process may have to act as a surrogate measure for good outcome, although the latter is clearly most significant. By good process is meant, for example, that information is fairly and accurately expressed, that groups are well facilitated (e.g., all participants get to have their say, not just the most dogmatic or verbose), and that group opinion is accurately recorded and reported (e.g., Renn *et al.*, 1995; Rowe and Frewer, 2000). If the process is poor, then a good outcome seems less likely: for example, if there is biased debate, then the recommendations that emerge from the end are liable to be poorly attained and not truly representative of opinions, and, if used as the basis for informing policy change, might ultimately prove contentious and disastrous. Concepts and measures related to fairness, and in particular, *representativeness* (i.e., that the participants are truly representative of the affected population) thus inform the choice of evaluation criteria in many of the few empirical evaluations that have been conducted (Rowe and Frewer, 2004). Outcome measures, where these are sought, tend to focus on whether the evaluated exercise has had any real effect on policy.

27.4.2   Evaluation results

No grand conclusions can be drawn on the relevance of participation approaches at a general level, given the few studies conducted, the wide variety of evaluation criteria and measures used, and the variability of mechanism types studied and contexts in which these have been employed. In terms of participation in the food domain, there are very few evaluations of relevance. Participation exercises reported in the academic literature *do* tend to be accompanied by commentary upon their quality, even when there is an absence of clearly stated *a priori* evaluation criteria and structured methods for acquiring and analysing data (Rowe *et al.*, 2005, suggest such descriptive case studies be termed 'assessments' to distinguish them from formal 'evaluations') – but often this is the only evidence we have. The lack of evaluative rigour in these cases, perhaps in accompaniment with the vested interests of authors who often believe in the participation message, means that we can have little faith in the generally positive assessments that tend to emerge. In terms of more rigorously acquired evidence, it is possible to comment upon the effectiveness of a couple of the food-policy related exercises discussed in the previous section.

The UK 'GM Nation?' event of 2003 was subject to an extensive evaluation, focusing upon *process* aspects (see Horlick-Jones *et al.*, 2006a,b). A wide variety of evaluation criteria were considered in this, including normative criteria derived from the literature, criteria implied in the stated objectives of the event sponsors, and criteria elicited from event participants. Data was collected through a variety of methods, including participant questionnaires, observation schedules, and document analysis. The result from this detailed analysis was a considerably negative evaluation of the event. Although the participants generally felt the various events were fairly and independently facilitated and run, there were problems in terms of the transfer and use of the supplied information (for example, in the conference events, few participants actually read the supplied information booklets, in part because there was no time set aside specifically for this). The most serious criticism, however, concerned the representativeness of the participants, who were demonstrated to be atypical of the general population in terms of socio-economic and demographic factors, but perhaps more importantly, in terms of their attitudes to GM technology (Horlick-Jones *et al.*, 2006a,b). In particular, participants were considerably more negative/less positive than a comparative nationally representative sample (Pidgeon *et al.*, 2005). In essence, the debate report to government expressed the views of a highly biased sample as though this reflected the views of the public at large.

As discussed earlier, participation outcomes are more difficult to evaluate than processes, but the *absence* of clear impacts is often most telling. With respect to Consensus Conferences (in general, and not only in relation to food policy), there is little evidence of these having much impact on policy at all. One possible exception is where such conferences have been conducted in Denmark, in which the use of this mechanism is more directly linked to policy makers who have influence and power to act upon their conclusions (in this case, the Danish Board of Technology, which is funded by the Danish Parliament), and which country arguably has a more consensual (amenable) political culture. For example, Einsiedel *et al.* (2001) linked a 1987 Consensus Conference in Denmark, which recommended against genetic engineering on animals, with a Danish Parliament decision not to fund such research in a subsequent research and development programme. However, Einsiedel *et al.* (2001) also noted that 'substantive impacts' of this type are less clear in the case of other Consensus Conferences held in other countries in which the sponsors have had no direct say on policy. One possible modern example concerns the 'GM Nation?' event, in which the extremely negative recommendations that emerged appear to have had no clear influence on government policy (which has largely been directed instead by, for example, EU policy – Horlick-Jones *et al.*, 2006a,b), arguably because the sponsors (the AEBC) had no significant political power. In this case, the sole pre-commitment of the event paymasters (the national government) to the event organisers (hence participants) was to formally respond to the final debate report – a commitment that made no great demands of them! Thus it would seem that both the *ability* and *will* (e.g., indicated by a significant pre-commitment) of sponsors to act on recommendations that emerge from such exercises is crucial for their influence on affairs.

In the absence of clear evidence of participation effectiveness, is there other research that might shed some light on whether participatory approaches are likely to be beneficial to policy making? There is a substantial body of work on the psychology of group processes that may be relevant here, and the results from this leave an uncomfortable suspicion that group-based participation processes may not provide the solutions to policy management dilemmas that their supporters hope. One well-documented finding is that of group 'polarisation', which is the tendency for groups to shift their attitudes and judgements after discussion towards the *initially dominant position* within the group (Moscovici and Zavalloni, 1969). Assuming, for example, that individuals have an exaggerated perception of the riskiness of some technology to begin with (as is arguably the case, for example, regarding genetically modified foods and crops), research suggests that polarisation processes may make *group* views even more exaggerated, as people become more *persuaded* of risks or amend their views for *social comparative* purposes – persuasion and social comparison being the two dominant explanations for attitude shifts in these scenarios. (With regards risk assessments, this phenomenon was initially termed the 'risky shift' – because groups appeared to have a tendency to make riskier decisions – until 'cautious shifts' were also observed in cases where the predominant initial position was against risk, and these phenomena are now seen as examples of the more general polarisation phenomenon.) Likewise, underestimates of the riskiness of activities, if pervasive, might be reinforced and exaggerated further after group debate – not only within public groups, but within expert committees too. Thus, rather than decreasing the space between competing parties (e.g., an expert advisory group and public assemblages, as in a Consensus Conference) typical group processes might well increase it. Furthermore, difficulties might be exacerbated by finding that the social sanctions are often placed on those with minority opposing views – who could be the lone public representatives co-opted onto scientific advisory committees, or minority 'experts' within larger groups of concerned citizens (though factors such as *consistency* and *moderacy* might enable a degree of influence for these minorities, e.g. Wood *et al.*, 1994). Other authors are also beginning to turn to wider literatures to raise concerns about the likely success of participatory approaches (e.g., Ryfe, 2005; Sunstein, 2005). At present, our knowledge of processes in real world groups in the participation domain is limited, but is clearly in need of much greater study.

27.5   Future trends

At the present time, and in the near future, the experiment of public involvement in the determination of food policy is likely to continue. There are signs, however, that initial enthusiasm for involvement is being tempered by realisation in various circles of a need to answer a number of practical questions – such as *when* should the public be involved, and *how* should this be done. As indicated previously, research funders are coming to appreciate the need to establish a suitable *evidence base* to support the optimistic contentions often expressed about this new way of doing things (e.g., the EU, and see also RCUK, 2005); and hard evidence is unlikely to find that involvement approaches (particularly public participation) will solve all of the policy-making problems. Perhaps the public involvement trend simply reflects a very typical human way of coping with difficulties, lurching from one extreme to the next, before gradually coming to rest on some sensible middle ground. Of course, the 'old way' of policy setting had its problems by taking the public lightly, and by overlooking the uncertainty inherent in risk analyses and the hidden values that lay behind many decisions. But do the people of Europe (for example) – a large percentage of whom cannot even be bothered to vote at elections – really want to be regularly involved in the affairs of policy makers? And even given that such motivation exists, are people able to deal with the complex facts often involved in food policy cases? (As a case in point, in the UK there has recently been a move to remove the right to trial-by-jury in fraud cases, as these currently tend to fall apart without conviction due to juries' difficulties in handling copious complex facts.) Maybe the key solution is simply to create more 'aware' scientists and policy makers, who avoid hyperbole, express opinions with caution, and so on? Perhaps institutional learning about past difficulties will suitably inform aspects such as the scientific assessment of food risks so that future crises may be prevented, trust in policy makers may be restored, and the need to involve the public through participatory mechanisms may quietly fade away? Even if this is the case, there will clearly still be a need to continue social science research into how and why people believe what they do, in order to inform the policy community as to what the public are thinking, and hence to inform their communication strategies.

In all likelihood, the most appropriate method of public engagement in policy making will depend upon the specifics of the particular situation. There is likely to be a place for public participation in developing food policy, but it is equally likely that 'communication' (or *non-participation*), which has negative connotations in certain circles because it may imply a subservient relationship of the public to policy makers, will also have its place. Indeed, learning how to effectively communicate complex ideas to laypeople is important: as Frewer and Shepherd (1998) point out, by developing the public understanding of science more generally, the public's capacity to actually enter scientific debate about issues such as risk and food policy will be enhanced. Achieving such public enlightenment would, in itself, appear to be a worthy goal.

27.6   Sources of further information and advice

For interesting analyses of food management and policy-making issues, see the work of Jasanoff (e.g., 1990; 1997). The edited book by Irwin and Wynne (1996) provides a number of perspectives on the 'public understanding of science' issue. To find out more about current policy of various national and international food safety organisations, the reader should consider the relevant websites, such as:

- http://efsa.eu.int/ (EFSA);
- www.afssa.fr (AFSA);
- www.verbraucherministerium.de (BMVEL);
- http://www.food.gov.uk/ (FSA).

Generic information on conducting evaluations of social programmes, with some applicability to evaluating public engagement exercises, can be found in a number of good textbooks, including those by: Clarke (1999); Patton (1990); Robson (2002); Rossi *et al.* (1999). Specific works looking at public engagement include the edited volume by Joss and Durrant (1995), which focuses on Consensus Conferences (which seem to have become the favoured one-off participation mechanism), and Renn *et al.* (1995), which takes one evaluation framework that is then applied by various authors to assess a number of different engagement mechanisms. The reader might also keep a look out for a forthcoming book by Horlick-Jones, Rowe, Walls, Pidgeon, Poortinga, O'Riordan and Murdock (2006a), which will describe the complex evaluation of the 'GM Nation?' event. Finally, one book worth referring to for details on how people perceive risks, how this may differ from experts, and the issue of trust and risk communication, is that of Slovic (2000), which contains reprints of many of that author's seminal papers.

27.7   Acknowledgement

Much of the work reported in this paper was supported by the Programme on Understanding Risk funded by the Leverhulme Trust (RSK990021).

27.8   References

BIER V M (2001), 'On the state of the art: Risk communication to the public', *Reliability Engineering and System Safety*, 71: 139–150.

BYRNE D (2002), *EFSA: Excellence, Integrity and Openness*, Inaugural Meeting of the Management Board of the European Food Safety Authority, Brussels (18 September 2002).

CHESS C and PURCELL K (1999), 'Public participation and the environment: Do we know what works?', *Environmental Science and Technology*, 33 (16): 2685–2692.

CHIPMAN H, KENDALL P, SLATER M and AULD G (1996), 'Audience responses to a risk communication message in 4 media formats', *Journal of Nutrition Education*, 28 (3), 133–139.

CLARKE A (1999), *Evaluation research: an introduction to principles, methods and practice*, London, Sage.

EFSA (2003), *EFSA plans greater public involvement in its work*, Press release, 3 December 2003.

EINSIEDEL E F, JELSOE E and BRECK T (2001), 'Publics at the technology table: The consensus conference in Denmark, Canada, and Australia', *Public Understanding of Science*, 10 (1), 83–98.

ELLAHI B (1995), 'UK National Consensus Conference on Plant Biotechnology', *Trends in Food Science and Technology*, 6 (2), 35–41.

FIORINO D J (1990), 'Citizen participation and environmental risk: a survey of institutional mechanisms', *Science, Technology, & Human Values*, 15 (2), 226–243.

FOOD STANDARDS AGENCY (2000), *Statement of General Objectives and Practices*, London, Food Standards Agency.

FREWER L J (1999), 'Risk perception, social trust, and public participation into strategic decision-making: implications for emerging technologies', *Ambio*, 28, 569–574.

FREWER L J and SHEPHERD R (1998), 'Consumer Perceptions of Modern Food Bio-technology', in Roller S and Harlander S, *Genetic engineering for the food industry: A strategy for food quality improvement*, New York, Blackie Academic, 27–46.

FREWER L J, HOWARD C, HEDDERLEY D and SHEPHERD R (1996), 'What determines trust in information about food-related risks? Underlying psychological constructs', *Risk Analysis*, 16, 473–486.

FUNTOWICZ S and RAVETZ J (1992), 'Risk management as a post-normal science', *Risk Analysis*, 12 (1), 95–97.

GOLDING D, KRIMSKY S and PLOUGH A (1992), 'Evaluating risk communication: Narrative vs technical presentations of information about radon', *Risk Analysis*, 12 (1), 27–35.

GREGORY N G (2000), 'Consumer concerns about food', *Outlook on Agriculture*, 29 (4), 251–257.

HILGARTNER S (1990), 'The dominant view of popularization – Conceptual problems, political uses', *Social Studies of Science*, 20 (3), 519–539.

HORLICK-JONES T (1998), 'Meaning and contextualisation in risk assessment', *Reliability Engineering & System Safety*, 59, 79–89.

HORLICK-JONES T (2004), 'Experts in risk? ... do they exist?', *Health, Risk & Society*, 6 (2), 107–114.

HORLICK-JONES T, WALLS J, ROWE G, PIDGEON N, POORTINGA W and O'RIORDAN T (2004), *A Deliberative Future? An Independent Evaluation of the GM Nation? Public Debate about the Possible Commercialisation of Transgenic Crops in Britain, 2003*, Programme on Understanding Risk Working Paper 04-02, Norwich, University of East Anglia.

HORLICK-JONES T, WALLS J, ROWE G, PIDGEON N, POORTINGA W, MURDOCK G and O'RIORDAN T (2006a), *The GM Debate: Risk, Politics and Public Engagement*, London, Routledge (in press).

HORLICK-JONES T, WALLS J, ROWE G, PIDGEON N, POORTINGA W and O'RIORDAN T (2006b), 'On evaluating the *GM Nation?* public debate about the commercialisation of transgenic crops in Britain', *New Genetics and Society* (in press).

IRWIN A and WYNNE B (1996), *Misunderstanding Science? The Public Reconstruction of Science and Technology* (edited), Cambridge, Cambridge University Press.

JASANOFF S (1990), *The 5th branch – Science advisers as policy-makers*, Cambridge, MA, Harvard University Press.

JASANOFF S (1997), 'Civilization and madness: The great BSE scare of 1996', *Public Understanding of Science*, 6 (3), 221–232.

JOSS S and DURANT J (1995), *Public Participation in Science: The Role of Consensus Conferences in Europe* (edited), London, The Science Museum.

KÜNAST R (2001), *Government policy statement on the new consumer protection and agricultural policies*, speech by Federal Minister of Consumer Protection, Food and Agriculture, 8 February 2001 (www.verbraucherministerium.de).

MOSCOVICI S and ZAVALLONI M (1969), 'The group as a polariser of attitudes', *Journal of Personality and Social Psychology*, 12, 125–135.

NATIONAL RESEARCH COUNCIL (1996), *Understanding Risk: Informing Decisions in a Democratic Society*, Washington DC, National Academy Press.

PATTON M (1990), *Qualitative evaluation and research methods*, 2nd edn, London, Sage.

PIDGEON N F, POORTINGA W, ROWE G, HORLICK-JONES T, WALLS J and O'RIORDAN T (2005), 'Using surveys in public participation processes for risk decision-making: The case of the 2003 British *GM Nation?* public debate', *Risk Analysis*, 25 (2), 467–479.

POST (2001), *Open Channels: Public Dialogue in Science and Technology*, London, Parliamentary Office of Science and Technology, Report 152.

PUBLIC DEBATE STEERING BOARD (PDSB) (2003), *GM Nation? The Findings of the Public Debate*, London, Department of Trade and Industry (www.gmnation.org.uk).

RCUK (2005), *Evaluating science and society programmes*, London, Research Councils UK.

RENN O (1992), 'Risk communication: Toward a rational discourse with the public', *Journal of Hazardous Materials*, 20, 465–519.

RENN O, WEBLER T and WIEDEMANN P (1995), *Fairness and competence in citizen participation: Evaluating models for environmental discourse*, Dordrecht, Netherlands, Kluwer Academic Publishers.

ROBSON C (2002), *Real world research*, 2nd edn, Oxford, Blackwell Publishing.

ROSSI P H, FREEMAN H E and LIPSEY M W (1999), *Evaluation: A Systematic Approach*, 6th edn, London, Sage Publications.

ROWE G and FREWER L J (2000), 'Public participation methods: A framework for evaluation', *Science, Technology, & Human Values*, 25 (1), 3–29.

ROWE G and FREWER L J (2004), 'Evaluating public participation exercises: A research agenda', *Science, Technology, & Human Values*, 29 (4), 512–556.

ROWE G and FREWER L J (2005), 'A typology of public engagement mechanisms', *Science, Technology, & Human Values*, 30 (2), 251–290.

ROWE G, REYNOLDS C and FREWER L J (2001), 'Public participation in developing policy related to food issues', in Frewer L J, Risvik E, Schifferstein H N J and von Alvensleben R, *Food Choice in Europe*, Berlin, Springer-Verlag, 415–432.

ROWE G, HORLICK-JONES T, WALLS J and PIDGEON N F (2005), 'Difficulties in evaluating public engagement initiatives: Reflections on an evaluation of the UK "GM Nation?" public debate', *Public Understanding of Science*, 14 (4), 331–352.

ROWE G, POORTINGA W and PIDGEON N F (2006), 'A comparison of responses to internet and postal surveys in a public engagement context', *Science Communication*, 27 (3), 352–375.

RYFE D M (2005), 'Does deliberative democracy work?', *Annual Review of Political Science*, 8, 49–71.

SLOVIC P (2000), *The perception of risk*, London, Earthscan.

SUDMAN S and BRADBURN N M (1974), *Response Effects in Surveys*, Chicago, Aldine.

SUNSTEIN C R (2005), 'Group judgments: Statistical means, deliberation, and information markets', *New York University Law Review*, 80 (3), 962–1049.

VERBEKE W (2001), 'Beliefs, attitude and behaviour towards fresh meat revisited after the Belgian dioxin crisis', *Food Quality and Preference*, 12 (8), 489–498.

WALLS J, PIDGEON N, WEYMAN A and HORLICK-JONES T (2004), 'Critical trust: understanding lay perceptions of health and safety risk regulation', *Health, Risk & Society*, 6 (2), 133–150.

WOOD W, LUNDGREN S, OUELLETTE J A, BUSCEME S and BLACKSTONE T (1994), 'Minority influence: A metaanalytic review of social-influence processes', *Psychological Bulletin*, 115 (3), 323–345.

WYNNE B (1991), 'Knowledges in context', *Science, Technology, & Human Values*, 16 (1), 111–121.

28

Food, citizens, and market: the quest for responsible consuming

F. W. A. Brom, T. Visak and F. Meijboom, Wageningen
University, The Netherlands

28.1   Introduction

Is there a relation between consumer concerns about the food market, consumer
behaviour and their ideas about the future of agriculture and food production?
When setting the title for this chapter we experienced difficulties in bringing the
words 'food', 'citizens', 'consumers', 'market' and 'state' together. This difficulty reflects changes in the relation between individuals in their roles of
consumer and citizen, and collective co-ordination systems such as the state and
the market. These changes become even more complicated when we realise that
food is linked with the identity and culture of the consumer.

In modern liberal democracies, we normally distinguish between the *market*
as co-ordination system for the exchange of commodities, and the *state* as co-
ordination mechanism for the creation and maintenance of public goods. Current
food practices are inextricably bound up with world trade. It is hard to imagine
today's world without the export and import of food products among countries
all over the world. The economic impact of food trade is immense. In a publication of the World Bank on international trade and agriculture it is calculated
that, together, the top 20 food exporters export approximately US$80.26 billion
a year (Diaz-Bonilla and Thomas, 2003, p. 233). From this perspective 'Food,
consumer, and market' seems an adequate title.

Food, however, is not just a commodity. Next to being important in
international trade, it is a scarce good in some places in the world, resulting in
massive undernourishment in some parts of the world. FAO estimates that, at the
present time, the number of undernourished people is around 800 million. This
implies that food trade touches upon basic questions of *global justice* and the

internationally recognised *Right to Adequate Food*. Food-security is therefore often 'state-business'. Food policy is directed at the public good of enough, safe food to feed the country. From that perspective the title 'Food, citizens, and state' is highly appropriate.

However, food represents more than just nourishment alone. Food can be important for peoples' identity, for example within (religious) views of life. This is clear for some religious doctrines including Jewish, Islamic and Hindu. It is, however, not limited to organised religion. In contemporary Western societies – where more and more people have not committed themselves to one specific organised religion – food is one of the strongest ways to demonstrate symbolic interaction with nature, other living beings and the universe. The growth of people making food choices according to certain self-chosen food laws (vegan, vegetarian, piscenarian, free range, organic and so on) makes this clear. The social and cultural meanings of food preparation, sharing food and the way food is part of our communication patterns, is a study in itself. A first example might be that on special occasions, such as weddings and funerals, in certain cultures specific dishes are served. By complying with those food rules or ignoring them, people communicate about their identity. Another example might be eating vegetarian or vegan. In doing so a perspective on life, nature and identity is often communicated. This is also reflected in the discussion regarding the Right to Adequate Food; it is asserted by the UN Committee on Economic, Social and Cultural Rights (CESCR) that the Right to Adequate Food implies that food should not only be safe ('free from adverse substances') but also acceptable within a given culture. From that perspective we even contemplated the title 'Food, identity and culture'.

The possible titles for this chapter demonstrate that food is of interest to different communities. This becomes even clearer when we look at current issues in international trade relations. In current food trade, food as a commodity can collide with the cultural and social meanings of food. This is the case in two recent transatlantic trade conflicts over food: the use of artificial growth hormones in beef production and the use of (modern) biotechnology in food production (Brom, 2004). In these discussions not only the acceptability of certain products is at stake, but also the future of food production and agriculture. A central way of looking at the questions brought forward by consumers is to label these as consumer concerns (see Chapter 5 by De Jonge *et al.*, this volume). This, however, seems to eliminate the political and moral message that lies behind these concerns. Sometimes citizens use the market for sending political and moral messages; that is why this chapter is about food, citizens and the market. However, the messages from consumers may not always be clear. In addition, consumers – as a group – may fail to behave according to what these messages actually say.

28.2   Background of consumer concerns

Consumers are wary; they express 'consumer concerns' about agriculture and food production. Before we analyse these concerns, (in the next section), it is important to contextualise them against developments in the agricultural and food sector. These changes can be summarised in three interlinked developments.

The first development is the growth of agricultural efficiency. In the second half of the 20th century governmental policy in the field of agriculture and food in Western countries was successful in providing enough and safe food. One of the driving forces of the growth of efficiency was intensification. Intensive farming, however, creates some externalities, that means consequences that are not intended by the stakeholders, but that follow from the way the practice of intensive farming is organised. Intensive agricultural food production systems are organised upon economic rationality; therefore, unwanted consequences that can only be prevented by making costs, are often seen as being inevitable by those that would have to pay those costs. Environmental damage caused by some approaches to plant protection, or compromised animal health and welfare in (intensive) animal production systems are examples of these externalities. These externalities result in consumer concerns, and agricultural policy has reached a point where a conflict has appeared between striving for more economic efficiency in food-production on the one hand, and satisfying societal concerns for a sustainable agriculture on the other hand.

The second development is the growth of the gap between producers and consumers. As a result of the growing agricultural efficiency and of urbanisation, a physical and intellectual distance between food production and food consumption has developed and grown. In the emerging global market for food and other agricultural products, the gap between the consumer and the farm has widened. In Western society, most consumers have no direct contact with the farms where their food is produced. Nearly all food, in the city as well as in villages, is purchased in supermarkets. Often food is imported from exotic or distant countries and many consumers have little real-life experience with modern farming. Both the physical and the intellectual distance between producers and consumers has grown. Farmers often consider critical questions raised by consumers about agricultural practices as typical 'city-issues'. This physical and mental gap between food production and consumption has important consequences for the way consumers perceive products, and for the way trust is built. Indeed, trust is no longer shaped by direct human interaction. While I generally trust the shopkeepers where I buy my food, I also know that they cannot guarantee the way the product was produced or how safe it is, as they have neither produced it nor are they in direct contact with the producer. Consumers have, generally speaking, a romantic, and artisan picture of food production that often is re-enforced by food advertisements. When – mostly in situations where food safety is presented in a crisis context – they are confronted with the reality of food production, they feel alienated, which leads to problems for governments and actors in the food chain, such as distrust in (the producers of) food. Technical, economic and scientific approaches to food and food safety seem to be out of touch with the role of food in the real world. Consumer concerns with regard to the safety of industrialised agriculture, as well as the consumer reactions and food scares in food crises, can be understood against this background. This gap also partly clarifies the political and moral resistance against the introduction of modern biotechnology in agriculture and food production in Europe.

The third development is the quest for transparency and traceability. It is generally acknowledged that in order to bridge the physical and intellectual gap between consumption and production it is necessary that the food sector 'opens up'; transparency and traceability are at the moment buzz-words in the food sector. The idea is that by creating transparency and improved systems of traceability, consumers become more empowered in their decisions regarding consumption choices. The problem, however, is how much transparency and traceability should the government and companies strive for. It is clear that you cannot show everything to everybody. And this also applies to traceability. Specifically, decisions need to be taken regarding which properties of food production are so important that they have to be traceable throughout the food chain. For example, should the origin, production method, the environmental consequences of the production method, and the type of labour used to produce the food (e.g., chocolate or meat) also be subjected to traceability? Transparency and traceability presuppose clarity about the importance of what has to be shown and what has to be traceable. It is not evident that such a shared system is feasible. Societal pluralism has entered the discussions in the food chain; different consumers have different concerns. This has raised the question of how governments and firms in the food chain could and should react to the societal value pluralism.

28.3   The distinction between consumer and citizen

One way of reacting to consumer concerns is to leave them to the market. In other words, consumers should be able to choose whether or not to buy particular products. One could claim that consumer behaviour is directed at the market and therefore ethically and politically irrelevant. This would reflect the classical distinction between consumers and citizens. In this classic analytical distinction consumers are rational actors who make choices in order to maximise their personal utility. Can citizens make political choices according to their values? This distinction goes back to the distinction between '*bourgeois*' and '*citoyen*' and between '*Homo Economicus*' and '*Homo Politicus*'. This distinction is brought forward by Marc Sagoff who in his *Economy of the Earth* states (1998, 8): 'As a citizen, I am concerned with the public interest, rather than my own interest; with the good of the community, rather than simply the well-being of my own family. (...) In my role as a consumer, (...) I pursue the goals I have as an individual.'

This classic distinction between the consumer and citizen is seen as empirically untenable by many authors. People do not live in two distinct worlds; they (sometimes) bring their civic values into the shop and it is clear that in their consumer preferences (sometimes) play a decisive role. The same holds for the relation between market and state: the market is not a natural phenomenon but its scope and form are created by social institutions such as legislative frameworks (see also Chapter 29 by Korthals, this volume).

One indication that the classic distinction does not hold anymore, and that citizens communicate political and moral messages in the market, is the rise of co-ordinated collective consumer action. We see, for instance, that for multi-national companies an agreement with the national government no longer guarantees the continuity of business operations. We see that a company's social licence to produce depends on the support of civil society in a country. And a way for civil society to communicate with companies is via the market. Multi-nationals have become dependent on the support of internationally operating single-issue organisations.

28.4   Different consumer concerns

If the classic distinction does not hold anymore, and if citizens communicate political and moral messages in the market, how can we interpret these messages? In the genetically modified food debate we see that European consumers seem to distrust (the producers of) GM-food. (See also Chapter 10 by Siegrist, this volume.) Tabloid newspapers write horror stories about 'Franken-stein food', non-governmental organisations (e.g., Greenpeace) campaign against the use of biotechnology in agriculture and food production, and, last but not least, governmental responses to these concerns could give rise to trade-disputes about barriers against genetically modified food products (Brom, 2004).

> Consumer Concern has recently been heightened in some countries by active campaigning against genetically modified organisms, particularly in food products. It is an emotive debate, with science caught in the middle of it. A clear political lead is therefore needed. The trouble is that short-term political pressures do not always influence policies for the better. They can lead to ad hoc regulatory interventions, which focus on and stigmatize new techniques, duplicate existing systems and lead to needless bureaucracy and the occasional trade dispute.
>
> Donald J. Johnston, secretary-general of the OECD (1999).

What *are* these consumer concerns, and how should governments cope with them? The term 'consumer concern' is often used as a 'container notion', which includes consumer concerns about food safety, environmental and animal welfare consequences of food production systems, and consumers intrinsic moral objections associated with different food technologies like genetic modification. The complexity of the relation between consumer and citizen becomes clear if we look at the different concerns consumers bring into the food market.

Concerns that matter in principle to all consumers
Certain consumer concerns matter equally to everybody in their role as a consumers. Food safety is a key issue in this field. Food safety is important to all

consumers. and it is clear that food safety problems require a governmental response; it is beyond the possibilities of individual consumers to assess these questions. Here we see that consumer concern asks for collective action in order to be taken seriously.

Concerns that matter to specific groups of consumers
Other consumer concerns matters to specific groups of consumers, because of the way these individuals express preferences for how they want to live their lives. It is important for *citizens* to be able to live according to their own life plans. Respect for consumer autonomy implies that consumers have the *prima facie* right to live their lives according to their own value systems. This implies that consumers ought to have the option of choosing products that fit in with their view of life. Vegetarians, for instance, can only live according to their own value system when they know whether or not their food contains animal products. In so far as vegetarianism is a lifestyle, we see that *personal* values enter the market. If vegetarianism transcends lifestyle and represents a moral choice that directs an appeal to others, it goes beyond *consumer* concerns. The concern does not relate to their personal consumption choices, but to broader ideas about how people should treat each other and animals or more generally, what constitutes a good society.

Concerns that have a public message
Finally, there are concerns brought forward in the market (consumer concerns), that find their origin in the political and moral views held by different individuals. These concerns are related to ideas about what constitutes a 'good' society. These concerns are not consumer-oriented in a technical sense, but they do entail a public message that goes beyond the sphere of consumption and are about how people should live in general. People are concerned about certain products because of the wider impact these products have, or potentially have, on their society and beyond. Take, as an example, meat that is produced by veal calves that are individually raised in confinement with severe animal welfare consequences. People are against this way of producing meat, not just because they don't want to eat meat produced by crated calves, but because they think that the way crated calves are treated is immoral and should be banned. Crating calves is problematic because it is not compatible with what consumers believe should be done in a good society. Here we see how *civic* values enter the market.

To sum up, consumer concerns create mixed messages for producers, retailers and government: it is a political and moral message in the market that is, however, not always consistent with the political and moral message in the political arena. The views that are expressed in societal discussions are often more demanding than what people communicate by their actual purchase choices. Consumer concerns reflect a plurality of voices, roles and messages, and are frequently incorporated into the agendas of pressure groups and NGOs.

Consumer concerns reflect public uneasiness with specific food issues, but this uneasiness is not always translated into a clear message.

28.5   Trust: the need for a reliable answer

A part of the 'mixed message' resulting from consumer concerns relates to consumer trust. Maintaining trust in the food sector is not only important for retailers, food industry, and the agricultural sector. The establishment of trust is also important for government, because public trust, in general, is of importance for society. Without people trusting each other, no co-operation seems possible and without co-operation no society can survive. This is no different for the food sector.

Since trust is related to situations of uncertainty and lack of control, we make ourselves vulnerable when we trust other persons or institutions. Trust is a risky enterprise (cf. Luhmann, 2000, 31). Therefore, it only seems to be a small step to equate trust with taking a risk of harm. However, it is important to note that trusting another is not the same as taking a risk. When a person trusts, it will appear that he or she does not perceive the situation as risky or as a gamble, although he or she certainly will run a risk. When we trust another 'we do not consider the possibility that she might deliberately let us down' (Lagerspetz, 1998, 48). Risk taking and trusting are on different levels.

This difference can be explained with the help of the distinction between a first-person and a third-person perspective. From a third-person perspective, trust is certainly a risky matter: a trustor takes a risk. In acting as if only one state of affairs were to be expected, one runs a risk and makes trust close to a gamble. Nevertheless, from a first-person perspective the picture is quite different. As a trustor one is not aware of taking this risk. If so, he would be a risk taker not a trustor. Hence from the perspective of the trustor risks are not the main element of trust. Only as an observer may one notice that another runs a risk.

Trust is primarily a concept defining human relations. Having trust in one another means that one takes each other, and each other's concerns, interests and wishes, seriously. The difference between risk taking and trust is the difference between a kind of relation and a certain act. Risk taking is an act upon a decision in which pros and cons are weighed; trust is a relation built up through time. Therefore building trust costs time and trust is easily lost.

The big differences between 'trusting' and 'risk-taking' are important for the way we build trust in the agro-food market. Rational and scientific procedures applied to the assessment and minimisation of risk are still necessary, but no longer adequate in themselves to ensure consumer trust in the food supply. Similarly, rational and scientific procedures that substantiate or counter health claims are also necessary but not enough in themselves to maintain consumer trust. Scientific analysis alone cannot build trust. Trust is not merely based upon risk assessment and management. Communication and transparency are necessary, too. This is more than just effective risk communication. It requires a dialogue with stakeholders (including consumers) about potential risks, but also the acknowledgement that our relationship with those whose trust we want to develop and maintain represents a moral relationship. In this relationship trust can only be developed if the moral concerns of people are taken seriously, that is if moral responses are given to moral questions.

This illustrates that trustworthiness plays an important role. Lack of trust is not simply the problem of having to rely on other parties, but implies also the question whether the other person is worth being trusted. Who wants to be trusted should be trustworthy: this means that he who wants to be trusted should not only be competent but also have goodwill towards the trustor.

These two elements are crucial with regard to addressing problems of trust and distrust. Trust can only grow if the reasons for consumer distrust are also addressed. As trustful expectations are formed based upon another's competence and good-will, distrust can be based upon 'well-meant but ill-judged or incompetent attempts to care for what is entrusted and ill-meant and cleverly disguised abuses of discretionary powers' (Baier, 1994, 104). We can, therefore, distinguish between four ideal-type reasons for distrust:

1. General distrust on grounds of motive.
2. Distrust on grounds of motive in a specific context.
3. General distrust on grounds of competence.
4. Distrust on grounds of competence in a specific situation.

MacLagan (1998, 57).

The relevance of these points for earning trust in the context of the agro-food sector is different for each point. In the first point, distrust is based upon the idea that the motives of the potential trustee, are self-interested or deceitful. For example, people may distrust food industries because they think these industries only want to make money at the expense of other factors, for example consumer protection. To earn trust against this criticism is to show over time that you are not merely driven by opportunistic motives, but that you take the other's interests and moral concerns seriously, i.e., to show your (moral) integrity.

In the second type of reason for distrust, the suspicion with regard to the moral sincerity of the motives of the other party is not as strong as in (1). With this type, people do not assume that the other agent acts merely upon opportunistic motives, but they distrust others, because they fear that the latter adhere to different values in comparison to themselves. This, for instance, could be the case with food safety regulations. People fear that those in charge of consumer protection or food production have different values about food safety to those held by consumers. To earn trust where it has been lost as a consequence of differences in motive may not necessarily imply that the distrusted actor must change their value system, but it is important for the distrusted actor to demonstrate that they are prepared to take due account of alternative value systems.

In points three and four, distrust is not based upon moral criticism, but upon the fear that those who want to be trusted are not able to do what they are expected to do. In order to earn trust, this fear implies that those who want to be trusted should be critical about their own abilities. Trust is often lost, not because of incompetence (3) but because of unrealistic expectations of abilities.

Parts of the debate about genetically modified food can best be analysed according to the second form of distrust. People will not trust those who state that their concerns are, at least in part, 'nonsensical', and that they do not need to reckon with these concerns. Thus, for example, if governments, retailers or other actors in the food chain state that public concerns for justice, animal welfare, sustainability and biodiversity should be kept out of the debate regarding genetically modified foods, then this is likely to result in societal distrust in food chain actors.

28.6   Do consumers have a responsibility for public goods?

Can we take consumer concerns seriously? Consumer concerns are not predictable, and the political and moral messages from consumers and citizens seem mixed and unclear. If we look at the 'citizen' in order to understand the 'consumer' we run into trouble. The recent Dutch debate about the distribution of responsibilities for realising a more nature- and animal-friendly agro-food sector makes this clear. In this debate the consumer is attacked by politicians because consumer behaviour is not in line with the preferences citizens bring to the public debate. Citizens want a change of agricultural practices towards more sustainability while consumers only incidentally buy environmental and animal-friendly products. This seems inconsistent.

Cees Veerman, the Dutch Minister of Agriculture, Nature Management and Food Quality, therefore wants consumers to take their responsibilities seriously: 'If they won't do this, I shall not hesitate to call them bluntly hypocritical'. This critical attitude towards the consumers comes from a great number of players in the sector: producers, retailers, government and NGOs. It is based on the presumption that the consumers' willingness to pay for more 'responsible' products is necessary in order to achieve the aim of developing more sustainable agricultural production practices. It is a common understanding in policy making that it is pointless to reform the food and agricultural sector as long as citizen demands for animal and environmentally friendly production systems are not translated into consumer behaviour in the market. However, just that willingness of the consumer appears to be problematic. One of the members of the Dutch parliament acknowledges that the citizen is becoming more and more demanding regarding questions of animal welfare, but that the consumer still does not pay for animal welfare oriented products. His conclusion is that, 'due to these double moral standards the market takes little initiative' to change in a more sustainable direction. This state of affairs leads to the reproach of double moral standards towards the consumer.

The double moral standard seems to result in a vicious circle which blocks any successful implementation of sustainable and animal-friendly practices in the agro-food sector. The citizen directs concerns about animal welfare and environmental protection towards government. The government acknowledges these concerns, but claims to depend on the collaboration of the producers and retailers in order to realise an adequate policy in this area. The food industry, in turn, cite consumer demand for cheap food as the underlying factor which maintains the status quo. These same consumers, in their role as citizens, subsequently direct their concerns back to the government. This completes the circle that seems to be built around the reproach by politicians of double moral standards. For these reasons, Platform Biologica, the Dutch organic producers' organisation, states in a letter to the minister that the breaking through the consumers' double moral standard is essential if any advances in structural developments towards a more sustainable animal husbandry are to be made. Lack of consumer support has become a huge problem for farmers pioneering new sustainable agricultural practices, as they have difficulties in competing with ordinary farmers.

The discrepancy between what people say they find important, and the behaviour they exhibit in the supermarket, in other words: the double moral standards of the citizen versus the consumer seems to be the main obstacle creating a barrier to responsible consumption and sustainable production. All players in the agro-food sector, including government, seem to be willing to strive for animal and environmentally friendly production, with the exception of the consumer. Is this, however, the whole story? Is the reproach of double moral standards only about consistency between a person's attitudes and behaviour? Is the problem of responsible consumption a problem of double moral standards?

The reason the Dutch government, as well as sustainable producers, criticise the lack of consumer support regarding changes in agricultural practices is that government wants to influence the market *via* consumer behaviour, and that some of those producers have already adopted a more sustainable method of production. An example is Dutch organic pig farmers that have to quit because of the lack of actual consumer demand. The reproach of double standards could be interpreted as a sign of frustration about the lack of consumer co-operation.

28.7   Looking behind the double standards

Is 'double standard' the best way to characterise the lack of consumer co-operation in changing the agricultural system? In order to assess this, we need to look more closely at the relation between moral preferences and consumer behaviour. Most statistics and interviews about consumption preferences and patterns only give a general picture and do not allow conclusions about any relevant discrepancies *within* individual persons (but see de Jonge *et al.*, Chapter 5, this volume). Besides, not all existing discrepancies in attitudes are equally relevant to the problem of responsible consumption. An individual might have several different preferences, concerning sustainability, price and availability which might conflict. Although an individual may exhibit a pro-environmental preference, for example, other preferences might be stronger and thus influence actual consumer behaviour.

Furthermore, it should be made clear that not all consumer preferences are moral preferences. Moral preferences originate from people's moral ideas. People can, however, find something valuable in senses other than those which can be described as moral. If, for instance, someone states that piano music is important for him, but he seldom or never goes to piano concerts, we talk about an aesthetic preference, but not about double moral standards. In the same way, we need not talk about double moral standards if someone likes to have certain products made available in the supermarket, but doesn't buy them. Talking about double moral standards in case of animal welfare and environment presupposes that people have moral preferences concerning animals and the environment. It would therefore be interesting to analyse further whether the preferences that are on the basis of the relevant consumer concern are indeed *moral* preferences.

Although it is not always clear whether measured discrepancies are related to moral preferences, moral statements concerning animals and the environment are indeed made by citizens. Animal welfare issues attract high levels of attention from citizens. For example, there has been strong public criticism of intensive husbandry practices, or regarding the culling of animals during recent outbreaks of animal diseases, like swine fever. In recent EU elections the *Animal Party* (which supports animal welfare causes) attracted over 3.4% of the votes of the Dutch electorate, almost enough to gain a seat in the European Parliament. The question remains, however, whether the voices of citizens in favour of more animal welfare voices are representative of society overall? Is it meaningful to talk about a societal double standard, or does the term 'double standard' make sense only at the level of the individual?

Suppose consumers would agree that the transformation of agriculture such that more sustainable and animal-friendly practices are adopted is a goal worth pursuing, and that they have a (strong) moral preference for this goal. Even under these circumstances, it is not clear that the reproach of 'double standards' would be justified, since the link between moral preferences and behaviour is not straightforward. The reasons why this is the case are outlined below.

Action is more than 'concerns'
Consumer behaviour is determined by behavioural intentions, which are determined by attitudes. However, this is not necessarily a simple causal relationship. Social influences may mediate the relationship between attitudes and behavioural intention. If, for instance, a consumer holds a positive attitude towards promotion of animal welfare, but it is socially accepted that meat from intensive farms is a normal food product, the consumer does not consider himself particularly unfriendly towards animals when buying those products. Consequently, the consumer does not form the intention to buy animal-friendly products, even though his attitude is in favour of positive animal welfare practices.

Efficacy
The step from attitude to behavioural intention is also influenced by ideas concerning one's efficacy, that means the degree to which one thinks one's behaviour will make a difference. Consider a consumer, who is in favour of animal welfare, but thinks that his own individual consumption pattern has only a small influence on the welfare of animals (Diederen, 2003). From that perspective, refraining from buying animal-friendly meat or environmentally-friendly bread is not a case of 'double standards' but a choice to explicitly refrain from an ineffective symbolic action.

Evaporating responsibility on the part of the consumer
'The consumer perceives that the impact of his or her behaviour on animal welfare practices or the environment is mediated *via* a long chain of actions. One cannot (simply) assume responsibility when one is a small part in a long chain of actions; the responsibility for the impact disappears in the "system".'

Although the frustration of policy-makers and producers in favour of sustainable and/or animal-friendly production practices regarding the lack of consumer co-operation seems understandable, the question is whether the problem is one of double moral standards in a strict sense. When consumers share the goal regarding the transformation of agriculture to sustainable practices, but do not act accordingly, this should not necessarily be attributed to double moral standards. Maybe the moral standard is clear, but is not realised in practice because of other, non-moral considerations. Furthermore, even if consumers share the goal regarding the transformation of agricultural practice, it is not evident that this ought to lead automatically to changes in consumption behaviour. The question whether, and to what extent, moral considerations may, and indeed should, influence consumer behaviour is complex and needs further reflection.

28.8   A quest for responsible consuming

People express all kind of concerns regarding the agricultural system; sometimes these concerns are – in part – translated into consumer behaviour and thereby explicitly or implicitly directed at influencing the structure of the agricultural system. Sometimes pressure groups and NGOs manage consumer behaviour as a political instrument by asking consumers to boycott specific items. Consumer behaviour is one of the most important factors which steer the production system. Meat produced using animal-friendly production systems will not be produced if consumers do not buy it. Transforming agriculture in a more sustainable direction needs consumer support. Therefore, consumer behaviour is politically and morally relevant in two ways: (1) it sends relevant messages, sometimes strong ones, (2) it seems necessary for governments to respond to consumer demand if they are to adopt the goal of a more sustainable and animal-friendly agriculture. In the notion of responsible consumption, both aspects of consumption are brought together in an ideal: in consumption consumers signal a direction and at the same time they support transformation into that direction. For example, they want a certain kind of production system (e.g., organic) and by buying certain products (e.g., organic) they support that system.

Ideals, however, cannot be directly morally binding. Ideals function as a perspective, a compass that gives direction to one's deliberation. And if one recognises an ideal, it is clear that one wants to live up to it. The policy impact of this ideal, therefore, cannot be to reproach consumers with adopting 'double standards' if they do not live up to the ideal through expressing particular consumption choices; the impact of the ideal should be in finding ways to stimulate consumers into recognising this ideal, as well as its practical consequences. A first step could be that governments make clear that they have accepted a more sustainable and animal-friendly agriculture as an important goal. If this step is really taken, governments cannot characterise non-animal-friendly and non-sustainable products as normal products. And if they want to stimulate this transfer *via* the market, they need to make clear that animal and environmentally friendly products really have an impact on animals and the environment. Finally governments cannot rely on changes in consumer behaviour alone; if a more sustainable and animal-friendly agriculture is an important goal, then governments should encourage other actors in the food chain to take responsibility as well. Only then, can governments expect responsible consumption from citizens.

28.9   Acknowledgements

This chapter is partly based on the results of a research project, funded by the Netherlands Organisation for Scientific Research (NWO) on 'Sustainable consumption and the double moral standards of consumers'. Thanks are also due to Annemarie Kalis who has participated in this NWO project.

28.10   References

BAIER, A. (1994), 'Trust and antitrust', in: *Moral Prejudices, essays on ethics,* Cambridge: Harvard University Press, pp. 95–129.

BROM, F.W.A. (2004), WTO, public reason and food. Public reasoning in the 'trade conflict' on GM-food, *Ethical Theory and Moral Practice*, 7 (4).

DIAZ-BONILLA, E. and M. THOMAS (2003), 'Trade liberalization, the World Trade Organization, and food security', in: M.D. Ingco (ed.), *Creating a Trading Environment for Development*, Washington, DC: The World Bank, pp. 225–246.

DIEDEREN, P. (2003), 'Burger, laat die consument met rust!' in: H. Dagevos, L. Sterrenberg, *Burgers en consumenten – tussen tweedeling en twee-eenheid*, Wageningen: Wageningen AP.

JOHNSTON, D. J. (1999), 'Editorial: a defence of modern biotechnology', *OECD Observer*: March.

LAGERSPETZ, O. (1998), *Trust: The Tacit Demand*, Dordrecht: Kluwer Academic Publishers.
LUHMANN, N. (2000), *Vertrauen, ein Mechanismus der Reduktion der sozialer Komplexität,* 4th edn, Stuttgart: Lucius and Lucius
MACLAGAN, P. (1998), *Management and Morality. A developmental perspective*. London: Sage.
SAGOFF, M. (1988). *The Economy of the Earth*, Cambridge: Cambridge University Press.

29

The ethics of food production and consumption

M. Korthals, Wageningen University, The Netherlands

29.1   Introduction: the importance of ethical considerations in food choice

Modern Western consumers are no longer involved in food production, and have less and less knowledge of, and trust in, production processes. Food, however, is and continues to be an intrinsic good for consumers; rice, for example, in some cultures has not only monetary value, but cultural, social and ethical value as well, because it has an intrinsic role to play in individuals experiencing life according to their ideas of what constitutes a good life (Visser, 1986; Watson, 1998). It is not clear how consumers' preferences can be communicated to food producers if the gap between consumer preferences and what is actually being done by producers remains in place. After discussing some political and ethical positions that do not align with consumers' values and responsibilities, I will go into details of current consumer attitudes towards ethics of food, as well as discussing recent trends, such as the increasing diversification into various food styles and corresponding farming and production styles. Many consumers complain about barriers which prevent them from realising ethically conscious food choices. As a consequence of many social scientific studies framing the buying, cooking and eating person as both a *citizen* concerned with ethical issues related to food production, and as a (materialist, profit maximising) *consumer*, researchers may neglect the fact that consumers are confronted with various difficulties, such as what and who to believe regarding the information about food and ingredients provided on product labels and by the media. In this context, where conflicting and potentially untrustworthy information is being presented, consumers often decide to buy the cheapest food stuffs available.

Only recently have these potential barriers to optimal consumer choice started to be addressed, with remarkable results. However, we can identify barriers from the producers' side as well: the 'productionist paradigm' applied to food production practices during the last sixty years actively militates against the producer taking heed of ethical concerns.

The co-responsibility of consumers with producers regarding their food choices has significant implications for food product development, labelling and advertising. It implies that consumers do not have uniform beliefs, which explains why emerging trends and food 'movements' such as food, farming and production styles, (for example, fast food, slow food, international food and health food) are attaining more importance and prominence. This again makes it necessary to discuss procedures to regulate the coexistence of these styles and the criteria of coexistence from an ethics point of view. The final section of the chapter discusses future trends within the food sector, and why the food sector has to learn to live with diversity and social contextualisation through increased consumer involvement and participation in activities through the entire food chain.

29.2   Current consumer attitudes towards ethics of food: some trends

In Western Europe, from the 1980s onwards, production and consumption of food has become increasingly politicised. In the 1950s, 1960s and 1970s, one can say that, at least with respect to the ethical values and goals of the food system, there was a large, implicit consensus across various stakeholder communities, including consumers: food was not seen to be a political and ethically controversial issue. Nothing political could happen with food; the only ethical issue that was at stake was food shortages in various parts of the world mostly due to misdistribution of food. Food was essentially seen as 'fuel' that could be made available for consumption in larger or smaller quantities, and could be unsafe to eat, but consideration was generally given to other issues. This consensus was a mainly result of the food security problems facing Europe in the first half of the 20th century.

One of the first reports on the genetic modification of food products is still written with this paradigmatic background in mind (Polkinghorne report (1994): Ministry of Agriculture: Report of the committee on the Ethics of Genetic Modification and Food Use, London). It is therefore no wonder that the Polkinghorne report only recommends with respect to genetic modification that these food stuffs 'require notification by those seeking to market a novel food of why a copy gene of human origin had been used rather than an alternative'.

Next to the total neglect of the ethical issues that could be addressed with genetically modified food, it is also remarkable that this report clearly subscribes to a concept of the consumer which was at that time prevalent: consumers are seen as to be protected with respect to food safety, but in other aspects consumer protection, or at least the provision of information needed by consumers to make an informed choice, is not seen to be necessary. Food is framed as politically and ideologically neutral, and *quality* is not an issue. There is a very strict division of responsibilities between companies, governments and consumer organisations: the food industry is responsible for food production and organising food choices, the authorities are responsible for guaranteeing the safety of the food, and consumer organisations lobby for food availability and fair access to the food supply for all.

However, since the 1980s, food has become more and more an item on the *political* agenda. Food catastrophes like BSE, Dioxin, Foot and Mouth Disease and other food safety incidents cause social crises which extend beyond straightforward matters of food safety. They demonstrate the gap which has developed between the locations where consumers shop for, prepare and and consume a meal, and the distant places where (parts or ingredients of) the final food stuffs are produced. This gap between production and consumption not only determines various kinds of ethically unacceptable production practices but also contributes to an increasing feeling of consumer alienation, and a lack of trust by consumers, in the motives of various actors in the food sector.

Policy measures and marketing strategies have contributed to the new awakening of ethical concerns with respect to food production. These phenomena have influenced the emergence of new ethical issues and intuitions, arguments or perspectives. Some (perhaps more cynical) observers would argue that the emergence of food ethics is correlated with the rise of the affluent, middle-class consumer, and has increasingly become the focus of societal debate in order to appease the moral unrest of this group of consumers. Ethics is partly constructed by, and a marketing tool for, organisations which promote specific ethical standards or political agendas, or non-government organisations which protest against the activities of particular multinational companies or methods of food production. Of course, the way these ethical orientations have emerged is dependent on how consumers and their values are conceptualised by press, communication and marketing activities (Miller and Rose, 1997). However, this comment is made from an outside perspective and has no constructive solution to the disturbances resulting from living with these ethical issues.

Consumer protests have often been limited to some of the usual ethical concerns (for example, animal welfare or fair trade), but at the same time, were sometimes effective. Via boycotts and other protests, consumers have ensured that certain products were taken off the shelves (for example, oranges from apartheid South Africa) and others were put on the shelf (for example, products produced using fair trade practices; Friedman, 1999).

An interesting description of consumer ethics trends is given by the Eurobarometer that is published every three years regarding the attitudes of European consumers to technology, including the medical, agricultural and food uses of genetic modification (GM). According to the Eurobarometer, consumers differentiate between different types of applications of biotechnology, particularly medical applications in contrast to agri-food applications. They also make a distinction between GM crops and GM foods, the latter being the least supported by European consumers. (Exceptionally, the majority of consumers in Spain, Portugal and Finland support GM food.) Perceptions regarding the risks for society, and potential usefulness of applications play the most important role in the consumer rejection or acceptance of GM foods and crops. This implies that consumer benefits are the most important factors in determining whether GM crops are accepted or not. Price is not often mentioned as a factor contributing to consumer decision-making. In addition, less than 50% of Europeans report high levels of trust in governments (Eurobarometer, 2002).

One interesting trend is the requirement of localisation of global developments, which implies that *local* food production and distribution (*terroir* as it is called in France) has gained importance in both food production and consumer policies (Winter, 2004). This trend of preference for food supplied locally is probably connected with the broader trend of increasing diversification of various food styles and the corresponding farming and production styles. The emergence of GM food, at least in Europe, gave rise to the distinction between GM and non-GM crops, foods and food ingredients, and resulted in all kinds of regulations relating to their coexistence in the food chain. Coexistence policies already existed between organic and non-organic productions styles, and it can be predicted that more types of styles (like healthy eating) need to be included by coexistence schemes (Kriflik and Yeatman, 2005).

Another trend within the food sector is the phenomenon of mergers of smaller food companies into larger ones, and the formation of global food chains with the concomitant development of the globalisation of markets. Outsourcing, seeking international sources of food ingredients, and implementing control of production processes (even if national legislation in the country of production on, for example, food safety is insufficient), is quite normal for the larger European retailers (Reardon *et al.*, 2001). Longer supply chains and connections, the rapid fragmentation of ingredient sourcing (e.g., herbs from Kenya, conservation stuffs from Canada, soy sauce from India and so on being used in the same product), and increased processing of ingredients make these chains increasingly vulnerable to various kinds of contamination (Lang and Heasman, 2004; Nestle, 2002).

Last but not least, the technologies applied to food production and conservation are rapidly progressing (Busch, 2003), resulting in an increase in novel processed foods, about which consumers are insufficiently informed. In addition, there is increased uncertainty regarding the extent to which producers respond consumer concerns and preferences.

29.3   Ethical arguments against and in favour of consumers' responsibilities

There are at least three positions that militate against consumers having a voice in the food market. On the one hand, we have the position promoted by the Chicago school of economy that postulates that the market should be *value free* and the consumer always has sufficient information and skills to make appropriate consumption choices. On the other hand, there is the alternative position that the consumer must always to be protected against negative or inappropriate choices by the state. In the latter case, the consumer is seen as a passive person with insufficient knowledge to make up his or her own mind, a person often in debt and as a consequence in need of protection from greedy producers (Reisch, 2004). The third argument which militates against consumer sovereignty mirrors the first position and stresses the need of governmental interventions in markets. It proposes consumer sovereignty a 'dead end street,' as consumers will always be utilitarian maximisers of their own private utility (for example, by buying cheaper foods) and therefore will always follow their own private interests and preferences, which means that the protection of political values like animal welfare and sustainability should only be conducted by governments. In all of these three cases sharp distinction is made between the citizen and the consumer: in this case, the citizen should be the main actor that influences politics by voting, thus contributing to the political issues that are left over by the markets and consumers.

The empirical evidence for these three views of consumers is not very impressive (Korthals, 2004). Firstly, markets are *never* value free, because norms of trust and decency (like keeping to an agreed contract) are always more or less upheld by markets. Secondly, although some consumers (for example, children) are particularly vulnerable, many consumers are able to shape their opinions regarding products, in particular given the rise of new knowledge systems such as the Internet and widespread education. However, knowledge is always incomplete, both for consumers and for producers and regulators. Thirdly, many consumer NGOs have noted that consumers are collectively mobilising on public interest issues over and beyond their private, short-term interests. This is also demonstrated by recent governmental and industrial interest in consumer concerns regarding food production. So the concept of the rational, profit maximising, egoistic, economic consumer is losing ground as a description of consumer behaviour and thought, but *also* as a theoretical construct. Fourthly, the distinction between the consumer, who is buying goods, *versus* citizens, who are voting for policies, in the field of food consumption is rather problematic. Empirically, there is only one human being that shops and prepares his or her food, and votes or contributes in other ways to the political process. The preferences in shopping cannot be disconnected from political preferences. Moreover, from a conceptual point of view, this distinction between consumer and citizen is not useful in the field of food because the existence of consumer concerns makes it clear that consumers think that the existing political process of regulating and enabling food production is insufficient to take into account consumer views on animal welfare and other concerns.

As early as 1962, the Kennedy government appealed to the rights of consumers in a rather broad way in the Bill of Consumers Rights (Reisch, 2004), which was incorporated into the EU consumer policy programme. These rights were: the right to safety; the right to be informed; the right to choose, the right to be heard; the right to representation; and the right to adequate and legal protection. After the Rio Convention (1992), in which the overall importance of sustainable production was agreed upon by most nations, and the formation of the European single market, the ethical consumer and diverse consumer concerns came to prominence. However, their concerns are multiple and often ambiguous.

In ethics, consumers' rights can be justified from at least three different perspectives that frame, in different ways, consumer sovereignty. A *deontological* position, that strongly advocates undeniable sovereignty, can be traced back to the German philosopher Kant. He states:

> Laziness and cowardice are the reasons why such a large part of humanity, even long after nature has liberated it from foreign control (*naturaliter maiorennes*), is still happy to remain infantile during its entire life, making it so easy for others to act as its keeper. It is so easy to be infantile. If I have a book that is wisdom for me, a therapist or preacher who serves as my conscience, a doctor who prescribes my diet, then I do not need to worry about these myself. I do not need to think, as long as I am willing to pay.
>
> (Immanuel Kant, *Was heisst Aufklärung*, 1785)

As consumption choices are included in one's autonomy, consumers should determine their own food (diet); as a consequence, the markets should follow these consumer preferences. In fact, this argument is one of the strongest arguments against the conceptual distinction between consumer and citizen, because it makes it clear that in the market, the autonomy of *consumer, not producers,* should prevail. As is clear from Kant's quotation, he presupposes that an adult is *educated*, has *capabilities*, and gets (reliable) *information* on the diets with which he or she wants to comply. Moreover, it presupposes also that production systems and markets deliver the goods and services such an autonomous person prefers.

However, consumer sovereignty can be justified from a *utilitarian* perspective also, although in a different way, as is clear from John Stuart Mill's, statement on freedom:

> The only freedom which deserves the name, is that of pursuing our own good in our own way, so long as we do not attempt to deprive others of theirs, or impede their efforts to obtain it. Each is the proper guardian of his own health, whether bodily, or mental or spiritual.
>
> (John Stuart Mill, *On Liberty*, 1863)

Again, from this perspective, the autonomous person should be enabled to strive for his own good through education, regulation, reliable information and responsive markets. However, from a utilitarian perspective, one is justified in balancing the overall costs of letting consumers choose and of letting experts on healthy food decide what actually constitutes healthy food and nutrition. There is not an inherent principle of consumer sovereignty which applies here.

The third perspective that I want to discuss here in more detail is the *pragmatist* perspective, because it pays a lot of attention to the fact that these ethical principles apply to what are essentially social developments (Korthals, 2004). Food is produced, prepared and consumed (and enjoyed, I hope) in social contexts. Food sovereignty can only be upheld when this is taken into account. Without social regulations, and encouragements, no one can exercise rights of autonomy. Consumer sovereignty can only have meaning in the context of markets, production sectors, governance, policy, and civil society. As a result of food having cultural and social functions, collectives in the sphere of civil societies, be it cultural or quasi-political, (such as NGOs or consumer organisations,) have an important role to play in shaping and exercising ones food choices. This implies that purely economic competition (through purchasing power) on markets should not be the only consideration with respect to the continuation or not of certain types of food or agriculture. With the production and marketing of food, competition for profits does not always means that the best win, because there can be so many costs that are not paid for, like environmental, health, animal welfare, loss of employment and livelihood, disrespect for human rights (food sovereignty). For food production and marketing not everything is allowed, and private sins do not always lead to public virtues, as the fierce proponents of free market believe.

29.4   Ethics of consumer concerns

Consumers of food products have concerns that differentiate according to at least three levels, which accordingly result in thee types of concern (Korthals, 2001). Consumers have substantive concerns about certain ethically questionable structural traits of the food chain, such as lack of animal welfare. Secondly, they complain about the lack of trustworthy information, or even partisan or distorted information, and lack of objectivity on the part of information sources. They also complain about lack of involvement with the food chain, and an increasing gap between the food chain and consumers, which treats them as complete outsiders and does not involve them in decisions made about the food supply.

The most common substantive consumer concerns that are mentioned by European consumers focus on seven ethical issues. These include the safety of food (e.g., the use of hormones and antibiotics in animal feed), the quality of the food, the healthiness of the food, issues relating to animal welfare (with criteria like the five freedoms or transport of animals and slaughtering, import/export of animals and animal products, and local *versus* trans-local production), the impact of food production on the quality of the landscape, the environmental effects of food production, and the fair treatment of farmers (implying good working conditions both in the developed and developing world).

These values are subject to lots of detailed specifications, depending on the circumstances (Korthals, 2001). All of these values can be specified as innumerable items and the concrete tasks and contexts associated with the different items are also innumerable; animal welfare can mean intact horns on farm animals, absence of lesions and injuries, good conditions to maintain the condition of feet and limbs and so forth. Good working conditions can mean that men and women get equal pay, that men and women have opportunities for child care provision, etc.

A second set of concerns focuses on the reliability of the information provided by producers and regulators, but also deals with the relevance of the information as making a contribution to balanced ethical decisions in food choices. This set of concerns also covers the issue of pluralism of preferences and value orientations: the information should not/cannot necessarily be neutral, but at least it should take account of consumer differences in information needs (for example, consumers with preference for organic meat products look for different information about the food chain and want different advice compared to consumers with other preferences, for example, low fat products).

The large range of food claims is, in particular, very confusing (as well as being potentially misleading in many cases). Martijn Katan, a well-known food scientist complains:

> However, the Food and Drug Administration's oversight over health claims has eroded, and the United States now allows 'qualified health claims' for which there is hardly any evidence, as long as a disclaimer is included. In the European Union the safety of novel foods is thoroughly regulated but health claims are not – EU legislation for nutrition claims is complex, fragmented, and poorly enforced. Paradoxically, current EU regulations prohibit claims that a food ingredient prevents a disease even when the claim is true – for example, that folic acid prevents neural tube defects.
>
> (Katan, 2004, p. 181)

A third type of consumer concern covers the widespread feeling of alienation from the food chain, and consumer assessment of efforts being applied to bridge the gap between producers and consumers. Issues of involvement and participation are connected with these concerns. A large minority of consumers do not feel at ease with this gap and seek to overcome it by having more voice in the food chain, e.g. by forms of participation in food policy (Rowe and Frewer, 2005; Rowe, Chapter 27, this volume).

29.4.1 Representation and misrepresentation

With respect to all these concerns it should be borne in mind that consumers differ in their ethical orientations, attitudes and purchasing behaviours. There are different types of consumers, and their choice between potentially conflicting values differs accordingly. Different weighing models and types of information are used for making choices. The same applies to producers: their value orientations and attitudes differ enormously across Europe. Attempts to re-establish trust should at least take into account the pluralism of consumers vis-à-vis their different ethical orientations, viewpoints, and way of balancing their preferred values.

I want to make a plea for considering an evil in the food and agricultural sector that is very much entwined with hunger, but has also some features of its own: lack of pluralism and representation. Ethically, hunger is a phenomenon that is rather easy to identify: the food is simply not equally distributed and fairly divided across population groups, which means that the principles of equality and fairness are distorted, resulting in a global situation where obesity is endemic in some population groups, whereas others are undernourished or even starving. Although many different interpretations of these principles can be identified, the issue of food misdistribution and nutritional inequalities tends to be viewed consensually as something which must be dealt with within the food chain.

However, lack of pluralism and representation is a lot more difficult to identify. It could mean something like *misrepresentation*, which in the food sector would mean that not all food styles are represented on the market and in research: some styles have no voice and some have more voice. This implies that the right on food choice of collectives or individuals is not respected.

Lack of pluralism is undeniably connected with the concept of *food quality*. This multi-interpretable concept is defined by different cultures in various ways. Quality of food is for a Moslem different for a Hindu or a Jewish person, to name only the largest and broadest lifestyle groups. There are also differences between different European nations and cultures (Rozin *et al.*, 1999).

The question of what type of food to choose, and why, is at first instance (*prima facie*) amenable to the decision of the individual citizen/consumer. As Kant says, it is so easy to let someone else decide, and ethically seen there is no justification to let someone else be your keeper in choosing your food (on the basis of paternalism). However, in fact, consumers strive for commonalities, and understanding the three types of consumer concerns offers a first step into discerning what these commonalities are. So, food choices are not only individual choices and autonomy has some collective aspects.

29.5   Dilemmas and barriers which prevent the food sector restructuring itself according to ethically acceptable measures

29.5.1   Dilemmas that block ethically conscious consumers' food choice

Recent Eurobarometers have made it clear that many consumers have experienced difficulties in finding the food that they want, or identifying the food that they prefer. Often, there is a lack of labelling (Gaskell *et al.*, 2005), or irrelevant things are labelled.

Consumers also identify barriers to ethical food choices such as availability and the lack of trustworthy information. As barriers to ethical food choices, consumer dilemmas can be categorised in two ways. First, I will list dilemmas

originating with the individual, and subsequently take into account more
increasingly inclusive social circles; then I will summarise dilemmas connected
with the various functions consumers perform in consuming food.

Individual dilemmas
- The individual consumer: do you choose products produced by the dieting
  industry, like weight watchers and 'Atkins' products, or by the fast food
  industry, like McDonald's?
- When eating with friends and relatives: do you eat your recommended or
  preferred diet alone, or are you social and do you adapt to their preferences?
  (See also Chapter 14 by Ueland, this volume.)
- In the context of local food purchasing: do you buy cheap food stuffs from
  cheaper, larger retail outlets and potentially destroy the local retail economy,
  or spend a little more money in the local vegetable shop, butcher, and bakery
  (if these are available)? Do you take into account the long-term effects of
  these purchases?
- Can you buy what is produced locally? Can you get information about local
  production?
- Can the consumer buy favourite products produced in other (possibly distant)
  areas? Should consumers choose between selecting products on the basis of
  low food miles *versus* quality and taste?
- Can the consumer buy from markets and small shops and do they realise that
  such retail outlets are not encouraged by regulations? Country level: should
  the consumer buy cheap food stuffs from abroad and not support the food
  production industry of their own country? Should the consumer buy expensive foodstuffs from abroad (for example, those produced using ecological
  production methods such as bananas), or cheaper foods from their own
  country?
- European level: *idem.*
- Global level: North-South divide and buying from poor farmers: is the food
  safe? Can the consumer trust that their purchases indeed help the poor? Can
  consumers trust the vendor?
- Global level: should consumers eat meat and fish (and contribute to the
  deterioration of nature), the farming of which has a negative environmental
  impact in terms of the food pyramid?
- Future generations: do consumers consider them or not in deciding upon the
  use of non-renewable resources in food production?

Consumer dilemmas relating to preparing, cooking and eating food
- When preparing foods: should you cook without using pre-packaged and non-
  frozen food. What if there are no shops providing alternative products
  accessible by consumers?
- When buying meat and meat products: should the consumer choose between
  locally produced meat, organically produced meat or free range meat?
- When buying fruit: should consumers buy from organic or ecological shop

'traceable' products which may be mislabelled regarding country of origin and production method?

- Should consumers purchase cheap food or that which is produced ethically but may be more expensive (short-term profit for the consumer *versus* long-term profit for society and the environment).
- Do consumers choose from one or more of these enormous amounts of labels, e.g. meat from pigs with or without teeth, with or without tails, or do you believe the critics that argue that these labels can't be trusted?
- Do consumers get more confidence in labelling because of intensive regulation and monitoring of the labelling companies, or do they lose sight and trust because of these complicated regulations?
- Do consumers trust the health claims of light food or do they believe the critics that these health claims are only partially valid and often neglect unhealthy ingredients (like acids) of the products?

(Böcker and Hanf, 2002; Brinkmann, 2004; Kriflik and Yeatman, 2005; Schroder and McEachem, 2004.)

29.5.2   The productionist paradigm of the food sector frames problems in an ethically unacceptable way

The productionist paradigm that still permeates the whole food sector (including regulatory activities) emphasises the importance of high levels of food production, together with a limited conception of food safety, which can be summarised as food free from biological and chemical contaminants. Food-related diseases, such as obesity, cardiovascular diseases and intestinal cancers (in particular caused by red meat), are neglected within this paradigm. More-over, food portions, and the amount of calories and of salt, saturated fat and sugar in foods, have increased in recent years (Nielsen and Popkin, 2003). WHO has published various reports on the connection between food intake and these diseases (FAO/WHO, 2003). There are strong positive associations between consumption of foods high in fat and sugar (associated with the products of the fast food industry), weight gain and insulin resistance, which increases the risk of obesity and type 2 diabetes (Pereira *et al.*, 2005). The food industry has mostly reacted very angrily in response to these reports and findings and has attacked organisations critical of its products, and threatened them with juridical and other sanctions (Nestle, 2002; Shell, 2002). In particular large American companies apparently do not feel responsible for the diseases and environmental costs related to processed food.

However, the advertising budgets for unhealthy food are enormous. A case in point is the high level of advertising promoting processed foods aimed at children which is broadcast during television schedules aimed at this same population group. Moreover, the direct and indirect costs to the individual and the economy associated with food-related diseases have escalated over recent decades, as have environmental costs and the damage done to human rights, i.e. the destruction of livelihood of poor farmers in the South (Thiele and Ashcroft, 2004).

It seems as if only reluctantly, under great social pressure, the food industry is willing to start to tackle the problem of obesity in more affluent countries. Many companies are hampered by the old productionist paradigm that frames consumer protection in terms of *safe* food free from biological and chemical contaminants, and which formulates food innovation as a *technology push* process, in which there is no room for the voice of the consumer.

29.6   Implications for food product development: representativeness, transparency (labelling), fair taxation and pricing

As was made clear in the first sections of this chapter, the existing situation within the food sector is far from ethically acceptable. Consumers are frequently confronted with unreliable and biased information, and with supplies of food that are largely unhealthy, animal and environmentally unfriendly, disrespectful of human rights and so forth. Although these ethically unacceptable activities cannot be attributed to the practices of large companies alone, the latter still have a large stake in continuation of the existing ethically unacceptable situation. A case in point, food advertisements aimed at children are worth annually US$12.7 billion, and do not promote fruit and vegetable consumption but do promote consumption of fatty and sugary food stuffs (Nestle, 2002). Widespread obesity is the result (Critser, 2003).

In this section, I will discuss four issues that can make the food production sector more ethically acceptable: the need for diversification of production and food styles, the need for greater transparency focused towards the consumer (ethical traceability), the need for taxation of unhealthy food stuffs, and the need for sufficient prices.

The need for greater and more representative *diversification* towards a multi-tier food system (intensive, extensive, organic, GM, non-GM, health food, fun food, etc.) is directly justifiable from the concept of respect for the cultural diversity of food choices. Before World War II, food was seen in most countries and cultures to be an important factor applied both to self-sufficiency of states as to the self-identity of a culture.

Historically, the short period after World War II stands alone as the time in which food was not framed in terms of cultural and emotional identity, although this does not appear to be the case at the present time. As a consequence of the increasing politicisation and culturalisation of food, food is again seen by many as an ethical, social and cultural commodity. In correspondence with the pluralism of cultures, we encounter in addition to the dominant food style, fast food, increased societal emphasis on different varieties of food, farming and production styles, such as slow food, international food, and health food, alternative food (urban community) networks. For example, organic food is the fastest growing agricultural sector in the United States. In response to its rapid growth, the United States Department of Agriculture implemented the National

Organic Program (NOP) in October of 2001 (2003). The NOP set the standard to which all food sold in the United States as 'organic' must be produced. In European countries we can see the same picture emerging. Coexistence of different systems (pluralism) requires procedures to regulate the peaceful coexistence of these diverse styles. Although the debate on the formulation of criteria of coexistence from an ethics point of view is still in its infancy, some comments can be made at this stage. First, very generally, the recognition of food choices and their representative and collective organisations is in line with the deontological, utilitarian and pragmatist arguments given earlier. Secondly, coexistence should take into account the representativeness of a food style, not its monetary value or market share. Although these last two criteria are to be taken into account, there are other means to find out what styles are representing food choices, like consultations and deliberations.

Transparency is still in its infancy in the food production sector (see also Chapter 5 by de Jonge *et al.*, this volume). For example, many subsidies in Europe and USA are not open to societal scrutiny and not made public. What has been made public until now (e.g., in UK) shows that large companies get a substantial part of the EU subsidies (in 2004 the sugar company Tate and Lyle got 192 million Euro in subsidy). Moreover, the names of companies that are fined because of lack of hygiene or because they didn't live up to certification rules, are not published; the inspection reports of the European Food Standard Agencies are not open to the public. It is important that the implementation of traceability systems should ensure the transparent provision of information regarding the origin of ingredients included in food products Under current legislation this is still rather vague, and does not include the traceability of ingredients provided by fringe suppliers (for example, within the animal feed chain) and waste companies (Lees, 2003). Traceability is mostly organised as a recall system for risk management purposes and not as an information system that keeps the consumer informed about ethical concerns. Traceability systems are only framed as a safety tool, not as a way to promote ethical food choices. This is a problem particularly because the longer the chain, the greater the waste will be. For example, longer food chains mean that more packaging of food ingredients will be used, and, as a consequence, the production system will be less sustainable. This is not considered in the current traceability framework. Ethical traceability schemas that do not reflect management concerns but consumer interests are required.

Taxation of unhealthy food ingredients (for example, polysaturated fats, salt and sugar) is a third requirement that could make the food system more ethically acceptable (American Public Health Association, 2002). As a result of the increasing costs of food-related diseases such as cancer, cardiovascular diseases and obesity, which are to some extent the result of the consumption of unhealthy food ingredients, the food sector has to take its responsibility and pay for these externalised costs. Some say that there are no unhealthy or healthy food stuffs, or that other factors also contribute to these diseases. It seems that the arguments for and against what constitutes healthy eating are dependent on the circumstances and the product to be sold. Of course, a whole range of factors determine obesity (for example, overnutrition in combination with a sedentary lifestyle) but it is easily observable that everywhere where fast food outlets (and processed food) are dominant, obesity is an increasing problem (see also Chapter 17 by Mela, this volume). Although industry is against a tax on fatty foods, it does appear to accept responsibility for poor consumer health through inappropriate nutrition, as is clear from the final outcome of the Trans Fat Lawsuits, where McDonald's settled with a payment of US$8.5 million to the American Heart Association. (One wonders why other nations didn't receive compensation, see www.bantransfat.com.) The idea of a tax on (saturated) fat needs further elaboration and experimentation, e.g. in the direction of taxing only certain food stuffs (like chips or burgers), but given the societal costs of the associated diseases and the potential effects of higher prices deterring consumption it is also ethically justifiable because it implies a modest type of protection of vulnerable (young and un-informed) consumers. This could also justify subsidising healthy food choices (for example, fresh fruits and vegetables), although this seems more difficult because of the wide variety in health requirements of consumers.

The need for *sufficient prices* is a fourth ethical requirement, whereby the term sufficient should express the ethical requirements discussed earlier. It is established that when food prices decrease below a certain threshold, animal welfare and other ethical values will be damaged. Although availability of food for all is an ethical requirement, this requirement must be balanced with others that undoubtedly will cost some money and effort. Increasingly cheaper prices of chicken in Europe are paid for by the inhumane management and housing systems of broiler chickens. Increasingly low food prices can often only be produced because producers pay employees' salaries which are too low or insufficient to provide a living wage; landscapes are destroyed or other non-monetarised values are compromised. It is, of course, not an easy task to determine what constitutes a sufficient and fair price for commodities and foods, and the topic merits further study and debate; however, the trend of making food stuffs (in particular those which are unhealthy, like fast foods) still cheaper, is endangering the implementation of values like animal welfare, environmental protection and the quality of the landscape. Producers and consumers, who are demanding cheaper food stuffs, are doing a dubious job: they are not only moral hazards (Reisch, 2004), but are compromising the well being of poor farmers, inarticulate animals, silent landscapes and their own long-term interests (Appleby et al., 2003).

29.7   Future trends: diversification of food and farming styles

The food sector has to learn to live with diversity and social contextualisation through consumer involvement and participation in the food chain. Particularly in Europe, the landscape will become more diversified, with increased involvement of consumers, in particular when the complex connection between food consumption in one place and certain types of production in another place will become clearer to them. For example, health messages from nutritionists to eat more fish because of the healthy omega 3 fatty acid, will undoubtedly give rise to more concerns with the rapid decline of fish resources, as well as the presence of certain toxins in fish products (see also Chapter 5 by de Jonge *et al.*, this volume).

The food sector will be confronted with more political conflicts over food, in particular associated with the trend of outsourcing, using controversial technologies, environmental impact of some production methods, and the increasing gap between poor and rich people. Given the uneasiness and even fear that many feel when confronted with technological globalisation, where the food industry represents one of the most globalised players, the food industry will feel the impact.

Moreover, coexistence schemes, and steps to rebuild trust will be necessary. These are only possible as a consequence of developing transparent and integrated ways of involving consumers in fundamental decisions concerning research, management and food styles. Of course, many consumers do not have the time or interest to be engaged in the food sector and they will try to follow at distance by trusting their more involved consumer colleagues. But an increasingly large minority is interested in activities in the food sector, and has strong opinions on what is happening with their food, which they will want to voice.

29.8   Implications for research and development

The food sector, in all its aspects, is one of the most controversial areas for research; both from a biological and from a social science or ethical perspective, problems continually arise. Experiments with different types of producer ethics and their systematic evaluation are necessary, as is also the case with coexistence schemes, and various mechanisms for deliberation and participation of consumers. Evaluation of ethical schemes, from consumer, producer, and governmental points of view will be needed, as will be the measures to rebuild relationships, trust and involvement. It is not known which types of participation are effective with respect to the various targets of ethical involvement. In addition, in the case of producers and regulators, it would be good to develop and apply better methods of inquiry regarding the ethical preferences of consumers with respect to food, labelling, and packaging.

Within the food sector itself, the new situation of diversification, consumer participation, and ethical justification, may be a period of immense creativity and entrepreneurship. Trying to satisfy ethical requirements may be difficult, but not impossible, and in the long term will pay for itself. A hundred and fifty years ago, many entrepreneurs were confronted with the situation that the large majority of Western nations did not find it ethically acceptable for human slave labour to be used in food production. Companies reacted by responding positively to new regulations outlawing slave labour, abolished these unethical practices, and prospered. At the present time, ethical requirements have been focusing, on new issues, in response to the novel situations evoked by globalisation and lack of consumer trust in the activities of different actors in the agrifood sector. Again, food companies are faced with new ethical challenges which demand 're-engineering' of the food chain (Trienekens and Hvolby, 2001). The worst thing that can happen is that the challenges are dismissed and consumers' trust is not regained (Brinkmann, 2004).

29.9   Conclusions

For a long time, the ethics of food was only concerned with food security and consequently with distribution and misdistribution, assessed against the criteria of fair distribution. However, since the food wars (on, e.g., genetically modified food) and the food scares which have occurred in the Western world over the last three decades, the whole social, cultural and political structure has been changed radically, providing new challenges not only to the food industry and policy makers but also to food ethicists and the discipline of food ethics. Nowadays, with the gap between consumers and producers increasing consumer alienation, it seems clear that the lack of food is not the only morally unacceptable issue, but that the lack of representation of the voices of the consumers in the food chain is also an ethical concern. Consumers voice concerns regarding at least three types of issues: substantive issues, (such as animal welfare), sustainability (environment, justice towards future generations) and landscape (aesthetic values, the use of the countryside as a recreational resource), information issues, (such as reliable labelling and branding, and transparent traceability systems) and procedural issues, (such as meaningful consumer involvement and participation, or some kind of reliable consumer representation regarding activities within the food chain). Moreover, these concerns are voiced in a pluralist way, which means that diversification of food production is necessary, accompanied by policies of coexistence of the various food, farming and production styles. A large minority of consumers do not want to be protected, but do want to be heard, and as long as the food sector is willing to change towards consumer pull, and indeed acknowledges the disadvantages of being a push sector, it is necessary to experiment and evaluate new types of societal participation. Participation is often promoted as a panacea for all societal ills, but here it is argued that to become important and effective, it should offer more than giving information and being representative and a form of reciprocal communication. It is a special task to find out what the role of NGOs in the debate about ethics and food should be, as they should contribute to the balance of countervailing powers in the food sector.

29.10   Acknowledgment

The author has no interest whatsoever in food production companies; the ethics of production and consumption of food should be conducted without any prior commitment in increasing or decreasing production, and only with reasonable ethical principles and values in mind.

29.11   Sources of further information

CRITSER, G. (2003), *Fat Land: How Americans Became the Fattest People in the World*, Boston: Houghton Mifflin.

FAO/WHO (2003), *Report: Diet, Nutrition and the Prevention of Chronic Diseases*, Geneva (WHO TRS 916).

KORTHALS, M. (2001), Taking consumers seriously: Two concepts of consumer sovereignty, *Journal of Agricultural and Environmental Ethics*, **14**, 2, 201–215.

KORTHALS, M. (2004), *Before Dinner: Philosophy and Ethics of Food*, Dordrecht: Springer.

LANG, T. and E. MILLSTONE (eds) (2002), *The Atlas of Food*, London: Earthscan Books.

LANG, T. and M. HEASMAN (2004), *Food Wars*, London: Earthscan Books.

NESTLE, M. (2002), *Food Politics: How food industry influences nutrition and health*, Berkeley: University of California Press.

POTTHAST, TH., C. BAUMGARTNER and E.-M. ENGELS (eds) (2005), *Die richtigen Maße für die Nahrung? Biotechnologie, Landwirtschaft und Lebensmittel in ethischer Perspektive*, Ethik in den Wissenschaften, Vol. 17, Tübingen: Francke Verlag.

SHELL, E.R. (2002), *The Hungry Gene*, London: Atlantic Books.

THIELE, F. and R. ASHCROFT (eds) (2004), *Bioethics in a Small World*, Dordrecht: Springer.

WATSON, R. (1998), *The Philosopher's Diet. How to lose weight and change the world*, Boston: Non-pareil.

29.12   References and further reading

AMERICAN PUBLIC HEALTH ASSOCIATION (2002), Reducing sodium content in the American diet, *Association News*, **4**, 5–6.

APPLEBY, M., N. CUTLER, J. GAZZARD, P. GODDARD, J.A. MILNE, C. MORGAN and A. REDFERN (2003), What price cheap food? *Journal of Agricultural and Environmental Ethics*, **16**, 4, 395–408.

BÖCKER, A. and C. HANF (2002), Confidence lost and – partially – regained: consumer response to food scares, *Journal of Economic Behavior and Organization*, **43**, 471–485.

BRINKMANN, J. (2004), Looking at consumer behavior in a moral perspective, *Journal of Business Ethics*, **51**, 129–141.

BUSCH, L. (2003), Virgil, vigilance, and voice: agrifood ethics in an age of globalization, *Journal of Agricultural and Environmental Ethics*, **16**, 459–477.

CRITSER, G. (2003), *Fat Land: How Americans Became the Fattest People in the World*, Boston: Houghton Mifflin.

EUROBAROMETER 58. *Europeans and Biotechnology in 2002. http://ec.europa.eu/public_opinion/archives/ebs/ebs_177_en.pdf*.

EUROBAROMETER (2005), *Risk Issues*, http://ec.europa.eu/public_opinion/archives/ebs/ebs_238_en.pdf.

EUROPEAN COMMISSION (2000), White Paper on Food Safety, 12 January 2000.

FAO/WHO (2003), *Report: Diet, Nutrition and the Prevention of Chronic Diseases*, Geneva (WHO TRS 916).

FRIEDMAN, M. (1999), *Consumer Boycotts. Effecting change through the marketplace and the media*, New York: Routledge

GASKELL, G., N. ALLUM and S. STARES (2003), *Europeans and Biotechnology in 2002*, Eurobarometer 58.0.

GASKELL, G., N. ALLUM and S. STARES (2005), *Europeans and Biotechnology in 2004*, Eurobarometer.

KANT, I. (1785/1995), *Was heisst Aufklärung, idem*, Werke, Darmstadt.

KATAN, M. (2004), Health claims for functional foods, *British Medical Journal*, **328**, 180–181.

KORTHALS, M. (2001), Taking consumers seriously: Two concepts of consumer sovereignty, *Journal of Agricultural and Environmental Ethics*, **14**, 2, 201–215.

KORTHALS, M. (2004), *Before Dinner: Philosophy and Ethics of Food*, Dordrecht: Springer.

KRIFLIK, L.S. and H. YEATMAN (2005), Food scares and sustainability: A consumer perspective, *Health, Risk and Society*, **7**, 1, 11–24.

LANG, T. and E. MILLSTONE (eds) (2002), *The Atlas of Food*, London: Earthscan Books.

LANG, T. and M. HEASMAN (2004), *Food Wars*, London: Earthscan Books.

LEES, M. (ed.) (2003), *Food Authenticity and Traceability*. Eurofins, France.

MILL, J.S. (1863/1975), On Liberty, in: Mill, J.S., *Three Essays*, Oxford: Oxford University Press.

MILLER, P. and N. ROSE (1997), Mobilizing the consumer: assembling the subject of consumption, *Theory, Culture and Society*, **14**, 1–36.

NESTLE, M. (2002), *Food Politics: How food industry influences nutrition and health*, Berkeley: University of California Press.

NIELSEN, S. and B.M. POPKIN (2003), Patterns and trends in food portion sizes, 1977–1998, *JAMA*, **289**, 450–453.

PEREIRA, M.A., A.I. KARTASHOV, C.B. EBBELING, L. VAN HORN, M.L. SLATTERY, D.R. JACOBS, JR and D.S. LUDWIG (2005), Fast-food habits, weight gain, and insulin resistance, *The Lancet*, **365**, Jan., 36–42.

POLKINGHORNE REPORT (1993), *Report of the Committee on the Ethics of Genetic Modification and Food Use*, HMSO, London.

REARDON, TH., J-M. CODRON and L. BUSCH (2001), Global change in agrifood grades and standards: agribusiness strategic responses in developing countries, *International Food and Agribusiness Management Review*, **2**, 3/4, 421–435.

REISCH, L. (2004), Principles and visions of a new consumer policy, *Journal of Consumer Policy*, **27**, 1–42.

ROWE, G. and L. FREWER (2005), A typology of public engagement mechanisms, *Science, Technology and Human Values*, **30**, 251–290.

ROZIN, P., C. FISCHLER, S. IMADA, A. SARUBIN and A. WRZESNIEWSKI (1999), Attitudes to food and the role of food in life in the U.S.A., Japan, Flemish Belgium and France: Possible omplications for the diet–health debate. *Appetite*, **33**, 163–180.

SCHRODER, M. and M. McEACHEM (2004), Consumer value conflicts surrounding ethical food purchase decisions: a focus on animal welfare, *International Journal of Consumer Studies*, **28**, 2, 168–178.

SHELL, E.R. (2002), *The Hungry Gene*, London: Atlantic Books.

THIELE, F. and R. ASHCROFT (eds) (2004), *Bioethics in a Small World*, Dordrecht: Springer.

TRIENEKENS J.H. and H.H. HVOLBY (2001), Models for supply chain reengineering, *Production Planning and Control*, **12**, 3, 254–264.

US DEPARTMENT OF AGRICULTURE, ECONOMIC RESEARCH SERVICE (2003), *United States Fact Sheet*.

VISSER, M. (1986), *Much Depends on Dinner*, Macmillan.

WATSON, R. (1998), *The Philosopher's Diet. How to lose weight and change the world*, Boston: Non-pareil.

WINTER, M. (2004), Geographies of food: agro food geographies – farming, food and politics, *Progress in Human Geography*, **28**, 5, 664–670.

30

Looking to the future

L. J. Frewer and H. van Trijp, Wageningen University,
The Netherlands

30.1 Summary

Reflecting on the diverse chapters included in this volume, it can be seen that consumer food choices will play an important role in fuelling the body, delivering sensory enjoyment from eating and improved health. In addition, food has a profound influence on other aspects of the lives (and preferred lifestyles) of consumers as well. The determinants of food choices are influenced by many different factors, and understanding issues related to food security, food safety, consumer health and well-being are important in the context of food choice. The earlier sections focus on key influences on food choice – for example, sensory factors are extremely influential in determining food preferences, but cannot be isolated from the social and cultural spaces which consumers inhabit. Whilst risk perceptions associated with food hazards may play a large part in determining consumer responses, much of choice behaviour is grounded in consumer desire to acquire some kind of benefit – whether to themselves, to the environment, or to society more generally. A key challenge for the future will be how consumers trade off the risks and benefits inherent in food choices. This will require the further integration and synthesis of various disciplines; some traditionally more focused on risk (such as risk psychology), others more on benefits (sensory science and various sub-disciplines in psychology and sociology). Other psychological factors relevant to food choice have been discussed – and, although limitations of space prohibit an extensive review of this research, it is also important to acknowledge that rapid developments in this area are providing opportunities for greater application to consumer health, well-being and lifestyle preferences.

Throughout the field of understanding the food consumer, we have witnessed a broadening of both its scope and area of application. Increasingly, we see that the theories of food choice have found their application and refinement in various applied fields of research, and in each of these application areas new theories have emerged. This has been reflected in the content of this book. One way of looking at this is by analogy to the Maslovian 'Hierarchy of Needs', which states that consumer needs can be classified in an overlapping hierarchical structure. At the lowest end there is a need for ensuring food security and food safety. Once these needs are adequately satisfied, consumer attention regarding foods moves toward the more experience-type consumption benefits that are derived from it, such as good taste, together with high convenience and affordability. Social motivations for food choice play an important role, too. Credence attributes such as healthiness, sustainability and moral consideration of food choice are positioned higher in the consumer need hierarchy as they are not related to just satisfaction, but require a more forward-looking perspective on future health and well-being, as well as consideration of welfare and well-being of others, including workers in food production, environmental factors, and animal husbandry practices in response to personal consumption decisions.

The field of consumer behaviour and how people make decisions about food products, as reflected in this volume, echoes different facets of these developments. Understanding consumer responses to food safety issues has developed strongly in recent decades, and has developed its own theoretical accounts. The traditional focus on sensory quality of food is increasingly being extended to include attitudinal determinants of overall food perceptions, as well as food liking and preferences. The latter include those related to convenience, price and cognitive aspects of food liking and preference. Increasingly, research in the area of consumer food choices has moved from an analysis of consumer food perceptions to a more behavioural approach where the challenge has been to understand purchase and consumption behaviours in a broader context. For food choices, it is important to realise that many actual consumer decisions are made under conditions of low consumer involvement, with little cognitive elaboration on the part of the consumer. This will increasingly require models that combine the more implicit and automated responses of consumers with the more rational, information-based aspects of decision making.

Consumer decisions regarding healthy and unhealthy consumption behaviours are still poorly understood, and available models tend to have low predictive validity. This is a serious limitation of the field and an area that requires more attention in the future, and is particularly pertinent as policy makers increasingly realise the importance of food consumer understanding in the design and execution of their intervention strategies, yet we seem only poorly equipped to give them the right answers. The same holds for applications in new product development, where we also still see low success rates, similar to the limited effects of health intervention studies. In addition, there is a need to develop more predictive, yet actionable models of consumer choice behaviour that provide guidance for enhancing success in behavioural modification efforts;

research which is relevant to the development of effective health interventions as well as new product development.

For most consumers, the point at which decisions are made regarding what food products are actually bought into the home is the retail environment. Decisions must therefore be made regarding the acceptability (or otherwise) of products and their attributes. These include, for example, branding and labelling of foods and consumer perceptions of quality and convenience, as well as attitudes towards production methods and processing technologies. Of course, consumers are not homogenous, and vary according to demographic and cultural factors, as well as at different stages in their lives. The food choices of an adolescent may not be the same as an elderly person, or someone with responsibility for the care of young children. Whilst it is impossible to cover all aspects of individual and group variation in the limited space available in this volume, some key determinants of food choices, such as cultural differences, gender and life experiences have been addressed. As populations in the developed world continue to age, then further research into the needs of the elderly will emerge in order to develop informed and effective policy pertinent to the needs of changing populations. In sub-Saharan Africa, where diseases such as malaria, tuberculosis and increasingly AIDS compromise consumer health status, healthy food choices are important but are also problematic given local food shortages and problems associated with food safety. Thus increasingly there is public debate about how to optimise consumer health and well-being through food choices, although both problems and solutions are likely to vary across different regions. Obesity continues to emerge as one of the most serious health concerns in affluent countries, and those with thriving economies and newly affluent groups, whilst in other parts of the world food security still represents a major problem. The section on consumers, food and health includes different chapters which cover many of the pertinent issues, ranging from discussion of the determinants of obesity, consumer attitudes towards novel foods which confer specific health benefits, and changing unhealthy food choices to those that promote health and well-being. Avoiding health risks (for example, those caused by allergic responses to food ingredients or microbial risks) is also important if consumer health is to be optimised. Research in this area must remain a priority if problems related to consumer food choices are to be solved.

The development of effective food policies must take due account of consumer attitudes, beliefs, and cultural and individual diversity regarding preferences, particularly as food chains become ever longer, even pan-global. Food policy must address not only issues of consumer health, but also some of the broader societal issues associated with production and consumption. In the final section of this volume, full discussion of the range of issues associated with developing effective food policy were discussed, ranging from analysis of the ethical issues associated with food production to promotion of healthy eating, ethical consumption and food safety governance. In the future, there will be increased discussion of how the results of research into consumers and food choices can be incorporated into effective policy frameworks and food governance systems.

   Despite the existing and ongoing research into food choice, there is, nonetheless, a list of emerging problems which continue to confront society. Consumer health associated with food choice is an outcome of a complex set of interactions between what is scientifically possible, what is acceptable to consumers, what is needed by society, and market forces. Solving problems of food security does not depend on scientific advances to improve crop yields in isolation, but also developing an effective infrastructure within particular regions or boundaries to facilitate movement of food to areas where it is needed. For example, water shortages are compromising food security and food safety in many parts of the world, but solving the problems may require changes in social attitudes to climate change and sustainability. Research into understanding consumers and food products must be interdisciplinary, focusing on combining research from the natural and social sciences in order to address key research questions.

30.2   Future challenges for regulators

One future challenge relates to developing governance systems which emphasise that consumer needs include not just protection against risk, but also acquisition of benefit. As described in several chapters in this volume, traditional approaches to consumer protection have been enshrined in the evaluation and communication of risks associated with foods. For example, activities within risk *assessment* focus on estimating the risk that a hazardous event or factor will negatively affect a population or subpopulation. Against this, risk *management* is defined as the process of weighing policy alternatives which emerge as a consequence of the results of risk assessment, as well as selecting and implementing appropriate control options, including regulatory measures. Risk communication is represented as the interactive exchange of information and opinions concerning risk and risk management among risk assessors, risk managers, consumers and other interested parties and, theoretically, interacts with both assessment and management. If the issues of potential benefit are to be included in the evaluation regarding governance activities, then it is also important to discuss *socio-economic factors* and *ethical issues* as a formal part of risk–benefit evaluation. Even when considering evaluation of health and environmental impact, the debate about how to evaluate the effect of a particular hazard on quality of life must specifically address socio-economic factors (for example, application of QUALYS, or quality adjusted life years, as opposed to DALYS (disability adjusted life years) requires systematic evaluation of socio-economic impacts as well as health effects. If benefits are to be systematically incorporated into the framework, then communication must also focus on the issue of consumer benefit, and (as is the case for risk perception) what psychological factors are important for consumers as part of their decision-making processes. Further research into this area is warranted. In addition, much has been written about how to develop effective public participation in the decision-making process, how best to take up the outputs of public consultation into the policy context is still a controversial point. In particular, how best to handle lack of consensus resulting from a public consultation about food safety priorities is an issue requiring further clarification, although methodologies adopted from the foresight literature (for example, Delphi methodologies) may prove fruitful.

It is also important to acknowledge that individuals are unlikely to process *all* information about *all* risks (both those originating in food and beyond this), as the amount of information received would be overwhelming. In designing and implementing appropriate risk management strategies, it is important to examine how both members of the public as well as other key stakeholders (experts and decision-makers) perceive both the practice and effectiveness of food risk management. How to integrate these views into governance practices represents an important challenge for the future. Policy initiatives also need to focus on public health, and thus address how people make decisions about healthy food choices, including those associated with technological inputs into innovative new products aimed to improve health (for example, in the area of nutri-genomics). Developing effective policy to deal with emerging health issues such as obesity or healthy eating also represents a challenge for the future.

The food industry has traditionally been one of the important end-users of knowledge in the area of food consumer understanding, particularly in the fields of food product optimisation and new product development. Today, the food industry is increasingly being confronted with a responsibility for the external effects that purchase and consumption of their food products may bring about. This is particularly evident in the current debates on safety and healthiness of food products. Also, owing to globalisation and increasingly competitiveness in the food industry, food companies are increasingly faced with issues of differentiation *vis-a-vis* competition and effectiveness, and the efficiency of cross-cultural and even global strategies. Many of the issues associated with safety, healthiness and sustainability of food consumption do not relate directly to the 'quality' of individual food products, but merely to the 'total food basket' that the consumer consumes. For the food industry this means increased reliance on the integration of food quality strategies (ensuring that individual food products themselves respond to requirements in safety, healthiness and sustain-ability). It also means that they take a co-responsibility, together with governments and consumer organisations, to ensure that consumers balance the overall portfolio of food products they consume. For the food industry, this also implies that there should be a healthy balance between company profit and the social and ecological consequences of their business performance, as highlighted in the people, planet and profit concepts targeting sustainability.

There are ample opportunities for the food industry in the balancing of product quality and consumer choice behaviour. This is probably best exemplified by the WHO challenge 'to make the healthy choice the easy choice'. As consumers are generally not willing to compromise direct benefits (such as taste and convenience) for credence qualities such as healthiness, the challenge is to make it as easy as possible for them to make the more desirable food choices. By better understanding the total context of consumer purchase and consumption behaviour, the food industry can develop healthier foods which are positioned in the market place, and which are associated with few barriers to consumers choosing them. Food retailers will also play an important role in making this possible.

Interestingly, several authors in this book plead for a better understanding of the motivational aspects of food choice ('wanting'), beyond the pure hedonics ('liking'). It has been well established that preferences (a motivational/ behavioural construct) do not necessarily coincide with liking (a hedonics construct). One of the challenges for future research will be to understand the contexts and underlying mechanisms that explain when liking and wanting associate and dissociate as it may help us identify more efficient and effective routes for food choice interventions from which both industry and public policy can benefit.

30.3   New technologies, new foods

Within this book, we have described how the attitudes of consumers towards emerging food technologies may compromise consumer acceptance and commercialisation. Scientific advances, for example in the area of nanotechnology, promise to deliver profound benefits in the area of food safety and consumer health. For example, nanotechnology has the potential to improve the bioavailability of nutrients, deliver them to specific sites in the body, or to determine the nutritional profile of the individual and to use this profile to optimise the nutritional intake, providing solutions to some of the health problems which are emerging internationally. Other consumer benefits may lie in the sensory area – for example, controlled release of flavours and fragrances for sensory enhancement is feasible through application of nanotechnology. Furthermore, the development of novel packaging to monitor and control product quality through controlled ripening and extended shelf life are all possible. However, consumer acceptance of nanotechnology applications in the agri-food sector is by no means automatic, despite obvious consumer benefits which are being developed. The attitudes of European consumers towards genetically modified organisms, particularly applied to agriculture and food production, is the example frequently cited as being of greatest relevance to commercialisation of nanotechnology applications. There are, of course, conceptual differences between GM technology and its applications, and the somewhat broader and more diverse scientific basis and commercialisation possibilities associated with developments in nanotechnology. Nanotechnology is not a single technology, but reflects various combinations and convergences of emerging technologies (for example, chemistry, biotechnology, information technology, cognitive science and engineering). Each area of application may be associated with differing public concerns and ethical issues, as well as perceptions of benefit, varying in both profundity and complexity, which need to be understood if governance, regulation, and commercialisation strategies are to match the needs of society.

30.4   Conclusions

Some key challenges for future research have been identified which follow on from this volume, and have international resonance as well as local application. One of the most important is the development of effective methodologies to communicate about, and influence, consumer food choices under conditions where risk, cost and benefit may be associated with specific food choices. Communication with consumers about uncertainty associated with risk-benefit evaluation, as well as developing targeted communication strategies to vulnerable groups, are becoming research priorities. The need to restructure risk and health governance practices to incorporate socio-economic and ethical evaluation is recognised internationally, but concrete guidelines regarding how this might be operationalised and harmonised globally are not yet available. Sustainable production is high on the agendas of many national and international bodies, but implementation of a successful sustainability policy is contingent on consumers buying foods produced using sustainable production methods. The potential for bioterroristic incidents has a negative effect on food safety is another area worthy of research, and needs to be included in evaluation of the effectiveness of crisis management activities. How best to combine effective detection of contaminants with ingredient traceability and crisis communication with consumers needs to be understood in this context.

Optimising food choice must take account of the potential impact of food on consumer health and wellness, whilst at the same time ensuring both food safety and food security. Novel products which have specific properties beneficial to consumer health (low fat products with improved sensory qualities, for example, or nutrigenomic products) may confer profound benefits to consumers if they are acceptable from other consumer perspectives. Optimising consumer health and well-being can only occur if a multidisciplinary perspective to understanding consumer food choices and product preferences is adopted, and this will increasingly be the focus of future research activities in the area of *understanding the consumers of food products.*

Index